Student Solutions Manual

Katy Murphy • Michael Sullivan III • Michael Sullivan

College Algebra

FIFTH EDITION

Michael Sullivan

PRENTICE HALL, Upper Saddle River, NJ 07458

Executive Editor: Sally Simpson
Special Projects Manager: Barbara A. Murray
Production Editor: Michele Wells
Supplement Cover Manager: Paul Gourhan
Supplement Cover Designer: Liz Nemeth
Manufacturing Buyer: Alan Fischer

Simon & Schuster / A Viacom Company
Upper Saddle River, NJ 07458

Printed in the United States of America

10 9 8 7 6 5 4 3

ISBN 0-13-081012-6

Prentice-Hall International (UK) Limited, *London*
Prentice-Hall of Australia Pty. Limited, *Sydney*
Prentice-Hall Canada, Inc., *London*
Prentice-Hall Hispanoamericana, S.A., *Mexico*
Prentice-Hall of India Private Limited, *New Delhi*
Prentice-Hall of Japan, Inc., *Tokyo*
Simon & Schuster Asia Pte. Ltd., *Singapore*
Editora Prentice-Hall do Brazil, Ltda., *Rio de Janeiro*

Preface

To the Student

This manual contains detailed solutions to all the odd-numbered problems in ***College Algebra***, Fifth Edition, by Michael Sullivan. Preceding the solutions to some of the groups of problems we have listed step-by-step procedures which may be applied to each exercise in the group. Hopefully, these will enable you to develop a systematic approach to solving certain types of exercises. Our desire is that after seeing several examples, you will be able to solve problems without referring to this manual.

We wish to thank everyone who has helped with this project. The enormous job of typing was done by Brenda Dobson, and the art work was prepared by Kelly Evans. A special thank you to our families for their support during this project.

Finally, we would be grateful to hear from readers who discover any errors in this solutions manual.

Katy Murphy
Michael Sullivan, III
and
Michael Sullivan

For Pat and Yola

Contents

CHAPTER 5 POLYNOMIAL AND RATIONAL FUNCTIONS 193

CHAPTER 6 EXPONENTIAL AND LOGARITHMIC FUNCTIONS 270

CHAPTER 7 THE CONICS 306

CHAPTER 8 SYSTEMS OF EQUATIONS AND INEQUALITIES 342

CHAPTER 9 SEQUENCES; INDUCTION; COUNTING; PROBABILITY 479

APPENDIX: GRAPHING UTILITIES 508

Chapter 1

PRELIMINARIES

1.1 Real Numbers

1. $\frac{1}{3} = 0.333 \ldots$

3. $\frac{1}{8} = 0.125$

5. $-\frac{8}{5} = -1.6$

7. $\frac{1}{9} = 0.111 \ldots$

9. $\frac{4}{25} = 0.16$

11. $-\frac{3}{7} = -0.428571428571 \ldots$

13. $9 - 4 + 2 = 5 + 2 = 7$

15. $-6 + 4 \cdot 3 = -6 + 12 = 6$

17. $4 + 5 - 8 = 9 - 8 = 1$

19. $4 + \frac{1}{3} = \frac{4}{1} + \frac{1}{3} = \frac{4 \cdot 3 + 1 \cdot 1}{1 \cdot 3}$ by Equation (13)

$= \frac{12 + 1}{3} = \frac{13}{3}$

21. $6 - [3 \cdot 5 + 2 \cdot (3 - 2)]$
$6 - [3 \cdot 5 + 2 \cdot 1]$
$6 - [15 + 2]$
$6 - 17$
-11

23. $2 \cdot (3 - 5) + 8 \cdot 2 - 1$
$2 \cdot (-2) + 16 - 1$
$-4 + 16 - 1$
$12 - 1$
11

25. $10 - [6 - 2 \cdot 2 + (8 - 3)] \cdot 2$
$10 - [6 - 4 + 5] \cdot 2$
$10 - [2 + 5] \cdot 2$
$10 - 7 \cdot 2$
$10 - 14$
-4

27. $5 - 3 \cdot \frac{1}{2} = 5 - \frac{3 \cdot 1}{1 \cdot 2}$ by Equation (14)

$= \frac{5}{1} - \frac{3}{2}$

$= \frac{5}{1} + \frac{-3}{2}$ by Equation (10)

$= \frac{5 \cdot 2 + 1 \cdot (-3)}{1 \cdot 2}$ by Equation (13)

$= \frac{10 + (-3)}{2} = \frac{7}{2}$

29. $$\frac{4}{5} + \frac{8}{3} = \frac{4 \cdot 3 + 5 \cdot 8}{5 \cdot 3} \quad \text{by Equation (13)}$$
$$= \frac{12 + 40}{15} = \frac{52}{15}$$

31. $$\frac{\frac{16}{3} - \frac{1}{2}}{\frac{1}{3}} = \frac{\frac{16}{3} + \frac{-1}{2}}{\frac{1}{3}} \quad \text{by Equation (10)}$$
$$= \frac{\frac{16 \cdot 2 + 3 \cdot (-1)}{3 \cdot 2}}{\frac{1}{3}} \quad \text{by Equation (13)}$$
$$= \frac{\frac{32 + (-3)}{6}}{\frac{1}{3}}$$
$$= \frac{\frac{29}{6}}{\frac{1}{3}} = \frac{29}{6} \cdot \frac{3}{1} \quad \text{by Equation (15)}$$
$$= \frac{29 \cdot 3}{6 \cdot 1} = \frac{87}{6} = \frac{29}{2}$$

33. $$\begin{aligned} 6(x + 4) &= 6 \cdot x + 6 \cdot 4 && \text{by Equation (3a)} \\ &= 6x + 24 \end{aligned}$$

35. $$\begin{aligned} x(x - 4) &= x(x + (-4)) && \text{by Equation (6)} \\ &= x \cdot x + x \cdot (-4) && \text{by Equation (3a)} \\ &= x^2 - 4x \end{aligned}$$

37. $$\begin{aligned} (x + 2)(x + 4) &= (x + 2) \cdot x + (x + 2)4 && \text{by Equation (3a)} \\ &= x \cdot x + 2 \cdot x + x \cdot 4 + 2 \cdot 4 && \text{by Equation (3b)} \\ &= x^2 + 2x + 4x + 8 \\ &= x^2 + 6x + 8 \end{aligned}$$

39. $$\begin{aligned} (x - 2)(x + 1) &= (x - 2) \cdot x + (x - 2) \cdot 1 && \text{by Equation (3a)} \\ &= (x + (-2)) \cdot x + (x + (-2)) \cdot 1 \\ &= x \cdot x + (-2)x + x \cdot 1 + (-2) \cdot 1 && \text{by Equation (3b)} \\ &= x^2 - 2x + x - 2 \\ &= x^2 - x - 2 \end{aligned}$$

41. $$\begin{aligned} (x - 8)(x - 2) &= (x - 8)(x + (-2)) \\ &= (x - 8) \cdot x + (x - 8) \cdot (-2) && \text{by Equation (3a)} \\ &= (x + (-8)) \cdot x + (x + (-8)) \cdot (-2) \\ &= x \cdot x + (-8) \cdot x + x \cdot (-2) + (-8)(-2) && \text{by Equation (3b)} \\ &= x^2 - 8x - 2x + 16 \\ &= x^2 - 10x + 16 \end{aligned}$$

43. $(x + 2)(x - 2) = (x + 2)(x + (-2))$
$= (x + 2) \cdot x + (x + 2) \cdot (-2)$ by Equation (3a)
$= x \cdot x + 2 \cdot x + x \cdot (-2) + 2 \cdot (-2)$ by Equation (3b)
$= x^2 + 2x - 2x - 4$
$= x^2 - 4$

49. No. The fraction $\frac{1}{3} = 0.3333 \ldots$, with 3 repeating. Thus, $\frac{1}{3}$ is larger than 0.333 by 0.000333 ... .

51. No. $5 - 3 \neq 3 - 5$

53. No. $20 \div 4 \neq 4 \div 20$ or $\frac{20}{4} \neq \frac{4}{20}$ because $5 \neq \frac{1}{5}$

55. Symmetric property

57.

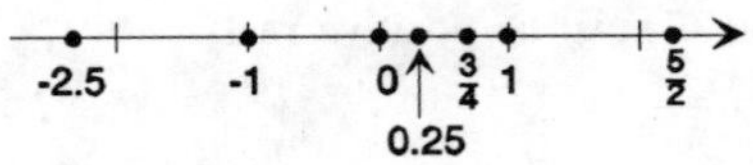

59. $x > 0$

61. $x < 3$

63. $|x| \leq 1$

65. Since $b < 0$, then $|b| = -b$, which is positive.

67. Since $b^2 = b \cdot b$ and $b < 0$, then b^2 is positive. (The product of two negative numbers is a positive number.)

69. If $a > 0$ and $b < 0$, then $-b > 0$ and $a - b = a + (-b)$ is positive.

71. $|a| + |b|$ is positive because both $|a|$ and $|b|$ are positive.

73. $b + |b| = b + (-b) = 0$

75. $|x + y| = |3 + (-2)| = |3 - 2|$
$= |1| = 1$

77. $|x| + |y| = |3| + |-2|$
$= 3 + 2 = 5$

79. $\frac{|x|}{x} = \frac{|3|}{3} = \frac{3}{3} = 1$

81. $|4x - 5y| = |4(3) - 5(-2)| = |12 + 10| = |22| = 22$

83. $||4x| - |5y|| = ||4(3)| - |5(-2)|| = ||12| - |-10|| = |12 - 10| = |2| = 2$

85. The domain of the variable x in the expression $\frac{4}{x - 5}$ is $\{x \mid x \neq 5\}$ since, if $x = 5$, the denominator becomes zero, which is not allowed.

87. The domain of the variable x in the expression $\frac{x}{x + 4}$ is $\{x \mid x \neq -4\}$ since, if $x = -4$, the denominator becomes zero.

89. The domain of the variable x in the expression $\frac{1}{x} + \frac{1}{x - 1}$ is $\{x \mid x \neq 0, x \neq 1\}$ since, if $x = 0$ or $x = 1$, the denominator becomes zero.

91. The domain of the variable x in the expression $\frac{x + 1}{x(x + 2)}$ is $\{x \mid x \neq 0, x \neq -2\}$ since, if $x = 0$ or $x = -2$, the denominator becomes zero.

93. The domain of the variable x in the expression $\frac{x+1}{x^2+4}$ is the set of all real numbers, since there is no x which causes the denominator to equal zero.

95. Area A is the product of length ℓ times width w

$A = \ell \cdot w$

$A = \ell w$

The variables A, ℓ, w must be positive real numbers.

97. C is π times d

$C = \pi \cdot d$

$C = \pi d$

The variables d and C must be positive real numbers.

99. A is $\frac{\sqrt{3}}{4}$ times the square of x.

$A = \frac{\sqrt{3}}{4} \cdot x^2$

$A = \frac{\sqrt{3}}{4}x^2$

The variables x and A must be positive real numbers.

101. V is $\frac{4}{3}$ times π times cube of r.

$V = \frac{4}{3} \cdot \pi \cdot r^3$

$V = \frac{4}{3}\pi r^3$

The variables r and V must be positive real numbers.

103. V is cube of x.

$V = x^3$

$V = x^3$

The variables x and V must be positive real numbers.

105. $C = \frac{5}{9}(F - 32)$

If $F = 32°$,

$C = \frac{5}{9}(32 - 32)$

$C = \frac{5}{9}(0)$

$C = 0°$

107. $C = \frac{5}{9}(F - 32)$

If $F = 77°$,

$C = \frac{5}{9}(77 - 32)$

$C = \frac{5}{9}(45)$

$C = 25°$

109. $C = 4000 + 2x$

(a) If $x = 1000$

$C = 4000 + 2(1000)$
$= 4000 + 2000$
$= \$6000$

(b) If $x = 2000$

$C = 4000 + 2(2000)$
$= 4000 + 4000$
$= \$8000$

111. $1 + 2 + 3 + 4 + \dots 99$

$= (1 + 99) + (2 + 98) + (3 + 97) \dots + (49 + 51) + 50$

$= \underbrace{100 + 100 + \dots + 100}_{49 \text{ times}} + 50$

$= 49 \cdot 100 + 50 = 4950$

113. $-2 - 4 - 6 - 8 \ldots - 98$

$$= (-2 - 98) + (-4 - 96) + (-6 - 94) + \ldots + (-48 - 52) - 50$$

$$= \underbrace{(-100) + (-100) + (-100) + \ldots + (-100)}_{24 \text{ times}} + (-50)$$

$$= 24 \cdot (-100) - 50 = -2450$$

115. $\frac{2}{3} \cdot 300 = 200$

117. Rational numbers are numbers that can be written in the form of $\frac{a}{b}$ where a and b are integers. Irrational numbers are non-terminating and non-repeating decimals. No, there are no real numbers that are both rational and irrational. No, there are no real numbers that are neither rational nor irrational.

119. $0.9999 \ldots = 1$

121. $(3.15)^2 = 9.9225$ or $(3.16)^2 = 9.9856$

1.2 Approximations; Calculators

1. (a) 18.953
   (b) 18.952

3. (a) 28.653
   (b) 28.653

5. (a) 0.063
   (b) 0.062

7. (a) 9.999
   (b) 9.998

9. 3/7 = 0.428571428
   (a) 0.429
   (b) 0.428

11. 521/15 = 34.7333 ...
   (a) 34.733
   (b) 34.733

13. Enter [8.51] Press [x^2]

Display: 8.51 72.4201

Thus, $(8.51)^2 \approx 72.42$

15. Enter [4.1] Press [+] Enter [3.2] Press [×]

Display 4.1 3.2

Enter [8.3] Press [=]

Display: 8.3 30.66

Thus, $4.1 + (3.2)(8.3) = 30.66$

17. Enter $\boxed{9.6}$ Press $\boxed{x^2}$ Press $\boxed{+}$ Enter $\boxed{6.4}$

Display: 9.6 92.16 6.4

Press $\boxed{x^2}$ Press $\boxed{=}$

Display: 40.96 133.12

Thus, $(9.6)^2 + (6.4)^2 = 133.12$

19. $8.6 + 10.2/4.2 \approx 11.02857143 \approx 11.03$

21. $\dfrac{2.3 - 9.25}{8.91 + 5.4} = \dfrac{6.95}{14.31} = -0.4856744 \approx -0.49$

23. $\dfrac{\pi + 8}{10.2 + 8.6} \approx 0.592637907 \approx 0.59$

25. $(22.6 + 8.5)/81.3 + 21.2 \approx 21.58253383 \approx 21.58$

27. $22.6 + 8.5/81.3 + 21.2 \approx 44.396556 \approx 43.90$

29. $(22.6 + 8.5)/(81.3 + 21.2) \approx 0.303414634 \approx 0.30$

31. $22.6 + 8.5/(81.3 + 21.2) \approx 22.68292683 \approx 22.68$

33. $[42.1(4.11 + 8.6) - 25.3]/7.6 \approx 67.07776 \approx 67.08$

35. $[(4.2 - 8.7)^2 + (5.1 - 2.77)^2]^2 \approx 659.40590521 \approx 659.41$

37. $\dfrac{5}{9} = 0.55555 \ldots \boxed{>} 0.5555$

39. $\sqrt{3} = 1.7320508 \ldots \boxed{>} 1.732$

41. $\dfrac{290}{1{,}000{,}000} = 0.00029$

43. $\dfrac{4}{5} \cdot 1000 = 800$

1.3 Integer Exponents

1. $4^2 = 16$

3. $4^{-2} = \dfrac{1}{4^2} = \dfrac{1}{16}$

5. $-4^{-2} = -\dfrac{1}{4^2} = -\dfrac{1}{16}$

7. $4^0 \cdot 2^{-3} = 1 \cdot \dfrac{1}{2^3} = 1 \cdot \dfrac{1}{8} = \dfrac{1}{8}$

9. $2^{-3} + \left(\dfrac{1}{2}\right)^3 = \dfrac{1}{2^3} = \dfrac{1^3}{2^3} = \dfrac{1}{8} + \dfrac{1}{8} = \dfrac{2}{8} = \dfrac{1}{4}$

11. $3^{-6} \cdot 3^4 = 3^{-6+4} = 3^{-2} = \dfrac{1}{3^2} = \dfrac{1}{9}$

13. $\dfrac{(3^2)^2}{(2^3)^2} = \dfrac{9^2}{8^2} = \dfrac{81}{64}$

15. $\left(\frac{2}{3}\right)^{-3} = \frac{1}{\left(\frac{2}{3}\right)^3} = \frac{1}{\frac{2^3}{3^3}} = \frac{3^3}{2^3} = \frac{27}{8}$

17. $\frac{2^3 \cdot 3^2}{2^4 \cdot 3^{-2}} = \frac{2^3}{2^4} \cdot \frac{3^2}{3^{-2}} = \frac{3^{2-(-2)}}{2^{4-3}} = \frac{3^4}{2^1} = \frac{81}{2}$

19. $\left(\frac{9}{2}\right)^{-2} = \frac{1}{\left(\frac{9}{2}\right)^2} = \frac{1}{\frac{9^2}{2^2}} = \frac{2^2}{9^2} = \frac{4}{81}$

21. $\frac{2^{-2}}{3} = \frac{\frac{1}{2^2}}{3} = \frac{\frac{1}{4}}{3} = \frac{1}{4} \cdot \frac{1}{3} = \frac{1}{12}$

23. $\frac{-3^{-1}}{2^{-1}} = \frac{-\frac{1}{3}}{\frac{1}{2}} = -\frac{1}{3} \cdot \frac{2}{1} = \frac{-2}{3}$

25. $x^0y^2 = 1 \cdot y^2 = y^2$

27. $xy^{-2} = x \cdot \frac{1}{y^2} = \frac{x}{y^2}$

29. $(8x^3)^{-2} = \frac{1}{(8x^3)^2} = \frac{1}{8^2 \cdot x^6} = \frac{1}{64x^6}$

31. $-4x^{-1} = -4 \cdot \frac{1}{x^1} = \frac{-4}{x}$

33. $3x^0 = 3 \cdot 1 = 3$

35. $\frac{x^{-2}y^3}{xy^4} = \frac{x^{-2}}{x} = \frac{y^3}{y^4} = x^{-2-1}y^{3-4} = x^{-3}y^{-1} = \frac{1}{x^3}\frac{1}{y} = \frac{1}{x^3y}$

37. $x^{-1}y^{-1} = \frac{1}{x}\frac{1}{y} = \frac{1}{xy}$

39. $\frac{x^{-1}}{y^{-1}} = \frac{\frac{1}{x}}{\frac{1}{y}} = \frac{1}{x} \cdot \frac{y}{1} = \frac{y}{x}$

41. $\left(\frac{4y}{5x}\right)^{-2} = \frac{1}{\left(\frac{4y}{5x}\right)^2} = \frac{1}{\frac{(4y)^2}{(5x)^2}} = \frac{(5x)^2}{(4y)^2} = \frac{5^2 \cdot x^2}{4^2 \cdot y^2} = \frac{25x^2}{16y^2}$

43. $x^{-2}y^{-2} = \frac{1}{x^2y^2}$

45. $\frac{x^{-1}y^{-2}z^3}{x^2yz^3} = \frac{x^{-1}}{x^2}\frac{y^{-2}}{y}\frac{z^3}{z^3} = x^{-1-2}y^{-2-1} = x^{-3}y^{-3} = \frac{1}{x^3}\frac{1}{y^3} = \frac{1}{x^3y^3}$

47. $\frac{(-2)^3x^4(yz)^2}{3^2xy^3z^4} = \frac{-8}{9}\frac{x^4}{x}\frac{y^2}{y^3}\frac{z^2}{z^4} = \frac{-8}{9}x^{4-1}y^{2-3}z^{2-4} = \frac{-8}{9}x^3y^{-1}z^{-2} = \frac{-8}{9}x^3\frac{1}{y}\frac{1}{z^2} = \frac{-8x^3}{9yz^2}$

49. $\frac{\left(\frac{x}{y}\right)^{-2} \cdot \left(\frac{y}{x}\right)^4}{x^2y^3} = \frac{\frac{x^{-2}}{y^{-2}} \cdot \frac{y^4}{x^4}}{x^2y^3} = \frac{\frac{x^{-2}}{x^4}\frac{y^4}{y^{-2}}}{x^2y^3} = \frac{x^{-2-4}y^{4-(-2)}}{x^2y^3} = \frac{x^{-6}}{x^2}\frac{y^6}{y^3} = x^{-6-2}y^{6-3}$

$= x^{-8}y^3 = \frac{y^3}{x^8}$

51. $\left(\dfrac{3x^{-1}}{4y^{-1}}\right)^{-2} = \dfrac{1}{\left(\dfrac{3x^{-1}}{4y^{-1}}\right)^{2}} = \dfrac{1}{\dfrac{(3x^{-1})^2}{(4y^{-1})^2}} = \dfrac{(4y^{-1})^2}{(3x^{-1})^2} = \dfrac{4^2y^{-2}}{3^2x^{-2}} = \dfrac{16}{9}\dfrac{\frac{1}{y^2}}{\frac{1}{x^2}} = \dfrac{16}{9}\dfrac{x^2}{y^2} = \dfrac{16x^2}{9y^2}$

53. $\dfrac{(xy^{-1})^{-2}}{xy^3} = \dfrac{x^{-2}y^2}{xy^3} = \dfrac{x^{-2}}{x}\dfrac{y^2}{y^3} = x^{-2-1}y^{2-3} = x^{-3}y^{-1} = \dfrac{1}{x^3y}$

55. $\left(\dfrac{x}{y^2}\right)^{-2} \cdot (y^2)^{-1} = \dfrac{(x)^{-2}}{(y^2)^{-2}}\dfrac{1}{y^2} = \dfrac{x^{-2}}{y^{-4}y^2} = \dfrac{x^{-2}}{y^{-2}} = \dfrac{\frac{1}{x^2}}{\frac{1}{y^2}} = \dfrac{y^2}{x^2}$

57. If $x = 2$, $2x^3 - 3x^2 + 5x - 4 = 2(2)^3 - 3(2)^2 + 5(2) - 4 = 2 \cdot 8 - 3 \cdot 4 + 10 - 4 = 10$
If $x = 1$, $2x^3 - 3x^2 + 5x - 4 = 2(1)^3 - 3(1)^2 + 5(1) - 4 = 2(1) - 3 \cdot 1 + 5 - 4 = 0$

59. $\dfrac{(666)^4}{(222)^4} = \left(\dfrac{666}{222}\right)^4 = 3^4 = 81$

61. $(8.2)^6 \approx 304{,}006.671$

63. $(6.1)^{-3} \approx 0.004$

65. $(-2.8)^6 \approx 481.890$

67. $(-8.11)^{-4} \approx 0.000$

69. $454.2 = 4.542 \times 10^2$

71. $0.013 = 1.3 \times 10^{-2}$

73. $32{,}155 = 3.2155 \times 10^4$

75. $0.000423 = 4.23 \times 10^{-4}$

77. $6.15 \times 10^4 = 6.1500 \times 10^4 = 61{,}500$

79. $1.214 \times 10^{-3} = 0001.214 \times 10^{-3}$
$= 0.001214$

81. $1.1 \times 10^8 = 1.10000000 \times 10^8$
$= 110{,}000{,}000$

83. $8.1 \times 10^{-2} = 008.1 \times 10^{-2} = 0.081$

85. $\dfrac{60 \text{ sec}}{1 \text{ min}} \cdot \dfrac{60 \text{ min}}{1 \text{ hr}} \cdot \dfrac{24 \text{ hr}}{1 \text{ day}} \cdot \dfrac{365 \text{ days}}{1 \text{ year}} \cdot \dfrac{186000 \text{ miles}}{1 \text{ sec}} = 5.8657 \times 10^{12}$ miles

1.4 Polynomials

1. $3x^2 - 5$ is a polynomial of degree 2.

3. 5 is a polynomial of degree 0.

5. $3x^2 - \dfrac{5}{x}$ is not a polynomial due to the $5x^{-1}$ term.

7. $2y^3 - \sqrt{2}$ is a polynomial of degree 3.

9. $\dfrac{x^2 + 5}{x^3 - 1}$ is not a polynomial since it is the quotient of two polynomials and the denominator polynomial has degree greater than one.

11. $(x^2 + 4x + 5) + (3x - 3)$
$= x^2 + (4x + 3x) + (5 - 3)$
$= x^2 + 7x + 2$

13. $(x^3 - 2x^2 + 5x + 10) - (2x^2 - 4x + 3) = x^3 - 2x^2 + 5x + 10 - 2x^2 + 4x - 3$
$= x^3 + (-2x^2 - 2x^2) + (5x + 4x) + (10 - 3)$
$= x^3 - 4x^2 + 9x + 7$

15. $(6x^5 + x^3 + x) + (5x^4 - x^3 + 3x^2) = 6x^5 + 5x^4 + (x^3 - x^3) + 3x^2 + x$
$= 6x^5 + 5x^4 + 3x^2 + x$

17. $3(x^2 - 3x + 1) + 2(3x^2 + x - 4)$
$3x^2 - 9x + 3 + 6x^2 + 2x - 8$
$(3x^2 + 6x^2) + (-9x + 2x) + (3 - 8)$
$9x^2 - 7x - 5$

19. $6(x^3 + x^2 - 3) - 4(2x^3 - 3x^2)$
$6x^3 + 6x^2 - 18 - 8x^3 + 12x^2$
$(6x^3 - 8x^3) + (6x^2 + 12x^2) - 18$
$-2x^3 + 18x^2 - 18$

21. $(x^2 - x + 2) + (2x^2 - 3x + 5) - (x^2 + 1)$
$x^2 - x + 2 + 2x^2 - 3x + 5 - x^2 - 1$
$(x^2 + 2x^2 - x^2) + (-x - 3x) + (2 + 5 - 1)$
$2x^2 - 4x + 6$

23. $9(y^2 - 3y + 4) - 6(1 - y^2)$
$9y^2 - 27y + 36 - 6 + 6y^2$
$(9y^2 + 6y^2) + (-27y) + (36 - 6)$
$15y^2 - 27y + 30$

25. $(x + a)^2 - x^2 = x^2 + 2xa + a^2 - x^2$ by (3a)
$= (x^2 - x^2) + 2ax + a^2$
$= 2ax + a^2$

27. $(x + a)^3 - x^3 = x^3 + 3ax^2 + 3a^2x + a^3 - x^3$ by (5a)
$= x^3 - x^3 + 3ax^2 + 3a^2x + a^3$
$= 3ax^2 + 3a^2x + a^3$

29. $(x + 5)(x - 3) = x^2 + 2x - 15$ by (4a)

31. $(x + 8)(2x + 1) = 2x^2 + 17x + 8$ by (4b)

33. $(-3x + 1)(x + 4) = -3x^2 - 11x + 4$ by (4b)

35. $(1 - 4x)(2 - 3x) = (-4x + 1)(-3x + 2) = 12x^2 - 11x + 2$ by (4b)

37. $(2x - 5)(2x + 5) = (2x)^2 - 5^2 = 4x^2 - 25$

39. $(2x - 5)^2 = (2x)^2 - 2(2x)(5) + 5^2 = 4x^2 - 20x + 25$

41. $(x - 1)(x^2 + x + 1) = x^3 - 1$ by (6)

43. $(2x + 5)(x^2 - 2) = 2x(x^2 - 2) + 5(x^2 - 2)$ by distributive law

$= 2x \cdot x^2 - 2x \cdot 2 + 5 \cdot x^2 - 5 \cdot 2$ by distributive law

$= 2x^3 - 4x + 5x^2 - 10$

$= 2x^3 + 5x^2 - 4x - 10$

45. $(2x - 3)(3x^2 - 2x + 1) = 2x(3x^2 - 2x + 1) - 3(3x^2 - 2x + 1)$

$= 6x^3 - 4x^2 + 2x - 9x^2 + 6x - 3$

$= 6x^3 - 13x^2 + 8x - 3$

47. $(3x + 4y)(3x - 4y) = (3x)^2 - (4y)^2 = 9x^2 - 16y^2$

49. $(x + y)(x - 2y) = x(x - 2y) + y(x - 2y)$

$= x \cdot x + x \cdot (-2y) + y \cdot x + y \cdot (-2y)$

$= x^2 - 2xy + xy - 2y^2$

$= x^2 - xy - 2y^2$

51. $(x^2 + 2x + y^2) + (x + y - 3y^2) = x^2 + (2x + x) + y + (y^2 - 3y^2)$

$= x^2 + 3x + y - 2y^2$

$= x^2 - 2y^2 + 3x + y$

53. $(x - y)^2 - (x + y)^2 = x^2 - 2xy + y^2 - (x^2 + 2xy + y^2)$ by (3a) and (3b)

$= x^2 - 2xy + y^2 - x^2 - 2xy - y^2$

$= (x^2 - x^2) + (-2xy - 2xy) + (y^2 - y^2)$

$= -4xy$

55. $(x + 2y)^2 + (x - 3y)^2 = x^2 + 2x(2y) + (2y)^2 + x^2 - 2x(3y) + (3y)^2$ by (3a) and (3b)

$= x^2 + x^2 + 4xy - 6xy + 4y^2 + 9y^2$

$= 2x^2 - 2xy + 13y^2$

57. $[(x + y)^2 + z^2] + [x^2 + (y + z)^2] = [x^2 + 2xy + y^2 + z^2] + [x^2 + y^2 + 2yz + z^2]$

$= 2x^2 + 2xy + 2y^2 + 2z^2 + 2yz$

$= 2x^2 + 2y^2 + 2z^2 + 2xy + 2yz$

59. $(x - y)(x^2 + xy + y^2) = x(x^2 + xy + y^2) - y(x^2 + xy + y^2)$

$= x \cdot x^2 + x \cdot xy + x \cdot y^2 - y \cdot x^2 - y \cdot xy - y \cdot y^2$

$= x^3 + x^2y + xy^2 - x^2y - xy^2 - y^3$

$= x^3 - y^3$

61. $(x + 2)^2(x - 2) = (x^2 + 4x + 4)(x - 2)$ by 3a

$= x^2(x - 2) + 4x(x - 2) + 4(x - 2)$

$= x^2 \cdot x + x^2 \cdot (-2) + 4x \cdot x + 4x \cdot (-2) + 4 \cdot x + 4 \cdot (-2)$

$= x^3 - 2x^2 + 4x^2 - 8x + 4x - 8$

$= x^3 + 2x^2 - 4x - 8$

63. $(x-1)^2(2x+3) = (x^2-2x+1)(2x+3)$ by (3b)

$= x^2(2x+3) - 2x(2x+3) + 1(2x+3)$

$= x^2 \cdot 2x + x^2 \cdot 3 - 2x \cdot 2x - 2x \cdot 3 + 1 \cdot 2x + 1 \cdot 3$

$= 2x^3 + 3x^2 - 4x^2 - 6x + 2x + 3$

$= 2x^3 - x^2 - 4x + 3$

65. $(x-1)(x-2)(x-3) = (x^2-3x+2)(x-3)$ by (4a)

$= x^2(x-3) - 3x(x-3) + 2(x-3)$

$= x^2 \cdot x + x^2 \cdot (-3) - 3x \cdot x - 3x \cdot (-3) + 2 \cdot x + 2 \cdot (-3)$

$= x^3 - 3x^2 - 3x^2 + 9x + 2x - 6$

$= x^3 - 6x^2 + 11x - 6$

67. $(x+1)^3 - (x-1)^3 = (x^3+3x^2+3x+1) - (x^3-3x^2+3x-1)$ by (5a) and (5b)

$= x^3 + 3x^2 + 3x + 1 - x^3 + 3x^2 - 3x + 1$

$= 6x^2 + 2$

69. $(x-1)^2(x+1)^2 = (x^2-2x+1)(x^2+2x+1)$ by (3a) and (3b)

$= x^2(x^2+2x+1) - 2x(x^2+2x+1) + 1(x^2+2x+1)$

$= x^2 \cdot x^2 + x \cdot 2x + x^2 \cdot 1 - 2x \cdot x^2 - 2x \cdot 2x - 2x$

$\cdot 1 + 1 \cdot x^2 + 1 \cdot 2x + 1 \cdot 1$

$= x^4 + 2x^3 + x^2 - 2x^3 - 4x^2 - 2x + x^2 + 2x + 1$

$= x^4 - 2x^2 + 1$

71. $(2x+3)^3 = 8x^3 + 3 \cdot 3(2x)^2 + 3 \cdot 9(2x) + 27$

$= 8x^3 + 36x^2 + 54x + 27$

73. $(2x-3a)^3 = 8x^3 - 3 \cdot (3a)(2x)^2 + 3(3a)^2(2x) - 27a^3$

$= 8x^3 - 36ax^2 + 54a^2x - 27a^3$

75. $(x+y+z)(x-y-z)$

$= x(x-y-z) + y(x-y-z) + z(x-y-z)$

$= x \cdot x + x \cdot (-y) + x \cdot (-z) + y \cdot x + y \cdot (-y) + y \cdot (-z) + z \cdot x + z \cdot (-y)$

$+ z \cdot (-z)$

$= x^2 - xy - xz + xy - y^2 - yz + xz - yz - z^2$

$= x^2 - y^2 - z^2 - 2yz$

77.

$$\begin{array}{r|l} & 4x^2 - 3x + 1 \\ \hline x & 4x^3 - 3x^2 + x + 1 \\ & \underline{4x^3} \\ & -3x^2 \\ & \underline{-3x^2} \\ & x \\ & \underline{x} \\ & 1 \end{array}$$

The quotient is $4x^2 - 3x + 1$; the remainder is 1.

Check: $x(4x^2 - 3x + 1) + 1 = 4x^3 - 3x + x + 1$

79.

$$\begin{array}{r|l} & 4x^2 - 11x + 23 \\ \hline x + 2 & 4x^3 - 3x^2 + x + 1 \\ & \underline{4x^3 + 8x^2} \\ & \quad -11x^2 + x \\ & \quad \underline{-11x^2 - 22x} \\ & \qquad 23x + 1 \\ & \qquad \underline{23x + 46} \\ & \qquad\quad -45 \end{array}$$

The quotient is $4x^2 - 11x + 23$; the remainder is -45.

Check: $(x + 2)(4x^2 - 11x + 23) + (-45)$
$= 4x^3 - 11x^2 + 23x + 8x^2 - 22x + 46 + (-45)$
$= 4x^3 - 3x^2 + x + 1$

81.

$$\begin{array}{r|l} & 4x^2 + 13x + 53 \\ \hline x - 4 & 4x^3 - 3x^2 + x + 1 \\ & \underline{4x^3 - 16x^2} \\ & \quad 13x^2 + x \\ & \quad \underline{13x^2 - 52x} \\ & \qquad 53x + 1 \\ & \qquad \underline{53x - 212} \\ & \qquad\quad 213 \end{array}$$

The quotient is $4x^2 + 13x + 53$; the remainder is 213.

Check: $(x - 4)(4x^2 + 13x + 53) + 213$
$= 4x^3 + 13x^2 + 53x - 16x^2 - 52x - 212 + 213$
$= 4x^3 - 3x^2 + x + 1$

83.

$$\begin{array}{r|l} & 4x - 3 \\ \hline x^2 & 4x^3 - 3x^2 + x + 1 \\ & \underline{4x^3} \\ & \quad -3x^2 \\ & \quad \underline{-3x^2} \\ & \qquad x + 1 \end{array}$$

The quotient is $4x - 3$; the remainder is $x + 1$.

Check: $x^2(4x - 3) + (x + 1) = 4x^3 - 3x^2 + x + 1$

85.

$$\begin{array}{r|l} & 4x - 3 \\ \hline x^2 + 2 & 4x^3 - 3x^2 + x + 1 \\ & \underline{4x^3 \qquad + 8x} \\ & \quad -3x^2 - 7x + 1 \\ & \quad \underline{-3x^2 \qquad - 6} \\ & \qquad -7x + 7 \end{array}$$

The quotient is $4x - 3$; the remainder is $-7x + 7$.

Check: $(x^2 + 2)(4x - 3) + (-7x + 7)$
$= 4x^3 - 3x + 8x - 6 - 7x + 7$
$= 4x^3 - 3x^2 + x + 1$

87.

$$\begin{array}{r|l} & 2 \\ \hline 2x^3 - 1 & 4x^3 - 3x^2 + x + 1 \\ & \underline{4x^3 \qquad\qquad - 2} \\ & \quad -3x^2 + x + 3 \end{array}$$

The quotient is 2; the remainder is $-3x^2 + x + 3$.

Check: $(2x^3 - 1)(2) + (-3x^2 + x + 3)$
$= 4x^3 - 2 - 3x^2 + x + 3 = 4x^3 - 3x^2 + x + 1$

89.

$$\begin{array}{r|l} & 2x - \frac{5}{2} \\ \hline 2x^2 + x + 1 & 4x^3 - 3x^2 + x + 1 \\ & \underline{4x^3 + 2x^2 + 2x} \\ & \quad -5x^2 - x + 1 \\ & \quad \underline{-5x^2 - \frac{5}{2}x - \frac{5}{2}} \\ & \qquad \frac{3}{2}x + \frac{7}{2} \end{array}$$

The quotient is $2x - \frac{5}{2}$; the remainder is $\frac{3}{2}x + \frac{7}{2}$

Check: $\left(2x - \frac{5}{2}\right)(2x^2 + x + 1) + \frac{3}{2}x + \frac{7}{2}$
$= 4x^3 + 2x^2 + 2x - 5x^2 - \frac{5}{2}x - \frac{5}{2} + \frac{3}{2}x + \frac{7}{2}$
$= 4x^3 - 3x^2 + x + 1$

91. $$\begin{array}{r} x - \frac{3}{4} \\ 4x^2 + 1\overline{)4x^3 - 3x^2 + x + 1} \\ \underline{4x^3 \qquad\quad + x \quad} \\ -3x^2 \qquad + 1 \\ \underline{-3x^2 \qquad - \frac{3}{4}} \\ \frac{7}{4} \end{array}$$

The quotient is $x - \frac{3}{4}$; the remainder is $\frac{7}{4}$.

Check: $(4x^2 + 1)\left[x - \frac{3}{4}\right] + \frac{7}{4}$
$= 4x^3 - 3x^2 + x - \frac{3}{4} + \frac{7}{4}$
$= 4x^3 - 3x^2 + x + 1$

93. $$\begin{array}{r} x^3 + x^2 + x + 1 \\ x - 1\overline{)x^4 + 0x^3 + 0x^2 + 0x - 1} \\ \underline{x^4 - x^3 \qquad\qquad\qquad\qquad} \\ x^3 \qquad\qquad\qquad\quad \\ \underline{x^3 - x^2 \qquad\qquad\quad} \\ x^2 \qquad\qquad \\ \underline{x^2 - x \qquad} \\ x - 1 \\ \underline{x - 1} \\ 0 \end{array}$$

The quotient is $x^3 + x^2 + x + 1$; the remainder is 0.

Check: $(x - 1)(x^3 + x^2 + x + 1)$
$= x^4 + x^3 + x^2 + x - x^3 - x^2 - x - 1$
$= x^4 - 1$

95. $$\begin{array}{r} x^2 + 1 \\ x^2 - 1\overline{)x^4 + 0x^3 + 0x^2 + 0x - 1} \\ \underline{x^4 \qquad\quad - x^2 \qquad\qquad} \\ x^2 \qquad\quad - 1 \\ \underline{x^2 \qquad\quad - 1} \\ 0 \end{array}$$

The quotient is $x^2 + 1$; the remainder is 0.

Check: $(x^2 - 1)(x^2 + 1) + 0 = x^4 + x^2 - x^2 - 1$
$= x^4 - 1$

97. $$\begin{array}{r} -4x^2 - 3x - 3 \\ x - 1\overline{)-4x^3 + x^2 + 0x - 4} \\ \underline{-4x^3 + 4x^2 \qquad\qquad\quad} \\ -3x^2 + 0x \qquad \\ \underline{-3x^2 + 3x \qquad} \\ -3x - 4 \\ \underline{-3x + 3} \\ -7 \end{array}$$

The quotient is $-4x^2 - 3x - 3$; the remainder is -7.

Check: $(x - 1)(-4x^2 - 3x - 3) + (-7)$
$= -4x^3 - 3x^2 - 3x + 4x^2 + 3x + 3 - 7$
$= -4x^3 + x^2 - 4$

99. $1 - x^2 + x^4 = x^4 - x^2 + 1$

$$\begin{array}{r} x^2 - x - 1 \\ x^2 + x + 1\overline{)x^4 + 0x^3 - x^2 + 0x + 1} \\ \underline{x^4 + x^3 + x^2 \qquad\qquad\qquad} \\ -x^3 - 2x^2 + 0x \qquad \\ \underline{-x^3 - x^2 - x \qquad} \\ x^2 + x + 1 \\ \underline{x^2 - x - 1} \\ 2x + 2 \end{array}$$

The quotient is $x^2 - x - 1$; the remainder is $2x + 2$.

Check: $(x^2 + x + 1)(x^2 - x - 1) + 2x + 2$
$= x^4 - x^3 - x^2 + x^3 - x^2 - x + x^2 - x - 1$
$+ 2x + 2$
$= x^4 - x^2 + 1$

101. $1 - x^2 = -x^2 + 1$;
$1 - x^2 + x^4 = x^4 + 0x^3 - x^2 + 0x + 1$

$$\begin{array}{r|l} & -x^2 \\ \hline -x^2 + 1 & x^4 + 0x^3 - x^2 + 0x + 1 \\ & \underline{x^4 \qquad\quad - x^2} \\ & \qquad\qquad\qquad\qquad 1 \end{array}$$

The quotient is $-x^2$; the remainder is 1.

Check: $-x^2(-x^2 + 1) + 1 = x^4 - x^2 + 1$

103.
$$\begin{array}{r|l} & x^2 + ax + a^2 \\ \hline x - a & x^3 + 0x^2 + 0x - a^3 \\ & \underline{x^3 - ax^2} \\ & \quad ax^2 + 0x \\ & \quad \underline{ax^2 - a^2x} \\ & \qquad a^2x - a^3 \\ & \qquad \underline{a^2x - a^3} \\ & \qquad\qquad 0 \end{array}$$

The quotient is $x^2 + ax + a^2$; the remainder is 0.

Check: $(x - a)(x^2 + ax + a^2)$
$= x^3 + ax^2 + a^2x - ax^2 - a^2x - a^3$
$= x^3 - a^3$

105.
$$\begin{array}{r|l} & x^3 + ax^2 + a^2x + a^3 \\ \hline x - a & x^4 + 0x^3 + 0x^2 + 0x - a^4 \\ & \underline{x^4 - ax^3} \\ & \quad ax^3 + 0x^2 \\ & \quad \underline{ax^3 - a^2x^2} \\ & \qquad a^2x^2 + 0x \\ & \qquad \underline{a^2x^2 - a^3x} \\ & \qquad\quad a^3x - a^4 \\ & \qquad\quad \underline{a^3x - a^4} \\ & \qquad\qquad 0 \end{array}$$

The quotient is $x^3 + ax^2 + a^2x + a^3$; the remainder is 0.

Check: $(x - a)(x^3 + ax^2 + a^2x + a^3)$
$= x^4 + ax^3 + a^2x^2 - a^3x - ax^3 - a^2x^2$
$- a^3x - a^4$
$= x^4 - a^4$

107. The degree of the product of two polynomials equals the degree of the product of the leading terms:
$$(a_nx^n + a_{n-1}x^{n-1} + \dots + a_1x + a_0)(b_mx^m + b_{m-1}x^{m-1} + \dots + b_1x + b_0)$$
$$= a_nb_mx^{n+m} + (a_nb_{m-1} + a_{n-1}b_m)x^{n+m-1} + \dots + (a_1b_0 + a_0b_1)x + a_0b_0$$
The largest exponent will be $n + m$ where n is the degree of the first polynomial and m is the degree of the second polynomial.

109. Derive (2): $(x - a)(x + a) = x(x + a) - a(x + a)$
$= x^2 + ax - ax - a^2$
$= x^2 - a^2$

Derive (3a): $(x + a)^2 = (x + a)(x + a)$
$= x(x + a) + a(x + a)$
$= x^2 + ax + ax + a^2$
$= x^2 + 2ax + a^2$

Derive (3b): $(x - a)^2 = (x - a)(x - a)$
$= x(x - a) - a(x - a)$
$= x^2 - ax - ax - a(-a)$
$= x^2 - 2ax + a^2$

Derive (6): $(x - a)(x^2 + ax + a^2) = x(x^2 + ax + a^2) - a(x^2 + ax + a^2)$
$= x^3 + ax^2 + a^2x - ax^2 - a^2x - a^3$
$= x^3 + (ax^2 - ax^2) + (a^2x - a^2x) - a^3$
$= x^3 - a^3$

111. $(x + a)^4 = (x + a)^2(x + a)^2$
$$= (x^2 + 2ax + a^2)(x^2 + 2ax + a^2)$$
$$= x^2(x^2 + 2ax + a^2) + 2ax(x^2 + 2ax + a^2) + a^2(x^2 + 2ax + a^2)$$
$$= x^4 + 2ax^3 + a^2x^2 + 2ax^3 + 4a^2x^2 + 2a^3x + a^2x^2 + 2a^3x + a^4$$
$$= x^4 + 4ax^3 + 6a^2x^2 + 4a^3x + a^4$$

115.
$$\begin{array}{r|l} & x^2 - 4x + 11 \\ \hline x + 2 & x^3 - 2x^2 + 3x + 5 \\ & \underline{x^3 + 2x^2} \\ & \quad -4x^2 + 3x \\ & \quad \underline{-4x^2 - 8x} \\ & \qquad 11x + 5 \\ & \qquad \underline{11x + 22} \\ & \qquad\quad -17 \end{array}$$

$$\frac{x^3 - 2x^2 + 3x + 5}{x + 2} = x^2 - 4x + 11 - \frac{17}{x + 2}$$
$$x^2 - 4x + 11 - \frac{17}{x + 2} = ax^2 + bx + c + \frac{d}{x + 2}$$
$a = 1$, $b = -4$, $c = 11$, $d = -17$
The sum of $a + b + c + d = 1 - 4 + 11 - 17 = -9$

1.5 Factoring Polynomials

1. $3x + 6 = 3(x + 2)$

3. $ax^2 + a = a(x^2 + 1)$

5. $x^3 + x^2 + x = x(x^2 + x + 1)$

7. $2x^2 - 2x = 2x(x - 1)$

9. $3x^2y - 6xy^2 + 12xy = 3xy(x - 2y + 4)$

11. $x^2 - 1 = x^2 - 1^2 = (x - 1)(x + 1)$ by (2)

13. $1 - 4x^2 = 1^2 - (2x)^2 = (1 - 2x)(1 + 2x) = -(2x - 1)(2x + 1)$ by (2)

15.

Factors of 10	1,10	−1,−10	2,5	−2,−5
Sum	11	−11	7	−7

$x^2 + 7x + 10 = (x + 2)(x + 5)$

17.

Factors of 21	1,21	−1,−21	3,7	−3,−7
Sum	22	−22	10	−10

$x^2 - 10x + 21 = (x - 7)(x - 3)$

19.

Factors of −8	1,−8	−1,8	2,−4	−2,4
Sum	−7	7	−2	2

$x^2 - 2x - 8 = (x + 2)(x - 4)$

21. $x^2 + 2x + 1 = (x + 1)^2$ by (3a)

23. $x^2 - 4x + 4 = x^2 - 2(2x) + 2^2 = (x - 2)^2$ by (3a)

25. $$15 + 2x - x^2 = \begin{cases} (1 \quad x)(15 \quad x) \\ (3 \quad x)(5 \quad x) \end{cases}$$

$$\text{(signs must alternate)} = \begin{cases} (1 - x)(15 + x) \\ (1 + x)(15 - x) \\ (3 - x)(5 + x) \\ (3 + x)(5 - x) \end{cases}$$

$$= \begin{cases} 15 - 14x - x^2 \\ 15 + 14x - x^2 \\ 15 - 2x - x^2 \\ 15 + 2x - x^2 \end{cases}$$

Thus, $15 + 2x - x^2 = (3 + x)(5 - x) = -(x - 5)(x + 3)$

27. $3x^2 - 12x - 36 = 3(x^2 - 4x - 12)$

$$\text{where } x^2 - 4x - 12 = (x \quad)(x \quad) = \begin{cases} (x \quad 1)(x \quad 12) \\ (x \quad 2)(x \quad 6) \\ (x \quad 3)(x \quad 4) \end{cases}$$

$$\text{(signs must alternate)} = \begin{cases} (x + 1)(x - 12) \\ (x - 1)(x + 12) \\ (x + 2)(x - 6) \\ (x - 2)(x + 6) \\ (x + 3)(x - 4) \\ (x - 3)(x + 4) \end{cases}$$

$$= \begin{cases} x^2 - 11x - 12 \\ x^2 + 11x - 12 \\ x^2 - 4x - 12 \\ x^2 + 4x - 12 \\ x^2 - x - 12 \\ x^2 + x - 12 \end{cases}$$

Thus, $x^2 - 4x - 12 = (x + 2)(x - 6)$ and
$3x^2 - 12x - 36 = 3(x^2 - 4x - 12) = 3(x + 2)(x - 6)$

29.

Factors of 30	1,30	−1,−30	2,15	−2,−15	3,10	−3,−10	5,6	−5,−6
Sum	31	−31	17	−17	13	−13	11	−11

$y^4 + 11y^3 + 30y^2 = y^2(y + 11y + 30) = y^2(y + 5)(y + 6)$

31. $16x^2 + 8x + 1 = (4x)^2 + 2(1)(4x) + 1^2 = (4x + 1)^2$ by (3a)

33. $4x^2 + 12x + 9 = (2x)^2 + 2(3)(2x) + 3^2 = (2x + 3)^2$ by (3a)

35.

Factors of $-45a^2$	$1a,-45a$	$-1a,45a$	$3a,-15a$	$-3a,15a$	$5a,-9a$	$-5a,9a$
Sum	$-44a$	$44a$	$-12a$	$12a$	$-4a$	$4a$

$ax^2 - 4a^2x - 45a^3 = a(x^2 - 4ax - 45a^2) = a(x + 5a)(x - 9a)$

37. $x^3 - 125 = x^3 - 5^3 = (x - 5)(x^2 + 5x + 25)$ by (6)

39. $27 - 8x^3 = (3)^3 - (2x)^3 = (3 - 2x)(9 + 6x + 4x^2) = -(2x - 3)(4x^2 + 6x + 9)$

41. $2x^4 + 16x = 2x(x^3 + 8) = 2x(x + 2)(x^2 - 2x + 4)$ by (7)

43. $3x^2 + 4x + 1 = (3x + 1)(x + 1)$

45. $x^4 - 81 = (x^2)^2 - 9^2 = (x^2 - 9)(x^2 + 9)$ by (2)
$= (x + 3)(x - 3)(x^2 + 9)$ by (2)

47. $x^6 - 2x^3 + 1 = (x^3)^2 - 2(x^3) + 1 = (x^3 - 1)^2$ by (3b)
$= [(x - 1)(x^2 + x + 1)]^2$
$= (x - 1)^2(x^2 + x + 1)^2$

49. $x^7 - x^5 = x^5(x^2 - 1) = x^5(x + 1)(x - 1)$ by (2)

51. $2z^2 + 5z + 3 = (2z \quad)(z \quad) = \begin{cases} (2z \quad 1)(z \quad 3) \\ (2z \quad 3)(z \quad 1) \end{cases}$

(all signs positive) $= \begin{cases} (2z + 1)(z + 3) \\ (2z + 3)(z + 1) \end{cases}$

$= \begin{cases} 2z^2 + 7z + 3 \\ 2z^2 + 5z + 3 \end{cases}$

Thus, $2z^2 + 5z + 3 = (2z + 3)(z + 1)$

53. $16x^2 + 24x + 9 = (4x)^2 + 2(3)(4x) + 3^2 = (4x + 3)^2$ by (3b)

55. $5 + 16x - 16x^2 = \begin{cases} (1 \quad x)(5 \quad 16x) \\ (1 \quad 16x)(5 \quad x) \\ (1 \quad 2x)(5 \quad 8x) \\ (1 \quad 8x)(5 \quad 2x) \\ (1 \quad 4x)(5 \quad 4x) \end{cases}$

(signs must alternate) $= \begin{cases} (1 + x)(5 - 16x) \\ (1 - x)(5 + 16x) \\ (1 + 16x)(5 - x) \\ (1 - 16x)(5 + x) \\ (1 + 2x)(5 - 8x) \\ (1 - 2x)(5 + 8x) \\ (1 + 8x)(5 - 2x) \\ (1 - 8x)(5 + 2x) \\ (1 + 4x)(5 - 4x) \\ (1 - 4x)(5 + 4x) \end{cases}$

$= \begin{cases} 5 - 11x - 16x^2 \\ 5 + 11x - 16x^2 \\ 5 + 79x - 16x^2 \\ 5 - 79x - 16x^2 \\ 5 + 2x - 16x^2 \\ 5 - 2x - 16x^2 \\ 5 + 38x - 16x^2 \\ 5 - 38x - 16x^2 \\ 5 + 16x - 16x^2 \\ 5 - 16x - 16x^2 \end{cases}$

Thus, $5 + 16x - 16x^2 = (5 - 4x)(1 + 4x) = -(4x - 5)(4x + 1)$

57. $$4y^2 - 16y + 15 = \begin{cases} (y \quad\;)(4y \quad\;) \\ (2y \quad\;)(2y \quad\;) \end{cases}$$

$$\text{(all signs negative)} = \begin{cases} (\;y - 1)(4y - 15) \\ (\;y - 15)(4y - 1) \\ (\;y - 3)(4y - 5) \\ (\;y - 5)(4y - 3) \\ (2y - 1)(2y - 15) \\ (2y - 3)(2y - 5) \end{cases}$$

$$= \begin{cases} 4y^2 - 19y + 15 \\ 4y^2 - 61y + 15 \\ 4y^2 - 17y + 15 \\ 4y^2 - 23y + 15 \\ 4y^2 - 32y + 15 \\ 4y^2 - 16y + 15 \end{cases}$$

Thus, $4y^2 - 16y + 15 = (2y - 3)(2y - 5)$

59. $18x^2 - 9x - 27 = 9(2x^2 - x - 3)$
where $2x^2 - x - 3 = (2x \quad\;)(x \quad\;)$

$$= \begin{cases} (2x \quad 1)(x \quad 3) \\ (2x \quad 3)(x \quad 1) \end{cases}$$

$$\text{(signs must alternate)} = \begin{cases} (2x + 1)(x - 3) \\ (2x - 1)(x + 3) \\ (2x + 3)(x - 1) \\ (2x - 3)(x + 1) \end{cases}$$

$$= \begin{cases} 2x^2 - 5x - 3 \\ 2x^2 + 5x - 3 \\ 2x^2 + x - 3 \\ 2x^2 - x - 3 \end{cases}$$

Thus, $2x^2 - x - 3 = (2x - 3)(x + 1)$ and $18x^2 - 9x - 27 = 9(2x - 3)(x + 1)$

61. $8x^2 + 2x + 6 = 2(4x^2 + x + 3)$

$$\text{where } 4x^2 + x + 3 = \begin{cases} (4x \quad\;)(x \quad\;) \\ (2x \quad\;)(2x \quad\;) \end{cases}$$

$$\text{(all signs positive)} = \begin{cases} (4x + 1)(x + 3) \\ (4x + 3)(x + 1) \\ (2x + 1)(2x + 3) \end{cases}$$

$$= \begin{cases} 4x^2 + 13x + 3 \\ 4x^2 + 7x + 3 \\ 4x^2 + 8x + 3 \end{cases}$$

Thus, $4x^2 + x + 3$ is prime and $8x^2 + 2x + 6 = 2(4x^2 + x + 3)$

63. $x^2 - x + 4$

Factors of 4	1,4	−1,−4	2,2	−2,−2
Sum	5	−5	4	−4

Since none of the sums equal -1, $x^2 - x + 4$ is prime.

65. $4x^3 - 10x^2 - 6x = 2x(2x^2 - 5x - 3)$

$$\text{where } 2x^2 - 5x - 3 = (2x \quad)(x \quad)$$

$$= \begin{cases} (2x \quad 1)(x \quad 3) \\ (2x \quad 3)(x \quad 1) \end{cases}$$

$$\text{(signs must alternate)} = \begin{cases} (2x + 1)(x - 3) \\ (2x - 1)(x + 3) \\ (2x + 3)(x - 1) \\ (2x - 3)(x + 1) \end{cases}$$

$$= \begin{cases} 2x^2 - 5x - 3 \\ 2x^2 + 5x - 3 \\ 2x^2 + x - 3 \\ 2x^2 - x - 3 \end{cases}$$

Thus, $2x^2 - 5x - 3 = (2x + 1)(x - 3)$ and $4x^3 - 10x^2 - 6x = 2x(2x + 1)(x - 3)$

67. $x^4 + 2x^2 + 1 = (x^2)^2 + 2(1)(x^2) + 1 = (x^2 + 1)^2$

69. $$1 - 8x^2 - 9x^4 = \begin{cases} (1 \quad x^2)(1 \quad 9x^2) \\ (1 \quad 3x^2)(1 \quad 3x^2) \end{cases}$$

$$\text{(signs must alternate)} = \begin{cases} (1 + x^2)(1 - 9x^2) \\ (1 - x^2)(1 + 9x^2) \\ (1 + 3x^2)(1 - 3x^2) \\ (1 - 3x^2)(1 + 3x^2) \end{cases}$$

$$= \begin{cases} 1 - 8x^2 - 9x^2 \\ 1 + 8x^2 - 9x^2 \\ 1 - 9x^2 \\ 1 - 9x^2 \end{cases}$$

$$\begin{aligned} \text{Thus, } 1 - 8x^2 - 9x^4 &= (1 + x^2)(1 - 9x^2) \\ &= (1 + x^2)[1^2 - (3x)^2] \\ &= (1 + x^2)(1 - 3x)(1 + 3x) \\ &= (x^2 + 1)[-1(3x - 1)(3x + 1) \\ &= -(3x - 1)(3x + 1)(x^2 + 1) \end{aligned}$$

71. $$\begin{aligned} 8 - 64x^3 = 8(1 - 8x^3) &= 8[1^3 - (2x)^3] \\ &= 8(1 - 2x)(1^2 + (2x) + (2x)^2] \\ &= 8(1 - 2x)(1 + 2x + 4x^2) \\ &= 8[-1(2x - 1)](4x^2 + 2x + 1) \\ &= -8(2x - 1)(4x^2 + 2x + 1) \end{aligned}$$

73. $x(x + 3) - 6(x + 3) = (x + 3)(x - 6)$

75. $(x + 2)^2 - 5(x + 2) = (x + 2)[(x + 2) - 5] = (x + 2)(x - 3)$

77. $$\begin{aligned} (2x - 1)^2 - 25 = (2x - 1)^2 - 5^2 &= [(2x - 1) - 5][(2x - 1) + 5] \\ &= (2x - 6)(2x + 4) \\ &= [2(x - 3)][2(x + 2)] \\ &= 4(x - 3)(x + 2) \end{aligned}$$

79. $(3x-2)^3 - 27 = (3x-2)^3 - 3^3 = [(3x-2)-3][(3x-2)^2 + 3(3x-2) + 3^2]$
$= (3x-5)(9x^2 - 12x + 4 + 9x - 6 + 9)$
$= (3x-5)(9x^2 - 3x + 7)$

81. $3(x^2 + 10x + 25) - 4(x+5) = 3(x^2 + 2(5)(x) + 5^2) - 4(x+5)$
$= 3(x+5)^2 - 4(x+5)$
$= (x+5)[3(x+5) - 4]$
$= (x+5)[3x + 15 - 4]$
$= (x+5)(3x+11)$

83. $x^3 + 2x^2 - x - 2 = x^2(x+2) - 1(x+2)$
$= (x+2)[x^2 - 1]$
$= (x+2)(x+1)(x-1)$ by (2)

85. $x^4 - x^3 + x - 1 = x^3(x-1) + (x-1)$
$= (x-1)[x^3 + 1]$
$= (x-1)(x+1)(x^2 - x + 1)$ by (7)

87. $x^5 + x^3 + 8x^2 + 8 = x^3(x^2+1) + 8(x^2+1)$
$= (x^2+1)[x^3 + 8]$
$= (x^2+1)(x^3 + 2^3)$
$= (x^2+1)(x+2)(x^2 - 2x + 4)$ by (7)

89.

Factors of 4	1,4	−1,−4	2,2	−2,−2
Sum	5	−5	4	−4

The possibilities are $(x \pm 1)(x \pm 4) = x^2 \pm 5x + 4$ or $(x \pm 2)(x \pm 2) = x^2 \pm 4x + 4$, none of which equal $x^2 + 4$.

1.6 Rational Expressions

1. $\dfrac{3x+9}{x^2-9} = \dfrac{3(x+3)}{(x-3)(x+3)} = \dfrac{3}{x-3}$

3. $\dfrac{x^2-2x}{3x-6} = \dfrac{x(x-2)}{3(x-2)} = \dfrac{x}{3}$

5. $\dfrac{24x^2}{12x^2-6x} = \dfrac{24x^2}{6x(2x-1)} = \dfrac{4x}{2x-1}$

7. $\dfrac{y^2-25}{2y^2-8y-10} = \dfrac{(y+5)(y-5)}{2(y^2-4y-5)} = \dfrac{(y+5)(y-5)}{2(y+1)(y-5)} = \dfrac{y+5}{2(y+1)}$

9. $\dfrac{x^2+4x-5}{x^2-2x+1} = \dfrac{(x+5)(x-1)}{(x-1)(x-1)} = \dfrac{x+5}{x-1}$

11. $\dfrac{x^2-4}{x^2+5x+6} = \dfrac{(x+2)(x-2)}{(x+2)(x+3)} = \dfrac{x-2}{x+3}$

13. $\dfrac{x^2+5x-14}{2-x} = \dfrac{(x+7)(x-2)}{-(x-2)} = -(x+7)$

15. $\dfrac{2x^3-x^2-10x}{x^3-2x^2-8x} = \dfrac{x(2x^2-x-10)}{x(x^2-2x-8)} = \dfrac{x(2x-5)(x+2)}{x(x-4)(x+2)} = \dfrac{2x-5}{x-4}$

17. $\dfrac{(x-4)^2-9}{(x+3)^2-16} = \dfrac{[(x-4)+3][(x-4)-3]}{[(x+3)+4][(x+3)-4]} = \dfrac{(x-1)(x-7)}{(x+7)(x-1)} = \dfrac{x-7}{x+7}$

19. $\dfrac{6x(x-1)-12}{x^3-8} = \dfrac{6[x(x-1)-2]}{(x-2)(x^2+2x+4)} = \dfrac{6(x^2-x-2)}{(x-2)(x^2+2x+4)} = \dfrac{6(x+1)(x-2)}{(x-2)(x^2+2x+4)}$

$= \dfrac{6(x+1)}{(x^2+2x+4)}$

21. $\dfrac{3x+6}{5x^2} \cdot \dfrac{x^2-x-6}{x^2-4} = \dfrac{3(x+2)}{5x^2} \cdot \dfrac{(x-3)(x+2)}{(x+2)(x-2)} = \dfrac{3(x-3)(x+2)}{5x^2(x-2)}$

23. $\dfrac{4x^2-1}{x^2-16} \cdot \dfrac{x^2-4x}{2x+1} = \dfrac{(2x+1)(2x-1)}{(x+4)(x-4)} \cdot \dfrac{x(x-4)}{(2x+1)} = \dfrac{x(2x-1)}{x+4}$

25. $\dfrac{4x-8}{-3x} \cdot \dfrac{12}{12-6x} = \dfrac{4(x-2)}{-3x} \cdot \dfrac{12}{6(2-x)} = \dfrac{4(x-2)}{-3x} \cdot \dfrac{12}{-6(x-2)} = \dfrac{8}{3x}$

27. $\dfrac{x-3x-10}{x^2+2x-35} \cdot \dfrac{x^2+4x-21}{x^2+9x+14} = \dfrac{(x-5)(x+2)}{(x+7)(x-5)} \cdot \dfrac{(x+7)(x-3)}{(x+7)(x+2)} = \dfrac{x-3}{x+7}$

29. $\dfrac{\dfrac{2x^2-x-28}{3x^2-x-2}}{\dfrac{4x^2+16x+7}{3x^2+11x+6}} = \dfrac{\dfrac{(2x+7)(x-4)}{(3x+2)(x-1)}}{\dfrac{(2x+1)(2x+7)}{(3x+2)(x+3)}} = \dfrac{(2x+7)(x-4)(3x+2)(x+3)}{(3x+2)(x-1)(2x+1)(2x+7)} = \dfrac{(x-4)(x+3)}{(x-1)(2x+1)}$

31. $\dfrac{\dfrac{8x^2-6x+1}{4x^2-1}}{\dfrac{12x^2-5x-2}{6x^2-x-2}} = \dfrac{\dfrac{(2x-1)(4x-1)}{(2x+1)(2x-1)}}{\dfrac{(3x-2)(4x+1)}{(3x-2)(2x+1)}} = \dfrac{(2x-1)(4x-1)(3x-2)(2x+1)}{(2x+1)(2x-1)(3x-2)(4x+1)} = \dfrac{4x-1}{4x+1}$

33. $\dfrac{5x^2-x}{3x+2} \cdot \dfrac{2x^2-x}{2x^2-x-1} \cdot \dfrac{10x^2+3x-1}{6x^2+x-2} = \dfrac{x(5x-1)}{3x+2} \cdot \dfrac{x(2x-1)}{(2x+1)(x-1)} \cdot \dfrac{(2x+1)(5x-1)}{(3x+2)(2x-1)}$

$= \dfrac{x^2(5x-1)^2}{(3x+2)^2(x-1)}$

35. $\dfrac{x}{2} + \dfrac{5}{2} = \dfrac{x+5}{2}$

37. $\dfrac{x^2}{2x-3} - \dfrac{4}{2x-3} = \dfrac{x^2-4}{2x-3} = \dfrac{(x-2)(x+2)}{2x-3}$

39. $\dfrac{x+1}{x-3} + \dfrac{2x-3}{x-3} = \dfrac{(x+1)+(2x-3)}{x-3} = \dfrac{3x-2}{x-3}$

41. $\dfrac{3x+5}{2x-1} - \dfrac{2x-4}{2x-1} = \dfrac{(3x+5)-(2x-4)}{2x-1} = \dfrac{3x+5-2x+4}{2x-1} = \dfrac{x+9}{2x-1}$

43. $\dfrac{4}{x-2} + \dfrac{x}{2-x} = \dfrac{4}{x-2} - \dfrac{x}{x-2} = \dfrac{4-x}{x-2}$

45. $\dfrac{4}{x-1} - \dfrac{2}{x+2} = \dfrac{4(x+2)}{(x-1)(x+2)} - \dfrac{2(x-1)}{(x+2)(x-1)} = \dfrac{4x+8-(2x-2)}{(x-1)(x+2)} = \dfrac{2x+10}{(x-1)(x+2)}$

$= \dfrac{2(x+5)}{(x-1)(x+2)}$

47. $\dfrac{x}{x+1} + \dfrac{2x-3}{x-1} = \dfrac{x(x-1)}{(x+1)(x-1)} + \dfrac{(2x-3)(x+1)}{(x-1)(x+1)} = \dfrac{x^2-x+2x^2-x-3}{(x+1)(x-1)}$

$= \dfrac{3x^2-2x-3}{(x+1)(x-1)}$

49. $\dfrac{x-3}{x+2} - \dfrac{x+4}{x-2} = \dfrac{(x-3)(x-2)}{(x+2)(x-2)} - \dfrac{(x+4)(x+2)}{(x-2)(x+2)} = \dfrac{x^2-5x+6-(x^2+6x+8)}{(x+2)(x-2)}$

$= \dfrac{x^2-5x+6-x^2-6x-8}{(x+2)(x-2)} = \dfrac{-11x-2}{(x+2)(x-2)}$

51. $\dfrac{x}{x^2-4} + \dfrac{1}{x} = \dfrac{x}{(x+2)(x-2)} + \dfrac{1}{x} = \dfrac{x(x)}{(x+2)(x-2)(x)} + \dfrac{1(x+2)(x-2)}{(x)(x+2)(x-2)}$

$= \dfrac{x^2+(x^2-4)}{(x+2)(x-2)(x)} = \dfrac{2x^2-4}{x(x+2)(x-2)} = \dfrac{2(x^2-2)}{x(x+2)(x-2)}$

53. $\dfrac{x^3}{(x-1)^2} - \dfrac{x^2+1}{x} = \dfrac{x^3(x)}{(x-1)^2x} - \dfrac{(x^2+1)(x-1)^2}{x(x-1)^2} = \dfrac{x^4-(x^2+1)(x^2-2x+1)}{x(x-1)^2}$

$= \dfrac{x^4-(x^4-2x^3+2x^2-2x+1)}{x(x-1)^2} = \dfrac{2x^3-2x^2+2x-1}{x(x-1)^2}$

55. $\dfrac{x}{x+1} + \dfrac{x-2}{x-1} - \dfrac{x+1}{x-2} = x(x-1)(x-2) + (x-2)(x-1)$

$= \dfrac{x(x-1)(x-2)+(x-2)(x+1)(x-2)-(x+1)(x+1)(x-1)}{(x+1)(x-1)(x-2)}$

$= \dfrac{x(x^2-3x+2)+(x^2-x-2)(x-2)-(x^2+2x+1)(x-1)}{(x+1)(x-1)(x-2)}$

$= \dfrac{(x^3-3x^2+2x)+(x^3-x^2-2x-2x^2+2x+4)-(x^3+2x^2+x-x^2-2x-1)}{(x+1)(x-1)(x-2)}$

$= \dfrac{x^3-7x^2+3x+5}{(x+1)(x-1)(x-2)}$

57. $\dfrac{1}{x} + \dfrac{1}{x+1} - \dfrac{1}{x-1} = \dfrac{(x+1)(x-1)+x(x-1)-x(x+1)}{x(x+1)(x-1)}$

$= \dfrac{x^2-1+x^2-x-x^2-x}{x(x+1)(x-1)} = \dfrac{x^2-2x-1}{x(x+1)(x-1)}$

59. $x^2 - 4 = (x+2)(x-2)$
$x^2 - x - 2 = (x+1)(x-2)$
The LCM is $(x+2)(x-2)(x+1)$

61. $x^3 - x = x(x^2-1) = x(x+1)(x-1)$
$x^2 - x = x(x-1)$
The LCM is $x(x+1)(x-1)$

63. $4x3 - 4x^2 + x = x(4x^2 - 4x + 1)$
$= x(2x - 1)^2$
$2x^3 - x^2 = x^2(2x - 1)$
$x^3 = x^3$
The LCM is $x^3(2x - 1)^2$

65. $x^3 - x = x(x^2 - 1) = x(x + 1)(x - 1)$
$x^3 - 2x^2 + x = x(x^2 - 2x + 1)$
$= x(x - 1)^2$
$x^3 - 1 = (x - 1)(x^2 + x + 1)$
The LCM is $x(x + 1)(x - 1)^2(x^2 + x + 1)$

67. $$\frac{x}{x^2 - 7x + 6} - \frac{x}{x^2 - 2x - 24} = \frac{x}{(x - 1)(x - 6)} - \frac{x}{(x + 4)(x - 6)} = \frac{x(x + 4) - x(x - 1)}{(x - 1)(x + 4)(x - 6)}$$
$$= \frac{x^2 + 4x - (x^2 - x)}{(x - 1)(x + 4)(x - 6)} = \frac{5x}{(x - 1)(x + 4)(x - 6)}$$

69. $$\frac{4x}{x^2 - 4} - \frac{2}{x^2 + x - 6} = \frac{4x}{(x + 2)(x - 2)} - \frac{2}{(x + 3)(x - 2)} = \frac{4x(x + 3) - 2(x + 2)}{(x + 2)(x - 2)(x + 3)}$$
$$= \frac{4x^2 + 12x - (2x + 4)}{(x + 2)(x - 2)(x + 3)} = \frac{4x^2 + 10x - 4}{(x + 2)(x - 2)(x + 3)}$$
$$= \frac{2(2x^2 + 5x - 2)}{(x - 2)(x + 2)(x + 3)}$$

71. $$\frac{3}{(x - 1)^2(x + 1)} + \frac{2}{(x - 1)(x + 1)^2} = \frac{3(x + 1) + 2(x - 1)}{(x - 1)^2(x + 1)^2} = \frac{3x + 3 + 2x - 2}{(x - 1)^2(x + 1)^2}$$
$$= \frac{5x + 1}{(x - 1)^2(x + 1)^2}$$

73. $$\frac{x + 4}{x^2 - x - 2} - \frac{2x + 3}{2x^2 + 2x - 8} = \frac{x + 4}{(x - 2(x + 1)} - \frac{2x + 3}{(x + 4)(x - 2)}$$
$$= \frac{(x + 4)(x + 4) - (2x + 3)(x + 1)}{(x + 1)(x - 2)(x + 4)}$$
$$= \frac{x^2 + 8x + 16 - (2x^2 + 5x + 3)}{(x + 1)(x - 2)(x + 4)} = \frac{-x^2 + 3x + 13}{(x + 1)(x - 2)(x + 4)}$$

75. $$\frac{1}{x} - \frac{2}{x^2 + x} + \frac{3}{x^3 - x^2} = \frac{1}{x} - \frac{2}{x(x + 1)} + \frac{3}{x^2(x - 1)}$$
$$= \frac{x(x + 1)(x - 1) - 2x(x - 1) + 3(x + 1)}{x^2(x + 1)(x - 1)}$$
$$= \frac{x^3 - x - 2x^2 + 2x + 3x + 3}{x^2(x + 1)(x - 1)} = \frac{x^3 - 2x^2 + 4x + 3}{x^2(x + 1)(x - 1)}$$

77. $$\frac{1}{h}\left[\frac{1}{x + h} - \frac{1}{x}\right] = \frac{1}{h}\left[\frac{x - (x + h)}{x(x + h)}\right] = \frac{1}{h}\left[\frac{-h}{x(x + h)}\right] = \frac{-1}{x(x + h)}$$

79. $$\frac{1 + \frac{1}{x}}{1 - \frac{1}{x}} = \frac{\frac{x}{x} + \frac{1}{x}}{\frac{x}{x} - \frac{1}{x}} = \frac{\frac{x + 1}{x}}{\frac{x - 1}{x}} = \frac{x + 1}{x} \cdot \frac{x}{x - 1} = \frac{x + 1}{x - 1}$$

81. $$\frac{x-\frac{1}{x}}{x+\frac{1}{x}} = \frac{\frac{x^2}{x}-\frac{1}{x}}{\frac{x^2}{x}+\frac{1}{x}} = \frac{\frac{x^2-1}{x}}{\frac{x^2+1}{x}} = \frac{x^2-1}{x} \cdot \frac{x}{x^2+1} = \frac{x^2-1}{x^2+1} = \frac{(x+1)(x-1)}{x^2+1}$$

83. $$\frac{\frac{x+4}{x-2}-\frac{x-3}{x+1}}{x+1} = \frac{\frac{(x+4)(x+1)}{(x-2)(x+1)}-\frac{(x-3)(x-2)}{(x+1)(x-2)}}{x+1} = \frac{\frac{x^2+5x+4-(x^2-5x+6)}{(x-2)(x+1)}}{\frac{x+1}{1}}$$

$$= \frac{10x-2}{(x-2)(x+1)} \cdot \frac{1}{x+1} = \frac{2(5x-1)}{(x-2)(x+1)} \cdot \frac{1}{x+1} = \frac{2(5x-1)}{(x-2)(x+1)^2}$$

85. $$\frac{\frac{x-2}{x+2}+\frac{x-1}{x+1}}{\frac{x}{x+1}-\frac{2x-3}{x}} = \frac{\frac{(x-2)(x+1)}{(x+2)(x+1)}+\frac{(x-1)(x+2)}{(x+1)(x+2)}}{\frac{x^2}{x(x+1)}-\frac{(2x-3)(x+1)}{x(x+1)}} = \frac{\frac{(x^2-x-2)+(x^2+x-2)}{(x+2)(x+1)}}{\frac{x^2-(2x^2-x-3)}{x(x+1)}}$$

$$= \frac{\frac{2x^2-4}{(x+2)(x+1)}}{\frac{-x^2+x+3}{x(x+1)}} = \frac{2(x^2-2)}{(x+2)(x+1)} \cdot \frac{x(x+1)}{-(x^2-x-3)}$$

$$= \frac{-2x(x^2-2)}{(x+2)(x^2-x-3)}$$

87. $$1-\frac{1}{1-\frac{1}{x}} = 1-\frac{1}{\frac{x}{x}-\frac{1}{x}} = 1-\frac{1}{\frac{x-1}{x}} = 1-\frac{x}{x-1} = \frac{x-1}{x-1}-\frac{x}{x-1}$$

$$= \frac{x-1-x}{x-1} = \frac{-1}{x-1}$$

89. $$\frac{\frac{x+h-2}{x+h+2}-\frac{x-2}{x+2}}{h} = \frac{\frac{(x+h-2)(x+2)}{(x+h+2)(x+2)}-\frac{(x-2)(x+h+2)}{(x+2)(x+h+2)}}{h}$$

$$= \frac{\frac{(x^2+hx+2h-4)-(x^2+hx-2h-4)}{(x+2)(x+h+2)}}{h}$$

$$= \frac{\frac{4h}{(x+2)(x+h+2)}}{\frac{h}{1}} = \frac{4h}{(x+2)(x+h+2)} \cdot \frac{1}{h}$$

$$= \frac{4}{(x+2)(x+h+2)}$$

91. $$1 + \frac{1}{x} = \frac{x+1}{x},\quad 1 + \frac{1}{1 + \frac{1}{x}} = 1 + \frac{1}{\frac{x+1}{x}} = 1 + \frac{x}{x+1} = \frac{x+1+x}{x+1}$$

$$= \frac{2x+1}{x+1},\quad 1 + \frac{1}{1 + \frac{1}{1 + \frac{1}{x}}} = 1 + \frac{1}{1 + \frac{1}{\frac{x+1}{x}}} = 1 + \frac{1}{1 + \frac{x}{x+1}}$$

$$= 1 + \frac{1}{\frac{x+1+x}{x+1}} = 1 + \frac{x+1}{2x+1} = \frac{2x+1+x+1}{2x+1} = \frac{3x+2}{2x+1}$$

$$1 + \frac{1}{1 + \frac{1}{1 + \frac{1}{1 + \frac{1}{x}}}} = 1 + \frac{1}{1 + \frac{1}{1 + \frac{1}{\frac{x+1}{x}}}} = 1 + \frac{1}{1 + \frac{1}{1 + \frac{x}{x+1}}}$$

$$= 1 + \frac{1}{1 + \frac{1}{\frac{x+1+x}{x+1}}} = 1 + \frac{1}{1 + \frac{x+1}{2x+1}} = 1 + \frac{1}{\frac{2x+1+x+1}{2x+1}}$$

$$= 1 + \frac{2x+1}{3x+2} = \frac{3x+2+2x+1}{3x+2} = \frac{5x+3}{3x+2}$$

Thus, a: 1, 2, 3, 5, 8, 13, ... These are the numbers of a Fibonacci sequence.
b: 1, 1, 2, 3, 5, 8, ...
c: 0, 1, 1, 2, 3, 5, ...

1.7 Square Roots; Radicals

Historical Problem

1. (a) √C8 = $\sqrt[3]{8}$

(b) √C√C512 = $\sqrt[3]{\sqrt[3]{512}}$

(c) √√C64 = $\sqrt[3]{\sqrt[3]{64}}$

(d) √C√64 = $\sqrt[3]{\sqrt{64}}$

(e) √√81 = $\sqrt{\sqrt{81}}$

(f) √√256x^8 = $\sqrt{\sqrt{256x^8}}$

(g) √$\overline{C - 4 + \sqrt{8}}$ = $\sqrt[3]{-4 + \sqrt{8}}$

Exercises

1. $\sqrt{25} = 5$

3. $\sqrt[3]{27} = 3$

5. $\sqrt[3]{-64} = -4$

7. $\sqrt{\frac{1}{9}} = \frac{\sqrt{1}}{\sqrt{9}} = \frac{1}{3}$

9. $\sqrt{25x^4} = 5x^2$

11. $\sqrt[3]{8(1 + x)^3} = 2(1 + x)$

13. $\sqrt{8} = \sqrt{4 \cdot 2} = \sqrt{4}\sqrt{2} = 2\sqrt{2}$

15. $\sqrt[3]{16x^4} = \sqrt[3]{8x^3 \cdot 2x} = \sqrt[3]{8x^3}\sqrt[3]{2x} = 2x\sqrt[3]{2x}$

17. $\sqrt[3]{\sqrt{x^6}} = \sqrt[3]{\sqrt{(x^3)^2}} = \sqrt[3]{x^3} = x$

19. $\sqrt{\frac{25x^3}{9x}} = \sqrt{\frac{25x^2}{9}} = \frac{5}{3}x$

21. $\sqrt[4]{x^{12}y^8} = \sqrt[4]{(x^3)^4(y^2)^4} = x^3y^2$

23. $\sqrt[4]{\frac{x^9y^8}{xy^{12}}} = \sqrt[4]{\frac{x^8}{y^4}} = \sqrt[4]{\frac{(x^2)^4}{y^4}} = \frac{x^2}{y}$

25. $\sqrt{36x} = \sqrt{36}\sqrt{x} = 6\sqrt{x}$

27. $\sqrt{3x^2}\sqrt{12x} = \sqrt{36x^3} = \sqrt{36x^2}\sqrt{x} = 6x\sqrt{x}$

29. $\frac{\sqrt{3xy^3}\sqrt{2x^2y}}{\sqrt{6x^3y^4}} = \frac{\sqrt{6x^3y^4}}{\sqrt{6x^3y^4}} = 1$

31. $\sqrt{\frac{16y^4}{9x^2}} = \frac{\sqrt{16x^4}}{\sqrt{9x^2}} = \frac{4y^2}{3|x|}$

33. $\left(\sqrt{5}\,\sqrt[3]{9}\right)^2 = \sqrt{5^2}\,\sqrt[3]{9^2} = 5\sqrt[3]{81} = 5\sqrt[3]{3^4} = 5\sqrt[3]{3^3 \cdot 3} = 15\sqrt[3]{3}$

35.
$$\sqrt{\frac{2x - 3}{2x^4 + 3x^3}}\sqrt{\frac{x}{4x^2 - 9}} = \sqrt{\frac{x(2x - 3)}{(2x^4 + 3x^3)(4x^2 - 9}} = \sqrt{\frac{x(2x - 3)}{x^3(2x + 3)(2x + 3)(2x - 3)}}$$
$$= \sqrt{\frac{1}{x^2(2x + 3)^2}} = \frac{\sqrt{1}}{\sqrt{x^2(2x + 3)^2}} = \frac{1}{x(2x + 3)}$$

37. $\sqrt{\sqrt[3]{\sqrt[4]{x}}} = \sqrt{\sqrt[12]{x}} = \sqrt[24]{x}$

39. $\sqrt{\frac{x - 1}{x + 1}}\sqrt{\frac{x^2 + 2x + 1}{x^2 - 1}} = \sqrt{\frac{(x - 1)(x^2 + 2x + 1)}{(x + 1)(x^2 - 1)}} = \sqrt{\frac{(x - 1)(x + 1)(x + 1)}{(x + 1)(x - 1)(x + 1)}} = \sqrt{1} = 1$

41. $3\sqrt{2} + 4\sqrt{2} - \sqrt{2} = (3 + 4 - 1)\sqrt{2} = 6\sqrt{2}$

43.
$$\begin{aligned}3\sqrt{2} - \sqrt{18} + 2\sqrt{8} &= 3\sqrt{2} - \sqrt{9 \cdot 2} + 2\sqrt{4 \cdot 2}\\ &= 3\sqrt{2} - \sqrt{9}\sqrt{2} + 2\sqrt{4}\sqrt{2}\\ &= 3\sqrt{2} - 3\sqrt{2} + 4\sqrt{2}\\ &= (3 - 3 + 4)\sqrt{2}\\ &= 4\sqrt{2}\end{aligned}$$

45. $\sqrt[3]{16} + 5\sqrt[3]{2} - 2\sqrt[3]{54} = \sqrt[3]{8 \cdot 2} + 5\sqrt[3]{2} - 2\sqrt[3]{27 \cdot 2}$
$= \sqrt[3]{8}\sqrt[3]{2} + 5\sqrt[3]{2} - 2\sqrt[3]{27}\sqrt[3]{2}$
$= 2\sqrt[3]{2} + 5\sqrt[3]{2} - 6\sqrt[3]{2}$
$= (2 + 5 - 6)\sqrt[3]{2}$
$= \sqrt[3]{2}$

47. $\sqrt{8x^3} - 3\sqrt{50x} + \sqrt{2x^5} = \sqrt{4x^2 \cdot 2x} - 3\sqrt{25 \cdot 2x} + \sqrt{x^4 \cdot 2x}$
$= 2x\sqrt{2x} - 15\sqrt{2x} + x^2\sqrt{2x}$
$= (2x - 15 + x^2)\sqrt{2x}$
$= (x^2 + 2x - 15)\sqrt{2x}$
$= (x + 5)(x - 3)\sqrt{2x}$

49. $\sqrt[3]{16x^4y} - 3x\sqrt[3]{2xy} + 5\sqrt[3]{-2xy^4} = \sqrt[3]{8x^3 \cdot 2xy} - 3x\sqrt[3]{2xy} + 5\sqrt[3]{(-y)^3 \cdot 2xy}$
$= 2x\sqrt[3]{2xy} - 3x\sqrt[3]{2xy} - 5y\sqrt[3]{2xy}$
$= (2x - 3x - 5y)\sqrt[3]{2xy}$
$= (-x - 5y)\sqrt[3]{2xy}$

51. $(3\sqrt{6})(4\sqrt{3}) = 12\sqrt{18}$
$= 12\sqrt{9 \cdot 2}$
$= 12\sqrt{9} \cdot \sqrt{2}$
$= 36\sqrt{2}$

53. $(\sqrt{3} + 3)(\sqrt{3} - 4) = (\sqrt{3})^2 - \sqrt{3} - 12$
$= 3 - \sqrt{3} - 12$
$= -9 - \sqrt{3}$

55. $(3\sqrt{7} + 4)(2\sqrt{7} + 3) = 6(\sqrt{7})^2 + 17\sqrt{7} + 12$
$= 6(7) + 17\sqrt{7} + 12$
$= 54 + 17\sqrt{7}$

57. $(\sqrt{x} - 1)^2 = (\sqrt{x})^2 - 2\sqrt{x} + 1$
$= x - 2\sqrt{x} + 1$

59. $(\sqrt[3]{x} - 1)^3 = (\sqrt[3]{x})^3 - (3\sqrt[3]{x})^2 + 3\sqrt[3]{x} - 1 = x - 3\sqrt[3]{x^2} + 3\sqrt[3]{x} - 1$

61. $(2\sqrt{x} - 3)(2\sqrt{x} + 5) = 4(\sqrt{x})^2 + 4\sqrt{x} - 15 = 4x + 4\sqrt{x} - 15$

63. $\sqrt{1 - x^2} - \dfrac{1}{\sqrt{1 - x^2}} = \sqrt{1 - x^2} \cdot \dfrac{\sqrt{1 - x^2}}{\sqrt{1 - x^2}} - \dfrac{1}{\sqrt{1 - x^2}}$

$$= \frac{\left(\sqrt{1 - x^2}\right)^2}{\sqrt{1 - x^2}} - \frac{1}{\sqrt{1 - x^2}} = \frac{1 - x^2}{\sqrt{1 - x^2}} - \frac{1}{\sqrt{1 - x^2}}$$

$$= \frac{1 - x^2 - 1}{\sqrt{1 - x^2}} = \frac{-x^2}{\sqrt{1 - x^2}}$$

65. $\dfrac{2}{\sqrt{5}} = \dfrac{2}{\sqrt{5}} \cdot \dfrac{\sqrt{5}}{\sqrt{5}} = \dfrac{2\sqrt{5}}{5}$

67. $\dfrac{8}{\sqrt{6}} = \dfrac{8}{\sqrt{6}} \cdot \dfrac{\sqrt{6}}{\sqrt{6}} = \dfrac{8\sqrt{6}}{6} = \dfrac{4\sqrt{6}}{3}$

69. $\dfrac{1}{\sqrt{x}} = \dfrac{1}{\sqrt{x}} \cdot \dfrac{\sqrt{x}}{\sqrt{x}} = \dfrac{\sqrt{x}}{x}, x > 0$

71. $\dfrac{3}{5 + \sqrt{2}} = \dfrac{3}{5 + \sqrt{2}} \cdot \dfrac{5 - \sqrt{2}}{5 - \sqrt{2}} = \dfrac{3\left(5 - \sqrt{2}\right)}{25 - 2} = \dfrac{3\left(5 - \sqrt{2}\right)}{23} = \dfrac{15 - 3\sqrt{2}}{23}$

73. $\dfrac{3}{4 + \sqrt{7}} = \dfrac{3}{4 + \sqrt{7}} \cdot \dfrac{4 - \sqrt{7}}{4 - \sqrt{7}} = \dfrac{3\left(4 - \sqrt{7}\right)}{16 - 7} = \dfrac{3\left(4 - \sqrt{7}\right)}{9} = \dfrac{4 - \sqrt{7}}{3}$

75. $\dfrac{\sqrt{5}}{2 + 3\sqrt{5}} = \dfrac{\sqrt{5}}{2 + 3\sqrt{5}} \cdot \dfrac{2 - 3\sqrt{5}}{2 - 3\sqrt{5}} = \dfrac{\sqrt{5}\left(2 - 3\sqrt{5}\right)}{4 - 9(5)} = \dfrac{\sqrt{5}\left(2 - 3\sqrt{5}\right)}{-41} = \dfrac{2\sqrt{5} - 3(5)}{-41}$

$= \dfrac{-15 + 2\sqrt{5}}{-41} = \dfrac{-\left(15 - 2\sqrt{5}\right)}{-41} = \dfrac{15 - 2\sqrt{5}}{41}$

77. $\dfrac{\sqrt{3} - \sqrt{2}}{\sqrt{3} + \sqrt{2}} = \dfrac{\sqrt{3} - \sqrt{2}}{\sqrt{3} + \sqrt{2}} \cdot \dfrac{\sqrt{3} - \sqrt{2}}{\sqrt{3} - \sqrt{2}} = \dfrac{\left(\sqrt{3} - \sqrt{2}\right)^2}{\left(\sqrt{3}\right)^2 - \left(\sqrt{2}\right)^2} = \dfrac{\left(\sqrt{3}\right)^2 - 2\sqrt{3}\sqrt{2} + \left(\sqrt{2}\right)^2}{3 - 2}$

$= \dfrac{5 - 2\sqrt{6}}{1} = 5 - 2\sqrt{6}$

79. $\dfrac{1}{\sqrt{x} + 2} = \dfrac{1}{\sqrt{x} + 2} \cdot \dfrac{\sqrt{x} - 2}{\sqrt{x} - 2} = \dfrac{\sqrt{x} - 2}{x - 4}$

81. $\dfrac{\sqrt{x + h} - \sqrt{x}}{\sqrt{x + h} + \sqrt{h}} = \dfrac{\sqrt{x + h} - \sqrt{x}}{\sqrt{x + h} + \sqrt{x}} \cdot \dfrac{\sqrt{x + h} - \sqrt{x}}{\sqrt{x + h} - \sqrt{x}} = \dfrac{\left(\sqrt{x + h} - \sqrt{x}\right)^2}{\left(\sqrt{x + h}\right)^2 - \left(\sqrt{x}\right)^2}$

$= \dfrac{\left(\sqrt{x + h}\right)^2 - 2\sqrt{x + h}\sqrt{x} + \left(\sqrt{x}\right)^2}{x + h - x}$

$= \dfrac{x + h - 2\sqrt{x(x + h)} + x}{h} = \dfrac{2x + h - 2\sqrt{x(x + h)}}{h}$

83. $\dfrac{2 - \sqrt{5}}{3 + 2\sqrt{5}} = \dfrac{\left(2 - \sqrt{5}\right)\left(2 + \sqrt{5}\right)}{\left(3 + 2\sqrt{5}\right)\left(2 + \sqrt{5}\right)} = \dfrac{4 - 5}{6 + 4\sqrt{5} + 3\sqrt{5} + 2 \cdot 5} = \dfrac{-1}{16 + 7\sqrt{5}}$

85. $\dfrac{\sqrt{x + h} - \sqrt{x}}{h} = \dfrac{\sqrt{x + h} - \sqrt{x}}{h} \cdot \dfrac{\sqrt{x + h} + \sqrt{x}}{\sqrt{x + h} + \sqrt{x}} = \dfrac{\left(\sqrt{x + h}\right)^2 - \left(\sqrt{x}\right)^2}{h\left(\sqrt{x + h} + \sqrt{x}\right)}$

$= \dfrac{x + h - x}{h\left(\sqrt{x + h} + \sqrt{x}\right)} = \dfrac{1}{\sqrt{x + h} + \sqrt{x}}$

87. $\sqrt{2} \approx 1.41$ Keystrokes: [2] [$\sqrt{\ }$]

89. $\sqrt[3]{4} \approx 1.59$ Keystrokes: [4] [SHIFT] [x^y] [3] [=]

91. $\dfrac{2+\sqrt{3}}{3-\sqrt{5}} \approx 4.89$ Keystrokes: [2] [+] [3] [√] [=] [÷] [(] [3] [−] [5] [√] [)] [=]

93. $\dfrac{3\sqrt[3]{5}-\sqrt{2}}{\sqrt{3}} \approx 2.15$

Keystrokes: [3] [×] [5] [SHIFT] [x^y] [3] [=] [−] [2] [√] [=] [÷] [3] [√] [=]

1.8 Rational Exponents

1. $8^{2/3} = \left(\sqrt[3]{8}\right)^2 = 2^2 = 4$

3. $(-27)^{\frac{2}{3}} = \sqrt[3]{-27^2} = (-3)^2 = 9$

5. $4^{-3/2} = \left(\sqrt{4}\right)^{-3} = 2^{-3} = \dfrac{1}{2^3} = \dfrac{1}{8}$

7. $9^{-3/2} = \left(\sqrt{9}\right)^{-3} = 3^{-3} = \dfrac{1}{3^3} = \dfrac{1}{27}$

9. $\left(\dfrac{9}{4}\right)^{3/2} = \left(\sqrt{\dfrac{9}{4}}\right)^3 = \left(\dfrac{\sqrt{9}}{\sqrt{4}}\right)^3 = \left(\dfrac{3}{2}\right)^3 = \dfrac{3^3}{2} = \dfrac{27}{8}$

11. $\left(\dfrac{4}{9}\right)^{-3/2} = \left(\sqrt{\dfrac{4}{9}}\right)^{-3} = \left(\dfrac{\sqrt{4}}{\sqrt{9}}\right)^{-3} = \left(\dfrac{2}{3}\right)^{-3} = \dfrac{1}{\left(\dfrac{2}{3}\right)^3} = \dfrac{1}{\dfrac{2^3}{3^3}} = \dfrac{3^3}{2^3} = \dfrac{27}{8}$

13. $4^{1.5} = 4^{3/2} = \left(\sqrt{4}\right)^3 = 2^3 = 8$

15. $\left(\dfrac{1}{4}\right)^{-1.5} = \left(\dfrac{1}{4}\right)^{-3/2} = \left(\sqrt{\dfrac{1}{4}}\right)^{-3} = \left(\dfrac{\sqrt{1}}{\sqrt{4}}\right)^{-3} = \left(\dfrac{1}{2}\right)^{-3} = \dfrac{1}{\left(\dfrac{1}{2}\right)^3} = \dfrac{1}{\dfrac{1}{8}} = 8$

17. $\left(\sqrt{3}\right)^6 = 3^{6/2} = 3^3 = 27$

19. $\left(\sqrt{5}\right)^{-2} = 5^{(-2/2)} = 5^{-1} = \dfrac{1}{5}$

21. $x^{2/3}x^{-1/2} = x^{(2/3-1/2)} = x^{(4/6)-(3/6)} = x^{1/6}$

23. $(x^3y^6)^{2/3} = x^{6/3}y^{12/3} = x^2y^4$

25. $(x^2y)^{1/3}(xy^2)^{2/3} = x^{2/3}y^{1/3} \cdot x^{2/3}y^{4/3} = x^{(2/3)+(2/3)}y^{(1/3)+(4/3)} = x^{4/3}y^{5/3}$

27. $(16x^2y^{-1/3})^{3/4} = 16^{3/4}x^{6/4}y^{-3/12} = (2^4)^{3/4}x^{3/2}y^{-1/4} = 2^3x^{3/2}\dfrac{1}{y^{1/4}} = \dfrac{8x^{3/2}}{y^{1/4}}$

29. $$\left[\frac{x^{2/5}y^{-1/5}}{x^{-1/3}}\right]^{15} = \left(x^{(2/5)-(-1/3)}y^{-(1/5)}\right)^{15} = \left(x^{11/15}y^{-1/5}\right)^{15} = x^{11}y^{-3} = \frac{x^{11}}{y^3}$$

31. $$\frac{x}{(1+x)^{1/2}} + 2(1+x)^{1/2} = \frac{x + 2(1+x)^{1/2}(1+x)^{1/2}}{(1+x)^{1/2}} = \frac{x + 2(1+x)}{(1+x)^{1/2}} = \frac{3x+2}{(1+x)^{1/2}}$$

33. $$2x(x^2+1)^{1/2} + x^2 \cdot \frac{1}{2}(x^2+1)^{-1/2} \cdot 2x$$
$$= 2x(x^2+1)^{1/2} + \frac{x^2}{(x^2+1)^{1/2}} = \frac{2x(x^2+1)^{1/2}(x^2+1)^{1/2} + x^3}{(x^2+1)^{1/2}}$$
$$= \frac{2x(x^2+1) + x^3}{(x^2+1)^{1/2}} = \frac{2x^3 + 2x + x^3}{(x^2+1)^{1/2}} = \frac{3x^3 + 2x}{(x^2+1)^{1/2}} = \frac{x(3x^2+2)}{(x^2+1)^{1/2}}$$

35. $$\sqrt{4x+3} \cdot \frac{1}{2\sqrt{x-5}} + \sqrt{x-5} \cdot \frac{1}{5\sqrt{4x+3}}$$
$$= \frac{\sqrt{4x+3}}{2\sqrt{x-5}} + \frac{\sqrt{x-5}}{5\sqrt{4x+3}} = \frac{\sqrt{4x+3}\left(5\sqrt{4x+3}\right) + \sqrt{x-5}\left(2\sqrt{x-5}\right)}{10\sqrt{x-5}\sqrt{4x+3}}$$
$$= \frac{5(4x+3) + 2(x-5)}{10\sqrt{x-5}\sqrt{4x+3}} = \frac{22x+5}{10\sqrt{x-5}\sqrt{4x+3}}$$

37. $$\frac{\sqrt{1+x} - x \cdot \dfrac{1}{2\sqrt{1+x}}}{1+x} = \frac{\sqrt{1+x} - \dfrac{x}{2\sqrt{1+x}}}{1+x} = \frac{\dfrac{\sqrt{1+x}\left(2\sqrt{1+x}\right) - x}{2\sqrt{1+x}}}{\dfrac{1+x}{1}}$$
$$= \frac{\dfrac{2(1+x) - x}{2(1+x)^{1/2}}}{\dfrac{1+x}{1}} = \frac{2+x}{2(1+x)^{1/2}(1+x)} = \frac{2+x}{2(1+x)^{3/2}}$$

39. $$\frac{(x+4)^{1/2} - 2x(x+4)^{-1/2}}{x+4} = \frac{(x+4)^{1/2} - \dfrac{2x}{(x+4)^{1/2}}}{x+4} = \frac{\dfrac{(x+4)(x+4)^{1/2} - 2x}{(x+4)^{1/2}}}{x+4}$$
$$= \frac{\dfrac{x+4-2x}{(x+4)^{1/2}}}{\dfrac{x+4}{1}} = \frac{4-x}{(x+4)^{3/2}}$$

41. $$\frac{\dfrac{x^2}{(x^2-1)^{1/2}} - (x^2-1)^{1/2}}{x^2} = \frac{\dfrac{x^2 - (x^2-1)^{1/2}(x^2-1)^{1/2}}{(x^2-1)^{1/2}}}{x^2} = \frac{\dfrac{x^2 - (x^2-1)}{(x^2-1)^{1/2}}}{\dfrac{x^2}{1}}$$
$$= \frac{1}{(x^2-1)^{1/2}} \cdot \frac{1}{x^2} = \frac{1}{x^2(x^2-1)^{1/2}}$$

43. $$\frac{\dfrac{1 + x^2}{2\sqrt{x}} - 2x\sqrt{x}}{(1 + x^2)^2} = \frac{\dfrac{1 + x^2 - 2x\sqrt{x}\left(2\sqrt{x}\right)}{2\sqrt{x}}}{(1 + x^2)^2} = \frac{\dfrac{1 + x^2 - 4x^2}{2\sqrt{x}}}{\dfrac{(1 + x^2)^2}{1}} = \frac{1 - 3x^2}{2\sqrt{x}\,(1 + x^2)^2}$$

45. $$(x + 1)^{3/2} + x \cdot \frac{3}{2}(x + 1)^{1/2} = (x + 1)^{1/2}\left[(x + 1)^{2/2} + \frac{3x}{2}\right] = (x + 1)^{1/2}\left[x + 1 + \frac{3}{2}x\right]$$
$$= (x + 1)^{1/2}\left[\frac{3}{2}x + 1\right] = \frac{1}{2}(x + 1)^{1/2}(5x + 2)$$

47. $$\begin{aligned} 6x^{1/2}(x^2 + x) - 8x^{3/2} - 8x^{1/2} &= 2x^{1/2}(3(x^2 + x) - 4x^{2/2} - 4) \\ &= 2x^{1/2}(3x^2 + 3x - 4x - 4) \\ &= 2x^{1/2}(3x^2 - x - 4) \\ &= 2x^{1/2}(3x - 4)(x + 1) \end{aligned}$$

49. $$\begin{aligned} 3(x^2 + 4)^{4/3} + x \cdot 4(x^2 + 4)^{1/3} \cdot 2x &= (x^2 + 4)^{1/3}[3(x^2 + 4)^{3/3} + 8x^2] \\ &= (x^2 + 4)^{1/3}(3x^2 + 12 + 8x^2) \\ &= (x^2 + 4)^{1/3}(11x^2 + 12) \end{aligned}$$

51. $$\begin{aligned} &4(3x + 5)^{1/3}(2x + 3)^{3/2} + 3(3x + 5)^{4/3}(2x + 3)^{1/2} \\ &\quad = (3x + 5)^{1/3}(2x + 3)^{1/2}[4(2x + 3)^{2/2} + 3(3x + 5)^{3/3}] \\ &\quad = (3x + 5)^{1/3}(2x + 3^{1/2}(8x + 12 + 9x + 15) \\ &\quad = (3x + 5)^{1/3}(2x + 3)^{1/2}(17x + 27) \end{aligned}$$

53. If $u = \frac{1}{2}\left(x^3 - \frac{1}{x^3}\right)$,

$$\begin{aligned} \sqrt{1 + u^2} &= \sqrt{1 + \left[\frac{1}{2}\left(x^3 - \frac{1}{x^3}\right)\right]^2} = \sqrt{1 + \frac{1}{4}\left(x^6 - 2(x^3) \cdot \frac{1}{x^3} + \frac{1}{x^6}\right)} \\ &= \sqrt{1 + \frac{1}{4}\left(x^6 - 2 + \frac{1}{x^6}\right)} = \sqrt{1 + \frac{1}{4} \cdot \frac{x^{12} - 2x^6 + 1}{x^6}} \\ &= \sqrt{\frac{4x^6 + x^{12} - 2x^6 + 1}{4x^6}} = \frac{1}{2}\sqrt{\frac{4x^6}{x^6} + \frac{x^{12}}{x^6} - \frac{2x^6}{x^6} + \frac{1}{x^6}} \\ &= \frac{1}{2}\sqrt{x^6 + 2 + \frac{1}{x^6}} = \frac{1}{2}\sqrt{\left(x^3 + \frac{1}{x^3}\right)^2} = \frac{1}{2}\left(x^3 + \frac{1}{x^3}\right) \end{aligned}$$

55. If $u = \frac{1}{2}\left(x^2 - \frac{1}{x^2}\right)$,

$$1 + u^2 = 1 + \left[\frac{1}{2}\left(x^2 - \frac{1}{x^2}\right)\right]^2 = 1 + \frac{1}{4}\left(x^4 - 2(x^2)\left(\frac{1}{x^2}\right) + \frac{1}{x^4}\right)$$

$$= 1 + \frac{1}{4}\left(x^4 - 2 - \frac{1}{x^4}\right) = 1 + \frac{1}{4} \cdot \frac{x^8 - 2x^4 + 1}{x^4}$$

$$= 1 + \frac{x^8 - 2x^4 + 1}{4x^4} = \frac{4x^4 + x^8 - 2x^4 + 1}{4x^4} = \frac{x^8 + 2x^4 + 1}{4x^4}$$

$$= \frac{1}{4}\left(\frac{x^8}{x^4} + \frac{2x^4}{x^4} + \frac{1}{x^4}\right) = \frac{1}{4}\left(x^4 + 2 + \frac{1}{x^4}\right) = \frac{1}{4}\left(x^2 + \frac{1}{x^2}\right)^2$$

1.9 Geometry Topics

For legs a and b of a right triangle in Problems 1-5, we use $c^2 = a^2 + b^2$ to find the hypotenuse c:

1. For $a = 5$ and $b = 12$, $c^2 = a^2 + b^2$
$= 5^2 + 12^2$
$= 25 + 144$
$c^2 = 169$
then $c = 13$

3. For $a = 10$ and $b = 24$, $c^2 = 10^2 + 24^2$
$= 100 + 576$
$c^2 = 676$
then $c = 26$

5. For $a = 7$ and $b = 24$, $c^2 = 7^2 + 24^2$
$= 49 + 576$
$c^2 = 625$
then $c = 25$

In Problems 7–13, we will test whether the given triangles are right triangles by using $c^2 = a^2 + b^2$. The hypotenuse must be the longest side in any case:

7. For sides 3, 4, 5, let $c = 5$: $c^2 = a^2 + b^2$
$5^2 = 3^2 + 4^2$
$25 = 9 + 16$
$25 = 25$
The given triangle is a right triangle with hypotenuse of length 5.

9. For sides 4, 5, and 6, let $c = 6$: but $6^2 \neq 4^2 + 5^2$
since $36 \neq 16 + 25 = 41$
The triangle is not a right triangle.

11. For sides 7, 24, and 25, let $c = 25$: $25^2 = 7^2 + 24^2$
$625 = 49 + 576$
$625 = 625$
The triangle is a right triangle with hypotenuse of length 25.

13. For sides 6, 4, and 3, let $c = 6$: but $6^2 \neq 4^2 + 3^2$
since $36 \neq 16 + 9 = 25$
The triangle is not a right triangle.

15. $A = \ell w = 5 \cdot 3 = 15$; $P = 2(\ell + w) = 2(5 + 3) = 16$

17. $A = \ell w = \frac{1}{2} \cdot \frac{1}{3} = \frac{1}{6}$; $P = 2(\ell + w) = 2\left[\frac{1}{2} + \frac{1}{3}\right] = 2\left[\frac{3+2}{6}\right] = \frac{10}{6} = \frac{5}{3}$

19. $A = \frac{1}{2}bh = \frac{1}{2}(4)(3) = 6$

21. $A = \frac{1}{2}bh = \frac{1}{2}\left[\frac{1}{2}\right]\left[\frac{3}{4}\right] = \frac{3}{16}$

23. $A = \pi r^2 = \pi(1)^2 = \pi \approx 3.14$; $C = 2\pi r = 2\pi(1) = 2\pi \approx 6.28$

25. $A = \pi r^2 = \pi\left[\frac{3}{2}\right]^2 = \frac{9}{4}\pi \approx 7.065$; $C = 2\pi r = 2\pi\left[\frac{3}{2}\right] = 3\pi \approx 9.42$

27. $V = \ell wh = (4)(2)(6) = 48$

29. $V = \ell wh = \left[\frac{1}{3}\right]\left[\frac{5}{2}\right](4) = \frac{10}{3}$

31. After 4 revolutions, the wheel rolls a distance of 4 times its circumference. Thus,
If diameter = 16 inches, then radius = 8 inches
Total Distance $= 4C = 4(2\pi r) = 8\pi r = 8\pi(8) = 64\pi = 201.1$ inches $= 16.8$ feet

33. Area of border = Area of EFGH − Area of ABCD
= (10)(10) − (6)(6)
= 100 − 36
= 64 square feet

35. The area of the opening can be found by adding the area of the rectangle and the area of the semicircle.

$$(6)(4) + \frac{\pi(2)^2}{2} = 24 + 2\pi \approx 30.28 \text{ ft.}^2$$

The amount of wood frame needed to enclose the window can be found by adding the perimeter of three sides of the rectangle and the perimeter of the semicircle.

$$6 + 4 + 6 + \frac{2\pi(2)}{2} = 16 + 2\pi \approx 22.28 \text{ ft.}$$

37. We have the triangle with side 3960 miles and hypotenuse $3960 + \frac{1450}{5280} = 3960.2746$ miles.
Let side a be the distance in miles one can see from the Sears Tower:
Here $c = 3960.3$ and $b = 3960$:

$(3960.2746)^2 = a^2 + (3960)^2$
$15683775 = a^2 + 15681600$
$2172 = a^2$
$46.6 = a$
46.6 miles

39. We have the triangle with hypotenuse c representing the distance $3960 + \dfrac{6}{5280} = 3960.0011$ miles.

Let a be the distance in miles the 6-foot–tall person can see the ship.

$$\begin{aligned} c^2 &= a^2 + b^2 \\ (3960.0011)^2 &= a^2 + (3960)^2 \\ 15681609 &= a^2 + 15681600 \\ 9 &= a^2 \\ 3 &= a \\ 3 \text{ miles} &= a \end{aligned}$$

a

3960

3960.0011

41. $a = m^2 - n^2,\ b = 2mn,\ c = m^2 + n^2$

$$\begin{aligned} \text{Then, } a^2 + b^2 &= (m^2 - n^2)^2 + (2mn)^2 \\ &= m^4 - 2m^2n^2 + n^4 + 4m^2n^2 \\ &= m^4 + 2m^2n^2 + n^4 \\ &= (m^2 + n^2)^2 = c^2 \end{aligned}$$

The conclusion follows from the converse of the Pythagorean Theorem.

1 Chapter Review

1. $3 - 4 \cdot 5 + 6 = 3 - 20 + 6 = -11$

3. $\dfrac{3}{7} - \dfrac{7}{12} = \dfrac{3}{4} \cdot \dfrac{3}{3} - \dfrac{7}{12} = \dfrac{9}{12} - \dfrac{7}{12} = \dfrac{9 - 7}{12} = \dfrac{2}{12} = \dfrac{1}{6}$

5. $\dfrac{\dfrac{15}{2} + \dfrac{1}{4}}{\dfrac{2}{3}} = \dfrac{\dfrac{30}{4} + \dfrac{1}{4}}{\dfrac{2}{3}} = \dfrac{\dfrac{30 + 1}{4}}{\dfrac{2}{3}} = \dfrac{31}{4} \cdot \dfrac{3}{2} = \dfrac{93}{8}$

7. $5^2 - 3^3 \cdot 2 = 25 - 27 \cdot 2 = 25 - 54 = -29$

9. $\dfrac{2^{-3} \cdot 5^0}{4^2} = \dfrac{\dfrac{1}{2^3} \cdot 1}{16} = \dfrac{\dfrac{1}{8}}{\dfrac{16}{1}} = \dfrac{1}{8} \cdot \dfrac{1}{16} = \dfrac{1}{128}$

11. $\left(2\sqrt{5} - 2\right)\left(2\sqrt{5} + 2\right) = 2\sqrt{5}\left(2\sqrt{5} + 2\right) - 2\left(2\sqrt{5} + 2\right) = 4\left(\sqrt{5}\right)^2 + 4\sqrt{5} - 4\sqrt{5} - 4 = 20 - 4 = 16$

13. $\left(\dfrac{8}{27}\right)^{-2/3} = \left(\dfrac{27}{8}\right)^{2/3} = \left(\sqrt[3]{\dfrac{27}{8}}\right)^2 = \left(\dfrac{\sqrt[3]{27}}{\sqrt[3]{8}}\right)^2 = \dfrac{3^2}{2^2} = \dfrac{9}{4}$

15. $\left(\sqrt[3]{2}\right)^{-3} = \dfrac{1}{\left(\sqrt[3]{2}\right)^3} = \dfrac{1}{2}$

17. $|6 - 8^{1/3}| = |6 - \sqrt[3]{8}| = |6 - 2| = |4| = 4$

19. $\sqrt{|3^2 - 5^2|} = \sqrt{|9 - 25|} = \sqrt{|-16|} = \sqrt{16} = 4$

21. $$\frac{x^{-2}}{y^{-2}} = \frac{\frac{1}{x^2}}{\frac{1}{y^2}} = \frac{1}{x^2} \cdot \frac{y^2}{1} = \frac{y^2}{x^2}$$

23. $$\frac{(x^2y)^{-4}}{(xy)^{-3}} = \frac{\frac{1}{(x^2y)^4}}{\frac{1}{(xy)^3}} = \frac{(xy)^3}{(x^2y)^4} = \frac{x^3y^3}{x^8y^4} = x^{3-8}y^{3-4} = x^{-5}y^{-1} = \frac{1}{x^5}\frac{1}{y} = \frac{1}{x^5y}$$

25. $$\frac{\left[\frac{x^2}{y}\right]^2}{\left[\frac{x}{y^2}\right]^3} = \frac{\frac{x^4}{y^2}}{\frac{x^3}{y^6}} = \frac{x^4}{y^2} \cdot \frac{y^6}{x^3} = xy^4$$

27. $$\frac{x^{-2}}{x^{-2} + y^{-2}} = \frac{\frac{1}{x^2}}{\frac{1}{x^2} + \frac{1}{y^2}} = \frac{\frac{1}{x^2}}{\frac{y^2 + x^2}{x^2y^2}} = \frac{1}{x^2} \cdot \frac{x^2y^2}{x^2 + y^2} = \frac{y^2}{(x^2 + y^2)}$$

29. $$\left(25x^{-4/3}y^{-2/3}\right)^{3/2} = 25^{3/2}x^{-12/6}y^{-6/6} = \left(\sqrt{25}\right)^3 x^{-2}y^{-1} = 5^3\frac{1}{x^2}\frac{1}{y} = \frac{125}{x^2y}$$

31. $$\left[\frac{2x^{-1/2}}{y^{-3/4}}\right]^{-4} = \frac{(2x^{1-2})^{-4}}{(y^{-3/4})^{-4}} = \frac{2^{-4}x^{4/2}}{y^{12/4}} = \frac{\frac{1}{2^4} \cdot x^2}{y^3} = \frac{\frac{x^2}{16}}{\frac{y^3}{1}} = \frac{x^2}{16y^3}$$

33. $$\begin{aligned}(2x - 3)(-4x + 2) &= 2x(-4x + 2) - 3(-4x + 2)\\ &= -8x^2 + 4x + 12x - 6\\ &= -8x^2 + 16x - 6\end{aligned}$$

35. $$\begin{aligned}3(3x^3 - 2x^2 + 1) - 3(x^3 + 4x^2 - 2x - 3) &= 12x^3 - 8x^2 + 4 - 3x^3 - 12x^2 + 6x + 9\\ &= 9x^3 - 20x^2 + 6x + 13\end{aligned}$$

37. $$\begin{aligned}(2x - 5)(3x^2 + 2) &= 2x(3x^2 + 2) - 5(3x^2 + 2)\\ &= 6x^3 + 4x - 15x^2 - 10\\ &= 6x^3 - 15x^2 + 4x - 10\end{aligned}$$

39. $(x+1)(x+2)(x-3) = (x+1)[x(x-3)+2(x-3)]$

$$\begin{aligned}
&= (x+1)[x^2-3x+2x-6] \\
&= (x+1)(x^2-x-6) \\
&= x(x^2-x-6)+1(x^2-x-6) \\
&= x^3-x^2-6x+x^2-x-6 \\
&= x^3-7x-6
\end{aligned}$$

41.

$$\begin{array}{r}
3x^2 + 8x + 25 \\
x-3\overline{)3x^3 - x^2 + x + 4} \\
\underline{3x^3 - 9x^2} \qquad\qquad \\
8x^2 + x \qquad \\
\underline{8x - 24x} \qquad \\
25x + 4 \\
\underline{25x - 75} \\
79
\end{array}$$

The quotient is $3x^2 + 8x + 25$; the remainder is 79.

Check: $(x-3)(3x^2+8x+25)+79$
$= 3x^2 + 8x^2 + 25x - 9x^2 - 24x - 75 + 79$
$= 3x^3 - x^2 + x + 4$

43.

$$\begin{array}{r}
-3x^2 + 4 \qquad\qquad\qquad \\
x^2+1\overline{)-3x^4 + 0x^3 + x^2 + 0x + 2} \\
\underline{-3x^4 \qquad\quad -3x^2} \qquad\qquad \\
4x^2 + 0x + 2 \\
\underline{4x^2 \qquad\quad + 4} \\
-2
\end{array}$$

The quotient is $-3x^2 + 4$; the remainder is -2.

Check: $(x^2+1)(-3x^2+4)-2$
$= -3x^4 + 4x^2 - 3x^2 + 4 - 2$
$= -3x^4 + x^2 + 2$

45.

$$\begin{array}{r}
8x^2 + 24x + 62 \qquad\qquad \\
x^2-3x+1\overline{)8x^4 + 0x^3 - 2x^2 + 5x + 1} \\
\underline{8x^4 - 24x^3 + 8x^2} \qquad\qquad\quad \\
24x^3 - 10x^2 + 5x \qquad \\
\underline{24x^3 - 72x^2 + 24x} \qquad \\
62x^2 - 19x + 1 \\
\underline{62x^2 - 186x + 62} \\
167x - 61
\end{array}$$

The quotient is $8x^2 + 24x + 62$; the remainder is $167x - 61$.

Check: $(x^2-3x+1)(8x^2+24x+62)$
$+ 167x - 61$
$= (8x^4 + 24x^3 + 62x^2) - 24x^3$
$- 72x^2 - 186x + 8x^2$
$+ 24x + 62 + 167x - 61$
$= 8x^4 - 2x^2 + 5x + 1$

47.

$$\begin{array}{r}
x^4 - x^3 + x^2 - x + 1 \qquad \\
x+1\overline{)x^5 + 0x^4 + 0x^3 + 0x^2 + 0x + 1} \\
\underline{x^5 + x^4} \qquad\qquad\qquad\qquad\qquad \\
-x^4 + 0x^3 \qquad\qquad\qquad\quad \\
\underline{-x^4 - x^3} \qquad\qquad\qquad\quad \\
x^3 + 0x^2 \qquad\qquad\quad \\
\underline{x^3 + x} \qquad\qquad\qquad \\
-x^2 + 0x \qquad\quad \\
\underline{-x^2 - x} \qquad\quad \\
x + 1 \\
\underline{x + 1} \\
0
\end{array}$$

The quotient is $x^4 - x^3 + x^2 - x + 1$; the remainder is 0.

Check: $(x+1)(x^4 - x^3 + x^2 - x + 1)$
$= x^5 - x^4 + x^3 - x^2 + x + x^4$
$- x^3 + x^2 - x + 1$
$= x^5 + 1$

49.

$$\begin{array}{r} 3x^4 - 2x^2 + 1 \\ 2x + 1 \overline{) 6x^5 + 3x^4 - 4x^3 - 2x^2 + 2x + 1} \\ \underline{6x^5 + 3x^4} \qquad\qquad\qquad\qquad\quad \\ -4x^3 - 2x^2 \qquad\qquad \\ \underline{-4x^3 - 2x^2} \qquad\qquad \\ 2x + 1 \\ \underline{2x + 1} \\ 0 \end{array}$$

The quotient is $3x^4 - 2x^2 + 1$; the remainder is 0.

Check: $(2x + 1)(3x^4 - 2x^2 + 1)$
$= 6x^5 - 4x^3 + 2x + 3x^4 - 2x^2 + 1$
$= 6x^5 + 3x^4 - 4x^3 - 2x^2 + 2x + 1$

51.

Factors of -14	$1,-14$	$-1,14$	$2,-7,$	$-2,7$
Sum	-13	13	-5	5

$x^2 + 5x - 14 = (x - 2)(x + 7)$

53.

$$6x^2 - 5x - 6 = \begin{cases} (6x \quad)(x \quad) \\ (3x \quad)(2x \quad) \end{cases}$$

$$= \begin{cases} (6x \quad 6)(x \quad 1) \\ (6x \quad 1)(x \quad 6) \\ (6x \quad 2)(x \quad 3) \\ (6x \quad 3)(x \quad 2) \\ (3x \quad 6)(2x \quad 1) \\ (3x \quad 1)(2x \quad 6) \\ (3x \quad 2)(2x \quad 3) \\ (3x \quad 3)(2x \quad 2) \end{cases}$$

$$\text{(signs must alternate)} = \begin{cases} (6x + 6)(x - 1) \\ (6x - 6)(x + 1) \\ (6x + 1)(x - 6) \\ (6x - 1)(x + 6) \\ (6x + 2)(x - 3) \\ (6x - 2)(x + 3) \\ (6x + 3)(x - 2) \\ (6x - 3)(x + 2) \\ (3x + 6)(2x - 1) \\ (3x - 6)(2x + 1) \\ (3x + 1)(2x - 6) \\ (3x - 1)(2x + 6) \\ (3x + 2)(2x - 3) \\ (3x - 2)(2x + 3) \\ (3x + 3)(2x - 2) \\ (3x - 3)(2x + 2) \end{cases}$$

$$= \begin{cases} 6x^2 - 6 \\ 6x^2 - 6 \\ 6x^2 - 35x - 6 \\ 6x^2 + 35x - 6 \\ 6x^2 - 16x - 6 \\ 6x^2 + 16x - 6 \\ 6x^2 - 9x - 6 \\ 6x^2 + 9x - 6 \\ 6x^2 + 9x - 6 \\ 6x^2 - 9x - 6 \\ 6x^2 - 16x - 6 \\ 6x^2 + 18x - 6 \\ 6x^2 - 5x - 6 \\ 6x^2 + 5x - 6 \\ 6x^2 - 6 \\ 6x^2 - 6 \end{cases}$$

Thus, $6x^2 - 5x - 6 = (3x + 2)(2x - 3)$

55. Factors of -14 $1,-14$ $-1,14$ $2,-7$ $-2,7$

Sum -13 13 -5 5

$$3x^2 - 15x - 42 = 3(x^2 - 5x - 14) = 3(x + 2)(x - 7)$$

57. $8x^3 + 1 = (2x)^3 + 1 = (2x + 1)((2x)^2 - 2x + 1) = (2x + 1)(4x^2 - 2x + 1)$

59.
$$\begin{aligned} 2x^3 + 3x^2 - 2x - 3 &= x^2(2x + 3) - 1(2x + 3) \\ &= (2x + 3)[x^2 - 1] \\ &= (2x + 3)(x + 1)(x - 1) \end{aligned}$$

61. $25x^2 - 4 = (5x)^2 - 2^2 = (5x + 2)(5x - 2)$

63. $9x^2 + 1$ is prime.

65. $$\frac{2x^2 + 11x + 14}{x^2 - 4} = \frac{(2x + 7)(x + 2)}{(x + 2)(x - 2)} = \frac{2x + 7}{x - 2}$$

67. $$\frac{9x^2 - 1}{x^2 - 9} \cdot \frac{3x - 9}{9x^2 + 6x + 1} = \frac{(3x + 1)(3x - 1)}{(x + 3)(x - 3)} \cdot \frac{3(x - 3)}{(3x + 1)^2} = \frac{3(3x - 1)}{(x + 3)(3x + 1)}$$

69. $$\frac{x + 1}{x - 1} - \frac{x - 1}{x + 1} = \frac{(x + 1)(x + 1)}{(x - 1)(x + 1)} - \frac{(x - 1)(x - 1)}{(x + 1)(x - 1)} = \frac{(x + 1)^2 - (x - 1)^2}{(x - 1)(x + 1)}$$

$$= \frac{x^2 + 2x + 1 - (x^2 - 2x + 1)}{(x - 1)(x + 1)} = \frac{4x}{(x - 1)(x + 1)}$$

71. $$\frac{3x + 4}{x^2 - 4} - \frac{2x - 3}{x^2 + 4x + 4} = \frac{3x + 4}{(x + 2)(x - 2)} - \frac{2x - 3}{(x + 2)^2} = \frac{(3x + 4)(x + 2)}{(x + 2)^2(x - 2)} - \frac{(2x - 3)(x - 2)}{(x + 2)^2(x - 2)}$$

$$= \frac{3x^2 + 10x + 8 - (2x^2 - 7x + 6)}{(x + 2)^2(x - 2)} = \frac{x^2 + 17x + 2}{(x + 2)^2(x - 2)}$$

73. $$1 + \frac{1}{1 + \frac{1}{x}} = 1 + \frac{1}{\frac{x}{x} + \frac{1}{x}} = 1 + \frac{1}{\frac{x + 1}{x}} = 1 + \frac{x}{x + 1} = \frac{x + 1}{x + 1} + \frac{x}{x + 1}$$

$$= \frac{x + 1 + x}{x + 1} = \frac{2x + 1}{x + 1}$$

75. $$\frac{4}{\sqrt{5}} = \frac{4}{\sqrt{5}} \cdot \frac{\sqrt{5}}{\sqrt{5}} = \frac{4\sqrt{5}}{\left(\sqrt{5}\right)^2} = \frac{4\sqrt{5}}{5}$$

77. $$\frac{2}{1 - \sqrt{2}} = \frac{2}{1 - \sqrt{2}} \cdot \frac{1 + \sqrt{2}}{1 + \sqrt{2}} = \frac{2\left(1 + \sqrt{2}\right)}{\left(1 - \sqrt{2}\right)\left(1 + \sqrt{2}\right)} = \frac{2\left(1 + \sqrt{2}\right)}{1 - \left(\sqrt{2}\right)^2} = \frac{2\left(1 + \sqrt{2}\right)}{1 - 2}$$

$$= \frac{2\left(1 + \sqrt{2}\right)}{-1} = -2\left(1 + \sqrt{2}\right)$$

79. $\dfrac{1+\sqrt{5}}{1-\sqrt{5}} = \dfrac{1+\sqrt{5}}{1-\sqrt{5}} \cdot \dfrac{1+\sqrt{5}}{1+\sqrt{5}} = \dfrac{\left(1+\sqrt{5}\right)^2}{\left(1-\sqrt{5}\right)\left(1+\sqrt{5}\right)} = \dfrac{1+2\sqrt{5}+\left(\sqrt{5}\right)^2}{1-\left(\sqrt{5}\right)^2}$

$= \dfrac{1+2\sqrt{5}+5}{1-5} = \dfrac{6+2\sqrt{5}}{-4} = \dfrac{-2\left(3+\sqrt{5}\right)}{4} = \dfrac{-\left(3+\sqrt{5}\right)}{2}$

81. $(2+x^2)^{1/2} + x \cdot \dfrac{1}{2}(2+x^2)^{-1/2} \cdot 2x = (2+x^2)^{1/2} + \dfrac{x^2}{(2+x^2)^{1/2}}$

$= \dfrac{(2+x^2)^{1/2}(2+x^2)^{1/2} + x^2}{(2+x^2)^{1/2}} = \dfrac{(2+x^2)+x^2}{(2+x^2)^{1/2}}$

$= \dfrac{2(1+x^2)}{(2+x^2)^{1/2}}$

83. $\dfrac{(x+4)^{1/2} \cdot 2x - x^2 \cdot \dfrac{1}{2}(x+4)^{-1/2}}{x+4}$

$= \dfrac{2x(x+4)^{1/2} - \dfrac{x^2}{2(x+4)^{1/2}}}{x+4} = \dfrac{\dfrac{2x(x+4)^{1/2}(2(x+4)^{1/2}) - x^2}{2(x+4)^{1/2}}}{x+4} = \dfrac{\dfrac{4x(x+4) - x^2}{2(x+4)^{1/2}}}{\dfrac{x+4}{1}}$

$= \dfrac{\dfrac{4x^2+16x-x^2}{2(x+4)^{1/2}}}{\dfrac{x+4}{1}} = \dfrac{3x^2+16x}{2(x+4)^{1/2}(x+4)} = \dfrac{x(3x+16)}{2(x+4)^{3/2}}$

85. $C = 3000 + 6x - \dfrac{x^2}{1000}$

(a) If $x = 1000$, then $C = 3000 + 6(1000) - \dfrac{(1000)^2}{1000}$
$= 3000 + 6000 - 1000$
$= 8000$
The cost of producing 1000 hand calculators is \$8000.

(b) If $x = 3000$, then $C = 3000 + 6(3000) - \dfrac{(3000)^2}{1000}$
$= 3000 + 18{,}000 - \dfrac{9{,}000{,}000}{1{,}000}$
$= 3000 + 18{,}000 - 9000$
$= 12{,}000$
The cost of producing 3000 hand calculators is \$12,000.

87. The area of the opening of the window is found by adding the area of the rectangle and the area of the triangle.

$$(6)(5) + \frac{1}{2}(5)\left(\sqrt{16 - 6.25}\right) = 30 + \frac{5}{2}\sqrt{9.75}$$
$$\approx 30 + \frac{5}{2}(3.1)$$
$$\approx 37.81 \text{ ft.}^2$$

The amount of wood frame needed to enclose the window is found by adding the perimeter of two sides of the triangle and three sides of the rectangle.

$$4 + 4 + 6 + 5 + 6 = 25 \text{ ft.}$$

89. The area of the pond is found by subtracting the area of the smaller circle from the area of the larger circle.

$$\pi(5)^2 - \pi(3)^2 = 25\pi - 9\pi = 16\pi \text{ ft}^2 \approx 50.24 \text{ ft.}^2$$

The amount of fence required to enclose the pond is found by evaluating the perimeter of the larger circle.

$$2\pi(5) = 10\pi \approx 31.4 \text{ ft.}$$

6/19/00

Chapter 2

EQUATIONS AND INEQUALITIES

2.1 Equations

Solving mentally in Problems 1–7:

1. For $7x = 21$
$7(3) = 21$
so that $x = 3$

3. For $5x + 15 = 0$
we want $-15 + 15 = 0$
or $5(-3) + 15 = 0$
so that $x = -3$

5. For $2x - 3 = 0$
we want $3 - 3 = 0$
or $2\left[\frac{3}{2}\right] - 3 = 0$
so that $x = \frac{3}{2}$

7. For $\frac{1}{3}x = \frac{5}{12}$
$\frac{1}{3} \cdot \frac{5}{4} = \frac{5}{12}$
so that $x = \frac{5}{4}$

9.
$$\begin{aligned} 3x + 2 &= x \\ (3x + 2) - 2 &= x - 2 \\ 3x &= x - 2 \\ 3x - x &= (x - 2) - x \\ 2x &= -2 \\ \frac{2x}{2} &= \frac{-2}{2} \\ x &= -1 \end{aligned}$$

Check:
$$\begin{aligned} 3x + 2 &= x \\ 3(-1) + 2 &\stackrel{?}{=} -1 \\ -3 + 2 &\stackrel{?}{=} -1 \\ -1 &= -1 \end{aligned}$$
The solution checks.

11.
$$\begin{aligned} 2t - 6 &= 3 - t \\ (2t - 6) + 6 &= (3 - t) + 6 \\ 2t &= 9 - t \\ 2t + t &= (9 - t) + t \\ 3t &= 9 \\ \frac{3t}{3} &= \frac{9}{3} \\ t &= 3 \end{aligned}$$

Check:
$$\begin{aligned} 2t - 6 &= 3 - t \\ 2(3) - 6 &\stackrel{?}{=} 3 - 3 \\ 6 - 6 &\stackrel{?}{=} 0 \\ 0 &= 0 \end{aligned}$$
The solution checks.

13.
$$\begin{aligned} 6 - x &= 2x + 9 \\ (6 - x) - 6 &= (2x + 9) - 6 \\ -x &= 2x + 3 \\ -x - 2x &= 2x + 3 - 2x \\ -3x &= 3 \\ \frac{-3x}{-3} &= \frac{3}{-3} \\ x &= -1 \end{aligned}$$

Check:
$$\begin{aligned} 6 - x &= 2x + 9 \\ -(-1) &\stackrel{?}{=} 2(-1) + 9 \\ 6 + 1 &\stackrel{?}{=} -2 + 9 \\ 7 &= 7 \end{aligned}$$

15.
$$\begin{aligned} 3 + 2n &= 5n + 7 \\ (3 + 2n) - 3 &= (5n + 7) - 3 \\ 2n &= 5n + 4 \\ 2n - 5n &= 5n + 4 - 5n \\ -3n &= 4 \\ \frac{-3n}{-3} &= \frac{4}{-3} \\ n &= -\frac{4}{3} \end{aligned}$$

Check:
$$\begin{aligned} 3 + 2n &= 5n + 7 \\ 3 + 2\left[\frac{-4}{3}\right] &\stackrel{?}{=} 5\left[\frac{-4}{3}\right] + 7 \\ 3 - \frac{8}{3} &\stackrel{?}{=} \frac{-20}{3} + 7 \\ \frac{9 - 8}{3} &\stackrel{?}{=} \frac{-20 + 21}{3} \\ \frac{1}{3} &= \frac{1}{3} \end{aligned}$$

17.
$$\begin{aligned} 2(3 + 2x) &= 3(x - 4) \\ 6 + 4x &= 3x - 12 \\ (6 + 4x) - 6 &= (3x - 12) - 6 \\ 4x &= 3x - 18 \\ 4x - 3x &= 3x - 18 - 3x \\ x &= -18 \end{aligned}$$

Check:
$$\begin{aligned} 2(3 + 2x) &= 3(x - 4) \\ 2(3 + 2(-18)) &\stackrel{?}{=} 3(-18 - 4) \\ 2(3 - 36) &\stackrel{?}{=} 3(-22) \\ 2(-33) &\stackrel{?}{=} -66 \\ -66 &= -66 \end{aligned}$$

19.
$$\begin{aligned} 8x - (3x + 2) &= 3x - 10 \\ 8x - 3x - 2 &= 3x - 10 \\ 5x - 2 &= 3x - 10 \\ (5x - 2) + 2 &= (3x - 10) + 2 \\ 5x &= 3x - 8 \\ 5x - 3x &= (3x - 8) - 3x \\ 2x &= -8 \\ \frac{2x}{2} &= \frac{-8}{2} \\ x &= -4 \end{aligned}$$

Check:
$$\begin{aligned} 8x - (3x + 2) &= 3x - 10 \\ 8(-4) - (3(-4) + 2) &= 3(-4) - 10 \\ -32 - (-12 + 2) &= -12 - 10 \\ -32 - (-10) &= -22 \\ -22 &= -22 \end{aligned}$$

21.
$$\begin{aligned} \frac{3}{2}x + 2 &= \frac{1}{2} - \frac{1}{2}x \\ 2\left[\frac{3}{2}x + 2\right] &= 2\left[\frac{1}{2} - \frac{1}{2}x\right] \\ 3x + 4 &= 1 - x \\ (3x + 6) - 4 &= (1 - x) - 4 \\ 3x &= -3 - x \\ 3x + x &= -3 - x + x \\ 4x &= -3 \\ \frac{4x}{4} &= \frac{-3}{4} \\ x &= -\frac{3}{4} \end{aligned}$$

Check:
$$\begin{aligned} \frac{3}{2}x + 2 &= \frac{1}{2} - \frac{1}{2}x \\ \frac{3}{2}\left(-\frac{3}{4}\right) + 2 &\stackrel{?}{=} \frac{1}{2} - \frac{1}{2}\left(\frac{-3}{4}\right) \\ \frac{-9}{8} + 2 &\stackrel{?}{=} \frac{1}{2} + \frac{3}{8} \\ \frac{-9 + 16}{8} &\stackrel{?}{=} \frac{4 + 3}{8} \\ \frac{7}{8} &= \frac{7}{8} \end{aligned}$$

23.

$$\begin{aligned}
\frac{1}{2}x - 5 &= \frac{3}{4}x \\
4\left[\frac{1}{2}x - 5\right] &= 4\left[\frac{3}{4}x\right] \\
2x - 20 &= 3x \\
(2x - 20) + 20 &= 3x + 20 \\
2x &= 3x + 20 \\
2x - 3x &= 3x + 20 - 3x \\
-x &= 20 \\
\frac{-x}{-1} &= \frac{20}{-1} \\
x &= -20
\end{aligned}$$

Check:

$$\begin{aligned}
\frac{1}{2}x - 5 &= \frac{3}{4}x \\
\frac{1}{2}(-20) - 5 &= \frac{3}{4}(-20) \\
-10 - 5 &= -15 \\
-15 &= -15
\end{aligned}$$

25.

$$\begin{aligned}
\frac{2}{3}p &= \frac{1}{2}p + \frac{1}{3} \\
6\left[\frac{2}{3}p\right] &= 6\left[\frac{1}{2}p + \frac{1}{3}\right] \\
4p &= 3p + 2 \\
4p - 3p &= (3p + 2) - 3p \\
p &= 2
\end{aligned}$$

Check:

$$\begin{aligned}
2p &= \frac{1}{2}p + \frac{1}{3} \\
\frac{2}{3}(2) &\stackrel{?}{=} \frac{1}{2}(2) + \frac{1}{3} \\
\frac{4}{3} &\stackrel{?}{=} 1 + \frac{1}{3} \\
\frac{4}{3} &= \frac{4}{3}
\end{aligned}$$

27.

$$\begin{aligned}
0.9t &= 0.4 + 0.1t \\
0.9t - 0.1t &= (0.4 + 0.1t) - 0.1t \\
0.8t &= 0.4 \\
\frac{0.8t}{0.8} &= \frac{0.4}{0.8} \\
t &= 0.5
\end{aligned}$$

Check:

$$\begin{aligned}
0.9t &= 0.4 + 0.1t \\
0.9(0.5) &\stackrel{?}{=} 0.4 + 0.01(0.5) \\
0.45 &\stackrel{?}{=} 0.4 + 0.05 \\
0.45 &= 0.45
\end{aligned}$$

29.

$$\begin{aligned}
\frac{x + 1}{3} + \frac{x + 2}{7} &= 5 \\
21\left[\frac{x + 1}{3} + \frac{x + 2}{7}\right] &= 21[5] \\
7(x + 1) + 3(x + 2) &= 105 \\
7x + 7 + 3x + 6 &= 105 \\
10x + 13 &= 105 \\
(10x + 13) - 13 &= 105 - 13 \\
10x &= 92 \\
\frac{10x}{10} &= \frac{92}{10} \\
x &= \frac{46}{5}
\end{aligned}$$

Check:

$$\begin{aligned}
\frac{x + 1}{3} + \frac{x + 2}{7} &= 5 \\
\frac{\frac{46}{5} + 1}{3} + \frac{\frac{46}{5} + 2}{7} &\stackrel{?}{=} 5 \\
\frac{\frac{46 + 5}{5}}{3} + \frac{\frac{46 + 10}{5}}{7} &\stackrel{?}{=} 5 \\
\frac{51}{15} + \frac{56}{35} &\stackrel{?}{=} 5 \\
\frac{357 + 168}{105} &\stackrel{?}{=} 5 \\
\frac{525}{105} &\stackrel{?}{=} 5 \\
5 &= 5
\end{aligned}$$

31. The domain is $\{y \mid y \neq 0\}$

$$\begin{aligned} \frac{2}{y} + \frac{4}{y} &= 3 \\ y\left[\frac{2}{y} + \frac{4}{y}\right] &= y[3] \\ 2 + 4 &= 3y \\ 6 &= 3y \\ \frac{6}{3} &= \frac{3y}{3} \\ 2 &= y \\ y &= 2 \end{aligned}$$

Check:
$$\begin{aligned} \frac{2}{y} + \frac{4}{y} &= 3 \\ \frac{2}{2} + \frac{4}{2} &\stackrel{?}{=} 3 \\ 1 + 2 &\stackrel{?}{=} 3 \\ 3 &= 3 \end{aligned}$$

33.
$$\begin{aligned} \frac{1}{2} + \frac{2}{x} &= \frac{3}{4} \\ 4x\left[\frac{1}{2} + \frac{2}{x}\right] &= \frac{3}{4}(4x) \\ 2x + 8 &= 3x \\ 2x + 8 - 8 &= 3x - 8 \\ 2x &= 3x - 8 \\ 2x - 3x &= 3x - 8 - 3x \\ \frac{-x}{-1} &= \frac{-8}{-1} \\ x &= 8 \end{aligned}$$

Check:
$$\begin{aligned} \frac{1}{2} + \frac{2}{x} &= \frac{3}{4} \\ \frac{1}{2} + \frac{2}{8} &= \frac{3}{4} \\ \frac{1}{2} + \frac{1}{4} &= \frac{3}{4} \\ 4\left[\frac{1}{2} + \frac{1}{4}\right] &= \frac{3}{4}(4) \\ 2 + 1 &= 3 \\ 3 &= 3 \end{aligned}$$

35.
$$\begin{aligned} (x + 7)(x - 1) &= (x + 1)^2 \\ x^2 + 6x - 7 &= x^2 + 2x + 1 \\ (x^2 + 6x - 7) - x^2 &= (x^2 + 2x + 1) - x^2 \\ 6x - 7 &= 2x + 1 \\ (6x - 7) + 7 &= (2x + 1) + 7 \\ 6x &= 2x + 8 \\ 6x - 2x &= (2x + 8) - 2x \\ 4x &= 8 \\ \frac{4x}{4} &= \frac{8}{4} \\ x &= 2 \end{aligned}$$

Check:
$$\begin{aligned} (x + 7)(x - 1) &= (x + 1)^2 \\ (2 + 7)(2 - 1) &\stackrel{?}{=} (2 + 1)^2 \\ (9)(1) &\stackrel{?}{=} 3^2 \\ 9 &= 9 \end{aligned}$$

37.
$$\begin{aligned} x(2x - 3) &= (2x + 1)(x - 4) \\ 2x^2 - 3x &= 2x^2 - 7x - 4 \\ (2x - 3x) - 2x^2 &= (2x^2 - 7x - 4) - 2x^2 \\ -3x &= -7x - 4 \\ -3x + 7x &= (-7x - 4) + 7x \\ 4x &= -4 \\ \frac{4x}{4} &= \frac{-4}{4} \\ x &= -1 \end{aligned}$$

Check:
$$\begin{aligned} (2x - 3) &= (2x + 1)(x - 4) \\ -1(2(-1) - 3) &\stackrel{?}{=} (2(-1) + 1)(-1 - 4) \\ -1(-2 - 3) &\stackrel{?}{=} (-2 + 1)(-5) \\ -1(-5) &\stackrel{?}{=} (-1)(-5) \\ 5 &= 5 \end{aligned}$$

39.
$$\begin{aligned} z(z^2 + 1) &= 3 + z^3 \\ z^3 + z &= 3 + z^3 \\ (z^3 + z) - z^3 &= (3 + z^3) - z^3 \\ z &= 3 \end{aligned}$$

Check:
$$\begin{aligned} z(z^2 + 1) &= 3 + z^3 \\ 3(3^2 + 1) &\stackrel{?}{=} 3 + 3^3 \\ 3(9 + 1) &\stackrel{?}{=} 30 \\ 3(10) &\stackrel{?}{=} 30 \\ 30 &= 30 \end{aligned}$$

41. $\dfrac{x}{x-2} + 3 = \dfrac{2}{x-2}$

Note that $x - 2$ cannot equal zero so $x = 2$ is ***not*** in the domain of the variable.

$$(x-2)\left[\frac{x}{x-2} + 3\right] = \frac{2}{x-2}(x-2)$$
$$x + 3(x-2) = 2$$
$$x + 3x - 6 = 2$$
$$4x = 8$$
$$x = 2$$

But $x = 2$ is not in the domain of the variable. Hence, the equation has no solution.

43.
$$x^2 = 9x$$
$$x^2 - 9x = 0$$
$$x(x-9) = 0$$
$$x = 0, x = 9$$

The solution set is $\{0, 9\}$

45.
$$t^3 - 9t^2 = 0$$
$$t^2(t-9) = 0$$
$$t^2 = 0, \quad t - 9 = 0$$
$$t = 0, \qquad t = 9$$

The solution set is $\{0, 9\}$

47. $\dfrac{2x}{x^2-4} = \dfrac{4}{x^2-4} - \dfrac{3}{x+2}$

Note that $x^2 - 4$ cannot equal zero and $x + 2$ cannot equal zero. Therefore, $x = -2$ and $x = 2$ are not in the domain of the variable.

$$(x^2-4)\frac{2x}{x^2-4} = (x^2-4)\left[\frac{4}{x^2-4} - \frac{3}{x+2}\right]$$
$$2x = 4 - \frac{3(x^2-4)}{x+2}$$
$$2x = 4 - \frac{3(x-2)(x+2)}{x+2}$$
$$2x = 4 - 3(x-2)$$
$$2x = 4 - 3x + 6$$
$$2x = 10 - 3x$$
$$5x = 10$$
$$x = 2$$

But $x = 2$ is not in the domain of the variable. The equation has no solution.

49. The domain is $\{x \mid x \neq -2\}$

$$\frac{x}{x+2} = \frac{1}{2}$$
$$2(x+2)\left[\frac{x}{x+2}\right] = 2(x+2)\left[\frac{1}{2}\right]$$
$$2x = x + 2$$
$$2x - x = (x+2) - x$$
$$x = 2$$

Check:
$$\frac{x}{x+2} = \frac{1}{2}$$
$$\frac{2}{2+2} \stackrel{?}{=} \frac{1}{2}$$
$$\frac{2}{4} \stackrel{?}{=} \frac{1}{2}$$
$$\frac{1}{2} = \frac{1}{2}$$

51. The domain is $\left\{x \mid x \neq \frac{3}{2}, x \neq -5\right\}$

$$
\begin{aligned}
\frac{5}{2x-3} &= \frac{2}{x+5} \\
(2x-3)(x+5)\left[\frac{5}{2x-3}\right] &= (2x-3)(x+5)\left[\frac{2}{x+5}\right] \\
5(x+5) &= 2(2x-3) \\
5x+25 &= 4x-6 \\
(5x+25)-25 &= (4x-6)-25 \\
5x &= 4x-31 \\
5x-4x &= (4x-31)-4x \\
x &= -31
\end{aligned}
$$

Check:

$$
\begin{aligned}
\frac{5}{2x-3} &= \frac{2}{x+5} \\
\frac{5}{2(-31)-3} &= \frac{2}{-31+5} \\
\frac{5}{-62-3} &= \frac{2}{-26} \\
\frac{5}{-65} &= -\frac{1}{13} \\
-\frac{1}{13} &= -\frac{1}{13}
\end{aligned}
$$

53.

$$
\begin{aligned}
\frac{6t+7}{4t-1} &= \frac{3t+8}{2t-4} \\
(4t-1)(2t-4)\left[\frac{6t+7}{4t-1}\right] &= (4t-1)(2t-4)\left[\frac{3t+8}{2t-4}\right] \\
(2t-4)(6t+7) &= (4t-1)(3t+8) \\
(12t^2-10t-28 &= 12t^2+29t-8 \\
(12t^2-10t-28)-12t^2 &= (12t^2+29t-8)-12t^2 \\
-10t-28 &= 29t-8 \\
(-10t-28)+28 &= (29t-8)+28 \\
-10t &= 29t+20 \\
-10t-29t &= 29t+20-29t \\
-39t &= 20 \\
\frac{-30t}{-39} &= \frac{20}{-39} \\
t &= \frac{-20}{39}
\end{aligned}
$$

Check:

$$
\begin{aligned}
\frac{6t+7}{4t-1} &= \frac{3t+8}{2t-4} \\
\frac{6\left(\frac{-20}{39}\right)+7}{4\left(\frac{-20}{39}\right)-1} &\stackrel{?}{=} \frac{3\left(\frac{-20}{39}\right)+8}{2\left(\frac{-20}{39}\right)-4} \\
\frac{\frac{-120}{39}+7}{\frac{-80}{39}-1} &\stackrel{?}{=} \frac{\frac{-60}{39}+8}{-\frac{40}{39}-4} \\
\frac{\frac{-120+273}{39}}{\frac{-80-39}{39}} &\stackrel{?}{=} \frac{\frac{-60+312}{39}}{\frac{-40-156}{39}}
\end{aligned}
$$

$$
\begin{aligned}
\frac{\frac{153}{39}}{\frac{-119}{39}} &\stackrel{?}{=} \frac{\frac{252}{39}}{\frac{-196}{39}} \\
\frac{153}{39} \cdot \frac{39}{-119} &\stackrel{?}{=} \frac{252}{39} \cdot \frac{39}{-196} \\
\frac{153}{119} - \frac{252}{196} &= -\frac{252}{39} \cdot \frac{39}{-196} \\
-\frac{153}{119} &= -\frac{252}{196} \\
-\frac{9 \cdot 17}{7 \cdot 17} &= -\frac{4 \cdot 7 \cdot 9}{4 \cdot 7 \cdot 9} \\
-\frac{9}{7} &= -\frac{9}{7}
\end{aligned}
$$

55. The domain is $\{x \mid x \neq 2, x \neq -5\}$.

$$\begin{aligned}
\frac{4}{x-2} &= \frac{-3}{x+5} + \frac{7}{(x+5)(x-2)} \\
(x+5)(x-2)\left[\frac{4}{x-2}\right] &= (x+5)(x-2)\left[\frac{-3}{x+5} + \frac{7}{(x+5)(x-2)}\right] \\
4(x+5) &= -3(x-2) + 7 \\
4x + 20 &= -3x + 6 + 7 \\
4x + 20 &= -3x + 13 \\
4x + 20 + 3x &= -3x + 13 + 3x \\
7x + 20 &= 13 \\
7x + 20 - 20 &= 13 - 20 \\
\frac{7x}{7} &= \frac{-7}{7} \\
x &= -1
\end{aligned}$$

Check:

$$\begin{aligned}
\frac{4}{x-2} &= \frac{-3}{x+5} + \frac{7}{(x+5)(x-2)} \\
\frac{4}{-1-2} &= \frac{-3}{-1+5} + \frac{7}{(-1+5)(-1-2)} \\
\frac{4}{-3} &= \frac{-3}{4} + \frac{7}{-12} \\
-12\left[\frac{4}{-3}\right] &= -12\left[\frac{-3}{4} + \frac{7}{-12}\right] \\
16 &= 9 + 7 \\
16 &= 16
\end{aligned}$$

57. The domain is $\{y \mid y \neq -3, y \neq 4, y \neq -6\}$.

$$\begin{aligned}
\frac{2}{y+3} + \frac{3}{y-4} &= \frac{5}{y+6} \\
(y+3)(y-4)(y+6)\left[\frac{2}{y+3} + \frac{3}{y-4}\right] &= (y+3)(y-4)(y+6\left[\frac{5}{y+6}\right] \\
2(y-4)(y+6) + 3(y+3)(y+6) &= 5(y+3)(y-4) \\
2(y^2 + 2y - 24) + 3(y^2 + 9y + 18) &= 5(y^2 - y - 12) \\
2y^2 + 4y - 48 + 3y^2 + 27y + 54 &= 5y^2 - 5y - 60 \\
5y^2 + 31y + 6 &= 5y^2 - 5y - 60 \\
(5y^2 + 31y + 6) - 5y^2 &= (5y^2 - 5y - 60) - 5y^2 \\
31y + 6 &= -5y - 60 \\
(31y + 6) - 6 &= (-5y - 60) - 6 \\
31y &= -5y - 66 \\
31y + 5y &= (-5y - 66) + 5y \\
36y &= -66 \\
\frac{36y}{36} &= \frac{-66}{36} \\
y &= \frac{-11}{6}
\end{aligned}$$

Check:

$$\frac{2}{y+3} + \frac{3}{y-4} = \frac{5}{y+6}$$

$$\frac{2}{\frac{-11}{6}+3} + \frac{3}{\frac{-11}{6}-4} \stackrel{?}{=} \frac{5}{\frac{-11}{6}+6}$$

$$\frac{2}{\frac{-11+18}{6}} + \frac{3}{\frac{-11-24}{6}} \stackrel{?}{=} \frac{5}{\frac{-11+36}{6}}$$

$$\frac{2}{\frac{7}{6}} + \frac{3}{\frac{-35}{6}} \stackrel{?}{=} \frac{5}{\frac{25}{6}}$$

$$\frac{12}{7} + \frac{18}{-35} \stackrel{?}{=} \frac{3}{2}$$

$$\frac{6- - 18}{35} \stackrel{?}{=} \frac{6}{5}$$

$$\frac{42}{35} \stackrel{?}{=} \frac{6}{5}$$

$$\frac{6}{5} = \frac{6}{5}$$

59.

$$\frac{x}{x^2-1} - \frac{x+3}{x^2-x} = \frac{-3}{x^2+x}$$

$$\frac{x}{(x-1)(x+1)} - \frac{x+3}{x(x-1)} = \frac{-3}{x(x+1)}$$

The domain is $\{x \mid x \neq -1, x \neq 0, x \neq 1\}$.

$$x(x-1)(x+1)\left[\frac{x}{(x-1)(x+1)} - \frac{x+3}{x(x-1)}\right] = \left[\frac{-3}{x(x+1)}\right]x(x-1)(x+1)$$

$$x^2 - (x+3)(x+1) = -3(x-1)$$

$$x^2 - (x^2+4x+3) = -3x+3$$

$$-4x-3 = -3x+3$$

$$-x = 6$$

$$x = -6$$

Check:

$$\frac{x}{x^2-1} - \frac{x+3}{x^2-x} = \frac{-3}{x^2+x}$$

$$\frac{-6}{(-6)^2-1} - \frac{(-6)+3}{(-6)^2-(-6)} = \frac{-3}{(-6)^2+(-6)}$$

$$\frac{-6}{35} - \frac{-3}{42} = \frac{-3}{30}$$

$$\frac{-6}{35} + \frac{1}{14} = \frac{-1}{10}$$

$$\frac{-12}{70} + \frac{5}{70} = \frac{-7}{70}$$

$$\frac{-7}{70} = \frac{-7}{70}$$

61.

$$3.2x + \frac{21.3}{65.871} = 19.23$$
$$3.2x + \frac{21.3}{65.871} - \frac{21.3}{65.871} = 19.23 - \frac{21.3}{65.871}$$
$$3.2x = 19.23 - \frac{21.3}{65.871}$$
$$\frac{3.2x}{3.2} = \frac{19.23 - \frac{21.3}{65.871}}{3.2}$$
$$x = 5.9083252$$
$$x \approx 5.91$$

63.

$$14.72 - 21.58x = \frac{18}{2.11}x + 2.4$$
$$14.72 - 21.58x - 2.4 = \frac{18}{2.11}x + 2.4 - 2.4$$
$$14.72 - 21.58x - 2.4 + 21.58x = \frac{18}{2.11}x + 21.58x$$
$$14.72 - 2.4 = x\left(\frac{18}{2.11} + 21.58\right)$$
$$\frac{14.72 - 2.4}{\frac{18}{2.11} + 21.58} = x$$
$$0.4091554 = x$$
$$x \approx 0.41$$

65.

$$x^2 - 7x + 12 = 0$$
$$(x - 3)(x - 4) = 0$$
$$x - 3 = 0 \quad \text{or} \quad x - 4 = 0$$
$$x = 3 \quad \text{or} \quad x = 4$$

The solution set is $\{3, 4\}$.

Check: $x = 3$:
$$3^2 - 7 \cdot 3 + 12 \stackrel{?}{=} 0$$
$$9 - 21 + 12 \stackrel{?}{=} 0$$
$$-12 + 12 \stackrel{?}{=} 0$$
$$0 \stackrel{?}{=} 0$$

So 3 is a solution.

$x = 4$:
$$4^2 - 7 \cdot 4 + 12 \stackrel{?}{=} 0$$
$$16 - 28 + 12 \stackrel{?}{=} 0$$
$$-12 + 12 \stackrel{?}{=} 0$$
$$0 \stackrel{?}{=} 0$$

So 4 is a solution.

67.

$$
\begin{aligned}
2x^2 + 5x - 3 &= 0 \\
(x + 3)(2x - 1) &= 0 \\
x + 3 = 0 \quad &\text{or} \quad 2x - 1 = 0 \\
x = -3 \quad &\text{or} \quad 2x = 1 \\
& \quad x = \frac{1}{2}
\end{aligned}
$$

The solution set is $\left\{-3, \frac{1}{2}\right\}$

Check: $x = -3$:

$$
\begin{aligned}
2(-3)^2 + 5(-3) - 3 &\overset{?}{=} 0 \\
2(9) + 5(-3) - 3 &\overset{?}{=} 0 \\
18 - 15 - 3 &\overset{?}{=} 0 \\
3 - 3 &\overset{?}{=} 0 \\
0 &= 0
\end{aligned}
$$

So -3 is a solution.

$x = \frac{1}{2}$:

$$
\begin{aligned}
2\left(\frac{1}{2}\right)^2 + 5\left(\frac{1}{2}\right) - 3 &\overset{?}{=} 0 \\
2\left(\frac{1}{4}\right) + \frac{5}{2} - 3 &\overset{?}{=} 0 \\
\frac{1}{2} + \frac{5}{2} - 3 &\overset{?}{=} 0 \\
3 - 3 &\overset{?}{=} 0 \\
0 &= 0
\end{aligned}
$$

So $\frac{1}{2}$ is a solution.

69.

$$
\begin{aligned}
x^3 &= 9x \\
x^3 - 9x &= 0 \\
x(x^2 - 9) &= 0 \\
x(x - 3)(x + 3) &= 0 \\
x = 0 \text{ or } x - 3 = 0 &\text{ or } x + 3 = 0 \\
x = 3 \quad & \quad x = -3
\end{aligned}
$$

The solution set is $\{-3, 0, 3\}$.

Check: $x = -3$: $(-3)^3 \overset{?}{=} 9(-3) \Rightarrow -27 = -27$

$x = 0$: $0^3 \overset{?}{=} 9(0) \Rightarrow 0 = 0$

$x = 3$: $3^3 \overset{?}{=} 9(3) \Rightarrow 27 = 27$

So $-3, 0, 3$ are solutions.

71.

$$
\begin{aligned}
x^3 + x^2 - 20x &= 0 \\
x(x^2 + x - 20) &= 0 \\
x(x + 5)(x - 4) &= 0 \\
x = 0 \quad \text{or} \quad x + 5 = 0 &\quad \text{or} \quad x - 4 = 0 \\
x = -5 \quad & \quad x = 4
\end{aligned}
$$

The solution set is $\{-5, 0, 4\}$.

Check: $x = 0$: $0(0 + 5)() - 4) = 0(5)(-4) = 0$

So 0 is a solution.

$x = -5$: $-5(-5 + 5)(-5 - 4) = -5(0)(-9) = 0$

So -5 is a solution.

$x = 4$: $4(4 + 5)(4 - 4) = 4(9)(0) = 0$

So 4 is a solution.

73.

$$
\begin{aligned}
x^3 + x^2 - x - 1 &= 0 \\
(x^3 + x^2) - 1(x + 1) &= 0 \\
x^2(x + 1) - 1(x + 1) &= 0 \\
(x^2 - 1)(x + 1) &= 0 \\
(x - 1)(x + 1)(x + 1) &= 0 \\
x - 1 = 0 \quad &\text{or} \quad x + 1 = 0 \\
x = 1 \quad &\text{or} \quad x = -1
\end{aligned}
$$

The solution set is $\{-1, 1\}$.

Check: $x = -1$: $-1^3 + (-1)^2 - (-1) - 1 = -1 + 1 + 1 - 1 = 0$

$x = 1$: $1^3 + 1^2 - 1 - 1 = 1 + 1 - 1 - 1 = 0$

So -1 and 1 are solutions.

75.
$$\begin{aligned}
x^3 - 3x^2 - 4x + 12 &= 0\\
(x^3 - 3x^2) - 4(x - 3) &= 0\\
x^2(x - 3) - 4(x - 3) &= 0\\
(x^2 - 4)(x - 3) &= 0\\
(x - 2)(x + 2)(x - 3) &= 0
\end{aligned}$$
$$x - 2 = 0 \text{ or } x + 2 = 0 \text{ or } x - 3 = 0$$
$$x = 2 \text{ or } x = -2 \text{ or } x = 3$$
The solution set is $\{-2, 2, 3\}$.

Check: $x = 2$: $2^3 - 3(2)^2 - 4(2) + 12 = 8 - 12 - 8 + 12 = 0$

$x = -2$: $(-2)^3 - 3(-2)^2 - 4(-2) + 12 = -8 - 12 + 8 + 12 = 0$

$x = 3$: $3^3 - 3(3)^2 - 4(3) + 12 = 27 - 27 - 12 + 12 = 0$

All solutions check.

77.
$$\begin{aligned}
ax - b &= c \text{ where } a \neq 0\\
(ax - b) + b &= c + b\\
ax &= c + b\\
\frac{ax}{a} &= \frac{c + b}{a} \text{ since } a \neq 0\\
x &= \frac{c + b}{a} = \frac{b + c}{a}
\end{aligned}$$

Check:
$$\begin{aligned}
ax - b &= c\\
a\frac{c + b}{a} - b &\stackrel{?}{=} c\\
(c + b) - b &\stackrel{?}{=} c\\
c &= c
\end{aligned}$$

79.
$$\begin{aligned}
\frac{x}{a} + \frac{x}{b} &= c \quad \text{where } a \neq 0,\ b \neq 0\\
(ax - b) + b &= c + b\\
ab\left[\frac{x}{a} + \frac{x}{b}\right] &= ac[c]\\
bx + ax &= abc\\
(b + a)x &= abc\\
\frac{(b + a)x}{b + a} &= \frac{abc}{b + a}\\
x &= \frac{abc}{b + a} = \frac{abc}{a + b}
\end{aligned}$$

Check:
$$\begin{aligned}
\frac{x}{a} + \frac{x}{b} &= c\\
\frac{\frac{abc}{a + b}}{a} + \frac{\frac{abc}{a + b}}{b} &\stackrel{?}{=} c\\
\frac{abc}{a(a + b)} + \frac{abc}{b(a + b)} &\stackrel{?}{=} c\\
\frac{ab^2c + a^2bc}{ab(a + b)} &\stackrel{?}{=} c\\
\frac{abc[b + a]}{ab(a + b)} &\stackrel{?}{=} c\\
c &= c
\end{aligned}$$

81.
$$\begin{aligned}
\frac{1}{x - a} + \frac{1}{x + a} &= \frac{2}{x - 1}\\
(x - a)(x + a)(x - 1)\left[\frac{1}{x - a} + \frac{1}{x + a}\right] &= (x - a)(x + a)(x - 1)\left[\frac{2}{x - 1}\right]\\
(x + a)(x - 1) + (x - a)(x - 1) &= 2(x - a)(x + a)\\
x^2 - x + ax - a + x^2 - x - ax + a &= 2(x^2 - a^2)\\
2x^2 - 2x &= 2x^2 - 2a^2\\
(2x^2 - 2x) - 2x^2 &= (2x^2 - 2a^2) - 2x^2\\
-2x &= -2a^2\\
\frac{-2x}{-2} &= \frac{-2a^2}{-2}\\
x &= a^2
\end{aligned}$$

Check:

$$\begin{aligned}
\frac{1}{x-a} + \frac{1}{x+a} &= \frac{2}{x-1} \\
\frac{1}{a^2-a} + \frac{1}{a^2+a} &\stackrel{?}{=} \frac{2}{a^2-1} \\
\frac{1}{a(a-1)} + \frac{1}{a(a+1)} &\stackrel{?}{=} \frac{2}{a^2-1} \\
\frac{a+1+a-1}{a(a-1)(a+1)} &\stackrel{?}{=} \frac{2}{a^2-1} \\
\frac{2a}{a(a-1)(a+1)} &\stackrel{?}{=} \frac{2}{a^2-1} \\
\frac{2}{(a-1)(a+1)} &= \frac{2}{(a-1)(a+1)}
\end{aligned}$$

83. Find a if $x = 4$ in $x + 2a = 16 + ax - 6a$

$$\begin{aligned}
4 + 2a &= 16 + a(4) - 6a \\
4 + 2a &= 16 + 4a - 6a \\
4 + 2a &= 16 - 2a \\
(4 + 2a) - 4 &= (16 - 2a) - 4 \\
2a &= 12 - 2a \\
2a + 2a &= 12 - 2a + 2a \\
4a &= 12 \\
\frac{4a}{4} &= \frac{12}{4} \\
a &= 3
\end{aligned}$$

Check:

$$\begin{aligned}
x + 2a &= 16 + ax - 6a \\
4 + 2(3) &\stackrel{?}{=} 16 + 3(4) - 6(3) \\
4 + 6 &\stackrel{?}{=} 16 + 12 - 18 \\
10 &= 10
\end{aligned}$$

85. Solve for R in $\frac{1}{R} = \frac{1}{R_1} + \frac{1}{R_2}$

$$\begin{aligned}
RR_1R_2\left[\frac{1}{R}\right] &= RR_1R_2\left[\frac{1}{R_1} + \frac{1}{R_2}\right] \\
R_1R_2 &= RR_2 + RR_1 \\
R_1R_2 &= R(R_2 + R_1) \\
\frac{R_1R_2}{R_2 + R_1} &= \frac{R(R_2 + R_1)}{R_2 + R_1} \\
\frac{R_1R_2}{R_1 + R_2} &= R \\
R &= \frac{R_1R_2}{R_1 + R_2}
\end{aligned}$$

Check:

$$\begin{aligned}
\frac{1}{R} &= \frac{1}{R_1} + \frac{1}{R_2} \\
\frac{1}{\frac{R_1R_2}{R_1 + R_2}} &\stackrel{?}{=} \frac{1}{R_1} + \frac{1}{R_2} \\
\frac{R_1 + R_2}{R_1R_2} &\stackrel{?}{=} \frac{R_2 + R_1}{R_1R_2} \\
\frac{R_1 + R_2}{R_1R_2} &= \frac{R_1 + R_2}{R_1R_2}
\end{aligned}$$

87. Solve for R in $F = \frac{mv^2}{R}$

$$\begin{aligned}
R[F] &= R\left[\frac{mv^2}{R}\right] \\
RF &= mv \\
\frac{RF}{F} &= \frac{mv^2}{F} \\
R &= \frac{mv^2}{F}
\end{aligned}$$

Check:

$$\begin{aligned}
F &= \frac{mv^2}{R} \\
F &\stackrel{?}{=} \frac{mv^2}{\frac{mv^2}{F}} \\
F &\stackrel{?}{=} \frac{mv^2}{1} \cdot \frac{F}{mv^2} \\
F &= F
\end{aligned}$$

89. Solve for r in $S = \dfrac{a}{1 - r}$

$$(1 - r)[S] = 1 - r\left[\frac{a}{1 - r}\right]$$
$$S - rS = a$$
$$(S - rS) - S = a - S$$
$$-rS = a - S$$
$$\frac{-rS}{-S} = \frac{a - S}{-S}$$
$$r = \frac{a - S}{-S}$$
$$r = \frac{-(a - S)}{S}$$
$$r = \frac{S - a}{S}$$

Check:
$$S = \frac{a}{1 - r}$$
$$S \stackrel{?}{=} \frac{a}{1 - \dfrac{S - a}{S}}$$
$$S \stackrel{?}{=} \frac{a}{\dfrac{S - (S - a)}{S}}$$
$$S \stackrel{?}{=} \frac{a}{\dfrac{S - S + a}{S}}$$
$$S \stackrel{?}{=} \frac{a}{\dfrac{a}{S}}$$
$$S \stackrel{?}{=} a \cdot \frac{S}{a}$$
$$S = S$$

91. In step (6), we have $(x - 2)(x + 5) = (x - 2)(x + 4)$. In step (7), we cannot divide by $x - 2$ because $x = 2$ from step (1) which means we actually divided by 0. Step (7) should read:

$$(x - 2)(x + 5) - (x - 2)(x + 4) = 0$$

Then,
$$(x - 2)([(x + 5) - (x + 4)] = 0$$
$$(x - 2)(1) = 0$$
$$x - 2 = 0$$
$$x = 2$$

2.2 Setting Up Equations: Applications

1. Let A represent area of the circle and r the radius:
Area of circle is the product of π times the square of the radius.
$A \quad = \quad \pi \quad \cdot \quad r^2$
$$A = \pi r^2$$

3. Let A represent the area of the square and s the length of a side:
Area of a square is the square of the length of a side.
$A \quad = \quad s^2$
$$A = s^2$$

5. Let F represent the force, m the mass, and a the acceleration:
Force equals the product of mass times acceleration.
$F \quad = \quad m \quad \cdot \quad a$
$$F = ma$$

7. Let W represent the work, F the force, and d the distance:
Work equals force times distance.
$W \quad = \quad F \quad \cdot \quad d$
$$W = Fd$$

9. C = total variable cost, x = number of dishwashers manufactured.
$C = 150x$

11.

Amount in Bonds	Amount in CD's	Total
x	$x - 2000$	20,000

$$\begin{aligned} x + (x - 2000) &= 20000 \\ 2x - 2000 &= 20000 \\ 2x &= 22000 \\ x &= 11000 \end{aligned}$$

\$11,000 will be invested in bonds. \$9,000 will be invested in CD's.

13.

Scott	Alice	Tricia	Total
x	$\frac{3}{4}x$	$\frac{1}{2}x$	900,000

$$\begin{aligned} x + \frac{3}{4}x + \frac{1}{2}x &= 900{,}000 \\ \left(1 + \frac{3}{4} + \frac{1}{2}\right)x &= 900{,}000 \\ \frac{9}{4}x &= 900{,}000 \\ x &= \frac{4}{9}(900{,}000) \\ x &= 400{,}000 \end{aligned}$$

Scott receives \$400,000. Alice receives \$300,000. Tricia receives \$200,000.

15. **Step 1:** We are being asked for an hourly wage in dollars per hour.
Step 2: Let x represent the hourly wage.
Step 3: We set up a table:

	Hourly Wage	Salary
Regular hours, 40	x	$40x$
Overtime hours, 8	$1.5x$	$8(1.5x) = 12x$

The total of regular salary plus overtime is \$442.00, then

Step 4:

$$\begin{aligned} 40x + 12x &= 442 \\ 52x &= 442 \\ x &= 8.50 \end{aligned}$$

The regular hourly wage is \$8.50 per hour.

Step 5: Forty hours yields a salary of $40(8.50) = \$340$, and 8 hours of overtime yields a salary of $8(1.5)(8.50) = \$102$, for a total of \$442.

17. **Step 1:** We are being asked to find the number of touchdowns scored by the Bears.
Step 2: Let x represent the number of touchdowns scored.

Step 3: We set up a table:

	Point Value	Points Earned
Safeties, 1	2	$(1)(2) = 2$
Field goals, 2	3	$(2)(6) = 6$
Touchdowns without extra points, 2	6	$(2)(6) = 12$
Touchdowns with extra points, $x - 2$	7	$(x - 2)(7) = 7x - 14$

The total points scored is 41; thus

$$2 + 6 + 12 + 7x - 14 = 41$$

Step 4:

$$7x + 6 = 41$$
$$7x = 35$$
$$x = 5$$

There were two touchdowns without extra points and three touchdowns with extra points for a total of 5 touchdowns.

Step 5: Two safeties (for 2 points) and two field goals (for 6 points) and two touchdowns without extra points (for 12 points) and three touchdowns with extra points (for 21 points) give a total of $2 + 6 + 12 + 21 = 41$ points.

19. ℓ = length, w = width

$$2\ell + 2w = 60 \qquad \text{Perimeter} = 2\ell + 2w$$
$$\ell = w + 8 \qquad \text{The length is 8 more than the width.}$$
$$2(w + 8) + 2w = 60$$
$$2w + 16 + 2w = 60$$
$$4w + 16 = 60$$
$$4w = 44$$
$$w = 11 \text{ feet}$$
$$\ell = 19 \text{ feet}$$

The length is 19 ft.; the width is 11 ft.

21. **Step 1:** We want to find two dollar amounts, the principle to invest in B-rated bonds pay 15% per year and the principle to invest in a CD paying 7% per year.

Step 2: Let x represent the amount invested in bonds at 15%. Then $50{,}000 - x$ is the amount that will be invested in a CD at 7%.

Step 3: We make a table:

	Principle	Rate	Time (yr)	Interest
Bonds at 15%	x	0.15	1	0.15
CD at 7%	$50{,}000 - x$	0.07	1	$0.07(50{,}000 - x)$

Since the total interest is to be $6000, we have

$$0.15x + 0.07\,(50{,}000 - x) = 6000$$

Step 4:

$$15x + 7(50{,}000 - x) = 600{,}000$$
$$15x + 350{,}000 - 7x = 600{,}000$$
$$350{,}000 + 8x = 600{,}000$$
$$8x = 250{,}000$$
$$x = 31{,}250$$

Thus, $31,250 will be invested in bonds at 15% and $18,750 in a CD at 7%.

Step 5: The interest on the bond after one year is $0.15(31{,}250) = \$4687.50$, and the interest on the CD is $0.07(18{,}750) = \$1312.50$, for a total interest of $\$4687.50 + \$1312.50 = \$6000.00$.

23. **Step 1:** We want to find the dollar amount loaned at 8%.

Step 2: Let x represent the amount invested at 8%. Then the amount invested at 18% is $12{,}000 - x$.

Step 3: We make a table:

	Principle	Rate	Time (yr.)	Interest
Loan at 8%	x	0.08	1	$0.08x$
Loan at 18%	$12{,}000 - x$	0.18	1	$0.18(12{,}000 - x)$

Since the total interest is to be $1000, we have

$$0.8x + 0.18(12{,}000 - x) = 1000$$

Step 4:

$$\begin{aligned} 8x + 18(12{,}000 - x) &= 100{,}000 \\ 8x + 216{,}000 - 18x &= 100{,}000 \\ -10x &= -116{,}000 \\ x &= 11{,}600 \end{aligned}$$

Thus, $11,600 will be loaned at 8% and $400 at 18%.

Step 5: The interest on the loan at 8% is 0.08(11,600) = $928, and the interest on the loan at 18% is 0.18(400) = $72; thus the total interest is $1000.

25. **Step 1:** We want to find the amounts in cubic centimeters of 15% and 5% HCl for the mixture.

Step 2: Let x represent the amount of 15% HCl. Then the amount of 5% HCl would be $100 - x$.

Step 3: We can make a table:

	Amount of solution	Percent	Amount HCl
15% HCl	x	0.15	$0.15x$
5% HCl	$100 - x$	0.05	$0.05(100 - x)$
8% HCl	100	0.08	$0.08(100) = 8$

The amount of HCl in the 8% solution must equal the amounts of HCl in the 15% and the 5% solutions:

$$0.15x + .05(100 - x) = 8$$

Step 4:

$$\begin{aligned} 15x + 500 - 5x &= 800 \\ 10x &= 300 \\ x &= 30 \end{aligned}$$

Thus, we mix 30 cubic centimeters of the 15% HCl solution with 70 cubic centimeters of the 5% HCl solution.

Step 5: Computing the amounts of HCl, we mix 30(0.15) = 4.5 with 70(0.05) = 3.5 cubic centimeters to get 100(0.08) = 8 cubic centimeters.

27. **Step 1:** We are asked to find the number of pounds of cashews.

Step 2: Let x be the weight in pounds of the cashews. Then the total weight of the mixture will be $60 + x$.

Step 3: We make a table:

	Price Per Pound	Total Value
Peanuts, 60	$1.50	1.50(60)
Cashews, x	$4.00	$4.00(x)$
Mixture, $60 + x$	$2.50	$2.50(60 + x)$

Since the profit is to remain the same, we have

$$1.50(60) + 4.00(x) = 2.50(60 + x)$$

Step 4:

$$90 + 4x = 150 + 2.5x$$
$$.5x = 60$$
$$x = 40$$

Thus, 40 pounds of cashews must be added to 60 pounds of peanuts.

Step 5: The peanuts are worth 60(1.50) = \$90, and the cashews are worth 40(4.00) = \$160. Thus the mixture is worth 100(2.50) = \$250, and the profit does not change.

29.

	Amount	Concentration of Water	Pure Water
Water	x	100% = 1.0	x
40% Solution	20	60% = .6	.6(20)
30% Solution	$x + 20$	70% = .7	$.7(x + 20)$

$$x + .6(20) = .7(x + 20)$$
$$x + 12 = .7x + 14$$
$$.3x = 2$$
$$x = \frac{20}{3} = 6\frac{2}{3}$$

$\frac{20}{3}$ ounces of pure water should be added.

31. **Step 1:** We want to find the number of adult tickets sold.

Step 2: Let x be the number of adult tickets sold. Then the number of children's tickets is $5200 - x$.

Step 3: We make a table:

Tickets	Price	Receipts
Adults, x	\$4.75	$4.75x$
Children's, $5200 - x$	\$2.50	$2.50(5200 - x)$
Total, 5200	--	\$20,335

Equating the receipts, we have

$$4.75x + 2.50(5200 - x) = 20,335$$

Step 4:

$$4.75x + 13000 - 2.5x = 20,335$$
$$2.25x = 7335$$
$$x = 3260$$

Thus, there were 3260 adults and 5200 − 3260 = 1940 children's tickets sold.

Step 5: Receipts of 4.75(3260) = \$15,485 (Adults) and 2.50(1940) = \$4850 (Children). The total is \$15,485 + \$4850 = \$20,335.

33. **Step 1:** We are asked to find the original price and *difference* between the original price and the new price.

Step 2: Let x be the *original price* of the house.

Step 3: Note that

$$\text{original price} = x$$
$$\text{amount reduced} = 0.15x$$
$$\text{new price} = \$125,000$$
$$\text{Original price} - \text{amount reduced} = \text{new price}$$
$$x - 0.15x = \$125,000$$

Step 4:

$$.85x = \$125,000$$
$$x = \$147,058.82$$

Thus, the original price is \$147,058.82, and the amount of the savings is 0.15(\$147,058.82) = \$22,058.82.

Step 5: The original price \$147,058.82 less the 15% reduction 0.15(147,058.82) gives the new price \$125,000.

35. **Step 1:** We are asked to find the amount the bookstore paid for the book.
Step 2: Let x be the bookstore's price for the book.
Step 3: Note that

$$\begin{aligned} \text{bookstore price} &= x \\ \text{percent increase} &= 0.25x \\ \text{selling price} &= \$56.00 \\ \text{bookstore price} + \text{percent increase} &= \text{selling price} \\ x + 0.25x &= \$56.00 \end{aligned}$$

Step 4:

$$\begin{aligned} 1.25x &= \$56.00 \\ x &= \$44.80 \end{aligned}$$

Step 5: The amount that the bookstore paid for the book was \$44.80. The 25% markup of the price increases the cost of the book \$11.20, making the selling price of the book \$56.00.

37. **Step 1:** We want to find a time in minutes.
Step 2: Let t be the time in minutes it takes for them to do the job together. Then in one minute Trent can do $\frac{1}{30}$ of the job, Lois can do $\frac{1}{20}$, and together they do $\frac{1}{t}$ of the job.
Step 3: We make a table:

	Minutes to do job	Part of job done in one minute
Trent	30	$\frac{1}{30}$
Lois	20	$\frac{1}{20}$
Together	t	$\frac{1}{t}$

Thus, $\frac{1}{30} + \frac{1}{20} = \frac{1}{t}$

Step 4:

$$\begin{aligned} \frac{2 + 3}{60} &= \frac{1}{t} \\ \frac{5}{60} &= \frac{1}{t} \\ \frac{1}{12} &= \frac{1}{t} \\ t &= 12 \end{aligned}$$

Thus, working together, the job can be done in 12 minutes.

Step 5: Trent does $\frac{1}{30}$ of the job in one minute.

Lois does $\frac{1}{20}$ of the job in one minute.

In one minute working together they do $\frac{1}{30} + \frac{1}{20}$ of the job.

In 12 minutes they do $12\left(\frac{1}{30} + \frac{1}{20}\right) = 12\left(\frac{5}{60}\right) = 1$ of the job, or ***all*** of it.

39. **Step 1:** We want to find a final exam score.
Step 2: Let x be the final exam score to yield an average of 80 when counted twice and combined with the other scores.
Step 3: The total points scored with the exam counted twice is:

$$80 + 83 + 71 + 61 + 95 + 2x = 390 + 2x$$

Averaging, we get $\frac{390 + 2x}{7} = 80$

Step 4: $390 + 2x = 560$

$$2x = 170$$
$$x = 85$$

Thus, Sarah needs a score of 85 to get an average of 80 if the final counts as two tests.

Step 5: $$\frac{80 + 83 + 71 + 61 + 95 + 2(85)}{7} = 80$$

41. **Step 1:** We want to find a speed in miles per hour.
Step 2: Let v be the speed of the current in miles per hour.
Step 3: We make a table (using $s = vt$):

	Velocity of boat	Time (hr.)	Distance (mi.)
Upstream	$16 - v$	$\frac{20}{60} = \frac{1}{3}$	$\frac{16 - v}{3}$
Downstream	$16 + v$	$\frac{15}{60} = \frac{1}{4}$	$\frac{16 + v}{4}$

Since the distance is the same in each direction:

$$\frac{16 - v}{3} = \frac{16 + v}{4}$$

Step 4:
$$4(16 - v) = 3(16 + v)$$
$$64 - 4v = 48 + 3v$$
$$16 = 7v$$
$$v = \frac{16}{7} \approx 2.286$$

Thus, the speed of the current is 2.286 miles per hour.

Step 5: The distance is the same in each direction:

$$\frac{16 - v}{3} = \frac{16 - 2.286}{3} = 4.57 \text{ miles}$$

$$\frac{16 + v}{4} = \frac{16 + 2.286}{4} = 4.57 \text{ miles}$$

43. ℓ = length of garden
w = width of garden

(a) If the length of garden is to be twice its width, then the width is to be half the length. Thus, $w = \frac{1}{2}\ell$

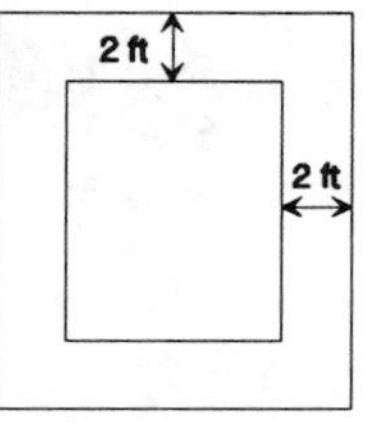

The dimensions of the fence are $\ell + 4$ and $w + 4$, which means the dimensions are $\ell + 4$ and $\frac{1}{2}\ell + 4$.

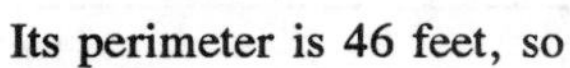

Its perimeter is 46 feet, so

$$2[(\ell + 4) + \frac{1}{2}\ell + 4] = 46$$
$$2\ell + 8 + \ell + 8 = 46$$
$$3\ell = 30$$
$$\ell = 10$$

The dimensions of the garden are 10 feet by 5 feet.

(b) Area $= \ell \cdot w = 10 \cdot 5 = 50$ square feet

(c) If the dimensions of the fence are the same, then the dimensions are $\ell + 4$ and $\ell + 4$. Its perimeter is 46 feet, so

$$
\begin{aligned}
2[(\ell + 4) + (\ell + 4)] &= 46 \\
2\ell + 8 + 2\ell + 8 &= 46 \\
4\ell &= 30 \\
\ell &= 7.5
\end{aligned}
$$

The dimensions of the garden are 7.5 feet by 7.5 feet.

(d) The area of this square garden is $\ell \cdot w\ (7.5)(7.5) = 56.25$ square feet.

45. **Step 1:** We want to find a position on a football field.

Step 2: Let s be the distance the tight end runs after catching the ball.

Step 3: We can make a table (using $v = \frac{s}{t}$ and $t = \frac{s}{v}$):

	Velocity	Time (seconds)	Distance (yards)
Tight end	$\frac{100}{12} = \frac{25}{3}$	$\frac{s}{\frac{25}{3}} = \frac{3s}{25}$	s
Defensive back	$\frac{100}{10} = 10$	$\frac{s + 5}{10}$	$s + 5$

Since the time is the same for both runners:

$$\frac{3s}{25} = \frac{s + 5}{10}$$

Step 4:

$$
\begin{aligned}
30s &= 25s + 125 \\
5s &= 125 \\
s &= 25
\end{aligned}
$$

Thus, the tight end goes 25 yards from his 20 yard line, or to his 45 yard line.

Step 5: Checking the times for the appropriate distances:

Tight end: $\dfrac{25 \text{ yds}}{\frac{25}{3} \text{ yds/sec}} = 3 \text{ sec}$

Defensive back: $\dfrac{30 \text{ yds}}{10 \text{ yds/sec}} = 3 \text{ sec}$

47. **Step 1:** We want to find an amount of water in gallons.

Step 2: Let x be the amount of water added in gallons.

Step 3: We can make a table:

	Amount	Percent antifreeze	Amount of antifreeze
Antifreeze	1	100	1
Water	x	0	0
Mixture	$1 + x$	60	$0.60(1 + x)$

The amount of antifreeze remains the same:

$$1 = .60(1 + x)$$

Step 4:

$$
\begin{aligned}
10 &= 6 + 6x \\
4 &= 6x \\
\frac{2}{3} &= x
\end{aligned}
$$

Thus, $\frac{2}{3}$ gallon of water must be added.

Step 5: We check by calculating the percentage of antifreeze:

$$1 \div 1\frac{2}{3} = 1 \div \frac{5}{3} = 1 \cdot \frac{3}{5} = .60 = 60\%$$

49.

	Amount	Concentration of Water	Pure Water
Water	x	$100\% = 1.0$	x
4% salt	32	$96\% = .96$	$.96(32)$
6% salt	$32 - x$	$94\% = .94$	$.94(32 - x)$

$$x + .94(32 - x) = .96(32)$$
$$x + 30.08 - .94x = 3.072$$
$$0.06x = 0.64$$
$$x = 10.67$$

10.67 ounces of water must be evaporated from the 4% salt solution.

51.

	Rate	Time	Distance
Metra commuter train	r	3	d
Amtrak train	$r + 50$	1	$d - 10$

$$\text{Rate} \times \text{Time} = \text{Distance}$$
$$(r + 50) \times 1 = d - 10$$
$$r + 50 = 3r - 10 \quad (d = 3r)$$
$$60 = 2r$$
$$30 = r$$

The Metra commuter train travels at a rate of 30 miles per hour and the Amtrak train travels at a rate of 80 miles per hour.

53.

	Amount	% Gold	Pure Gold
12 karat	x	$\frac{12}{24} = 50\% = 0.5$	$.5x$
Pure	$60 - x$	$\frac{24}{24} = 100\% = 1.0$	$50 - x$
16 karat	60	$\frac{16}{24} = 66\frac{2}{3}\% = 0.67$	$.67(60)$

$$.5x + (60 - x) = .67(60)$$
$$.5x + 60 - x = 40$$
$$-.5x = -20$$
$$x = 40$$

Mix 40 grams of 12 karat gold with 20 grams of pure gold.

55. **Step 1:** We want to find a time in minutes and a distance in miles.
Step 2: Let t be the time in minutes to run the mile.
Step 3: We can make a table:

	Minutes to run the mile	Time	Part of mile run in one minute	Distance
Mike	6	t	$\frac{1}{6}$	$\frac{1}{6}t$
Dan	9	$t + 1$	$\frac{1}{9}$	$\frac{1}{9}(t + 1)$

$$\frac{1}{6}t = \frac{1}{9}(t + 1)$$

Step 4:

$$\frac{1}{6}t = \frac{1}{9}t + \frac{1}{9}$$

$$\frac{3t - 2t}{18} = \frac{1}{9}$$

$$\frac{1}{18}t = \frac{1}{9}$$

$$t = 2$$

Thus, after 2 minutes, Mike will pass Dan.

Step 5: If Mike gives Dan a headstart of 1 minute, Mike will pass Dan $\frac{1}{3}$ of a mile from the start in 2 minutes.

57. **Step 1:** We want to find a time.
Step 2: Let t be the length of time the auxiliary pump must run.
Step 3: We can make a table:

	Hours to do job	Part of job done in 1 hour	Part of job done in 3 hours
Main pump	4	$\frac{1}{4}$	$\frac{3}{4}$
Auxiliary pump	9	$\frac{1}{9}$	--

$\frac{1}{4}$ of the job must be done by the auxiliary pump.

Step 4:

$$\left(\frac{1}{9}\right)(t) = \frac{1}{4}$$

$$\frac{t}{9} = \frac{1}{4}$$

$$4t = 9$$

$$t = \frac{9}{4} = 2\frac{1}{4}$$

Thus, the auxiliary pump must run $2\frac{1}{4}$ hours, and to finish at noon it must be started at 9:45 A.M.

Step 5: In 3 hours the main pump does $3\left(\frac{1}{4}\right) = \frac{3}{4}$ of the job.

In $2\frac{1}{4}$ hours the auxiliary pump does $2\frac{1}{4}\left(\frac{1}{9}\right) = \frac{1}{4}$ of the job.

59. **Step 1:** We want to find a time in minutes.
Step 2: Let t be the time in minutes to fill the tub with both faucets open and the stopper removed.

Step 3: We can make a table:

	Minutes to do job	Part of job done in one minute
Both faucets open	15	$\frac{1}{15}$
Stopper removed	20	$\frac{-1}{20}$
Both faucets open and stopper removed	t	$\frac{1}{t}$

$$\frac{1}{15} - \frac{1}{20} = \frac{1}{t}$$

Step 4:
$$\frac{4-3}{60} = \frac{1}{t}$$

$$\frac{1}{60} = \frac{1}{t}$$

$$t = 60$$

Thus, one hour is required to fill the tub.

Step 5: In one minute $\frac{1}{15} - \frac{1}{20}$ of the tub is filled.

In 60 minutes $60\left(\frac{1}{15} - \frac{1}{20}\right) = 1$ of the tub is filled.

61.

	Distance	Velocity	Time
Lewis	100	$\frac{100}{9.99}$	9.99
Burke	$\frac{100}{12}(9.99)$	$\frac{100}{12}$	9.99

In 9.99 seconds, Burke ran a distance of $\frac{100}{12}(9.99) = 83.25$ meters. Lewis would win by 16.75 meters.

2.3 Quadratic Equations

1.
$$x^2 = 9x$$
$$x^2 - 9x = 0$$
$$x(x-9) = 0$$
$$x = 0 \quad \text{or} \quad x - 9 = 0$$
$$x = 9$$
$$\{0, 9\}$$

3.
$$x^2 - 25 = 0$$
$$(x+5)(x-5) = 0$$
$$x + 5 = 0 \quad \text{or} \quad x - 5 = 0$$
$$x = -5 \qquad x = 5$$
$$\{-5, 5\}$$

5.
$$z^2 + z - 6 = 0$$
$$(z + 3)(z - 2) = 0$$
$$z + 3 = 0 \quad \text{or} \quad z - 2 = 0$$
$$z = -3 \qquad z = 2$$
$$\{-3, 2\}$$

7.
$$2x^2 - 5x - 3 = 0$$
$$(2x + 1)(x - 3) = 0$$
$$2x + 1 = 0 \quad \text{or} \quad x - 3 = 0$$
$$x = -\frac{1}{2} \qquad x = 3$$
$$\left\{-\frac{1}{2}, 3\right\}$$

9.
$$3t^2 - 48 = 0$$
$$3(t^2 - 16) = 0$$
$$3(t + 4)(t - 4) = 0$$
$$t + 4 = 0 \quad \text{or} \quad t - 4 = 0$$
$$t = -4 \qquad t = 4$$
$$\{-4, 4\}$$

11.
$$x(x - 8) + 12 = 0$$
$$x^2 - 8x + 12 = 0$$
$$(x - 2)(x - 6) = 0$$
$$x - 2 = 0 \quad \text{or} \quad x - 6 = 0$$
$$x = 2 \qquad x = 6$$
$$\{2, 6\}$$

13.
$$4x^2 + 9 = 12x$$
$$4x^2 - 12x + 9 = 0$$
$$(2x - 3)(2x - 3) = 0$$
$$2x - 3 = 0$$
$$x = \frac{3}{2}$$
$$\left\{\frac{3}{2}\right\}$$

15.
$$6(p^2 - 1) = 5p$$
$$6p^2 - 6 = 5p$$
$$6p^2 - 5p - 6 = 0$$
$$(2p - 3)(3p + 2) = 0$$
$$2p - 3 = 0 \quad \text{or} \quad 3p + 2 = 0$$
$$p = \frac{3}{2} \qquad p = -\frac{2}{3}$$
$$\left\{-\frac{2}{3}, \frac{3}{2}\right\}$$

17.
$$6x - 5 = \frac{6}{x}$$
$$6x^2 - 5x = 6$$
$$6x^2 - 5x - 6 = 0$$
$$(2x - 3)(3x + 2) = 0$$
$$2x - 3 = 0 \quad \text{or} \quad 3x + 2 = 0$$
$$x = \frac{3}{2} \qquad x = -\frac{2}{3}$$
$$\left\{-\frac{2}{3}, \frac{3}{2}\right\}$$

19.
$$\frac{4(x - 2)}{x - 3} + \frac{3}{x} = \frac{-3}{x(x - 3)}$$
$$4x(x - 2) + 3(x - 3) = -3$$
$$4x^2 - 8x + 3x - 9 = -3$$
$$4x^2 - 5x - 6 = 0$$
$$(4x + 3)(x - 2) = 0$$
$$4x + 3 = 0 \quad \text{or} \quad x - 2 = 0$$
$$x = -\frac{3}{4} \qquad x = 2$$
$$\left\{-\frac{3}{4}, 2\right\}$$

For Problems 21–39, we get each equation into standard form $ax^2 + bx + c = 0$ and use the quadratic formula,

$$x = \frac{-b \pm \sqrt{b^2 - 4ac}}{2a}$$

21. $x^2 - 4x + 2 = 0$

Here, $a = 1, b = -4, c = 2.$

$$x = \frac{-(-4) \pm \sqrt{(-4)^2 - 4(1)(2)}}{2(1)}$$
$$= \frac{-4 \pm \sqrt{16 - 8}}{2}$$
$$= \frac{4 \pm 2\sqrt{2}}{2}$$
$$= 2 \pm \sqrt{2}$$
$$\{2 - \sqrt{2}, 2 + \sqrt{2}\}$$

23. $x^2 - 4x - 1 = 0$

Here $a = 1, b = -4, c = -1.$

$$x = \frac{-(-4) \pm \sqrt{(-4)^2 - 4(1)(-1)}}{2(1)}$$
$$= \frac{4 \pm \sqrt{16 + 4}}{2}$$
$$= \frac{4 \pm \sqrt{20}}{2}$$
$$= \frac{4 \pm 2\sqrt{5}}{2}$$
$$= 2 \pm \sqrt{5}$$
$$\{2 - \sqrt{5}, 2 + \sqrt{5}\}$$

25. $2x^2 - 5x + 3 = 0$

Here $a = 2, b = -5, c = 3.$

$$x = \frac{-(-5) \pm \sqrt{(-5)^2 - 4(2)(3)}}{2(2)}$$
$$= \frac{-5 \pm \sqrt{25 - 24}}{4}$$
$$= \frac{5 \pm 1}{4}$$
$$\left\{1, \frac{3}{2}\right\}$$

27. $4y^2 - y + 2 = 0$

Here $a = 4, b = -1, c = 2.$

$$x = \frac{-(-1) \pm \sqrt{(-1)^2 - 4(4)(2)}}{2(4)}$$
$$= \frac{1 \pm \sqrt{1 - 32}}{8}$$
$$= \frac{1 \pm \sqrt{-31}}{8}$$

No real solution.

29.
$$4x^2 = 1 - 2x$$
$$4x^2 + 2x - 1 = 0$$

Here, $a = 4, b = 2, c = -1.$

$$x = \frac{-2 \pm \sqrt{2^2 - 4(4)(-1)}}{2(4)}$$
$$= \frac{-2 \pm \sqrt{4 + 16}}{8}$$
$$= \frac{-2 \pm 2\sqrt{5}}{8}$$
$$= \frac{-1 \pm \sqrt{5}}{4}$$
$$\left\{\frac{-1 - \sqrt{5}}{4}, \frac{-1 + \sqrt{5}}{4}\right\}$$

31.
$$4x^2 = 9x$$
$$4x^2 - 9x = 0$$

Here $a = 4, b = -9, c = 0.$

$$x = \frac{-9 \pm \sqrt{(-9)^2 - 4(4)(0)}}{2(4)}$$
$$= \frac{9 \pm \sqrt{81}}{8} = \frac{9 \pm 9}{8}$$
$$\left\{0, \frac{9}{4}\right\}$$

33. $9t^2 - 6t + 1 = 0$

Here $a = 9$, $b = -6$, $c = 1$.

$$x = \frac{-(-6) \pm \sqrt{(-6)^2 - 4(9)(1)}}{2(9)}$$

$$= \frac{6 \pm \sqrt{36 - 36}}{18} = \frac{1}{3}$$

$$\left\{\frac{1}{3}\right\}$$

35. $\frac{3}{4}x^2 - \frac{1}{4}x - \frac{1}{2} = 0$

Here $a = \frac{3}{4}$, $b = \frac{-1}{4}$, $c = \frac{-1}{2}$.

$$x = \frac{-\left(-\frac{1}{4}\right) \pm \sqrt{\left(-\frac{1}{4}\right)^2 - 4\left(-\frac{3}{8}\right)}}{2\left(\frac{3}{4}\right)}$$

$$= \frac{\frac{1}{4} \pm \sqrt{\frac{1}{16} + \frac{24}{16}}}{\frac{3}{2}}$$

$$= \left(\frac{1}{4} \pm \frac{5}{4}\right)\frac{2}{3}$$

$$= \frac{1 \pm 5}{6}$$

$$\{1, \frac{-2}{3}\} = \left\{\frac{-2}{3}, 1\right\}$$

37. $4 - \frac{1}{x} - \frac{2}{x^2} = 0$

$4x^2 - x - 2 = 0$

Here $a = 4$, $b = -1$, $c = -2$.

$$x = \frac{-(-1) \pm \sqrt{(-1)^2 - 4(4)(-2)}}{2(4)}$$

$$= \frac{1 \pm \sqrt{1 + 32}}{8}$$

$$= \frac{1 \pm \sqrt{33}}{8}$$

$$\left\{\frac{1 - \sqrt{33}}{8}, \frac{1 + \sqrt{33}}{8}\right\}$$

39. $3x = 1 - \frac{1}{x}$

$3x^2 = x - 1$

$3x^2 - x + 1 = 0$

Here $a = 3$, $b = -1$, $c = 1$.

$$x = \frac{-1(1) \pm \sqrt{(-1)^2 - 4(3)(1)}}{2(3)}$$

$$= \frac{1 \pm \sqrt{1 - 12}}{6}$$

$$= \frac{1 \pm \sqrt{-11}}{6}$$

No real solution.

41. $x^2 - 4.1x + 2.2 = 0$

Here $a = 1$, $b = -4.1$, $c = 2.2$.

$$x = \frac{-(-4.1) \pm \sqrt{(-4.1)^2 - 4(1)(2.2)}}{2(1)}$$

$$= \frac{4.1 \pm \sqrt{16.81 - 8.8}}{2}$$

$$= \frac{4.1 \pm \sqrt{8.01}}{2}$$

$$= \frac{4.1 \pm 3\sqrt{.89}}{2}$$

$$= \frac{4.1 \pm 3(.94)}{2}$$

$$\{0.63, 3.47\}$$

43. $x^2 + \sqrt{3}x - 3 = 0$

Here, $a = 1$, $b = \sqrt{3}$, $c = -3$.

$$x = \frac{-\sqrt{3} \pm \sqrt{\left(\sqrt{3}\right)^2 - 4(1)(-3)}}{2(1)}$$

$$= \frac{-\sqrt{3} \pm \sqrt{3 + 12}}{2}$$

$$= \frac{-\sqrt{3} \pm \sqrt{15}}{2} = \frac{-1.73 \pm 3.87}{2}$$

$$\{-2.80, 1.07\}$$

45. $\pi x^2 - x - \pi = 0$

Here $a = \pi$, $b = -1$, $c = -\pi$.

$$x = \frac{-(-1) \pm \sqrt{(-1)^2 - 4(\pi)(-\pi)}}{2(\pi)}$$
$$= \frac{1 \pm \sqrt{1 + 4\pi^2}}{2\pi}$$
$$= \frac{1 + 6.36}{6.28}$$

$\{-0.85, 1.17\}$

47. $3x^2 + 8\pi x + \sqrt{29} = 0$

Here, $a = 3$, $b = 8\pi$, $c = \sqrt{29}$.

$$x = \frac{-8\pi \pm \sqrt{64\pi^2 - 4(3)(\sqrt{29})}}{2(3)}$$
$$= \frac{-8\pi \pm \sqrt{64\pi^2 - 12\sqrt{29}}}{6}$$
$$= \frac{-8\pi \pm \sqrt{567.03}}{6}$$
$$= \frac{-8\pi \pm 23.81}{6}$$
$$x = -0.22, -8.16$$

$\{-8.16, -0.22\}$

49. $x^2 - 5 = 0$

$x^2 = 5$

$x = \pm\sqrt{5}$

$\{-\sqrt{5}, \sqrt{5}\}$

51. $16x^2 - 8x + 1 = 0$

$(4x - 1)(4x - 1) = 0$

$4x - 1 = 0$

$x = \frac{1}{4}$

$\frac{1}{4}$

53. $10x^2 - 19x - 15 = 0$

$(5x + 3)(2x - 5) = 0$

$5x + 3 = 0$ or $2x - 5 = 0$

$x = \frac{-3}{5}$ $\quad x = \frac{5}{2}$

$\left\{\frac{-3}{5}, \frac{5}{2}\right\}$

55. $2 + z = 6z^2$

$6z^2 - z - 2 = 0$

$(2z + 1)(3z - 2) = 0$

$2z + 1 = 0$ or $3z - 2 = 0$

$z = -\frac{1}{2}$ $\quad z = \frac{2}{3}$

$\left\{-\frac{1}{2}, \frac{2}{3}\right\}$

57. $x^2 + \sqrt{2}x = \frac{1}{2}$

$x^2 + \sqrt{2}x - \frac{1}{2} = 0$

$2x^2 + 2\sqrt{2}x - 1 = 0$

Here $a = 2$, $b = 2\sqrt{2}$, $c = -1$.

$$x = \frac{-2\sqrt{2} \pm \sqrt{(2\sqrt{2})^2 - 4(2)(-1)}}{2(2)}$$
$$= \frac{-2\sqrt{2} \pm \sqrt{16}}{4}$$
$$= \frac{-\sqrt{2} \pm 2}{2}$$

$\left\{\frac{-\sqrt{2} - 2}{2}, \frac{-\sqrt{2} + 2}{2}\right\}$

59. $x^2 + x = 4$

$x^2 + x - 4 = 0$

Here $a = 1$, $b = 1$, $c = -4$.

$$x = \frac{-1 \pm \sqrt{1^2 - 4(1)(-4)}}{2(1)}$$
$$= \frac{-1 \pm \sqrt{17}}{2}$$

$\left\{\frac{-1 - \sqrt{17}}{2}, \frac{-1 + \sqrt{17}}{2}\right\}$

In Problems 61-65, we use the discriminant $b^2 - 4ac$:

61. $2x^2 - 6x + 7 = 0$
    Here $a = 2$, $b = -6$, $c = 7$.
    $b^2 - 4ac = (-6)^2 - 4(2)(7)$
    $= -20 < 0$
    No real solution.

63. $9x^2 - 30x + 25 = 0$
    Here $a = 9$, $b = -30$, $c = 25$.
    $b^2 - 4ac = (-30)^2 - 4(9)(25)$
    $= 0$
    Repeated real solution.

65. $3x^2 + 5x - 8 = 0$
    Here $a = 3$, $b = 5$, $c = -8$.
    $b^2 - 4ac = 5^2 - 4(3)(-8) = 121$
    Two unequal real solutions.

In Problems 67-71, tell what number to add to complete square:

67. $x^2 + 8x$
    $x^2 + 8x + \left(\frac{8}{2}\right)^2$
    $x^2 + 8x + 16$
    $(x + 4)^2$
    i.e., add 16

69. $x^2 + \frac{1}{2}x$
    $x^2 + \frac{1}{2}x + \left(\frac{\frac{1}{2}}{2}\right)^2$
    $x^2 + \frac{1}{2}x + \frac{1}{16}$
    $\left(x + \frac{1}{4}\right)^2$
    i.e., add $\frac{1}{16}$

71. $x^2 - \frac{2}{3}x$
    $x^2 - \frac{2}{3}x + \left(\frac{\frac{2}{3}}{2}\right)^2$
    $x^2 - \frac{2}{3}x + \frac{1}{9}$
    $\left(x - \frac{1}{3}\right)^2$
    i.e., add $\frac{1}{9}$

73. $x^2 + 4x - 21 = 0$
    $x^2 + 4x = 21$
    We add $\left(\frac{4}{2}\right)^2$ or 4 to both sides:
    $x^2 + 4x + 4 = 21 + 4$
    $(x + 2)^2 = 25$
    $x + 2 = \pm\sqrt{25}$
    $x = -2 \pm \sqrt{25}$
    $= -2 \pm 5$
    $\{-7, 3\}$

75. $x^2 - \frac{1}{2}x = \frac{3}{16}$

We add $\left(\frac{\frac{1}{2}}{2}\right)^2$ or $\frac{1}{16}$

$$x^2 - \frac{1}{2}x + \frac{1}{16} = \frac{3}{16} + \frac{1}{16}$$
$$\left(x - \frac{1}{4}\right)^2 = \frac{4}{16}$$
$$\left(x - \frac{1}{4}\right)^2 = \frac{1}{4}$$
$$x - \frac{1}{4} = \pm\sqrt{\frac{1}{4}}$$
$$x = \frac{1}{4} \pm \sqrt{\frac{1}{4}}$$
$$= \frac{1}{4} \pm \frac{1}{2}$$
$$\left\{-\frac{1}{4}, \frac{3}{4}\right\}$$

77. $3x^2 + x - \frac{1}{2} = 0$

$$x^2 + \frac{1}{3}x = \frac{1}{6}$$

We add $\left(\frac{\frac{1}{3}}{2}\right)^2$ or $\frac{1}{36}$

$$x^2 + \frac{1}{3}x + \frac{1}{36} = \frac{1}{6} + \frac{1}{36}$$
$$\left(x + \frac{1}{6}\right)^2 = \frac{7}{36}$$
$$x + \frac{1}{6} = \pm\sqrt{\frac{7}{36}}$$
$$x + \frac{1}{6} = \frac{\pm\sqrt{7}}{6}$$
$$x = -\frac{1}{6} \pm \frac{\sqrt{7}}{6}$$
$$x = \frac{-1 \pm \sqrt{7}}{6}$$
$$\left\{\frac{-1 - \sqrt{7}}{6}, \frac{-1 + \sqrt{7}}{6}\right\}$$

79. Let w = width of opening and ℓ = length of opening.

$\ell = w + 2$

Area of opening $= w(w + 2) = 143$

$$w^2 + 2w = 143$$
$$w^2 + 2w - 143 = 0$$
$$(w + 13)(w - 11) = 0$$
$$w + 13 = 0 \qquad w - 11 = 0$$
$$w = -13 \qquad w = 11$$

Since measurements must be positive, the dimensions of the opening are 11 feet by 13 feet.

81. Let ℓ = length, w = width,

Perimeter $= 2\ell + 2w$.

(1) $\quad 2\ell + 2w = 26$

$$\ell w = 40$$

Simplifying (1), we get,

$$\ell + w = 13$$
$$w = 13 - \ell$$
$$\ell(13 - \ell) = 40$$
$$13\ell - \ell^2 = 40$$
$$\ell^2 - 13\ell + 40 = 0$$
$$(\ell - 5)(\ell - 8) = 0$$
$$\ell = 5 \qquad \ell = 8$$
$$w = 8 \qquad w = 5$$

The dimensions of the rectangle are 5 m by 8 m.

83. We want the dimensions of a box with square base and volume of 4 cubic feet. Let x be the side of the sheet of metal. Then $v = \ell w$ gives

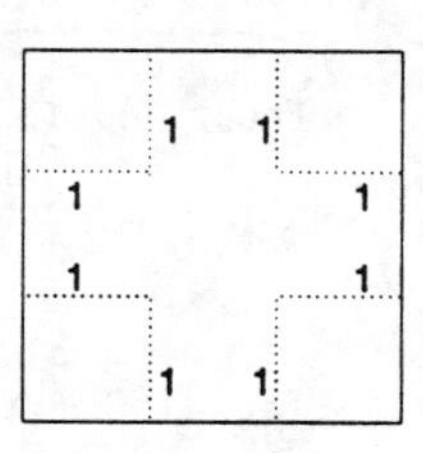

$$4 = (x - 2)(x - 2)$$
$$4 = x^2 - 4x + 4$$
$$0 = x^2 - 4x$$
$$0 = x(x - 4)$$
$$x = 0 \text{ or } x - 4 = 0$$
$$x = 4$$

Thus, the sheet should be 4 ft. $\times$ 4 ft. to give a box of $2 \times 2 = 4$ cu. ft.

85. (a) For $s = 96 + 80t - 16t^2$, $s = 0$ when ball strikes ground:

$$0 = 96 + 80t - 16t^2$$
$$0 = 16(6 + 5t - t^2)$$
$$0 = 6 + 5t - t^2$$
$$0 = t^2 - 5t - 6$$
$$0 = (t - 6)(t + 1)$$
$$t - 6 = 0 \quad \text{or} \quad t + 1 = 0$$
$$t = 6 \qquad\qquad t = -1$$

Thus, after 6 seconds, the ball strikes the ground.

(b) $s = 96$ when the ball is at the level of the top of the building:

$$96 = 96 + 80t - 16t^2$$
$$0 = 80t - 16t^2$$
$$0 = -16t(t - 5)$$
$$-16t = 0 \quad \text{or} \quad t - 5 = 0$$
$$t = 0 \qquad\qquad t = 5$$

Thus, the ball passes the top of the building after 5 seconds.

87. We want the time in days for Mike to do the job. Let t be this number. Then Dan takes $t + 5$ days.

	Days to do job	Part of job done in 1 day
Mike	t	$\frac{1}{t}$
Danny	$t + 5$	$\frac{1}{(t + 5)}$
Together	6	$\frac{1}{6}$

Then,

$$\frac{1}{t} + \frac{1}{t + 5} = \frac{1}{6}$$
$$6(t + 5) + 6t = t(t + 5)$$
$$6t + 30 + 6t = t^2 + 5t$$
$$0 = t^2 - 7t - 30$$
$$0 = (t - 10)(t + 3)$$
$$t - 10 = 0 \quad \text{or} \quad t + 3 = 0$$
$$t = 10 \qquad\qquad t = -3$$

Thus, Mike can do the job alone in 10 days, i.e., $(10)\frac{1}{10} = 1$ whole part of the job; and Danny can do the job alone in 15 days, i.e., $15\left(\frac{1}{15}\right) = 1$ whole part of the job.

89. We want to find the number x of boxes over 150 ordered to make a total of \$30,625. We make a table:

Number of boxes	Price each	Total Cost
$150 + x$	$\$200 - x$	$(150 + x)(200 - x)$

Thus,

$$(150 + x)(200 - x) = 30{,}625$$
$$30{,}000 - 50x - x^2 = 30{,}625$$
$$0 = x^2 + 50x + 625$$
$$0 = (x + 25)(x + 25)$$
$$x = 25$$

Thus, $150 + x = 175$ boxes are ordered at $200 - 25 = 175$ per box and $(175)(175) = 30{,}625$.

91. We are looking for a width. Let x represent the width of the border in feet. It is best to convert all units to feet now:

$$1 \text{ cubic yard} = 27 \text{ cubic feet}$$

$$3 \text{ inches} = \frac{3}{12} = \frac{1}{4} \text{ ft.}$$

From the figure: The total area is $A_T = (6 + 2x)(10 + 2x)$
The area of the garden is $A_G = 6 \times 10 = 60$ sq ft.
Then, the area of the border, A_B, is $A_B = (6 + 2x)(10 + 2x) - 60$

The volume of the border is 27 cubic feet or $\left(\frac{1}{4}\text{ ft}\right)(A_B)$ so that

$$27 = \frac{1}{4}A_B$$

$$108 = A_B$$

Thus,

$$108 = (6 + 2x)(10 + 2x) - 60$$

$$108 = 60 + 12x + 20x + 4x^2 - 60$$

$$0 = 4x^2 + 32x - 108$$

$$0 = x^2 + 8x - 27$$

Here, $a = 1$, $b = 8$, $c = -27$ and $b^2 - 4ac = 8^2 - 4(1)(-27) = 172$

Then, $x = \dfrac{-8 \pm \sqrt{172}}{2(1)} = \dfrac{-8 \pm 2\sqrt{43}}{2} = -4 \pm \sqrt{43}$

so that $x = -4 - \sqrt{43}$ cannot be negative or $x = -4\sqrt{43}$

$$x = -4 + 6.56$$

$$x = 2.56 \text{ ft.}$$

The width of the border is 2.56 ft.

Check: Thus, the area of the border is:

$$A_B = (6 + 2(2.56))(10 + 2(2.56)) - (6)(10)$$

$= 108.13$ which is close to 108 sq. ft.

The given volume of cement is then $(108)\left(\frac{1}{4}\right) = 27$ cubic ft.

93. We want to find the new dimensions of length and width in centimeters.

Let x be the amount of the reduction of the length and width measured in centimeters.

The current bar has volume $V_c = (12)(7)(3) = 252$ cubic centimeters.

The new volume is to be $V_N = .90V_c = .9(252) = 226.8$ cubic centimeters.

Then

$$V_N = 226.8 = (3)(12 - x)(7 - x)$$

$$226.8 = 3(84 - 19x + x^2)$$

$$75.6 = 84 - 19x + x^2$$

$$0 = 8.4 - 19x + x^2$$

Here, $a = 1$, $b = -19$, $c = 8.4$ and $b^2 - 4ac = (-19)^2 - 4(1)(8.4) = 327.4$

Then, $x = \dfrac{-(-19) \pm \sqrt{327.4}}{2(1)} = \dfrac{19 \pm 18.09}{2}$

$x = 0.455$ or $x = 18.55$; but since 18.55 exceeds the measurements it would be subtracted from, it is not a practical solution.

The new dimensions are:

7 − 0.455, and 12 − 0.455, and 3 centimeters

or width = 6.55 cm, length = 11.55 cm, and thickness = 3 cm.

Check: $(6.55)(11.55)(3) = 226.9 \cong 226.8$.

95. We want to find the width of a concrete pool border. Let x represent the width in feet of the border. It is best to convert all units to feet now:

$$1 \text{ cubic yard} = 27 \text{ cubic feet}$$

$$3 \text{ inches} = \frac{3}{12} = \frac{1}{4} \text{ foot}$$

We will use $A = \pi r^2$.

We are given that the distance across the pool is 10 feet, which means the radius is $\frac{10}{2} = 5$ feet.

The total area, pool and border, is $A_T = \pi(5 + x)^2$.
The area of the pool alone is $A_p = \pi(5)^2 = 25\pi$.
The area of the border is $A_B = A_T - A_p = \pi(5 + x)^2 - 25\pi$.

The volume of the border, $\frac{1}{4}A_B = \frac{1}{4}(\pi(5 + x)^2 - 25\pi)$.

Then,
$$\frac{1}{4}(\pi(5 + x)^2 - 25\pi) = 27$$
$$\pi(25 + 10x + x^2 - 25) = 108$$
$$x^2 + 10x - \frac{108}{\pi} = 0$$
$$x^2 + 10x - 34.38 = 0$$

Here, $a = 1$, $b = 10$, $c = -34.38$, and $b^2 - 4ac = 10^2 - 4(1)(-34.38) = 237.52$

Then, $x = \dfrac{-10 \pm \sqrt{237.52}}{2(1)} = \dfrac{-10 \pm 15.41}{2}$

Thus, $x = 2.71$ or $x = -12.71$

We ignore -12.71 since a measurement must be positive. The width of the border is 2.71 ft.

Check: Area of border $= \pi[(5 + 2.71)^2 - 25] = 108.21$

Volume of border $= \frac{1}{4}(108.2) = 27.1$ cubic ft.

97. We want to find a speed in miles per hour. Let x represent the speed of the current in miles per hour. We can make a table using $s = vt$:

	Velocity	Time	Distance
Downstream	$15 + x$	$\frac{10}{15 + x}$	10
Upstream	$15 - x$	$\frac{10}{15 - x}$	10

Then the total time
$$\frac{10}{15 + x} + \frac{10}{15 - x} = 1.5$$
$$10(15 - x) + 10(15 + x) = 1.5(15 + x)(15 - x)$$
$$150 - 10x + 150 + 10x = 1.5(225 - x^2)$$
$$-1.5(225 - x^2) + 300 = 0$$
$$-225 + x^2 + 200 = 0$$
$$x^2 - 25 = 0$$
$$(x + 5)(x - 5) = 0$$
$$x = -5 \text{ or } x = 5$$

We choose the positive solution. The speed of the current is 5 miles per hour.

Check: Downstream $(15 + 5)\left(\frac{10}{15 + 5}\right) = 10$ miles

Upstream $(15 - 5)\left(\frac{10}{15 - 5}\right) = 10$ miles

99.

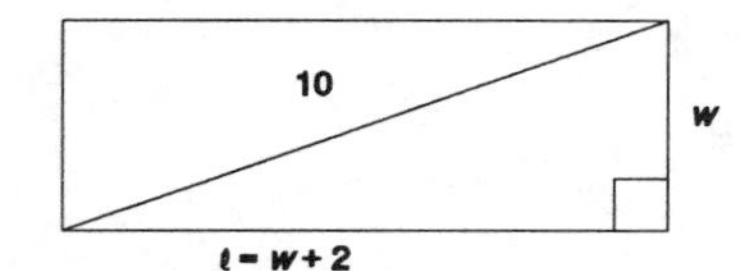

ℓ = length, w = width

Using the Pythagorean Theorem,

$$w^2 + (w + 2)^2 = 10^2$$
$$w^2 + w^2 + 4w + 4 = 100$$
$$2w^2 + 4w - 96 = 0$$
$$w^2 + 2w - 48 = 0$$
$$(w + 8)(w - 6) = 0$$
$$w = -8 \quad w = 6$$
$$\ell = 8$$

Impossible because width cannot be negative. The dimensions of the rectangle are 6×8 inches.

101. The roots of a quadratic equation are:

$$\frac{-b - \sqrt{b^2 - 4ac}}{2a} \text{ and } \frac{-b + \sqrt{b^2 - 4ac}}{2a}$$

The sum $\dfrac{-b - \sqrt{b^2 - 4ac}}{2a} + \dfrac{-b + \sqrt{b^2 - 4ac}}{2a}$

$$= \frac{-b}{2a} - \frac{\sqrt{b^2 - 4ac}}{2a} + \frac{-b}{2a} + \frac{\sqrt{b^2 - 4ac}}{2a} = \frac{-2b}{2a} = \frac{-b}{a}$$

103. $kx^2 + x + k = 0$

Here, $a = k$, $b = 1$, $c = k$ and $b^2 - 4ac = 1 - 4(k)(k) = 1 - 4k^2$. For repeated real roots, we want

$$1 - 4k^2 = 0$$
$$1 = 4k^2$$
$$\frac{1}{4} = k^2$$
$$\pm\frac{1}{2} = k$$

$k = -\dfrac{1}{2}$ or $k = \dfrac{1}{2}$ will give a repeated real solution.

105. For $ax^2 + bx + c = 0$, the real solutions are:

$$x = \frac{-b \pm \sqrt{b^2 - 4ac}}{2a} = \frac{-b + \sqrt{b^2 - 4ac}}{2a} \text{ or } \frac{-b + \sqrt{b^2 - 4ac}}{2a}$$

For $ax^2 - bx + c = 0$, the real solutions are:

$$x = \frac{-(-b) \pm \sqrt{(-b)^2 - 4ac}}{2a} = \frac{b + \sqrt{b^2 - 4ac}}{2a} \text{ or } \frac{b - \sqrt{b^2 - 4ac}}{2a}$$

$$\text{or } \frac{b + \sqrt{b^2 - 4ac}}{2a}$$

Since $\sqrt{b^2 - 4ac}$ is the same in both answers, we have

$$-\frac{-b - \sqrt{b^2 - 4ac}}{2a} = \frac{b + \sqrt{b^2 - 4ac}}{2a} \text{ and } -\frac{-b + \sqrt{b^2 - 4ac}}{2a} = \frac{b - \sqrt{b^2 - 4ac}}{2a}$$

so that the real solutions to $ax^2 - bx + c = 0$ are the negatives of those of $ax^2 + bx + c = 0$.

107. For $\frac{1}{2}n(n+1) = 666$, we solve for n to get the number of integers:

$$2\left[\frac{1}{2}n(n+1)\right] = (666)(2)$$
$$n(n+1) = 1332$$
$$n^2 + n - 1332 = 0$$
$$(n+37)(n-36) = 0$$

$n = -37$ or $n = 36$ so that 36 consecutive integers, starting with 1, add up to 666.

109. We want to find the average speed from Chicago to Miami.

	Velocity	Time	Distance
Chicago to Atlanta	45	t_1	$45t_1$
Atlanta to Miami	55	t_2	$55t_2$

If Atlanta is halfway between Chicago and Miami, then the distances from Chicago to Atlanta and Atlanta to Miami are equal.

$$45t_1 = 55t_2 \Rightarrow t_1 = \frac{55}{45}t_2$$

$$\text{Average speed} = \frac{\text{Distance}}{\text{Time}}$$
$$= \frac{45t_1 + 55t_2}{t_1 + t_2} = \frac{45\left[\frac{55}{45}t_2\right] + 55t_2}{\frac{55}{45}t_2 + t_2}$$
$$= \frac{55t_2 + 55t_2}{\frac{55t_2 + 45t_2}{45}}$$
$$\frac{100t_2}{\frac{100}{45}t_2} = \frac{110}{\frac{100}{45}}$$
$$\frac{(45)(110)}{100} = 49.5 \text{ miles per hour}$$

The average speed from Chicago to Miami is 49.5 mph.

2.4 Other Types of Equations

1.
$$\sqrt{2t - 1} = 1$$
$$\left(\sqrt{2t - 1}\right)^2 = 1^2$$
$$2t - 1 = 1$$
$$t = 1$$

Check: $\sqrt{2(1) - 1} = \sqrt{1} = 1$

3.
$$\sqrt{3t + 1} = -6$$
$$3t + 1 = 36$$
$$3t = 35$$
$$t = \frac{35}{3}$$

Check: $\sqrt{3\left[\frac{35}{3}\right] + 1} = \sqrt{36} = 6 \neq -6$

No real solution.

5.
$$\sqrt[3]{1 - 2x} - 3 = 0$$
$$\sqrt[3]{1 - 2x} = 3$$
$$\left(\sqrt[3]{1 - 2x}\right)^3 = 3^3$$
$$1 - 2x = 27$$
$$-26 = 2x$$
$$-13 = x$$
Check: $\sqrt[3]{1 - 2(-13)} - 3$
$$= \sqrt{27} - 3 = 0$$

7.
$$x = 6\sqrt{x}$$
$$(x)^2 = \left(6\sqrt{x}\right)^2$$
$$x^2 = 36x$$
$$x^2 - 36x = 0$$
$$x(x - 36) = 0$$
$$x = 0 \text{ or } x - 36 = 0$$
$$x = 36$$
Check: $0 = 6\sqrt{0} = 0$
$$36 = 6\sqrt{36} = 36$$
The solution set is $\{0, 36\}$.

9.
$$\sqrt{15 - 2x} = x$$
$$\left(\sqrt{15 - 2x}\right) = x^2$$
$$15 - 2x = x^2$$
$$-x^2 - 2x + 15 = 0$$
$$x^2 + 2x - 15 = 0$$
$$(x + 5)(x - 3) = 0$$
$$x = -5 \text{ or } x = 3$$
Check 3: $\sqrt{15 - 2(3)} \stackrel{?}{=} 3$
$$\sqrt{9} \stackrel{?}{=} 3$$
$$3 = 3$$
Check -5: $\sqrt{15 - 2(-5)} \neq -5$

Does not check.

The solution is $x = 3$.

11.
$$x = 2\sqrt{x - 1}$$
$$x^2 = \left(2\sqrt{x - 1}\right)^2$$
$$x^2 = 4(x - 1)$$
$$x = 4x - 4$$
$$x^2 - 4x + 4 = 0$$
$$(x - 2)^2 = 0$$
$$x = 2$$
Check: $2 \stackrel{?}{=} 2\sqrt{2 - 1}$
$$2 = 2$$

13.
$$\sqrt{x^2 - x - 4} = x + 2$$
$$\left(\sqrt{x^2 - x - 4}\right)^2 = (x + 2)^2$$
$$x^2 - x - 4 = x^2 + 4x + 4$$
$$-5x = 8$$
$$x = -\frac{8}{5}$$

Check:
$$\sqrt{\left(-\frac{8}{5}\right)^2 + \frac{8}{5} - 4} \stackrel{?}{=} -\frac{8}{5} + 2$$
$$\sqrt{\frac{64}{25} + \frac{40}{25} - \frac{100}{25}} \stackrel{?}{=} \frac{2}{5}$$
$$\sqrt{\frac{4}{25}} \stackrel{?}{=} \frac{2}{5}$$
$$\frac{2}{5} = \frac{2}{5}$$

15.
$$3 + \sqrt{3x + 1} = x$$
$$\sqrt{3x + 1} = x - 3$$
$$3x + 1 = (x - 3)^2$$
$$3x + 1 = x^2 - 6x + 9$$
$$0 = x^2 - 9x + 8$$
$$0 = (x - 1)(x - 8)$$
$$x = 1 \qquad x = 8$$

Check 1: $3 + \sqrt{3(1) + 1} \stackrel{?}{=} 1$
$$3 + 2 \stackrel{?}{=} 1$$
$$5 \neq 1$$
Check 8: $3 + \sqrt{3(8) + 1} \stackrel{?}{=} 8$
$$3 + 5 \stackrel{?}{=} 8$$
$$8 = 8$$
The solution is $x = 8$.

17.

$$\begin{aligned}
\sqrt{2x+3} - \sqrt{x+1} &= 1 \\
\left(\sqrt{2x+3}\right)^2 &= \left(1 + \sqrt{x+1}\right)^2 \\
2x + 3 &= 1 + 2\sqrt{x+1} + x + 1 \\
x + 1 &= 2\sqrt{x+1} \\
(x+1)^2 &= \left(2\sqrt{x+1}\right)^2 \\
x^2 + 2x + 1 &= 4(x+1) \\
x^2 + 2x + 1 &= 4x + 4 \\
x^2 - 2x - 3 &= 0 \\
(x+1)(x-3) &= 0 \\
x = -1 \quad &\text{or } x = 3
\end{aligned}$$

Check -1:

$$\begin{aligned}
\sqrt{2(-1)+3} - \sqrt{-1+1} &\stackrel{?}{=} 1 \\
\sqrt{1} - \sqrt{0} &\stackrel{?}{=} 1 \\
1 &= 1
\end{aligned}$$

Check 3:

$$\begin{aligned}
\sqrt{2(3)+3} - \sqrt{3+1} &\stackrel{?}{=} 1 \\
\sqrt{9} - \sqrt{4} &\stackrel{?}{=} 1 \\
3 - 2 &\stackrel{?}{=} 1 \\
1 &= 1
\end{aligned}$$

The solution set is $\{-1, 3\}$.

19.

$$\begin{aligned}
\sqrt{3+1} - \sqrt{x-1} &= 2 \\
\sqrt{3x+1} &= 2 + \sqrt{x-1} \\
\left(\sqrt{3x+1}\right)^2 &= \left(2 + \sqrt{x-1}\right)^2 \\
3x + 1 &= 4 + 4\sqrt{x-1} + x - 1 \\
2x - 2 &= 4\sqrt{x-1} \\
x - 1 &= 2\sqrt{x-1} \\
(x-1)^2 &= \left(2\sqrt{x-1}\right)^2 \\
x^2 - 2x + 1 &= 4(x-1) \\
x^2 - 6x + 5 &= 0 \\
(x-1)(x-5) &= 0 \\
x = 1 \quad \text{or} \quad &x = 5
\end{aligned}$$

Check 1:

$$\begin{aligned}
\sqrt{3(1)+1} - \sqrt{x-1} &\stackrel{?}{=} 2 \\
\sqrt{4} &\stackrel{?}{=} 2 \\
2 &= 2
\end{aligned}$$

Check 5:

$$\begin{aligned}
\sqrt{3(5)+1} - \sqrt{5-1} &\stackrel{?}{=} 2 \\
\sqrt{16} - \sqrt{4} &\stackrel{?}{=} 2 \\
4 - 2 &\stackrel{?}{=} 2 \\
2 &= 2
\end{aligned}$$

The solution set is $\{1, 5\}$.

21.

$$\begin{aligned}
\sqrt{3 - 2\sqrt{x}} &= \sqrt{x} \\
3 - 2\sqrt{x} &= x \\
-2\sqrt{x} &= x - 3 \\
4x &= x^2 - 6x + 9 \\
x^2 - 10x + 9 &= 0 \\
(x-9)(x-1) &= 0 \\
x &= 9,\ x = 1
\end{aligned}$$

Check 9:

$$\begin{aligned}
\sqrt{3 - 2\sqrt{9}} &\stackrel{?}{=} 9 \\
\sqrt{3 - 6} &\stackrel{?}{=} 3 \\
\sqrt{-3} &\neq 3
\end{aligned}$$

Extraneous Solution.

Check 1:

$$\begin{aligned}
\sqrt{3 - 2\sqrt{1}} &\stackrel{?}{=} \sqrt{1} \\
\sqrt{3 - 2} &\stackrel{?}{=} \sqrt{1} \\
\sqrt{1} &= \sqrt{1}
\end{aligned}$$

The solution is $x = 1$.

23.

$$\begin{aligned}
(3x+1)^{1/2} &= 4 \\
\sqrt{3x+1} &= 4 \\
3x + 1 &= 16 \\
3x &= 15 \\
x &= 5
\end{aligned}$$

Check:

$$\begin{aligned}
\left(3(5) + 1\right)^{1/2} &\stackrel{?}{=} 4 \\
16^{1/2} &\stackrel{?}{=} 4 \\
4 &= 4
\end{aligned}$$

The solution is $x = 5$.

25. $(5x - 2)^{1/3} = 2$

$\sqrt[3]{5x - 2} = 2$

$5x - 2 = 2^3$

$5x - 2 = 8$

$5x = 10$

$x = 2$

Check: $(5 \cdot 2 - 2)^{1/3} \stackrel{?}{=} 2$

$8^{1/3} = 2$

$2 = 2$

The solution is $x = 2$.

27. $(x^2 + 9)^{1/2} = 5$

$\sqrt{x^2 + 9} = 5$

$x^2 + 9 = 25$

$x^2 = 16$

$x = -4, 4$

Check -4: $((-4)^2 + 9)^{1/2} \stackrel{?}{=} 5$

$(25)^{1/2} \stackrel{?}{=} 5$

$5 = 5$

Check 4: $((4)^2 + 9)^{1/2} \stackrel{?}{=} 5$

$25^{1/2} \stackrel{?}{=} 5$

$5 = 5$

The solution set is $\{-4, 4\}$.

29. $x^{3/2} - 2x^{1/2} = 0$

Let $u = x^{1/2}$

$u^3 - 2u = 0$

$u(u^2 - 2) = 0$

$u = 0 \quad u^2 - 2 = 0$

$u^2 = 2$

$u = \pm\sqrt{2}$

$x^{1/2} = 0 \quad x^{1/2} = \pm\sqrt{2}$

$x = 0^2 \quad x = \left(\pm\sqrt{2}\right)^2$

$x = 0 \quad x = 2$

Check:

$x = 0$: $0^{3/2} - 2(0)^{1/2} = 0$

$x = 2$: $2^{3/2} - 2(2)^{1/2} = \left(\sqrt{2}\right)^3 - 2\left(\sqrt{2}\right)$

$= \sqrt{2}\left(\sqrt{2^2} - 2\right)$

$= \sqrt{2}(2 - 2)$

$= \sqrt{2}(0)$

$= 0$

The solution set is $\{0, 2\}$.

31. $x^4 - 5x^2 + 4 = 0$

$(x^2 - 4)(x^2 - 1) = 0$

$x^2 - 4 = 0 \qquad x^2 - 1 = 0$

$(x - 2)(x + 2) = 0 \qquad (x - 1)(x + 1) = 0$

$x - 2 = 0 \quad x + 2 = 0 \quad x - 1 = 0 \quad x + 1 = 0$

$x = 2 \quad x = -2 \quad x = 1 \quad x = -1$

$\{-2, -1, 1, 2\}$

33. $3x^4 - 2x^2 - 1 = 0$

$(3x^2 + 1)(x^2 - 1) = 0$

$x^2 + 1 = 0 \qquad x^2 - 1 = 0$

$x^2 = -1 \qquad (x - 1)(x + 1) = 0$

$x^2 = -\frac{1}{3} \qquad x - 1 = 0 \quad x + 1 = 0$

No real solution $\quad x = 1 \quad x = -1$

$\{-1, 1\}$

35. $x^6 + 7x^3 - 8 = 0$

$(x^3 + 8)(x^3 - 1) = 0$

$x^3 + 8 = 0 \qquad x^3 - 1 = 0$

$x^3 = -8 \qquad x^3 = 1$

$x = -2 \qquad x = 1$

$\{-2, 1\}$

37. $(x + 2)^2 + 7(x + 2) + 12 = 0$

Let $u = x + 2$

$u^2 + 7u + 12 = 0$

$(u + 4)(u + 3) = 0$

$u + 4 = 0 \qquad u + 3 = 0$

$u = -4 \qquad u = -3$

$x + 2 = -4 \qquad x + 2 = -3$

$x = -6 \qquad x = -5$

$\{-6, -5\}$

39. $(3x + 4)^2 - 6(3x + 4) + 9 = 0$

Let $u = 3x + 4$

$u^2 - 6u + 9 = 0$

$(u - 3)^2 = 0$

$u = 3$

$3x + 4 = 3$

$3x = -1$

$x = -\frac{1}{3}$

41. $2(s+1)^2 - 5(s+1) = 3$

Let $u = s + 1$.

Then $u^2 = (s+1)^2$ and

$$2u^2 - 5u = 3$$
$$2u^2 - 5u - 3 = 0$$
$$(2u+1)(u-3) = 0$$

$$u = -\frac{1}{2} \quad \text{or} \quad u = 3$$
$$s + 1 = -\frac{1}{2} \quad \text{or} \quad s + 1 = 3$$
$$s = \frac{-3}{2} \quad \text{or} \quad s = 2$$

Check $\frac{1}{2}$: $2\left[\frac{-3}{2} + 1\right]^2 - 5\left[\frac{-3}{2} + 1\right] \stackrel{?}{=} 3$

$$2\left[\frac{-1}{2}\right](^2 - 5\left[\frac{-1}{2}\right] \stackrel{?}{=} 3$$
$$\frac{1}{2} + \frac{5}{2} \stackrel{?}{=} 3$$
$$3 = 3$$

Check 2: $(2+1)^2 - 5(2+1) \stackrel{?}{=} 3$

$$18 - 15 \stackrel{?}{=} 3$$
$$3 = 3$$

The solution set is $\left\{-\frac{3}{2}, 2\right\}$.

43.
$$x - 4\sqrt{x} = 0$$
$$x = 4\sqrt{x}$$
$$x^2 = 16x$$
$$x^2 - 16x = 0$$
$$x(x - 16) = 0$$
$$x = 0 \text{ or } x = 16$$

Check 0: $0 - 4\sqrt{0} \stackrel{?}{=} 0$

$$0 = 0$$

16: $16 - 4\sqrt{16} \stackrel{?}{=} 0$

$$0 = 0$$

The solution set is $\{0, 16\}$.

45.
$$x + \sqrt{x} = 20$$
$$x + \sqrt{x} - 20 = 0$$

Let $u = x^{1/2}$. Then $u^2 = x$, and

$$u^2 + u - 20 = 0$$
$$(u-4)(u+5) = 0$$
$$u = 4 \quad \text{or} \quad u = -5$$
$$x^{1/2} = 4 \qquad x^{1/2} = -5$$

(impossible)

$$x = 16$$

Check: $16 + \sqrt{16} \stackrel{?}{=} 20$

$$16 + 4 \stackrel{?}{=} 20$$
$$20 = 20$$

The solution is $x = 16$.

47. $t^{1/2} - 2t^{1/4} + 1 = 0$

Let $u = t^{1/4}$. Then $u^2 = (t^{1/4})^2 = t^{1/2}$, and

$$u^2 - 2u + 1 = 0$$
$$(u-1)(u-1) = 0$$
$$u = 1$$
$$t^{1/4} = 1$$
$$t = 1^4 = 1$$

Check: $1^{1/2} - 2(1)^{1/4} + 1 \stackrel{?}{=} 0$

$$0 = 0$$

The solution is $t = 1$.

49. $4x^{1/2} - 9x^{1/4} + 4 = 0$

Let $u = x^{1/4}$; then $u^2 = x^{1/2}$, and $4u^2 - 9u + 4 = 0$

Here, $a = 4$, $b = -9$, $c = 4$, and $b^2 - 4ac = (-9)^2 - 4(4)(4) = 17$.

Then,

$$u = \frac{-(-9) \pm \sqrt{17}}{2(4)} = \frac{9 \pm \sqrt{17}}{8}$$

$$x^{1/4} = \frac{9 \pm \sqrt{17}}{8}$$

$$x = \left(\frac{9 \pm \sqrt{17}}{8}\right)^4$$

Check $\left(\frac{9 - \sqrt{17}}{8}\right)^4$: $4\left(\left(\frac{9 - \sqrt{17}}{8}\right)^4\right)^{1/2} - 9\left(\left(\frac{9 - \sqrt{17}}{8}\right)^4\right)^{1/4} + 4 \stackrel{?}{=} 0$

$$4\left(\frac{9 - \sqrt{17}}{8}\right)^2 - 9\left(\frac{9 - \sqrt{17}}{8}\right) + 4 \stackrel{?}{=} 0$$

$$\frac{81 - 18\sqrt{17} + 17}{16} - \frac{81 + 9\sqrt{17}}{8} + 4 \stackrel{?}{=} 0$$

$$81 - 18\sqrt{17} + 17 - 162 + 18\sqrt{17} + 64 \stackrel{?}{=} 0$$

$$0 = 0$$

Check $\left(\frac{9 + \sqrt{17}}{8}\right)^4$: $4\left(\left(\frac{9 + \sqrt{17}}{8}\right)^4\right)^{1/2} - 9\left(\left(\frac{9 + \sqrt{17}}{8}\right)^4\right)^{1/4} + 4 \stackrel{?}{=} 0$

$$4\left(\frac{9 + \sqrt{17}}{8}\right)^2 - 9\left(\frac{9 + \sqrt{17}}{8}\right) + 4 \stackrel{?}{=} 0$$

$$\frac{81 + 18\sqrt{17} + 17}{16} - \frac{81 + 9\sqrt{17}}{8} + 4 \stackrel{?}{=} 0$$

$$81 + 18\sqrt{17} + 17 - 162 - 18\sqrt{17} + 64 \stackrel{?}{=} 0$$

$$0 = 0$$

The solution set is $\left\{\left(\frac{9 - \sqrt{17}}{8}\right)^4, \left(\frac{9 + \sqrt{17}}{8}\right)^4\right\}$.

51. $\sqrt[4]{5x^2 - 6} = x$

$5x^2 - 6 = x^4$

$x^4 - 5x^2 + 6 = 0$

Let $u = x^2$. Then $u^2 = x^4$ and

$u^2 - 5u + 6 = 0$

$(u - 3)(u - 2) = 0$

$u = 2$ or $u = 3$

$x^2 = 2$ or $x^2 = 3$

$x = \pm\sqrt{2}$ or $x = \pm\sqrt{3}$

The solution set is $\{\sqrt{2}, \sqrt{3}\}$.

Check $-\sqrt{3}$: $\sqrt[4]{5(-\sqrt{3})^2 - 6} \neq -\sqrt{3}$

$\sqrt{3}$: $\sqrt[4]{5(\sqrt{3})^2 - 6} \stackrel{?}{=} \sqrt{3}$

$\sqrt[4]{9} \stackrel{?}{=} \sqrt{3}$

$(3^2)^{1/4} \stackrel{?}{=} \sqrt{3}$

$3^{1/2} = \sqrt{3}$

$-\sqrt{2}$: $\sqrt[4]{5(-\sqrt{2})^2 - 6} \neq -\sqrt{2}$

$\sqrt{2}$: $\sqrt[4]{5(\sqrt{2})^2 - 6} \stackrel{?}{=} \sqrt{2}$

$\sqrt[4]{4} \stackrel{?}{=} \sqrt{2}$

$(2^2)^{1/4} \stackrel{?}{=} \sqrt{2}$

$2^{1/2} = \sqrt{2}$

53. $x^2 + 3x + \sqrt{x^2 + 3x} = 6$

Let $u = \sqrt{x^2 + 3x}$. Then $u^2 = x^2 + 3x$ and

$$u^2 + u = 6$$
$$u^2 + u - 6 = 0$$
$$(u + 3)(u - 2) = 0$$
$$u = -3 \quad \text{or} \quad u = 2$$
$$\sqrt{x^2 + 3x} = -3 \quad \text{or} \quad \sqrt{x^2 + 3x} = 2$$

No solution.
$$x^2 + 3x = 4$$
$$x^2 + 3x - 4 = 0$$
$$(x + 4)(x - 1) = 0$$
$$x = -4 \text{ or } x = 1$$

Check -4: $(-4)^2 + 3(-4) + \sqrt{(-4)^2 + 3(-4)} \stackrel{?}{=} 6$
$$16 - 12 + \sqrt{4} \stackrel{?}{=} 6$$
$$6 = 6$$

1: $1^2 + 3(1) + \sqrt{1^2 + 3(1)} \stackrel{?}{=} 6$
$$6 = 6$$

The solution set is $\{-4, 1\}$.

55. $\dfrac{1}{(x + 1)^2} = \dfrac{1}{x + 1} + 2$

Let $u = \dfrac{1}{x + 1}$. Then $u^2 = \dfrac{1}{(x + 1)^2}$ and

$$u^2 = u + 2$$
$$u^2 - u - 2 = 0$$
$$(u + 1)(u - 2) = 0$$
$$u = -1 \quad \text{or} \quad u = 2$$

$$\frac{1}{x + 1} = -1 \quad \text{or} \quad \frac{1}{x + 1} = 2$$
$$x + 1 = -1 \quad \text{or} \quad x + 1 = \frac{1}{2}$$
$$x = -2 \quad \text{or} \quad x = -\frac{1}{2}$$

Check -2: $\dfrac{1}{(-2 + 1)^2} \stackrel{?}{=} \dfrac{1}{-2 + 1} + 2$

$-\dfrac{1}{2}$: $\dfrac{1}{\left(-\frac{1}{2} + 1\right)^2} \stackrel{?}{=} \dfrac{1}{-\frac{1}{2} + 1} + 2$
$$4 = 4$$

The solution set is $\left\{-2, -\dfrac{1}{2}\right\}$.

57. $3x^{-2} - 7x^{-1} - 6 = 0$

Let $u = x^{-1}$

$$3u^2 - 7u - 6 = 0$$
$$(3u + 2)(u - 3) = 0$$
$$3u = -2 \qquad u = 3$$
$$u = \frac{-2}{3} \quad \text{or} \quad u = 3$$
$$x = \left(\frac{-2}{3}\right)^{-1} \quad \text{or} \quad x = (3)^{-1}$$
$$x = \frac{-3}{2} \quad \text{or} \quad x = \frac{1}{3}$$

The solution set is $\left\{\frac{-3}{2}, \frac{1}{3}\right\}$.

59. $2x^{2/3} - 5x^{1/3} - 3 = 0$

Let $u = x^{1/3}$

$$2u^2 - 5u - 3 = 0$$
$$(2u + 1)(u - 3) = 0$$
$$2u + 1 = 0 \quad \text{or} \quad u - 3 = 0$$
$$u = \frac{-1}{2} \quad \text{or} \quad u = 3$$
$$x^{1/3} = \frac{-1}{2} \quad \text{or} \quad x^{1/3} = 3$$
$$x = \left(\frac{-1}{2}\right)^3 \quad \text{or} \quad x = (3)^3$$
$$x = \frac{-1}{8} \quad \text{or} \quad x = 27$$

The solution set is $\left\{\frac{-1}{8}, 27\right\}$.

61. $\left(\frac{v}{v+1}\right)^2 + \frac{2v}{v+1} = 8$

Let $u = \frac{v}{v+1}$. Then $u^2 = \left(\frac{v}{v+1}\right)^2$

and

$$u + 2u = 8$$
$$u^2 + 2u - 8 = 0$$
$$(u - 2)(u + 4) = 0$$
$$u = 2 \quad \text{or} \quad u = -4$$
$$v = 2(v + 1) \quad \text{or} \quad v = -4(v + 1)$$
$$v = 2v + 2 \quad \text{or} \quad v = -4v - 4$$
$$-2 = v \quad \text{or} \quad 5v = -4$$
$$v = -2 \quad \text{or} \quad v = -\frac{4}{5}$$

Check -2: $\left(\frac{-2}{-2+1}\right)^2 + \frac{2(-2)}{-2+1} \stackrel{?}{=} 8$

$$8 = 8$$

Check $-\frac{4}{5}$: $\left(\frac{-\frac{4}{5}}{-\frac{4}{5}+1}\right)^2 + \frac{2\left(-\frac{4}{5}\right)}{-\frac{4}{5}+1} \stackrel{?}{=} 8$

$$\left(\frac{-\frac{4}{5}}{\frac{1}{5}}\right)^2 + \frac{-\frac{8}{5}}{\frac{1}{5}} \stackrel{?}{=} 8$$
$$16 - 8 \stackrel{?}{=} 8$$
$$8 = 8$$

The solution set is $\left\{-2, -\frac{4}{5}\right\}$.

63. $x - 4x^{1/2} + 2 = 0$

Let $u = x^{1/2}$. Then $u^2 = x$ and $u^2 - 4u + 2 = 0$.

Here $a = 1$, $b = -4$, $c = 2$, and $b^2 - 4ac = (-4)^2 - 4(1)(2) = 8$.

Then, $u = \frac{-(-4) \pm \sqrt{8}}{2(1)} = \frac{4 \pm 2\sqrt{2}}{2} = 2 \pm \sqrt{2}$

so that

$$u = 2 - \sqrt{2} \quad \text{or} \quad u = 2 + \sqrt{2}$$
$$x^{1/2} = 2 - \sqrt{2} \qquad x^{1/2} = 2 + \sqrt{2}$$
$$(x^{1/2})^2 = (2 - \sqrt{2})^2 \qquad (x^{1/2})^2 = (2 + \sqrt{2})^2$$
$$x = 4 - 4\sqrt{2} + 2 \qquad x = 4 + 4^2 + 2$$
$$x = 6 - 4\sqrt{2} \qquad x = 6 + 4\sqrt{2}$$
$$x = 0.34 \qquad x = 11.66$$

Check 0.34: $0.34 - 4(0.34)^{1/2} + 2 \stackrel{?}{=} 0$

$$0.34 - 2.33 + 2 \stackrel{?}{=} 0$$
$$0 = 0$$

11.66: $11.66 - 4(11.66)^{1/2} + 2 \stackrel{?}{=} 0$

$$11.66 - 13.66 + 2 \stackrel{?}{=} 0$$
$$0 = 0$$

The solution set is $\{0.34, 11.66\}$.

65. $x^4 + \sqrt{3}x^2 - 3 = 0$

Let $u = x^2$. Then $u^2 = x^4$ and $u^2 + \sqrt{3}\ u - 3 = 0$.

Here $a = 1$, $b = \sqrt{3}$, $c = -3$, and $b^2 - 4ac = \left(\sqrt{3}\right)^2 - 4(1)(-3) = 15$.

Then, $u = \dfrac{-\left(\sqrt{3}\right) \pm \sqrt{15}}{2(1)} = \dfrac{-\sqrt{3} \pm \sqrt{15}}{2}$

so that

$$u = \frac{-\sqrt{3} - \sqrt{15}}{2} \quad \text{or} \quad u = \frac{-\sqrt{3} + \sqrt{15}}{2}$$

$$x^2 = \frac{-\sqrt{3} - \sqrt{15}}{2} \qquad x^2 = \frac{-\sqrt{3} + \sqrt{15}}{2}$$

No real solution.

$$x = \pm \left[\frac{-\sqrt{3} + \sqrt{15}}{2}\right]^{1/2}$$

$$x = \pm\ 1.03$$

Check -1.03: $(-1.03)^4 + \sqrt{3}(-1.03)^2 - 3 \stackrel{?}{=} 0$

$$1.13 + \sqrt{3}(1.06) - 3 \stackrel{?}{=} 0$$

$$1.13 + 1.84 - 3 \stackrel{?}{=} 0$$

$$0 = 0$$

1.03: $(1.03)^4 + \sqrt{3}(1.03)^2 - 3 \stackrel{?}{=} 0$

$$1.13 + \sqrt{3}(1.06) - 3 \stackrel{?}{=} 0$$

$$0 = 0$$

The solution set is $\{-1.03, 1.03\}$.

67. $\pi(1 + t)^2 = \pi + 1 + t$

Let $u = 1 + t$. Then $u^2 = (1 + t)^2$ and

$$\pi u^2 = \pi + u$$

$$\pi u^2 - u - \pi = 0$$

Here $a = \pi$, $b = -1$, $c = -\pi$, and $b^2 - 4ac = (-1)^2 - 4(\pi)(-\pi) = 1 + 4\pi^2$

Then, $u = \dfrac{-(-1) \pm \sqrt{1 + 4\pi^2}}{2(\pi)} = \dfrac{1 \pm \sqrt{1 + 4\pi^2}}{2\pi}$

so that

$$u = \frac{1 \pm \sqrt{1 + 4\pi^2}}{2\pi}$$

$$1 + t = \frac{1 \pm \sqrt{1 + 4\pi^2}}{2\pi}$$

$$t = \frac{1 \pm \sqrt{1 + 4\pi^2}}{2\pi} - 1$$

$$t = \frac{1 \pm \sqrt{1 + 4\pi^2}}{2\pi} - 1 \quad \text{or } t = \frac{1 + \sqrt{1 + 4\pi^2}}{2\pi} - 1$$

$$= -1.85 \qquad\qquad = 0.17$$

Check -1.85: $\pi(1 - 1.85)^2 \stackrel{?}{=} \pi + 1 - 1.85$

$$\pi(.72) \stackrel{?}{=} 2.29$$
$$2.27 \approx 2.29$$

0.17: $\pi(1 + 0.17)^2 \stackrel{?}{=} \pi + 1 + 0.17$

$$\pi(1.37) \stackrel{?}{=} 4.31$$
$$4.30 \approx 4.31$$

The solution set is $\{-1.85, 0.17\}$.

69. $k^2 - k = 12$ and $k = \dfrac{x+3}{x-3}$

$$k^2 - k - 12 = 0$$
$$(k - 4)(k + 3) = 0$$

$k - 4 = 0$	$k + 3 = 0$
$k = 4$	$k = -3$
$\dfrac{x+3}{x-3} = 4$	$\dfrac{x+3}{x-3} = -3$
$x + 3 = 4(x - 3)$	$x + 3 = -3(x - 3)$
$x + 3 = 4x - 12$	$x + 3 = -3x + 9$
$15 = 3x$	$4x = 6$
$5 = x$	$x = \dfrac{3}{2}$

$$\left\{\frac{3}{2}, 5\right\}$$

71. Total time elapsed $= 4 = \dfrac{\sqrt{s}}{4} + \dfrac{s}{1100}$

$$(1100)(4) = \left[\frac{\sqrt{s}}{4} + \frac{s}{1100}\right](1100)$$
$$4400 = 275\sqrt{s} + s$$

Let $u = \sqrt{s}$. Then $u^2 = s$ and $4400 = 275u + u^2$.

$$0 = u^2 + 275u - 4400$$

Here, $a = 1$, $b = 275$, $c = -4400$, and $b^2 - 4ac = (275)^2 - 4(1)(-4400) = 93225$.
Then,

$$u = \frac{-275 \pm \sqrt{93225}}{2(1)}$$
$$u = -1160.66 \text{ or } u = 15.164$$
$$\sqrt{s} = -1160 \text{ or } \sqrt{s} = 15.164$$

No real solution. $s = 229.95$

The depth of the well is 230 ft.

2.5 Inequalities

1. $[0, 2]$; $0 \le x \le 2$

3. $(-1, 2)$; $-1 < x < 2$

5. $(-\infty, 0]$ or $(2, \infty)$; $x \le 0$ or $x > 2$

7. $[0, 3)$; $0 \le x < 3$

9. If $x < 5$, then $x - 5 < 0$

11. If $x > -4$, then $x + 4 > 0$

13. If $x > -4$, then $3x > -12$

15. If $x < 6$, then $-2x > -12$
Note: When multiplying or dividing by a negative value, the inequality symbol changes to its opposite.

17. $x \geq -2$

19. $x \geq 4$ and $x < 6$

21. $x \leq 0$ or $x < 6$

23. $x \leq -2$ and $x > 1$: There are no numbers for which $x \leq -2$ and $x > 1$.

25. $x \leq -2$ or $x > 1$

27. $[0, 4]$

29. $[4, 6)$

31. $[4, \infty)$

33. $(-\infty, -4)$

35. $2 \leq x \leq 5$

37. $-3 < x < -2$

39. $x \geq 4$

41. $x < -3$

43. Since $a \leq b$, $c > 0$,

$$\begin{aligned} \text{then } a - b &\leq 0 \\ (a - b)c &\leq 0 \cdot c \\ ca - cb &\leq 0 \\ ca &\leq cb \text{ or } ac \leq bc \end{aligned}$$

Thus, if $a \leq b$ and $c > 0$, then $ac \leq bc$

45. Since $a < b$, $a \cdot \frac{1}{2} < b \cdot \frac{1}{2}$

$$\begin{aligned} \frac{a}{2} &< \frac{b}{2} \\ \frac{a}{2} + \frac{a}{2} &< \frac{b}{2} + \frac{a}{2} \\ a &< \frac{a + b}{2} \end{aligned}$$

Also,

$$\begin{aligned} \frac{a}{2} &< \frac{b}{2} \\ \frac{a}{2} + \frac{b}{2} &< \frac{b}{2} + \frac{b}{2} \\ \frac{a + b}{2} &< b \end{aligned}$$

Thus, $a < \frac{a + b}{2} < b$

47. If $0 < a < b$, then $\left(\sqrt{ab}\right)^2 - a^2 = ab - a^2 = a(b - a) > 0$.

Therefore, $\left(\sqrt{ab}\right)^2 > a^2$. Hence, $\sqrt{ab} > a$.

$$b^2 - \left(\sqrt{ab}\right)^2 = b^2 - ab = b(b - a) > 0$$

Therefore, $b^2 > \left(\sqrt{ab}\right)^2$. Hence, $b > \sqrt{ab}$.

Combining the inequalities, $a < \sqrt{ab} < b$ and $\sqrt{ab}$ is the geometric mean of a and b.

49. For $0 < a < b$, $\frac{1}{h} = \frac{1}{2}\left(\frac{1}{a} + \frac{1}{b}\right)$

$$h \cdot \frac{1}{h} = \frac{1}{2}\left(\frac{b+a}{ab}\right) \cdot h$$
$$1 = \frac{1}{2}\left(\frac{b+a}{ab}\right) \cdot h$$
$$2 = \left(\frac{b+a}{ab}\right) h$$
$$\frac{2ab}{a+b} = h$$

$$h - a = \frac{2ab}{a+b} - a = \frac{2ab - a(a+b)}{a+b} = \frac{2ab - a^2 - ab}{a+b} = \frac{ab - a^2}{a+b} = \frac{a(b-a)}{a+b} > 0$$

Therefore, $h > a$.

$$b - h = b - \frac{2ab}{a+b} = \frac{b(a+b) - 2ab}{a+b} = \frac{ab + b^2 - 2ab}{a+b} = \frac{b^2 - ab}{a+b} = \frac{b(b-a)}{a+b} > 0$$

Therefore, $h < b$.
Combining these inequalities, $a < h < b$.

51. $21 <$ young adult's age < 30

53. (a) An average 25-year-old male can expect to live at least 48.4 more years. $25 + 48.4 = 73.4$. Therefore, average male lives ≥ 73.4.
(b) An average 25-year-old female can expect to live at least 54.7 more years. $25 + 54.7 = 79.7$. Therefore, average female lives ≥ 79.7.
(c) By the above information, a female can expect to live longer by 6.3 years.

2.6 Linear Inequalities

1. (a)
$$3 < 5$$
$$3 + (-3) < 5 + (-3)$$
$$0 < 2$$

(b)
$$3 < 5$$
$$3 - 5 < 5 - 5$$
$$-2 < 0$$

(c)
$$3 < 5$$
$$3(3) < (3)(5)$$
$$9 < 15$$

(d)
$$3 < 5$$
$$(-2)(3) > (-2)(5)$$
$$-6 > -10$$

3. (a)
$$2x + 1 < 2$$
$$2x + 1 + (-3) < 2 + (-3)$$
$$2x - 2 < -1$$

(b)
$$2x + 1 < 2$$
$$2x + 1 - 5 < 2 - 5$$
$$2x - 4 < -3$$

(c)
$$2x + 1 < 2$$
$$3(2x + 1) < 3(2)$$
$$6x + 3 < 6$$

(d)
$$2x + 1 < 2$$
$$(-2)(2x + 1) > (-2)(2)$$
$$-4x - 2 > -4$$

5.
$$x + 1 < 5$$
$$x + 1 - 1 < 5 - 1$$
$$x < 4$$
$\{x \mid x < 4\}$ or $(-\infty, 4)$

4

7.
$$1 - 2x \leq 3$$
$$1 - 2x - 1 \leq 3 - 1$$
$$-2x \leq 2$$
$$\frac{-2x}{2} \geq \frac{2}{-2}$$
$$x \geq -1$$
$\{x \mid x \geq -1\}$ or $[-1, \infty)$

-1

9.
$$3x - 7 > 2$$
$$3x - 7 + 7 > 2 + 7$$
$$3x > 9$$
$$\frac{3x}{3} > \frac{9}{3}$$
$$x > 3$$
$\{x \mid x > 3\}$ or $(3, \infty)$

3

11.
$$3x - 1 \geq 3 + 5x$$
$$3x - 1 + 1 \geq 3 + 5x + 1$$
$$3x \geq 4 + 5x$$
$$3x - 5x \geq 4 + 5x - 5x$$
$$-2x \geq 4$$
$$\frac{-2x}{-2} \geq \frac{4}{-2}$$
$$x \leq -2$$
$\{x \mid x \leq -2\}$ or $(-\infty, -2]$

-2

13.
$$-2(x + 3) < 8$$
$$\frac{-2(x + 3)}{-2} > \frac{8}{-2}$$
$$x + 3 > -4$$
$$x + 3 - 3 > -4 - 3$$
$$x > -7$$
$\{x \mid x > -7\}$ or $(-7, \infty)$

-7

15.
$$4 - 3(1 - x) \leq 3$$
$$4 - 3 + 3x \leq 3$$
$$1 + 3x - 1 \leq 3 - 1$$
$$3x \leq 2$$
$$\frac{3x}{3} \leq \frac{2}{3}$$
$$x \leq \frac{2}{3}$$
$\left\{x \mid x \leq \frac{2}{3}\right\}$ or $\left[-\infty, \frac{2}{3}\right]$

$\frac{2}{3}$

17.
$$\frac{1}{2}(x - 4) > x + 8$$
$$\frac{1}{2}x - 2 > x + 8$$
$$\frac{1}{2}x - 2 - 8 > x + 8 - 8$$
$$\frac{1}{2}x - 10 > x$$
$$\frac{1}{2}x - 10 - \frac{1}{2}x > x - \frac{1}{2}x$$
$$-10 > \frac{1}{2}x$$
$$2(-10) > 2\left[\frac{1}{2}x\right]$$
$$-20 > x$$
$$x < -20$$
$\{x \mid x < -20\}$ or $(-\infty, -20)$

-20

19.
$$\frac{x}{2} \geq 1 - \frac{x}{4}$$
$$\frac{x}{2} + \frac{x}{4} \geq 1 - \frac{x}{4} + \frac{x}{4}$$
$$\frac{3x}{4} \geq 1$$
$$\frac{4}{3}\left(\frac{3x}{4}\right) \geq \frac{4}{3}(1)$$
$$x \geq \frac{4}{3}$$
$\left\{x \mid x \geq \frac{4}{3}\right\}$ or $\left[\frac{4}{3}, \infty\right)$

$\frac{4}{3}$

21.
$$0 \le 2x - 6 \le 4$$
$$0 + 6 \le 2x - 6 + 6 \le 4 + 6$$
$$6 \le 2x \le 10$$
$$\frac{6}{2} \le \frac{2x}{2} \le \frac{10}{2}$$
$$3 \le x \le 5$$
$\{x \mid 3 \le x \le 5\}$ or $[3, 5]$

3 5

23.
$$-5 \le 4 - 3x \le 2$$
$$-5 - 4 \le -3x \le 2 - 4$$
$$-9 \le -3x \le -2$$
$$\frac{-9}{-3} \le \frac{-3x}{-3} \le \frac{-2}{-3}$$
$$3 \ge x \ge \frac{2}{3}$$
$$\frac{2}{3} \le x \le 3$$
$\left\{x \mid \frac{2}{3} \le x \le 3\right\}$ or $\left[\frac{2}{3}, 3\right]$

-2/3 3

25.
$$-3 < \frac{2x - 1}{3} < 0$$
$$(3)(-3) < 3\left[\frac{2x - 1}{3}\right] < 3(0)$$
$$-9 < 2x - 1 < 0$$
$$-9 + 1 < 2x - 1 + 1 < 0 + 1$$
$$-8 < 2x < 1$$
$$\frac{-8}{2} < \frac{2x}{2} < \frac{1}{2}$$
$$-4 < x < \frac{1}{2}$$
$\left\{x \mid -4 < x < \frac{1}{2}\right\}$ or $\left(-4, \frac{1}{2}\right)$

-4 1/2

27.
$$1 < 1 - \frac{1}{2}x < 4$$
$$1 - 1 < 1 - \frac{1}{2}x - 1 < 4 - 1$$
$$0 < \frac{-1}{2}x < 3$$
$$(-2)(0) > (-2)\left[-\frac{1}{2}x\right] > (-2)(3)$$
$$0 > x > -6$$
$$-6 < x < 0$$
$\{x \mid -6 < x < 0\}$ or $(-6, 0)$

-6 0

29.
$$(x + 2)(x - 3) > (x - 1)(x + 1)$$
$$x^2 - x - 6 > x^2 - 1$$
$$x^2 - x - 6 - x^2 > x^2 - 1 - x^2$$
$$-x - 6 > -1$$
$$-x - 6 + 6 > -1 + 6$$
$$-x > 5$$
$$\frac{-x}{-1} < \frac{5}{-1}$$
$$x < -5$$
$\{x \mid x < -5\}$ or $(-\infty, -5)$

-5

31.
$$x(4x + 3) \le (2x + 1)^2$$
$$4x^2 + 3x \le 4x^2 + 4x + 1$$
$$4x^2 + 3x - 4x^2 \le 4x^2 + 4x + 1 - 4x^2$$
$$3x \le 4x + 1$$
$$3x - 3x \le 4x + 1 - 3x$$
$$0 \le x + 1$$
$$0 - 1 \le x + 1 - 1$$
$$-1 \le x$$
$$x \ge -1$$
$\{x \mid x \ge -1\}$ or $[-1, \infty)$

-1

33.
$$\frac{1}{2} \le \frac{x+1}{3} < \frac{3}{4}$$
$$3\left[\frac{1}{2}\right] \le 3\left[\frac{x+1}{3}\right] < 3\left[\frac{3}{4}\right]$$
$$\frac{3}{2} \le x + 1 < \frac{9}{4}$$
$$\frac{3}{2} - 1 \le x + 1 - 1 < \frac{9}{4} - 1$$
$$\frac{1}{2} \le x < \frac{5}{4}$$
$$\left\{x \middle| \frac{1}{2} \le x < \frac{5}{4}\right\} \text{ or } \left[\frac{1}{2}, \frac{5}{4}\right)$$

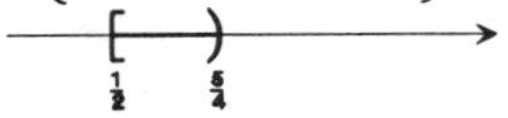

35.
$$(4x + 2)^{-1} < 0$$
$$4x + 2 < 0$$
$$\left\{x \middle| x < -\frac{1}{2}\right\} \text{ or } \left(-\infty, -\frac{1}{2}\right)$$

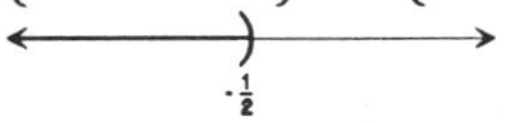

37.
$$0 < \frac{2}{x} < \frac{3}{5}$$
$$\frac{x}{2} > \frac{5}{3}$$
$$x > \frac{10}{3}$$
$$\left\{x \middle| x > \frac{10}{3}\right\} \text{ or } \left(\frac{10}{3}, \infty\right)$$

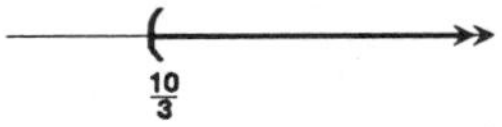

39.
$$0 < (2x - 4)^{-1} < \frac{1}{2}$$
$$2x - 4 > 2$$
$$2x > 6$$
$$x > 3$$
$$\{x | x > 3\} \text{ or } (3, \infty)$$

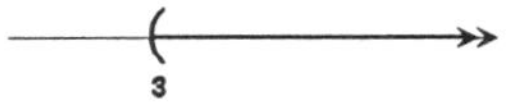

41.
$$-1 < x < 1$$
$$-1 + 4 < x + 4 < 1 + 4$$
$$3 < x + 4 < 5$$
$$a = 3, b = 5$$

43.
$$2 < x < 3$$
$$-4(2) > -4x > -4(3)$$
$$-8 > -4x > -12$$
$$-12 < -4x < -8$$
$$a = -12, b = -8$$

45.
$$0 < x < 4$$
$$2(0) < 2x < 2(4)$$
$$0 < 2x < 8$$
$$0 + 3 < 2x + 3 < 8 + 3$$
$$3 < 2x + 3 < 11$$
$$a = 3, b = 11$$

47.
$$-3 < x < 0$$
$$-3 + 4 < x + 4 < 0 + 4$$
$$1 < x + 4 < 4$$
$$\frac{1}{1} > \frac{1}{x+4} > \frac{1}{4}$$
$$\frac{1}{4} < \frac{1}{x+4} < 1$$
$$a = \frac{1}{4}, b = 1$$

49.
$$6 < 3x < 12$$
$$\frac{1}{3}(6) < \frac{1}{3}(3x) < \frac{1}{3}(12)$$
$$2 < x < 4$$
$$2^2 < x^2 < 4^2$$
$$4 < x^2 < 16$$
$$a = 4, b = 16$$

51.

$$-10 < x < 5$$

$-10 + 10 < x + 10 < 5 + 10$ and $-10 - 5 < x - 5 < 5 - 5$

$0 < x + 10 < 15$ and $-15 < x - 5 < 0$

$\therefore x + 10 > 0$ and $x - 5 < 0$

$x > -10$ and $x < 5$

$$2 > \frac{(-1)(x + 10)}{(x - 5)}$$

$$2(x - 5) < (-1)(x + 10)$$

$$2x - 10 < -x - 10$$

$$3x < 0$$

$$x < 0$$

$$\{x \mid -10 < x < 0\}$$

53. $\sqrt{3x + 6}$ is defined when

$$3x + 6 \geq 0$$

$$3x \geq -6$$

$$x \geq -2$$

For real numbers $\{x \mid x \geq -2\}$ or $[-2, \infty)$, $\sqrt{3x + 6}$ is defined.

55. Here $V = 20T$ and

$$80 \leq T \leq 120$$

$$(20)(80) \leq 20T \leq (20)(120)$$

$$1600 \leq V \leq 2400$$

The volume of the gas ranges from 1600 to 2400 cc, inclusive.

Check: $\frac{1600}{20} = 80$

$\frac{2400}{20} = 120$

57. Let P represent the selling price and C the commission. Then

$C = 45{,}000 + 0.25\,(P - 900{,}000) = 0.25P - 180{,}000$ and we are given that

$$900{,}000 \leq P \leq 1{,}100{,}000$$

$$(0.25)(900{,}000) \leq 0.25P \leq (0.25)(1{,}100{,}000)$$

$$225{,}000 \leq 0.25P \leq 275{,}000$$

$$225{,}000 - 180{,}000 \leq 0.25P - 180{,}000 \leq 275{,}000 - 180{,}000$$

$$45{,}000 \leq C \leq 95{,}000$$

The agent's commission ranges from \$45,000 to \$95,000, inclusive.

$$\frac{45{,}000}{900{,}000} = 0.05 = 5\% \text{ to } \frac{95{,}000}{1{,}100{,}000} = 0.086 = 8.6\%, \text{ inclusive}$$

As a percent of selling price, the commission ranges from 5% to 8.6%.

Check: For a \$900,000 complex, the commission is \$45,000 or 5%.

For a \$1,100,000 complex, the commission is

$\$45{,}000 + 0.25(1{,}100{,}000 - 900{,}000) = \$95{,}000$ or 8.6%.

59. Let W = weekly wage and T the withholding tax. Then

$T = 69.90 + .28(W - 517) = 69.90 + .28W - 144.76$ and since

$$525 \leq W \leq 600$$

$$(0.28)(525) \leq 0.28W \leq 0.28(600)$$

$$147 \leq 0.28W \leq 168$$

$$147 - 74.86 \leq 0.28W - 74.86 \leq 168 - 74.86$$

$$72.14 \leq T \leq 93.14$$

The amount of withholding ranges from \$72.14 to \$93.14, inclusive.

Check: For a wage of \$525, the withholding is

$69.90 + 0.28(525 - 517) = \$72.14$

For a wage of \$600, the withholding is

$69.90 + 0.28(600 - 517) = \$93.14$

61. Let K = the monthly usage in kilowatt hours and C = the total monthly charges per customer. Then $C = 0.10494K + 9.36$ and

$$
\begin{aligned}
80.24 \leq & \quad C & \leq 271.80 \\
80.24 \leq & \quad 0.10494K + 9.36 & \leq 271.80 \\
70.88 \leq & \quad 0.10494K & \leq 262.44 \\
675.43 \leq & \quad K & \leq 2500.86
\end{aligned}
$$

The range of usage in kilowatt hours varied from 675.43 to 2500.86.

Check: For 675.43 kilowatt hours, the charge is $0.10494(675.43) + 9.36 = \$80.24$.
For 2500.86 kilowatt hours, the charge is $0.10494(2500.86) + 9.36 = \$271.80$

63. Let C = the dealer cost and M = the markup over dealer's cost. If the price is \$8800, then $8800 = C + MC = C(1 + M)$ which gives

$$C = \frac{8000}{1 + M}$$

Also

$$
\begin{aligned}
0.12 \leq & \quad M & \leq 0.18 \\
1.12 \leq & \quad 1 + M & \leq 1.18 \\
\frac{1}{1.12} \geq & \quad \frac{1}{1 + M} & \geq \frac{1}{1.18} \\
\frac{800}{1.12} \geq & \quad \frac{8800}{1 + M} & \geq \frac{8800}{1.18} \\
7857.14 \geq & \quad C & \geq 7457.63 \\
7457.63 \leq & \quad C & \leq 7857.14
\end{aligned}
$$

The cost ranged from \$7457.63 to \$7857.14, inclusive.

Check: For a cost of \$7457.63 and markup of 18%, $7457.63 + 0.18(7457.63) = \$8800$.
For a cost of \$7857.14 and markup of 12%, $7857.14 + 0.12(7857.14) = \$8800$.

65. Let T = the score on the last test and G = the resulting grade. Then

$$
\begin{aligned}
& G = (68 + 82 + 87 + 89 + T) \div 5 \text{ so that} \\
& T = 5G - 326 \text{ and} \\
& 80 \leq \quad G \quad \leq 90 \\
& 400 \leq \quad 5G \quad \leq 450 \\
& 74 \leq \quad 5G - 326 \quad \leq 124
\end{aligned}
$$

The fifth test score must be 74 or greater (≥ 74).

Check: $G = \dfrac{70 + 82 + 85 + 89 + 74}{5} = 80$

67. Let G = the amount (in gallons) of gasoline in the car at the start of the trip, and let D = the distance covered in the trip. Then $D = 25G$ or $G = \dfrac{D}{25}$, and

$$
\begin{aligned}
& 300 \leq D \\
& 12 \leq \frac{D}{25} \\
& 12 \leq G
\end{aligned}
$$

The amount of gasoline ranged from 12 to 20 gallons, inclusive, at the beginning of the trip.

Check: For 12 gallons at 25 miles per gallon the car will go $(12)(25) = 300$ miles.

2.7 Polynomial and Rational Inequalities

1. $(x - 5)(x + 2) < 0$

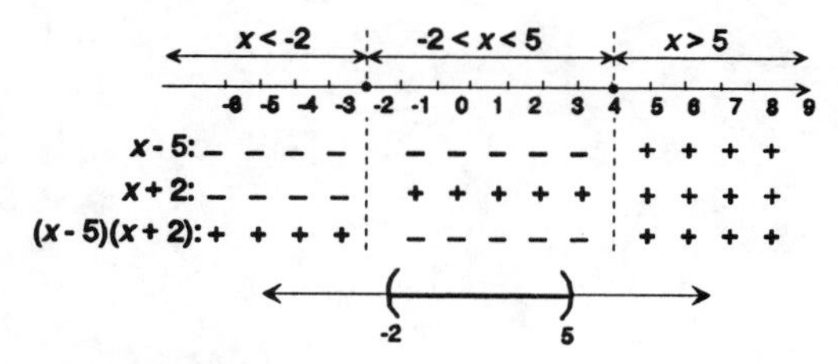

The solution set is $\{x \mid -2 < x < 5\}$.

3. $x^2 - 4x > 0$
$x(x - 4) > 0$

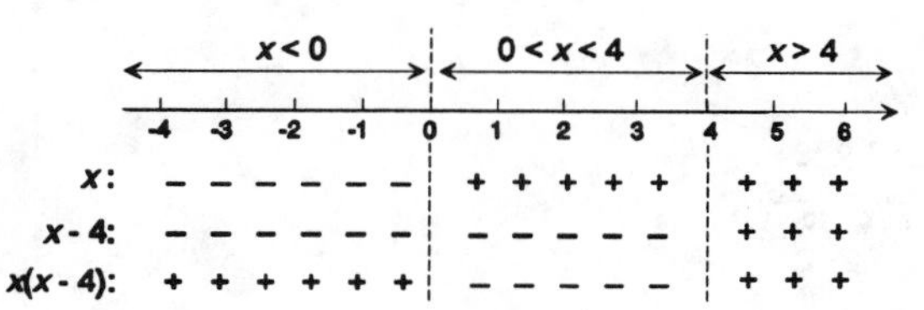

$x(x - 4) > 0$ if $x < 0$ or $x > 4$
The solution set is $\{x \mid x < 0 \text{ or } x > 4\}$.

5. $x^2 - 9 < 0$
$(x - 3)(x + 3) < 0$

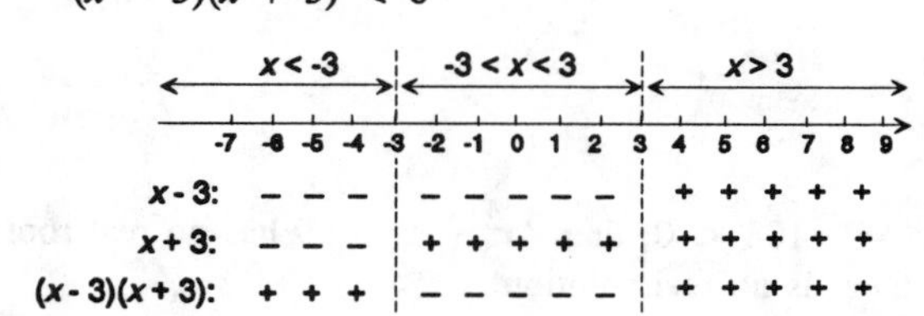

$(x - 3)(x + 3) < 0$ if $-3 < x < 3$
The solution set is $\{x \mid -3 < x < 3\}$.

7. $x^2 + x > 2$
$x^2 + x - 2 > 0$
$(x + 2)(x - 1) > 0$

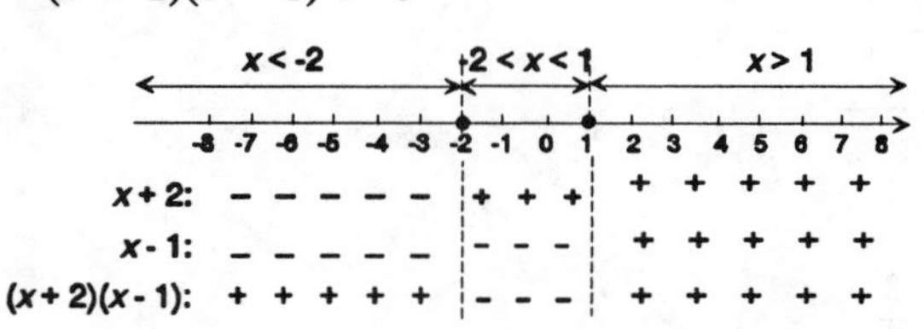

$(x + 2)(x - 1) > 0$ if $x < -2$ or $x > 1$
The solution set is $\{x \mid x < -2 \text{ or } x > 1\}$.

9.
$$2x^2 \le 5x + 3$$
$$2x^2 - 5x - 3 \le 0$$
$$(x - 3)(2x + 1) \le 0$$

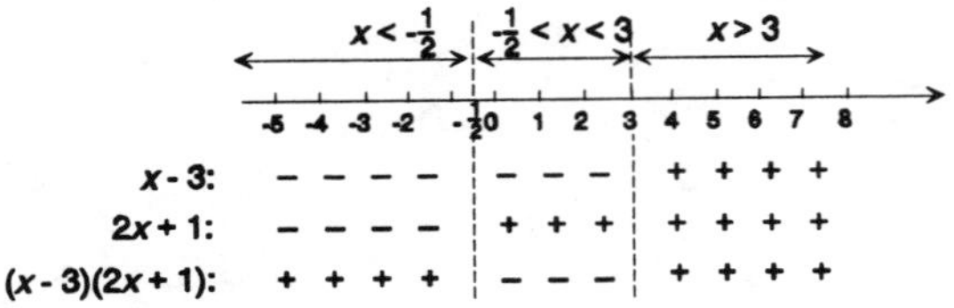

	$x < -\frac{1}{2}$	$-\frac{1}{2} < x < 3$	$x > 3$
$x - 3$:	− − − −	− − −	+ + + +
$2x + 1$:	− − − −	+ + +	+ + + +
$(x - 3)(2x + 1)$:	+ + + +	− − −	+ + + +

$(x - 3)(2x + 1) < 0$ if $-\frac{1}{2} \le x \le 3$

The solution set is $\left\{x \mid -\frac{1}{2} \le x \le 3\right\}$.

11.
$$x(x - 7) > 8$$
$$x^2 - 7x - 8 > 0$$
$$(x - 8)(x + 1) > 0$$

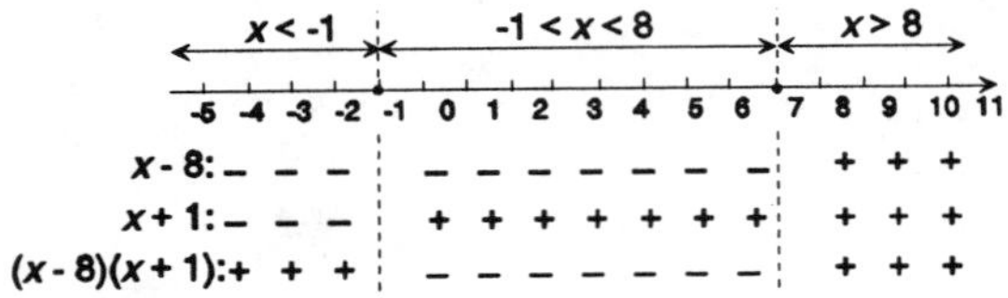

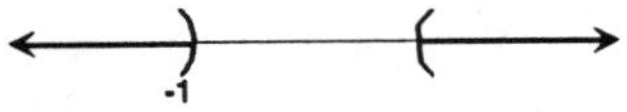

	$x < -1$	$-1 < x < 8$	$x > 8$
$x - 8$:	− − −	− − − − − − − −	+ + +
$x + 1$:	− − −	+ + + + + + +	+ + +
$(x - 8)(x + 1)$:	+ + +	− − − − − − − −	+ + +

-1

$(x - 8)(x + 1) > 0$ if $x < -1$ or $x > 8$
The solution set is $\{x \mid x < -1 \text{ or } x > 8\}$.

13.
$$4x^2 + 9 < 6x$$
$$4x^2 - 6x + 9 < 0$$
Since for $a = 4$, $b = -6$, $c = 9$, $b^2 - 4ac = -108 < 0$, then $4x^2 - 6x + 9$ has no real roots.
For $x = 0$, $4x^2 - 6 + 9 = 9 > 0$. Thus, there is no real solution.

15.
$$6(x^2 - 1) > 5x$$
$$6x^2 - 5x - 6 > 0$$
$$(2x - 3)(3x + 2) > 0$$

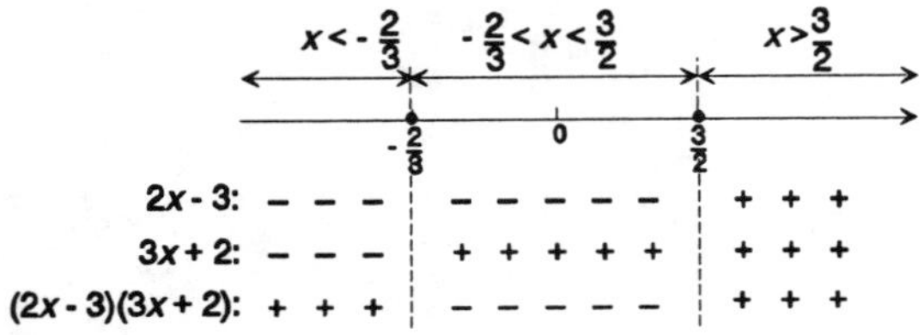

	$x < -\frac{2}{3}$	$-\frac{2}{3} < x < \frac{3}{2}$	$x > \frac{3}{2}$
$2x - 3$:	− − −	− − − − − −	+ + +
$3x + 2$:	− − −	+ + + + + +	+ + +
$(2x - 3)(3x + 2)$:	+ + +	− − − − − −	+ + +

$(2x - 3)(3x + 2) > 0$ if $x < -\frac{2}{3}$ or $x > \frac{3}{2}$

The solution set is $\left\{x \mid x < -\frac{2}{3} \text{ or } x > \frac{3}{2}\right\}$.

17. $(x - 1)(x^2 + x + 4) > 0$

	$x < 1$	$x > 1$
$x - 1$:	− − − − −	+ + + + + +
$x^2 + x + 4$:	+ + + + +	+ + + + + +
$(x - 1)(x^2 + x + 4)$:	− − − − −	+ + + + + +

$(x - 1)(x^2 + x + 4) > 0$ if $x > 1$

The solution set is $\{x \mid x > 1\}$.

19. $(x - 1)(x - 2)(x - 3) \le 0$

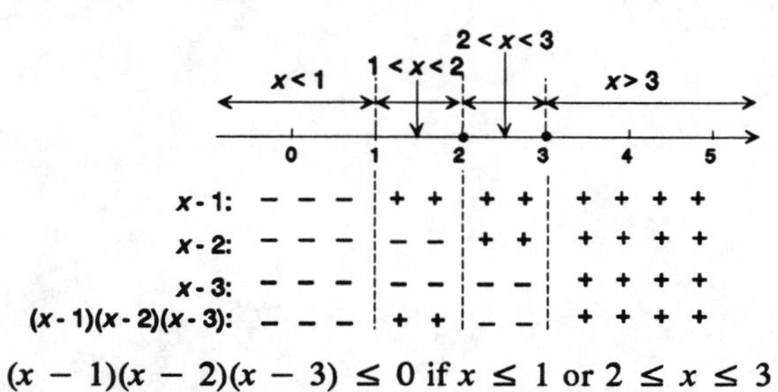

$(x - 1)(x - 2)(x - 3) \le 0$ if $x \le 1$ or $2 \le x \le 3$

The solution set is $\{x \mid x \le 1 \text{ or } 2 \le x \le 3\}$.

21.

$$x^3 - 2x^2 - 3x > 0$$
$$x(x^2 - 2x - 3) > 0$$
$$x(x - 3)(x + 1) > 0$$

	$x < -1$	$-1 < x < 0$	$0 < x < 3$	$x > 3$
x:	− − −	−	+ + + + + + +	+ +
$x - 3$:	− − −	−	− − − − − − −	+ +
$x + 1$:	− − −	+	+ + + + + + +	+ +
$x(x - 3)(x + 1)$:	− − −	+	− − − − − − −	+ +

The solution set is $\{x \mid -1 < x < 0 \text{ or } x > 3\}$.

23.

$$x^4 > x^2$$
$$x^4 - x^2 > 0$$
$$x^2(x^2 - 1) > 0$$
$$x^2(x + 1)(x - 1) > 0$$

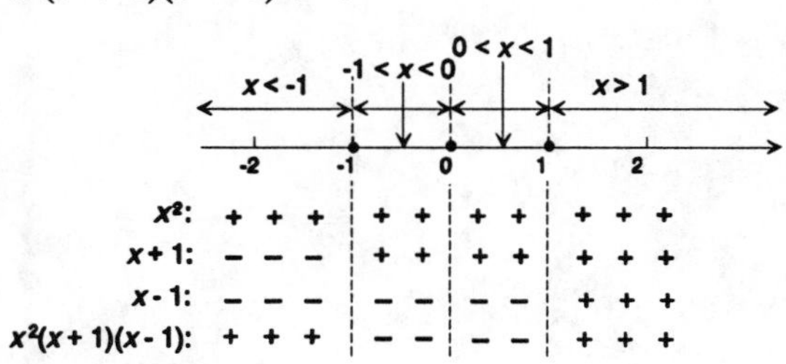

$x^2(x + 1)(x + 1) > 0$ if $x < -1$ or $x > 1$

The solution set is $\{x \mid x < -1 \text{ or } x > 1\}$.

25.
$$x^3 > 4x^2$$
$$x^3 - 4x^2 > 0$$
$$x^2(x - 4) > 0$$

	$x < 0$	$0 < x < 4$	$x > 4$
x^2:	+ + + +	+ + +	+ + +
$x - 4$:	− − − −	− − −	+ + +
$x^2(x - 4)$:	− − − −	− − −	+ + +

$x^2(x - 4) > 0$ if $x > 4$

The solution set is $\{x \mid x > 4\}$.

27.
$$x^4 > 1$$
$$x^4 - 1 > 0$$
$$(x^2 + 1)(x^2 - 1) > 0$$
$$(x^2 + 1)(x - 1)(x + 1) > 0$$

	$x < -1$	$-1 < x < 1$	$x > 1$
$x^2 + 1$:	+ + +	+ + + +	+ + + +
$x - 1$:	− − −	− − − −	+ + + +
$x + 1$:	− − −	+ + + +	+ + + +
$(x^2 + 1)(x - 1)(x + 1)$:	+ + +	− − − −	+ + + +

$(x^2 + 1)(x - 1)(x + 1) > 0$ if $x < -1$ or $x > 1$

The solution set is $\{x \mid x < -1 \text{ or } x > 1\}$.

29. $\dfrac{x + 1}{x - 1} > 0$

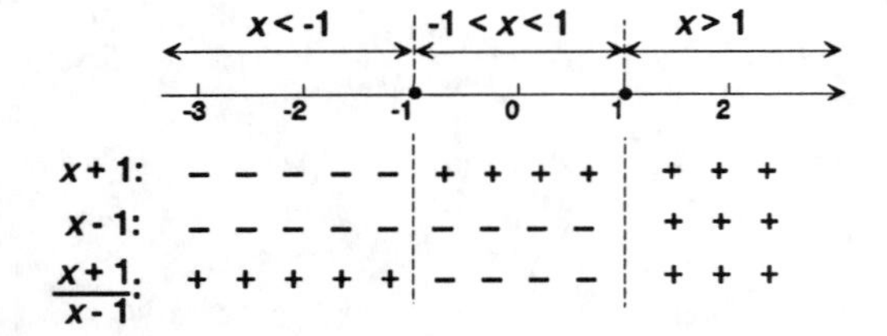

$\dfrac{x + 1}{x - 1} > 0$ if $x < -1$ or $x > 1$

The solution set is $\{x \mid x < -1 \text{ or } x > 1\}$.

31. $\dfrac{(x - 1)(x + 1)}{x} < 0$

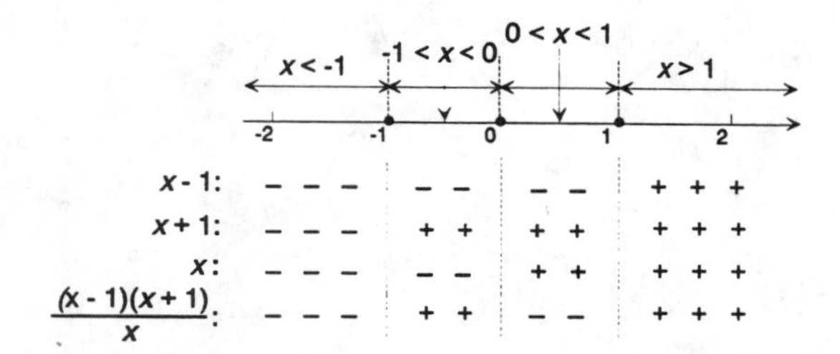

$\dfrac{(x - 1)(x + 1)}{x} < 0$ if $x < -1$ or $0 < x < 1$

The solution set is $\{x \mid x < -1 \text{ or } 0 < x < 1\}$.

33. $\dfrac{(x - 2)^2}{x^2 - 1} \geq 0$

$\dfrac{(x - 2)^2}{(x - 1)(x + 1)} \geq 0$

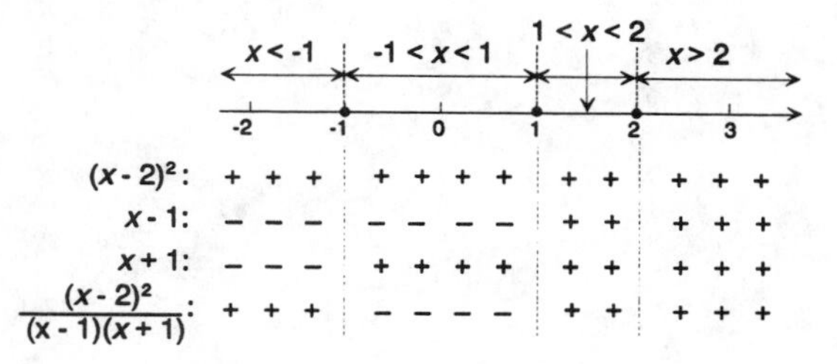

$\dfrac{(x - 2)^2}{(x - 1)(x + 1)} \geq 0$ if $x < -1$ or $x > 1$

The solution set is $\{x \mid x < -1 \text{ or } x > 1\}$.

35. $6x - 5 < \dfrac{6}{x}$

$6x - 5 - \dfrac{6}{x} < 0$

$\dfrac{6x^2 - 5x - 6}{x} < 0$

$\dfrac{(2x - 3)(3x + 2)}{x} < 0$

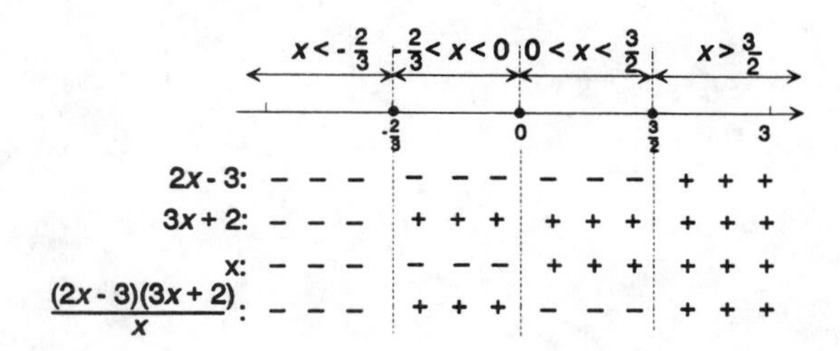

$$\frac{(2x - 3)(3x + 2)}{x} < 0 \text{ if } x < \frac{-2}{3} \text{ or } 0 < x < \frac{3}{2}$$

The solution set is $\left\{x \middle| x < \frac{-2}{3} \text{ or } 0 < x < \frac{3}{2}\right\}$.

37.
$$\frac{x + 4}{x - 2} \le 1$$
$$\frac{x + 4}{x - 2} - 1 \le 0$$
$$\frac{x + 4 - (x - 2)}{x - 2} \le 0$$
$$\frac{6}{x - 2} \le 0$$

	$x < 2$		$x > 2$
	0 1	2	3 4
$x - 2$:	− − − − − −		+ + + + +
$\frac{6}{x-2}$:	− − − − − −		+ + + + +

$\frac{6}{x - 2} \le 0$ if $x < 2$

The solution set is $\{x \mid x < 2\}$.

39.

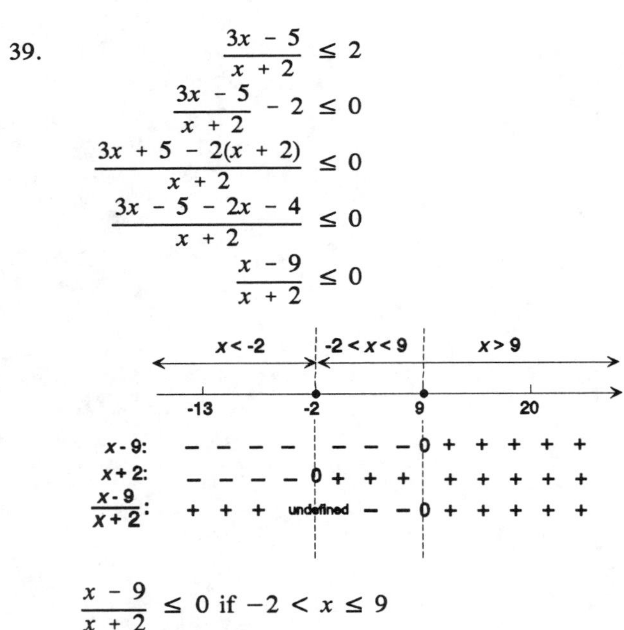

$$\frac{3x - 5}{x + 2} \le 2$$
$$\frac{3x - 5}{x + 2} - 2 \le 0$$
$$\frac{3x + 5 - 2(x + 2)}{x + 2} \le 0$$
$$\frac{3x - 5 - 2x - 4}{x + 2} \le 0$$
$$\frac{x - 9}{x + 2} \le 0$$

$\frac{x - 9}{x + 2} \le 0$ if $-2 < x \le 9$

The solution set is $\{x \mid -2 < x \le 9\}$.

41.

$$\frac{1}{x-2} < \frac{2}{3x-9}$$

$$\frac{1}{x-2} - \frac{2}{3x-9} < 0$$

$$\frac{3x-9-2x+4}{3(x-2)(x-3)} < 0$$

$$\frac{x-5}{3(x-2)(x-3)} < 0$$

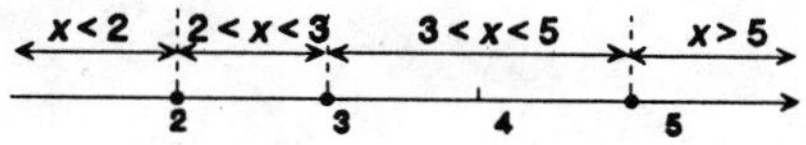

	$x<2$	$2<x<3$	$3<x<5$	$x>5$
$x-5$:	− − −	− −	− − − −	+ + +
$x-2$:	− − −	+ +	+ + + +	+ + +
$x-3$:	− − −	− −	+ + + +	+ + +
$\frac{x-5}{3(x-2)(x-3)}$:	− − −	+ +	− − − −	+ + +

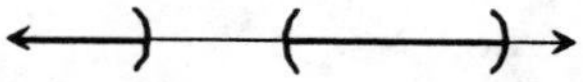

The solution set is $\{x \mid x < 2 \text{ or } 3 < x < 5\}$.

43.

$$\frac{2x+5}{x+1} > \frac{x+1}{x-1}$$

$$\frac{2x+5}{x+1} - \frac{x+1}{x-1} > 0$$

$$\frac{(2x+5)(x-1)-(x+1)(x+1)}{(x+1)(x-1)} > 0$$

$$\frac{2x^2+3x-5-x^2-2x-1}{(x+1)(x-1)} > 0$$

$$\frac{x^2+x-6}{(x+1)(x-1)} > 0$$

$$\frac{(x+3)(x-2)}{(x+1)(x-1)} > 0$$

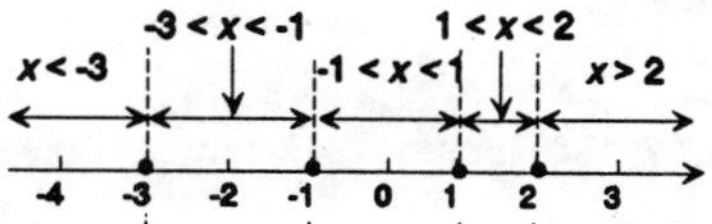

	$x<-3$	$-3<x<-1$	$-1<x<1$	$1<x<2$	$x>2$
$x+3$:	− −	+ +	+ +	+	+ +
$x-2$:	− −	− −	− −	−	+ +
$x+1$:	− −	− −	+ +	+	+ +
$x-1$:	− −	− −	− −	+	+ +
$\frac{(x+3)(x-2)}{(x+1)(x-1)}$:	+ +	− −	+ +	−	+ +

$$\frac{(x+3)(x-2)}{(x+1)(x-1)} > 0 \text{ if } x < -3 \text{ or } -1 < x < 1 \text{ or } x > 2$$

The solution set is $\{x \mid x < -3 \text{ or } -1 < x < 1 \text{ or } x > 2\}$.

45. $\dfrac{x^2(3 + x)(x + 4)}{(x + 5)(x - 1)} \geq 0$

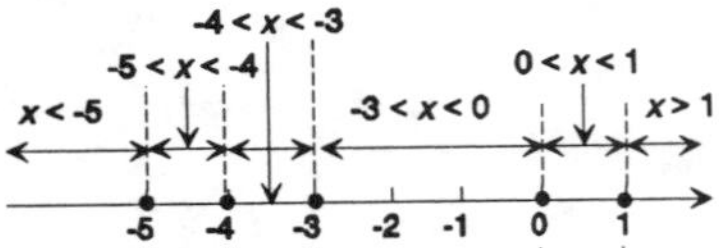

	$x<-5$	$-5<x<-4$	$-4<x<-3$	$-3<x<0$	$0<x<1$	$x>1$
x^2:	+ +	+	+	+ + +	+	+
$3 + x$:	− −	−	−	+ + +	+	+
$x + 4$:	− −	−	+	+ + +	+	+
$x + 5$:	− −	+	+	+ + +	+	+
$x - 1$:	− −	−	−	− − −	−	+
$\frac{x^2(3+x)(x+4)}{(x+5)(x-1)}$:	+ +	−	+	− − −	−	+

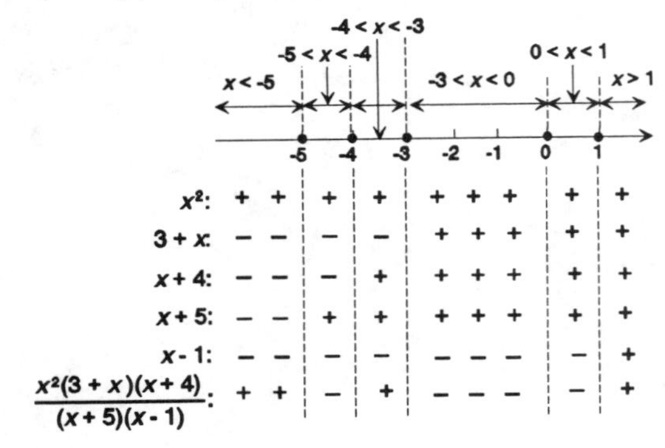

$\dfrac{x^2(3 + x)(x + 4)}{(x + 5)(x - 1)} > 0$ if $x < -5$ or $-4 \leq x \leq -3$ or $x > 1$

The solution set is $\{x \mid x < -5 \text{ or } -4 \leq x \leq -3 \text{ or } x > 1\}$.

47. Let x be the number. Then we want the set so that

$$x^3 > 4x^2$$
$$x^3 - 4x^2 > 0$$
$$x^2(x - 4) > 0$$

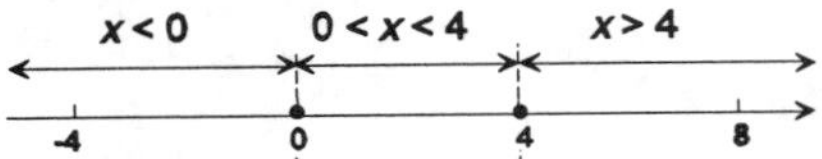

	$x<0$	$0<x<4$	$x>4$
x^2:	+ + + +	+ + +	+ + + +
$x - 4$:	− − − −	− − −	+ + + +
$x^2(x - 4)$:	− − − −	− − −	+ + + +

$x^2(x - 4) > 0$ if $x > 4$

The solution set is $\{x \mid x > 4\}$, all numbers larger than 4.

49. The domain of the variable in the expression $\sqrt{x^2 - 16}$ includes all values for which

$$x^2 - 16 \geq 0$$
$$(x - 4)(x + 4) \geq 0$$

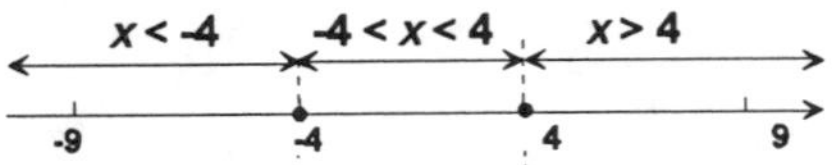

	$x<-4$	$-4<x<4$	$x>4$
$x - 4$:	− − − −	− − −	+ + + +
$x + 4$:	− − − −	+ + +	+ + + +
$(x - 4)(x + 4)$:	+ + + +	− − −	+ + + +

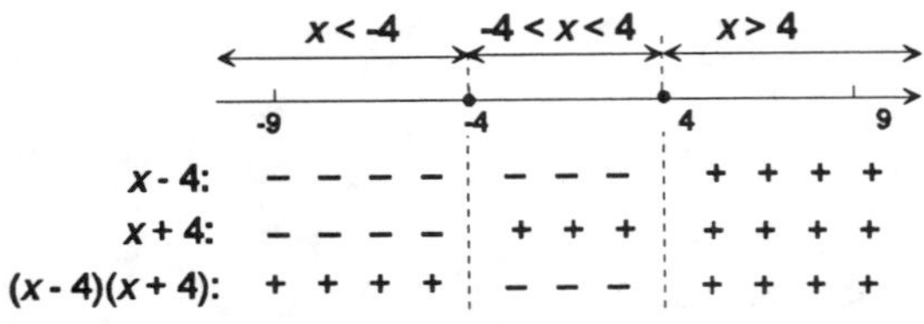

$(x - 4)(x + 4) \geq 0$ if $x \leq -4$ or $x \geq 4$

The domain is $\{x \mid x \leq -4 \text{ or } x \geq 4\}$.

51. The domain of the variable in the expression $\sqrt{\dfrac{x-2}{x+4}}$ includes all values for which $\dfrac{x-2}{x+4} \geq 0$.

	$x < -4$	$-4 < x < 2$	$x > 2$
Number line	-10, -4	2	8
$x - 2$:	− − − −	− − −	+ + + +
$x + 4$:	− − − −	+ + +	+ + + +
$\dfrac{x-2}{x+4}$:	+ + + +	− − −	+ + + +

$\dfrac{x-2}{x+4} \geq 0$ if $x < -4$ or $x \geq 2$

The domain is $\{x \mid x < -4 \text{ or } x \geq 2\}$.

53. We want to find the set of t values for which

$80t - 16t^2 > 96$

$-16t^2 + 80t - 96 > 0$

$+16t^2 - 80t + 96 < 0$ is an equivalent inequality

$16(t^2 - 5t + 6) < 0$

$16(t - 2)(t - 3) < 0$ gives the boundary points 2, 3:

	Test Number	$t - 2$	$t - 3$	$16(t - 2)(t - 3)$
$t < 2$	1	−	−	+
$2 < t < 3$	2.5	+	−	−
$t > 3$	4	+	+	+

The solution set is $2 < t < 3$. The ball is more than 96 feet above the ground for time t between 2 and 3 seconds, $2 < t < 3$

55. Profit = Revenue − Cost

$= x(40 - 0.2x) - 28x$

We want the values of x for which

$$x(40 - 0.2x) - 28x \geq 100$$
$$40x - 0.2x^2 - 28x \geq 100$$
$$-0.2x^2 + 12x - 100 \geq 0$$
$$-10(-0.2x^2 + 12x - 100) \leq 0(-10)$$
$$2x^2 - 120x + 1000 \leq 0$$
$$x^2 - 60x + 500 \leq 0$$

$(x - 10)(x - 50) \leq 0$ gives boundary points 10, 50:

	$-\infty < x < 10$	$10 < x < 50$	$50 < x <$
Number line	0, 10	20, 30, 40	50, 60
$x - 10$:	− −	+ + + + + +	+ +
$x - 50$:	− −	− − − − − −	+ +
$(x - 10)(x - 50)$:	+ +	− − − − − −	+ +

$(x - 10)(x - 50) \leq 0$ if $10 \leq x \leq 50$

Thus, for a profit of at least \$100, between 10 and 50 watches must be sold, $10 \leq x \leq 50$.

57. Prove that if a, b are real numbers and $a \geq 0$, $b \geq 0$, then $a \leq b$ is equivalent to $\sqrt{a} \leq \sqrt{b}$.

$a \leq b \Rightarrow b - a \geq 0$

$$\left(\sqrt{b} - \sqrt{a}\right)\left(\sqrt{b} + \sqrt{a}\right) \geq 0$$

Either (1) $\sqrt{b} - \sqrt{a} \geq 0$ and $\sqrt{b} + \sqrt{a} \geq 0$

or (2) $\sqrt{b} - \sqrt{a} \leq 0$ and $\sqrt{b} + \sqrt{a} \leq 0$

In (1) $\sqrt{b} \geq \sqrt{a}$ and $\sqrt{b} \geq -\sqrt{a}$

$\sqrt{a} \leq \sqrt{b}$ and $-\sqrt{a} \leq \sqrt{b}$

In (2) $\sqrt{b} \leq \sqrt{a}$ and $\sqrt{b} \leq -\sqrt{a}$

Case (2) is impossible since $\sqrt{a} \geq 0$ and $\sqrt{b} \geq 0$ which means $\sqrt{b} \leq -\sqrt{a}$ is impossible.

Case (1) is true and if $a \leq b$, then $\sqrt{a} \leq \sqrt{b}$

So $a \leq b$ is equivalent to $\sqrt{a} \leq \sqrt{b}$.

59. Given $kx^2 + 2x + 1 = 0$, $a = k$, $b = 2$, $c = 1$

We want a to be such that $b^2 - 4ac > 0$, so

$$2^2 - 4(k)(1) > 0$$
$$4 - 4k > 0$$
$$4 > 4k$$
$$1 > k$$
$$k < 1$$

For $kx^2 + 2x + 1 = 0$ to have two distinct real solutions, $k < 1$.

2.8 Equations and Inequalities Involving Absolute Value

1. $|2x| = 6$

$2x = 6$ or $2x = -6$

$x = 3$ or $x = -3$

The solution set is $\{-3, 3\}$.

3. $|2x + 3| = 5$

$2x + 3 = 5$ or $2x + 3 = -5$

$2x = 2$ or $2x = -8$

$x = 1$ $\quad x = -4$

The solution set is $\{-4, 1\}$.

5. $|1 - 4t| = 5$

$1 - 4t = 5$ or $1 - 4t = -5$

$-4t = 4$ or $-4t = -6$

$t = -1$ or $t = \frac{3}{2}$

The solution set is $\left\{-1, \frac{3}{2}\right\}$

7. $|-2x| = 8$

$-2x = 8$ or $-2x = -8$

$x = -4$ $\quad x = 4$

The solution set is $\{-4, 4\}$.

9. $|-2|x = 4$

$2x = 4$

$x = 2$

The solution is $x = 2$.

11. $\frac{2}{3}|x| = 8$

$|x| = 12$

$x = 12$ or $x = -12$

The solution set is $\{-12, 12\}$.

13. $\left|\frac{x}{3} + \frac{2}{5}\right| = 2$

$$\frac{x}{3} + \frac{2}{5} = 2 \quad \text{or} \quad \frac{x}{3} + \frac{2}{5} = -2$$
$$5x + 6 = 30 \quad \text{or} \quad 5x + 6 = -30$$
$$5x = 24 \quad \text{or} \quad 5x = -36$$
$$x = \frac{24}{5} \quad \text{or} \quad x = \frac{-36}{5}$$

The solution set is $\left\{\frac{-36}{5}, \frac{24}{5}\right\}$

15. $|u - 2| = -\frac{1}{2}$

It is not possible that $|u - 2| < 0$.
Therefore, there is no real solution.

17. $|x^2 - 9| = 0$

$$x^2 - 9 = 0$$
$$(x - 3)(x + 3) = 0$$
$$x = -3 \text{ or } x = 3$$

The solution set is $\{-3, 3\}$.

19. $|x^2 - 2x| = 3$

$$x^2 - 2x = 3 \text{ or } \qquad x^2 - 2x = -3$$
$$x^2 - 2x - 3 = 0 \text{ or } \qquad x^2 - 2x + 3 = 0$$
$$(x - 3)(x + 1) = 0 \text{ or } \qquad \text{No real solution (Note that } b^2 - 4ac = -8)$$
$$x = -1 \text{ or } x = 3$$

The solution set is $\{-1, 3\}$.

21. $|x^2 + x - 1| = 1$

$$x^2 + x - 1 = 1 \text{ or } \qquad x^2 + x - 1 = -1$$
$$x^2 + x - 2 = 0 \text{ or } \qquad x^2 + x = 0$$
$$(x + 2)(x - 1) = 0 \text{ or } \qquad x(x + 1) = 0$$
$$x = -2 \text{ or } x = 1 \text{ or } \quad x = -1 \text{ or } x = 0$$

The solution set is $\{-2, -1, 0, 1\}$.

23. $|2x| < 8$

$$-8 < 2x < 8$$
$$\frac{-8}{2} < \frac{2x}{2} < \frac{8}{2}$$
$$-4 < x < 4$$

The solutions set consists of all numbers x for which $\{x \mid -4 < x < 4\}$ or $(-4, 4)$.

-4 4

25. $|3x| > 12$

$$3x < -12 \quad \text{or} \quad 3x > 12$$
$$\frac{3x}{3} < \frac{-12}{3} \quad \text{or} \quad \frac{3x}{3} > \frac{12}{3}$$
$$x < -4 \quad \text{or} \quad x > 4$$

The solution set consists of all numbers x for which $\{x \mid x < -4 \text{ or } x > 4\}$ or $(-\infty, -4)$ or $(4, \infty)$.

-4 4

27. $|x - 2| < 1$

$$-1 < x - 2 < 1$$
$$-1 + 2 < x - 2 + 2 < 1 + 2$$
$$1 < x < 3$$

The solution set consists of all numbers x for which $\{x \mid 1 < x < 3\}$ or $(1, 3)$.

29. $|3t - 2| \leq 4$

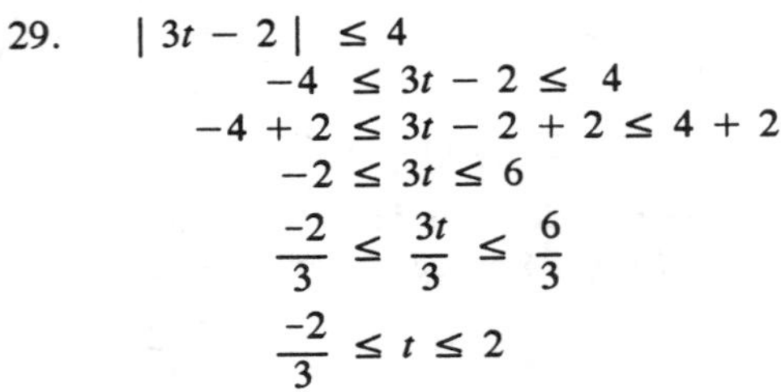

$$-4 \leq 3t - 2 \leq 4$$
$$-4 + 2 \leq 3t - 2 + 2 \leq 4 + 2$$
$$-2 \leq 3t \leq 6$$
$$\frac{-2}{3} \leq \frac{3t}{3} \leq \frac{6}{3}$$
$$\frac{-2}{3} \leq t \leq 2$$

The solution set consists of all numbers t for which

$\left\{t \mid -\frac{2}{3} \leq t \leq 2\right\}$ or $\left[-\frac{2}{3}, 2\right]$.

31. $|x - 3| \geq 2$

$x - 3 \leq -2$ or $x - 3 \geq 2$

$x \leq 1$ or $x \geq 5$

$(-\infty, 1]$ or $[5, \infty)$

33. $|1 - 4x| < 5$

$$-5 < 1 - 4x < 5$$
$$-5 - 1 < 1 - 4x - 1 < 5 - 1$$
$$-6 < -4x < 4$$
$$\frac{-6}{-4} > \frac{-4x}{-4} > \frac{4}{-4}$$
$$\frac{3}{2} > x > -1$$
$$-1 < x < \frac{3}{2}$$

The solution set consists of all numbers x for which $\left\{x \mid -1 < x < \frac{3}{2}\right\}$ or $\left(-1, \frac{3}{2}\right)$.

35. $|1 - 2x| > 3$

$1 - 2x < -3$ or $1 - 2x > 3$

$1 - 2x - 1 < -3 - 1$ or $1 - 2x - 1 > 3 - 1$

$-2x < -4$ or $-2x > 2$

$\frac{-2x}{-2} > \frac{4}{2}$ or $\frac{-2x}{-2} < \frac{2}{-2}$

$x > 2$ or $x < -1$

The solution set consists of all numbers x for which $\{x \mid x < -1 \text{ or } x > 2\}$ or $(-\infty, -1)$ or $(2, \infty)$.

37. $|x + 3| > 0$

$x + 3 < 0$ or $x + 3 > 0$

$x < -3$ or $x > -3$

$(-\infty, -3)$ or $(-3, \infty)$

39. $|x + 2| > -3$

Since $|x + 2| \geq 0$ for all x, we know $|x + 2| > -3$ for all real numbers x or $(-\infty, \infty)$.

41. $|2x - 1| < 0.02$

$$-0.02 < 2x - 1 < 0.02$$
$$-0.02 + 1 < 2x - 1 + 1 < 0.02 + 1$$
$$0.98 < 2x < 1.02$$
$$0.49 < x < 0.51$$

The solution set consists of all numbers x for which $\{x \mid 0.49 < x < 0.51\}$ or $(0.49, 0.51)$.

0.49 0.51

43. $|x - 1| < 3$

$$-3 < x - 1 < 3$$
$$-3 + 1 < x < 3 + 1$$
$$-2 < x < 4$$
$$2 < x + 4 < 8$$
$$a = 2, b = 8$$

45. $|x + 4| \leq 2$

$$-2 \leq x + 4 \leq 2$$
$$-2 - 4 \leq x \leq 2 - 4$$
$$-6 \leq x \leq -2$$
$$-12 \leq 2x \leq -4$$
$$-15 \leq 2x - 3 \leq -7$$
$$a = -15, b = -7$$

47. $|x - 2| \leq 7$

$$-7 \leq x - 2 \leq 7$$
$$-15 \leq x - 10 \leq -1$$
$$\frac{1}{15} \geq \frac{1}{x - 10} \geq \frac{-1}{1}$$
$$-1 \leq \frac{1}{x - 10} \leq \frac{-1}{15}$$
$$a = -1, b = -\frac{1}{15}$$

49. $$\left|\frac{a}{b}\right| = \sqrt{\left(\frac{a}{b}\right)^2} = \sqrt{\frac{a^2}{b^2}} = \frac{\sqrt{a^2}}{\sqrt{b^2}} = \frac{|a|}{|b|}$$

51. $(a + b)^2 = a^2 + 2ab + b^2 = |a|^2 + 2|a|\,|b| + |b|^2 = [\,|a| + |b|\,]^2$,

therefore $\sqrt{(a + b)^2} \leq \sqrt{(|a| + |b|)^2}$ or $|a + b| \leq |a| + |b|$

53. x differs from 3 by less than $\frac{1}{2}$

$$|x - 3| < \frac{1}{2}$$
$$|x - 3| < \frac{1}{2}$$
$$-\frac{1}{2} < x - 3 < \frac{1}{2}$$
$$-\frac{1}{2} + 3 < x - 3 + 3 < \frac{1}{2} + 3$$
$$\frac{5}{2} < x < \frac{7}{2}$$

The solution set consists of all numbers x for which $\frac{5}{2} < x < \frac{7}{2}$.

55. x differs from -3 by more than 2

$$|x - (-3)| > 2$$
$$|x + 3| > 2$$
$$x + 3 < -2 \text{ or } x + 3 > 2$$
$$x < -5 \text{ or } x > -1$$

The solution set consists of all numbers x for which $x < -5$ or $x > -1$.

57. A temperature x that differs from 98.6°F by at least 1.5°

$$|x - 98.6°| \geq 1.5°$$
$$x - 98.6° \leq -1.5° \text{ or } x - 98.6° \geq 1.5°$$
$$x \leq 97.1° \text{ or } x \geq 100.1°$$

The temperatures that are considered unhealthy are those that are less than 97.1°F or greater than 100.1°F inclusive.

59. $$x^2 < a$$
$$x^2 - a < 0$$

$(x + \sqrt{a})(x - \sqrt{a}) < 0$ gives boundary points $-\sqrt{a}$, $\sqrt{a}$:

	Test Number	$x + \sqrt{a}$	$x - \sqrt{a}$	$(x + \sqrt{a})(x - \sqrt{a})$
$x < -\sqrt{a}$	$-\sqrt{a} - 1$	$-$	$-$	$+$
$-\sqrt{a} < x < \sqrt{a}$	0	$+$	$-$	$-$
$x > \sqrt{a}$	$\sqrt{a} + 1$	$+$	$+$	$+$

The solution set is $-\sqrt{a} < x < \sqrt{a}$.

61. $x^2 < 1$

Using the results of Problem 55 with $a = 1$, we get the solution set

$-\sqrt{1} < x < \sqrt{1}$ or $-1 < x < 1$.

63. $x^2 \geq 9$

Using the result of Problem 56 with $a = 9$, we get the solution for $x^2 > 9$ to be $x > \sqrt{9}$ or $x < -\sqrt{9}$ which becomes $x > 3$ or $x < -3$. Also $x^2 = 9$ when $x = 3$ or $x = -3$.

The solution set is $x \geq 3$ or $x \leq -3$.

65. $x^2 \leq 16$

Using the result of Problem 55 with $a = 16$, we get the solution for $x^2 < 16$ to be $-\sqrt{16} < x < \sqrt{16}$ or $-4 < x < 4$. Also $x^2 = 16$ when $x = 4$ or $x = -4$.

The solution set is $-4 \leq x \leq 4$.

67. $x^2 > 4$

Using the result of Problem 56 with $a = 4$, we get the solution set $x > \sqrt{4}$ or $x < -\sqrt{4}$; i.e., the solution set is $x > 2$ or $x < -2$.

69. $$|\,3x - |2x + 1|\,| = 4$$

$3x - |2x + 1| = -4$ or $3x - |2x + 1| = 4$

$3x + 4 = |2x + 1|$ or $3x - 4 = |2x + 1|$

$2x + 1 = -(3x + 4)$ or $2x + 1 = 3x + 4$ or $2x + 1 = (-3x - 4)$ or $2x + 1 = 3x - 4$

$2x + 1 = -3x - 4$ | $-3 = x$ | $2x + 1 = -3x + 4$ | $5 = x$

$5x = -5$ | | $5x = 3$ |

$x = -1$ | $x = -3$ | $x = \dfrac{3}{5}$ | $x = 5$

Check:

$x = -1$	$|\,3(-1) - |2(-1) + 1|\,| = |-3 - 1| = 4$	Yes
$x = -3$	$|\,3(-3) - |2(-3) + 1|\,| = |-9 - 5| = 14$	No
$x = \dfrac{3}{5}$	$\left|\,3\left(\dfrac{3}{5}\right) - \left|2\left(\dfrac{3}{5}\right) + 1\right|\,\right| = \left|\dfrac{9}{5} - \dfrac{11}{5}\right| = \dfrac{2}{5}$	No
$x = 5$	$|\,3(5) - |2(5) + 1|\,| = |15 - 11| = 4$	Yes

The solution set is $\{-1, 5\}$.

2 Chapter Review

1.
$$2 - \frac{x}{3} = 8$$
$$2 - \frac{x}{3} - 2 = 8 - 2$$
$$-\frac{x}{3} = 6$$
$$(-3)\left(-\frac{x}{3}\right) = 6(-3)$$
$$x = -18$$
Check: $2 - \frac{-18}{3} \stackrel{?}{=} 8$
$$2 + 6 \stackrel{?}{=} 8$$
$$8 = 8$$

3.
$$-2(5 - 3x) + 8 = 4 + 5x$$
$$-10 + 6x + 8 = 4 + 5x$$
$$-2 + 6x - 4 = 4 + 5x - 4$$
$$6x - 6 = 5x$$
$$6x - 6 - 5x = 5x - 5x$$
$$x - 6 = 0$$
$$x = 6$$
Check: $-2(5 - 3 \cdot 6) + 8 \stackrel{?}{=} 4 + 5 \cdot 6$
$$-2(-13) + 8 \stackrel{?}{=} 4 + 30$$
$$26 + 8 \stackrel{?}{=} 34$$
$$34 = 34$$

5.
$$\frac{3x}{4} - \frac{x}{3} = \frac{1}{12}$$
$$12\left(\frac{3x}{4} - \frac{x}{3}\right) = \left(\frac{1}{12}\right)(12)$$
$$9x - 4x = 1$$
$$5x = 1$$
$$\frac{5x}{5} = \frac{1}{5}$$
$$x = \frac{1}{5}$$
Check: $\frac{3\left(\frac{1}{5}\right)}{4} - \frac{\left(\frac{1}{5}\right)}{3} \stackrel{?}{=} \frac{1}{12}$
$$\frac{\frac{9}{5}}{12} - \frac{\frac{4}{5}}{12} \stackrel{?}{=} \frac{1}{12}$$
$$\frac{1}{12} = \frac{1}{12}$$

7.
$$\frac{x}{x - 1} = \frac{6}{5}, \quad x \neq 1$$
$$5(x - 1)\frac{x}{x - 1} = \frac{6}{5} \cdot 5(x - 1)$$
$$5x = 6(x - 1)$$
$$5x = 6x - 6$$
$$5x + 6 = 6x - 6 + 6$$
$$5x + 6 - 5x = 6x - 5x$$
$$6 = x$$
$$x = 6$$
Check: $\frac{6}{6 - 1} \stackrel{?}{=} \frac{6}{5}$
$$\frac{6}{5} \stackrel{?}{=} \frac{6}{5}$$

9.
$$x(1 - x) = 6$$
$$x - x^2 = 6$$
$$0 = x^2 - x + 6$$
Here $a = 1$, $b = -1$, $c = 6$, and $b^2 - 4ac = -23$. No real solution.

11.

$$\frac{1}{2}\left(x - \frac{1}{3}\right) = \frac{3}{4} - \frac{x}{6}$$

$$12 \cdot \frac{1}{2}\left(x - \frac{1}{3}\right) = 12\left(\frac{3}{4} - \frac{x}{6}\right)$$

$$6x - 2 = 9 - 2x$$
$$6x - 2 + 2 = 9 - 2x + 2$$
$$6x = 11 - 2x$$
$$6x + 2x = 11 - 2x + 2x$$
$$8x = 11$$
$$\frac{8x}{8} = \frac{11}{8}$$
$$x = \frac{11}{8}$$

Check: $\frac{1}{2}\left(\frac{11}{8} - \frac{1}{3}\right) \overset{?}{=} \frac{3}{4} - \frac{\frac{11}{8}}{6}$

$$\frac{1}{2}\left(\frac{33}{24} - \frac{8}{24}\right) \overset{?}{=} \frac{3}{4} - \frac{11}{48}$$

$$\frac{1}{2}\left(\frac{25}{24}\right) \overset{?}{=} \frac{36}{48} - \frac{11}{48}$$

$$\frac{25}{48} = \frac{25}{48}$$

13.

$$(x - 1)(2x + 3) = 3$$
$$2x^2 + 3x - 2x - 3 = 3$$
$$2x^2 + x - 3 = 3$$
$$2x^2 + x - 3 - 3 = 3 - 3$$
$$2x^2 + x - 6 = 0$$
$$(2x - 3)(x + 2) = 0$$

$2x - 3 = 0$ or $x + 2 = 0$

$x = \frac{3}{2}$ or $x = -2$

$$\left\{-2, \frac{3}{2}\right\}$$

Check -2: $(-2 - 1)(2 \cdot -2 + 3) \overset{?}{=} 3$

$$(-3)(-1) \overset{?}{=} 3$$
$$3 = 3$$

Check $\frac{3}{2}$: $\left(\frac{3}{2} - 1\right)\left(2 \cdot \frac{3}{2} + 3\right) \overset{?}{=} 3$

$$\left(\frac{1}{2}\right)(6) \overset{?}{=} 3$$
$$3 = 3$$

15.

$$2x + 3 = 4x^2$$
$$2x + 3 - 2x = 4x^2 - 2x$$
$$3 = 4x^2 - 2x$$
$$3 - 3 = 4x^2 - 2x - 3$$
$$0 = 4x^2 - 2x - 3$$

Here $a = 4$, $b = -2$, $c = -3$ and $b^2 - 4ac = 52$.

Then $x = \frac{-(-2) \pm \sqrt{52}}{2(4)} = \frac{2 \pm 2\sqrt{13}}{8} = \frac{1 \pm \sqrt{13}}{4}$

$$\left\{\frac{1 - \sqrt{13}}{4}, \frac{1 + \sqrt{13}}{4}\right\}$$

Check: $\frac{1 - \sqrt{13}}{4}$: $2\left(\frac{1 - \sqrt{13}}{4}\right) + 3 \overset{?}{=} 4\left(\frac{1 - \sqrt{13}}{4}\right)^2$

$$\frac{1 - \sqrt{13}}{2} + \frac{6}{2} \overset{?}{=} \frac{\left(1 - \sqrt{13}\right)^2}{4}$$

$$\frac{7 - \sqrt{13}}{2} \overset{?}{=} \frac{1 - 2\sqrt{13} + 13}{4}$$

$$\frac{7 - \sqrt{13}}{2} \overset{?}{=} \frac{14 - 2\sqrt{13}}{4}$$

$$\frac{7 - \sqrt{13}}{2} \overset{?}{=} \frac{2\left(7 - \sqrt{13}\right)}{4}$$

$$\frac{7 - \sqrt{13}}{2} = \frac{7 - \sqrt{13}}{2}$$

Check: $\frac{1+\sqrt{13}}{4}$: $2\left[\frac{1+\sqrt{13}}{4}\right] + 3 \stackrel{?}{=} 4\left[\frac{1+\sqrt{13}}{4}\right]^2$

$$\frac{1+\sqrt{13}+6}{2} \stackrel{?}{=} \frac{\left(1+2\sqrt{13}\right)+13}{4}$$
$$\frac{7+\sqrt{13}}{2} \stackrel{?}{=} \frac{14+2\sqrt{13}}{4}$$
$$\frac{7+\sqrt{13}}{2} \stackrel{?}{=} \frac{2\left(7+\sqrt{13}\right)}{4}$$
$$\frac{7+\sqrt{13}}{2} = \frac{7+\sqrt{13}}{2}$$

17. $\sqrt[3]{x^2-1} = 2$

$$\left(\sqrt[3]{x^2-1}\right)^3 = 2^3$$
$$x^2 - 1 = 8$$
$$x^2 = 9$$
$$x = \pm\sqrt{9} = \pm 3$$
$\{-3, 3\}$

Check -3: $\sqrt[3]{(-3)^2 - 1} \stackrel{?}{=} 2$

$$\sqrt[3]{8} \stackrel{?}{=} 2$$
$$2 = 2$$

Check 3: $\sqrt[3]{3^2 - 1} \stackrel{?}{=} 2$

$$\sqrt[3]{8} \stackrel{?}{=} 2$$
$$2 = 2$$

19. $x(x+1) + 2 = 0$

$$x^2 + x + 2 = 0$$
Here, $a = 1, b = 1, c = 2$
and $b^2 - 4ac = -7$.
No real solution.

21. $x^4 - 5x^2 + 4 = 0$

Let $u = x^2$. Then $u^2 = x^4$

$$u^2 - 5u + 4 = 0$$
$$(u-4)(u-1) = 0$$
$u - 4 = 0$ or $u - 1 = 0$

$u = 4$ or $u = 1$

$x^2 = 4$ or $x^2 = 1$

$x = \pm 2$ or $x = \pm 1$

The solution set is $\{-2, -1, 1, 2\}$.

Check -2: $(-2)^4 - 5(-2)^2 + 4 \stackrel{?}{=} 0$

$$6 - 20 + 4 \stackrel{?}{=} 0$$
$$0 = 0$$

Check -1: $(-1)^4 - 5(-1)^2 + 4 \stackrel{?}{=} 0$

$$1 - 5 + 4 \stackrel{?}{=} 0$$
$$0 = 0$$

Check 1: $(1)^4 - 5(1)^2 + 4 \stackrel{?}{=} 0$

$$1 - 5 + 4 \stackrel{?}{=} 0$$
$$0 = 0$$

Check 2: $(2)^4 - 5(2)^2 + 4 \stackrel{?}{=} 0$

$$16 - 20 + 4 \stackrel{?}{=} 0$$
$$0 = 0$$

23. $\sqrt{2x-3} + x = 3$

$$\sqrt{2x-3} = 3 - x$$
$$\left(\sqrt{2x-3}\right)^2 = (3x - x)^2$$
$$2x - 3 = 9 - 6x + x^2$$
$$2x - 3 - 2x + 3 = x^2 - 6x + 9 - 2x + 3$$
$$0 = x^2 - 8x + 12$$
$$0 = (x-2)(x-6)$$
$x - 2 = 0$ or $x - 6 = 0$

$x = 2$ or $x = 6$

Check 6: $\sqrt{2(6) - 3} + 2 \stackrel{?}{=} 3$

$$5 \neq 3$$

Check 2: $\sqrt{2(2) - 3} + 2 \stackrel{?}{=} 3$

$$1 + 2 \stackrel{?}{=} 3$$
$$3 = 3$$

The solution is $x = 2$.

25.

$$x^{3/2} + 5x^{1/2} = 0$$
$$x^{1/2}(x^{2/2} + 5) = 0$$
$$x^{1/2}(x + 5) = 0$$
$$x^{1/2} = 0 \quad \text{or} \quad x + 5 = 0$$
$$x = 0 \qquad x = -5$$

Check 0: $0^{3/2} + 5 \cdot 0^{1/2} \stackrel{?}{=} 0$

$$0 = 0$$

Check -5: $-5^{3/2} + 5 \cdot -5^{1/2} \stackrel{?}{=} 0$

$$5\sqrt{-5} + 5\sqrt{-5} \stackrel{?}{=} 0$$
$$10\sqrt{-5} \stackrel{?}{=} 0$$

The solution is $x = 0$.

27.

$$\sqrt{x + 1} + \sqrt{x - 1} = \sqrt{2x + 1}$$
$$\left(\sqrt{x + 1} + \sqrt{x - 1}\right)^2 = \left(\sqrt{2x + 1}\right)^2$$
$$x + 1 + 2\sqrt{x + 1}\sqrt{x - 1} + x - 1 = 2x + 1$$
$$2\sqrt{x + 1}\sqrt{x - 1} + 2x = 2x + 1$$
$$2\sqrt{x + 1}\sqrt{x - 1} = 1$$
$$\left(2\sqrt{x + 1}\sqrt{x - 1}\right)^2 = 1^2$$
$$4(x + 1)(x - 1) = 1$$
$$4(x^2 - 1) = 1$$
$$4x^2 - 5 = 0$$
$$x^2 = \frac{5}{4}$$
$$x = \pm\sqrt{\frac{5}{4}} = \frac{\pm\sqrt{5}}{2}$$

Check $\frac{-\sqrt{5}}{2}$: $\sqrt{\frac{-\sqrt{5}}{2} + 1} + \sqrt{\frac{-\sqrt{5}}{2} - 1} = \sqrt{2\left(\frac{-\sqrt{5}}{2}\right) + 1}$

$$\sqrt{-0.118} + \sqrt{-2.118} = \sqrt{-0.118}$$

This is not defined.

Check $\frac{\sqrt{5}}{2}$: $\sqrt{\frac{\sqrt{5}}{2} + 1} + \sqrt{\frac{\sqrt{5}}{2} - 1} = \sqrt{2\left(\frac{\sqrt{5}}{2}\right) + 1}$

$$\sqrt{\frac{\sqrt{5} + 2}{2}} + \sqrt{\frac{\sqrt{5} - 2}{2}} \stackrel{?}{=} \sqrt{\sqrt{5} + 1}$$
$$\frac{\sqrt{5} + 2}{2} + 2\sqrt{\left(\frac{\sqrt{5} + 2}{2}\right)\left(\frac{\sqrt{5} - 2}{2}\right)} + \frac{\sqrt{5} - 2}{2} \stackrel{?}{=} \sqrt{5} + 1$$
$$\frac{2\sqrt{5}}{2} + 2\sqrt{\frac{5 - 4}{4}} \stackrel{?}{=} \sqrt{5} + 1$$
$$\sqrt{5} + 2\left(\frac{1}{2}\right) \stackrel{?}{=} \sqrt{5} + 1$$
$$\sqrt{5} + 1 = \sqrt{5} + 1$$

The solution set is $x = \dfrac{\sqrt{5}}{2}$.

29.
$$2\sqrt[3]{x^2} - \sqrt[3]{x} = 1$$
$$2x^{2/3} - x^{1/3} = 1$$
Let $u = x^{1/3}$
$$2u^2 - u = 1$$
$$2u^2 - u - 1 = 0$$
$$(2u + 1)(u - 1) = 0$$
$$2u + 1 = 0 \quad \text{or} \quad u - 1 = 0$$
$$u = -\frac{1}{2} \quad \text{or} \quad u = 1$$
$$x^{1/3} = -\frac{1}{2} \quad \text{or} \quad x^{1/3} = 1$$
$$x = -\frac{1}{8} \quad \text{or} \quad x = 1$$
The solution set is $\left\{\frac{-1}{8}, 1\right\}$.

Check $-\frac{1}{8}$: $2\sqrt[3]{\left(\frac{-1}{8}\right)^2} - \sqrt[3]{\frac{-1}{8}} \stackrel{?}{=} 1$
$$2\sqrt[3]{\frac{1}{64}} - \frac{-1}{2} \stackrel{?}{=} 1$$
$$2\left(\frac{1}{4}\right) + \frac{1}{2} \stackrel{?}{=} 1$$
$$1 = 1$$

Check 1: $2\sqrt[3]{1^2} - \sqrt[3]{1} \stackrel{?}{=} 1$
$$2 - 1 \stackrel{?}{=} 1$$
$$1 = 1$$

31.
$$x^{-6} - 7x^{-3} - 8 = 0$$
Let $u = x^{-3}$. Then, $u^2 = x^{-6}$
and $u^2 - 7u - 8 = 0$
$$(u - 8)(u + 1) = 0$$
$$u - 8 = 0 \quad \text{or} \quad u + 1 = 0$$
$$u = 8 \quad \text{or} \quad u = -1$$
$$x^{-3} = 8 \quad \text{or} \quad x^{-3} = -1$$
$$(x^{-3})^{-1} = 8^{-1} \quad \text{or} \quad (x^{-3})^{-1} = (-1)^{-1}$$
$$x^3 = \frac{1}{8} \quad \text{or} \quad x^3 = -1$$
$$x = \frac{1}{2} \quad \text{or} \quad x = -1$$
The solution set is $\left\{-1, \frac{1}{2}\right\}$.

Check -1: $(-1)^{-6} - 7(-1)^{-3} - 8 \stackrel{?}{=} 0$
$$1 + 7 - 8 \stackrel{?}{=} 0$$
$$0 = 0$$

Check $\frac{1}{2}$: $\left(\frac{1}{2}\right)^{-6} - 7\left(\frac{1}{2}\right)^{-3} - 8 \stackrel{?}{=} 0$
$$64 - 7(8) - 8 \stackrel{?}{=} 0$$
$$0 = 0$$

33.
$$x^2 + m^2 = 2mx + (nx)^2$$
$$x^2 - n^2x^2 - 2mx + m^2 = 0$$
$$(1 - n^2)x^2 - 2mx + m^2 = 0$$
Here $a = 1 - n^2$, $b = -2m$, $c = m^2$ and $b^2 - 4ac = 4n^2m^2$.

Then, $x = \dfrac{-(-2m) \pm \sqrt{4n^2m^2}}{2(1 - n^2)}$
$$= \frac{2m \pm 2nm}{2(1 - n^2)}$$
$$= \frac{2m(1 \pm n)}{2(1 - n^2)}$$
$$= \frac{m(1 \pm n)}{1 - n^2}$$

$$\text{Thus, } x = \frac{m(1 - n)}{1 - n^2} = \frac{m(1 - n)}{(1 + n)(1 - n)} = \frac{m}{1 + n}$$

$$\text{and } x = \frac{m(1 + n)}{1 - n^2} = \frac{m(1 + n)}{(1 + n)(1 - n)} = \frac{m}{1 - n}$$

The solution set is $\left\{\frac{m}{1 - n}, \frac{m}{1 + n}\right\}$.

Check $\frac{m}{1 - n}$:

$$\left(\frac{m}{1 - n}\right)^2 + m^2 \stackrel{?}{=} 2m\left(\frac{m}{1 - n}\right) + \left(n\left(\frac{m}{1 - n}\right)\right)^2$$

$$\frac{m^2}{(1 - n)^2} + \frac{m^2(1 - n)^2}{(1 - n)^2} \stackrel{?}{=} \frac{2m^2}{1 - n} + \frac{n^2m^2}{(1 - n)^2}$$

$$\frac{m^2 + m^2 - 2m^2n + m^2n^2}{(1 - n)^2} = \frac{2m^2(1 - n) + n^2m^2}{(1 - n)^2}$$

$$\frac{2m^2 - 2m^2n + m^2n^2}{(1 - n)^2} = \frac{2m^2 - 2m^2n + m^2n^2}{(1 - n)^2}$$

Check $\frac{m}{1 + n}$:

$$\left(\frac{m}{1 + n}\right)^2 + m^2 \stackrel{?}{=} 2m\left(\frac{m}{1 + n}\right) + \left(n\left(\frac{m}{1 + n}\right)\right)^2$$

$$\frac{m^2 + m^2(1 + n)^2}{(1 + n)^2} + \frac{2m^2}{1 + n} = \frac{n^2m^2}{(1 + n)^2}$$

$$\frac{2m^2 + 2m^2n + m^2n^2}{(1 + n)^2} = \frac{2m^2 + 2m^2n + m^2n^2}{(1 + n)^2}$$

35. $10a^2x^2 - 2abx - 36b^2 = 0$

Here $A = 10a^2$, $B = -2ab$, $C = -36b^2$ and

$$B^2 - 4AC = (-2ab)^2 - 4(10a^2)(-36b^2)$$
$$= 4a^2b^2 + 1440a^2b^2 = 1444a^2b^2$$

$$\text{Then, } x = \frac{-(-2ab) \pm \sqrt{148a^2b^2}}{2(10a^2)} = \frac{2ab \pm 38ab}{20a^2}$$

$$\text{or } x = \frac{2ab - 38ab}{20a^2} = \frac{-36ab}{20a^2} = \frac{-9b}{5a}$$

$$\text{and } x = \frac{2ab + 38ab}{20a^2} = \frac{40ab}{20a^2} = \frac{2b}{a}$$

The solution set is $\left\{\frac{-9b}{5a}, \frac{2b}{a}\right\}$.

Check $\frac{-9b}{5a}$: $10a^2\left(\frac{-9b}{5a}\right)^2 - 2ab\left(\frac{-9b}{5a}\right) - 36b^2 \stackrel{?}{=} 0$

$$10a^2\left(\frac{81b^2}{25a^2}\right) + \frac{18b^2}{5} - 36b^2 \stackrel{?}{=} 0$$

$$\frac{162b^2}{5} + \frac{18b^2}{5} - \frac{180b^2}{5} \stackrel{?}{=} 0$$

$$0 \stackrel{?}{=} 0$$

Check $\frac{2b}{a}$: $10a^2\left(\frac{2b}{a}\right)^2 - 2ab\left(\frac{2b}{a}\right) - 36b^2 \stackrel{?}{=} 0$

$$40b^2 - 4b^2 - 36b^2 \stackrel{?}{=} 0$$
$$0 \stackrel{?}{=} 0$$

37. $\sqrt{x^2 + 3x + 7} - \sqrt{x^2 - 3x + 9} + 2 = 0$

$$\sqrt{x^2 + 3x + 7} = \sqrt{x^2 - 3x + 9} - 2$$
$$\left(\sqrt{x^2 + 3x + 7}\right)^2 = \left(\sqrt{x^2 - 3x + 9} - 2\right)^2$$
$$x^2 + 3x + 7 = x^2 - 3x + 9 - 4\sqrt{x^2 - 3x + 9} + 4$$
$$6x - 6 = -4\sqrt{x^2 - 3x + 9}$$
$$3x - 3 = -2\sqrt{x^2 - 3x + 9}$$
$$(3x - 3)^2 = \left(-2\sqrt{x^2 - 3x + 9}\right)^2$$
$$9x^2 - 18x + 9 = 4(x^2 - 3x + 9)$$
$$9x^2 - 18x + 9 = 4x^2 - 12x + 36$$
$$5x^2 - 6x - 27 = 0$$
$$(5x + 9)(x - 3) = 0$$
$$5x + 9 = 0 \quad \text{or} \quad x - 3 = 0$$
$$x = -\frac{9}{5} \quad \text{or} \quad x = 3$$

Check $\frac{-9}{5}$: $\sqrt{\left(\frac{-9}{5}\right)^2 + 3\left(\frac{-9}{5}\right) + 7} - \sqrt{\left(\frac{-9}{5}\right)^2 - 3\left(\frac{-9}{5}\right) + 9} + 2 = 0$

$$\sqrt{\frac{81}{25} - \frac{135}{25} + \frac{175}{25}} - \sqrt{\frac{81}{25} + \frac{135}{25} + \frac{225}{25}} + 2 = 0$$
$$\sqrt{\frac{121}{25}} - \sqrt{\frac{441}{25}} + 2 = 0$$
$$\frac{11}{5} - \frac{21}{5} + 2 = 0$$
$$0 = 0$$

Check 3: $\sqrt{3^2 + 3(3) + 7} - \sqrt{3^2 - 3(3) + 9} + 2 = 0$

$$5 - 3 + 2 = 0$$
$$4 \neq 0$$

Does not check.

The solution set is $x = \frac{-9}{5}$.

39. $|2x + 3| = 7$

$2x + 3 = 7$ or $2x + 3 = -7$

$2x = 4$ $\qquad 2x = -10$

$x = 2$ $\qquad x = -5$

The solution set is $\{-5, 2\}$.

41. $|2 - 3x| = 7$

$2 - 3x = 7$ or $2 - 3x = -7$

$-3x = 5$ $\qquad -3x = -9$

$x = \frac{-5}{3}$ $\qquad x = 3$

The solution set is $\left\{\frac{-5}{3}, 3\right\}$

43.
$$\frac{2x-3}{5} + 2 \le \frac{x}{2}$$
$$10\left[\frac{2x-3}{5} + 2\right] \le \left[\frac{x}{2}\right]10$$
$$2(2x-3) + 20 \le 5x$$
$$4x - 6 + 20 \le 5x$$
$$-x \le -14$$
$$x \ge 14$$
The solution set is $\{x \mid x \ge 14\}$.

45.
$$-9 \le \frac{2x+3}{-4} \le 7$$
$$(-4)(-9) \ge (-4)\frac{2x+3}{-4} \ge 7(-4)$$
$$36 \ge 2x + 3 \ge -28$$
$$36 - 3 \ge 2x + 3 - 3 \ge -28 - 3$$
$$33 \ge 2x \ge -31$$
$$\frac{33}{2} \ge \frac{2x}{2} \ge \frac{-31}{2}$$
$$\left\{x \middle| \frac{-31}{2} \le x \le \frac{33}{2}\right\}$$

47.
$$6 > \frac{3-3x}{12} > 2$$
$$12(6) > 12\left[\frac{3-3x}{12}\right] > (12)(2)$$
$$72 > 3 - 3x > 24$$
$$72 - 3 > 3 - 3x - 3 > 24 - 3$$
$$69 > -3x > 21$$
$$\frac{69}{-3} < \frac{-3x}{-3} < \frac{21}{-3}$$
$$\{x \mid -23 < x < -7\}$$

49.
$$2x^2 + 5x - 12 < 0$$
$$(2x-3)(x+4) < 0$$

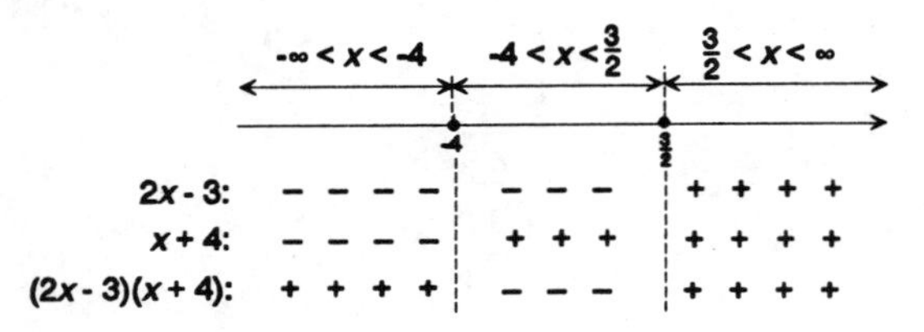

$(2x-3)(x+4) < 0$ if $-4 < x < \frac{3}{2}$. The solution set is $\left\{x \middle| -4 < x < \frac{3}{2}\right\}$.

51.
$$\frac{6}{x+3} \ge 1$$
$$\frac{6}{x+3} - 1 \ge 0$$
$$\frac{6-(x+3)}{x+3} \ge 0$$
$$\frac{3-x}{x+3} \ge 0$$

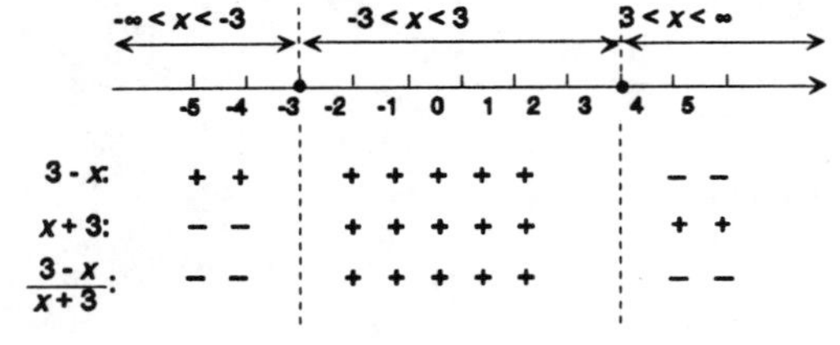

$\frac{3-x}{x+3} \ge 0$ if $-3 < x \le 3$. The solution set is $\{x \mid -3 < x \le 3\}$.

53.
$$\frac{2x - 6}{1 - x} < 2$$
$$\frac{2x - 6}{1 - x} - 2 < 0$$
$$\frac{2x - 6 - 2(1 - x)}{1 - x} < 0$$
$$\frac{2x - 6 - 2 + 2x}{1 - x} < 0$$
$$\frac{4x - 8}{1 - x} < 0$$
$$\frac{4(x - 2)}{1 - x} < 0$$

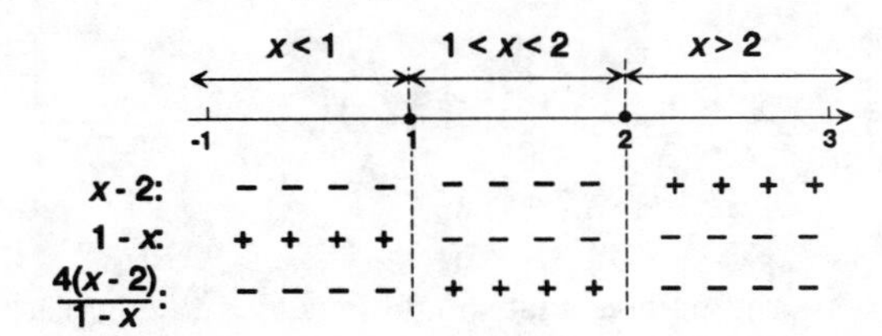

$\frac{4(x - 2)}{1 - x} < 0$ if $x < 1$ or $x > 2$. The solution set is $\{x \mid x < 1 \text{ or } x > 2\}$.

55. $$\frac{(x - 2)(x - 1)}{x - 3} > 0$$

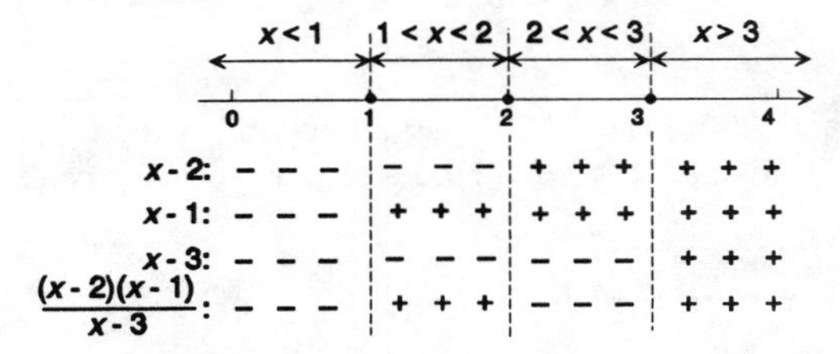

$\frac{(x - 2)(x - 1)}{x - 3} > 0$, if $1 < x < 2$ or $x > 3$. The solution set is $\{x \mid 1 < x < 2 \text{ or } x > 3\}$.

57.
$$\frac{x^2 - 8x + 12}{x^2 - 16} > 0$$
$$\frac{(x - 2)(x - 6)}{(x - 4)(x + 4)} > 0$$

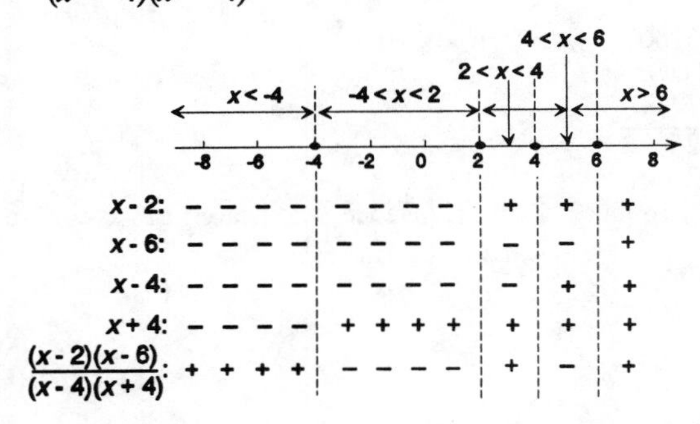

$\frac{(x - 2)(x - 6)}{(x - 4)(x + 4)} > 0$ if $x < -4$ or $2 < x < 4$ or $x > 6$

The solution set is $\{x \mid x < -4 \text{ or } 2 < x < 4 \text{ or } x > 6\}$.

59. $|3x + 4| < \frac{1}{2}$

$-\frac{1}{2} < 3x + 4 < \frac{1}{2}$

$-\frac{9}{2} < 3x < -\frac{7}{2}$

$-\frac{3}{2} < x < -\frac{7}{6}$

The solution set is $\left\{x \mid -\frac{3}{2} < x < -\frac{7}{6}\right\}$.

61. $|2x - 5| \geq 9$

$2x - 5 \leq -9$ or $2x - 5 \geq 9$

$2x \leq -4$ or $2x \geq 14$

$x \leq -2$ $x \geq 7$

The solution set is $\{x \mid x \leq -2 \text{ or } x \geq 7\}$.

63. $x^2 + 6x$

$\left[\frac{6}{2}\right]^2 = 9$

The number that should be added to complete the square is 9.

65. $x^2 - \frac{4}{3}x$

$\left[\frac{-4}{3} \cdot \frac{1}{2}\right]^2 = \frac{4}{9}$

The number that should be added to complete the square is $\frac{4}{9}$.

67. Using $s - vt$, we have $t = 3$ and $v = 1100$. We want the distance s in feet. Then,

$s = (1100)(3)$

$s = 3300$

The storm is 3300 feet away.

69. Using $s = vt$, we have a downwind speed of $v_d = 250 + 30 = 280$ and an upwind speed of $v_u = 250 - 30 = 220$ and the time total $t \leq 5$. We want to find the distance s the plane can travel.

	Velocity	Time	Distance
Downwind	280	$\frac{s/2}{280}$	$\frac{s}{2}$
Upwind	220	$\frac{s/2}{280}$	$\frac{s}{2}$

Then,

$$\frac{s/2}{280} + \frac{s/2}{220} \leq 5$$

$$(6160)\left[\frac{s}{560} + \frac{s}{440}\right] \leq (5)(6160)$$

$$11s + 14s \leq (5)(6160)$$

$$25s \leq 5(6160)$$

$$s \leq \frac{(5)(6160)}{25}$$

$$= 1232$$

The distance out is 616 miles, and the round trip is 1232 miles. Hence, the search plane can go as far as 616 miles.

Check: Downwind: time $= \dfrac{\frac{1232}{2}}{280} = 2.2$ hours

Upwind: time $= \dfrac{\frac{1232}{2}}{220} = 2.8$ hours

Roundtrip: time $= 2.2 + 2.8 = 5$ hours maximum.

71. We are asked to find a time in hours. Using $s = vt$, we can make a table as follows:

	Velocity	Time	Distance
Raft	5	t	$5t$
Helicopter	90	t	$90t$

Note that the times are equal. Also

$$5t + 90t = 150$$
$$95t = 150$$
$$t = 1.58 \text{ hours} \approx 1 \text{ hour, } 35 \text{ minutes}$$

The helicopter will reach the life raft in 1 hour, 35 minutes.

Check: $5(1.58) + 90(1.58) \approx 150$ miles

73. Let c = number of days it takes Clarissa to complete the job.
Let s = number of days its takes Shawna to complete the job.
Then $\frac{1}{c}$ and $\frac{1}{s}$ give the amount of the job done in one day.
If it takes 6 days to complete the job, then $\frac{1}{6}$ of the job is completed in one day.
Therefore,

$$\frac{1}{c} + \frac{1}{s} = \frac{1}{6}$$

We are told that Clarissa by herself can complete this job in 5 days less than Shawna. This means that $c = -5$.
We now have

$$\frac{1}{s-5} + \frac{1}{s} = \frac{1}{6}$$
$$\frac{s + s - 5}{s(s-5)} = \frac{1}{6}$$
$$6(2s - 5) = s(s - 5)$$
$$12s - 5 = s^2 - 5s$$
$$0 = s^2 - 17s + 30$$
$$0 = (s - 15)(s - 2)$$
$$s = 15 \text{ or } s = 2$$

If $s = 15$, then $c = 15 - 5 = 10$
If $s = 2$, then $c = 2 - 5 = -3$. This is impossible since the answer cannot be negative.
Therefore, it will take Clarissa 10 days to complete the job by herself.

75.

% HCl acid	Amount	Amount of Hcl acid
40%	60	.40(60)
15%	x	$.15x$
25%	$x + 60$	$.25(x + 60)$

$$.40(60) + .15x = .25(x + 60)$$
$$24 + .15x = .25x + 15$$
$$9 = .10x$$
$$90 = x$$

90 cc of 15% solution of Hcl acid should be mixed with the 60 cc of 40% acid to obtain a solution of 25% HCl. There is 150 cc of the 25% solution.

77.

% Salt	Amount	Amount of Salt
10%	64	.10(64)
0%	x	$0x$
2%	$64 + x$	$.02(64 + x)$

$$.10(64) + 0x = .02(64 + x)$$
$$6.4 = 1.28 + .02x$$
$$5.12 = .02x$$
$$\frac{512}{2} = x$$
$$256 = x$$

Add 256 ounces of water to the 64 ounces of a 10% salt solution to make 320 ounces of a 2% salt solution.

79. We want a length in feet. The effective speed of the train (i.e, relative to the man) is $30 - 4 = 26$ miles per hour. The time is 5 seconds $= \frac{5}{60}$ minutes $= \frac{5/60}{60}$ hours $= \frac{1}{720}$ hrs. Using $s = vt$, we get $s = (26)\left(\frac{1}{720}\right) = \frac{26}{720}$ miles $= \frac{26/720}{5280} = 190.67$ feet. The freight train is 190.67 feet long.

Check: $v = \frac{26}{720} \div \frac{5}{3600} = 26$

81. An 8 hp pump can fill a tank in eight hours and a 3 hp pump can fill a tank in twelve hours.

Let t = number of hours.

$$W = \frac{t}{8} + \frac{t}{12}$$

To find the amount of work done in 4 hours:

$$W = \frac{4}{8} + \frac{4}{12}$$
$$W = \frac{1}{2} + \frac{1}{3}$$
$$W = \frac{3 + 2}{6}$$
$$W = \frac{5}{6}$$

Hence, $\frac{5}{6}$ of the job is completed in 4 hours.

Therefore, the smaller pump needs to complete $\frac{1}{6}$ of the job.

If it takes 12 hours to complete the entire job, then it takes $\frac{5}{6}(12)$ to complete $\frac{5}{6}$ of the job. It takes the smaller pump two hours.

83. We want to find the number x of passengers over 20 to make a total of \$482.40. We make a table:

Number passengers	Fare Each	Total Cost
$20 + x$	$\$15 - 0.1x$	$(20 + x)(15 - 0.1x)$

Then,
$$\begin{aligned} (20 + x)(15 - 0.1x) &= 482.40 \\ 300 - 2x + 15x - 0.1x^2 &= 482.40 \\ -0.1x^2 + 13x - 182.40 &= 0 \\ x^2 - 130x + 1824 &= 0 \end{aligned}$$

Here $a = 1$, $b = -130$, $c = 1824$, and $b^2 - 4ac = (130)^2 - 4(1)(1824) = 9604$

Thus, $x = \dfrac{-(-130) \pm \sqrt{9604}}{2} = \dfrac{130 \pm 98}{2}$

$= \dfrac{130 + 98}{2}$ or $\dfrac{130 - 98}{2}$

$= 114$ or 16

Since the capacity of the bus is 44, we discard the 114. The total number of passengers is $20 + 16 = 36$, and the ticket price per passenger is $15 - 0.1(16) = \$13.40$.

Check: $(36)(15 - 0.1(16) = 482.40$

36 seniors went on the trip; each one paid \$13.40.

85. Distance = Velocity × Time

	Distance	Velocity	Time
Todd	100	v_m	$t \Rightarrow v_m = \dfrac{100}{t}$
Scott	95	v_d	$t \Rightarrow v_d = \dfrac{95}{t}$
Todd	105	$\dfrac{100}{t}$	$\dfrac{105}{100}t$
Scott	100	$\dfrac{95}{t}$	$\dfrac{100}{95}t$

$$\frac{105t}{100} \geq \frac{100}{95}t$$

$105t = 1.0526316t$

	Distance	Velocity	Time
Todd	100	$\dfrac{100}{t}$	t
Scott	95	$\dfrac{95}{t}$	t

(a) No
(b) Todd wins again.
(c) Todd wins by 0.25 meters.
(d) Todd should line up 5.26316 meters behind the start line.
(e) Yes.

Chapter 3

GRAPHS

3.1 Rectangular Coordinates; Scatter Diagrams

1. (a) Quadrant II
   (b) Positive x-axis
   (c) Quadrant III
   (d) Quadrant I
   (e) Negative y-axis
   (f) Quadrant IV

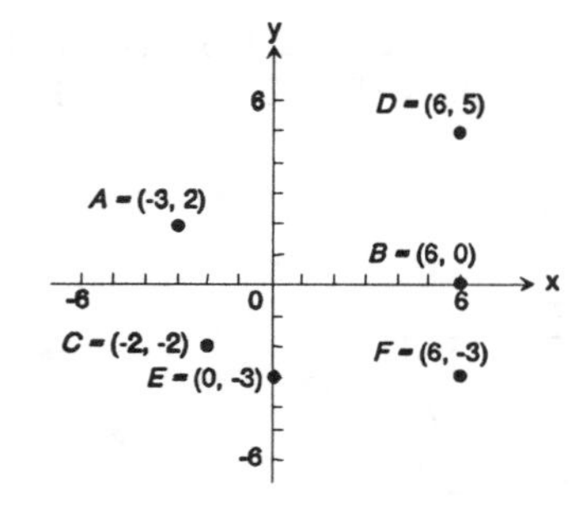

3. The points will be on a vertical line that is two units to the right of the y-axis.

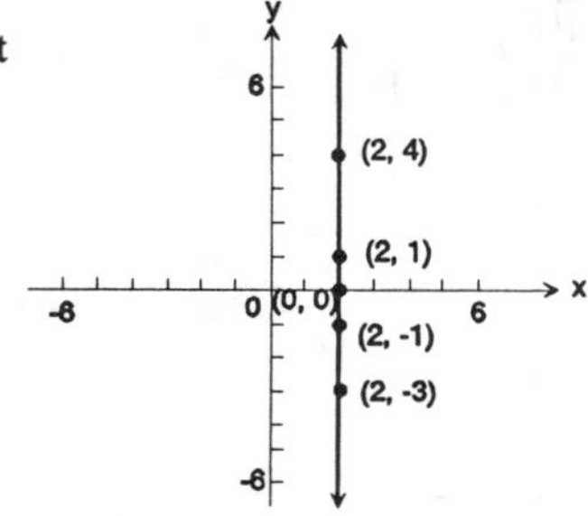

5. $d(P_1, P_2) = \sqrt{(2 - 0)^2 + (1 - 0)^2} = \sqrt{4 + 1} = \sqrt{5}$

7. $d(P_1, P_2) = \sqrt{(-2 - 1)^2 + (2 - 1)^2} = \sqrt{(-3)^2 + (1)^2} = \sqrt{0}$

9. $P_1 = (3, -4), P_2 = (5, 4)$

$d(P_1, P_2) = \sqrt{(5 - 3)^2 + [4 - (-4)]^2}$

$d(P_1, P_2) = \sqrt{(2)^2 + (8)^2}$

$d(P_1, P_2) = \sqrt{4 + 64}$

$d(P_1, P_2) = \sqrt{68} = 2\sqrt{17}$

11. $P_1 = (-3, 2), P_2 = (6, 0)$

$d(P_1, P_2) = \sqrt{[6 - (-3)]^2 + (0 - 2)^2}$

$d(P_1, P_2) = \sqrt{(9)^2 + (-2)^2}$

$d(P_1, P_2) = \sqrt{81 + 4}$

$d(P_1, P_2) = \sqrt{85}$

13. $P_1 = (4, -3), P_2 = (6, 4)$

$$d(P_1, P_2) = \sqrt{(6 - 4)^2 + [4 - (-3)]^2}$$
$$d(P_1, P_2) = \sqrt{(2)^2 + (7)^2}$$
$$d(P_1, P_2) = \sqrt{4 + 49}$$
$$d(P_1, P_2) = \sqrt{53}$$

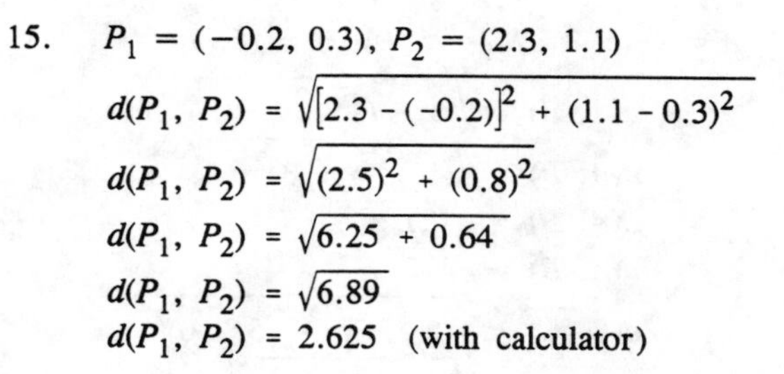

15. $P_1 = (-0.2, 0.3), P_2 = (2.3, 1.1)$

$$d(P_1, P_2) = \sqrt{[2.3 - (-0.2)]^2 + (1.1 - 0.3)^2}$$
$$d(P_1, P_2) = \sqrt{(2.5)^2 + (0.8)^2}$$
$$d(P_1, P_2) = \sqrt{6.25 + 0.64}$$
$$d(P_1, P_2) = \sqrt{6.89}$$
$$d(P_1, P_2) = 2.625 \quad \text{(with calculator)}$$

17. $P_1 = (a, b), P_2 = (0, 0)$

$$d(P_1, P_2) = \sqrt{(0 - a)^2 + (0 - b)^2}$$
$$d(P_1, P_2) = \sqrt{a^2 + b^2}$$

19. $A = (-2, 5), B = (1, 3), C = (-1, 0)$

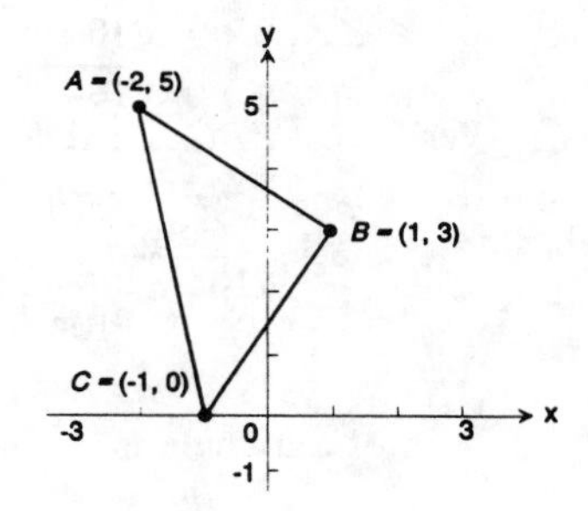

$$d(A, B) = \sqrt{[1 - (-2)^2 + (3 - 5)^2]}$$
$$d(A, B) = \sqrt{3^2 + (-2)^2}$$
$$d(A, B) = \sqrt{9 + 4}$$
$$d(A, B) = \sqrt{13}$$
$$d(B, C) = \sqrt{(-1 - 1)^2 + (0 - 3)^2}$$
$$d(B, C) = \sqrt{(-2)^2 + (-3)^2}$$
$$d(B, C) = \sqrt{4 + 9}$$
$$d(B, C) = \sqrt{13}$$
$$d(A, C) = \sqrt{[-1 - (-2)]^2 + (0 - 5)^2}$$
$$d(A, C) = \sqrt{(-1)^2 + (-5)^2}$$
$$d(A, C) = \sqrt{1 + 25}$$
$$d(A, C) = \sqrt{26}$$

Verify that ABC is a right triangle by the Pythagorean Theorem:

$$[d(A, B)]^2 + (d(B, C)]^2 = [d(A, C)]^2$$
$$\left(\sqrt{13}\right)^2 + \left(\sqrt{13}\right)^2 = \left(\sqrt{26}\right)^2$$
$$13 + 13 = 26$$
$$26 = 26$$

Area of a triangle is $A = \frac{1}{2}bh$. In this problem,

$$A = \frac{1}{2}[d(B, C)][d(A, B)]$$
$$A = \frac{1}{2}\left(\sqrt{13}\right)\left(\sqrt{13}\right)$$
$$A = \frac{1}{2}(13)$$
$$A = \frac{13}{2} \text{ square units}$$

21. $A = (-5, 3), B = (6, 0), C = (5, 5)$

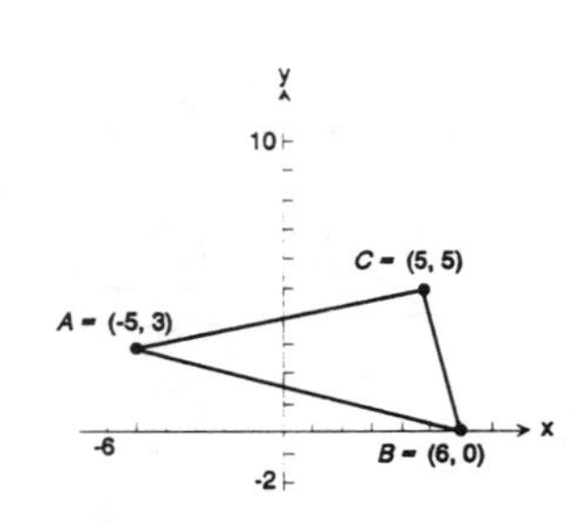

$$d(A, B) = \sqrt{[6 - (-5)]^2 + (0 - 3)^2}$$
$$d(A, B) = \sqrt{(11)^2 + (-3)^2}$$
$$d(A, B) = \sqrt{121 + 9}$$
$$d(A, B) = \sqrt{130}$$
$$d(B, C) = \sqrt{(5 - 6)^2 + (5 - 0)^2}$$
$$d(B, C) = \sqrt{(-1)^2 + (5)^2}$$
$$d(B, C) = \sqrt{1 + 25}$$
$$d(B, C) = \sqrt{26}$$
$$d(A, C) = \sqrt{[5 - (-5)]^2 + (5 - 3)^2}$$
$$d(A, C) = \sqrt{(10)^2 + (2)^2}$$
$$d(A, C) = \sqrt{100 + 4}$$
$$d(A, C) = \sqrt{104}$$

Verify that ABC is a right triangle by the Pythagorean Theorem:

$$[d(A, C)]^2 + (d(B, C)]^2 = [d(A, B)]^2$$
$$\left(\sqrt{104}\right)^2 + \left(\sqrt{26}\right)^2 = \left(\sqrt{130}\right)^2$$
$$104 + 26 = 130$$
$$130 = 130$$

Area of a triangle is $A = \frac{1}{2}bh$. In this problem,

$$A = \frac{1}{2}[d(A, C)][d(B, C)]$$
$$A = \frac{1}{2}\left(\sqrt{104}\right)\left(\sqrt{26}\right)$$
$$A = \frac{1}{2}\left(\sqrt{2704}\right)$$
$$A = \frac{1}{2}(52)$$
$$A = 26 \text{ square units}$$

23. $A = (4, -3), B = (0, -3), C = (4, 2)$

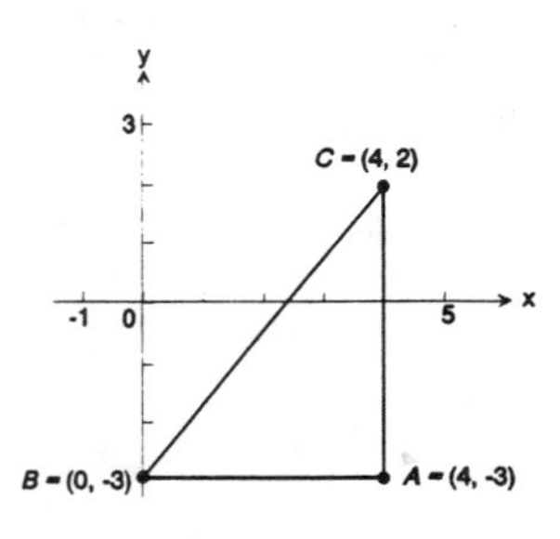

$$d(A, B) = \sqrt{(0 - 4)^2 + [-3 - (-3)]^2}$$
$$d(A, B) = \sqrt{(-4)^2 + (0)^2}$$
$$d(A, B) = \sqrt{16 + 0}$$
$$d(A, B) = 4$$
$$d(B, C) = \sqrt{(4 - 0)^2 + [(2 - (-3)]^2}$$
$$d(B, C) = \sqrt{(4)^2 + (5)^2}$$
$$d(B, C) = \sqrt{16 + 25}$$
$$d(B, C) = \sqrt{41}$$
$$d(A, C) = \sqrt{(4 - 4)^2 + [2 - (-3)]^2}$$
$$d(A, C) = \sqrt{(0)^2 + (5)^2}$$
$$d(A, C) = \sqrt{0 + 25}$$
$$d(A, C) = 5$$

Verify that ABC is a right triangle by the Pythagorean Theorem:

$$[d(A, C)]^2 + (d(A, B))^2 = [d(B, C)]^2$$
$$5^2 + 4^2 = \left(\sqrt{41}\right)^2$$
$$25 + 16 = 41$$
$$41 = 41$$

Area of a triangle is $A = \frac{1}{2}bh$. In this problem,

$$A = \frac{1}{2}[d(A, B)][d(A, C)]$$
$$A = \frac{1}{2}(4)(5)$$
$$A = 10 \text{ square units}$$

25. All points having an x-coordinate of 2 would be of the form $(2, y)$. Those which are 5 units from $(-2, -1)$ would be:

$$\sqrt{[2 - (-2)]^2 + (y - (-1))^2} = 5$$
$$\sqrt{(4)^2 + (y + 1)^2} = 5$$

Square both sides:

$$4^2 + (y + 1)^2 = 25$$
$$16 + y^2 + 2y + 1 = 25$$
$$y^2 + 2y + 17 = 25$$
$$y^2 + 2y - 8 = 0$$
$$(y + 4)(y - 2) = 0$$
$$y + 4 = 0 \quad \text{or} \quad y - 2 = 0$$
$$y = -4 \quad \text{or} \quad y = 2$$

Therefore, the points are $(2, -4)$, $(2, 2)$.

27. All points on the x-axis would be of the form $(x, 0)$. Those which are 5 units from $(4, -3)$ would be:

$$\sqrt{(x - 4)^2 + (0 - (-3))^2} = 5$$
$$\sqrt{(x - 4)^2 + (3)^2} = 5$$

Square both sides:

$$(x - 4)^2 + 9 = 25$$
$$x^2 - 8x + 16 + 9 = 25$$
$$x^2 - 8x + 25 = 25$$
$$x^2 - 8x = 0$$
$$x(x - 8) = 0$$
$$x = 0 \quad \text{or} \quad x = 8$$

Therefore, the points are $(0, 0)$ and $(8, 0)$.

29. $P_1 = (5, -4)$, $P_2 = (3, 2)$

Let $x_1 = 5$ $\quad y_1 = -4$
$x_2 = 3$ $\quad y_2 = 2$

Then, the coordinates (x, y) of the midpoint are:

$$x = \frac{x_1 + x_2}{2} = \frac{5 + 3}{2} = \frac{8}{2} = 4$$

$$y = \frac{y_1 + y_2}{2} = \frac{-4 + 2}{2} = -\frac{2}{2} = -1$$

Midpoint $= (4, -1)$

31. $P_1 = (-3, 2)$, $P_2 = (6, 0)$

Let $x_1 = -3$ $\quad y_1 = 2$
$x_2 = 6$ $\quad y_2 = 0$

Then, the coordinates (x, y) of the midpoint are:

$$x = \frac{x_1 + x_2}{2} = \frac{-3 + 6}{2} = \frac{3}{2}$$

$$y = \frac{y_1 + y_2}{2} = \frac{2 + 0}{2} = 1$$

Midpoint $= \left(\frac{3}{2}, 1\right)$

33. $P_1 = (4, -3), P_2 = (6, 1)$

Let $x_1 = 4 \quad y_1 = -3$
$x_2 = 6 \quad y_2 = 1$

Then, the coordinates (x, y) of the midpoint are:

$$x = \frac{x_1 + x_2}{2} = \frac{4 + 6}{2} = \frac{10}{2} = 5$$

$$y = \frac{y_1 + y_2}{2} = \frac{-3 + 1}{2} = \frac{-2}{2} = -1$$

Midpoint $= (5, -1)$

35. $P_1 = (-0.2, 0.3), P_2 = (2.3, 1.1)$

Let $x_1 = -0.2 \quad y_1 = 0.3$
$x_2 = 2.3 \quad y_2 = 1.1$

Then, the coordinates (x, y) of the midpoint are:

$$x = \frac{x_1 + x_2}{2} = \frac{-0.2 + 2.3}{2} = \frac{2.1}{2} = 1.05$$

$$y = \frac{y_1 + y_2}{2} = \frac{0.3 + 1.1}{2} = \frac{1.4}{2} = .7$$

Midpoint $= (1.05, 0.7)$

37. $P_1 = (a, b), P_2 = (0, 0)$

Let $x_1 = a \quad y_1 = b$
$x_2 = 0 \quad y_2 = 0$

Then, the coordinates (x, y) of the midpoint are:

$$x = \frac{x_1 + x_2}{2} = \frac{a + 0}{2} = \frac{a}{2}$$

$$y = \frac{y_1 + y_2}{2} = \frac{b + 0}{2} = \frac{b}{2}$$

Midpoint $= \left(\frac{a}{2}, \frac{b}{2}\right)$

39. $d[(0, 0) \,\&\, (2, 5)] = \sqrt{5^2 + 2^2} = \sqrt{29}$

$d[(0, 6) \,\&\, (2, 2)] = \sqrt{4^2 + 2^2} = \sqrt{20} = 2\sqrt{5}$

$d[(4, 4) \,\&\, (0, 3)] = \sqrt{4^2 + 1^2} = \sqrt{17}$

41. $d(P_1, P_2) = \sqrt{(-4 - 2)^2 + (1 - 1)^2} = 6$

$d(P_2, P_3) = \sqrt{(-4 + 4)^2 + (-3 - 1)^2} = 4$

$d(P_1, P_3) = \sqrt{(-4 - 2)^2 + (-3 - 1)^2} = \sqrt{52} = 2\sqrt{13}$

$4^2 + 6^2 = \left(2\sqrt{13}\right)^2$

Right Triangle

43. $d(P_1, P_2) = \sqrt{(-2 - 0)^2 + (-1 - 7)^2} = \sqrt{68}$

$d(P_2, P_3) = \sqrt{(3 + 0)^2 + (2 - 7)^2} = \sqrt{34}$

$d(P_1, P_3) = \sqrt{(-2 - 3)^2 + (-1 - 2)^2} = \sqrt{34}$

$\left(\sqrt{34}\right)^2 + \left(\sqrt{34}\right)^2 = \left(\sqrt{68}\right)^2$

Isosceles Right Triangle

45. $P_1 = (1, 3), P_2 = (5, 15)$

$d(P_1, P_2) = \sqrt{(5 - 1)^2 + (15 - 3)^2}$

$d(P_1, P_2) = \sqrt{(4)^2 + (12)^2}$

$d(P_1, P_2) = \sqrt{16 + 144}$

$d(P_1, P_2) = \sqrt{160}$

$d(P_1, P_2) = 4\sqrt{10}$

47. $P_1 = (-4, 6), P_2 = (4, -8)$

$d(P_1, P_2) = \sqrt{(4 - (-4))^2 + (-8 - 6)^2}$

$d(P_1, P_2) = \sqrt{(8)^2 + (-14)^2}$

$d(P_1, P_2) = \sqrt{64 + 196}$

$d(P_1, P_2) = \sqrt{260}$

$d(P_1, P_2) = 2\sqrt{65}$

49. (a)

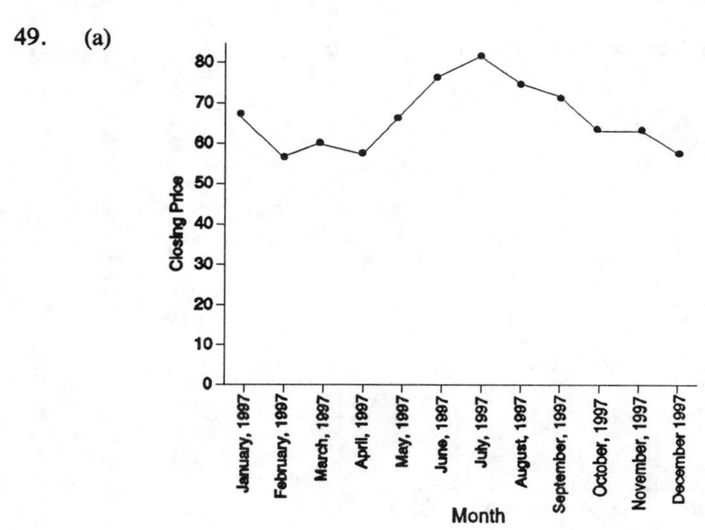

(b)

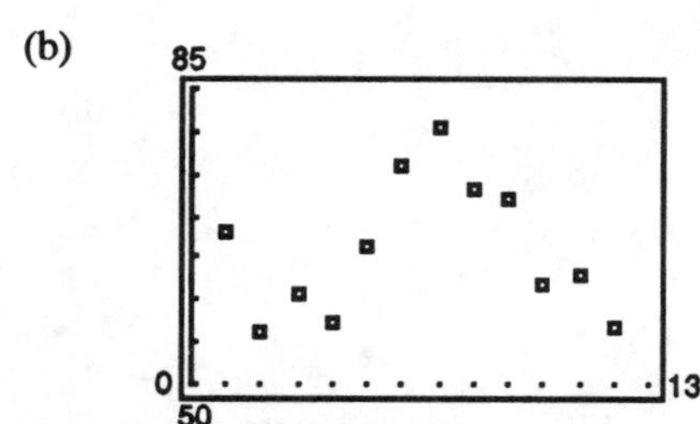

(c) The price of the stock is decreasing, increasing, and then decreasing over time.

51. (a)

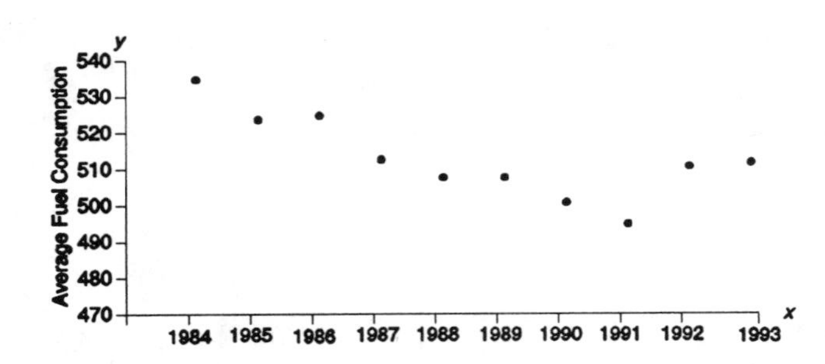

(b)

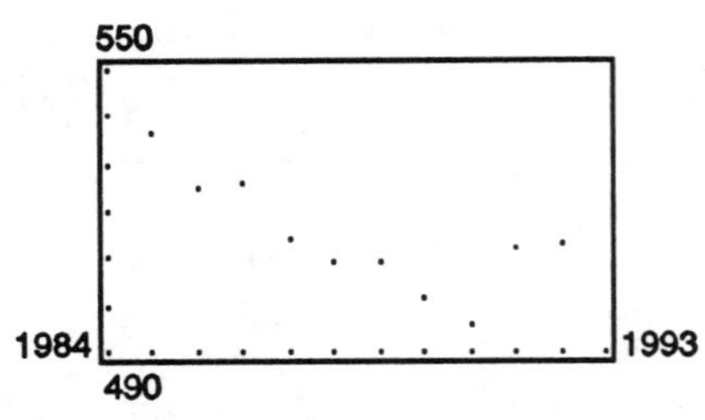

(c) The average fuel consumption decreases with time.

53. (a)

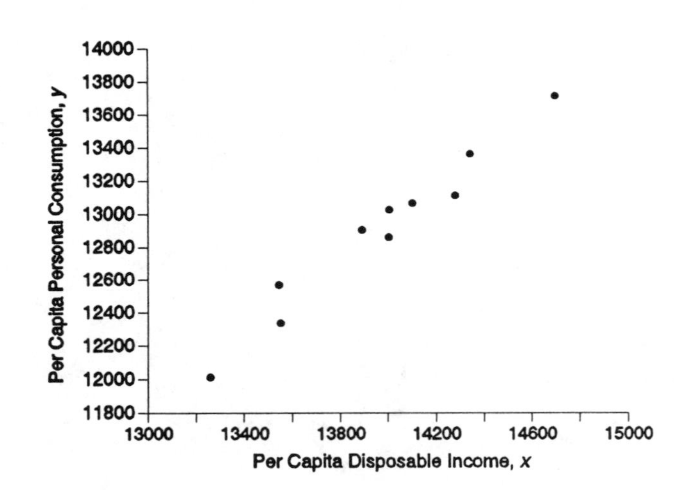

(b)

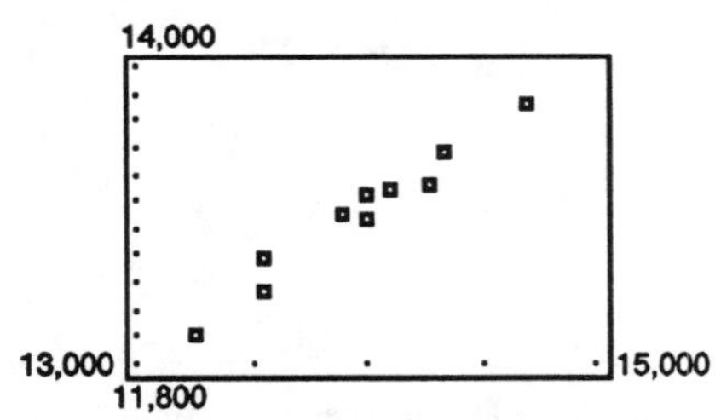

(c) Per capita personal income increases as per capita disposable income increases.

55. (a)

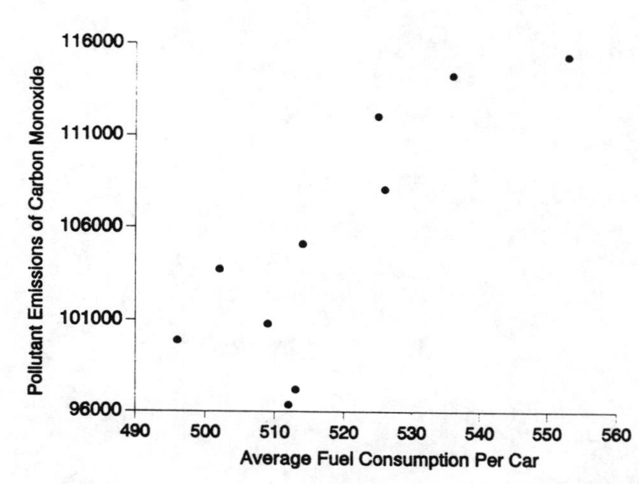

(b)

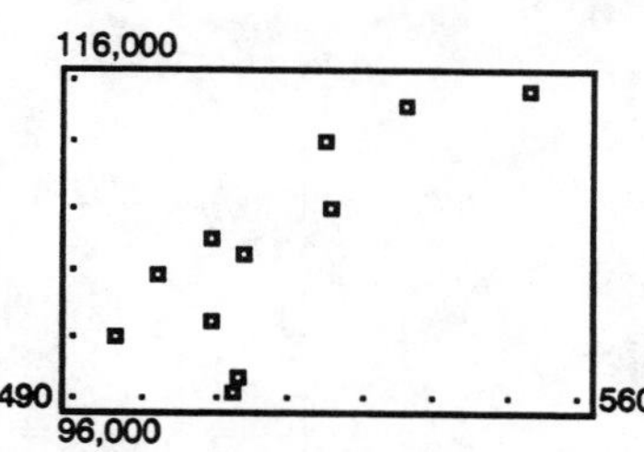

(c) The level of carbon monoxide increases as the average fuel consumption per car increases.

57.
$$\begin{aligned} 90^2 + 90^2 &= d^2 \\ 8100 + 8100 &= d^2 \\ 16200 &= d^2 \\ \sqrt{16200} &= d \\ 90\sqrt{2} &= d \\ 127.28 &= d \end{aligned}$$
127.28 ft.

59. (a)

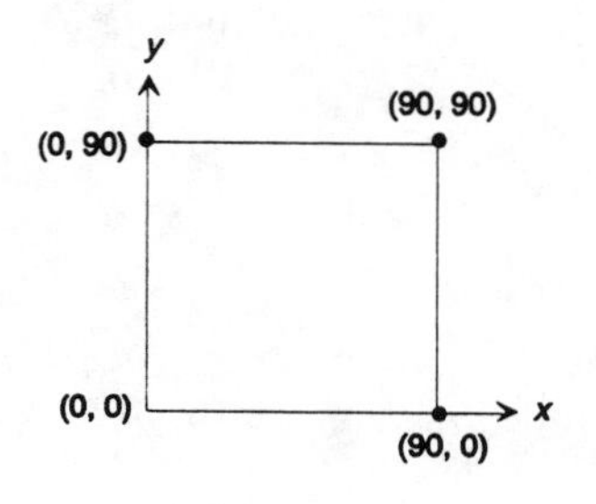

(b)
$$\begin{aligned} d &= \sqrt{(310 - 90)^2 + (15 - 90)^2} \\ &= \sqrt{(220)^2 + (-75)^2} \\ &= 232.4 \text{ feet} \end{aligned}$$

(c)
$$\begin{aligned} d &= \sqrt{(300 - 0)^2 + (300 - 90)^2} \\ &= \sqrt{(300)^2 + (210)^2} \\ &= 366.2 \text{ feet} \end{aligned}$$

61. The automobile heading east moves a distance of $40t$ after t hours. The truck heading south moves a distance of $30t$ after t hours. Their distance apart after 1 hour is:

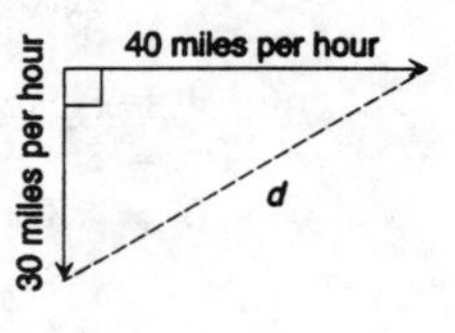

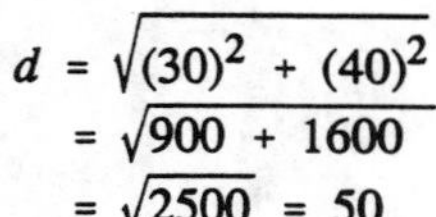

$$\begin{aligned} d &= \sqrt{(30)^2 + (40)^2} \\ &= \sqrt{900 + 1600} \\ &= \sqrt{2500} = 50 \end{aligned}$$

$d = 50t$ is the expression for their distance apart after t hours.

3.2 Graphs of Equations

1.

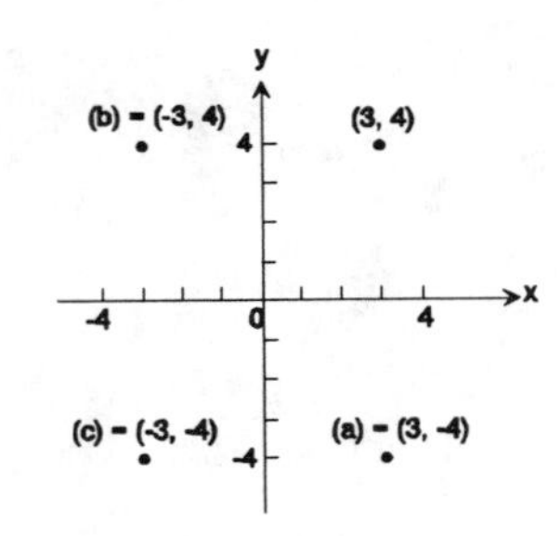

3.

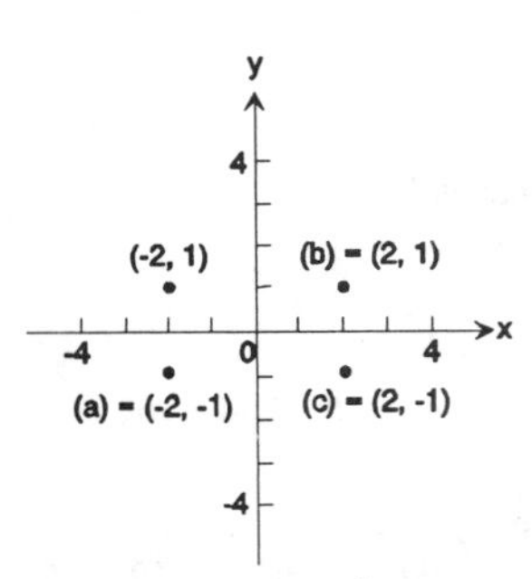

5.

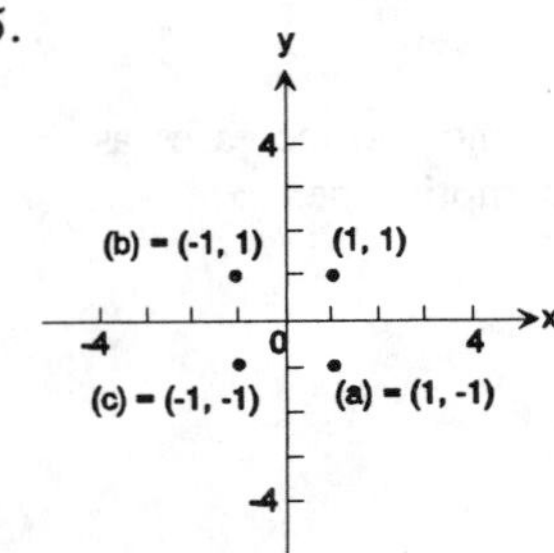

7.

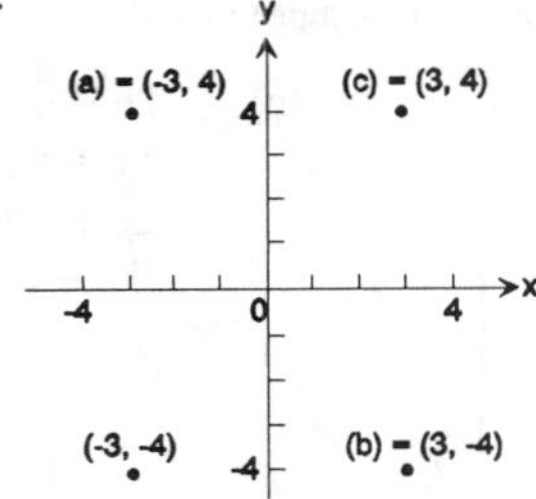

9.

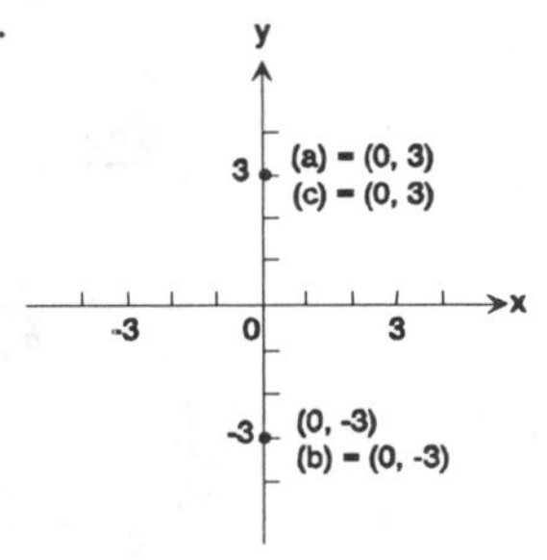

11. (a) $(-1, 0); (1, 0)$
 (b) x-axis, y-axis, origin

13. (a) $\left(-\frac{\pi}{2}, 0\right), \left(\frac{\pi}{2}, 0\right), (0, 1)$
 (b) y-axis

15. (a) $(0, 0)$
 (b) x-axis

17. (a) $(1, 0)$
 (b) none

19. (a) $(-3, 0), (0, 2), (3, 0)$
 (b) y-axis

21. (a) $(x, 0), 0 \le x < 2$
 (b) none

23. (a) $(-1.5, 0), (0, -2), (1.5, 0)$
 (b) y-axis

25. (a) none
 (b) origin

27.
$$\begin{aligned} y &= x^4 - \sqrt{x} \\ 0 &= 0^4 - \sqrt{0} \\ 0 &= 0 \\ 1 &= 1^4 - \sqrt{1} \\ 1 &\ne 0 \\ 0 &= (-1)^4 - \sqrt{-1} \\ 0 &\ne 1 - \sqrt{-1} \end{aligned}$$
$(0, 0)$ is on the graph.

29.
$$\begin{aligned} y^2 &= x^2 + 9 \\ 3^2 &= 0^2 + 9 \\ 9 &= 9 \\ 0^2 &= 3^2 + 9 \\ 0 &\ne 18 \\ 0^2 &= (-3)^2 + 9 \\ 0 &\ne 18 \end{aligned}$$
$(0, 3)$ is on the graph.

31.
$$\begin{aligned} x^2 + y^2 &= 4 \\ 0^2 + 2^2 &= 4 \\ 4 &= 4 \\ (-2)^2 + (2)^2 &= 4 \\ 8 &\ne 4 \\ \left(\sqrt{2}\right)^2 + \left(\sqrt{2}\right)^2 &= 4 \\ 4 &= 4 \end{aligned}$$
$(0, 2)$ and $\left(\sqrt{2}, \sqrt{2}\right)$ are on the graph.

33.
$$\begin{aligned} y &= 3x + 5 \\ 2 &= 3a + 5 \\ 3a &= -3 \\ a &= -1 \end{aligned}$$

35.
$$\begin{aligned} 2x + 3y &= 6 \\ 2a + 3b &= 6 \end{aligned}$$

37. Symmetry with respect to the x-axis means $(x, -y)$ is on the graph for every (x, y) on the graph. Therefore, given $(-4, 1)$, $(-2, 1)$, $(2, -1)$, $(4, 1)$; plot $(-4, -1)$, $(-2, -1)$, $(2, 1)$, $(4, -1)$.

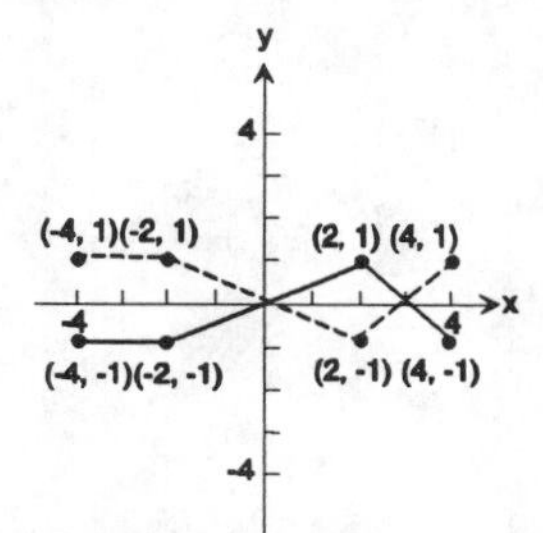

39. Symmetry with respect to the origin means $(-x, -y)$ is on the graph for every (x, y) on the graph. Therefore, given $(-4, 1)$, $(-2, 1)$, $(2, -1)$, $(4, 1)$; plot $(4, -1)$, $(2, -1)$, $(-2, 1)$, $(-4, -1)$.

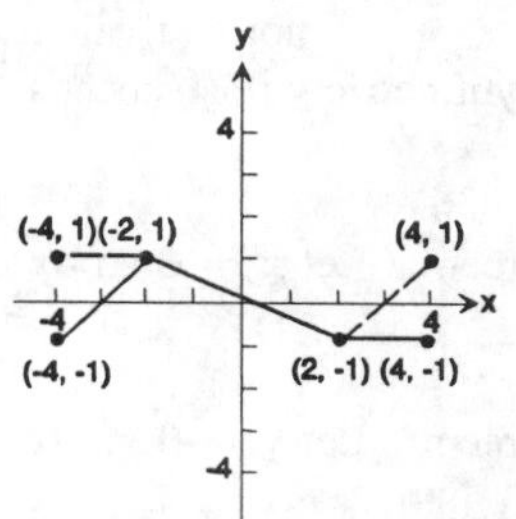

41. Symmetry with respect to the x-axis means $(x, -y)$ is on the graph for every (x, y) on the graph. Therefore, given $(-3, 1)$, $(0, 0)$, $(1, 2)$, $(4, 2)$; plot $(-3, -1)$, $(0, 0)$, $(1, -2)$, $(4, -2)$.

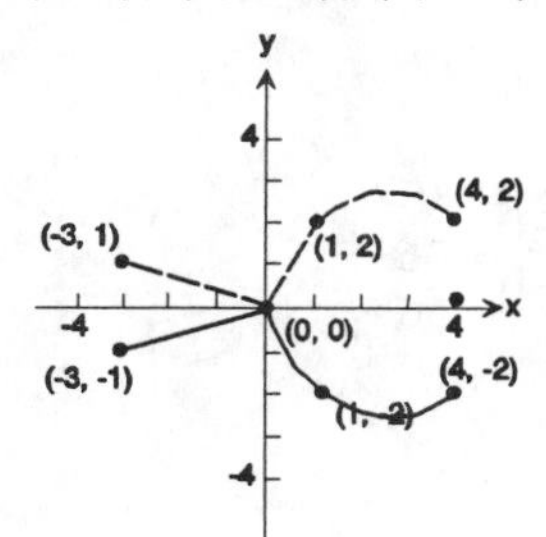

43. Symmetry with respect to the origin means $(-x, -y)$ is on the graph for every (x, y) on the graph. Therefore, given $(-3, 1)$, $(0, 0)$, $(1, 2)$, $(4, 2)$; plot $(3, -1)$, $(0, 0)$, $(-1, -2)$, $(-4, -2)$.

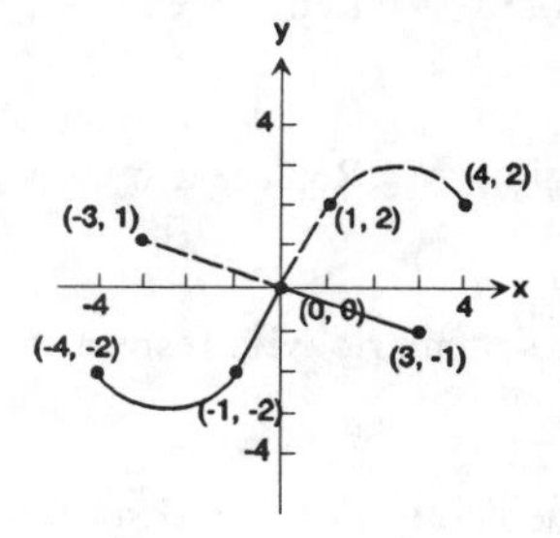

45. $x^2 = y$

y-intercept: Let $x = 0$ so $y = 0$ $(0, 0)$
x-intercept: Let $y = 0$ so $x = 0$ $(0, 0)$

Test for symmetry:

x-axis: Replace y by $-y$ so $x^2 = -y$, which is not equivalent to $x^2 = y$.
y-axis: Replace x by $-x$ so $(-x)^2 = y$ or $x^2 = y$ is equivalent to $x^2 = y$.
Origin: Replace x by $-x$ and y by $-y$ so $(-x)^2 = -y$ or $x^2 = -y$ is not equivalent to $x^2 = y$.

Therefore, symmetric with respect to the y-axis.

47. $y = 3x$

y-intercept: Let $x = 0$ so $y = 0$ $(0, 0)$
x-intercept: Let $y = 0$ so $x = 0$ $(0, 0)$

Test for symmetry:

x-axis: Replace y by $-y$ so $-y = 3x$ is not equivalent to $y = 3x$.
y-axis: Replace x by $-x$ so $y = -3x$ is not equivalent to $y = 3x$.
Origin: Replace x by $-x$ and y by $-y$ so $-y = -3x$ is $y = 3x$ which is equivalent to $y = 3x$.

Therefore, symmetric with respect to the origin.

49. $x^2 + y - 9 = 0$

y-intercept: Let $x = 0$ so $y = 9$ $(0, 9)$

x-intercept: Let $y = 0$ so $x^2 = 9$ $(-3, 0)$

$x = \pm 3$ $(3, 0)$

Test for symmetry:

x-axis: Replace y by $-y$ so $x^2 - y - 9 = 0$ is not equivalent to $x^2 + y - 9 = 0$.

y-axis: Replace x by $-x$ so $(-x)^2 + y - 9 = 0$ is $x^2 + y - 9 = 0$, which is equivalent to $x^2 + y - 9 = 0$.

Origin: Replace x by $-x$ and y by $-y$ so $(-x)^2 - y - 9 = 0$ is $x^2 - y - 9 = 0$, which is not equivalent to $x^2 + y - 9 = 0$.

Therefore, symmetric with respect to the y-axis.

51. $9x^2 + 4y^2 = 36$

y-intercept: Let $x = 0$ so $4y^2 = 36$ $(0, -3)$

$y^2 = 9$ $(0, 3)$

$y = \pm 3$

x-intercept: Let $y = 0$ so $9x^2 = 36$ $(-2, 0)$

$x^2 = 4$ $(2, 0)$

$x = \pm 2$

Test for symmetry:

x-axis: Replace y by $-y$ so $9x^2 + 4(-y)^2 = 36$

$9x^2 + 4y^2 = 36$

is equivalent to $9x^2 + 4y^2 = 36$

y-axis: Replace x by $-x$ so $9(-x)^2 + 4y^2 = 36$

$9x^2 + 4y^2 = 36$

is equivalent to $9x^2 + 4y^2 = 36$

Origin: Replace x by $-x$ and y by $-y$ so $9(-x)^2 + 4(-y)^2 = 36$

$9x^2 + 4y^2 = 36$

is equivalent to $9x^2 + 4y^2 = 36$

Therefore, symmetric with respect to the x-axis, y-axis, and origin.

53. $y = x^3 - 27$

y-intercept: Let $x = 0$ so $y = -27$ $(0, -27)$

x-intercept: Let $y = 0$ so $0 = x^3 - 27$

$27 = x^3$

$3 = x$ $(3, 0)$

Test for symmetry:

x-axis: Replace y by $-y$ so $-y = x^3 - 27$ is not equivalent to $y = x^3 - 27$

y-axis: Replace x by $-x$ so $y = (-x)^3 - 27$

$y = -x^3 - 27$ is not equivalent to $y = x^3 - 27$

Origin: Replace x by $-x$ and y by $-y$ so $-y = (-x)^3 - 27$

$-y = -x^3 - 27$

$y = x^3 + 27$ is not equivalent to $y = x^3 - 27$

Therefore, no symmetry.

55. $y = x^2 - 3x - 4$

y-intercept: Let $x = 0$ so $y = -4$ $(0, -4)$

x-intercept: Let $y = 0$ so $0 = x^2 - 3x - 4$ $(4, 0)$

$0 = (x - 4)(x + 1)$ $(-1, 0)$

$x - 4 = 0$ or $x + 1 = 0$

$x = 4$ or $x = -1$

Test for symmetry:

x-axis: Replace y by $-y$ so $-y = x^2 - 3x - 4$ is not equivalent to $y = x^2 - 3x - 4$.

y-axis: Replace x by $-x$ so $y = (-x)^2 - 3(-x) - 4$
$y = x^2 + 3x - 4$ is not equivalent to
$y = x^2 - 3x - 4$.

Origin: Replace x by $-x$ and y by $-y$ so
$-y = (-x)^2 - 3(-x) - 4$
$-y = x^2 + 3x - 4$
$y = -x^2 - 3x + 4$ is not equivalent to
$y = x^2 - 3x - 4$

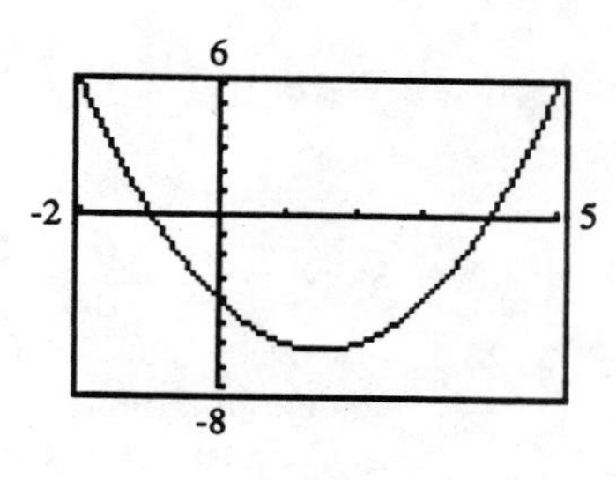

Therefore, no symmetry.

57. $y = \dfrac{x}{x^2 + 9}$

y-intercept: Let $x = 0$ so $y = 0$ (0, 0)

x-intercept: Let $y = 0$ so $0 = \dfrac{x}{x^2 + 9}$

$0 = x$ (0, 0)

Test for symmetry:

x-axis: Replace y by $-y$ so $-y = \dfrac{x}{x^2 + 9}$ is not equivalent to $y = \dfrac{x}{x^2 + 9}$

y-axis: Replace x by $-x$ so $y = \dfrac{-x}{(-x)^2 + 9}$

$y = \dfrac{-x}{x^2 + 9}$ is not equivalent to $y = \dfrac{x}{x^2 + 9}$

Origin: Replace x by $-x$ and y by $-y$ so $-y = \dfrac{-x}{(-x)^2 + 9}$

$-y = \dfrac{-x}{x^2 + 9}$

$y = \dfrac{x}{x^2 + 9}$ is equivalent to $y = \dfrac{x}{x^2 + 9}$

Therefore, symmetric with respect to the origin.

59. (a)

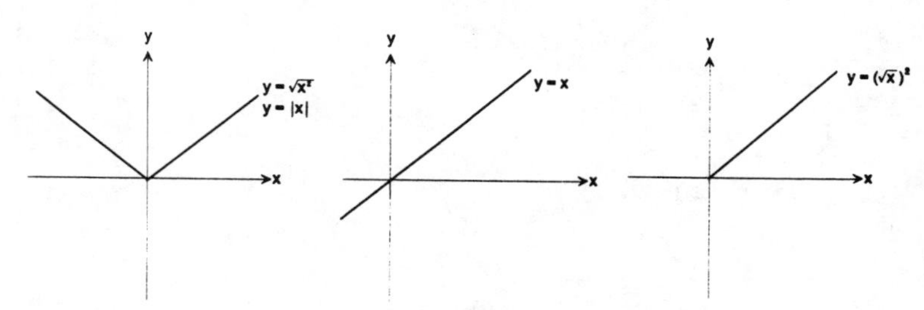

(b) Since $\sqrt{x^2} = |x|$, then for all x, the graphs of $y = \sqrt{x^2}$ and $y = |x|$ are the same.

(c) For $y = \left(\sqrt{x}\right)^2$, the domain of the variable x is $x \geq 0$; for $y = x$, the domain of the variable x is all real numbers. Thus, $\left(\sqrt{x}\right)^2 = x$ only for $x \geq 0$.

(d) For $y = \sqrt{x^2}$, the range of the variable y is $y \geq 0$; for $y = x$, the range of the variable y is all real numbers. Also, $\sqrt{x^2} = |x|$ equals x only if $x \geq 0$.

1. (a) Slope $= \dfrac{1-0}{2-0} = \dfrac{1}{2}$

   (b) For every 2 unit change in x, y will change by 1 unit; i.e., if x increases by 2 units, y will increase by 1 unit.

3. (a) Slope $= \dfrac{2-1}{-2-1} = \dfrac{1}{-3} = \dfrac{-1}{3}$

   (b) If x increases by 3 units, y will decrease by 1 unit.

5. (x_1, y_1) (x_2, y_2)
   (2, 3) (4, 0)

$$\text{Slope} = \frac{y_2 - y_1}{x_2 - x_1} = \frac{0-3}{4-2} = \frac{-3}{2}$$

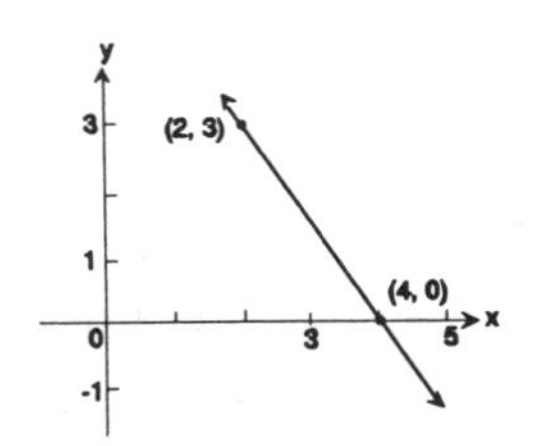

7. (x_1, y_1) (x_2, y_2)
   (−2, 3) (2, 1)

$$\text{Slope} = \frac{y_2 - y_1}{x_2 - x_1} = \frac{1-3}{2-(-2)} = \frac{-2}{4} = -\frac{1}{2}$$

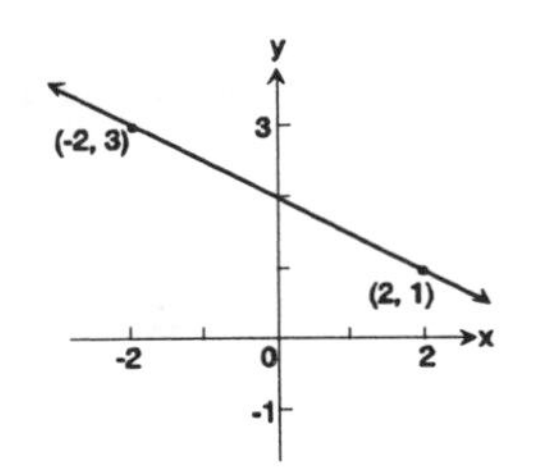

9. (x_1, y_1) (x_2, y_2)
   (−3, −1) (2, −1)

$$\text{Slope} = \frac{y_2 - y_1}{x_2 - x_1} = \frac{-1-(-1)}{2-(-3)} = \frac{0}{5} = 0$$

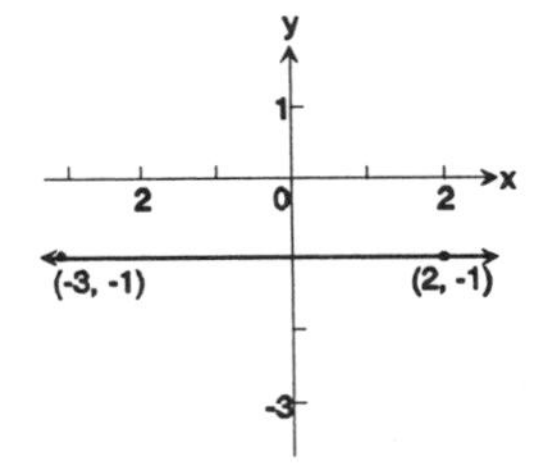

11. (x_1, y_1) (x_2, y_2)
    (−1, 2) (−1, −2)

$$\text{Slope} = \frac{y_2 - y_1}{x_2 - x_1} = \frac{-2-2}{1-(-1)} = \frac{-4}{0}$$

(slope undefined)

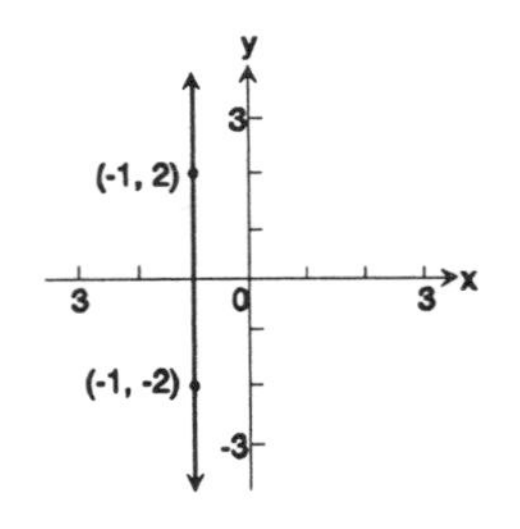

13. (x_1, y_1) (x_2, y_2)
    $(\sqrt{2}, 3)$ $(1, \sqrt{3})$

$$\text{Slope} = \frac{y_2 - y_1}{x_2 - x_1} = \frac{\sqrt{3} - 3}{1 - \sqrt{2}}$$

$$\approx \frac{1.732 - 3}{1 - 1.414} \approx \frac{-1.268}{-.414} \approx 3.06$$

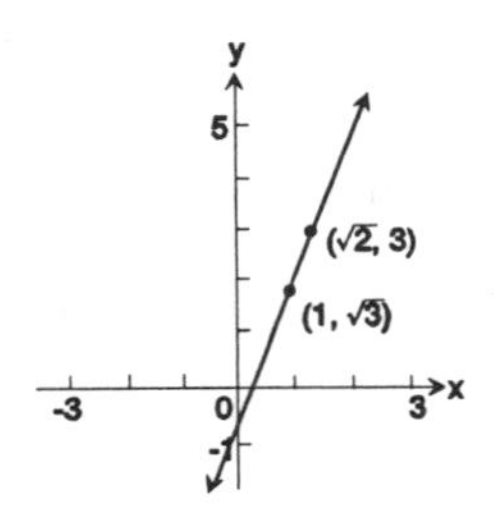

15.

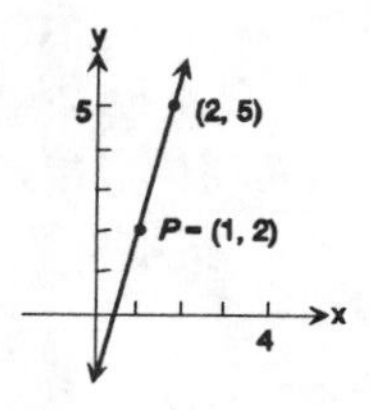

17.

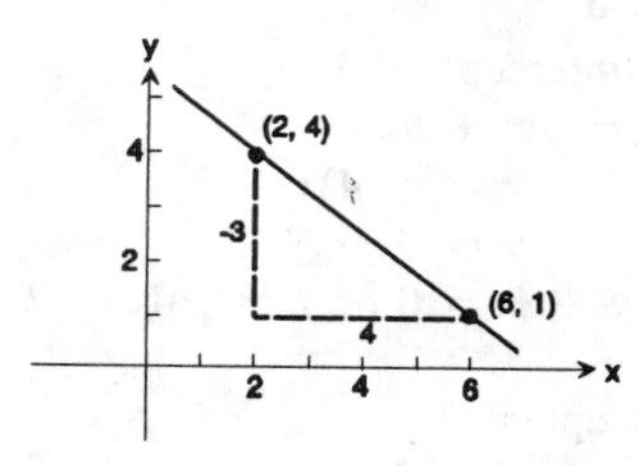

19.

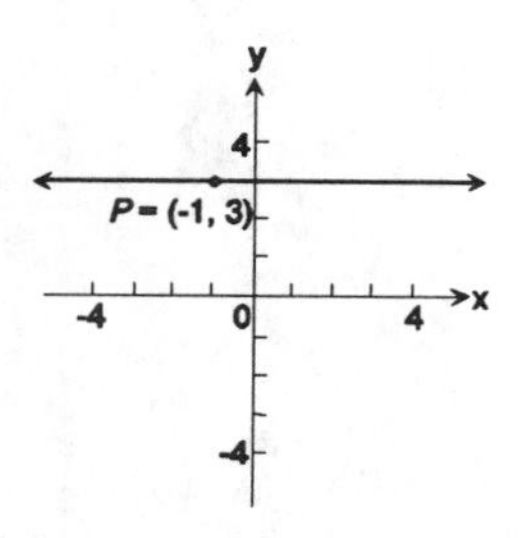

21.

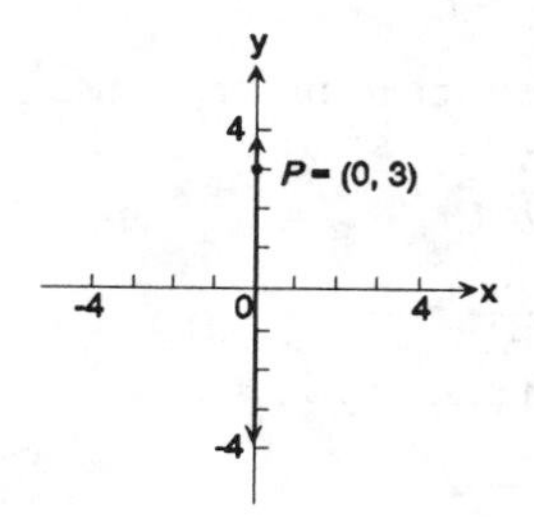

23. (0, 0) and (2, 1) are points on the line.

Slope $= \dfrac{1 - 0}{2 - 0} = \dfrac{1}{2}$

y-intercept is 0; so using $y = mx + b$, we get

$$y = \frac{1}{2}x + 0$$
$$2y = x$$
$$0 = x - 2y$$
$$x - 2y = 0 \quad \text{or} \quad y = \frac{1}{2}x$$

25. $(-2, 2)$ and $(1, 1)$ are points on the line.

Slope $= \dfrac{1 - 2}{1 - (-2)} = \dfrac{-1}{3} = \dfrac{-1}{3}$

Use $y - y_1 = m(x - x_1)$ with (x_1, y_1) being either point on the line.

$$y - 2 = \frac{-1}{3}(x - (-2))$$
$$y - 2 = \frac{-1}{3}x - \frac{2}{3}$$
$$-3y + 6 = x + 2$$
$$0 = x + 3y - 4$$
or
$$y = \frac{-1}{3}x + \frac{4}{3}$$

27. $(x_1, y_1) = (-2, 5)$

Slope = 4

$$y - y_1 = m(x - x_1)$$
$$y - 5 = 4[x - (-2)]$$
$$y - 5 = 4(x + 2)$$
$$y - 5 = 4x + 8$$
$$0 = 4x - y + 13$$
$$4x - y + 13 = 0 \quad \text{or} \quad y = 4x + 13$$

29. $(x_1, y_1) = (1, -1)$

Slope $= -\dfrac{2}{3}$

$$y - y_1 = m(x - x_1)$$
$$y - (-1) = -\frac{2}{3}(x - 1)$$
$$y + 1 = -\frac{2}{3}x + \frac{2}{3}$$

Multiply both sides by 3:

$$3y + 3 = -2x + 2$$
$$2x + 3y + 1 = 0 \text{ or } y = \frac{-2}{3}x - \frac{1}{3}$$

31. Passing through $(1, 3)$ and $(-1, 2)$.

Slope $= \dfrac{2 - 3}{-1 - 1} = \dfrac{-1}{-2} = \dfrac{1}{2}$

Use either point and slope.

$$y - 3 = \frac{1}{2}(x - 1)$$
$$y - 3 = \frac{1}{2}x - \frac{1}{2}$$

Multiply both sides by 2:

$$2y - 6 = x - 1$$
$$0 = x - 2y + 5$$
$$x - 2y + 5 = 0 \text{ or } y = \frac{1}{2}x + \frac{5}{2}$$

33. $m = -4$

$b = y\text{-intercept} = 3$

Use $y = mx + b$

$$y = (-4)x + 3$$

$$y = -4x + 3$$

$4x + y - 3 = 0$ or $y = -4x + 3$

35. x-intercept $= 2$

y-intercept $= -1$

Using the intercept form

$$\frac{x}{a} + \frac{y}{b} = 1$$

$$\frac{x}{2} + \frac{y}{-1} = 1$$

Multiply both sides by 2:

$$x - 2y = 2$$

$x - 2y - 2 = 0$ or $y = \frac{1}{2}x - 1$

or use points (2, 0) and (0, -1)

$$m = \frac{-1 - 0}{0 - 2} = \frac{1}{2}$$

$$b = -1$$

$$y = mx + b$$

$$y = \frac{1}{2}x - 1$$

$$2y = x - 2$$

$$0 = x - 2y - 2$$

$$x - 2y - 2 = 0$$

37. Slope undefined; passing through (2, 4).

Vertical line $x = a$ has undefined slope.

Thus, $x = 2$

$x - 2 = 0$; No slope-intercept form

39.

$$y = 2x + 3$$

$$y = mx + b$$

m: slope $= 2$

b: y-intercept $= 3$

Using intercepts, draw graph.

$$\left(-\frac{3}{2}, 0\right)$$

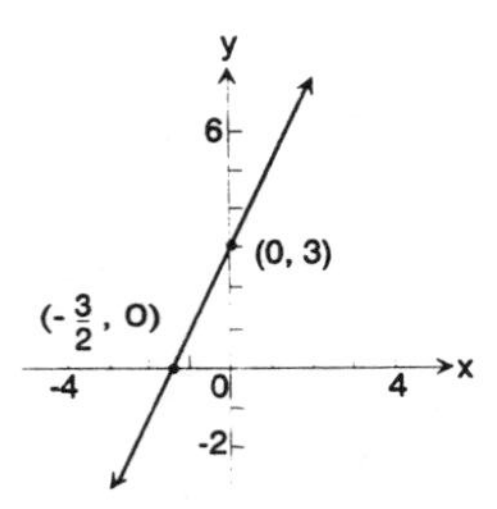

41.

$$\frac{1}{2}y = x - 1$$

$$y = 2x - 2$$

$$y = mx + b$$

m: slope $= 2$

b: y-intercept $= -2$

Using intercepts, draw graph.

(0, -2)

(1, 0)

43.

$$y = \frac{1}{2}x + 2$$

m: slope $= \frac{1}{2}$

b: y-intercept $= 2$

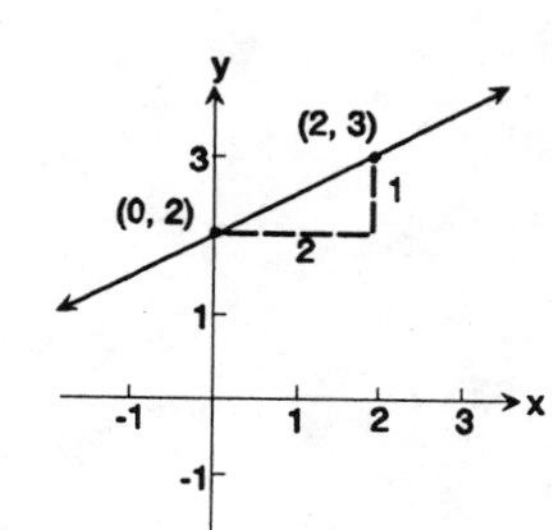

45.

$$x + 2y = 4$$
$$2y = -x + 4$$
$$y = -\frac{1}{2}x + 2$$

m: slope $= -\frac{1}{2}$

b: y-intercept $= 2$

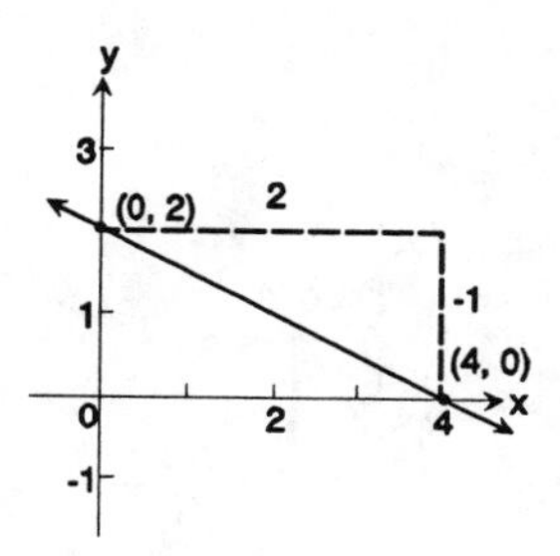

47.

$$2x - 3y = 6$$
$$-3y = -2x + 6$$
$$y = \frac{2}{3}x - 2$$
$$y = mx + b$$

m: slope $= \frac{2}{3}$

b: y-intercept $= -2$

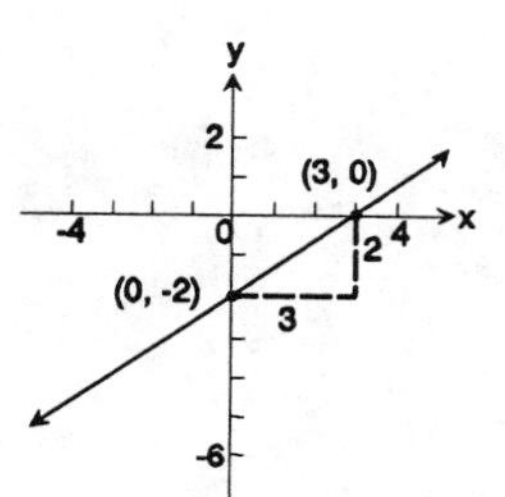

49.

$$x + y = 1$$
$$y = -x + 1$$

m: slope $= -1$

b: y-intercept $= 1$

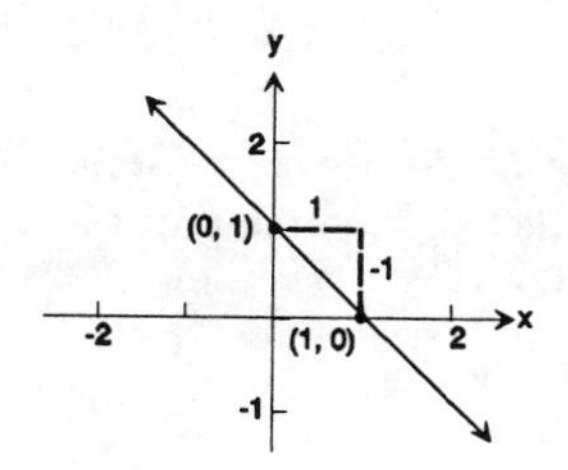

51. $x = -4$

of the form $x = a$, a vertical line
slope is undefined; no y-intercept.

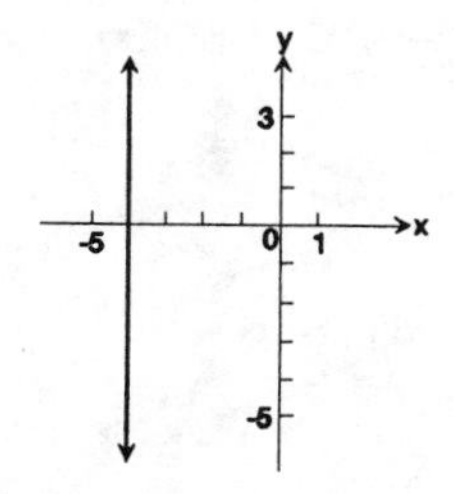

53. $y = 5$

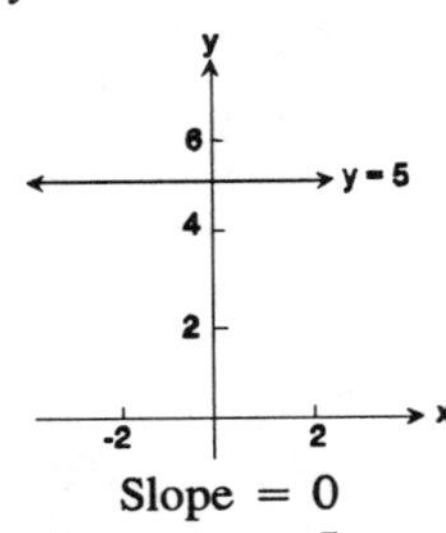

Slope = 0
y-intercept = 5

55. $y - x = 0$ or $y = x$

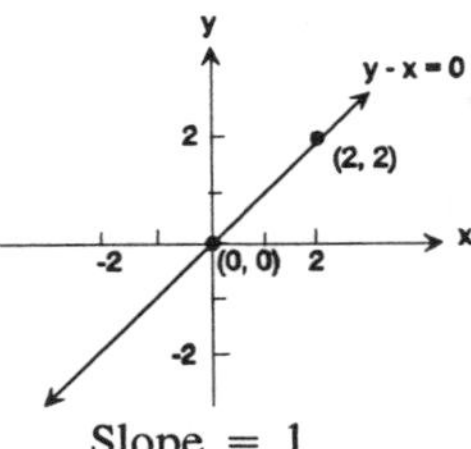

Slope = 1
y-intercept = 0

57.

$$2y - 3x = 0$$
$$2y = 3x$$
$$y = \frac{3}{2}x$$
$$y = mx + b$$
$$m:\ \text{slope} = \frac{3}{2}$$
$$b:\ y\text{-intercept} = 0$$

59. The general equation of the x-axis is $y = 0$ because the slope = 0 and x can be any real number on the number line but y must equal zero.

$$y = 0x + 0$$
$$y = 0$$

61.

$$(°F, °C) = (32, 0)$$
$$(°F, °C) = (212, 100)$$
$$\text{slope} = \frac{100 - 0}{212 - 32} = \frac{100}{180} = \frac{5}{9}$$
$$°C - 0 = \frac{5}{9}(°F - 32)$$
$$°C = \frac{5}{9}(°F - 32)$$

If °F = 70, then

$$°C = \frac{5}{9}(70 - 32) = \frac{5}{9}(38)$$
$$°C \approx 21°$$

63. (a) Since there is only a profit of \$0.50 per copy (\$1.00 − \$0.50) and the expense of \$100 must be deducted, then the profit is

$$P = (0.50)x - 100$$
$$\text{or } P = 0.5x - 100$$

Plot points: $(0, -100)$
$(200, 0)$

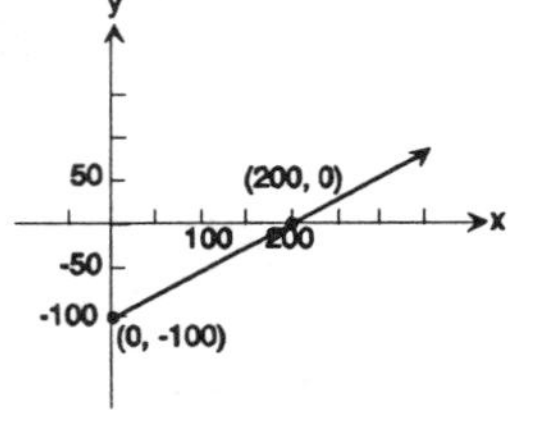

(b) $P = 0.5(1000) - 100 = 500 - 100 = \$400$

(c) $P = 0.5(5000) - 100 = 2500 - 100 = \$2400$

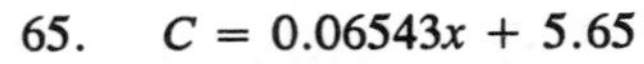

65. $C = 0.06543x + 5.65$

For 300 kWh, $C = 0.06543(300) + 5.65$
$= \$25.28$

For 750 kWh, $C = 0.06543(750) + 5.65$
$= \$54.72$

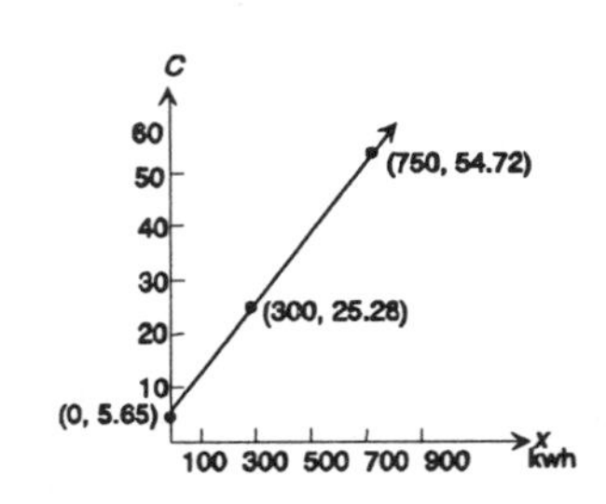

67. (b)

69. (d)

71.
$$y - 2 = x$$
$$y = x + 2$$
or
$$x - y + 2 = 0$$

73.
$$y - 0 = -\frac{1}{3}(x - 3)$$
$$y = -\frac{1}{3}x + 1$$
$$3y = -x + 3$$
$$x + 3y - 3 = 0 \text{ or } y = -\frac{1}{3}x + 1$$

75.
$$y - (-2) = -\frac{2}{3}(x - 3)$$
$$y + 2 = -\frac{2}{3}x + 2$$
$$3y + 6 = -2x + 6$$
$$2x + 3y = 0 \text{ or } y = -\frac{2}{3}x$$

79. No, if the intercepts are (0, 0). No. Every line crosses at least one axis. A horizontal line crosses the y-axis and a vertical line crosses the x-axis.

81. Their slopes are -1. They are the same line.

83. Yes, if the y-intercept $= 0$.

3.4 Parallel and Perpendicular Lines; Circles

1. $y = 6x$
   (a) 6
   (b) $-\frac{1}{6}$

3. $y = \frac{-1}{2}x + 2$
   (a) $\frac{-1}{2}$
   (b) 2

5. $2x - 4y + 5 = 0$
$$4y = 2x + 5$$
$$y = \frac{1}{2}x + \frac{5}{4}$$
   (a) $\frac{1}{2}$
   (b) -2

7. $3x + 5y - 10 = 0$
$$5y = -3x + 10$$
$$y = \frac{-3}{5}x + 2$$
   (a) $\frac{-3}{5}$
   (b) $\frac{5}{3}$

9. $x = 7$ is a vertical line; slope is undefined.
   (a) Slope is undefined.
   (b) Since $x = 7$ is vertical, a line perpendicular would be horizontal; slope $= 0$.

11.
$$y - y_1 = m(x - x_1),\ m = 2$$
$$y - 3 = 2(x - 3)$$
$$y - 3 = 2x - 6$$
$$2x - y - 3 = 0 \text{ or } y = 2x - 3$$

13. $y - y_1 = m(x - x_1),\ m = -\frac{1}{2}$

$$y - 2 = -\frac{1}{2}(x - 1)$$
$$2y - 4 = -x + 1$$
$$x + 2y - 5 = 0 \text{ or } y = -\frac{1}{2}x + \frac{5}{2}$$

15. Parallel to $y = 4x$;
passing through $(-1, 2)$.
Slope $= 4$ and parallel line has same slope.

$$y - 2 = 4(x - (-1))$$
$$y - 2 = 4(x + 1)$$
$$y - 2 = 4x + 4$$
$$0 = 4x - y + 6$$
$$4x - y + 6 = 0 \text{ or } y = 4x + 6$$

17. Parallel to $2x - y + 2 = 0$; passing through $(0, 0)$.

$$2x + 2 = y$$
$$y = 2x + 2$$

Slope $= 2$ and parallel line has same slope.

$$y - 0 = 2(x - 0)$$
$$y = 2x$$
$$0 = 2x - y$$
$$2x - y = 0 \text{ or } y = 2x$$

19. Parallel to $x = 5$; passing through $(4, 2)$.
$x = 5$ is a vertical line;
slope is undefined.
Therefore, $x = 4$ is parallel to $x = 5$.
$x - 4 = 0$; no slope-intercept form

21. Perpendicular to $y = \frac{1}{2}x + 4$; passing through $(1, -2)$.

Slope of given line $= \frac{1}{2}$

Slope of perpendicular line $= -2$

$$[y - (-2)] = -2(x - 1)$$
$$y + 2 = -2x + 2$$
$$2x + y = 0 \text{ or } y = -2x$$

23. Perpendicular to $2x + y - 2 = 0$; passing through $(-3, 0)$.

$$y = -2x + 2$$

Slope of given line $= -2$

Slope of perpendicular line $= \frac{1}{2}$

$$y - 0 = \frac{1}{2}[x - (-3)]$$
$$y = \frac{1}{2}(x + 3)$$
$$y = \frac{1}{2}x + \frac{3}{2}$$
$$2y = x + 3$$
$$0 = x - 2y + 3$$
$$x - 2y + 3 = 0 \text{ or } y = \frac{1}{2}x + \frac{3}{2}$$

25. Perpendicular to $x = 8$; passing through $(3, 4)$.
Since $x = 8$ is vertical, a line perpendicular would be horizontal of the form $y = b$.
Therefore, $y = 4$ is the line perpendicular to $x = 8$.

$$y - 4 = 0 \text{ or } y = 4$$

27. Center $= (2, 1)$
Radius $=$ Distance from $(0, 1)$ to $(2, 1)$

$$= \sqrt{(2 - 0)^2 + (1 - 1)^2}$$
$$= \sqrt{4} = 2$$
$$(x - 2)^2 + (y - 1)^2 = 4$$

29. Center $=$ Midpoint of $(1, 2)$ and $(4, 2)$

$$= \left(\frac{1 + 4}{2}, \frac{2 + 2}{2}\right)$$
$$= \left(\frac{5}{2}, 2\right)$$

Radius $=$ Distance from $\left(\frac{5}{2}, 2\right)$ to $(4, 2)$

$$= \sqrt{\left(4 - \frac{5}{2}\right)^2 + (2 - 2)^2}$$
$$= \sqrt{\frac{9}{4}} = \frac{3}{2}$$
$$\left(x - \frac{5}{2}\right)^2 + (y - 2)^2 = \frac{9}{4}$$

31. Use $(x - h)^2 + (y - k)^2 = r^2$, where $r = 1$; $(h, k) = (1, -1)$

$$(x - 1)^2 + [y - (-1)]^2 = 1^2$$
$$(x - 1)^2 + (y + 1)^2 = 1$$

Square each part and gather up terms:

$$(x^2 - 2x + 1) + (y^2 + 2y + 1) = 1$$
$$x^2 + y^2 - 2x + 2y + 2 = 1$$
$$x^2 + y^2 - 2x + 2y + 1 = 0$$

33. Use $(x - h)^2 + (y - k)^2 = r^2$, where $r = 2$; $(h, k) = (0, 2)$

$$(x - 0)^2 + (y - 2)^2 = 2^2$$
$$x^2 + (y - 2)^2 = 4$$
$$x^2 + y^2 - 4y + 4 = 4$$
$$x^2 + y^2 - 4y = 0$$

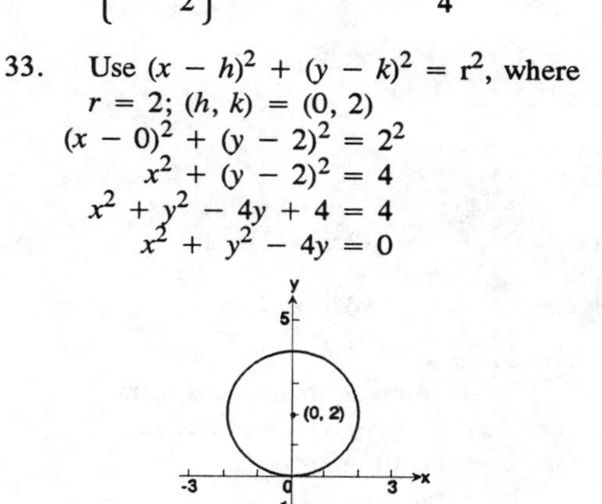

35. Use $(x - h)^2 + (y - k)^2 = r^2$, where $r = 5$; $(h, k) = (4, -3)$

$$(x - 4)^2 + [y - (-3)]^2 = (5)^2$$
$$(x - 4)^2 + (y + 3)^2 = 25$$

Square each part and gather up terms:

$$(x^2 - 8x + 16) + (y^2 + 6y + 9) = 25$$
$$x^2 + y^2 - 8x + 6y + 25 = 25$$
$$x^2 + y^2 - 8x + 6y = 0$$

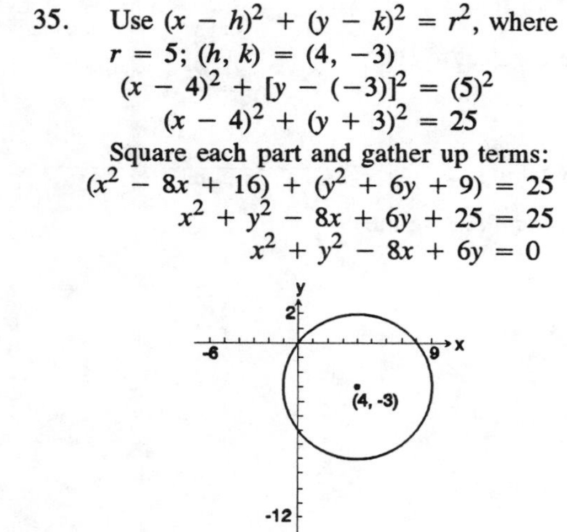

37. Since $(h, k) = (0, 0)$, the center is at the origin so use:

$$x^2 + y^2 = r^2 \text{ when } r = 2$$
$$x^2 + y^2 = 2^2$$
$$x^2 + y^2 = 4$$
$$x^2 + y^2 - 4 = 0$$

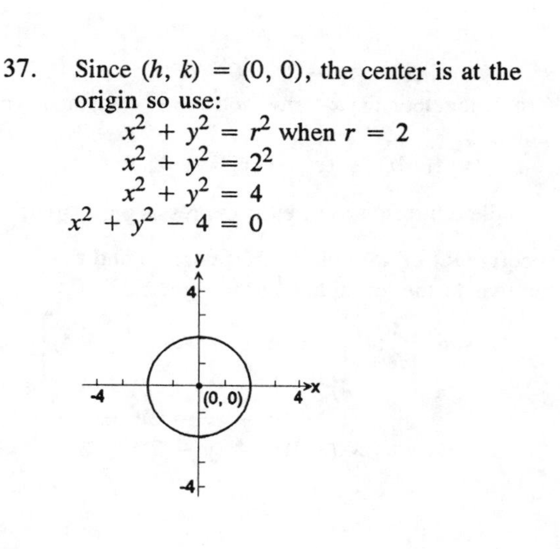

39. $r = \frac{1}{2}$; $(h, k) = \left(\frac{1}{2}, 0\right)$

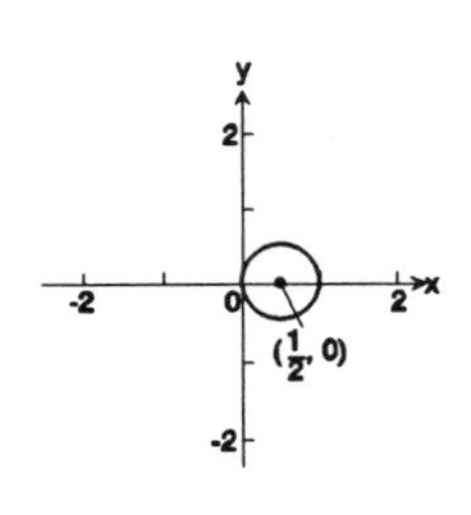

$$\left(x - \frac{1}{2}\right)^2 + (y - 0)^2 = \left(\frac{1}{2}\right)^2$$

$$\left(x - \frac{1}{2}\right)^2 + y^2 = \frac{1}{4} \quad \text{standard form}$$

$$x^2 - x + \frac{1}{4} + y^2 = \frac{1}{4}$$

$$x^2 + y^2 - x = 0 \quad \text{general form}$$

41. Since $x^2 + y^2 = 4$ is of the form $x^2 + y^2 = r^2$, it can be written as $x^2 + y^2 = (2)^2$, so $(h, k) = (0, 0)$ and $r = 2$.

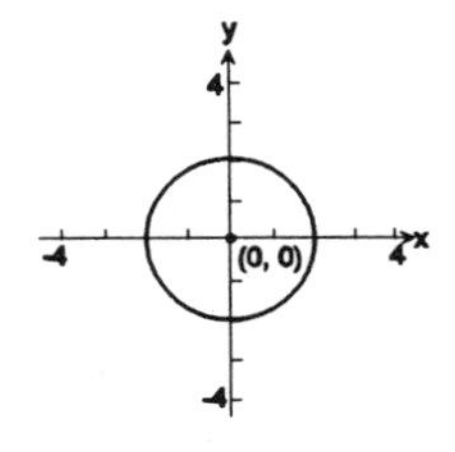

43. $2(x - 3)^2 + 2y^2 = 8$ can be written as:

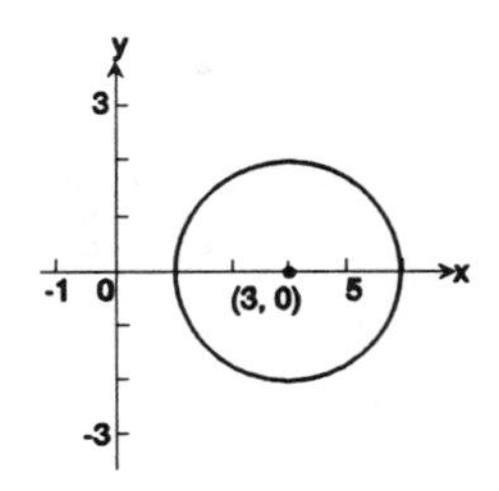

$$2[(x - 3)^2 + (y - 0)^2] = 2 \cdot 2^2$$

$$\frac{2}{2}[(x - 3)^2 + (y - 0)^2] = \frac{2 \cdot 2^2}{2}$$

$$(x - 3)^2 + (y - 0)^2 = 2^2$$

so compare to standard form
$(x - h)^2 + (y - k)^2 = r^2$ and get
$(h, k) = (3, 0)$ and $r = 2$.

45. $x^2 + y^2 + 4x - 4y - 1 = 0$

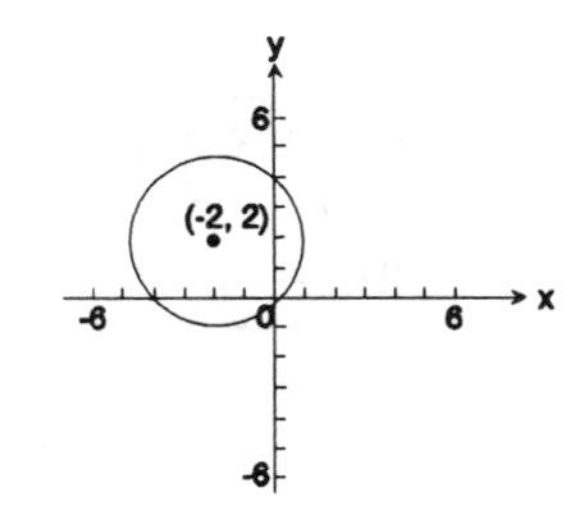

Group together the x-terms and the y-terms and rearrange constant to right side:

$$(x^2 + 4x) + (y^2 - 4y) = 1$$

Complete the square of each expression in parentheses by taking $\frac{1}{2}$ (coefficient of variable to first degree) and then squaring it and adding the number to the left side and the right side.

Add $\left[\frac{1}{2}(4)\right]^2 = 4$ Add $\left[\frac{1}{2}(-4)\right]^2 = 4$

$$(x^2 + 4x + 4) + (y^2 - 4y + 4) = 1 + 4 + 4$$

$$(x + 2)^2 + (y - 2)^2 = 9$$

$$[x-(-2)]^2 + (y - 2)^2 = 3^2$$

$$(h, k) = (-2, 2)$$

$$r = 3$$

47. $$x^2 + y^2 - x + 2y + 1 = 0$$

Proceed as in 45:

$$(x^2 - x) + (y^2 + 2y) = -1$$

$$\text{Add } \left[\frac{1}{2}(-1)\right]^2 = \frac{1}{4} \quad \text{Add } \left[\frac{1}{2}(2)\right]^2 = 1$$

$$\left(x^2 - x + \frac{1}{4}\right) + (y^2 + 2y + 1) = -1 + \frac{1}{4} + 1$$

$$\left(x - \frac{1}{2}\right)^2 + (y + 1)^2 = \frac{1}{4}$$

$$\left(x - \frac{1}{2}\right)^2 + [y - (-1)]^2 = \left(\frac{1}{2}\right)^2$$

$$(h, k) = \left(\frac{1}{2}, -1\right)$$

$$r = \frac{1}{2}$$

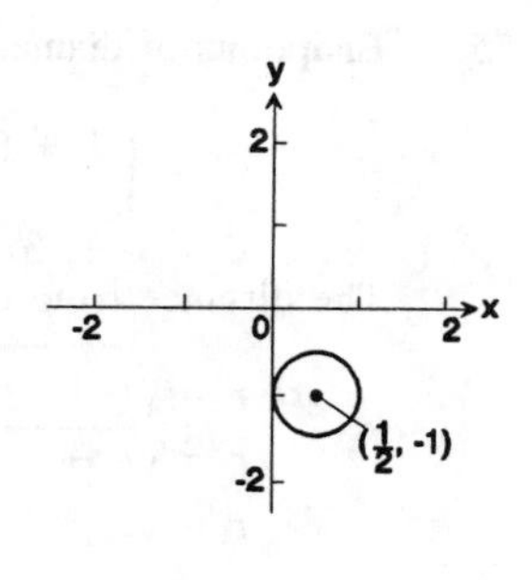

49. $2x^2 + 2y^2 - 12x + 8y - 24 = 0$

The coefficients of x^2 and y^2 should be 1 in order to put the equation in standard form, so divide each term by 2.

$$x^2 + y^2 - 6x + 4y - 12 = 0$$

$$(x^2 - 6x) + (y^2 + 4y) = 12$$

$$\text{Add } \left[\frac{1}{2}(-6)\right]^2 = 9 \qquad \text{Add } \left[\frac{1}{2}(4)\right]^2 = 4$$

$$(x^2 - 6x + 9) + (y^2 + 4y + 4) = 12 + 9 + 4$$

$$(x - 3)^2 + (y + 2)^2 = 25$$

$$(x - 3)^2 + [y - (-2)]^2 = 5^2$$

$$(h, k) = (3, -2)$$

$$r = 5$$

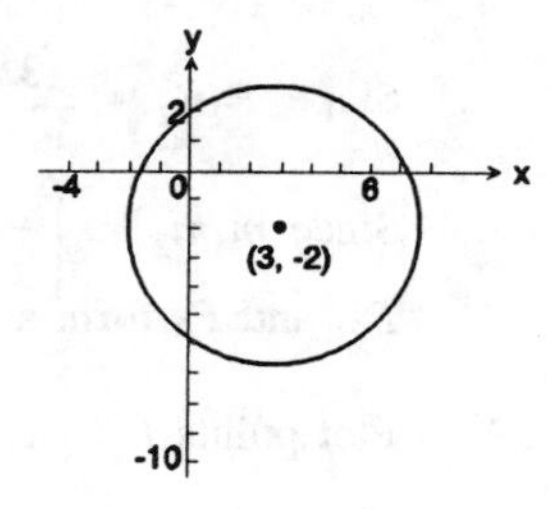

51. Center at origin and containing the point $(-3, 2)$.

$$r = d(C, P) = \sqrt{(-3 - 0)^2 + (2 - 0)^2}$$

$$= \sqrt{9 + 4}$$

$$= \sqrt{13}$$

Use $x^2 + y^2 = r^2$ since center is at origin.

$$x^2 + y^2 = \left(\sqrt{13}\right)^2$$

$$x^2 + y^2 = 13$$

$$x^2 + y^2 - 13 = 0$$

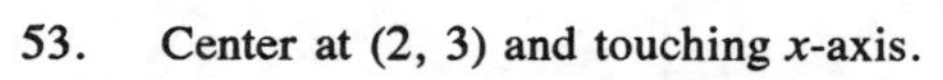

53. Center at (2, 3) and touching x-axis.

$$r = d(C, P) = \sqrt{(2 - 2)^2 + (3 - 0)^2} = \sqrt{9} = 3$$

[r could have been obtained by carefully examining the graph.]

$$(x - 2)^2 + (y - 3)^2 = 3^2$$

$$x^2 - 4x + 4 + y^2 - 6y + 9 = 9$$

$$x^2 + y^2 - 4x - 6y + 4 = 0$$

r is perpendicular to x-axis at point of tangency so P has coordinate (2, 0).

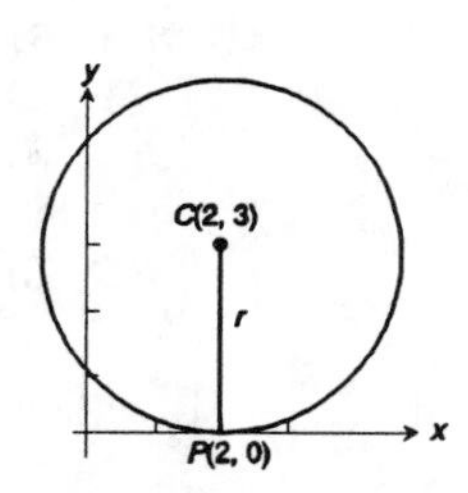

55. Endpoints of diameter (1, 4) and (−3, 2); midpoint of diameter is center of circle.

$$C = \left[\frac{1 + (-3)}{2}, \frac{4 + 2}{2}\right]$$
$$C = (-1, 3)$$

The distance from the center to either endpoint is the radius.

$$r = \sqrt{(-1 - 1)^2 + (3 - 4)^2}$$
$$r = \sqrt{4 + 1}$$
$$r = \sqrt{5}$$

$$[x - (-1)]^2 + (y - 3)^2 = \left(\sqrt{5}\right)^2$$
$$(x + 1)^2 + (y - 3)^2 = 5$$
$$x^2 + 2x + 1 + y^2 - 6y + 9 = 5$$
$$x^2 + y^2 + 2x - 6y + 5 = 0$$

57. Plot points $P_1(-2, 5)$, $P_2(1, 3)$, and $P_3(-1, 0)$.

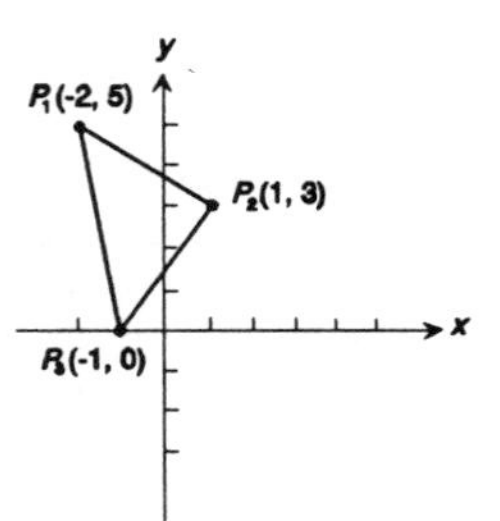

Slope $P_1P_2 = \dfrac{3 - 5}{1 + 2} = \dfrac{-2}{3} = m_1$

Slope $P_2P_3 = \dfrac{3 - 0}{1 - (-1)} = \dfrac{3}{2} = m_2$

Since $m_1m_2 = \left[-\dfrac{2}{3}\right]\left[\dfrac{3}{2}\right] = -1$, the lines are perpendicular so points P_1, P_2, and P_3 form a right triangle.

59. Plot points $P_1(-1, 0)$, $P_2(2, 3)$, and $P_3(1, -2)$, and $P_4(4, 1)$.

Slope $P_1P_2 = m_1 = \dfrac{3 - 0}{2 - (-1)} = \dfrac{3}{3} = 1$

Slope $P_3P_4 = m_2 = \dfrac{1 - (-2)}{4 - 1} = \dfrac{3}{3} = 1$

Slope $P_1P_3 = m_3 = \dfrac{-2 - 0}{1 - (-1)} = \dfrac{-2}{2} = -1$

Slope $P_2P_4 = m_4 = \dfrac{1 - 3}{4 - 2} = \dfrac{-2}{2} = -1$

Since the opposite sides are parallel and the adjacent sides are perpendicular, the points form a rectangle.

61. (c)

63. (b)

65. $(x + 3)^2 + (y - 1)^2 = 16$

67. $(x - 2)^2 + (y - 2)^2 = 9$

69. Refer to Fig. 53 in the text.
If $m_1m_2 = -1$, then $m_1,m_2 = -1$, then

$$d(A, B) = \sqrt{(1 - 1)^2 + (m_2 - m_1)^2} = \sqrt{(m_2 - m_1)^2}$$
$$d(O, A) = \sqrt{(1 - 0)^2 + (m_2 - 0)^2} = \sqrt{1 + m_2^2}$$
$$d(O, B) = \sqrt{(1 - 0)^2 + (m_1 - 0)^2} = \sqrt{1 + m_1^2}$$

Show: $[d(O, B)]^2 + [d(O, A)]^2 = [d(A, B)]^2$

$$\left[\sqrt{1 + m_1^2}\right]^2 + \left[\sqrt{1 + m_2^2}\right]^2 = \left[\sqrt{(m_2 - m_1)^2}\right]^2$$
$$1 + m_1^2 + 1 + m_2^2 = (m_2 - m_1)^2$$
$$m_1^2 + m_2^2 + 2 = m_2^2 - 2m_1m_2 + m_1^2$$

Since $m_1m_2 = -1$, $m_1^2 + m_2^2 + 2 = m_2^2 - 2(-1) + m_1^2$
$$m_1^2 + m_2^2 + 2 = m_1^2 + m_2^2 + 2$$

71. (a)
$$x^2 + (mx + b)^2 = r^2$$
$$(1 + m^2)x^2 + 2mbx + b^2 - r^2 = 0$$
One solution if and only if discriminant = 0
$$(2mb)^2 - 4(1 + m^2)(b^2 - r^2) = 0$$
$$-4b^2 + 4r^2 + 4m^2r^2 = 0$$
$$r^2(1 + m^2) = b^2$$

(b) $$x = \frac{-2mb}{2(1 + m^2)} = \frac{-2mb}{2b^2/r^2} = \frac{-r^2m}{b}$$

$$y = m\left[\frac{-r^2m}{b}\right] + b = \frac{-r^2m^2}{b} + b = \frac{-r^2m^2 + b^2}{b} = \frac{r^2}{b}$$

(c) Slope of tangent line = m

Slope of line joining center to point of tangency = $\dfrac{r^2/b}{-r^2m/b} = \dfrac{-1}{m}$

73.
$$x^2 - 4x + y^2 + 6y = -4$$
$$(x - 2)^2 + (y + 3)^2 = 9$$

Center $(2, -3)$

Slope from center to $\left(3, 2\sqrt{2} - 3\right)$ is $\dfrac{2\sqrt{2} - 3 + 3}{3 - 2} = 2\sqrt{2}$

Slope of tangent line is $\dfrac{-1}{2\sqrt{2}} = \dfrac{-2\sqrt{2}}{4}$

$$y - \left(2\sqrt{2} - 3\right) = \frac{-\sqrt{2}}{4}(x - 3)$$
$$\sqrt{2}x + 4y - 11\sqrt{2} + 12 = 0$$

75.
$$x^2 + y^2 - 4x + 6y + 4 = 0$$
$$x^2 - 4x + y^2 + 6y = -4$$
$$(x - 2)^2 + (y + 3)^2 = 9$$
Center $(2, -3)$

$$x^2 + y^2 + 6x + 4y + 9 = 0$$
$$x^2 + 6x + y^2 + 4y = -9$$
$$(x + 3)^2 + (y + 2)^2 = 4$$
Center $(-3, -2)$

Slope of line joining centers is $\dfrac{-2 + 3}{-3 - 2} = \dfrac{1}{-5}$

$$y + 3 = \frac{-1}{5}(x - 2)$$
$$x + 5y + 13 = 0$$

77. $2x - y + c = 0$

If $c = -4$, $2x - y - 4 = 0$

Plot points $(0, -4)$
$(1, -2)$
$(2, 0)$

If $c = 0$, $2x - y + 0 = 0$ or $2x = y$

Plot points $(0, 0)$
$(1, 2)$

If $c = 2$, $2x - y + 2 = 0$

Plot points $(0, 2)$
$(1, 4)$

All have the same slope, 2. The lines are parallel.

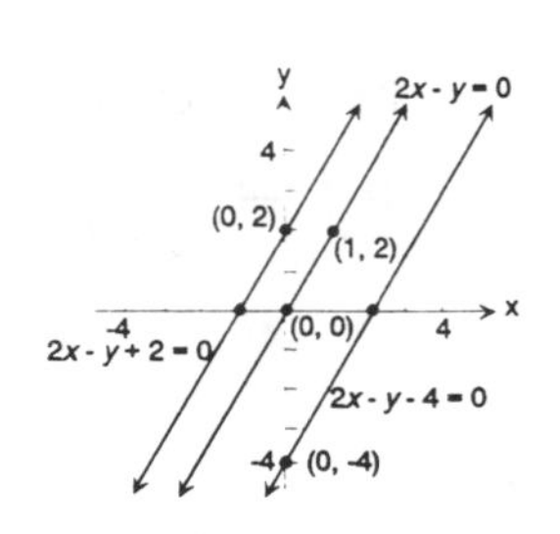

79. $y = 2$

3.5 Linear Curve Fitting

1. Linear relation

3. Linear relation

5. Nonlinear relation

7. (a)

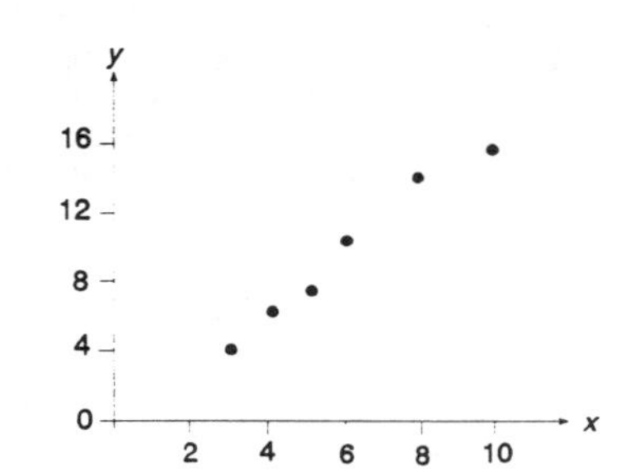

(b) Answers will vary. We select (4, 6) and (8, 14). The slope of the line joining the points (4, 6) and (8, 14) is

$$m = \frac{14 - 6}{8 - 4} = \frac{8}{4} = 2$$

The equation of the line with slope 2 passing through (4, 6) is found using the point-slope form with $m = 2$, $x_1 = 4$, $y_1 = 6$:

$$\begin{aligned} y - y_1 &= m(x - x_1) \\ y - 6 &= 2(x - 4) \\ y - 6 &= 2x - 8 \\ y &= 2x - 2 \end{aligned}$$

(c)

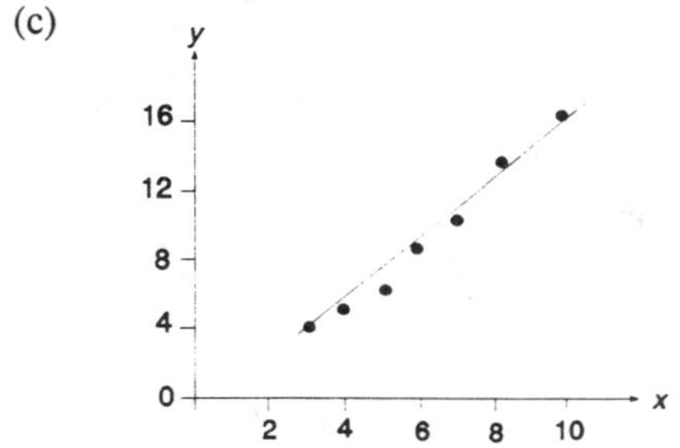

(d) Using the LINear REGression program, we obtain $y = 2.0357x - 2.3571$.

(e)

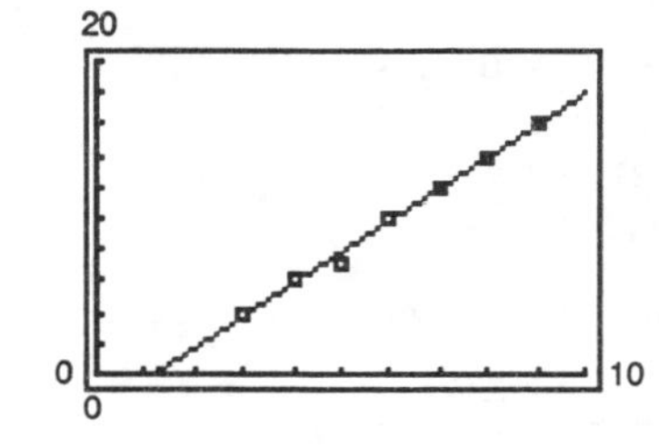

9. (a)

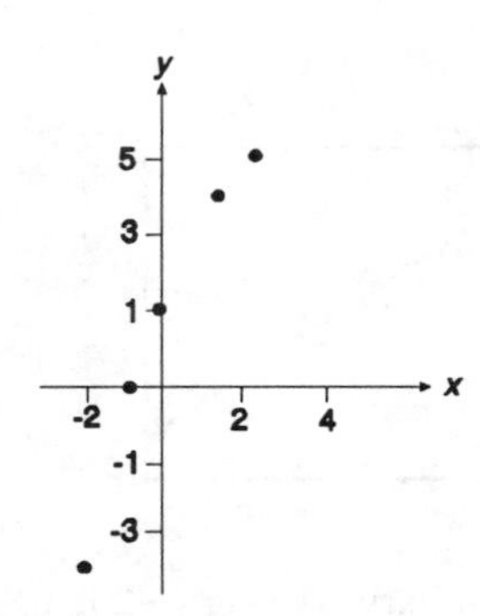

(b) Answers will vary. Using $(-1, 0)$ and $(2, 5)$, the slope of the line is

$$m = \frac{5 - 0}{2 - (-1)} = \frac{5}{3}$$

The equation of the line with slope $\frac{5}{3}$ passing through $(2, 5)$ is found using the point-slope form with $m = \frac{5}{3}$, $x_1 = 2$, $y_1 = 5$:

$$\begin{aligned} y - y_1 &= m(x - x_1) \\ y - 5 &= \frac{5}{3}(x - 2) \\ y - 5 &= \frac{5}{3}x - \frac{10}{3} \\ y &= \frac{5}{3}x + \frac{5}{3} \\ y &= \frac{5}{3}(x + 1) \end{aligned}$$

(c)

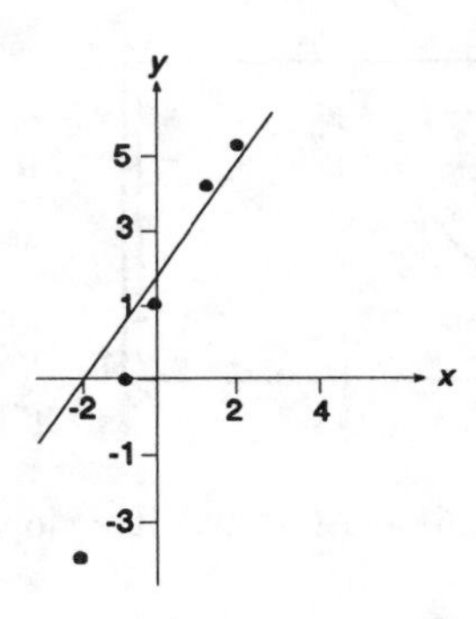

(d) $y = 2.2x + 1.2$

(e)

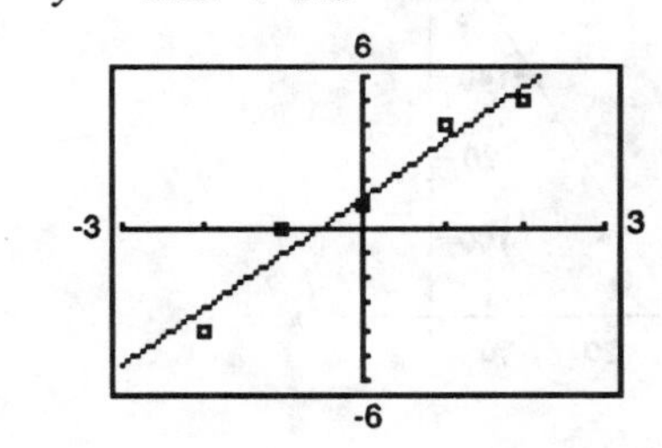

11. (a)

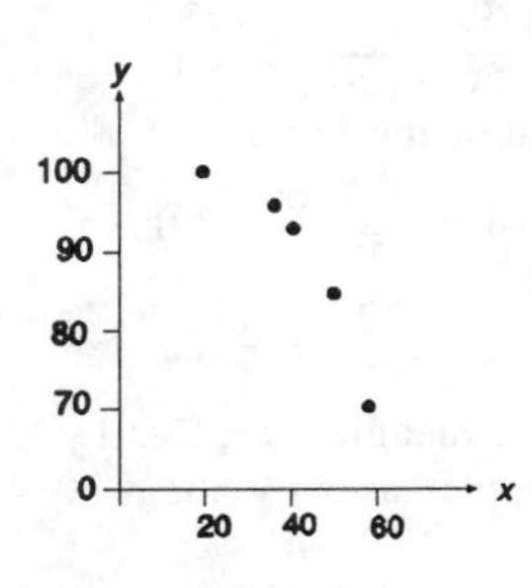

(b) Answers will vary. Using $(20, 100)$ and $(50, 83)$, the slope of the line is

$$m = \frac{100 - 83}{20 - 50} = \frac{17}{-30} = \frac{-17}{30}$$

The equation of the line with slope $\frac{-17}{30}$ passing through $(20, 100)$ is found using the point-slope form with $m = \frac{-17}{30}$, $x_1 = 20$, $y_1 = 100$:

$$\begin{aligned} y - y_1 &= m(x - x_1) \\ y - 100 &= \frac{-17}{30}(x - 20) \\ y &= 100 - \frac{17}{30}(x - 20) \end{aligned}$$

(c)

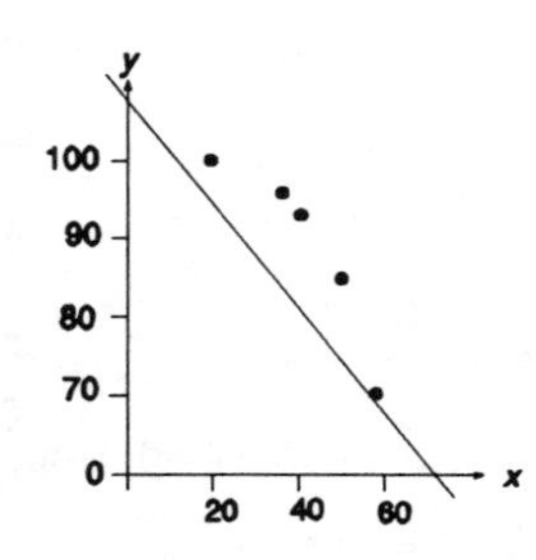

(d) $y = -0.72x + 116.6$

(e)

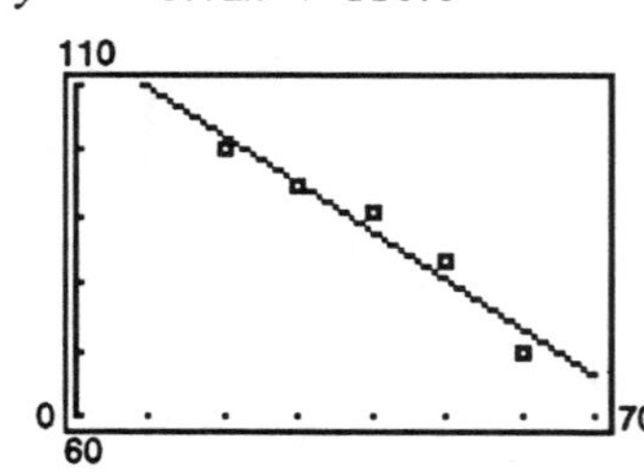

13. (a)

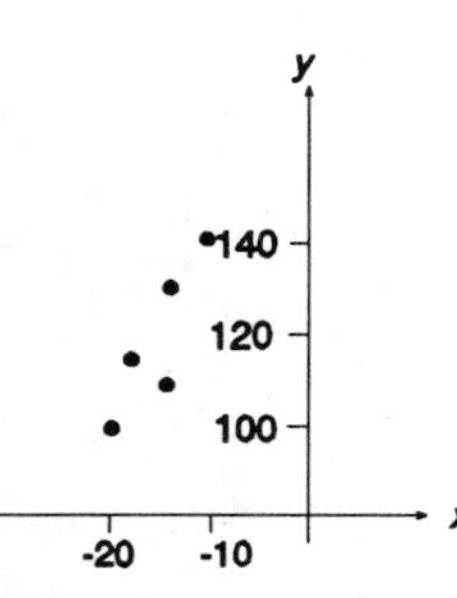

(b) Answers will vary. Using $(-20, 100)$ and $(-10, 140)$, the slope of the line is

$$m = \frac{140 - 100}{-10 - (-20)} = \frac{40}{10} = 4$$

The equation of the line with slope 4 passing through $(-20, 100)$ is

$$y - y_1 = m(x - x_1)$$
$$y - 100 = 4(x - (-20))$$
$$y - 100 = 4x + 80$$
$$y = 4x + 180$$

(c)

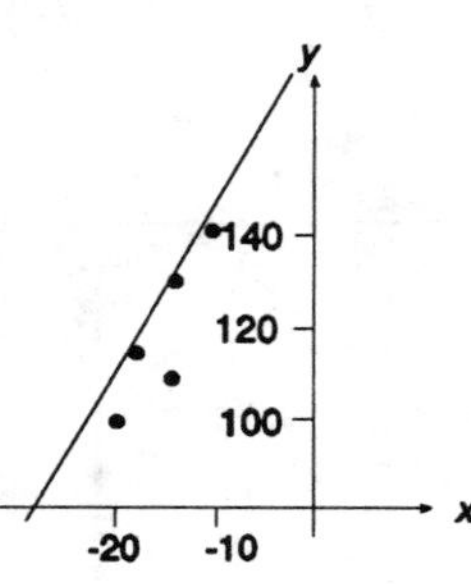

(d) $y = 3.8613x + 180.292$

(e)

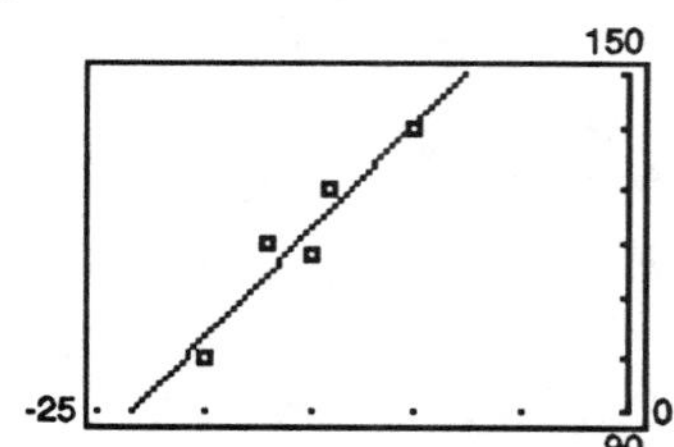

15. (a)

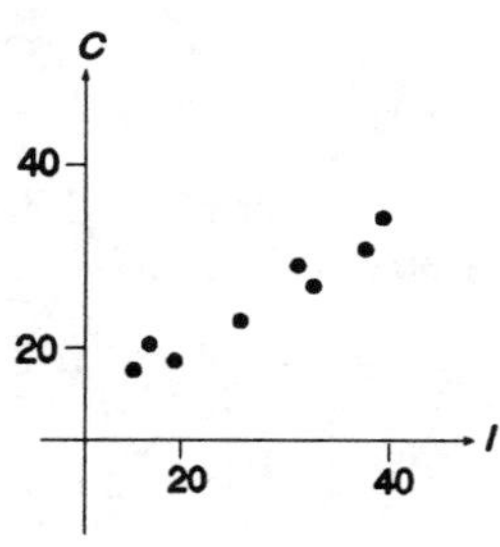

(b) Using $(20, 16)$ and $(50, 39)$, the slope of the line is:

$$m = \frac{39 - 16}{50 - 20} = \frac{23}{30}$$

The equation of the line is

$$y - 16 = \frac{23}{30}(I - 20)$$
$$C - 16 = \frac{23}{30}(I - 20)$$

(c) $C = 0.755I + 0.6266$

(d) As disposable income increases by \$1, consumption increases by about \$0.755.

(e) $C = 0.755(\$42) + 0.6266 = \$32.337$
A family with disposable income of \$42,000 will consume about \$32,337.

17. (a)

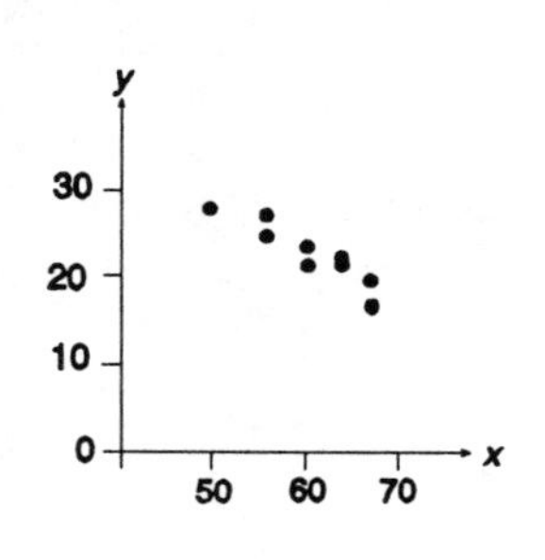

(b) Using $(50, 28)$ and $(65, 15)$, the slope is

$$m = \frac{28 - 15}{50 - 65} = \frac{13}{-15} = \frac{-13}{15}$$

$$y - 28 = \frac{-13}{15}(x - 50)$$

(c) $y = -0.8265x + 70.3903$

(d) As speed increases by 1 mph, mpg decreases by 0.8265.

(e) $y = -0.8265(61) + 70.3903 \approx 20$ mpg

19. (a)

Net Income

150, 130, 110, 90, 70, 0

2000, 2400, 2800 Sales

(b) Using (1912.8, 92.6) and (2378.2, 115.6), $y - 92.6 = 0.05(x - 1912.8)$

(c) NIBT = 0.0842 sales − 88.5776

(d) As sales increase by \$1, NIBT increases by \$0.0842.

(e) NIBT = 0.0842(2456.4) − 88.5776 ≈ \$118.3

21. (a) $y = 1.13x - 2841.69$

(b) If per capita disposable income increases \$1, per capita personal consumption increases \$1.13.

(c) $y = 1.13(14989) - 2841.69 = \$14{,}095.88$

23. (a) $y = 323.42x - 62080.03$

(b) If average fuel consumption per car increases by 1 gallon, pollutant emissions of carbon monoxide increase by 323.42 thousand tons.

(c) $y = 323.42(505) - 62080.03 \approx 101{,}247$ thousand tons

3.6 Variation

1. $y = kx$, $y = 2$ when $x = 10$

$$2 = k(10)$$
$$\frac{1}{5} = k$$
$$y = \frac{1}{5}x$$

3. $A = kx^2$, $A = 4\pi$ when $x = 2$

$$4\pi = k(2)^2$$
$$4\pi = 4k$$
$$\pi = k$$
$$A = \pi x^2$$

5. $F = \dfrac{K}{d^2}$, $F = 10$ when $d = 5$

$$10 = \frac{k}{25}$$
$$250 = k$$
$$F = \frac{250}{d^2}$$

7. $z = k(x^2 + y^2)$, $z = 5$ when $x = 3$, $y = 4$

$$5 = k(9 + 16)$$
$$5 = k(25)$$
$$\frac{1}{5} = k$$
$$z = \frac{1}{5}(x^2 + y^2)$$

9. $M = k\dfrac{d^2}{\sqrt{x}}$, $M = 24$ when $x = 9$, $d = 4$

$$24 = k\left[\frac{16}{\sqrt{9}}\right]$$
$$24 = k\left[\frac{16}{3}\right]$$
$$\frac{9}{2} = k$$
$$M = \frac{9}{2}\left[\frac{d^2}{\sqrt{x}}\right] = \frac{9d^2}{2\sqrt{x}}$$

11. $T^2 = k\dfrac{a^3}{d^2}$, $T = 2$ when $a = 2$, $d = 4$

$$4 = k\frac{8}{16}$$
$$8 = k$$
$$T^2 = 8\frac{a^3}{d^2}$$

13. $V = kr^3$, $k = \dfrac{4\pi}{3}$

$$V = \frac{4\pi}{3}r^3$$

15. $A = k(bh)$, $k = \dfrac{1}{2}$

$$A = \frac{1}{2}bh$$

17. $V = k(r^2h)$, $k = \pi$

$$V = \pi r^2 h$$

19. $F = G\left[\dfrac{mM}{d^2}\right]$, $G = 6.67 \times 10^{-11}$

$$F = 6.67 \times 10^{-11}\left[\frac{mM}{d^2}\right]$$

21.
$$s = kt^2$$
$$16 = k(1^2)$$
$$16 = k$$
$$s = 16t^2$$
$$s = 16(3)^2$$
$$s = 16(9)$$
$$s = 144 \text{ ft.}$$

$$64 = 16t^2$$
$$4 = t^2$$
$$\pm 2 = t$$

time is positive, so $t = 2$ seconds

23. $E = kW$, $E = 3$ when $W = 20$

$$3 = k(20)$$
$$\frac{3}{20} = k$$
$$E = \frac{3}{20}W, \text{ if } W = 15$$
$$E = \frac{3}{20}(15)$$
$$E = \frac{9}{4} = 2\frac{1}{4} = 2.25$$

25. $W = \dfrac{k}{d^2}$, $W = 55$ when $d = 4 \times 10^3$

$$55 = \frac{k}{(4 \times 10^3)^2}$$
$$(4 \times 10^3)^2(55) = k$$
$$(16 \times 10^6)(55) = k$$
$$(880)(10^6) = k$$
$$(8.8)(10^8) = k$$
$$W = \frac{8.8 \times 10^8}{(4.4 \times 10^3)^2}$$

when $d = 4.4 \times 10^3$

$$W = \frac{8.8 \times 10^8}{19.36 \times 10^6}$$
$$W = 0.4545 \times 10^2$$
$$W = 45.45 \text{ pounds}$$

27. $h = k(sd^3), d = 2$

$$h = 36$$
$$s = 75$$
$$36 = k(75)(8)$$
$$36 = 600k$$
$$\frac{36}{600} = k$$
$$0.06 = k$$
$$h = 0.06(sd^3), \qquad d = ? \text{ when } h = 45,\ s = 125$$
$$45 = 0.06(125)d^3$$
$$45 = 7.5d^3$$
$$6 = d^3$$
$$\sqrt[3]{6} = d$$
$$1.8171 = d$$
$$1.82 \text{ inches} \approx d$$

29. $K = k(mv^2) \qquad m = 25 \text{ pounds},\ v = 100 \text{ ft./sec.},\ K = 400 \text{ foot-pounds}$

$$400 = k(25)(100)^2$$
$$400 = 250{,}000k$$
$$0.0016 = k$$
$$K = 0.0016mv^2, \qquad m = 25 \text{ pounds},\ v = 150 \text{ ft./sec.},\ K = ?$$
$$K = 0.0016(25)(150)^2$$
$$K = 900 \text{ ft.-lb.}$$

31. $S = k\frac{(pd)}{t}, \qquad S = 100,\ d = 5,\ t = .75,\ p = 25$

$$100 = k\frac{(25)(5)}{(.75)}$$
$$100 = 166.6667k$$
$$0.5999 = k$$
$$S = (0.5999)\frac{pd}{t} \qquad S = ?,\ p = 48,\ d = 8,\ t = 0.5$$
$$S = (0.5999)\frac{(40)(8)}{0.5}$$
$$S = 383.999$$
$$S = 384 \text{ psi}$$

33. $R = \frac{k\ell}{r^2} \quad \ell = 50,\ r = 6 \times 10^{-3},\ R = 10$

$$10 = \frac{k(50)}{(6 \times 10^{-3})}$$
$$10 = \frac{50}{36 \times 10^{-6}}k$$
$$10 = (1.3888 \times 10^6)k$$
$$7.3 \times 10^{-6} = k$$

$R = \frac{(7.2 \times 10^{-6}\ell)}{r^2} \quad R = ?,\ \ell = 100,\ r = 7 \times 10^{-3}$

$$R = \frac{(7.2 \times 10^{-6})(100)}{(7 \times 10^{-3})^2}$$
$$R = \frac{7.2 \times 10^{-4}}{49 \times 10^{-6}}$$
$$R = 0.1469388 \times 10^2$$
$$R = 14.69 \text{ ohms}$$

35. $v = k\sqrt{r} = \sqrt{g}$

$v = \sqrt{g} \cdot \sqrt{r}$

$v = \sqrt{gr}$

37. $v = \sqrt{g}\sqrt{(3960 + r)}$

$v = \sqrt{g(3960 + r)}$

$r = 100 \qquad g = 79{,}036$

$v = \sqrt{79{,}036(3960 + 100)} = 18{,}001$ mph

39. $v = \sqrt{gr}$

diameter $= 2(3960 + r)$, distance $d =$ diameter $\cdot\ \pi$

$$v = \frac{d}{t} = \frac{2(3960 + r)\pi}{1.5} = 4.1888(3960 + r)$$

$v = 4.1888(3960 + r) = \sqrt{(79{,}036)(3960 + r)}$ (from Problem 37)

$$[4.1888(3960 + r)]^2 = 79{,}036(3960 + r)$$
$$17.546(3960 + r) = 79{,}036$$
$$3960 + r = \frac{79{,}036}{17.546}$$
$$3960 + r = 4504.5$$
$$r \approx 545 \text{ miles}$$

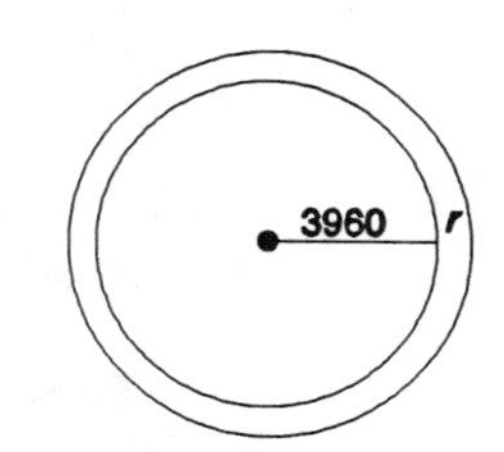

41. $F = \dfrac{mv^2}{r}$

43. $F = \dfrac{m(1.1v)^2}{r} = \dfrac{(1.21)mv^2}{r}$

21% increase

45. $F = \dfrac{m(3v)^2}{r} = \dfrac{9mv^2}{r}$

9 times

3 Chapter Review

1. Slope $= -2$; passing through $(3, -1)$

Use $y - y_1 = m(x - x_1)$

$$y - (-1) = -2(x - 3)$$
$$y + 1 = -2x + 6$$
$$2x + y - 5 = 0 \text{ or } y = -2x + 5$$

3. Slope undefined; passing through $(-3, 4)$.

Use $x = a$ so $x = -3$ or $x + 3 = 0$;

no y-intercept form.

5. y-intercept $= -2$; passing through $(5, -3)$

y-intercept: $(0, -2)$

$$m = \frac{-3 - (-2)}{5 - 0} = \frac{-3 + 2}{5} = -\frac{1}{5}$$
$$y = mx + b$$
$$y = -\frac{1}{5}x - 2$$
$$5y = -x - 10$$
$$x + 5y + 10 = 0 \text{ or } y = -\frac{1}{5}x - 2$$

7. **Parallel to $2x - 3y + 4 = 0$; passing through $(-5, 3)$.**

$$-3y = -2x - 4$$
$$y = \frac{2}{3}x + \frac{4}{3}$$
$$m = \frac{2}{3}$$

Line parallel has same slope.

Use
$$y - y_1 = m(x - x_1)$$
$$y - 3 = \frac{2}{3}[x - (-5)]$$
$$y - 3 = \frac{2}{3}(x + 5)$$
$$y - 3 = \frac{2}{3}x + \frac{10}{3}$$
$$3y - 9 = 2x + 10$$
$$0 = 2x - 3y + 19$$
$$2x - 3y + 19 = 0 \text{ or } y = \frac{2}{3}x + \frac{19}{3}$$

9. **Perpendicular to $x + y - 2 = 0$; passing through $(4, -3)$.**

$$x + y - 2 = 0$$
$$y = -x + 2$$
$$m = -1$$
$$m_{\perp} = 1$$
$$y - (-3) = 1(x - 4)$$
$$y + 3 = x - 4$$
$$0 = x - y - 7$$
$$x - y - 7 = 0$$
$$\text{or } -x + y + 7 = 0 \text{ or } y = x - 7$$

11.
$$4x - 5y + 20 = 0$$
$$-5y = -4x - 20$$
$$y = \frac{4}{5}x + 4$$

x-intercept: $(-5, 0)$
y-intercept: $(0, 4)$

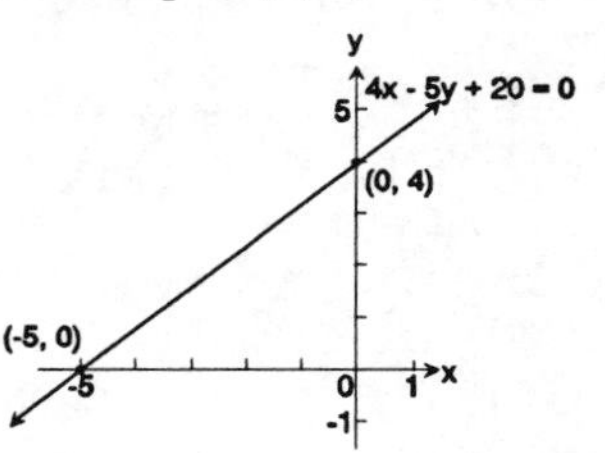

13.
$$\frac{1}{2}x - \frac{1}{3}y + \frac{1}{6} = 0$$
$$-\frac{1}{3}y = -\frac{1}{2}x - \frac{1}{6}$$
$$y = \frac{3}{2}x + \frac{1}{2}$$

Let $y = 0$, so x-intercept: $\left(-\frac{1}{3}, 0\right)$

Let $x = 0$, so y-intercept: $\left(0, \frac{1}{2}\right)$

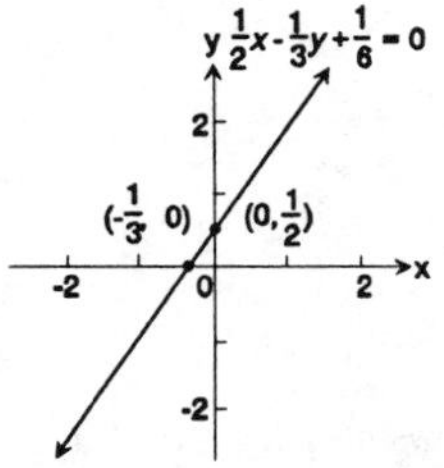

15.
$$\sqrt{2}x + \sqrt{3}y = \sqrt{6}$$
$$\sqrt{3}y = -\sqrt{2}x + \sqrt{6}$$
$$y = \frac{-\sqrt{6}}{3}x + \sqrt{2}$$

Let $y = 0$, so $\sqrt{2}x = \sqrt{6}$
$$x = \frac{\sqrt{6}}{\sqrt{2}}$$
$$= \sqrt{3}$$
x-intercept: $(\sqrt{3}, 0)$

Let $x = 0$, so $\sqrt{3}y = \sqrt{6}$
$$x = \frac{\sqrt{6}}{\sqrt{3}}$$
$$y = \sqrt{2}$$
y-intercept: $(0, \sqrt{2})$

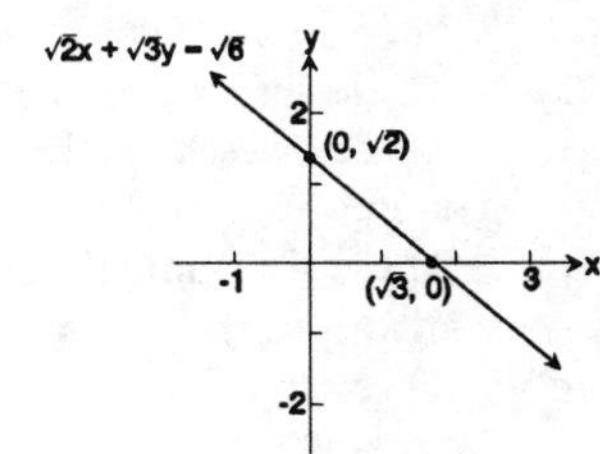

17. $x^2 + (y - 1)^2 = 4$
Center (0, 1)

Radius $= \sqrt{4} = 2$

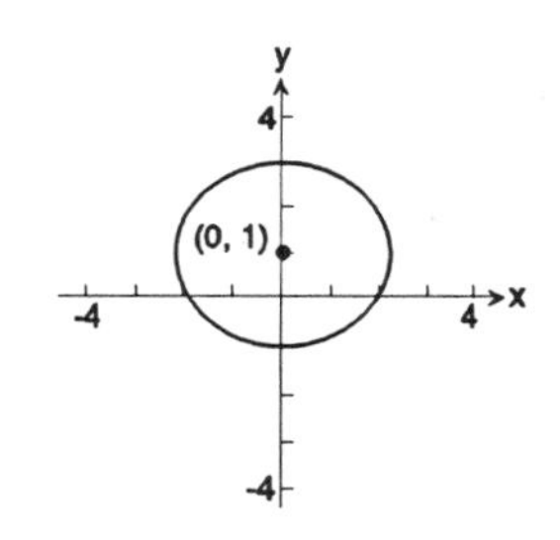

19. $$x^2 + y^2 - 2x + 4y - 4 = 0$$
$$(x^2 - 2x \quad) + (y^2 + 4y \quad) = 4$$
Complete the square twice:
$$(x^2 - 2x + 1) + (y^2 + 4y + 4) = 4 + 1 + 4$$
$$(x - 1)^2 + (y + 2)^2 = 9$$
Center $(1, -2)$

Radius $= \sqrt{9} = 3$

21. $3x^2 + 3y^2 - 6x + 12y = 0$
Divide by 3:
$$x^2 + y^2 - 2x + 4y = 0$$
$$(x^2 - 2x \quad) + (y^2 + 4y \quad) = 0$$
Complete the square twice:
$$(x^2 - 2x + 1) + (y^2 + 4y + 4) = 1 + 4$$
$$(x - 1)^2 + (y + 2)^2 = 5$$
Center $(1, -2)$

Radius $= \sqrt{5}$

23. (7, 4) and $(-3, 2)$

Slope $= \dfrac{2 - 4}{-3 - 7} = \dfrac{-2}{-10} = \dfrac{1}{5}$

Distance $= \sqrt{(-3 - 7)^2 + (2 - 4)^2} = \sqrt{(-10)^2 + (-2)^2} = \sqrt{100 + 4} = \sqrt{104} = 2\sqrt{26}$

Midpoint: $x = \dfrac{7 - 3}{2} = \dfrac{4}{2} = 2$

$y = \dfrac{4 + 2}{2} = \dfrac{6}{2} = 3$

Midpoint = (2, 3)

25. The lines shown are parallel, so they have the same slope. This slope must be a positive number. The y-intercepts of one is positive and the other one is negative.
(a) These lines have unequal slopes.
(b) These lines have the same slope, but it is negative (-1). The graph has lines with positive slope.
(c) These lines might be the ones graphed; each has slope 1; one has a positive y-intercept (2); the other y-intercept is -1.
(d) These lines have the same slope (1), but also have the same y-intercept (2); therefore, not parallel.
(e) These lines have equal slopes, but it is negative $\left(-\frac{1}{2}\right)$.

The only possibility is (c).

27. $2x = 3y^2$

x-intercept: (0, 0)

y-intercept: (0, 0)

Replace y by $(-y)$.

$2x = 3(-y)^2$

$2x = 3y^2$ so symmetric with respect to x-axis.

Replace x by $(-x)$.

$2(-x) = 3y^2$

$-2x = 3y^2$, ***not*** symmetric with respect to y-axis.

Replace x by $(-x)$ and y by $(-y)$.

$2(-x) = 3(-y)^2$

$-2x = 3y^2$, ***not*** symmetric with respect to origin.

29. $x^2 + 4y^2 = 16$

To find x-intercepts, let $y = 0$.

$x^2 = 16$

$x = \pm 4$ (4, 0) and (−4, 0)

To find y-intercepts, let $x = 0$.

$4y^2 = 16$

$y^2 = 4$

$y = \pm 2$ (0, 2) and (0, −2)

Replace y by $(-y)$.

$x^2 + 4(-y)^2 = 16$

$x^2 + 4y^2 = 16$ so symmetric with respect to x-axis

Replace x by $(-x)$.

$(-x)^2 + 4y^2 = 16$

$x^2 + 4y^2 = 16$ so symmetric with respect to y-axis

Replace x by $(-x)$ and y by $(-y)$.

$(-x)^2 + 4(-y)^2 = 16$

$x^2 + 4y^2 = 16$ so symmetric with respect to the origin

31. $y = x^4 + 2x^2 + 1$

To find x-intercept, let $y = 0$.

$x^4 + 2x^2 + 1 = 0$

$(x^2 + 1)^2 = 0$

$x^2 = -1$ no x-intercept

To find y-intercept, let $x = 0$.

$y = 0 + 0 + 1$

$y = 1$ (0, 1)

Replace y by $-y$.

$-y = x^4 + 2x^2 + 1$ so ***not*** symmetric with respect to x-axis

Replace x by $-x$.

$y = (-x)^4 + 2(-x)^2 + 1$

$y = x^4 + 2x^2 + 1$ so symmetric with respect to y-axis

Replace x by $(-x)$ and y by $(-y)$.

$-y = (-x)^4 + 2(-x^2) + 1$

$-y = x^4 + 2x^2 + 1$ so ***not*** symmetric with respect to the origin

33. $x^2 + x + y^2 + 2y = 0$

To find x-intercepts, let $y = 0$.

$$x^2 + x = 0$$
$$x(x + 1) = 0$$
$$x = 0 \text{ or } \quad x + 1 = 0$$
$$x = -1 \qquad (0, 0) \text{ and } (-1,0)$$

To find y-intercept, let $x = 0$

$$y^2 + 2y = 0$$
$$y(y + 2) = 0$$
$$y = 0 \quad \text{or} \quad y = -2 \qquad (0, 0) \text{ and } (0, -2)$$

$(-1, 0), (0, 0), (0, -2)$

Replace x by $-x$.

$$(-x)^2 - x + y^2 + 2y = 0$$
$x^2 - x + y^2 + 2y = 0$ so ***not*** symmetric with respect to x-axis

Replace y by $-y$.

$$x^2 + x + (-y)^2 + 2(-y) = 0$$
$x^2 + x + y^2 - 2y = 0$ so ***not*** symmetric with respect to y-axis

Replace x by $-x$ and y by $-y$.

$$(-x)^2 + (-x) + (-y)^2 + 2(-y) = 0$$
$x^2 - x + y^2 - 2y = 0$ so ***not*** symmetric with respect to the origin

35. $A = ks^2$, $A = \dfrac{\sqrt{3}}{4}$ if $s = 1$

$$\frac{\sqrt{3}}{4} = k(1)^2$$
$$k = \frac{\sqrt{3}}{4}$$
$$A = \frac{\sqrt{3}}{4}s^2, \ A = 16, \ s = ?$$
$$16 = \frac{\sqrt{3}}{4}s^2$$
$$64 = \sqrt{3}s^2$$
$$\frac{64}{\sqrt{3}} = s^2$$
$$\sqrt{\frac{64}{\sqrt{3}}} = s$$
$$\frac{\sqrt{64}}{\sqrt[4]{3}} = s$$
$$\frac{8}{\sqrt[4]{3}} = s$$

6.08 *cm* $\approx s$ (with calculator)

37. $T^2 = ka^3$

for Earth: $a = 93 \times 10^6$, $T = 365$ days

$$k = \frac{T^2}{a^3} = \frac{(365)^2}{(93 \times 10^6)^3}$$

for Mercury: $T = 88$ days

$$a^3 = \frac{T^2}{k} = \frac{(88)^2}{\dfrac{(365)^2}{(93 \times 10^6)^3}}$$
$$a = \sqrt[3]{\frac{(88)^2(93 \times 10^6)^3}{(365)^2}}$$
$$= 36{,}025{,}449 \text{ miles}$$
$$\approx 36 \text{ million miles}$$

39. $A = (3, 4) \quad B = (1, 1) \quad C = (-2, 3)$

$$d(B, C) = \sqrt{(-2 - 1)^2 + (3 - 1)^2} = \sqrt{9 + 4} = \sqrt{13}$$

$$d(A, B) = \sqrt{(3 - 1)^2 + (4 - 1)^2} = \sqrt{4 + 9} = \sqrt{13}$$

Since $BC = AB$, Triangle ABC is isosceles.

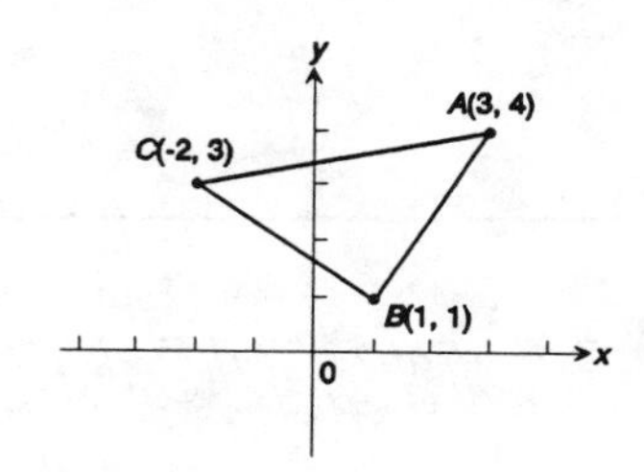

41. $A = (2, 5) \quad B = (6, 1) \quad C = (8, -1)$

$$m_{AB} = \frac{1 - 5}{6 - 2} = \frac{-4}{4} = -1$$

$$m_{BC} = \frac{-1 - 1}{8 - 6} = \frac{-2}{2} = -1$$

43. Endpoints of diameter: $A(-3, 2)$ and $B(5, -6)$.

Midpoint of diameter $= C = \left(\frac{-3 + 5}{2}, \frac{2 - 6}{2}\right) = (1, -2)$

radius $= d(A, C) = \sqrt{(-3 - 1)^2 + (2 - (-2)^2} = \sqrt{16 + 16} = 4\sqrt{2}$

equation:

$$(x - 1)^2 + (y + 2)^2 = \left(4\sqrt{2}\right)^2$$
$$x^2 - 2x + 1 + y^2 + 4y + 4 = 32$$
$$x^2 + y^2 - 2x + 4y - 27 = 0$$

45. (a)

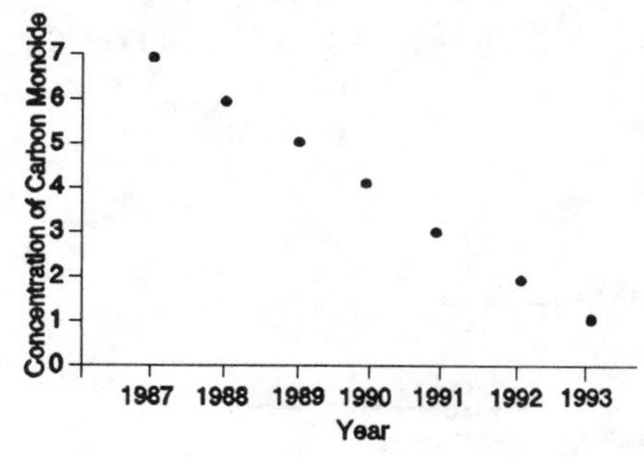

(b) $m = \frac{6.69 - 5.87}{1987 - 1990} = \frac{0.82}{-3}$
$= -0.27$

(c) Between the years 1987 and 1990, as x increases by 1 year, the level of carbon monoxide decreases by 0.27 ppm.

(d) $m = \frac{5.87 - 4.88}{1990 - 1993} = \frac{0.99}{-3} = -0.33$

(e) Between the years 1990 and 1993, as x increases by 1 year, the level of carbon monoxide decreases by 0.33 ppm.

(f) $m \approx -0.308$

(g) Between the years 1987 and 1993, as x increases by 1 year, the level of carbon monoxide decreases by 0.308 ppm.

Chapter 4

FUNCTIONS AND THEIR GRAPHS

4.1 Functions

1. Function
3. Not a function
5. Function
7. Function
9. Not a Function
11. Function

13. $f(x) = -3x^2 + 2x - 4$

(a) $f(0) = -3(0)^2 + 2(0) - 4 = -4$

(b) $f(1) = -3(1)^2 + 2(1) - 4 = -3 + 2 - 4 = -5$

(c) $f(-1) = -3(-1)^2 + 2(-1) - 4 = -3 - 2 - 4 = -9$

(d) $f(-x) = -3(-x)^2 + 2(-x) - 4 = -3x^2 - 2x - 4$

(e) $-f(x) = -(-3x^2 + 2x - 4) = 3x^2 - 2x + 4$

(f) $f(x + 1) = -3(x + 1)^2 + 2(x + 1) - 4 = -3(x^2 + 2x + 1) + 2x + 2 - 4 = -3x^2 - 6x - 3 + 2x + 2 - 4 = -3x^2 - 4x - 5$

15. $f(x) = \dfrac{x}{x^2 + 1}$

(a) $f(0) = \dfrac{0}{0^2 + 1} = 0$

(b) $f(1) = \dfrac{1}{1^2 + 1} = \dfrac{1}{2}$

(c) $f(-1) = \dfrac{-1}{(-1)^2 + 1} = -\dfrac{1}{2}$

(d) $f(-x) = \dfrac{-x}{(-x)^2 + 1} = \dfrac{-x}{x^2 + 1}$

(e) $-f(x) = -\dfrac{x}{x^2 + 1} = \dfrac{-x}{x^2 + 1}$

(f) $f(x + 1) = \dfrac{x + 1}{(x + 1)^2 + 1} = \dfrac{x + 1}{x^2 + 2x + 1 + 1} = \dfrac{x + 1}{x^2 + 2x + 2}$

17. $f(x) = |x| + 4$

(a) $f(0) = |0| + 4 = 4$

(b) $f(1) = |1| + 4 = 5$

(c) $f(-1) = |-1| + 4 = 1 + 4 = 5$

(d) $f(-x) = |-x| + 4 = |x| + 4$

(e) $-f(x) = -(|x| + 4) = -|x| - 4$

(f) $f(x + 1) = |x + 1| + 4$

19. $f(x) = \dfrac{2x + 1}{3x - 5}$

(a) $f(0) = \dfrac{2(0) + 1}{3(0) - 5} = \dfrac{1}{-5} = -\dfrac{1}{5}$

(b) $f(1) = \dfrac{2(1) + 1}{3(1) - 5} = \dfrac{3}{-2} = -\dfrac{3}{2}$

(c) $f(-1) = \dfrac{2(-1) + 1}{3(-1) - 5} = \dfrac{-2 + 1}{-3 - 5} = \dfrac{-1}{-8} = \dfrac{1}{8}$

(d) $f(-x) = \dfrac{2(-x) + 1}{3(-x) - 5} = \dfrac{-2x + 1}{-3x - 5}$

(e) $-f(x) = -\dfrac{2x + 1}{3x - 5} = \dfrac{-2x - 1}{3x - 5}$

(f) $f(x + 1) = \dfrac{2(x + 1) + 1}{3(x + 1) - 5} = \dfrac{2x + 2 + 1}{3x + 3 - 5} = \dfrac{2x + 3}{3x - 2}$

21. $f(0) = 3$ since $(0, 3)$ is on graph
$f(-6) = -3$ since $(-6, -3)$ is on graph

23. $f(2) =$ is positive since $f(2) = 4$

25. $f(x) = 0$ when $x = -3$
$x = 6$
$x = 10$

27. Domain: $[-6, 11]$ or $\{x \mid -6 \le x \le 11\}$

29. The x-intercepts are -3, 6, 10.

31. 3 times

33. $f(x) = \dfrac{x + 2}{x - 6}$

(a) $14 \stackrel{?}{=} \dfrac{3 + 2}{3 - 6}$

$14 \stackrel{?}{=} \dfrac{5}{-3}$

No, $(3, 14)$ is not on the graph of f.

(b) $f(4) = \dfrac{4 + 2}{4 - 6} = \dfrac{6}{-2} = -3$

$(4, -3)$ is on the graph.

(c) $2 = \dfrac{x + 2}{x - 6}$

$2x - 12 = x + 2$

$x = 14$

$(14, 2)$ is on the graph.

(d) Domain of $f = \{x \mid x \neq 6\}$

35. $f(x) = \dfrac{2x^2}{x^4 + 1}$

(a) $1 \stackrel{?}{=} \dfrac{2(-1)^2}{(-1)^4 + 1}$

$1 \stackrel{?}{=} \dfrac{2}{2}$

$1 = 1$

Yes, $(-1, 1)$ is on the graph of f.

(b) $f(2) = \dfrac{2(2)^2}{(2)^4 + 1} = \dfrac{8}{17}$

$\left(2, \dfrac{8}{17}\right)$ is on the graph.

(c) $1 = \dfrac{2x^2}{x^4 + 1}$

$x^4 + 1 = 2x^2$

$x^4 - 2x^2 + 1 = 0$

$(x^2 - 1)^2 = 0$

$x = \pm 1$

$(-1, 1)$ and $(1, 1)$ are on the graph.

(d) Domain of
$f = \{x \mid x \in \text{Real Numbers}\}$

37. Not a function since there are vertical lines that intersect the graph in more than one point.

39. Function (a) Domain: $\{x \mid -\pi \le x \le \pi\}$; Range: $\{y \mid -1 \le y \le 1\}$

(b) $\left(-\dfrac{\pi}{2}, 0\right)$, $\left(\dfrac{\pi}{2}, 0\right)$, $(0, 1)$

(c) y-axis

41. Not a function since there are vertical lines that intersect the graph in more than one point.

43. Function (a) Domain: $\{x \mid 0 < x < \infty\}$; Range: all real numbers

(b) $(1, 0)$ There are no y-intercepts

(c) None

45. Function (a) Domain: all real numbers; Range: $\{y \mid -\infty < y \leq 2\}$
(b) $(-3, 0)$, $(3, 0)$, $(0, 2)$
(c) y-axis

47. Function (a) Domain: $\{x \mid x \neq 2\}$; Range: $\{y \mid y \neq 1\}$
(b) $(0, 0)$
(c) None

49. $f(x) = 3x + 4$
all real numbers

51. $f(x) = \dfrac{x}{x^2 + 1}$
all real numbers

53. $g(x) = \dfrac{x}{x^2 - 1}$ $\{x \mid x \neq -1, x \neq 1\}$

$x^2 - 1 \neq 0$
$x^2 \neq 1$
$x \neq \pm 1$

55. $F(x) = \dfrac{x - 2}{x^3 + x}$ $\{x \mid x \neq 0\}$

$x^3 + x \neq 0$
$x(x^2 + 1) \neq 0$
$x \neq 0 \quad x^2 \neq -1$

57. $h(x) = \sqrt{3x - 12}$ $\{x \mid x \geq 4\}$

$3x - 12 \geq 0$
$3x \geq 12$
$x \geq 4$

59. $f(x) = \dfrac{4}{\sqrt{x - 9}}$

$x - 9 > 0$
$x > 9$
Hence, $(9, \infty)$.

61. $p(x) = \sqrt{\dfrac{x - 2}{x - 1}}$

$\dfrac{x - 2}{x - 1} \geq 0$

Interval	Test Number	$f(x) = \dfrac{x-2}{x-1}$	Positive/Negative
$x < 1$	0	$f(0) = 2$	+
$1 < x < 2$	1.5	$f(1.5) = -1$	−
$x > 2$	3	$\frac{1}{2}$	+

Domain: $\{x \mid x < 1 \text{ or } x \geq 2\}$ or $(-\infty, 1)$ or $[2, \infty)$

63. (a) III
(b) IV
(c) I
(d) V
(e) II

65.

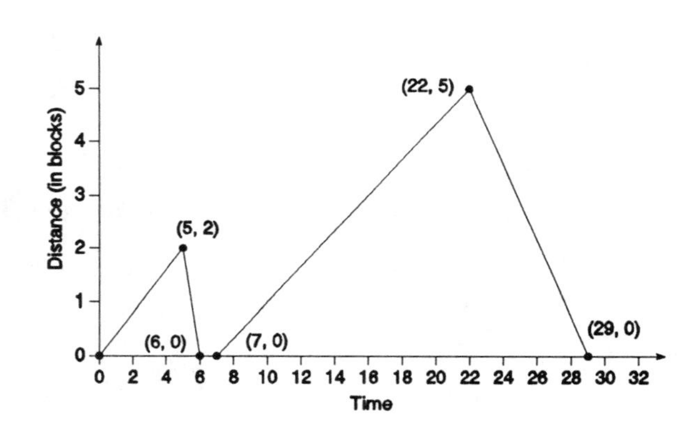

67. $f(x) = 2x^3 + Ax^2 + 4x - 5$ and $f(2) = 5$

$$f(2) = 2(2)^3 + A(2)^2 + 4(2) - 5 = 5$$
$$16 + 4A + 8 - 5 = 5$$
$$4A + 19 = 5$$
$$4A = -14$$
$$A = -\frac{7}{2}$$

69. $f(x) = \dfrac{3x + 8}{2x - A}$ and $f(0) = 2$

$$f(0) = \frac{3(0) + 8}{2(0) - A} = 2$$
$$\frac{8}{-A} = \frac{2}{1} \quad \text{(cross multiply)}$$
$$-2A = 8$$
$$A = -4$$

71. $f(x) = \dfrac{2x - A}{x - 3}$ and $f(4) = 0$

$$f(4) = \frac{2(4) - A}{4 - 3} = 0$$
$$\frac{8 - A}{1} = 0$$
$$8 - A = 0$$
$$8 = A$$

Since $x - 3 = 0$ when $x = 3$, then f is not defined at 3.

73. (a) No, because 23 corresponds to two different values in the range (56 and 53).

(b)

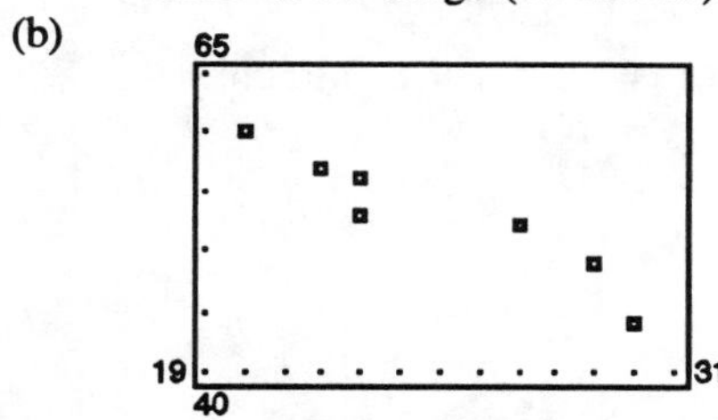

(c) $(p_1, D_1) = (20, 60)$; $(p_2, D_2) = (30, 44)$

$$m = \frac{D_2 - D_1}{p_2 - p_1} = \frac{44 - 60}{30 - 20} = \frac{-16}{10} = -1.6$$
$$D - D_1 = m(p - p_1)$$
$$D - 60 = -1.6(p - 20)$$
$$D - 60 = -1.6p + 32$$
$$D = -1.6p + 92$$

(d) If p increases \$1, quantity demanded decreases by 1.6.

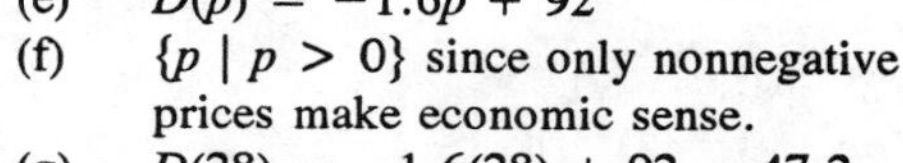

(e) $D(p) = -1.6p + 92$

(f) $\{p \mid p > 0\}$ since only nonnegative prices make economic sense.

(g) $D(28) = -1.6(28) + 92 = 47.2$. About 47 pairs of jeans.

(h) $D = -1.34p + 86.20$

75. (a) Yes, since each unique element in the domain corresponds to one element in the range.

(b)

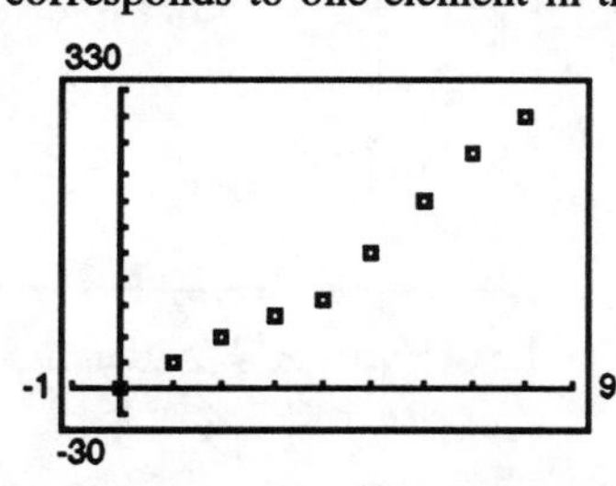

(c) Using (0, 0) and (8, 300)

$$m = \frac{300 - 0}{8 - 0} = \frac{300}{8} = 37.5$$

y-intercept: 0

$s = 37.5t + 0 = 37.5t$

(d) If t increases by 1 hour, distance increases by 37.5 miles.

(e) $s(t) = 37.5t$

(f) $\{t \mid t \geq 0\}$ since time must be non-negative.

(g) $s(11) = 37.5(11) = 412.5$ miles

(h) $s(t) = 37.78t - 19.13$

77. (a) $H(x) = 20 - 4.9x^2$

$x = 1$

$H(1) = 20 - 4.9(1)^2 = 20 - 4.9 = 15.1\ m$

$x = 1.1$

$H(1.1) = 20 - 4.9(1.1)^2 = 20 - 4.9(1.21) = 20 - 5.929 = 14.071 = 14.07\ m$

$x = 1.2$

$H(1.2) = 20 - 4.9(1.2)^2 = 20 - 4.9(1.44) = 20 - 7.056 = 12.944 = 12.94\ m$

$x = 1.3$

$H(1.3) = 20 - 4.9(1.3)^2 = 20 - 4.9(1.69) = 20 - 8.281 = 11.719 = 11.72\ m$

(b) $H(x) = 15$, $15 = 20 - 4.9x^2$, $-5 = -4.9x^2$, $x = 1.01$ seconds

$H(x) = 10$, $10 = 20 - 4.9x^2$, $-10 = -4.9x^2$, $x = 1.42$ seconds

$H(x) = 5$, $5 = 20 - 4.9x^2$, $-15 = -4.9x^2$, $x = 1.74$ seconds

(c) The rock strikes the ground when $H = 0$.

$$H(x) = 20 - 4.9x^2 = 0$$
$$20 = 4.9x^2$$
$$4.0816 = x^2$$
$$\sqrt{4.0816} = x$$
$$2.02 \text{ sec} = x$$

79. $\ell = x$

$x = 2w$ or $\left(\frac{x}{2}\right) = w$

$A = \ell \cdot w$

$A(x) = x\left(\frac{x}{2}\right) = \frac{x^2}{2} = \frac{1}{2}x^2$

81. $G(x) = (\text{amt. per hr.})(\text{no. of hrs.})$

$G(x) = 10x$

85. (a) The total cost of installing cable along the road is $10x$. If cable is installed x miles along the road, there are $5 - x$ miles left from the road to the house and where the cable ends. Therefore, using the Pythagorean Theorem, there is $\sqrt{(5 - x)^2 + 2^2} = \sqrt{25 - 10x + x^2 + 4} = \sqrt{x^2 - 10x + 29}$ miles of cable installed off the road. The total cost of installation is:

$$C(x) = 10x + 14\sqrt{x^2 - 10x + 29}$$

(b) $C(1) = 10(1) + 14\sqrt{(1)^2 - 10(1) + 29}$
$= 10 + 14\sqrt{20}$
$= 10 + 62.61$
$= \$72.61$

(c) $C(3) = 10(3) + 14\sqrt{(3)^2 - 10(3) + 29}$
$= 30 + 14\sqrt{8}$
$= 30 + 39.60$
$= \$69.60$

(d)

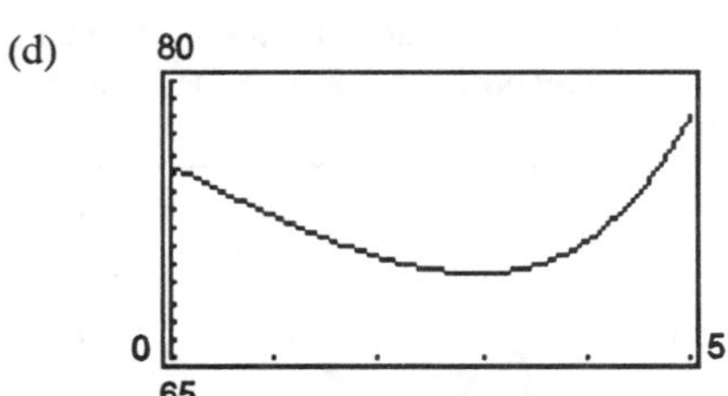

(e) Least cost: $x = 2.95$ miles

87. (a) $A(x) = (8.5 - 2x)(11 - 2x)$

(b) Domain: $0 \le x \le 4.25$

Range: $0 \le A \le 93.5$

(c) $A(1) = (8.5 - 2)(11 - 2)$
$= 58.5$ square inches

$A(1.2) = (8.5 - 2.4)(11 - 2.4)$
$= (6.1)(8.6)$
$= 52.46$ square inches

$A(1.5) = (8.5 - 3)(11 - 3)$
$= 44$ square inches

(d)

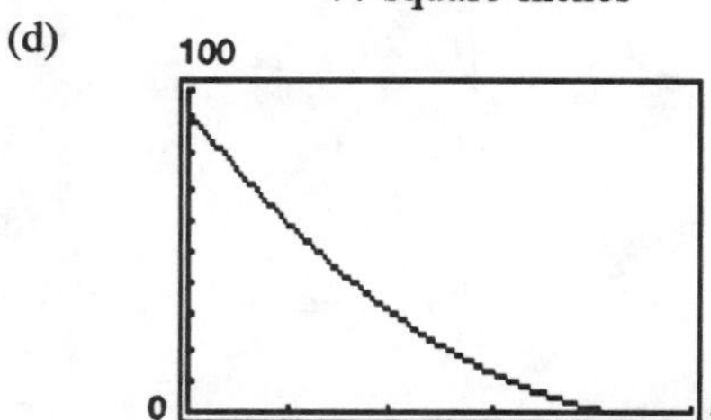

(e) $A(x) = 70$ when
$x = 0.64$ inches
$A(x) = 50$ when
$x = 1.28$ inches

89. (a)

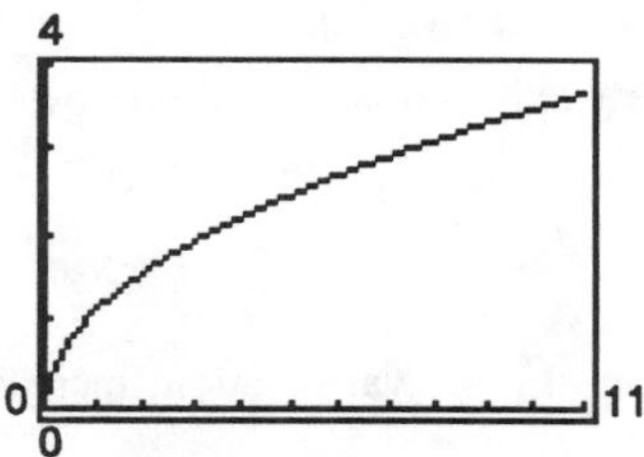

(b) As ℓ changes from 1 to 10, T varies from 1.1 to 3.5.

(c)
$$2\pi\sqrt{\frac{\ell}{32.2}} = 10$$
$$\sqrt{\frac{\ell}{32.2}} = \frac{5}{\pi}$$
$$\frac{\ell}{32.2} = \frac{25}{\pi^2}$$
$$\ell = \frac{25}{\pi^2} \cdot 32.2 \approx 81.56 \text{ feet}$$

91. $y = x^2 + 2x$

Graph $y_1 = x^2 + 2x$. The graph passes the vertical line test. Thus, the set (x, y) such that $y = x^2 + 2x$ represents a *function*.

93. $y = \dfrac{2}{x}$

The graph passes the vertical line test. Therefore, the set (x, y) such that $y = \dfrac{2}{x}$ represents a *function*.

95. $y^2 = 1 - x^2$

Solve for y: $y = \pm\sqrt{1 - x^2}$

For $x = 0$, $y = \pm 1$

Thus, $(0, 1)$ and $(0, -1)$ is on the graph. *Not a function* since a distinct x $(x = 0)$ corresponds to two different y's.

97. $x^2 + y = 1$

The graph passes the vertical line test. Therefore, the set (x, y) such that $x^2 + y = 1$ represents a *function*.

99. (a) $h(x) = 2x$

$h(a + b) \stackrel{?}{=} h(a) + h(b)$

$h(a + b) = 2(a + b) = 2a + 2b = h(a) + h(b)$

(b) $g(x) = x^2$

$g(a + b) \stackrel{?}{=} g(a) + g(b)$

$g(a + b) = (a + b)^2 + a^2 + 2ab + b^2 \neq a^2 + b^2 = g(a) + g(b)$

(c) $F(x) = 5x - 2$

$$F(a + b) \stackrel{?}{=} F(a) + F(b)$$

$$F(a + b) = 5(a + b) - 2 = 5a + 5b - 2 \neq 5a - 2 + 5b - 2 = 5a + 5b - 4$$
$$= F(a) + F(b)$$

(d) $G(x) = \dfrac{1}{x}$

$$G(a + b) \stackrel{?}{=} G(a) + G(b)$$

$$G(a + b) = \frac{1}{a + b} \neq \frac{1}{a} + \frac{1}{b} = G(a) + G(b)$$

4.2 More about Functions

1. C 3. E 5. B 7. F

9. (a) Domain: $\{x \mid -3 \leq x \leq 4\}$; Range: $\{y \mid 0 \leq y \leq 3\}$
   (b) In interval notation, increasing on $(-3, 0)$ and on $(2, 4)$; and decreasing on $(0, 2)$. In inequality notation, increasing on $-3 < x < 0$ and on $2 < x < 4$ and decreasing on $0 < x < 2$.
   (c) Since the graph is not symmetric with respect to the y-axis and is not symmetric with respect to the origin, it is NEITHER even nor odd.
   (d) The intercepts are $(-3, 0)$, $(0, 3)$, $(2, 0)$.

11. (a) Domain: all real numbers; Range: $\{y \mid 0 < y < \infty\}$
   (b) In inequality notation, increasing on $-\infty < x < \infty$. In interval notation, increasing on $(-\infty, \infty)$.
   (c) Since the graph is neither symmetric with respect to the y-axis nor the origin, it is NEITHER even nor odd.
   (d) The intercept is $(0, 1)$.

13. (a) Domain: $\{x \mid -\pi \leq x \leq \pi\}$; Range: $\{y \mid -1 \leq y \leq 1\}$
   (b) In interval notation, increasing on $\left(-\dfrac{\pi}{2}, \dfrac{\pi}{2}\right)$; and decreasing on $\left(-\pi, -\dfrac{\pi}{2}\right)$ and on $\left(\dfrac{\pi}{2}, \pi\right)$. In inequality notation, increasing on $-\dfrac{\pi}{2} < x < \dfrac{\pi}{2}$; and decreasing on $-\pi < x < -\dfrac{\pi}{2}$ and $\dfrac{\pi}{2} < x < \pi$.
   (c) Since the graph is symmetric with respect to the origin, the graph is ODD.
   (d) The intercepts are $(-\pi, 0)$, $(0, 0)$, $(\pi, 0)$.

15. (a) Domain: $\{x \mid x \neq 2\}$; Range: $\{y \mid y \neq 1\}$
   (b) In interval notation, decreasing on $(-\infty, 2)$ and on $(2, \infty)$. In inequality notation, decreasing on $-\infty < x < 2$ and on $2 < x < \infty$.
   (c) Since the graph is neither symmetric with respect to the y-axis nor symmetric with respect to the origin, it is NEITHER even nor odd.
   (d) The intercept is $(0, 0)$.

17. (a) Domain: $\{x \mid x \neq 0\}$; Range: all real numbers
   (b) In interval notation, increasing on $(-\infty, 0)$ and $(0, \infty)$. In inequality notation, increasing on $-\infty < x < 0$ and on $0 < x < \infty$.
   (c) Since the graph is symmetric with respect to the origin, the graph is ODD.

(d) The intercepts are $(-1, 0)$ and $(1, 0)$.

19. (a) Domain: $\{x \mid x \neq -2, x \neq 2\}$; Range: $\{y \mid -\infty < y \leq 0 \text{ and } 1 < y < \infty\}$.
(b) In interval notation, increasing on $(-\infty, -2)$ and on $(-2, 0)$, and decreasing on $(0, 2)$ and on $(2, \infty)$. In inequality notation, increasing on $-\infty < x < -2$ and on $-2 < x < 0$ and decreasing on $0 < x < 2$ and on $2 < x < \infty$.
(c) Since the graph is symmetric with respect to the y-axis, it is EVEN.
(d) The intercept is $(0, 0)$.

21. (a) Domain: $\{x \mid -4 \leq x \leq 4\}$; Range: $\{y \mid 0 \leq y \leq 2\}$
(b) In interval notation, increasing on $(-2, 0)$ and $(2, 4)$ and decreasing on $(-4, -2)$ and $(0, 2)$. In inequality notation, increasing on $-2 < x < 0$ and $2 < x < 4$ and decreasing on $-4 < x < -2$ and $0 < x < 2$.
(c) Since the graph is symmetric with respect to the y-axis, it is even.
(d) The intercepts are $(-2, 0)$, $(0, 2)$ and $(2, 0)$.

23. (a) $f(-2) = (-2)^2 = 4$ since $-2 < 0$
(b) $f(0) = 2$ since $x = 0$
(c) $f(2) = 2(2) + 1 = 5$ since $2 > 0$

25. (a) $f(1.2) = \text{int}(2(1.2)) = \text{int}(2.4) = 2$
(b) $f(1.6) = \text{int}(2(1.6)) = \text{int}(3.2) = 3$
(c) $f(-1.8) = \text{int}(2(-1.8)) = \text{int}(-3.6) = -4$

27. $$\frac{f(x) - f(1)}{x - 1} = \frac{3x - 3}{x - 1} = \frac{3(x - 1)}{x - 1} = 3$$

29. $$\frac{f(x) - f(1)}{x - 1} = \frac{(1 - 3x) - (-2)}{x - 1} = \frac{3 - 3x}{x - 1} = \frac{-3(x - 1)}{x - 1} = -3$$

31. $$\frac{f(x) - f(1)}{x - 1} = \frac{(3x^2 - 2x) - (1)}{x - 1} = \frac{3x^2 - 2x - 1}{x - 1} = \frac{(3x + 1)(x - 1)}{x - 1} = 3x + 1$$

33. $$\frac{f(x) - f(1)}{x - 1} = \frac{(x^3 - x) - (0)}{x - 1} = \frac{x^3 - x}{x - 1} = \frac{x(x - 1)(x + 1)}{x - 1} = x(x + 1)$$

35. $$\frac{f(x) - f(1)}{x - 1} = \frac{\frac{2}{x + 1} - 1}{x - 1} = \frac{\frac{2 - (x + 1)}{x + 1}}{x - 1} = \frac{1 - x}{(x + 1)(x - 1)} = \frac{-1(x - 1)}{(x + 1)(x - 1)} = \frac{-1}{x + 1}$$

37. $$\frac{f(x) - f(1)}{x - 1} = \frac{\sqrt{x} - 1}{x - 1} = \frac{\left(\sqrt{x} - 1\right)}{\left(\sqrt{x} - 1\right)\left(\sqrt{x} + 1\right)} = \frac{1}{\sqrt{x} + 1}$$

39. $f(x) = 4x^3$
odd: $f(-x) = -f(x)$
$4(-x)^3 = -(4x^3)$
$-4x^3 = -4x^3$

41. $g(x) = 2x^2 - 5$
even: $f(-x) = f(x)$
$2(-x)^2 - 5 = 2x^2 - 5$
$2x^2 - 5 = 2x^2 - 5$

43. $F(x) = \sqrt[3]{x}$

odd: $f(-x) = -f(x)$

$$\sqrt[3]{-x} = -\sqrt[3]{x}$$
$$-\sqrt[3]{x} = -\sqrt[3]{x}$$

45. $f(x) = x + |x|$

even: $f(-x) = f(x)$

$$-x + |-x| = x + |x|$$
$$-x + |x| \neq x + |x|$$

odd: $f(-x) = -f(x)$

$$-x + |-x| = -(x + |x|)$$
$$-x + |x| \neq -x - |x|$$

neither

47. $g(x) = \dfrac{1}{x^2}$

even: $f(-x) = f(x)$

$$\frac{1}{(-x)^2} = \frac{1}{x^2}$$
$$\frac{1}{x^2} = \frac{1}{x^2}$$

49. $h(x) = \dfrac{x^3}{3x^2 - 9}$

odd: $f(-x) = -f(x)$

$$\frac{(-x)^3}{3(-x)^2 - 9} = -\frac{x^3}{3x^2 - 9}$$
$$\frac{-x^3}{3x^2 - 9} = \frac{-x^3}{3x^2 - 9}$$

51. One at most because if f is increasing it could only cross the x-axis at most one time. It could not "turn" and cross it again or it would start to decrease.

53. $f(x) = \begin{cases} 2x & \text{if } x \neq 0 \\ 0 & \text{if } x = 0 \end{cases}$

(a) The domain is all real numbers.

(b) x-intercept(s): $0 = 2x$, $0 = x$ $\quad$ y-intercept: $y = 0$

The intercept is $(0, 0)$.

(c)

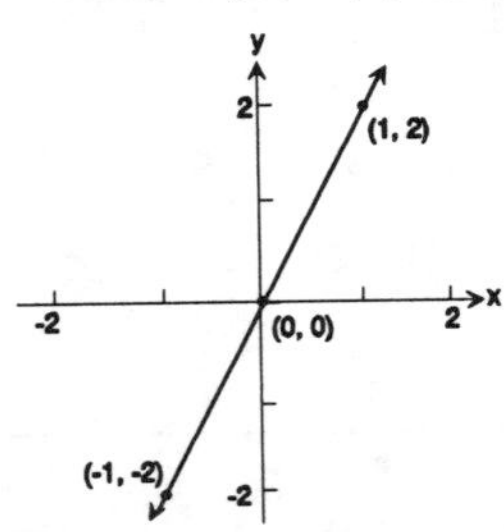

(d) The range is all real numbers.

55. $f(x) = \begin{cases} 1 + x & \text{if } x < 0 \\ x^2 & \text{if } x \geq 0 \end{cases}$

(a) The domain is all real numbers.

(b) x-intercept(s): $\quad$ y-intercept:

$0 = 1 + x$

or $0 = x^2$ $\qquad$ $y = 0^2$

$-1 = x$

or $x = 0$ $\qquad$ $y = 0$

The intercepts are $(-1, 0)$, $(0, 0)$.

(c)

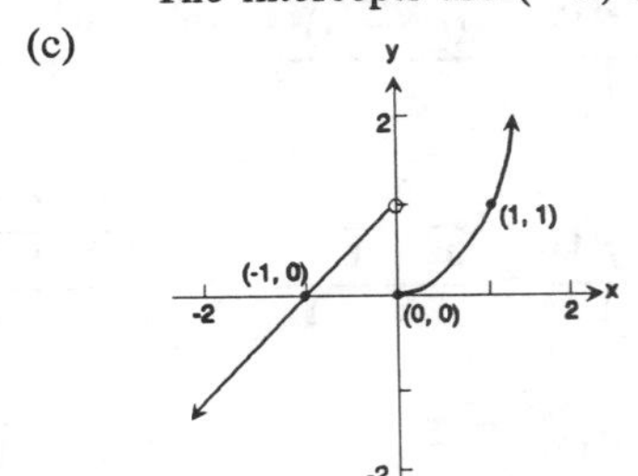

(d) The range is all real numbers.

57. $f(x) = \begin{cases} |x| & \text{if } -2 \leq x < 0 \\ 1 & \text{if } x = 0 \\ x^3 & \text{if } x > 0 \end{cases}$

(a) The domain is $\{x \mid -2 \leq x < \infty\}$

(b) x-intercept(s): None $\quad$ y-intercept: $y = 1$ if $x = 0$

The intercept is $(0, 1)$.

(c)

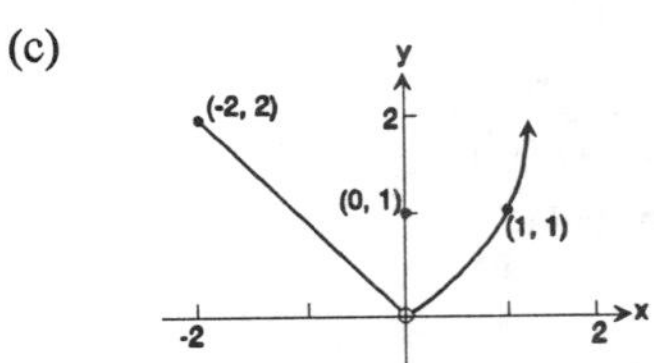

(d) The range is $\{y \mid 0 < y < \infty\}$.

59. $f(x) = 2x + 5$

(a) $f(-x) = 2(-x) + 5 = -2x + 5$ (b) $-f(x) = -(2x + 5) = -2x - 5$

(c) $f(2x) = 2(2x) + 5 = 4x + 5$

(d) $f(x - 3) = 2(x - 3) + 5 = 2x - 6 + 5 = 2x - 1$

(e) $f\left[\frac{1}{x}\right] = 2\left[\frac{1}{x}\right] + 5 = \frac{2}{x} + 5 = \frac{2}{x} + \frac{5x}{x} = \frac{5x + 2}{x}$

(f) $\frac{1}{f(x)} = \frac{1}{2x + 5}$

61. $f(x) = 2x^2 - 4$

(a) $f(-x) = 2(-x)^2 - 4 = 2x^2 - 4$ (b) $-f(x) = -(2x^2 - 4) = -2x^2 + 4$

(c) $f(2x) = 2(2x)^2 - 4 = 8x^2 - 4$

(d) $f(x - 3) = 2(x - 3)^2 - 4 = 2(x^2 - 6x + 9) - 4$
$= 2x^2 - 12x + 18 - 4 = 2x^2 - 12x + 14$

(e) $f\left[\frac{1}{x}\right] = 2\left[\frac{1}{x}\right]^2 - 4 = \frac{2}{x^2} - 4 = \frac{2}{x^2} - \frac{4x^2}{x^2} = \frac{2 - 4x^2}{x^2}$

(f) $\frac{1}{f(x)} = \frac{1}{2x^2 - 4}$

63. $f(x) = x^3 - 3x$

(a) $f(-x) = (-x)^3 - 3(-x) = -x^3 + 3x$ (b) $-f(x) = -(x^3 - 3x) = -x^3 + 3x$

(c) $f(2x) = (2x)^3 - 3(2x) = 8x^3 - 6x$

(d) $f(x - 3) = (x - 3)^3 - 3(x - 3) = x^3 - 9x^2 + 27x - 27 - 3x + 9$
$= x^3 - 9x^2 + 24x - 18$

(e) $f\left[\frac{1}{x}\right] = \left[\frac{1}{x}\right]^3 - 3\left[\frac{1}{x}\right] = \frac{1}{x^3} - \frac{3}{x} = \frac{1}{x^3} - \frac{3x^2}{x^3} = \frac{1 - 3x^2}{x^3}$

(f) $\frac{1}{f(x)} = \frac{1}{x^3 - 3x}$

65. $f(x) = |x|$

(a) $f(-x) = |-x| = |x|$

(b) $-f(x) = -|x|$

(c) $f(2x) = |2x| = 2|x|$

(d) $f(x - 3) = |x - 3|$

(e) $f\left(\frac{1}{x}\right) = \left|\frac{1}{x}\right| = \frac{1}{|x|}$

(f) $\frac{1}{f(x)} = \frac{1}{|x|}$

67. $\frac{f(x + h) - f(x)}{h} = \frac{2(x + h) + 5 - (2x + 5)}{h} = \frac{2x + 2h + 5 - 2x - 5}{h} = \frac{2h}{h} = 2$

69. $\dfrac{f(x+h)-f(x)}{h} = \dfrac{(x+h)^2 + 2(x+h) - (x^2+2x)}{h}$

$= \dfrac{x^2 + 2xh + h^2 + 2x + 2h - x^2 - 2x}{h} = \dfrac{2xh + h^2 + 2h}{h}$

$= 2x + h + 2$

71. $f(x) = \begin{cases} -x & \text{if } -1 \le x \le 0 \\ \frac{1}{2}x & \text{if } 0 < x \le 2 \end{cases}$

Other answers are possible.

73. $f(x) = \begin{cases} -x & \text{if } x \le 0 \\ 2 - x & \text{if } 0 < x \le 2 \end{cases}$

Other answers are possible.

75. $f(x) = \begin{cases} x^2 + 4 & \text{if } x \ne 2 \\ 6 & \text{if } x = 2 \end{cases}$

To see if f is even, we need to show that $f(x) = f(-x)$ for all possible values of x.

$f(-x) = (-x)^2 + 4 = x^2 + 4 = f(x)$

However, when $x = -2$,

$f(-2) = (-2)^2 + 4 = 8$, but

$f(2) = 6$

Because $f(-2) \ne f(2)$, the function is not even.

77. (a), (b), (e)

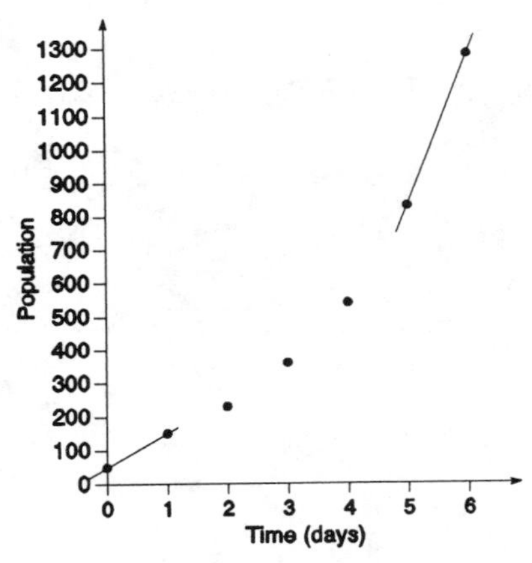

(c) Average Rate of Change $= \dfrac{153 - 50}{1 - 0} = \dfrac{103}{1} = 103$

(d) The average rate of change of the population between 0 and 1 day is 103. The population is increasing at a rate of 103 per day between 0 and 1 day.

(e) See (a)

(f) Average Rate of Change $= \dfrac{1280 - 839}{6 - 5} = \dfrac{441}{1} = 441$

(g) The average rate of change of the population between 5 and 6 days is 441. The population is increasing at a rate of 441 per day between the 5th and 6th days.

(h) As time passes, the average rate of change increases. Thus, the population is increasing at an increasing rate.

79. (a), (b), (e)

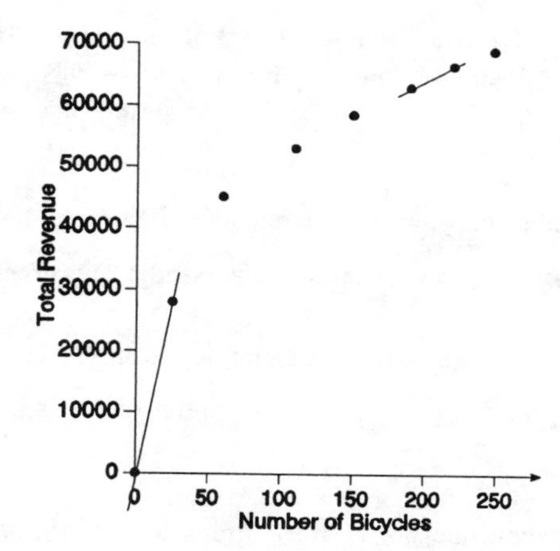

(c) Average Rate of Change $= \dfrac{28{,}000 - 0}{25 - 0} = \dfrac{28{,}000}{25} = 1120$

(d) For each additional bicycle sold between 0 and 25, total revenue increases by $1120.

(e) See (a)

(f) Average Rate of Change $= \dfrac{64{,}835 - 62{,}360}{223 - 190} = \dfrac{2475}{33} = 75$

(g) For each additional bicycle sold between 190 and 223, total revenue increases by $75.

(h) As the number of bicycles sold increases, marginal revenue decreases.

(i) Marginal revenue between 102 and 150 bicycles is $120. Marginal cost between 102 and 150 bicycles is $78.125. Therefore, profits are increasing in this interval. However, marginal revenue between 150 and 190 bicycles is $80 while marginal cost is $93.75. Thus, profits begin declining after the 150th bicycle is sold.

81. (a) For 50 therms, the charge $C = 9.00 + 0.36375(50) + 0.3256(50) = \$43.47$

(b) For 500 therms, the charge

$$C = 9.00 + 0.36375(50) + 0.11445(500 - 50) + 0.3256(500) = \$241.49$$

(c) If C is the monthly charge, then

$$C = \begin{cases} 9 + 0.36375x + 0.3256x & \text{for } 0 \le x \le 50 \\ 9 + 0.36375(50) + 0.11445(x - 50) + 0.3256x & \text{for } x > 50 \end{cases}$$

$$= \begin{cases} 9 + 0.68935x & \text{for } 0 \le x \le 50 \\ 9 + 18.1875 + 0.11445x - 5.7225 + 0.3256x & \text{for } > 50 \end{cases}$$

$$= \begin{cases} 9 + 0.68935x & \text{for } 0 \le x \le 50 \\ 21.465 + 0.44005x & \text{for } x > 50 \end{cases}$$

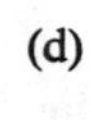

(d)

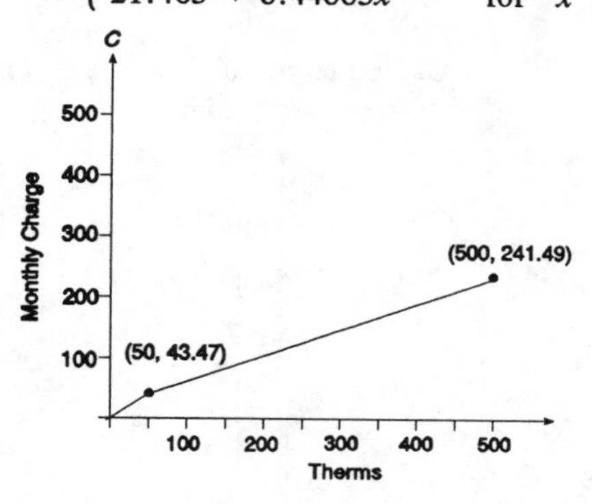

83. Each graph is that of $y = x^2$, but shifted vertically. If $y = x^2 + k$, $k > 0$, the shift is up k units; if $y = x^2 + k$, $k < 0$, the shift is down $|k|$ units. The graph of $y = x^2 - 4$ is the same as the graph of $y = x^2$ but shifted down 4. The graph of $y = x^2 + 5$ is the graph of $y = x^2$, but shifted up 5.

85. Each graph is that of $y = |x|$, but either compressed or stretched. If $y = k|x|$ and $k > 1$, the graph is stretched; if $y = k|x|$, $0 < k < 1$, the graph is compressed. The graph of $y = \frac{1}{4}|x|$ is the same as the graph of $y = |x|$, but compressed [for example, from (2, 2) to $\left(2, \frac{1}{2}\right)$]. The graph of $y = 5|x|$ is the same as the graph of $y = |x|$, but stretched [for example, from (2, 2) to (2, 10)].

87. The graph of $y = \sqrt{-x}$ is the reflection about the y-axis of the graph of $y = \sqrt{x}$. The same type of reflection occurs when graphing $y = 2x + 1$ and $y = 2(-x) + 1$. The conclusion is that the graph of $y = f(-x)$ is the reflection about the y-axis of the graph of $y = f(x)$.

89. For the graph $y = x^n$, n an even integer, as n increases, the graph of the function is narrower for $|x| > 1$ and flatter for $|x| < 1$.

4.3 Graphing Techniques; Transformations

1. B

3. H

5. I

7. L

9. F

11. G

13. $y = (x - 4)^3$

15. $y = x^3 + 4$

17. $y = -x^3$

19. $y = 4x^3$

21. (1) $y = \sqrt{x} + 2$
(2) $y = -\left(\sqrt{x} + 2\right)$
(3) $y = -\left(\sqrt{-x} + 2\right)$

23. (1) $y = -\sqrt{x}$
(2) $y = -\sqrt{x} + 2$
(3) $y = -\sqrt{x + 3} + 2$

25. $f(x) = x^2 - 1$
Using the graph of $y = x^2$, vertically shift downward 1 unit.

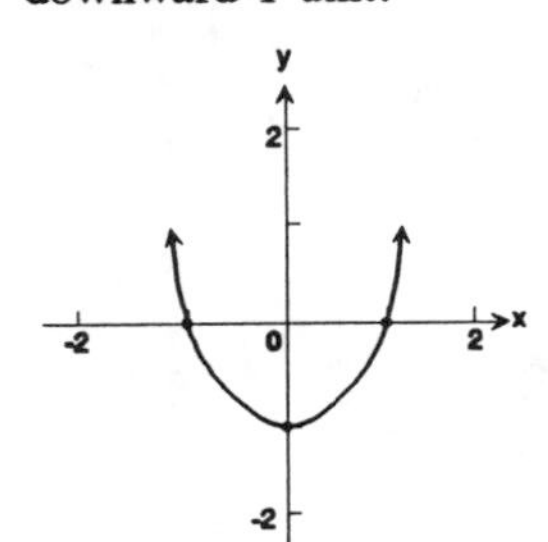

27. $g(x) = x^3 + 1$
Using the graph of $y = x^3$, vertically shift upward 1 unit.

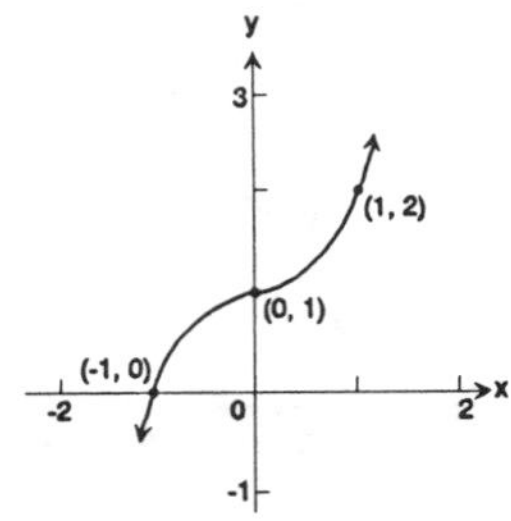

29. $h(x) = \sqrt{x - 2}$

Using the graph of $y = \sqrt{x}$, horizontally shift to the right 2 units.

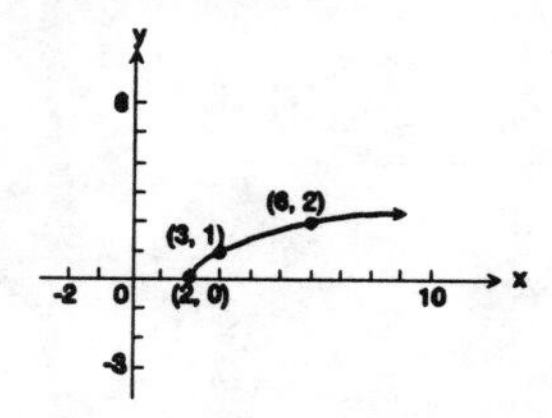

31. $f(x) = (x - 1)^3$

Using the graph of $y = x^3$, horizontally shift to the right 1 unit.

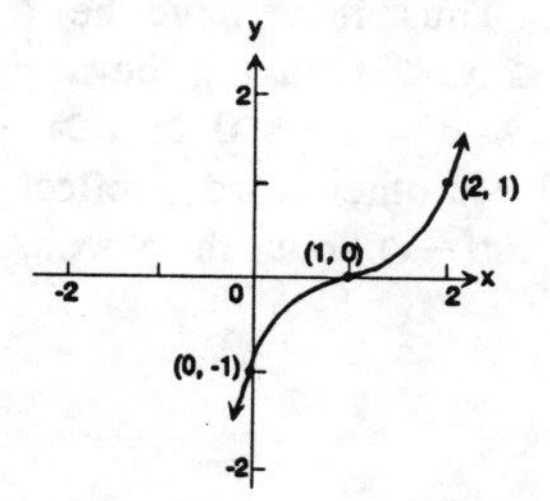

33. $g(x) = 4\sqrt{x}$

Using the graph of $y = \sqrt{x}$, vertically stretch so that (1, 1) becomes (1, 4).

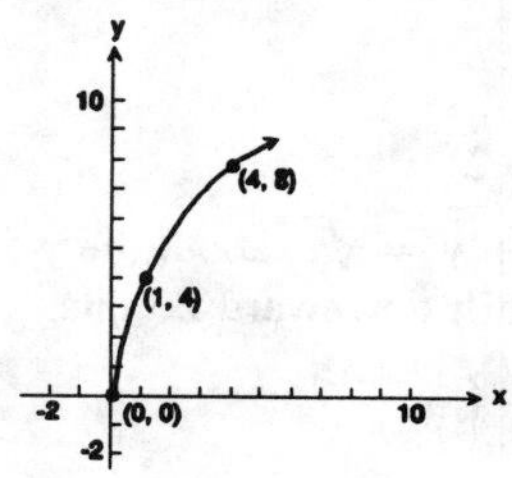

35. $h(x) = \dfrac{1}{2x}$

Using the graph of $y = \dfrac{1}{x}$, vertically compress so that (1, 1) becomes $\left(1, \dfrac{1}{2}\right)$.

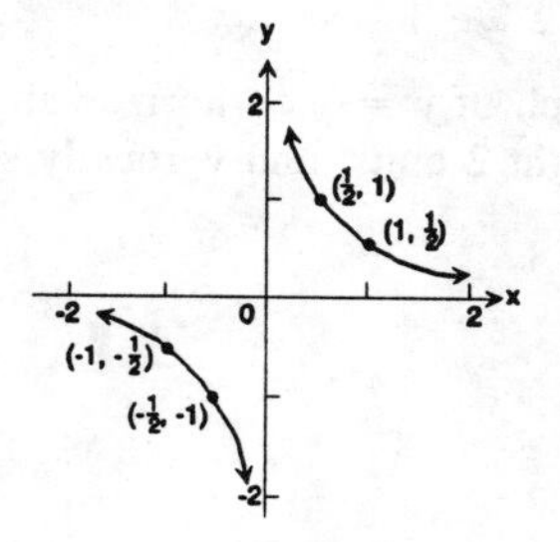

37. $f(x) = -|x|$

Reflect the graph of $y = |x|$ about the x-axis.

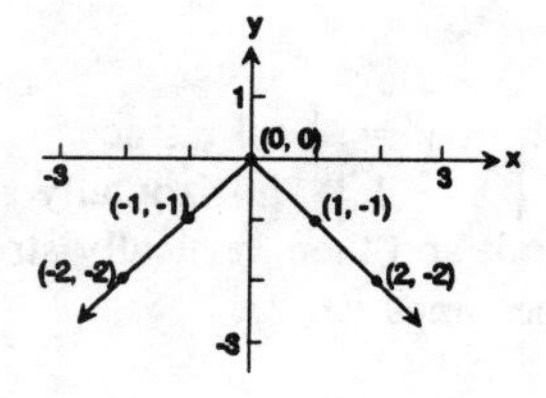

39. $g(x) = \dfrac{-1}{x}$

Reflect the graph of $y = \dfrac{1}{x}$ about the x-axis.

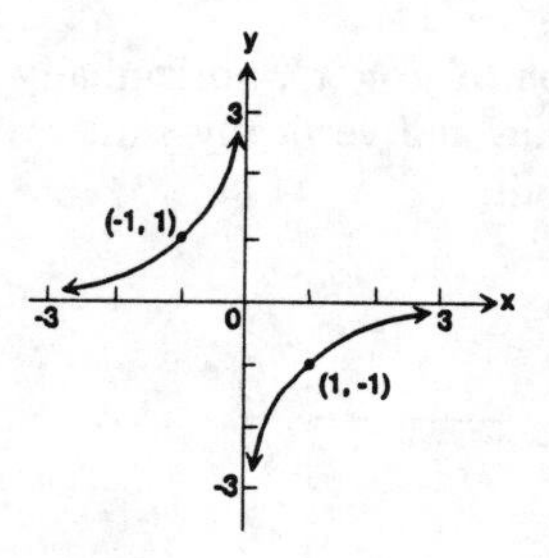

41. $h(x) = \text{int}(-x)$
The greatest integer function $y = \text{int}(-x)$ takes the integer values less than or equal to the given $-x$. Thus, if we have the inequality, $0 \le x < 1$, taking the negative, we have $-(0 \le x < 1) = 0 \ge x > -1$ or $-1 < x \le 0$. In other words, reflect the graph of $y = \text{int}(-x)$ about the y-axis.

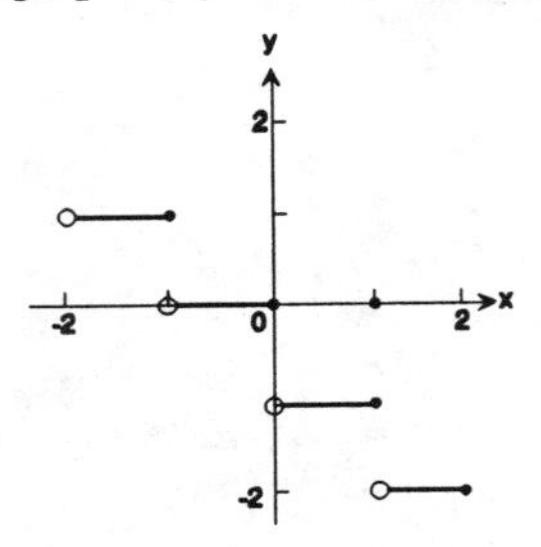

43. $f(x) = (x + 1)^2 - 3$
Using the graph of x^2, horizontally shift to the left 1 unit and vertically shift downward 3 units.

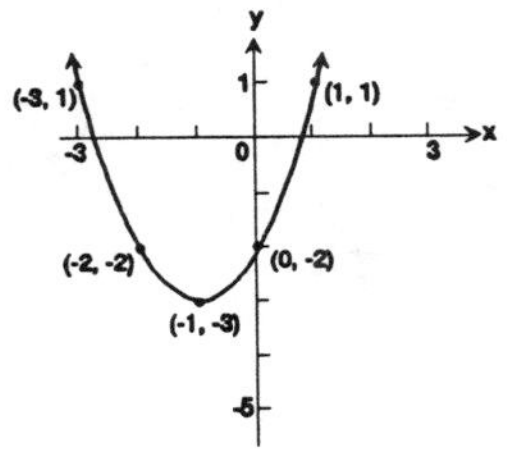

45. $g(x) = \sqrt{x - 2} + 1$
Using the graph of $y = \sqrt{x}$, horizontally shift to the right 2 units, and vertically shift upward 1 unit.

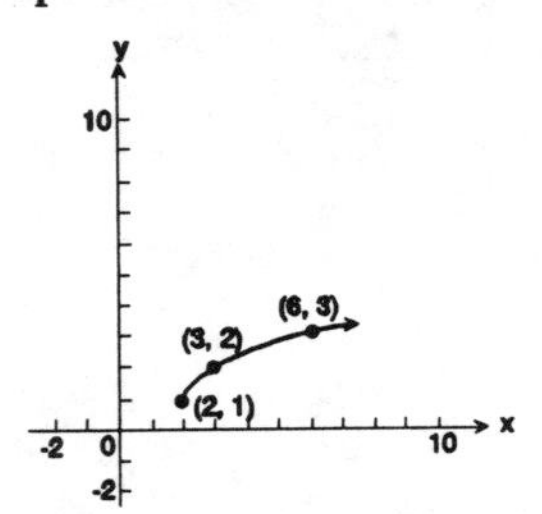

47. $h(x) = \sqrt{-x} - 2$
Reflect the graph $y = \sqrt{x}$ about the y-axis, and vertically shift downward 2 units.

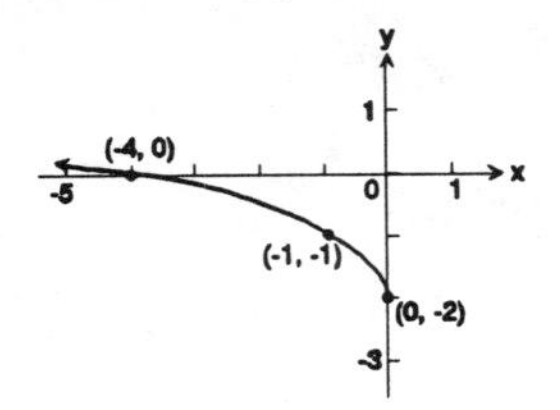

49. $f(x) = (x + 1)^3 - 1$
Using the graph of $y = x^3$, horizontally shift to the left 1 unit, and vertically shift downward 1 unit.

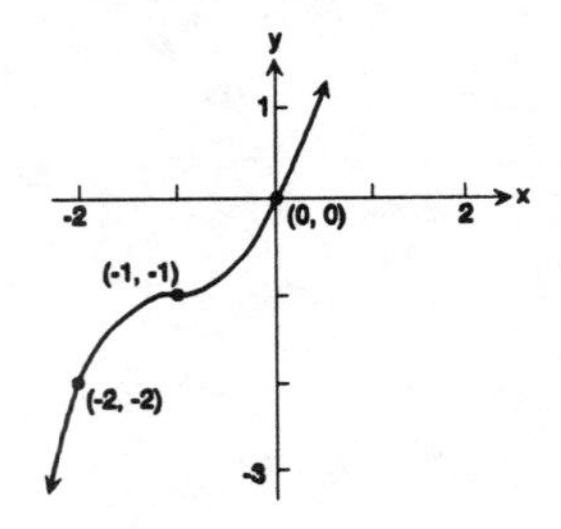

51. $g(x) = 2 \mid 1 - x \mid$
Using the graph of $y = \mid x \mid$, since $\mid 1 - x \mid = \mid x - 1 \mid$, horizontally shift to the right 1 unit and then vertically stretch so that (0, 1) becomes (0, 2).

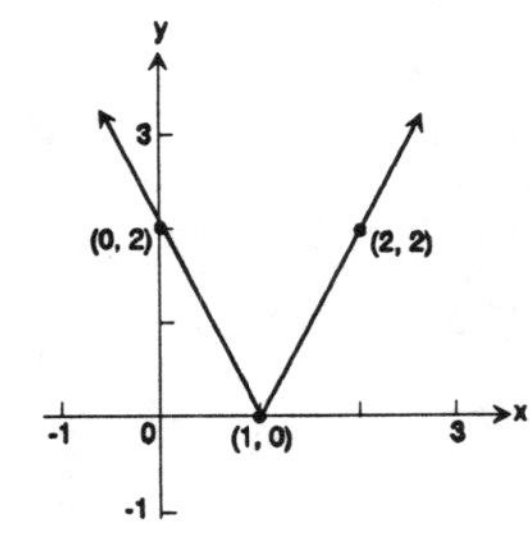

53. $h(x) = 2 \text{ int}(x - 1)$
Using the graph of $y = \text{int}(x)$, horizontally shift to the right 1 unit, then vertically stretch so that the range becomes even integers instead of all integers.

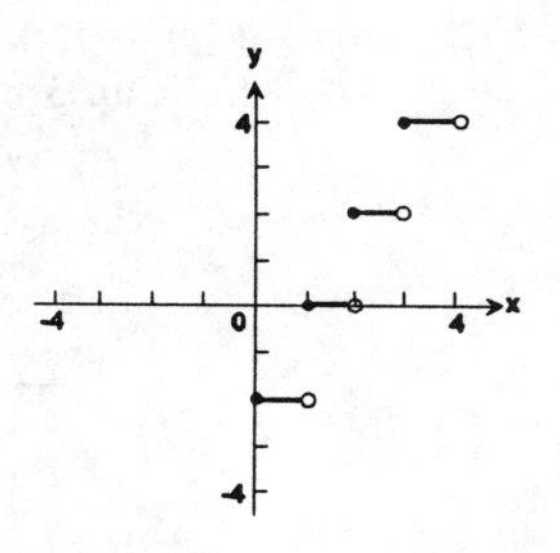

55. (a) $F(x) = f(x) + 3$
Shift up 3 units.

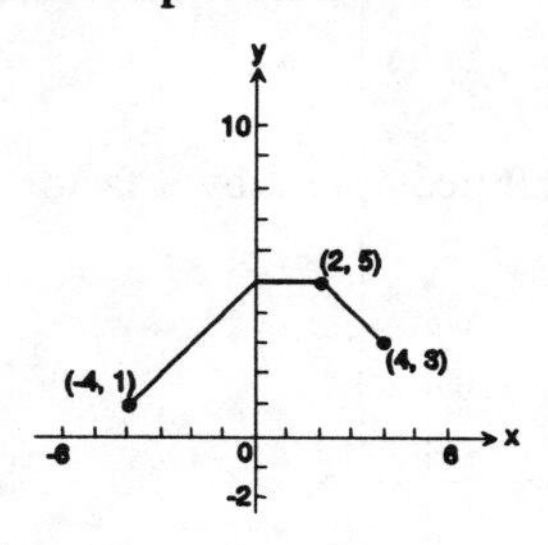

(b) $G(x) = f(x + 2)$
Shift left 2 units.

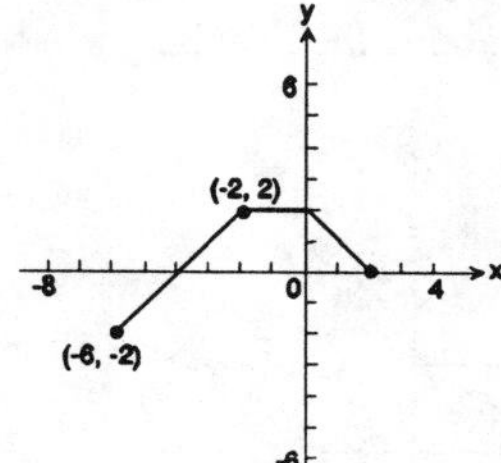

(c) $P(x) = -f(x)$
Reflect about the x-axis.

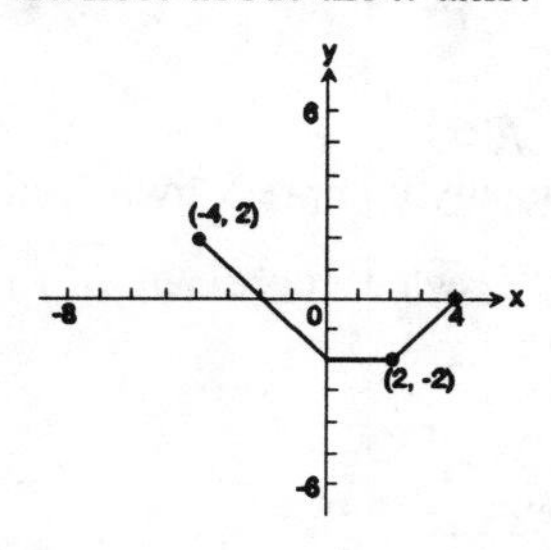

(d) $Q(x) = \frac{1}{2}f(x)$
Vertically compress so (2, 2) becomes (2, 1).

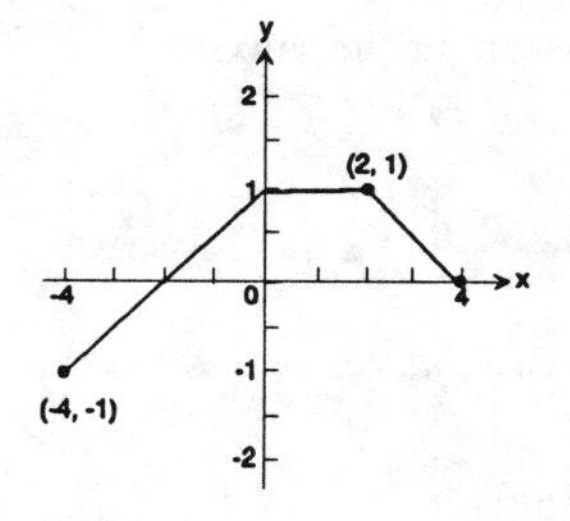

(e) $g(x) = f(-x)$
Reflect about the y-axis.

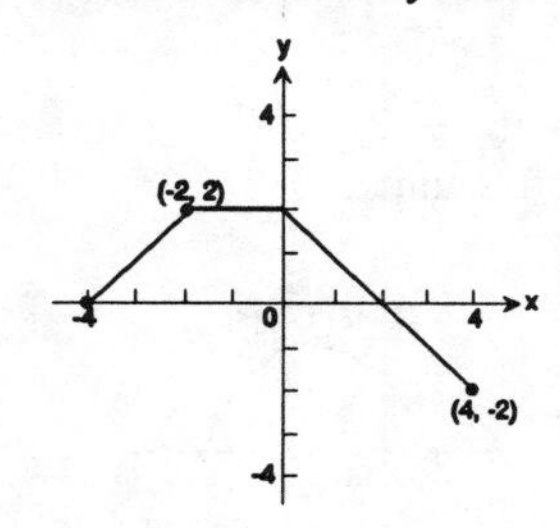

(f) $h(x) = f(2x)$
Horizontally compress the graph by a factor of 2. Multiply each x-coordinate of f by $\frac{1}{2}$.

57. (a) $F(x) = f(x) + 3$
Shift up 3 units.

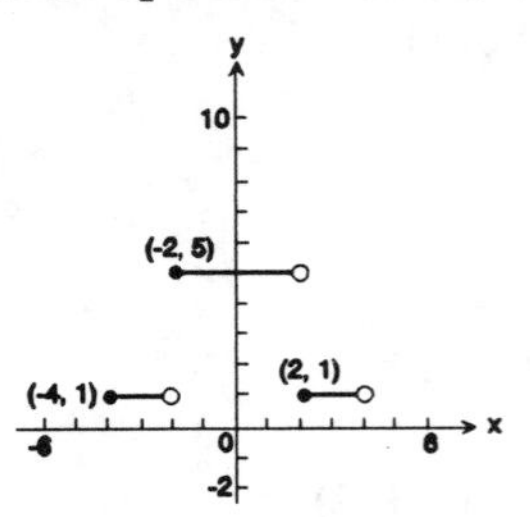

(b) $G(x) = f(x + 2)$
Shift left 2 units.

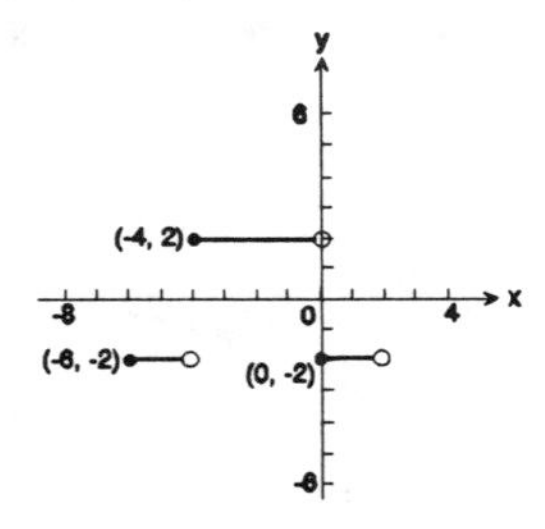

(c) $P(x) = -f(x)$
Reflect about the x-axis.

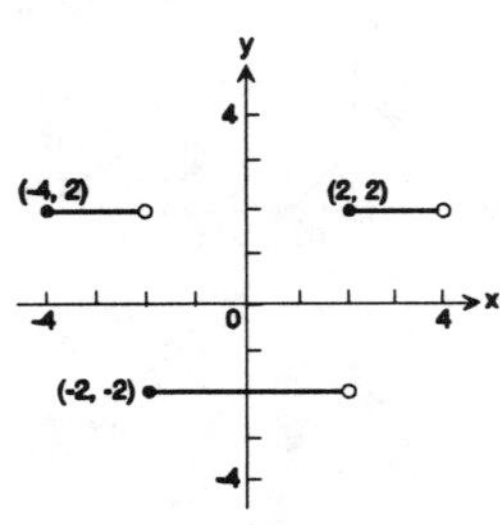

(d) $Q(x) = \frac{1}{2}f(x)$

Vertically compress by a factor of $\frac{1}{2}$.

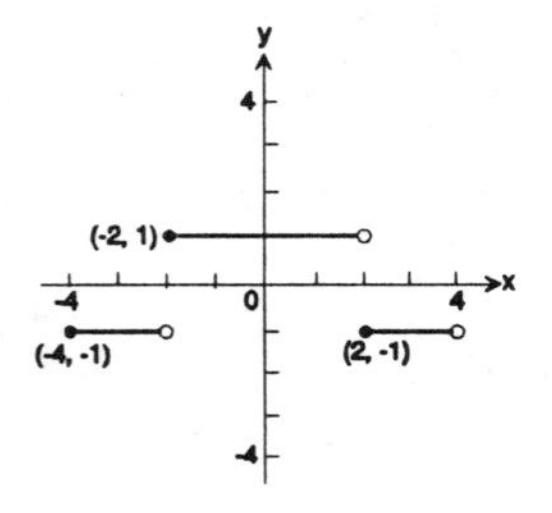

(e) $g(x) = f(-x)$
Reflect about the y-axis.

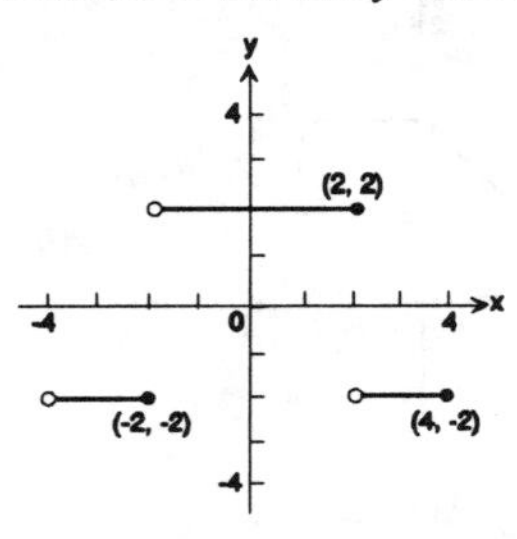

(f) $h(x) = f(2x)$
Horizontally compress by a factor of 2.
Multiply each x-coordinate of f by $\frac{1}{2}$.

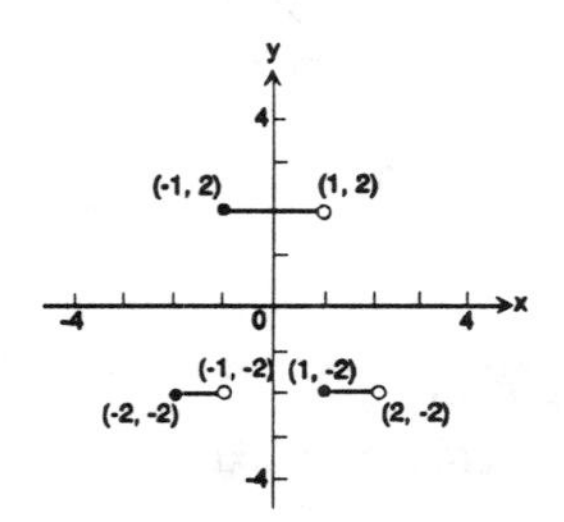

59. (a) $F(x) = f(x) + 3$
Shift up 3 units.

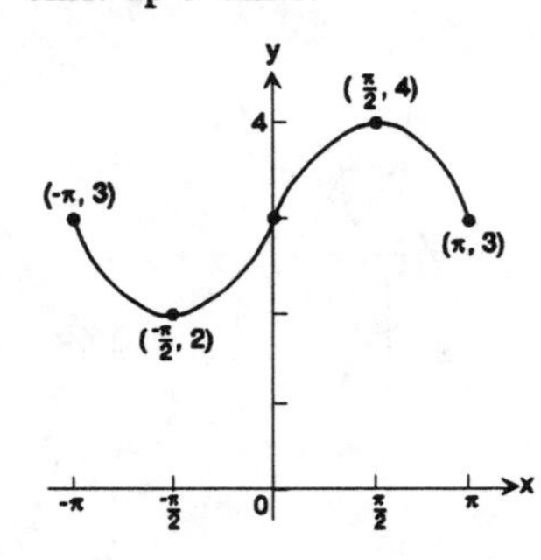

(b) $G(x) = f(x + 2)$
Shift left 2 units.

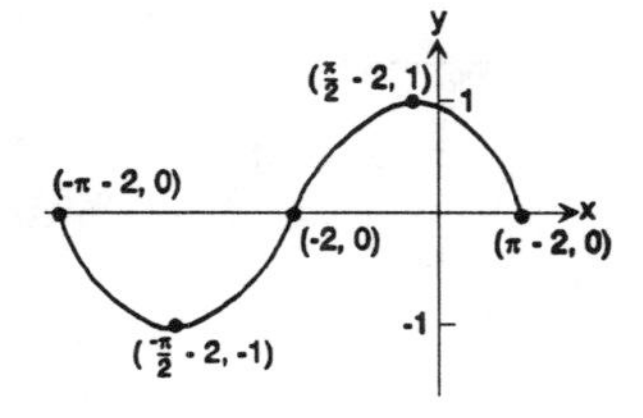

(c) $P(x) = -f(x)$
Reflect about the x-axis.

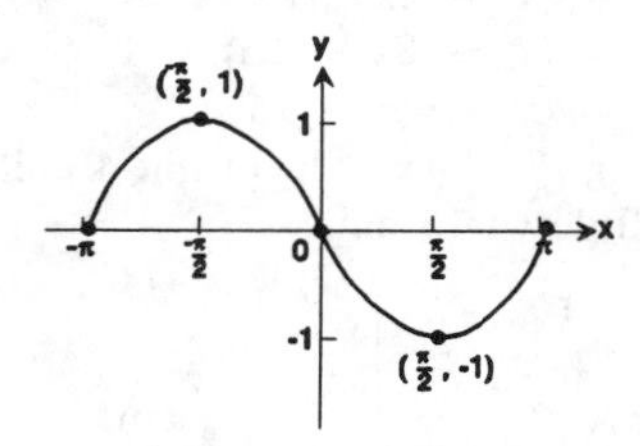

(d) $Q(x) = \frac{1}{2}f(x)$

Vertically compress by a factor of $\frac{1}{2}$.

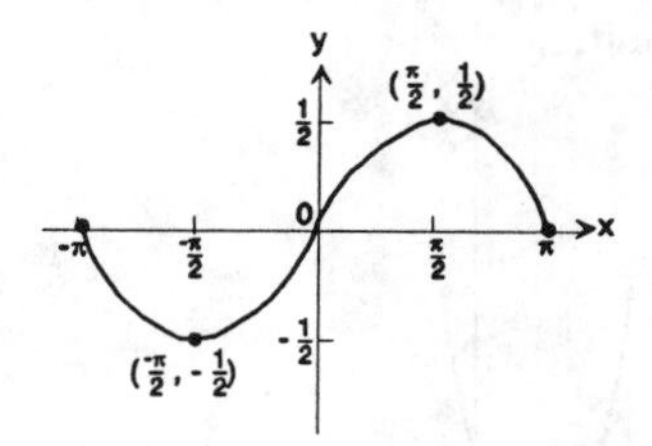

(e) $g(x) = f(-x)$
Reflect about the y-axis.

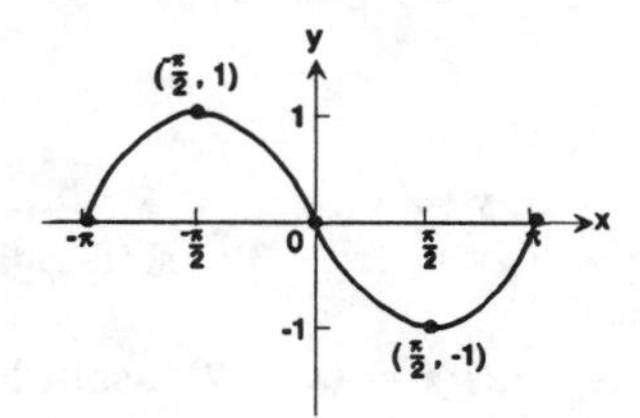

(f) $h(x) = f(2x)$
Horizontally compress by a factor of 2.
Multiply each x-coordinate of f by $\frac{1}{2}$.

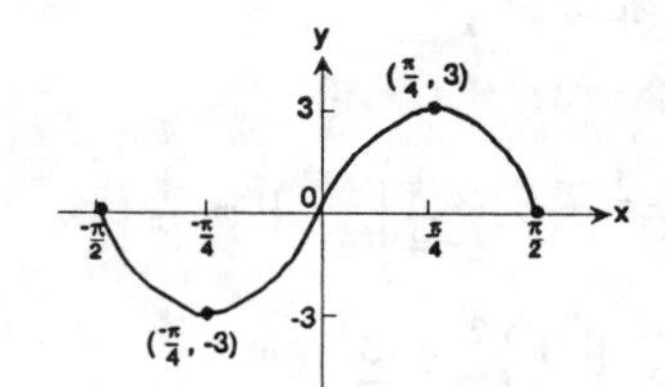

61. (a)

y = | x + 1 |

y = x + 1

10, -10, 10, -10

(b)

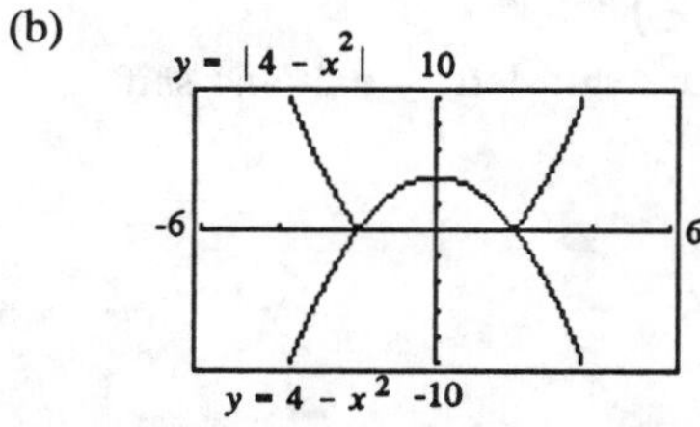

(c)

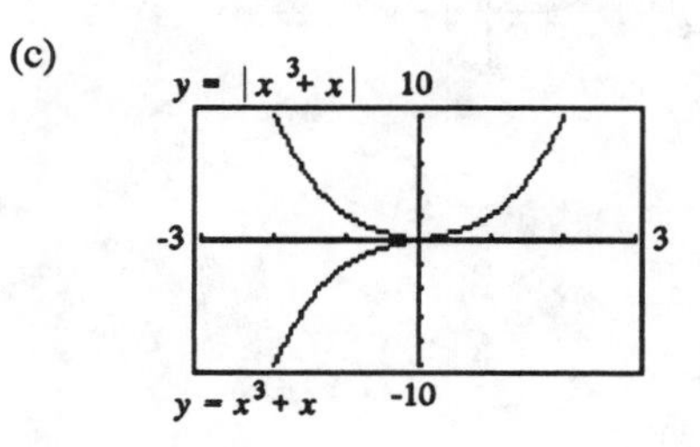

(d) Any part of the graph of $y = f(x)$ that lies below the x-axis is reflected about the x-axis to obtain the graph of $y = |f(x)|$

63. (a) Given the graph of $y = f(x)$, if $y = |f(x)|$, then all negative values for y become positive values for y. So the portion of the graph in quadrant III, where y coordinates are negative, become positive y coordinates in quadrant II. In other words, reflect the negative portion about the x-axis.

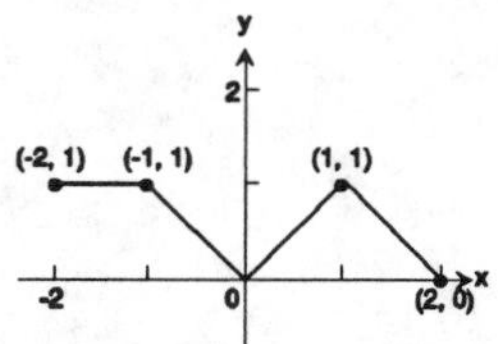

(b) Given the graph $y = f(x)$, if $y = f(|x|)$, then we must reflect about the y-axis, because $f(|+x|) = f(|-x|)$.

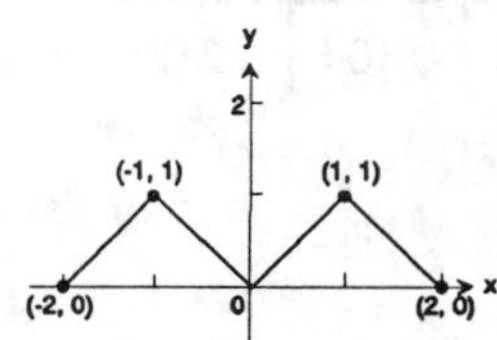

65. $f(x) = x^2 + 2x$
$f(x) = (x^2 + 2x + 1) - 1$
$f(x) = (x + 1)^2 - 1$
Using $f(x) = x^2$, shift left 1 unit and shift down 1 unit.

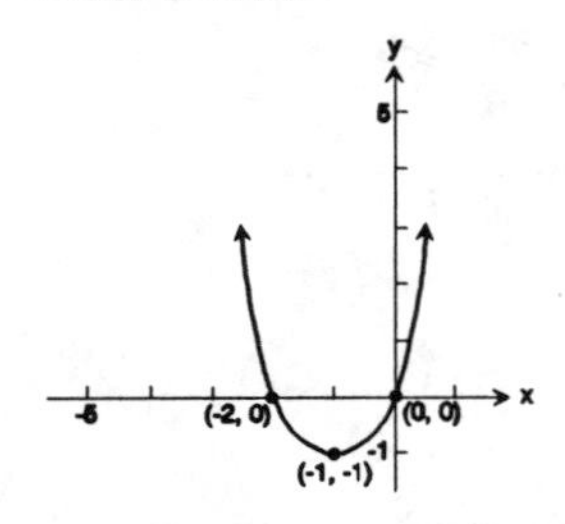

67. $f(x) = x^2 - 8x + 1$
$f(x) = (x^2 - 8x) + 1$
$f(x) = (x^2 - 8x + 16) + 1 - 16$
$f(x) = (x - 4)^2 - 15$
Using $f(x) = x^2$, shift right 4 units and shift down 15 units.

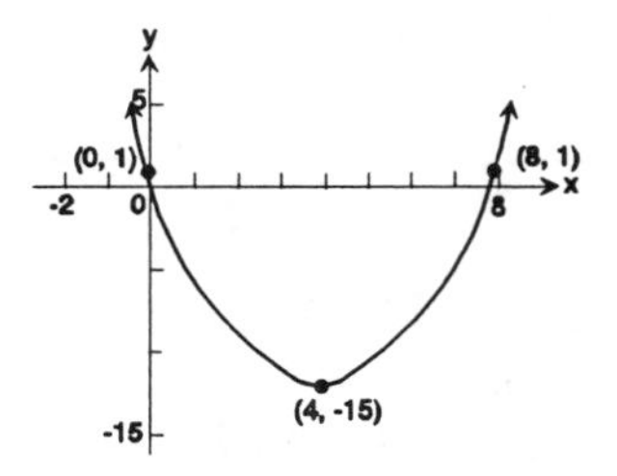

69. $f(x) = x^2 + x + 1$
$f(x) = (x^2 + x) + 1$

$$f(x) = \left(x^2 + x + \frac{1}{4}\right) + 1 - \frac{1}{4}$$

$$f(x) = \left(x + \frac{1}{2}\right)^2 + \frac{3}{4}$$

Using $f(x) = x^2$, shift left $\frac{1}{2}$ unit and shift up $\frac{3}{4}$ unit.

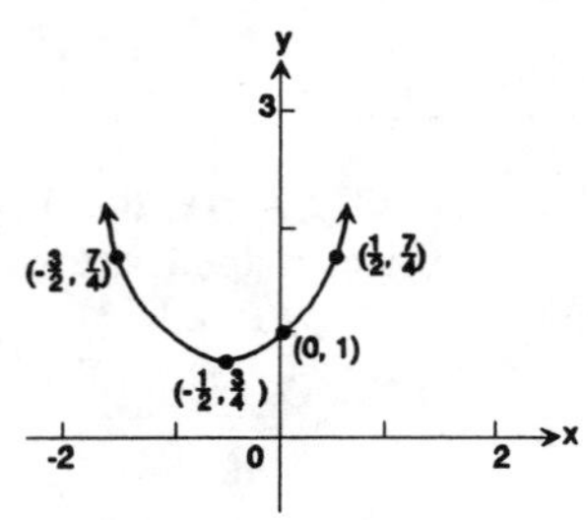

71. $y = (x - c)^2$
If $c = 0$, $y = x^2$.
If $c = 3$, $y = (x - 3)^2$, shift right 3 units.
If $c = -2$, $y = (x + 2)^2$, shift left 2 units.

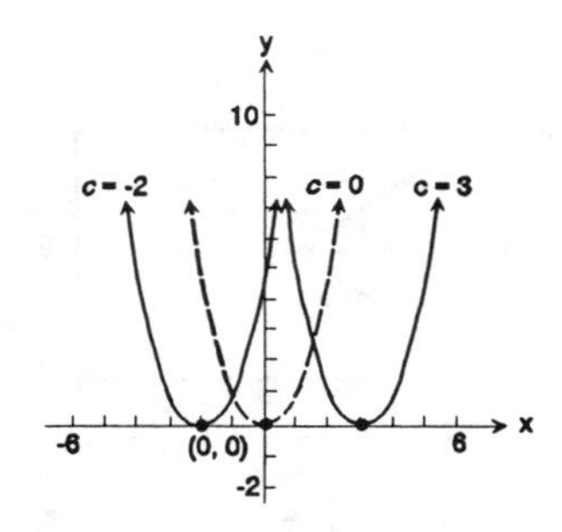

73. $F = \frac{9}{5}C + 32$
Graph:

C	0	40	100
F	32	104	212

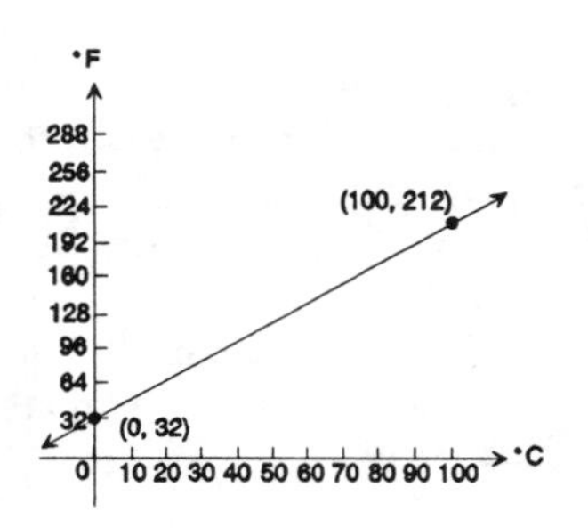

C	0	40	100
$K = C + 273$	273	313	373
F	32	104	212

or

$C = K - 273$

$F = \frac{9}{5}(K - 273) + 32$

Shift the graph of $F = F = \frac{9}{5}C + 32$, 273 units to the right.

75. (a)

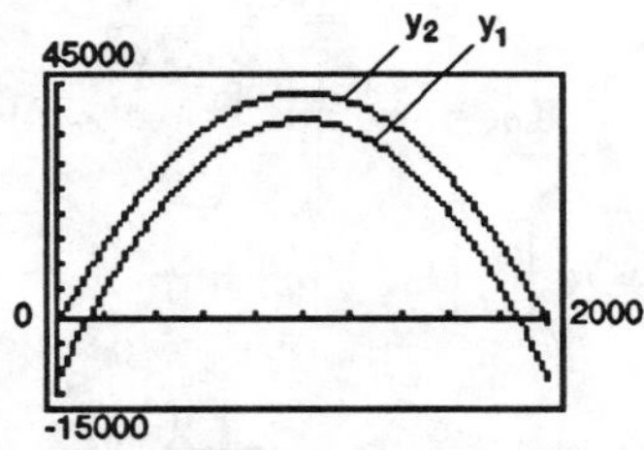

(b) Select the 10% tax since profits will be higher.

(c) The graph of y_1 is obtained by shifting the graph of $p(x)$ vertically down 10,000. The graph of y_2 is obtained by multiplying the y-coordinate of the graph of $p(x)$ by 0.9. Thus, y_2 is the graph of $p(x)$ vertically compressed by a factor of 0.9.

(d) Select the 10% tax since the graph of $y_1 = 0.9p(x) \geq y_2 = -0.5x^2 + 100x - 6800$ for all x in the domain.

4.4 Operations on Functions; Composite Functions

1. $f(x) = 3x + 4 \qquad g(x) = 2x - 3$

(a) $(f + g)(x) = 3x + 4 + 2x - 3$
$= 5x + 1$
The domain is all real numbers.

(b) $(f - g)(x) = (3x + 4) - (2x - 3)$
$= 3x + 4 - 2x + 3$
$= x + 7$
The domain is all real numbers.

(c) $(f \cdot g)(x) = (3x + 4)(2x - 3)$
$= 6x^2 - 9x + 8x - 12$
$= 6x^2 - x - 12$
The domain is all real numbers.

(d) $\left(\frac{f}{g}\right)(x) = \frac{3x + 4}{2x - 3}$
The domain is all real numbers except $\frac{3}{2}$.

3. $f(x) = x - 1 \qquad g(x) = 2x^2$

(a) $(f + g)(x) = (x - 1) + 2x^2$
$= 2x^2 + x - 1$
The domain is all real numbers.

(b) $(f - g)(x) = (x - 1) - (2x^2)$
$= -2x^2 + x - 1$
The domain is all real numbers.

(c) $(f \cdot g)(x) = (x - 1)(2x^2)$
$= 2x^3 - 2x^2$
The domain is all real numbers.

(d) $\left(\frac{f}{g}\right)(x) = \frac{x - 1}{2x^2}$
The domain is all real numbers except 0, $\{x \mid x \neq 0\}$.

5. $f(x) = \sqrt{x},\ x \geq 0 \quad g(x) = 3x - 5$

(a) $(f + g)(x) = \sqrt{x} + 3x - 5$
The domain is $\{x \mid 0 \leq x < \infty\}$.

(b) $(f - g)(x) = \sqrt{x} - (3x - 5)$
$= \sqrt{x} - 3x + 5$
The domain is $\{x \mid 0 \leq x < \infty\}$.

(c) $(f \cdot g)(x) = \sqrt{x}\,(3x - 5)$
$= 3x\sqrt{x} - 5\sqrt{x}$,
The domain is $\{x \mid 0 \leq x < \infty\}$.

(d) $\left(\dfrac{f}{g}\right)(x) = \dfrac{\sqrt{x}}{3x - 5}$
The domain is $\{x \mid 0 \leq x < \infty$ and $x \neq \frac{5}{3}\}$.

7. $f(x) = 1 + \dfrac{1}{x},\ x \neq 0 \quad g(x) = \dfrac{1}{x},\ x \neq 0$

(a) $(f + g)(x) = \left(1 + \dfrac{1}{x}\right) + \dfrac{1}{x}$
$= 1 + \dfrac{2}{x}$
The domain is $\{x \mid x \neq 0\}$.

(b) $(f - g)(x) = \left(1 + \dfrac{1}{x}\right) - \dfrac{1}{x} = 1$
The domain is $\{x \mid x \neq 0\}$.

(c) $(f \cdot g)(x) = \left(1 + \dfrac{1}{x}\right)\left(\dfrac{1}{x}\right)$
$= \dfrac{1}{x} + \dfrac{1}{x^2}$
The domain is $\{x \mid x \neq 0\}$.

(d) $\left(\dfrac{f}{g}\right)(x) = \dfrac{1 + \frac{1}{x}}{\frac{1}{x}} = \dfrac{\frac{x + 1}{x}}{\frac{1}{x}}$
$= \dfrac{x + 1}{x} \cdot \dfrac{x}{1} = x + 1$
The domain is $\{x \mid x \neq 0\}$.

9. $f(x) = \dfrac{2x + 3}{3x - 2},\ x \neq \dfrac{2}{3} \quad g(x) = \dfrac{4x}{3x - 2},\ x \neq \dfrac{2}{3}$

(a) $(f + g)(x) = \dfrac{2x + 3}{3x - 2} + \dfrac{4x}{3x - 2} = \dfrac{6x + 3}{3x - 2}$
The domain is $\left\{x \middle| x \neq \dfrac{2}{3}\right\}$.

(b) $(f - g)(x) = \dfrac{2x + 3}{3x - 2} - \dfrac{4x}{3x - 2} = \dfrac{-2x + 3}{3x - 2}$
The domain is $\left\{x \middle| x \neq \dfrac{2}{3}\right\}$.

(c) $(f \cdot g)(x) = \left(\dfrac{2x + 3}{3x - 2}\right)\left(\dfrac{4x}{3x - 2}\right) = \dfrac{8x^2 + 12x}{(3x - 2)^2}$
The domain is $\left\{x \middle| x \neq \dfrac{2}{3}\right\}$.

(d) $\left(\dfrac{f}{g}\right)(x) = \dfrac{\frac{2x + 3}{3x - 2}}{\frac{4x}{3x - 2}} = \dfrac{2x + 3}{3x - 2} \cdot \dfrac{3x - 2}{4x} = \dfrac{2x + 3}{4x}$
The domain is $\left\{x \middle| x \neq \dfrac{2}{3} \text{ and } x \neq 0\right\}$.

11. $f(x) = 3x + 1$
$(f + g)(x) = 6 - \dfrac{1}{2}x$
$6 - \dfrac{1}{2}x = (3x + 1) + g(x)$
$-\dfrac{7}{2}x + 5 = g(x)$
$g(x) = 5 - \dfrac{7}{2}x$

13. $f(x) = 2x \quad g(x) = 3x^2 + 1$

(a) $(f \circ g)(4) = f(g(4)) = f(49) = 2(49) = 98$
(b) $(g \circ f)(2) = g(f(2)) = g(4) = 3(4)^2 + 1 = 48 + 1 = 49$
(c) $(f \circ f)(1) = f(f(1)) = f(2) = 2(2) = 4$
(d) $(g \circ g)(0) = g(g(0)) = g(1) = 3(1)^2 + 1 = 4$

15. $f(x) = 4x^2 - 3 \qquad g(x) = 3 - \frac{1}{2}x^2$

(a) $(f \circ g)(4) = f(g(4)) = f(-5) = 4(-5)^2 - 3 = 97$

(b) $(g \circ f)(2) = g(f(2)) = g(13) = 3 - \frac{1}{2}(13)^2 = 3 - \frac{169}{2} = \frac{6 - 169}{2} = -\frac{163}{2}$

(c) $(f \circ f)(1) = f(f(1)) = f(1) = 1$

(d) $(g \circ g)(0) = g(g(0)) = g(3) = 3 - \frac{1}{2}(3)^2 = 3 - \frac{9}{2} = \frac{6 - 9}{2} = -\frac{3}{2}$

17. $f(x) = \sqrt{x} \qquad g(x) = 2x$

(a) $(f \circ g)(4) = f(g(4)) = f(8) = \sqrt{8} = 2\sqrt{2}$

(b) $(g \circ f)(2) = g(f(2)) = g\left(\sqrt{2}\right) = 2\sqrt{2}$

(c) $(f \circ f)(1) = f(f(1)) = f(1) = \sqrt{1} = 1$

(d) $(g \circ g)(0) = g(g(0)) = g(0) = 2(0) = 0$

19. $f(x) = |\, x \,| \qquad g(x) = \dfrac{1}{x^2 + 1}$

(a) $(f \circ g)(4) = f(g(4)) = f\left(\frac{1}{17}\right) = \left|\frac{1}{17}\right| = \frac{1}{17}$

(b) $(g \circ f)(2) = g(f(2)) = g(2) = \dfrac{1}{2^2 + 1} = \dfrac{1}{5}$

(c) $(f \circ f)(1) = f(f(1)) = f(1) = 1$

(d) $(g \circ g)(0) = g(g(0)) = g(1) = \dfrac{1}{1^2 + 1} = \dfrac{1}{2}$

21. $f(x) = \dfrac{3}{x^2 + 1} \qquad g(x) = \sqrt{x}$

(a) $(f \circ g)(4) = f(g(4)) = f(2) = \dfrac{3}{2^2 + 1} = \dfrac{3}{5}$

(b) $(g \circ f)(2) = g(f(2)) = g\left[\frac{3}{5}\right] = \sqrt{\frac{3}{5}} = \frac{\sqrt{15}}{5}$

(c) $(f \circ f)(1) = f(f(1)) = f\left[\frac{3}{2}\right] = \dfrac{3}{\left[\frac{3}{2}\right]^2 + 1} = \dfrac{3}{\frac{9}{4} + \frac{4}{4}} = \dfrac{3}{\frac{13}{4}} = \dfrac{3}{1} \cdot \dfrac{4}{13} = \dfrac{12}{13}$

(d) $(g \circ g)(0) = g(g(0)) = g(0) = \sqrt{0} = 0$

23. The domain of g is $\{x \mid x \neq 0\}$. The domain of f is $\{x \mid x \neq 1\}$.
Thus, $g(x) \neq 1$, so we solve:

$$g(x) = 1$$
$$\frac{2}{x} = 1$$
$$x = 2$$

Thus, $x \neq 2$, so the domain of $f \circ g$ is $\{x \mid x \neq 0, x \neq 2\}$.

25. The domain of g is $\{x \mid x \neq 0\}$. The domain of f is $\{x \mid x \neq 1\}$.
Thus, $g(x) \neq 1$, so we solve:

$$g(x) = 1$$
$$\frac{-4}{x} = 1$$
$$x = -4$$

Thus, $x \neq -4$, so the domain of $f \circ g$ is $\{x \mid x \neq -4, x \neq 0\}$.

27. The domain of $g(x)$ is all real numbers. The domain of f is $\{x \mid x \geq 0\}$.
Thus, $g(x) \geq 0$, so we solve:

$$g(x) \geq 0$$
$$2 + 3 \geq 0$$
$$2x \geq -3$$
$$x \geq \frac{-3}{2}$$

Thus, the domain of $f \circ g$ is $\left\{x \mid x \geq \frac{-3}{2}\right\}$.

29. The domain of $g(x)$ is $\{x \mid x \neq 1\}$. The domain of f is $\{x \mid x \geq -1\}$.
Thus, $g(x) \geq -1$.

$$g(x) \geq -1$$
$$\frac{2}{x - 1} \geq -1$$
$$\frac{2}{x - 1} + 1 \geq 0$$
$$\frac{2 + x - 1}{x - 1} \geq 0$$
$$\frac{x + 1}{x - 1} \geq 0$$

Interval	$\frac{x+1}{x-1}$
$x < -1$	Positive
$-1 < x < 1$	Negative
$x > 1$	Positive

The domain of $f \circ g$ is $\{x \mid x \leq -1 \text{ or } x > 1\}$.

31. $f(x) = 2x + 3$ $g(x) = 3x$. The domain of f and g is all real numbers.
(a) $(f \circ g)(x) = f(g(x)) = f(3x) = 2(3x) + 3 = 6x + 3$. Domain: All real numbers.
(b) $(g \circ f)(x) = g(f(x)) = g(2x + 3) = 3(2x + 3) = 6x + 9$. Domain: All real numbers.
(c) $(f \circ f)(x) = f(f(x)) = f(2x + 3) = 2(2x + 3) + 3 = 4x + 6 + 3 = 4x + 9$.
Domain: All real numbers.
(d) $(g \circ g)(x) = g(g(x)) = g(3x) = 3(3x) = 9x$. Domain: All real numbers.

33. $f(x) = 3x + 1$ $g(x) = x^2$. The domain of f and g is all real numbers.
(a) $(f \circ g)(x) = f(g(x)) = f(x^2) = 3x^2 + 1$. Domain: All real numbers.
(b) $(g \circ f)(x) = g(f(x)) = g(3x + 1) = (3x + 1)^2 = 9x^2 + 6x + 1$. Domain: All real numbers.
(c) $(f \circ f)(x) = f(f(x)) = f(3x + 1) = 3(3x + 1) + 1 = 9x + 3 + 1 = 9x + 4$.
Domain: All real numbers.
(d) $(g \circ g)(x) = g(g(x)) = g(x^2) = (x^2)^2 = x^4$. Domain: All real numbers.

35. $f(x) = x^2 \quad g(x) = x^2 + 4$. The domain of f is all real numbers. The domain of g is all real numbers.

(a) $(f \circ g) = f(g(x)) = f(x^2 + 4) = (x^2 + 4)^2 = x^4 + 8x^2 + 16$. The domain is all real numbers.

(b) $(g \circ f) = g(f(x)) = g(x^2) = (x^2)^2 + 4 = x^4 + 4$. The domain is all real numbers.

(c) $(f \circ f) = f(f(x)) = f(x^2) = (x^2)^2 = x^4$. The domain is all real numbers.

(d) $(g \circ g) = g(g(x)) = g(x^2 + 4) = (x^2 + 4)^2 + 4 = x^4 + 8x^2 + 16 + 4 = x^4 + 8x^2 + 20$. The domain is all real numbers.

37. $f(x) = \dfrac{3}{x - 1}$. The domain of f is $\{x \mid x \neq 1\}$.

$g(x) = \dfrac{2}{x}$. The domain of g is $\{x \mid x \neq 0\}$.

(a) $f(g(x)) = f\left(\dfrac{2}{x}\right) = \dfrac{3}{\dfrac{2}{x} - 1} = \dfrac{3}{\dfrac{2}{x} - \dfrac{x}{x}} = \dfrac{3}{\dfrac{2 - x}{x}} = \dfrac{3x}{2 - x}$

Domain of $f \circ g$ is $\{x \mid x \neq 0, x \neq 2\}$ since $x = 0$ is not in the domain of g and $x = 2$ is not in the domain of $f \circ g$.

(b) $(g \circ f) = g(f(x)) = g\left(\dfrac{3}{x - 1}\right) = \dfrac{2}{\dfrac{3}{x - 1}} = \dfrac{2(x - 1)}{3}$. Domain: $\{x \mid x \neq 1\}$ since $x = 1$ is not in the domain of f.

(c) $(f \circ f) = f(f(x)) = f\left(\dfrac{3}{x - 1}\right) = \dfrac{3}{\dfrac{3}{x - 1} - 1} = \dfrac{3}{\dfrac{3 - (x - 1)}{x - 1}} = \dfrac{3(x - 1)}{-x + 4}$.

Domain: $\{x \mid x \neq 1, x \neq 4\}$ since $x = 1$ is not in the domain of f and $x = 4$ is not in the domain of $f(f(x))$.

(d) $(g \circ g) = g(g(x)) = g\left(\dfrac{2}{x}\right) = \dfrac{2}{\dfrac{2}{x}} = \dfrac{2x}{2} = x$. Domain: $\{x \mid x \neq 0\}$ since $x = 0$ is not in the domain of g.

39. $f(x) = \dfrac{x}{x - 1}$. The domain of f is $\{x \mid x \neq 1\}$.

$g(x) = \dfrac{-4}{x}$. The domain of g is $\{x \mid x \neq 0\}$.

(a) $f(g(x)) = f\left(\dfrac{-4}{x}\right) = \dfrac{-\dfrac{4}{x}}{\dfrac{-4}{x} - 1} = \dfrac{-\dfrac{4}{x}}{\dfrac{-4 - x}{x}} = \dfrac{-4}{-4 - x}$.

Domain of $f \circ g$ is $\{x \mid x \neq -4, x \neq 0\}$ since $x = -4$ is not in the domain of $f \circ g$ and $x = 0$ is not in the domain of g.

(b) $g(f(x)) = g\left(\dfrac{x}{x - 1}\right) = \dfrac{-4}{\dfrac{x}{x - 1}} = \dfrac{-4(x - 1)}{x}$. Domain: $\{x \mid x \neq 0, x \neq 1\}$ since $x = 1$ is not in the domain of f.

(c) $(f \circ f) = f(f(x)) = f\left(\dfrac{x}{x-1}\right) = \dfrac{\dfrac{x}{x-1}}{\dfrac{x}{x-1} - 1} = \dfrac{\dfrac{x}{x-1}}{\dfrac{x-(x-1)}{x-1}} = \dfrac{x}{1} = x$

Domain: $\{x \mid x \neq 1\}$ since $x = 1$ is not in the domain of f.

(d) $(g \circ g) = g(g(x)) = g\left(\dfrac{-4}{x}\right) = \dfrac{-4}{\dfrac{-4}{x}} = x$. Domain: $\{x \mid x \neq 0\}$ since $x = 0$ is not in the domain of g.

41. $f(x) = \sqrt{x}$. The domain of f is $\{x \mid x \geq 0\}$.
$g(x) = 2x + 3$. The domain of g is all real numbers.

(a) $f(g(x)) = f(2x + 3) = \sqrt{2x+3}$. Domain of $f \circ g$ is $\left\{x \mid x \geq \dfrac{-3}{2}\right\}$ since $2x + 3 \geq 0$.

(b) $g(f(x)) = g(\sqrt{x}) = 2\sqrt{x} + 3$. Domain: $\{x \mid x \geq 0\}$ since the domain of $g(x)$ is $\{x \mid x \geq 0\}$.

(c) $f(f(x)) = f(\sqrt{x}) = \sqrt{\sqrt{x}} = x^{\frac{1}{4}} = \sqrt[4]{x}$. The domain is $\{x \mid x \geq 0\}$ since the domain of f is $\{x \mid x \geq 0\}$.

(d) $g(g(x)) = g(2x + 3) = 2(2x + 3) + 3 = 4x + 6 + 3 = 4x + 9$. The domain is the set of all real numbers.

43. $f(x) = \sqrt{x+1}$. The domain of f is $\{x \mid x \geq -1\}$.
$g(x) = \dfrac{2}{x-1}$. The domain of g is $\{x \mid x \neq 1\}$.

(a) $f(g(x)) = f\left(\dfrac{2}{x-1}\right) = \sqrt{\dfrac{2}{x-1} + 1} = \sqrt{\dfrac{2 + x - 1}{x-1}} = \dfrac{x+1}{x-1}$.

$\dfrac{x+1}{x-1} \geq 0$ when $x \leq -1$ or $x \geq 1$. Since $x \neq 1$, the domain of $f \circ g$ is $\{x \mid x \leq -1$ or $x > 1\}$.

(b) $g(f(x)) = g(\sqrt{x+1}) = \dfrac{2}{\sqrt{x+1} - 1}$. Domain: $\{x \mid x \geq -1, x \neq 0\}$. Since the domain of f is $\{x \mid x \geq -1\}$ and if $x = 0$, then $g(f(x))$ is not defined.

(c) $f(f(x)) = f(\sqrt{x+1}) = \sqrt{\sqrt{x+1} + 1}$. Domain: $\{x \mid x \geq -1\}$ since this is the domain of $f(x)$.

(d) $g(g(x)) = g\left(\dfrac{2}{x-1}\right) = \dfrac{2}{\dfrac{2}{x-1} - 1} = \dfrac{2}{\dfrac{2-(x-1)}{x-1}} = \dfrac{2(x-1)}{-x+3}$. Domain: $\{x \mid x \neq 1,$ $x \neq 3\}$ since the domain of g is $\{x \mid x \neq 1\}$.

45. $f(x) = ax + b \qquad g(x) = cx + d$
The domain of f and g is all real numbers.

(a) $(f \circ g)(x) = f(g(x)) = f(cx + d) = a(cx + d) + b = acx + ad + b$.
Domain: All real numbers.

(b) $(g \circ f)(x) = g(f(x)) = g(ax + b) = c(ax + b) + d = acx + bc + d$
Domain: All real numbers.

(c) $(f \circ f)(x) = f(f(x)) = f(ax + b) = a(ax + b) + b = a^2x + ab + b$
Domain: All real numbers.

(d) $(g \circ g)(x) = g(g(x)) = g(cx + d) = c(cx + d) = c^2x + cd + d$
Domain: All real numbers.

47. $(f \circ g)(x) = f(g(x)) = f\left[\frac{1}{2}x\right] = 2\left[\frac{1}{2}x\right] = x$

$(g \circ f)(x) = g(f(x)) = g(2x) = \frac{1}{2}(2x) = x$

49. $(f \circ g)(x) = f(g(x)) = f\left(\sqrt[3]{x}\right) = \left(\sqrt[3]{x}\right)^3 = x$

$(g \circ f)(x) = g(f(x)) = g(x^3) = \sqrt[3]{x^3} = x$

51. $(f \circ g)(x) = f(g(x)) = f\left[\frac{1}{2}(x + 6)\right] = 2\left[\frac{1}{2}(x + 6)\right] - 6 = x + 6 - 6 = x$

$(g \circ f)(x) = g(f(x)) = g(2x - 6) = \frac{1}{2}(2x - 6 + 6) = x$

53. $(f \circ g)(x) = f(g(x)) = f\left[\frac{1}{a}(x - b)\right] = a\left[\frac{1}{a}(x - b)\right] + b = a\left[\frac{x}{a} - \frac{b}{a}\right] + b = x - b + b = x$

$(g \circ f)(x) = g(f(x)) = g(ax + b) = \frac{1}{a}(ax + b - b) = x$

55. $f(x) = x^2,\ g(x) = 3x,\ h(x) = \sqrt{x} + 1$

$F(x) = 9x^2$
$F(x) = (3x)^2$
$F(x) = (g(x))^2$
$F(x) = f(g(x))$
$F = f \circ g$

57. $H(x) = |x| + 1$

$H(x) = \sqrt{x^2} + 1$
$H(x) = \sqrt{f(x)} + 1$
$H(x) = h(f(x))$
$H = h \circ f$

59. $q(x) = x + 2\sqrt{x} + 1$

$q(x) = \left(\sqrt{x} + 1\right)^2$
$q(x) = (h(x))^2$
$q(x) = f(h(x))$
$q = f \circ h$

61. $P(x) = x^4$

$P(x) = (x^2)^2$
$P(x) = (f(x))^2$
$P(x) = f(f(x))$
$P = f \circ f$

63. $H(x) = (2x + 3)^4 = f(g(x))$

$f(x) = x^4,\ g(x) = 2x + 3$

65. $H(x) = \sqrt{x^2 + x + 1} = f(g(x))$

$f(x) = \sqrt{x},\ g(x) = x^2 + x + 1$

67. $H(x) = \left[1 - \frac{1}{x^2}\right]^2 = f(g(x))$

$f(x) = x^2,\ g(x) = 1 - \frac{1}{x^2}$

69. $H(x) = \text{int}(x^2 + 1) = f(g(x))$

$f(x) = \text{int}(x),\ g(x) = x^2 + 1$

71. $f(x) = 2x^3 - 3x^2 + 4x - 1 \quad g(x) = 2$

$(f \circ g)(x) = f(g(x)) = f(2) = 2(2)^3 - 3(2)^2 + 4(2) - 1 = 16 - 12 + 8 - 1 = 11$

$(g \circ f)(x) = g(f(x)) = 2$

73. $(f \circ g)(x) = f(g(x)) = f(3x + a) = 2(3x + a)^2 + 5$
When $x = 0$, $(f \circ g)(x) = f \circ g)(0) = 23$
Then, $2(3 \cdot 0 + a)^2 + 5 = 23$
$2a^2 + 5 = 23$
$2a^2 = 18$
$a^2 = 9$
$a = -3, 3$

75. $S(r) = 4\pi r^2$
$r(t) = \frac{2}{3}t^3,\ t \geq 0$
$S(r(t)) = 4\pi\left[\frac{2}{3}t^3\right]^2$
$= 4\pi\left[\frac{4}{9}t^6\right]$
$= \frac{16}{9}\pi t^6$

77. $N(t) = 100\,t - 5t^2,\ 0 \leq t \leq 10$
$C(N) = 15000 + 8000N$
$C(N(t)) = 15000 + 8000(100t - 5t^2)$
$C(N(t)) = 15000 + 800{,}000t - 40{,}000t^2$

79. $p = -\frac{1}{4}x + 100 \qquad 0 \leq x \leq 400$
$\frac{1}{4}x = 100 - p$
$x = 4(100 - p)$
$C = \frac{\sqrt{x}}{25} + 600 = \frac{\sqrt{4(100 - p)}}{25} + 600$
$= \frac{2\sqrt{100 - p}}{25} + 600$

81. Since f and g are odd, $f(-x) = -f(x)$ and $g(-x) = -g(x)$.
$(f \circ g)(-x) = f(g(-x)) = f(-g(x)) = -f(g(x)) = -(f \circ g)(x)$.
Because $(f \circ g)(-x) = -(f \circ g)(x)$, by definition, $f \circ g$ is odd.

4.5 Mathematical Models: Constructing Functions

1. If $V = \pi r^2 h$ and $h = 2r$, then $V(r) = \pi r^2(2r) = 2\pi r^3$

3. (a) If $p = \frac{-1}{6}x + 100$ and $R = xp$, then $R(x) = x\left[\frac{-1}{6}x + 100\right] = -\frac{1}{6}x^2 + 100x$

(b) $R(200) = \frac{-1}{6}(200)^2 + 100(200) = \$13{,}333$

(c)

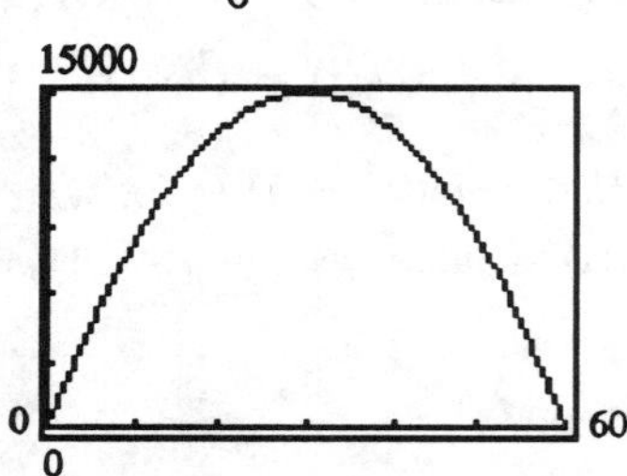

(d) 300; $15,000

(e) $p = -\frac{1}{6} \cdot 300 + 100 = \$50$

5. (a) If $x = -5p + 100$ and $R = xp$, then $p = \frac{100 - x}{5}$ and $R(x) = x\left[\frac{100 - x}{5}\right] = \frac{-1}{5}x^2 + 20x$

(b) $R(15) = \frac{-1}{5}(15)^2 + 20(15) = \$255$

(c)

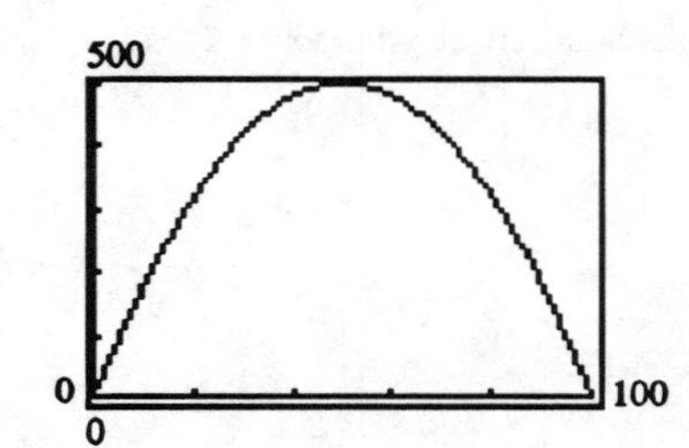

(d) 50; \$500

(e) $p = \dfrac{100 - 50}{5} = \dfrac{50}{5} = \$10$

7. (a) We know that width $= x$. In order to enclose a rectangular area, we need the perimeter, P. Let ℓ = length. $P = 2\ell + 2x = 400$.

Then $\ell = \dfrac{400 - 2x}{2} = 200 - x$. The area of the rectangle as a function of the width x, represented by $A(x) = \ell x = (200 - x)x = -x^2 + 200x$.

(b) $\{x \mid 0 < x < 200\}$

(c)

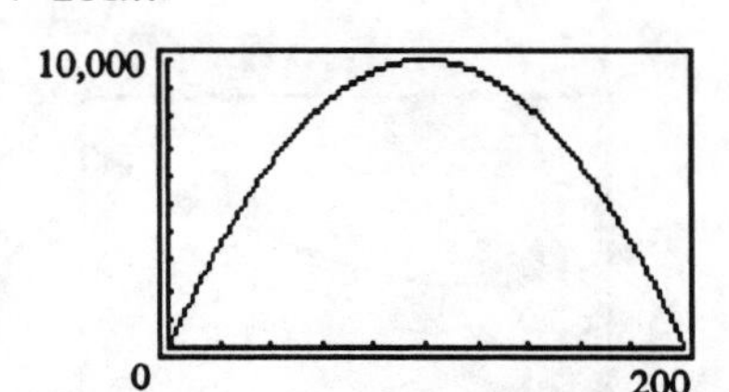

The value of a is largest when $x = 100$ yards.

9. (a) Let C = circumference, r = radius. We know that $C = 2\pi r$ by definition. If a wire of length x is bent into a circle, then x is the circumference, so $C = x = 2\pi r$. Writing the circumference as a function of x we have $C(x) = x$.

(b) We know that $C = x = 2\pi r$, so $r = \dfrac{x}{2\pi}$. By definition, the area of a circle is $A = \pi r^2$.

Expressing the area of the circle as a function of x, we have $A(x) = \pi\left[\dfrac{x}{2\pi}\right]^2 = \dfrac{x^2}{4\pi}$.

11. By definition, a triangle has area $A = \dfrac{1}{2}bh$, b = base, h = height. Because a vertex of the triangle is at the origin, we know that $b = x$ and $h = y$. Expressing the area of the triangle as a function of x, we have $A(x) = \dfrac{1}{2}xy = \dfrac{1}{2}x(x^3) = \dfrac{1}{2}x^4$.

13. (a) The distance d from P to the origin is $d = \sqrt{x^2 + y^2}$. Since P is a point on the graph of $y = x^2 - 8$, we have

$$d(x) = \sqrt{x^2 + (x^2 - 8)^2} = \sqrt{x^4 - 15x^2 + 64}$$

(b) If $x = 0$, the distance d is $d(0) = \sqrt{64} = 8$

(c) If $x = 1$, the distance d is $d(1) = \sqrt{1 - 15 + 64} = \sqrt{50} = 5\sqrt{2} \approx 7.07$

(d)

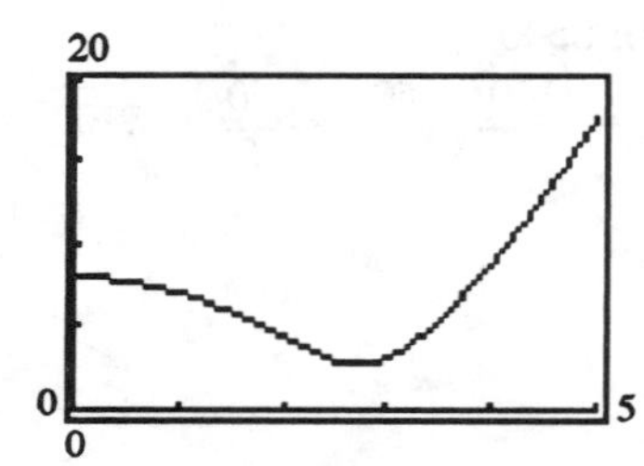

(e) d is smallest when x is 2.73.

15. (a) The distance d from P to the point (1, 0) is $d = \sqrt{(x - 1)^2 + y^2}$

Since P is a point on the graph of $y = \sqrt{x}$, we have

$$d(x) = \sqrt{(x - 1)^2 + \left(\sqrt{x}\right)^2} = \sqrt{x^2 - x + 1}$$

(b)

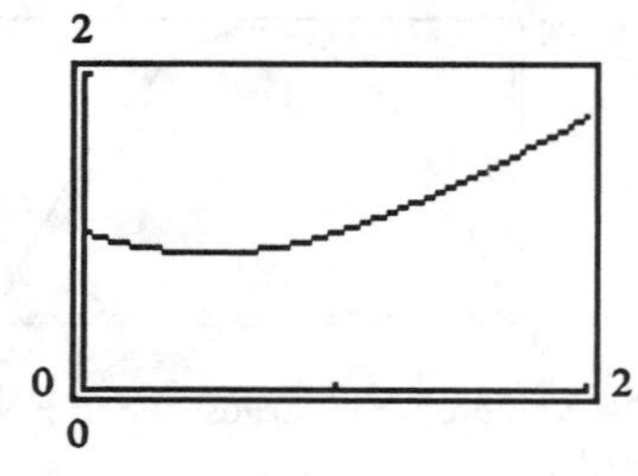

(c) d is smallest when x is 0.50.

17. We know that distance = (velocity)(time), $d = vt$. $d_1 = 25t$ and $d_2 = 40t$. By the Pythagorean Theorem,

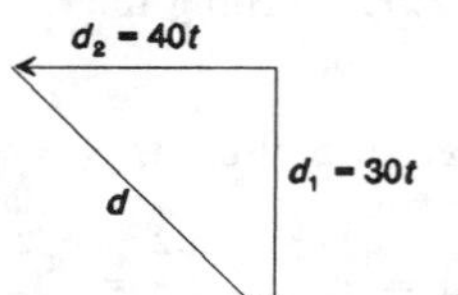

$$d^2 = d_1^2 + d_2^2$$

$$d^2 = (30t)^2 + (40t)^2$$

$$d(t) = \sqrt{900t^2 + 1600t^2}$$

$$d(t) = \sqrt{2500t^2} = 50t$$

19. (a) By definition, Volume,

V = (length)(width)(height)

length = $24 - 2x$, width = $24 - 2x$,

height = x

Therefore, $V(x) = x(24 - 2x)^2$

(b) $V(3) = 3(24 - 2(3))^2 = 972$ cubic inches

(c) $V(10) = 10(24 - 2(10))^2 = 160$ cubic inches

(d)

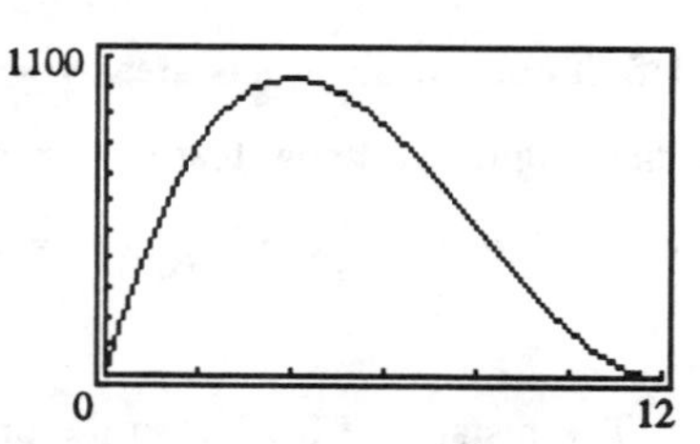

The volume is largest when x is about 4 inches.

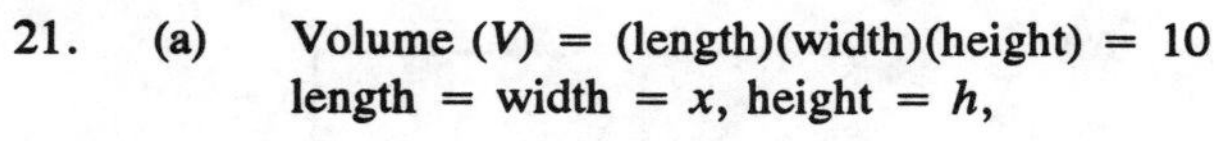

21. (a) Volume (V) = (length)(width)(height) = 10
length = width = x, height = h,
so $10 = x^2h$ and $h = \dfrac{10}{x^2}$
Area, $A = 2x^2 + 2xh + 2xh$

$$A(x) = 2x^2 + 4x\left(\frac{10}{x^2}\right) = 2x^2 + \frac{40}{x}$$

(b) $A(1) = 2(1)^2 + \dfrac{40}{1} = 2 + 40 = 42$ ft^2.

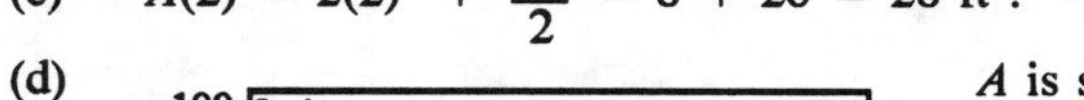

(c) $A(2) = 2(2)^2 + \dfrac{40}{2} = 8 + 20 = 28$ ft^2.

(d)

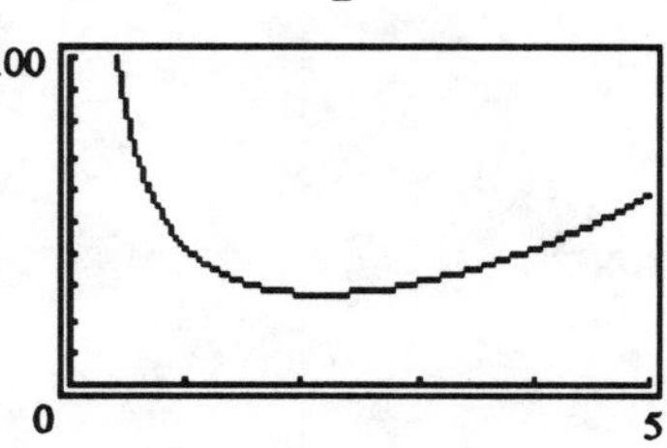

A is smallest when x is about 2.15 feet.

23. (a) A = Area
$A(x) = xy = x(16 - x^2)$
(b) Domain of $A = \{x \mid 0 < x < 4\}$ because $x > 0$ and $16 - x^2 > 0$

(c)

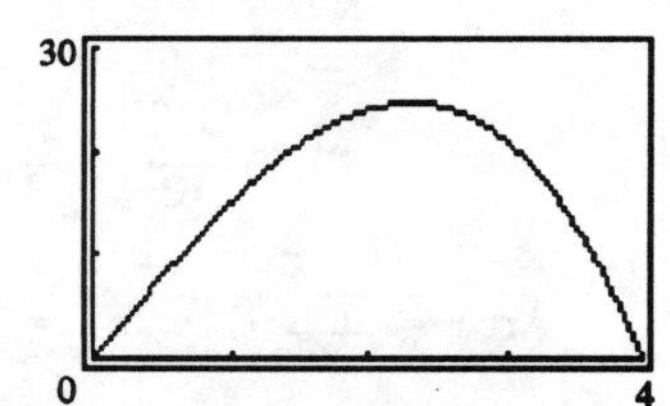

The area is largest for x about 2.31

25. A = Area, p = perimeter
(a) $A(x)$ = (length)(width) = $(2x)(2y) = 4x(4 - x^2)^{1/2}$
(b) $p(x)$ = 2 length + 2 width = $2(2x) = 2(2y) = 4x + 4(4 - x^2)^{1/2}$
(c)

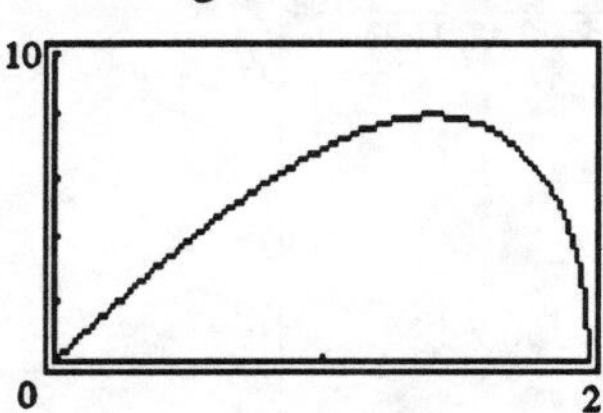

The area is largest for x about 1.41

(d)

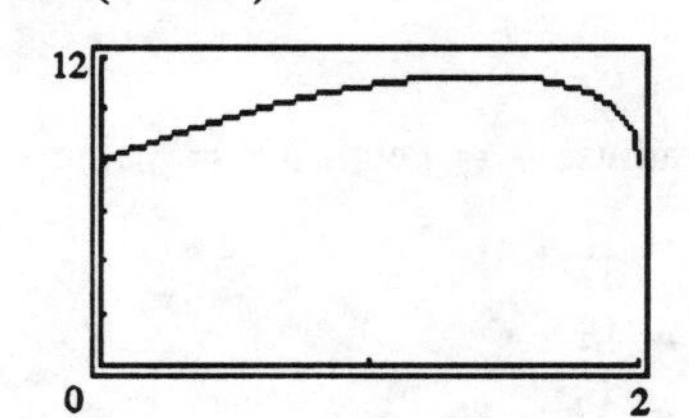

The perimeter is largest for x about 1.41

27. (a) C = Circumference, A = Area, r = radius, x = side of square

$$C = 2\pi r = 10 - 4x;\; r = \frac{5 - 2x}{\pi}$$

$$A(x) = x^2 + \pi r^2 = x^2 + \pi\left(\frac{5 - 2x}{\pi}\right)^2 = x^2 + \frac{25 - 20x + 4x^2}{\pi}$$

(b) Since all lengths must be positive, we have $x > 0$ and

$$10 - 4 > 0$$
$$-4x > -10$$
$$x < 2.5$$

Thus, domain

$$A = \{x \mid 0 < x < 2.5\}.$$

(c)

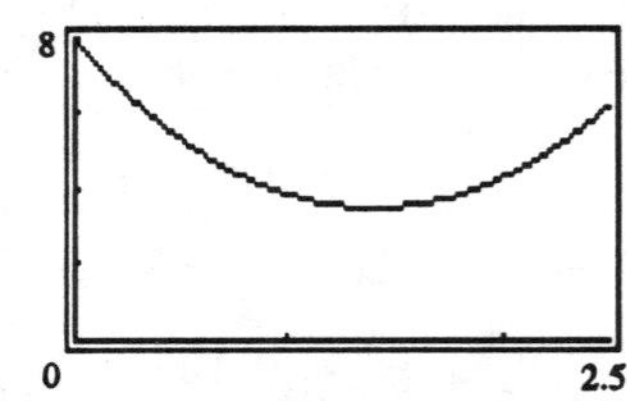

The area is smallest for x about 1.40 meters.

29. (a) A = Area, r = radius; diameter = $2r$

$A(r) = (2r)r = 2r^2$

(b) p = perimeter

$p(r) = 2(2r) + 2r = 6r$

31. Area of equilateral triangle $= \pi r^2 - \dfrac{\sqrt{3}}{4}x^2$

Area of equilateral triangle $= \dfrac{1}{2}x\sqrt{r^2 - \dfrac{x^2}{4}} = \dfrac{1}{3} \cdot \dfrac{\sqrt{3}}{4}x^2$

$$\sqrt{r^2 - \frac{x^2}{4}} = \frac{2}{3}\frac{\sqrt{3}}{4}x = \frac{x}{2\sqrt{3}}$$

$$r^2 - \frac{x^2}{4} = \frac{x^2}{12}$$

$$r^2 = \frac{4x^2}{12} = \frac{x^2}{3}$$

$$\text{Area} = \frac{\pi x^2}{3} - \frac{\sqrt{3}}{4}x^2 = \left(\frac{\pi}{3} - \frac{\sqrt{3}}{4}\right)x^2$$

33.

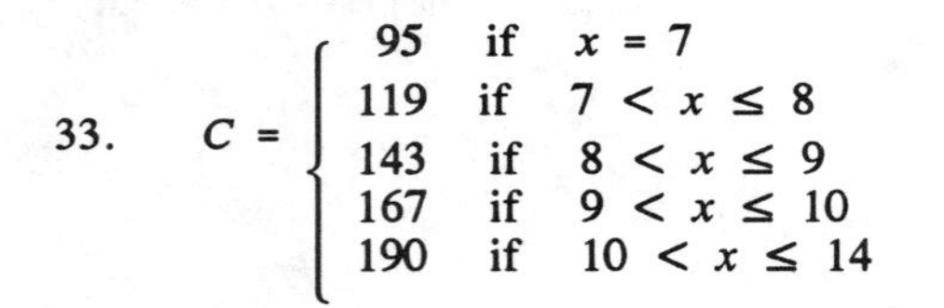

$$C = \begin{cases} 95 & \text{if } x = 7 \\ 119 & \text{if } 7 < x \le 8 \\ 143 & \text{if } 8 < x \le 9 \\ 167 & \text{if } 9 < x \le 10 \\ 190 & \text{if } 10 < x \le 14 \end{cases}$$

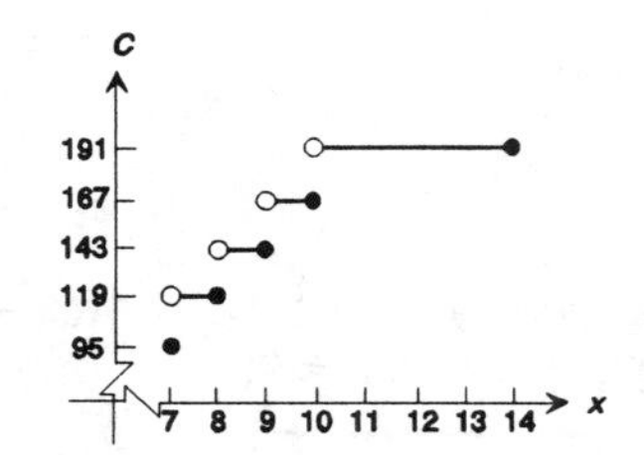

35. r = radius, h = height, V = volume of a cone

$$\frac{r}{h} = \frac{4}{16}$$
$$16r = 4h$$
$$r = \frac{1}{4}h$$
$$V = \frac{1}{3}\pi r^2 h$$
$$V(h) = \frac{1}{3}\pi\left(\frac{h}{4}\right)^2 h$$
$$= \frac{1}{48}\pi h^3$$

37. Since the height of the sphere is h, the length of the leg of the triangle is $\frac{h}{2}$. For the right triangle inside the sphere, we have

$$R^2 = \left(\frac{h}{2}\right)^2 + r^2$$
$$= \frac{h^2}{4} + r^2$$

Thus, $r^2 = R^2 - \frac{h^2}{4}$. The volume of the cylinder is

$$V = \pi r^2 h$$

Since $r^2 = R^2 - \frac{h^2}{4}$, we have

$$V(h) = \pi h \left(R^2 - \frac{h^2}{4}\right)$$

4 Chapter Review

1. $f(4) = -5$
$f(0) = 3$
$f(x) = mx + b$

$f(4)$: $4m + b = -5$
$f(0)$: $0m + b = 3$ } Solve two equations in two unknowns.

$$4m + b = -5$$
$$- b = -3$$
$$4m = -8$$
$$m = -2$$

$$-2(0) + b = 3$$
$$b = 3$$

Hence, $f(x) = -2x + 3$

3. $f(x) = \frac{Ax + 5}{6x - 2}$ and $f(1) = 4$

$f(1)$: $\frac{A(1) + 5}{6(1) - 2} = 4$

$$\frac{A + 5}{4} = 4$$
$$A + 5 = 16$$
$$A = 11$$

5. B, C, and D pass the vertical line test and therefore are functions.

7. $f(x) = \frac{3x}{x^2 - 4}$

(a) $f(-x) = \frac{-3x}{(-x)^2 - 4} = \frac{-3x}{x^2 - 4}$

(b) $-f(x) = -\left(\frac{3x}{x^2 - 4}\right) = -\frac{3x}{x^2 - 4}$

(c) $f(x + 2) = \frac{3(x + 2)}{(x + 2)^2 - 4} = \frac{3x + 6}{x^2 + 4x + 4 - 4} = \frac{3x + 6}{x^2 + 4x}$

(d) $f(x - 2) = \frac{3(x - 2)}{(x - 2)^2 - 4} = \frac{3x - 6}{x^2 - 4x + 4 - 4} = \frac{3x - 6}{x^2 - 4x}$

9. $f(x) = \sqrt{x^2 - 4}$

(a) $f(-x) = \sqrt{(-x)^2 - 4} = \sqrt{x^2 - 4}$

(b) $-f(x) = -\sqrt{x^2 - 4}$

(c) $f(x + 2) = \sqrt{(x + 2)^2 - 4} = \sqrt{x^2 + 4x + 4 - 4} = \sqrt{x^2 + 4x}$

(d) $f(x - 2) = \sqrt{(x - 2)^2 - 4} = \sqrt{x^2 - 4x + 4 - 4} = \sqrt{x^2 - 4x}$

11. $f(x) = \dfrac{x^2 - 4}{x^2}$

(a) $f(-x) = \dfrac{(-x)^2 - 4}{(-x)^2} = \dfrac{x^2 - 4}{x^2}$

(b) $-f(x) = -\left(\dfrac{x^2 - 4}{x^2}\right) = \dfrac{4 - x^2}{x^2} = -\dfrac{x^2 - 4}{x^2}$

(c) $f(x + 2) = \dfrac{(x + 2)^2 - 4}{(x + 2)^2} = \dfrac{x^2 + 4x + 4 - 4}{x^2 + 4x + 4} = \dfrac{x^2 + 4x}{x^2 + 4x + 4}$

(d) $f(x - 2) = \dfrac{(x - 2)^2 - 4}{(x - 2)^2} = \dfrac{x^2 - 4x + 4 - 4}{x^2 - 4x + 4} = \dfrac{x^2 - 4x}{x^2 - 4x + 4}$

13. $f(x) = x^3 - 4x$

if even: $f(-x) = f(x)$

$(-x)^3 - (-4x) = x^3 - 4x$

$-x^3 + 4x \neq x^3 - 4x$

if odd: $f(-x) = -f(x)$

$(-x)^3 - (-4x) = -(x^3 - 4x)$

$-x^3 + 4x = -x^3 + 4x$

Hence, function is odd.

15. $h(x) = \dfrac{1}{x^4} + \dfrac{1}{x^2} + 1$

if even: $h(-x) = h(x)$

$\dfrac{1}{(-x)^4} + \dfrac{1}{(-x)^2} + 1 = \dfrac{1}{x^4} + \dfrac{1}{x^2} + 1$

$\dfrac{1}{x^4} + \dfrac{1}{x^2} + 1 = \dfrac{1}{x^4} + \dfrac{1}{x^2} + 1$

Hence, $h(x)$ is even.

17. $G(x) = 1 - x + x^3$

if even: $G(-x) = G(x)$

$1 - (-x) + (-x)^3 = 1 - x + x^3$

$1 + x - x^3 \neq 1 - x + x^3$

if odd: $G(-x) = -G(x)$

$1 - (-x) + (-x)^3 = -(1 - x + x^3)$

$1 + x - x^3 \neq -1 + x - x^3$

Hence, $G(x)$ is neither even nor odd since $G(-x) \neq G(x)$ and $G(-x) \neq -G(x)$.

19. $f(x) = \dfrac{x}{x^2 - 9}$

The domain is the set of all values x such that

$x^2 - 9 \neq 0$

$(x - 3)(x + 3) \neq 0$

$x \neq 3, -3$

The domain is $\{x \mid x \neq -3, x \neq 3\}$.

21. $f(x) = \sqrt{2 - x}$

The domain consists of all values such that

$2 - x \geq 0\ x \leq 2$

The domain is $\{x \mid x \leq 2\}$ or $(-\infty, 2]$.

23. $h(x) = \dfrac{\sqrt{x}}{|x|}$

The domain is $\{x \mid x > 0\}$ or $(0, \infty)$.

25. $f(x) = \dfrac{x}{x^2 + 2x - 3}$

The domain consists of all values x such that

$$x^2 + 2x - 3 \neq 0$$
$$(x + 3)(x - 1) \neq 0$$
$$x \neq -3, 1$$

The domain is $\{x \mid x \neq -3, x \neq 1\}$.

27. $G(x) = \begin{cases} |x| & \text{if } -1 \leq x \leq 1 \\ \dfrac{1}{x} & \text{if } x > 1 \end{cases}$

The domain is $\{x \mid x \geq -1\}$ or $[-1, \infty)$.

29. $f(x) = \begin{cases} \dfrac{1}{x - 2} & \text{if } x > 2 \\ 0 & \text{if } x = 2 \\ 3x & \text{if } 0 \leq x \leq 2 \end{cases}$

The domain is $\{x \mid x \geq 0\}$ or $[0, \infty)$.

31. Average Rate of Change $= \dfrac{f(x) - f(0)}{x - 0} = \dfrac{2 - 5x - (2 - 5(0))}{x} = \dfrac{2 - 5x - 2}{x} = \dfrac{-5x}{x} = -5$

33. Average Rate of Change $= \dfrac{f(x) - f(0)}{x - 0} = \dfrac{3x - 4x^2 - (3(0) - 4(0)^2)}{x} = \dfrac{3x - 4x^2}{x} = \dfrac{x(3 - 4x)}{x}$

$= 3 - 4x$

35. $F(x) = |x| - 4$

Shift the graph of $y = |x|$ down 4 units.

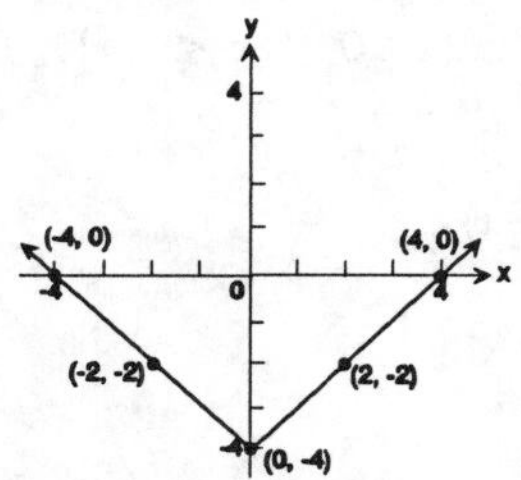

The domain is all real numbers. The range is $\{x \mid y \geq -4\}$.

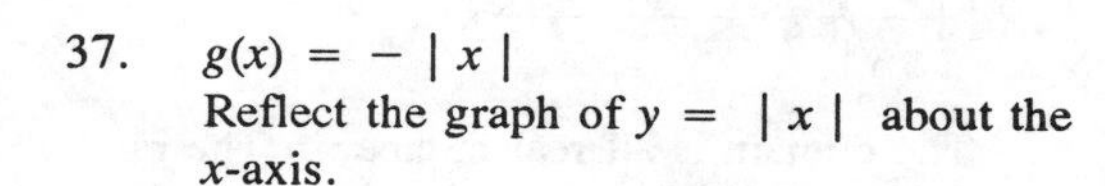

37. $g(x) = -|x|$

Reflect the graph of $y = |x|$ about the x-axis.

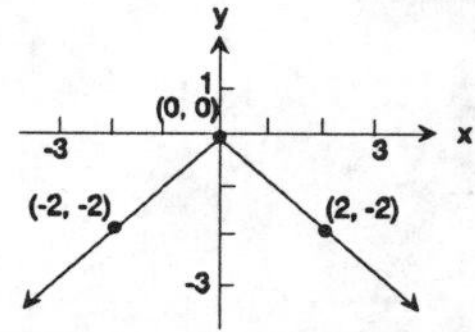

The domain is all real numbers. The range is $\{y \mid y \leq 0\}$.

39. $h(x) = \sqrt{x - 1}$

Shift the graph of $y = \sqrt{x}$ horizontally right one unit.

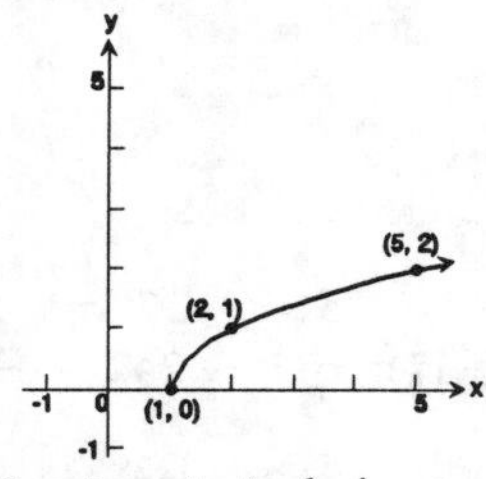

The domain is $\{x \mid x \geq 1\}$. The range is $\{y \mid y \geq 0\}$.

41. $f(x) = \sqrt{1 - x} = \sqrt{-x + 1}$

$= \sqrt{-(x - 1)}$

Reflect the graph of $y = \sqrt{x}$ about the y-axis, then shift right 1 unit.

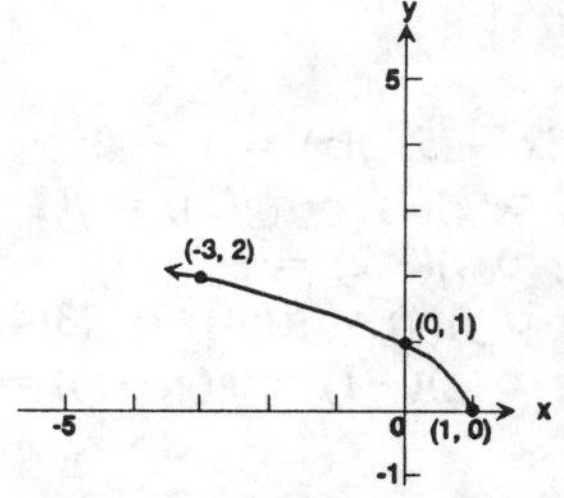

The domain is $\{x \mid x \leq 1\}$. The range is $\{y \mid y \geq 0\}$.

43. $F(x) = \begin{cases} x^2 + 4 & \text{if} \quad x < 0 \\ 4 - x^2 & \text{if} \quad x \geq 0 \end{cases}$

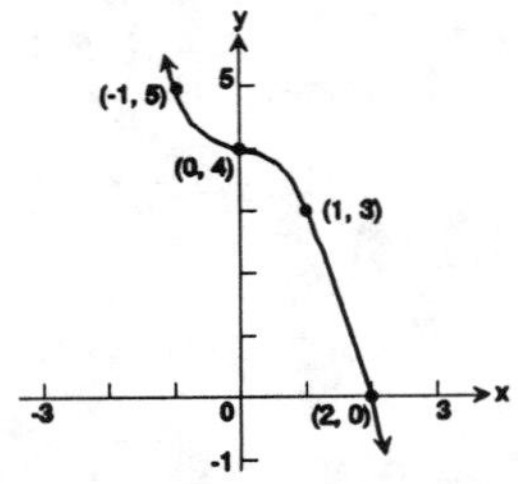

The domain is all real numbers. The range is all real numbers.

45. $h(x) = (x - 1)^2 + 2$

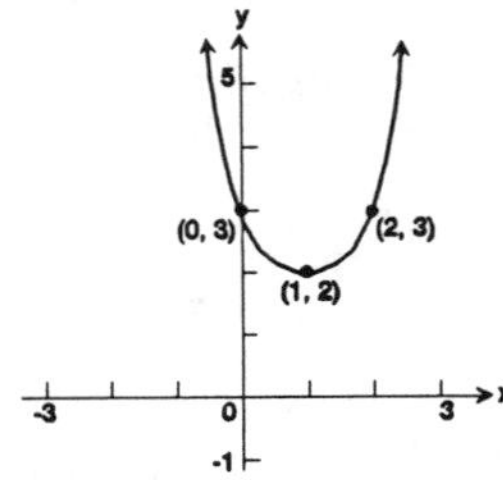

The domain is all real numbers. The range is $\{y \mid 2 \leq y < \infty\}$.

47. $g(x) = (x - 1)^3 + 1$

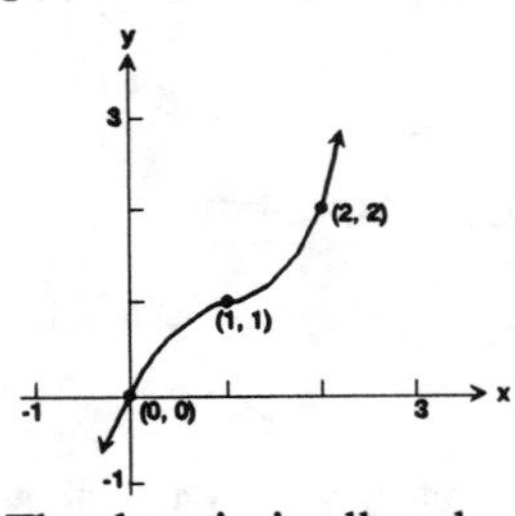

The domain is all real numbers. The range is all real numbers.

49. $f(x) = \begin{cases} 2\sqrt{x} & \text{if} \quad x \geq 4 \\ x & \text{if} \quad 0 < x < 4 \end{cases}$

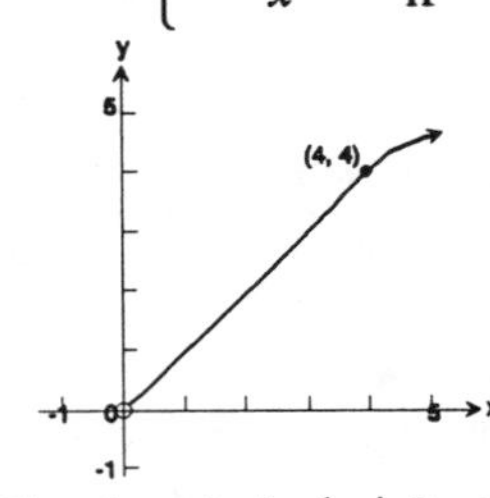

The domain is $\{x \mid 0 < x < \infty\}$. The range is $\{x \mid 0 < y < \infty\}$.

51. $g(x) = \dfrac{1}{x - 1} + 1$

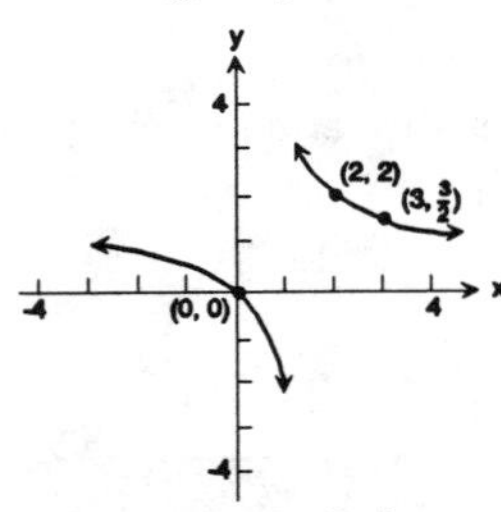

The domain is $\{x \mid x \neq 1\}$. Because $\dfrac{1}{x - 1}$ can never be zero, the range is $\{y \mid y \neq 1\}$.

53. $h(x) = \text{int}(-x)$

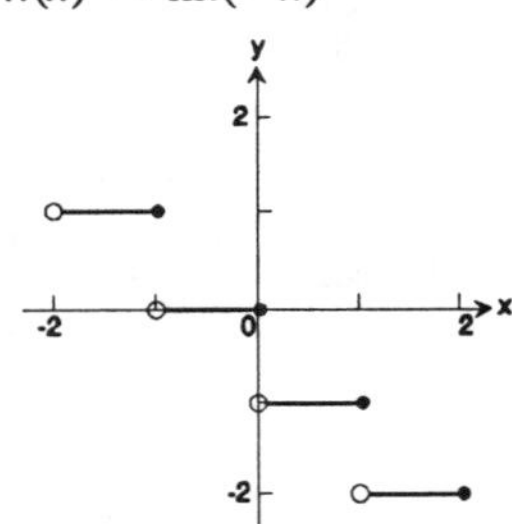

The domain is all real numbers. The range is set of all integers.

55. $f(x) = 3x - 5, \; g(x) = 1 - 2x^2$

(a) $(f \circ g)(2) = f(g(2)) = f(1 - 2(2)^2) = f(1 - 8) = f(-7) = 3(-7) - 5 = -26$

(b) $(g \circ f)(-2) = g(f(-2)) = g(3(-2) - 5) = g(-11) = 1 - 2(-11)^2 = 1 - 242 = -241$

(c) $(f \circ f)(4) = f(f(4)) = f(3(4) - 5) = f(7) = 3(7) - 5 = 16$

(d) $(g \circ g)(-1) = g(g(-1)) = g(1 - 2(-1)^2) = g(1 - 2) = g(-1) = 1 - 2(-1)^2 = 1 - 2 = -1$

57. $f(x) = \sqrt{x + 2}$, $g(x) = 2x^2 + 1$

(a) $(f \circ g)(2) = f(g(2)) = f(9) = \sqrt{11}$

(b) $(g \circ f)(-2) = g(f(-2)) = g(0) = 1$

(c) $(f \circ f)(4) = f(f(4)) = f(\sqrt{6}) = \sqrt{\sqrt{6} + 2}$

(d) $(g \circ g)(-1) = g(g(-1)) = g(3) = 19$

59. $f(x) = \dfrac{1}{x^2 + 4}$, $g(x) = 3x - 2$

(a) $(f \circ g)(2) = f(g(2)) = f(4) = \dfrac{1}{4^2 + 4} = \dfrac{1}{20}$

(b) $(g \circ f)(-2) = g(f(-2)) = g\left(\dfrac{1}{8}\right) = \dfrac{3}{8} - 2 = \dfrac{3}{8} - \dfrac{16}{8} = -\dfrac{13}{8}$

(c) $(f \circ f)(4) = f(f(4)) = f\left(\dfrac{1}{20}\right) = \dfrac{1}{\dfrac{1}{400} + 4} = \dfrac{1}{\dfrac{1601}{400}} = \dfrac{400}{1601}$

(d) $(g \circ g)(-1) = g(g(-1)) = g(-5) = -17$

61. $f(x) = 2 - x$, $g(x) = 3x + 1$.
The domain of f is all real numbers, and the domain of g is all real numbers.
$(f \circ g)(x) = f(g(x)) = f(3x + 1) = 2 - (3x + 1) + 2 - 3x - 1 = 1 - 3x.$
The domain of $f \circ g$ is all real numbers.
$(g \circ f)(x) = g(f(x)) = g(2 - x) = 3(2 - x) + 1 = 6 - 3x + 1 = 7 - 3x.$
The domain of $g \circ f$ is all real numbers.
$(f \circ f)(x) = f(f(x)) = f(2 - x) = 2 - (2 - x) = 2 - 2 + x = x$
The domain of $f \circ f$ is all real numbers.
$(g \circ g)(x) = g(g(x)) = g(3x + 1) = 3(3x + 1) + 1 = 9x + 3 + 1 = 9x + 4$
The domain of $g \circ g$ is all real numbers.

63. $f(x) = 3x^2 + x + 1$, $g(x) = |3x|$. The domain of f and g is all real numbers.
$(f \circ g)(x) = f(g(x)) = f(|3x|) = 3(|3x|)^2 + |3x| + 1 = 27x^2 + 3|x| + 1$
The domain of $f \circ g$ is all real numbers.
$(g \circ f)(x) = g(f(x)) = g(3x^3 + x + 1) = |3(3x^2 + x + 1)| = |9x^2 + 3x + 3|$
$= 3|3x^2 + x + 1|$. All real numbers.
$(f \circ f)(x) = f(f(x)) = f(3x^2 + x + 1) = 3(3x^2 + x + 1)^2 + (3x^2 + x + 1) + 1$
$= 3(3x^2 + x + 1)^2 + 3x^2 + x + 2$. All real numbers.
$(g \circ g)(x) = g(g(x)) = g(|3x|) = |3|3x|| = 9|x|$. All real numbers.

65. $f(x) = \dfrac{x + 1}{x - 1}$, $g(x) = \dfrac{1}{x}$. The domain of f is $\{x \mid x \neq 1\}$. The domain of g is $\{x \mid x \neq 0\}$.

$$(f \circ g)(x) = f(g(x)) = f\left(\frac{1}{x}\right) = \frac{\dfrac{1}{x} + 1}{\dfrac{1}{x} - 1} = \frac{\dfrac{1 + x}{x}}{\dfrac{1 - x}{x}} = \frac{1 + x}{x} \cdot \frac{x}{1 - x} = \frac{1 + x}{1 - x}.$$

Domain: $\{x \mid x \neq 0, x \neq 1\}$.

$(g \circ f)(x) = g(f(x)) = g\left(\dfrac{x + 1}{x - 1}\right) = \dfrac{1}{\dfrac{x + 1}{x - 1}} = \dfrac{x - 1}{x + 1}$. Domain is $\{x \mid x \neq -1, x \neq 1\}$.

$$(f \circ f)(x) = f(f(x)) = f\left[\frac{x+1}{x-1}\right] = \frac{\dfrac{x+1}{x-1}+1}{\dfrac{x+1}{x-1}-1} = \frac{\dfrac{x+1+x-1}{x-1}}{\dfrac{x+1-(x-1)}{x-1}} = \frac{\dfrac{2x}{x-1}}{\dfrac{2}{x-1}}$$

$$= \frac{2x}{x-1} \cdot \frac{x-1}{2} = x$$

Domain is $\{x \mid x \neq 1\}$.

$(g \circ g)(x) = g(g(x)) = g\left[\dfrac{1}{x}\right] = \dfrac{1}{\dfrac{1}{x}} = x$. Domain is $\{x \mid x \neq 0\}$.

67. (a) $y = f(-x)$
Reflect about the y-axis.

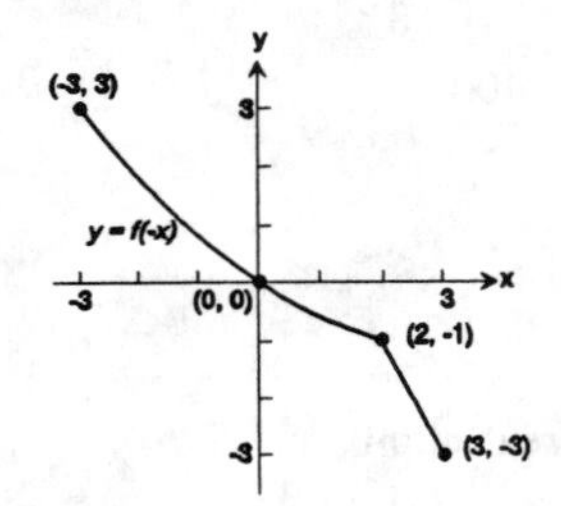

(b) $y = -f(x)$
Reflect about the x-axis.

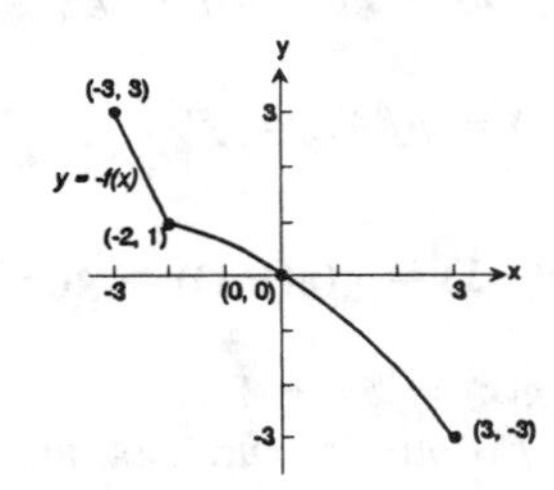

(c) $y = f(x + 2)$
Shift left 2 units.

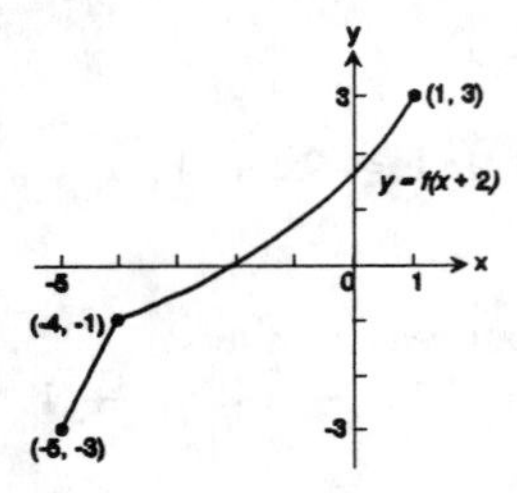

(d) $y = f(x) + 2$
Shift up 2 units.

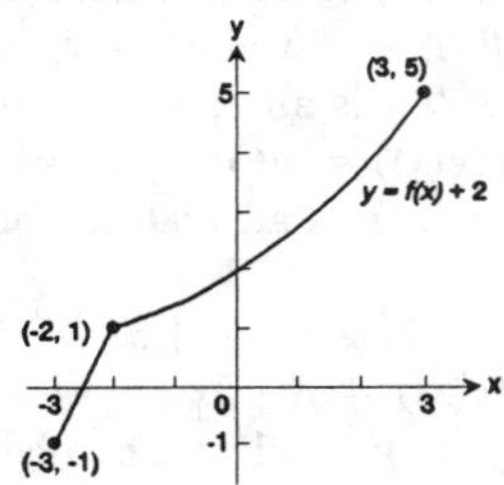

(e) $y = 2f(x)$
Multiply each y-coordinate by 2.

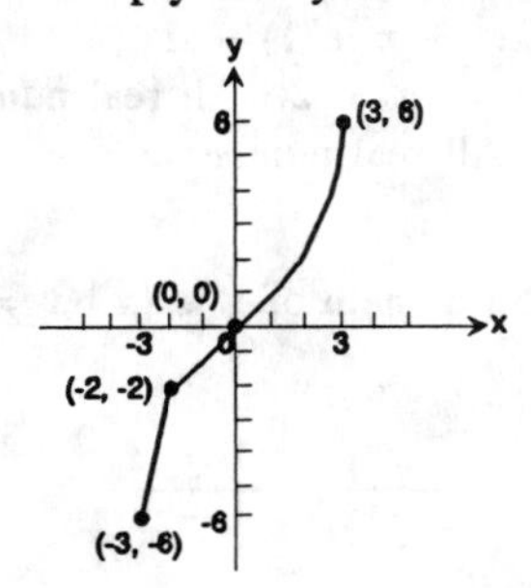

(f) $y = f(3x)$

Multiply each x-coordinate by $\dfrac{1}{3}$.

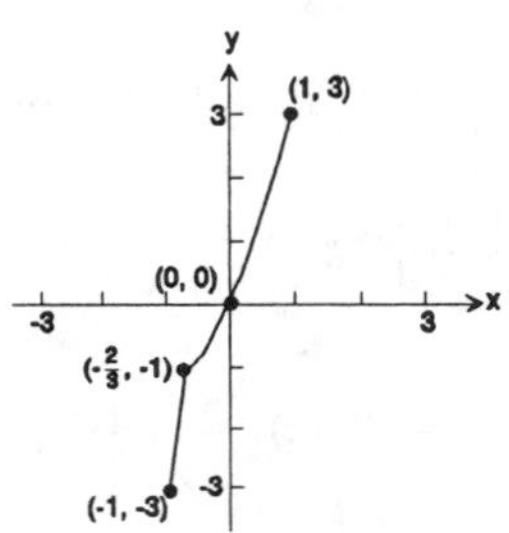

69. $T = T(h)$, $h = 0$ when $T = 30°$ and $h = 10{,}000$ when $T = 5°$
$T(h) = mh + b$ (linear function)

$$\left.\begin{array}{ll} T(0): & m(0) + b = 30° \\ T(10{,}000): & m(10{,}000) + b = 5° \end{array}\right\}$$

$$\begin{aligned} b &= 30° \\ 10{,}000m + 30 &= 5 \\ 10{,}000m &= -25 \\ m &= \frac{-25}{10{,}000} \\ m &= -0.0025 \end{aligned}$$

Hence, $T(h) = -0.0025h + 30$

71. Strength $= s = kxd^3$ where x = width, d = depth, and k is the constant of proportionality.

If the radius = 3 ft., then the diameter = 6 ft. and is a diagonal of the rectangle.

By the Pythagorean Theorem, we have $x^2 + d^2 = 6^2 \Rightarrow d = \sqrt{36 - x^2}$. Hence,

$$s(x) = kxd^3 = kx\left(\sqrt{36 - x^2}\right)^3 = kx(36 - x^2)^{3/2}.$$

Since all lengths must be positive, we have $x > 0$ and

$$\begin{aligned} 36 - x^2 &> 0 \\ (6 - x)(6 + x) &> 0 \\ 6 - x > 0 &\text{ and } 6 + x > 0 \\ x < 6 &\text{ and } x > -6 \end{aligned}$$

Thus, the domain of $s = \{x \mid 0 < x < 6\}$.

73. (a), (b), (e)

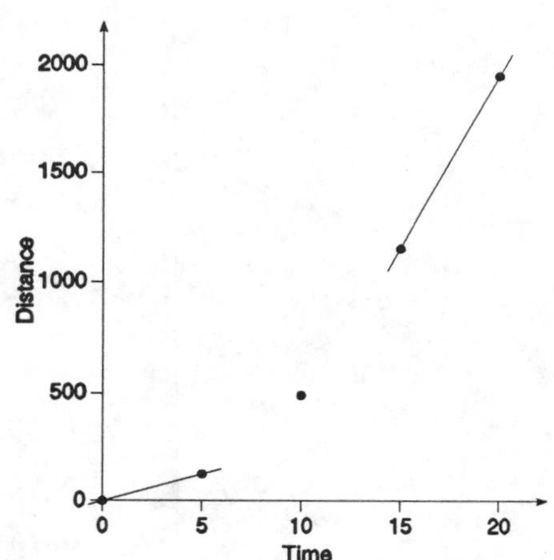

(c) Average Rate of Change $= \dfrac{112.5 - 0}{5 - 0} = 22.5$

(d) Between 0 and 5 seconds, the average speed of the parachutist is 22.5 ft/sec.

(e) See (a).

(f) Average Rate of Change $= \dfrac{1960 - 1102.5}{20 - 15} = \dfrac{857.5}{5} = 171.5$

(g) Between 15 and 20 seconds, the average speed of the parachutist is 171.5 ft/sec.

(h) As time passes, average speed increases.

75. (a) C = Cost of the material (in cents), r = radius, h = height

$$500 = \pi r^2 h, \; h = \frac{500}{\pi r^2}$$

$$C(r) = 6(2\pi r^2) + 4(2\pi r h)$$

$$= 12\pi r^2 + 8\pi r\left(\frac{500}{\pi r^2}\right)$$

$$= 12\pi r^2 + \frac{4000}{r}$$

(b) $C(4) = 12\pi(4)^2 + \dfrac{4000}{4} = 192\pi + 1000 = 1603$ cents or \$16.03.

(c) $C(8) = 12\pi(8)^2 + \dfrac{4000}{8} = 768\pi + 500 = 2913$ cents or \$19.13

(d) The cost is least for r about 3.75 cm.

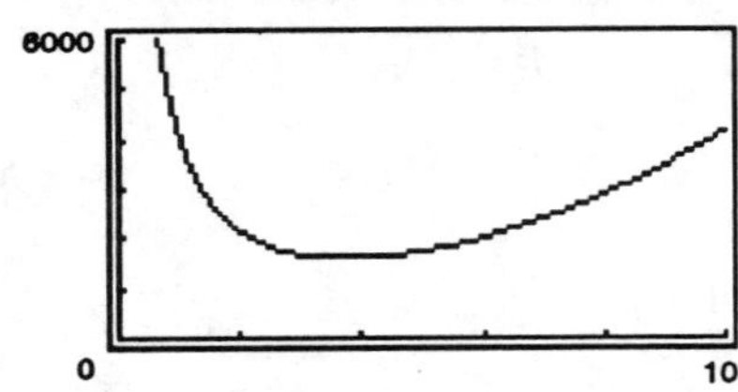

Chapter 5

POLYNOMIAL AND RATIONAL FUNCTIONS

5.1 Quadratic Functions; Curve Fitting

1. D
3. A
5. B
7. E

In each of Problems 9–27, we start with the graph of $y = x^2$, and apply shifting, compressions and stretches and/or reflections.

9. The graph of $f(x) = \frac{1}{4}x^2$ is a vertically compressed version of the graph of $y = x^2$. For each x, the y-coordinate of a point on the graph of $f(x) = \frac{1}{4}x^2$ is $\frac{1}{4}$ times as large as the corresponding y-coordinate on the graph of $y = x^2$.

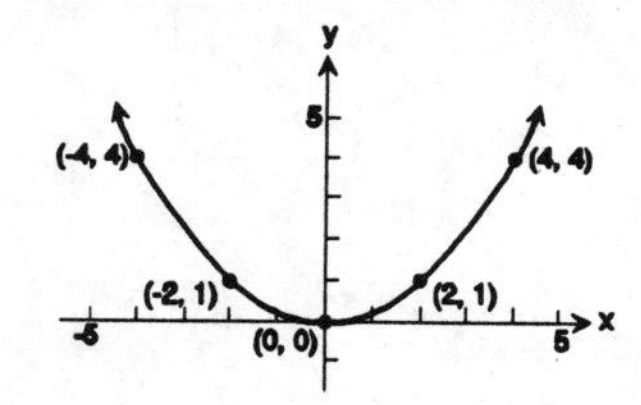

11. For $f(x) = \frac{1}{4}x^2 - 2$, we use the results of Problem 9, and then apply a vertical shift, moving the graph ***down*** 2 units.

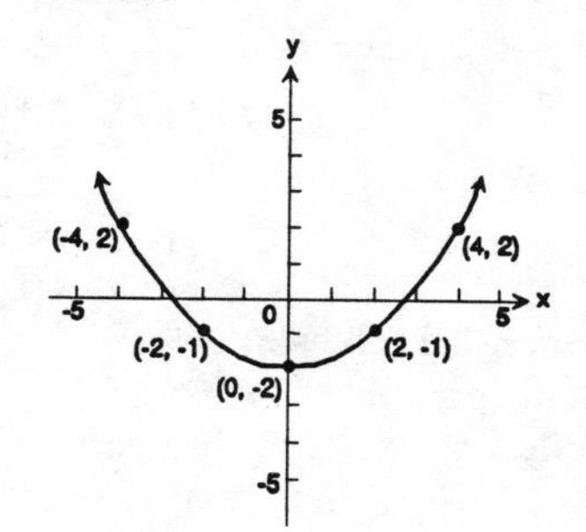

13. $f(x) = \frac{1}{4}x^2 + 2$. Use the results of Problem 9, followed by a vertical shift, 2 units ***upward***.

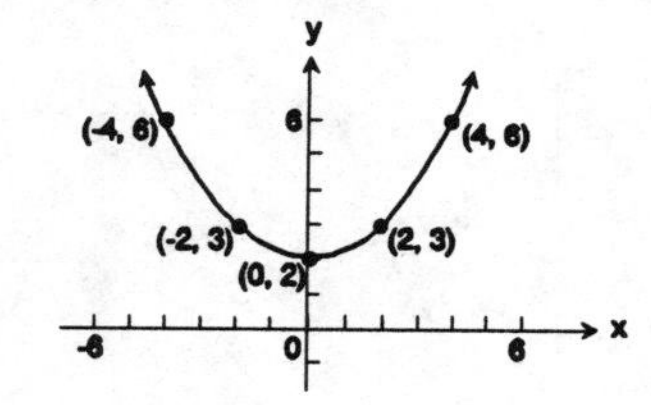

15. For $f(x) = \frac{1}{4}x^2 + 1$, use the results of Problem 9, followed by a vertical shift 1 unit ***upward***.

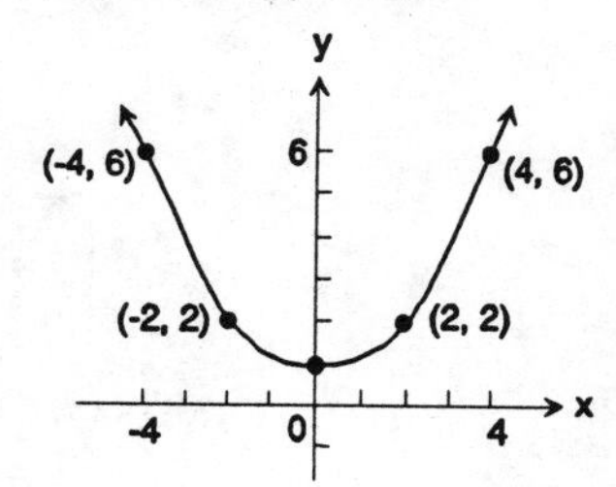

17. For $f(x) = x^2 + 4x + 2$, we must first complete the square in order to write the function in the form

$$f(x) = (x + h)^2 + k$$

We will then be able to perform horizontal and vertical shifts.

$$\begin{aligned} f(x) &= x^2 + 4x + 2 \\ &= (x^2 + 4x + 4) + 2 - 4 \\ &= (x + 2)^2 - 2 \end{aligned}$$

For $f(x) = (x + 2)^2$, shift the graph of $f(x) = x^2$ ***horizontally*** 2 units to the ***left***. Then, to obtain $f(x) = (x + 2)^2 - 2$, shift the graph ***down*** 2 units.

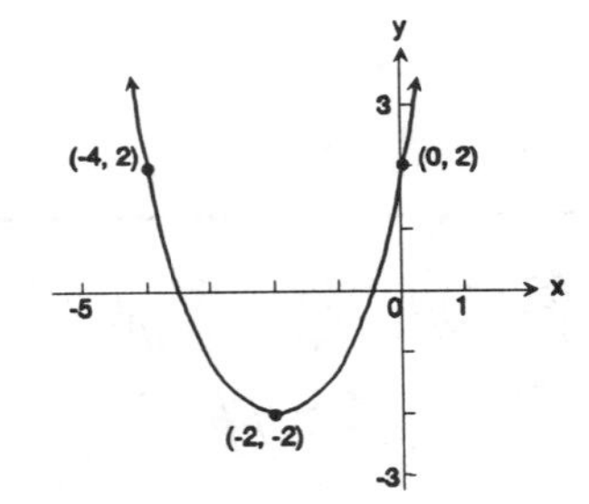

19. By completing the square, $f(x) = 2x^2 - 4x + 1$ can be put in the form $f(x) = 2(x + h)^2 + k$:

$$\begin{aligned} f(x) &= 2x^2 - 4x + 1 \\ &= 2(x^2 - 2x + \underline{\quad}) + 1 - \underline{\quad} \\ &= 2(x^2 - 2x + 1) + 1 - 2 \end{aligned}$$

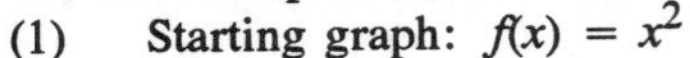

$$(2)(1) = 2$$

$$= 2(x - 1)^2 - 1$$

We can now perform the following steps:

(1) Starting graph: $f(x) = x^2$
(2) Vertical stretch of the graph of $f(x) = x^2$: $f(x) = 2x^2$
(3) Shift to the ***right*** one unit: $f(x) = 2(x - 1)^2$
(4) Shift ***down*** one unit: $f(x) = 2(x - 1)^2 - 1$

The final graph is shown at right.

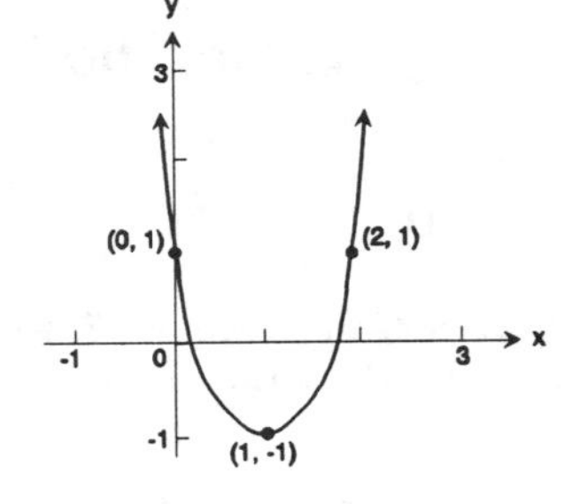

21. First complete the square:

$$\begin{aligned} f(x) &= -x^2 - 2x \\ &= -1(x^2 + 2x + \underline{\quad}) - \underline{\quad} \\ &= -1(x^2 + 2x + 1) - (-1) \end{aligned}$$

$$(-1)(1) = -1$$

$$= -1(x + 1)^2 + 1$$

Now perform the following steps:

(1) Starting graph: $f(x) = x^2$
(2) Reflect about the x-axis: $f(x) = -x^2$
(3) Shift ***left*** one unit: $f(x) = -(x + 1)^2$
(4) Shift ***up*** one unit: $f(x) = -(x + 1)^2 + 1$

The final graph is shown at right.

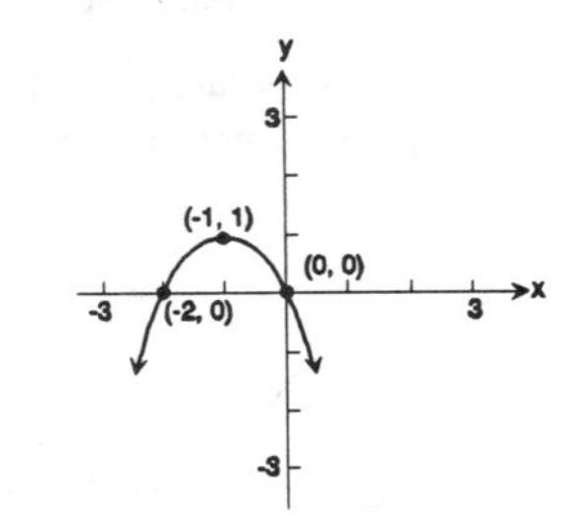

23. Complete the square:

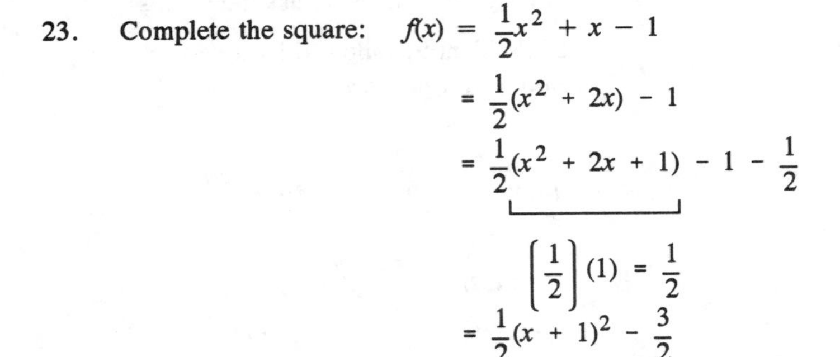

$$\begin{aligned} f(x) &= \frac{1}{2}x^2 + x - 1 \\ &= \frac{1}{2}(x^2 + 2x) - 1 \\ &= \frac{1}{2}(x^2 + 2x + 1) - 1 - \frac{1}{2} \end{aligned}$$

$$\left[\frac{1}{2}\right](1) = \frac{1}{2}$$

$$= \frac{1}{2}(x + 1)^2 - \frac{3}{2}$$

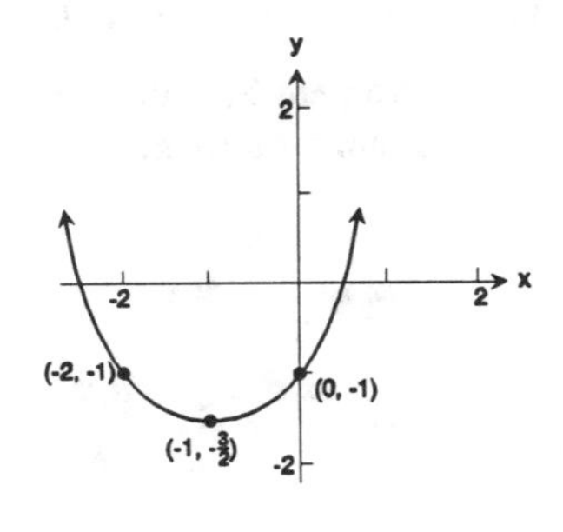

To graph $f(x) = \frac{1}{2}(x + 1)^2 - \frac{3}{2}$, apply a vertical compression of the graph of $y = x^2$, shift to the ***left*** one unit, and then shift down $\frac{3}{2}$ unit.

In Exercises 29–41, use the formulas in Display (2) to locate the vertex and axis. The vertex will be denoted (h, k).

25. $f(x) = x^2 + 2x - 8$. Here, $a = 1$, $b = 2$, $c = -8$, so that

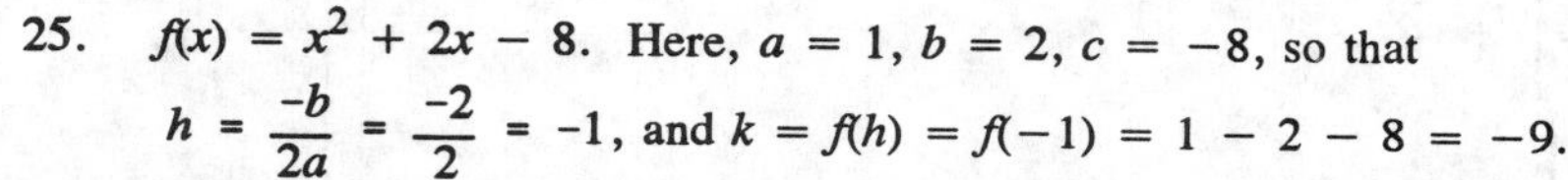

$h = \dfrac{-b}{2a} = \dfrac{-2}{2} = -1$, and $k = f(h) = f(-1) = 1 - 2 - 8 = -9$.

Therefore, the vertex is $(h, k) = (-1, -9)$, the axis is the line $x = h$ or $x = -1$, and the parabola opens upward since $a = 1 > 0$.

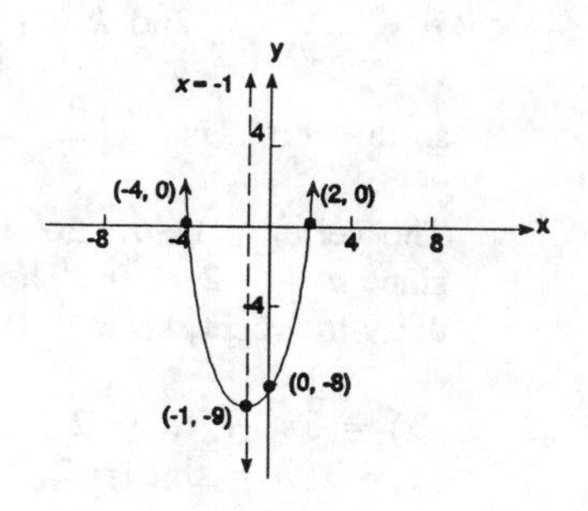

To find the y-intercepts, set $x = 0$: $y = f(0) = -8$. To find the x-intercepts, set $y = 0$ and solve for x:

$$y = x^2 + 2x - 8 = 0$$
$$(x + 4)(x - 2) = 0$$
$$x = -4 \text{ and } x = 2$$

27. $f(x) = -x^2 - 3x + 4$

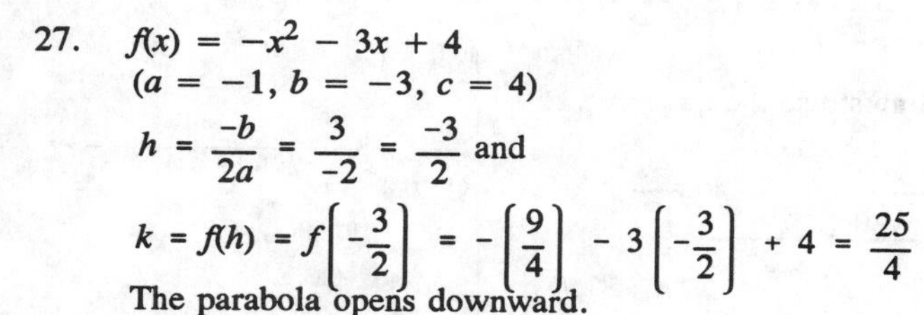

$(a = -1, b = -3, c = 4)$

$h = \dfrac{-b}{2a} = \dfrac{3}{-2} = \dfrac{-3}{2}$ and

$k = f(h) = f\left(-\dfrac{3}{2}\right) = -\left(\dfrac{9}{4}\right) - 3\left(-\dfrac{3}{2}\right) + 4 = \dfrac{25}{4}$

The parabola opens downward.

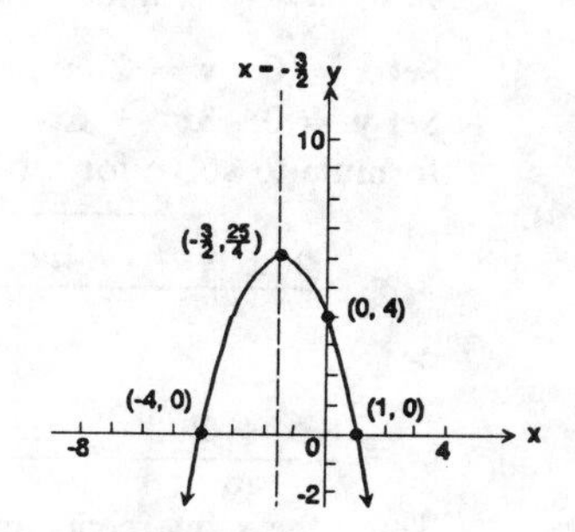

Set $x = 0$: $y = 4$ is the y-intercept.

Set $y = 0$: $y = -x^2 - 3x + 4 = 0$

$$x^2 + 3x - 4 = 0$$
$$(x + 4)(x - 1) = 0$$
$$x = -4, x = 1 \text{ (}x\text{-intercepts)}$$

29. $f(x) = x^2 + 2x + 1$

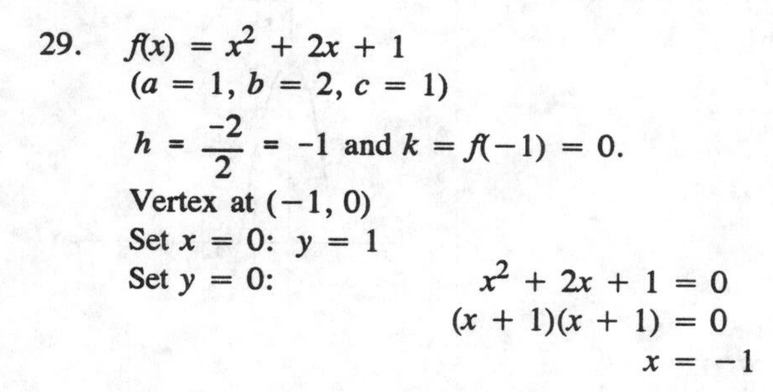

$(a = 1, b = 2, c = 1)$

$h = \dfrac{-2}{2} = -1$ and $k = f(-1) = 0$.

Vertex at $(-1, 0)$

Set $x = 0$: $y = 1$

Set $y = 0$:

$$x^2 + 2x + 1 = 0$$
$$(x + 1)(x + 1) = 0$$
$$x = -1$$

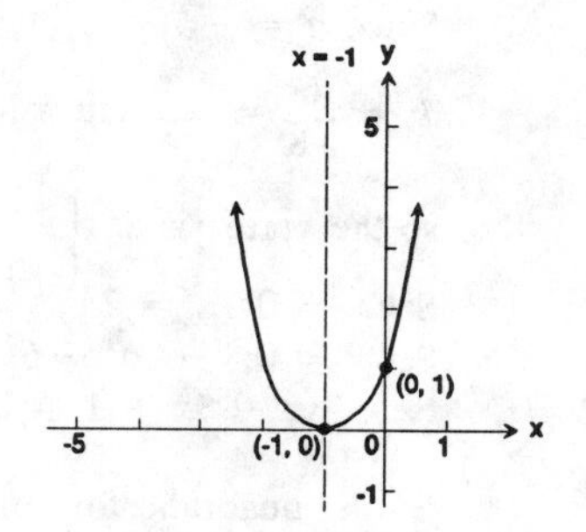

31. $f(x) = 2x^2 - x + 2$

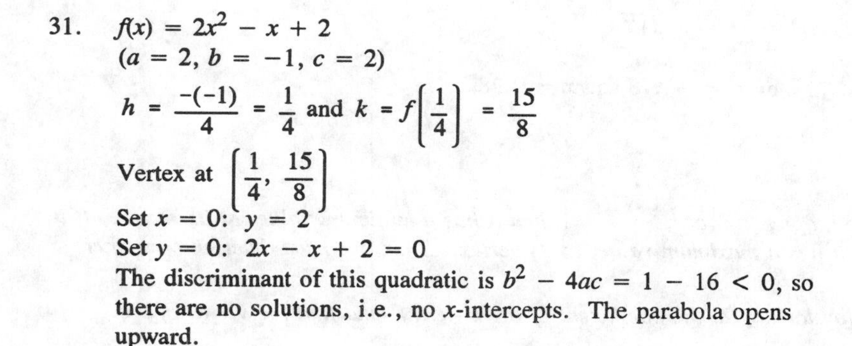

$(a = 2, b = -1, c = 2)$

$h = \dfrac{-(-1)}{4} = \dfrac{1}{4}$ and $k = f\left(\dfrac{1}{4}\right) = \dfrac{15}{8}$

Vertex at $\left(\dfrac{1}{4}, \dfrac{15}{8}\right)$

Set $x = 0$: $y = 2$

Set $y = 0$: $2x - x + 2 = 0$

The discriminant of this quadratic is $b^2 - 4ac = 1 - 16 < 0$, so there are no solutions, i.e., no x-intercepts. The parabola opens upward.

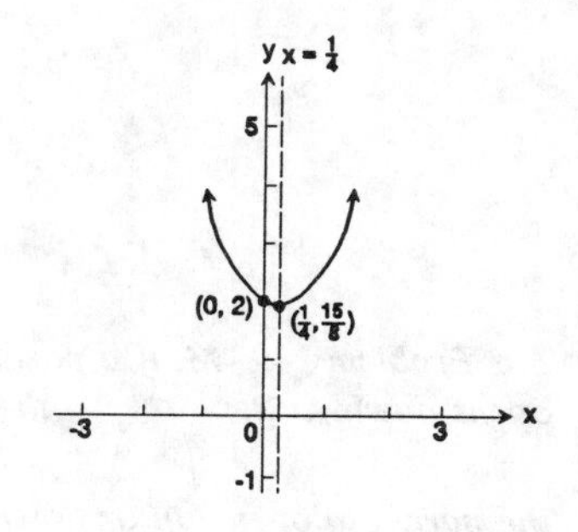

33. $f(x) = -2x^2 + 2x - 3$
Here, $a = -2$, $b = 2$, $c = -3$

$$h = \frac{-2}{-4} = \frac{1}{2} \text{ and } k = f\left(\frac{1}{2}\right) = \frac{-5}{2},$$

so the vertex is at $\left(\frac{1}{2}, \frac{-5}{2}\right)$.

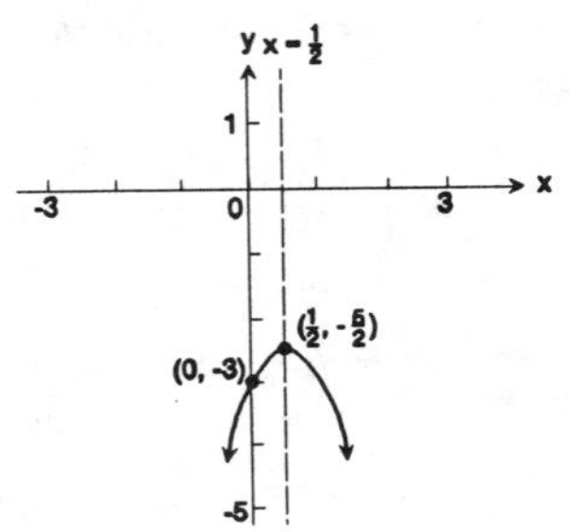

The vertex is below the x-axis, and the parabola opens downward, since $a = -2 < 0$. Therefore, there can be no x-intercepts. To find the y-intercept, set $x = 0$: $y = f(0) = -3$.

35. $f(x) = 3x^2 + 6x + 2$
($a = 3$, $b = 6$, $c = 2$)

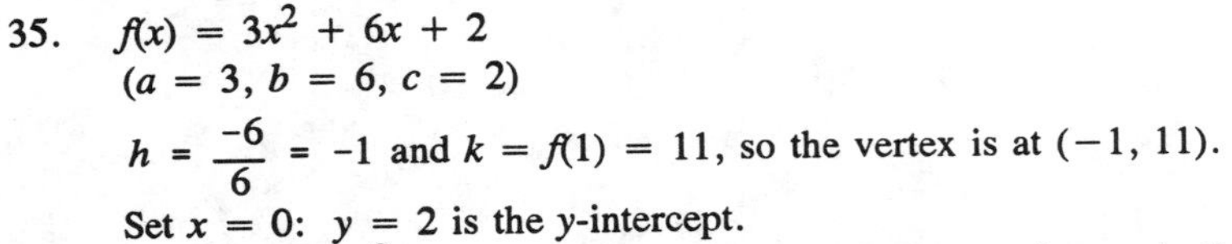

$h = \frac{-6}{6} = -1$ and $k = f(1) = 11$, so the vertex is at $(-1, 11)$.

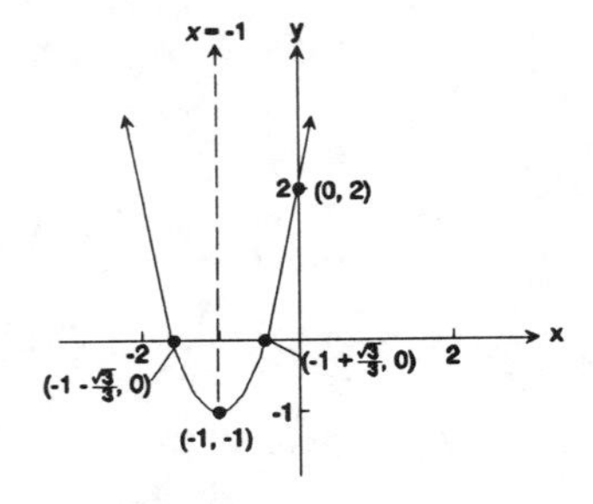

Set $x = 0$: $y = 2$ is the y-intercept.
Set $y = 0$: $3x^2 + 6x + 2 = 0$. Here we must use the quadratic formula to solve for x (the x-intercepts):

$$x = \frac{-b \pm \sqrt{b^2 - 4ac}}{2a} = \frac{-6 \pm \sqrt{36 - 4(6)}}{6} = \frac{-6 \pm \sqrt{12}}{6}$$

or

$$x = \frac{-6 \pm 2\sqrt{3}}{6} = \frac{-3 \pm \sqrt{3}}{3}, \; x \approx \frac{-3 \pm 1.732}{3}$$

Thus, the x-intercepts are at $x = -0.42$ and $x = -1.58$

37. $f(x) = -4x^2 - 6x + 2$
($a = -4$, $b = -6$, $c = 2$)
Since $a = -4 < 0$, the parabola opens downward.

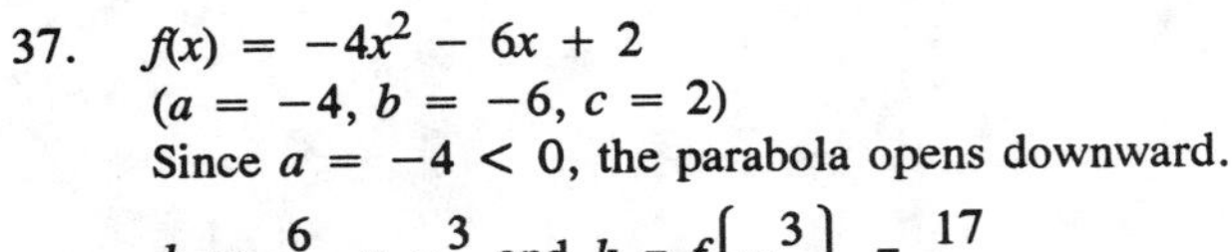

$$h = \frac{6}{-8} = -\frac{3}{4} \text{ and } k = f\left(-\frac{3}{4}\right) = \frac{17}{4}$$

so the vertex is at $\left(-\frac{3}{4}, \frac{17}{4}\right)$.

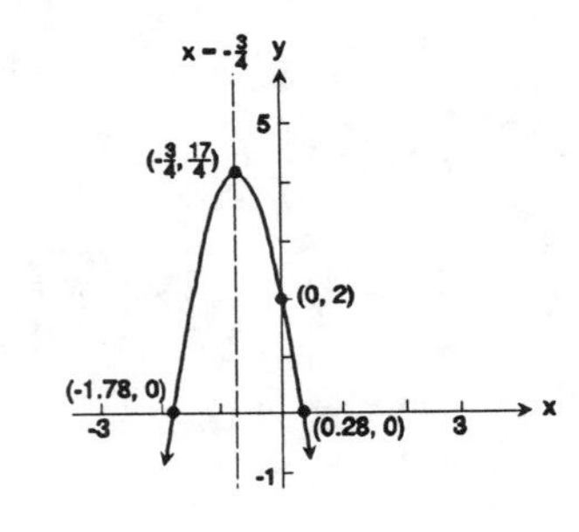

Set $x = 0$: $y = 2$
Set $y = 0$: $-4x^2 - 6x + 2 = 0$
or $-2x^2 - 3x + 1 = 0$.

By the quadratic formula,
$$x = \frac{3 \pm \sqrt{9 - 4(-2)}}{-4}$$
$$x = \frac{3 \pm \sqrt{17}}{-4}$$
or $x = -1.78$ and $x = 0.28$

*For Problems 39–43, if a parabola opens upward, ⋃, then it has a minimum value, **at its vertex.** If it opens downward, ⋂, then it has a maximum value, **at its vertex.** Thus, we must determine whether the parabola opens up or down (by looking at the coefficient of the x^2-term), and find the vertex of the parabola, using Display (2).*

39. $f(x) = 2x^2 + 12x - 3$ $(a = 2, b = 12, c = -3)$
Since $a = 2 > 0$, the parabola opens *upward*, and therefore has a *minimum* at the vertex (h, k).
By (3), $h = \dfrac{-b}{2a} = \dfrac{-12}{4} = -3$, and $k = f(h) = f(-3)$ or $k = -21$. Therefore the minimum value of $f(x)$ is $k = -21$.

41. $f(x) = -x^2 + 10x - 4$ $(a = -1, b = 10, c = -4)$
Here, $a = -1 < 0$, so the parabola opens downward and has a *maximum* at (h, k).
$h = \dfrac{-10}{-2} = 5$ and $k = f(5) = 21$. The maximum value of $f(x)$ is $k = 21$.

43. $f(x) = -3x^2 + 12x + 1$ $(a = -3, b = 12, c = 1)$
Here, $a = -3 < 0$, so the parabola opens downward and $f(x)$ has a maximum at (h, k).
$h = \dfrac{-12}{-6} = 2$ and $k = f(2) = 13$. So the maximum value of $f(x)$ is $k = 13$.

45. Since $f(x) = ax^2 + bx + c$ has vertex at $x = 0$, we have $\dfrac{-b}{2a}$ or $b = 0$. At $(0, 2)$, we have $2 = a(0)^2 + c$ or $c = 2$. At $(1, 8)$, we have $8 = a(1)^2 = 2$ or $a = 6$. Thus, $f(x) = 6x^2 + 2$.

47. We are given $R = -4p^2 + 4000p$, which represents a parabola that opens downward (since $a = -4 < 0$). Thus, R will be a maximum at the vertex. Now, $h = \dfrac{-b}{2a} = \dfrac{-4000}{-8} = 500$, and $k = f(h) = f(500) = -4(500)^2 + 4000(500) = 1{,}000{,}000$. The maximum revenue is $R = k = \$1{,}000{,}000$. It occurs when $p = h = 500$ dollars per dryer.

49. Let ℓ be the length and w be the width of the rectangle. Since the perimeter is 400 feet, we have $2\ell + 2w = 400$, so that $\ell = 200 - w$. We want to maximize the area, $A = \ell w = (200 - w)w$ $= -w^2 + 200w$. Now $A = -w^2 + 200w$ is the equation of a parabola that opens downward, and hence has a maximum value w at $w = \dfrac{-b}{2a} = \dfrac{-200}{-2} = 100$, and the value is $f(100) =$ $-(100)^2 + 200(100) = 10000$. The maximum area is 10000 sq. ft., and this occurs when $w = 100$ and $\ell = 200 - w = 200 - 100 = 100$. Thus, the dimensions of the rectangle are 100 ft. by 100 ft.

51. The area of the rectangular plot is given by width times length, or $A = x(4000 - 2x)$. (Refer to the figure.) Thus, $A = -2x^2 + 4000x$; hence, A will be a maximum at the vertex.
$h = \dfrac{-4000}{-4} = 1000$, and $k = f(1000) = -2{,}000{,}000 + 4{,}000{,}000 = 2{,}000{,}000$. Thus, the largest area that can be enclosed is $A = k = 2{,}000{,}000$ square meters.

53. Consider the figure: The total length of the fence is $x + x + x + y + y$, and this must equal 10,000 meters: $3x + 2y = 10{,}000$; or $2y = 10{,}000 - 3x$ so that $y = \dfrac{10{,}000 - 3x}{2}$. The total area enclosed is $A = xy = x\left(\dfrac{10{,}000 - 3x}{2}\right)$, or $A = \dfrac{-3}{2}x^2 + 5{,}000x$, which attains a *maximum* at the vertex (since $a = \dfrac{-3}{2} < 0$). Now, $h = \dfrac{-5{,}000}{-3} = 1666.67$. Thus, the maximum area is $A = k = f(1666.67)$ $= 4{,}166{,}666.67$ square meters.

55. (a) The maximum height of the projectile occurs at $x = \dfrac{-b}{2a}$ where $a = \dfrac{-32}{2500}$ and $b = 1$.

Therefore, the maximum height of the projectile occurs at $x = \dfrac{-1}{2\left(\dfrac{-32}{2500}\right)} = 39.0625$ feet.

(b) The maximum height is $f(39.0625) = \dfrac{-32(39.065)^2}{2500} + 39.06252 + 200 = 219.53$ feet.

(c) The projectile will strike the water when the height is zero. Therefore, solve the quadratic equation $\dfrac{-32}{2500}x^2 + x + 200 = 0$ using the quadratic formula with $a = \dfrac{-32}{2500}$, $b = 1$ and $c = 200$.

$$x = \frac{-1 \pm \sqrt{1^2 - 4\left(\dfrac{-32}{2500}\right)(200)}}{2\left(\dfrac{-32}{2500}\right)}$$

$$x = \frac{-1 \pm \sqrt{11.24}}{\dfrac{-64}{2500}}$$

$x \approx -91.90$ or $x \approx 170.02$

Since distance is not negative, the projectile will strike the water 170.02 feet from the base of the cliff.

(d)

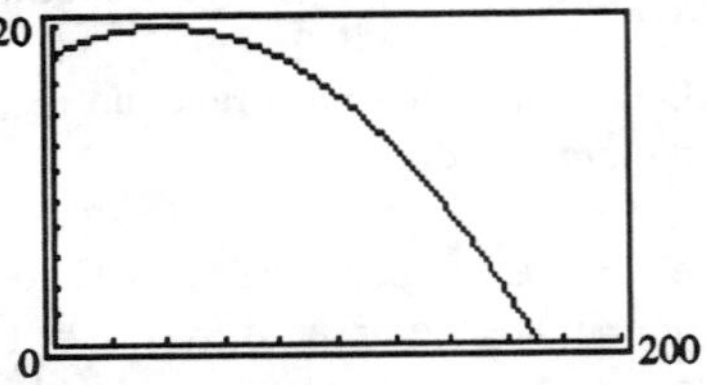

(f) When the height is 100 feet, the projectile is 135.69 feet from the cliff.

57. The situation at 4:00 P.M. is depicted below. 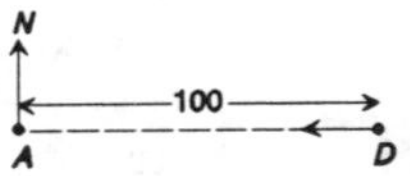
Let x denote the number of hours that have elapsed since 4:00 P.M. In x hours, the USS Independence (A) will have travelled $10x$ nautical miles, and the destroyer (D) will have gone $20x$ miles. The situation will then be as indicated:
If we let y denote the distance from A to D, then by the Pythagorean Theorem,

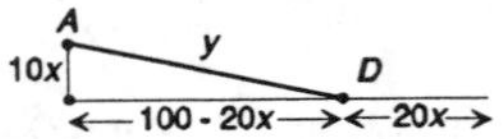

$$y^2 = (10x)^2 + (100 - 20x)^2$$
$$= 100x^2 + 10000 - 4000x + 400x^2, \text{ or}$$
$$y^2 = 500x^2 - 4000x + 10000$$

This equation is a parabola that opens upward, since $a = 500 > 0$, so it will have a minimum value at its vertex, (h, k). Now $h = \dfrac{-b}{2a} = \dfrac{4000}{1000} = 4$, and $k = f(h) = f(4) = 2000$. Thus, the smallest possible value of y^2 is 2000, so the minimum value for $y = \sqrt{2000} \approx 44.72$ nautical miles. But the question is to find what time it is. The minimum distance y occurs when $x = h = 4$ hours, i.e., at 8:00 P.M.

59. For simplicity, locate the origin at the point where the cable touches the road:

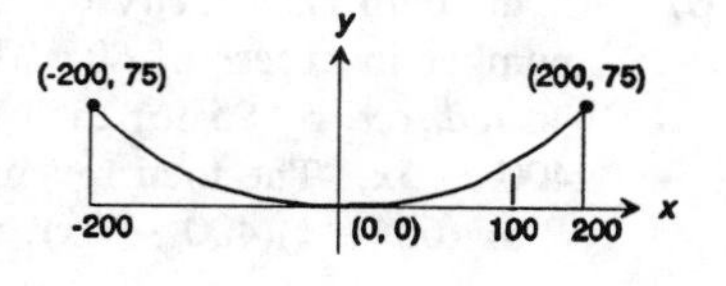

Then the equation of the parabola must be of the form:

$y = ax^2$, for some $a > 0$

Since the point (200, 75) is on the parabola, we can determine the constant a:

$$75 = a(200)^2, \text{ or } a = \frac{75}{(200)^2} = 0.001875 \ldots$$

Then when $x = 80$, we have $y = ax^2$, or

$$y = (0.001875)(100)^2 = 18.75 \text{ meters.}$$

61. The area A of the gutter of height x and length y is $A = xy$. Since $2x + y = 12$, we have $y = 12 - 2x$. We want to maximize the area $A = xy = x(12 - 2x)$ $= -2x^2 + 12x$. This is the equation of a parabola that opens downward, and hence the maximum area occurs when $x = \frac{-b}{2a} = \frac{-12}{-4} = 3$, and the value is $f(3) = -2(3)^2 + 12(3) = 18$. Thus, a depth of 3 inches will provide maximum cross-sectional area.

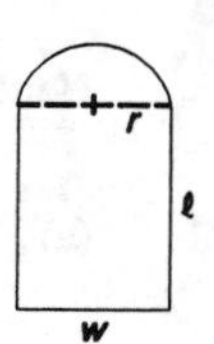

63. Since the diameter $(2r)$ equals the width of the rectangle, then $w = 2r$. The perimeter is

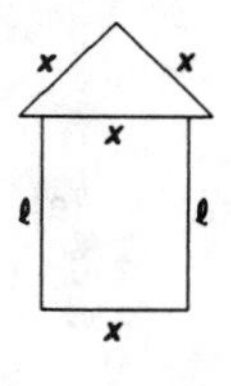

$$20 = 2\ell + w = \frac{1}{2}(2\pi r) \text{ or}$$

$$20 = 2\ell + 2r + \pi r \text{ so that}$$

$$2\ell = 20 - 2r - \pi r$$

$$A = \ell w + \frac{1}{2}(\pi r^2) = \ell(2r) + \frac{1}{2}(\pi r^2)$$

$$= r(20 - 2r - \pi r) + \frac{1}{2}\pi r^2$$

$$= \left[-2 - \frac{\pi}{2}\right] r^2 + 20r$$

The area is maximum at $r = \frac{-b}{2a} = \frac{-20}{-4 - \pi} \approx 2.80$.

Thus, $w = \frac{40}{\pi + 4} \approx 5.6$ ft., and $\ell = 10 - r - \frac{\pi}{2}r \approx 10 - 2.8 - \frac{\pi}{2}(2.8) \approx 2.8$ ft.

65. If ℓ is the length of the rectangle, and x is the width, the perimeter of the window is $16 = 2\ell + 3x$ so that $\ell = 8 - \frac{3}{2}x$. The area of the window is

$$A = \ell x + \frac{\sqrt{3}}{4}x^2 \text{ or}$$

$$A = \left[8 - \frac{3}{2}x\right] x + \frac{\sqrt{3}}{4}x^2$$

$$A = \left[\frac{-3}{2} + \frac{\sqrt{3}}{4}\right] x^2 + 8x$$

The area is maximum at $x = \frac{-b}{2a} - \frac{-8}{-3 + \frac{\sqrt{3}}{2}} \approx \frac{16}{6 - \sqrt{3}} \approx 3.75$ ft. and

$$\ell = 8 - \frac{3}{2}x \approx 8 - \frac{3}{2}(3.75) \approx 2.38 \text{ ft.}$$

67. If the club has exactly 60 members, its revenue would be $60(400) = 24{,}000$ dollars. Let x be the number in ***excess*** of 60. Then the total membership is $60 + x$. Now the price for ***every*** member will be ***reduced*** by \$5 for each member in excess of 60, i.e., by $5x$. Hence, the price per person will be $400 - 5x$. The total revenue is equal to the number of members times the charge per member: $R = (60 + x)(400 - 5x)$, or $R = -5x^2 + 100x + 24{,}000$. This represents a parabola with a ***maximum*** at its vertex. $h = \dfrac{-b}{2a} = \dfrac{-100}{-10} = 10$; $k = f(20) = -2000 + 500 + 24{,}000 = 22{,}500$. The maximum possible revenue is $R = k = 22{,}500$, and this occurs when $x = h = 10$, so that the number of members is $60 + 10 = 70$.

69. We have

$$\begin{aligned} ah^2 - bh + c &= y_0 \\ c &= y_1 \\ ah^2 + bh + c &= y_2 \end{aligned}$$

so that

$$\begin{aligned} y_0 + y_2 &= 2ah^2 + 2c \\ 4y_1 &= 4c \end{aligned}$$

Hence, Area $= \dfrac{h}{3}(2ah^2 + 6c) = \dfrac{h}{3}(y_0 + 4y_1 + y_2)$

71. (a)

50,000

15 … 75

15,000

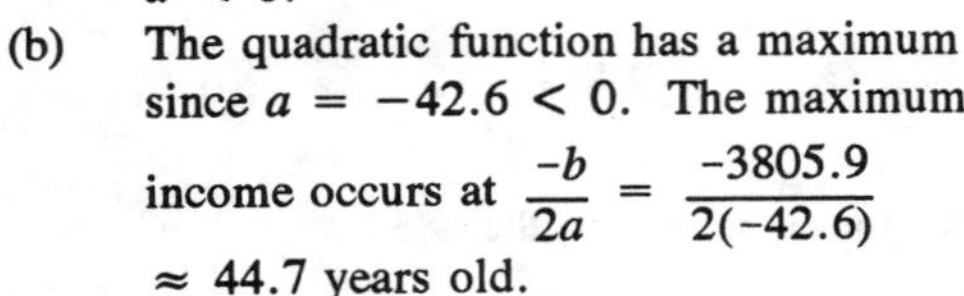

The data appears to be quadratic with $a < 0$.

(b) The quadratic function has a maximum since $a = -42.6 < 0$. The maximum income occurs at $\dfrac{-b}{2a} = \dfrac{-3805.9}{2(-42.6)}$ ≈ 44.7 years old.

(c) $I(44.7) = -42.6(44.7)^2 + 3805.5(44.7) - 38526 = \$46{,}461$

(e)

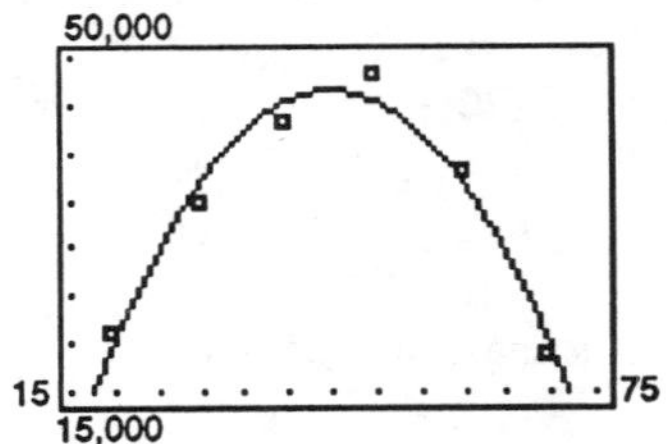

73. (a)

76

1979 … 1994

68

Quadratic with $a > 0$.

(b) The function has a minimum at its vertex since $a > 0$. The x-coordinate of the vertex is $\dfrac{-b}{2a} = \dfrac{-(-250)}{2(0.063)}$ ≈ 1984.

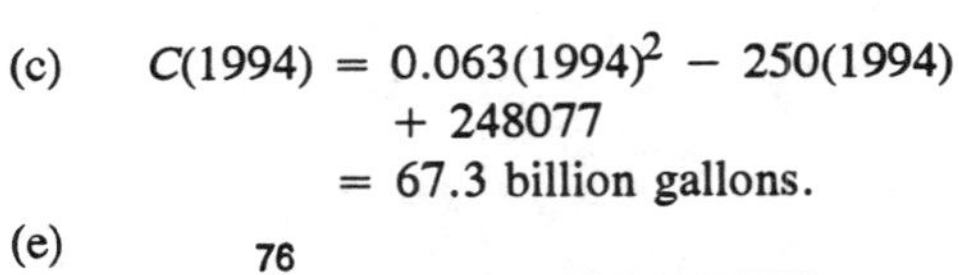

(c) $C(1994) = 0.063(1994)^2 - 250(1994) + 248077 = 67.3$ billion gallons.

(e)

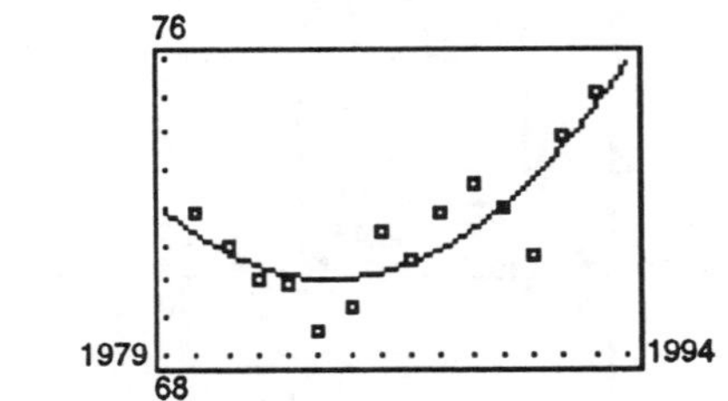

75. (a)

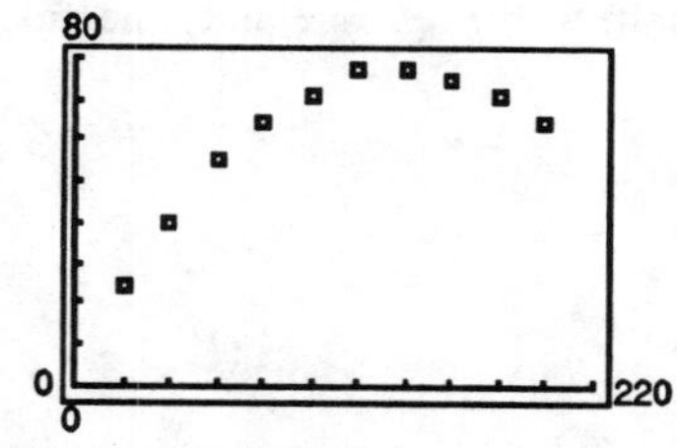

Quadratic with $a < 0$.

(b) The function has a maximum at its vertex since $a < 0$. The x-coordinate of the vertex is $\frac{-b}{2a} = \frac{-1.03}{2(-0.0037)}$ ≈ 139.2 feet.

(c) $f(139.2) = -0.0037(139.2)^2 + 1.03(139.2) + 5.7 = 77.4$ feet.

(e)

77. We are given $V(x) = kx(a - x) = -kx^2 + kax$

This is a maximum when $x = \frac{-ka}{-2k} = \frac{a}{2}$.

5.2 Polynomial Functions

1. $f(x) = 4x + x^3$ is a polynomial of degree 3.

3. $g(x) = \frac{1 - x^2}{2} = \frac{1}{2} - \frac{1}{2}x^2$ is a polynomial of degree 2.

5. $f(x) = 1 - \frac{1}{x} = 1 - x^{-1}$ is ***not*** a polynomial, since it contains x raised to a negative power.

7. $g(x) = x^{3/2} - x^2 + 2$ is ***not*** a polynomial since it contains x raised to a fractional power.

9. $F(x) = 5x^4 - \pi x^3 + \frac{1}{2}$ is a polynomial, of degree 4.

In Problems 11–17, start with the graph of $y = x^4$, and perform change of scale, shifting and reflection to obtain the graph of the given function.

11. For $f(x) = (x + 1)^4$, all that is needed is a horizontal shift, one unit to the left.

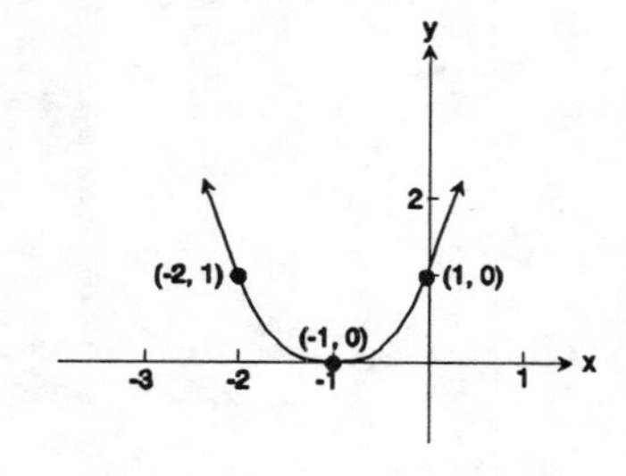

13. For $f(x) = \frac{1}{2}x^4$, vertically compress the graph of $y = x^4$. The graph will pass through $\left(1, \frac{1}{2}\right)$ instead of $(1, 1)$.

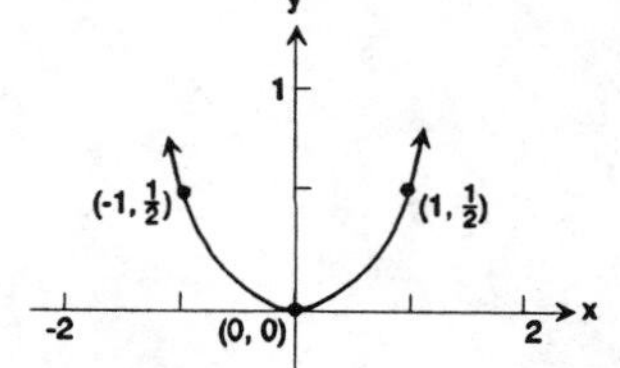

15. To graph $f(x) = 2(x+1)^4 + 1$, vertically stretch, then shift to the ***left*** one unit, and finally shift ***up*** one unit.

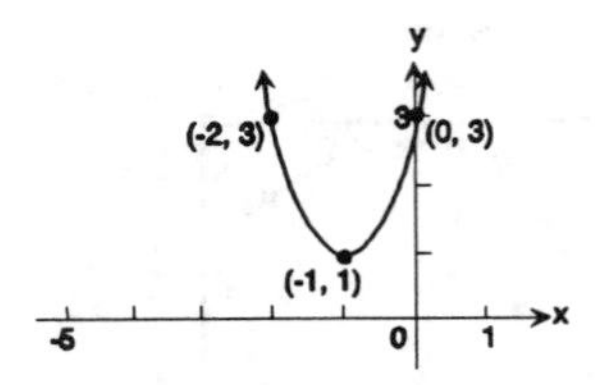

17. Perform the following steps:
(1) Starting graph: $y = x^4$
(2) Vertically compress: $y = \frac{1}{2}x^4$
(3) Reflect about the x-axis: $y = -\frac{1}{2}x^4$
(4) Shift ***right*** 2 units: $y = -\frac{1}{2}(x-2)^4$
(5) Shift ***down*** one unit: $y = -\frac{1}{2}(x-2)^4 - 1$

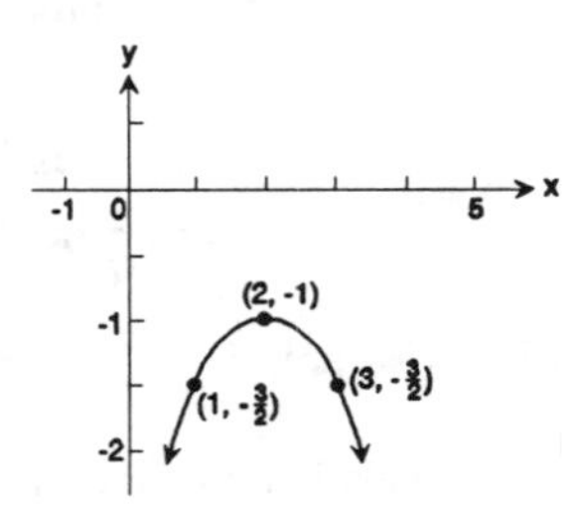

19. The zeros of $f(x) = 3(x-7)(x+3)^2$ are: 7, with multiplicity one; and -3, with multiplicity two. The graph touches the x-axis at -3 and crosses it at 7. Power Function: $y = 3 \cdot x \cdot x^2 = 3x^3$

21. The zeros of $f(x) = 4(x^2+1)(x-2)^3$ are: 2, with multiplicity three. Note: $x^2 + 1 = 0$ has no real solution. The graph crosses the x-axis at 2. Power Function: $y = 4 \cdot x^2 \cdot x^3 = 4x^5$

23. The zeros of $f(x) = -2\left[x + \frac{1}{2}\right]^2(x^2+4)^2$ are $\frac{-1}{2}$, with multiplicity two and $x^2 + 4 = 0$ has no real solutions. The graph touches the x-axis at $\frac{-1}{2}$. Power Function: $y = -2 \cdot x^2 \cdot (x^2)^2 = -2x^6$

25. The zeros of $f(x) = (x-5)^3(x+4)^2$ are: 5, with multiplicity three; and -4, with multiplicity two. The graph touches the x-axis at -4 and crosses it at 5. Power Function: $y = x^3 \cdot x^2 = x^5$

27. $f(x) = 3(x^2+8)(x^2+9)^2$ has no real zeros. The graph neither touches nor crosses the x-axis. Power Function: $y = 3 \cdot x^2 \cdot (x^2)^2 = 3x^6$

29. The zeros of $f(x) = -2x^2(x^2-2)$ are 0, with multiplicity 2; $\sqrt{2}$, with multiplicity 1; and $-\sqrt{2}$ with multiplicity 1.

Note: $x^2 - 2 = 0$

$x^2 = \sqrt{2}$

$x = \pm\sqrt{2}$

The graph touches the x-axis at $x = 0$ and crosses the x-axis at $-\sqrt{2}$ and $\sqrt{2}$.
Power Function: $y = -2 \cdot x^2 \cdot x^2 = -2x^4$

31. $f(x) = (x - 1)^2$

(a) x-intercept: 1; y-intercept: 1

(b) touches at 1

(c) $y = x^2$

(d) 1

(e)

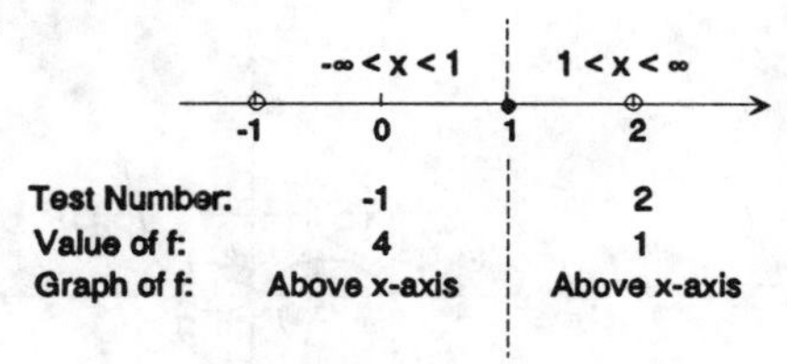

(f)

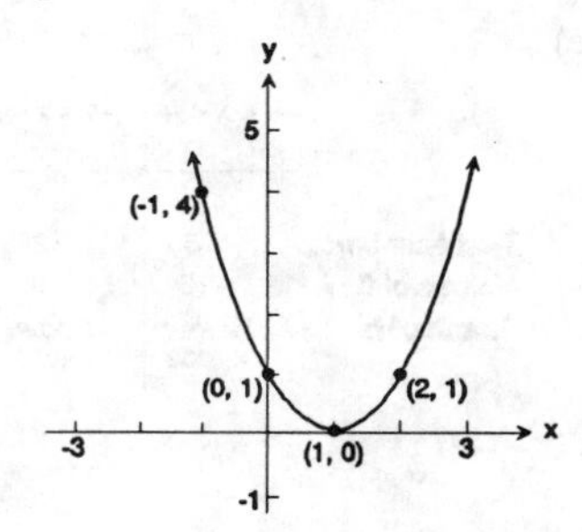

33. $f(x) = x^2(x - 3)$

(a) x-intercepts: 0, 3; y-intercept: 0

(b) touches at 0; crosses at 3

(c) $y = x^3$

(d) 2

(e)

	$-\infty < x < 0$	$0 < x < 3$	$3 < x < \infty$
Test Number:	-1	2	4
Values of f:	-4	-4	16
Graph of f:	Below x-axis	Below x-axis	Above x-axis

(f)

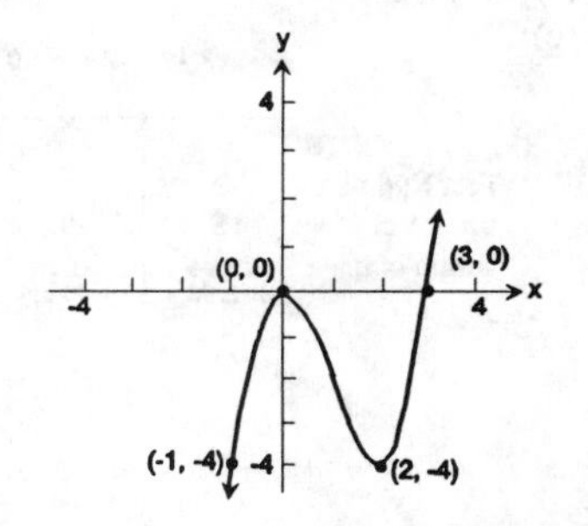

35. $f(x) = 6x^3(x + 4)$

(a) x-intercepts: -4, 0; y-intercept: 0

(b) crosses at -4 and 0

(c) $y = 6x^4$

(d) 3

(e)

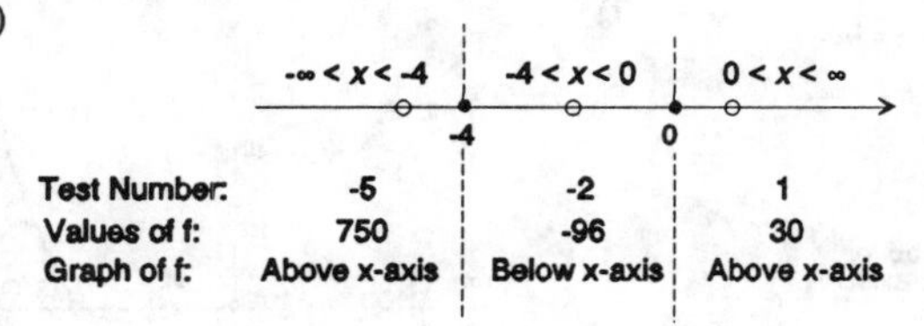

(f)

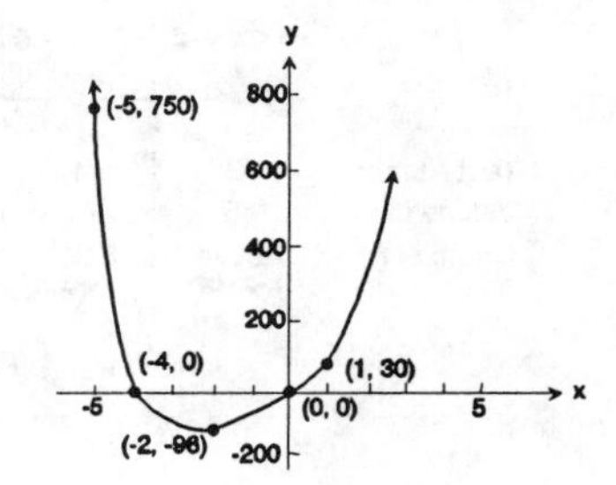

37. $f(x) = -4x^2(x + 2)$

(a) x-intercepts: -2, 0; y-intercept: 0

(b) crosses at -2; touches at 0

(c) $y = -4x^3$

(d) 2

(e)

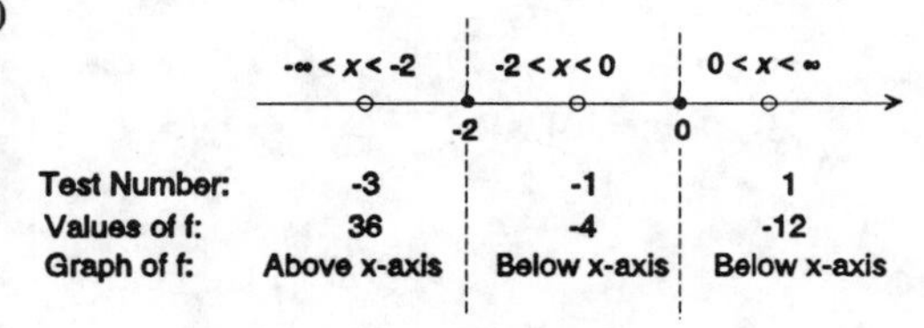

(f)

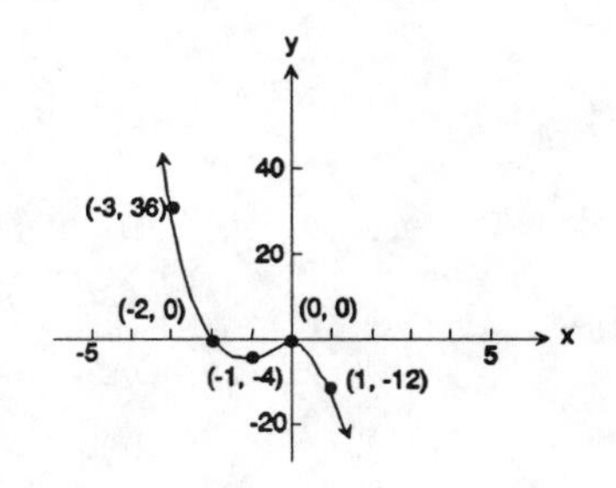

39. $f(x) = x(x - 2)(x + 4)$

(a) x-intercepts: -4, 0, 2; y-intercept: 0

(b) crosses at -4, 0, and 2

(c) $y = x^3$

(d) 2

(e)

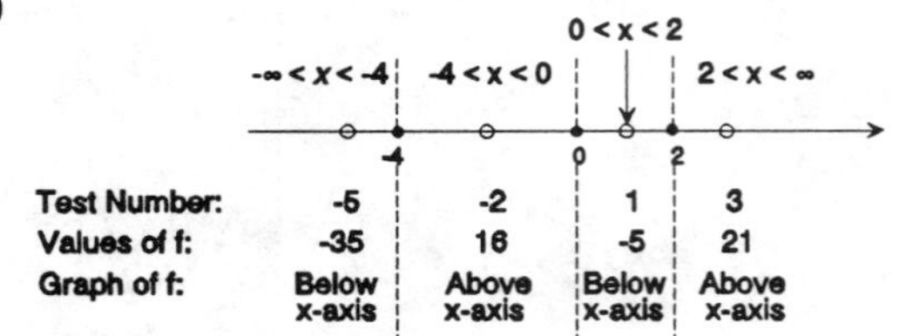

	$-\infty < x < -4$	$-4 < x < 0$	$0 < x < 2$	$2 < x < \infty$
Test Number:	-5	-2	1	3
Values of f:	-35	16	-5	21
Graph of f:	Below x-axis	Above x-axis	Below x-axis	Above x-axis

(f)

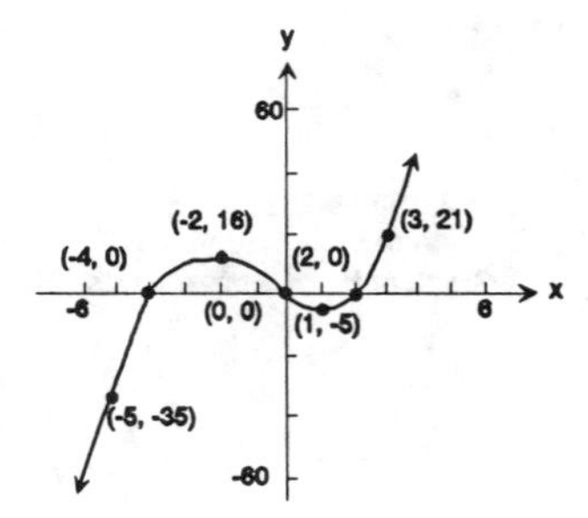

41. $f(x) = 4x - x^3 = -x(x^2 - 4) = -x(x + 2)(x - 2)$

(a) x-intercepts: -2, 0, 2; y-intercept: 0

(b) crosses at -2, 0, and 2

(c) $y = -x^3$

(d) 2

(e)

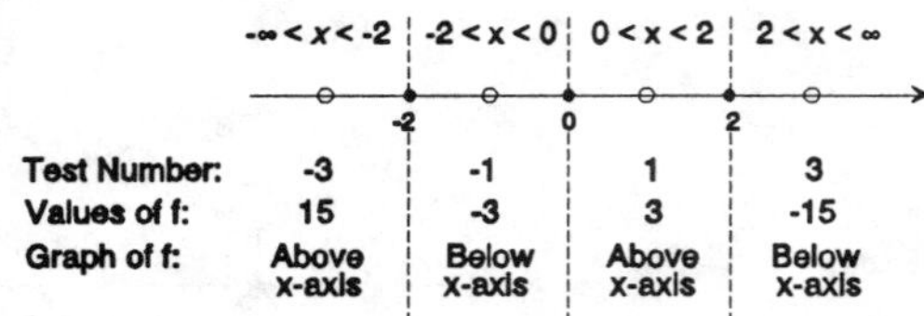

	$-\infty < x < -2$	$-2 < x < 0$	$0 < x < 2$	$2 < x < \infty$
Test Number:	-3	-1	1	3
Values of f:	15	-3	3	-15
Graph of f:	Above x-axis	Below x-axis	Above x-axis	Below x-axis

(f)

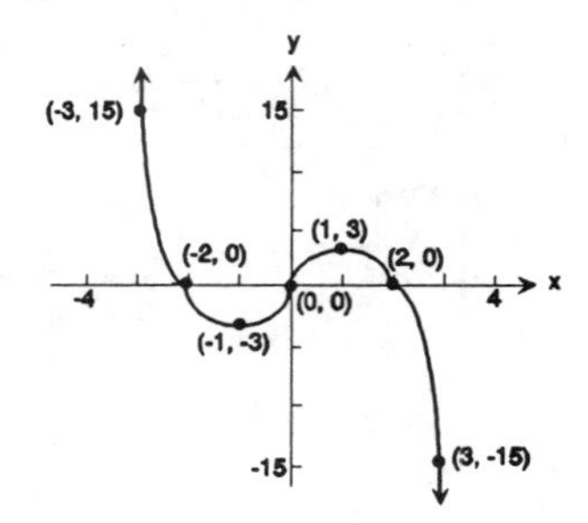

43. $f(x) = x^2(x - 2)(x + 2)$

(a) x-intercepts: -2, 0, 2; y-intercept: 0

(b) crosses at -2 and 2; touches at 0

(c) $y = x^4$

(d) 3

(e)

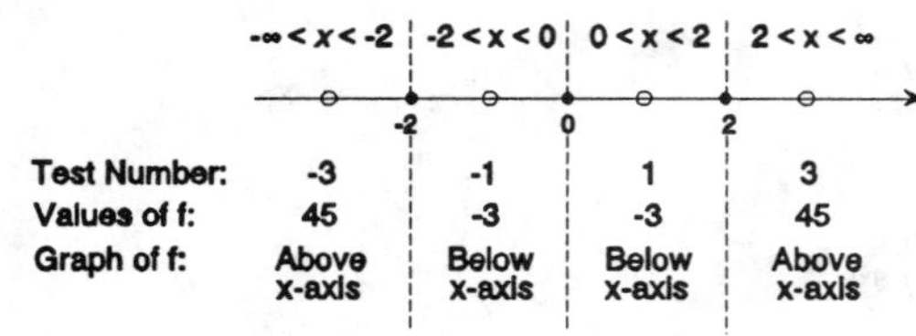

	$-\infty < x < -2$	$-2 < x < 0$	$0 < x < 2$	$2 < x < \infty$
Test Number:	-3	-1	1	3
Values of f:	45	-3	-3	45
Graph of f:	Above x-axis	Below x-axis	Below x-axis	Above x-axis

(f)

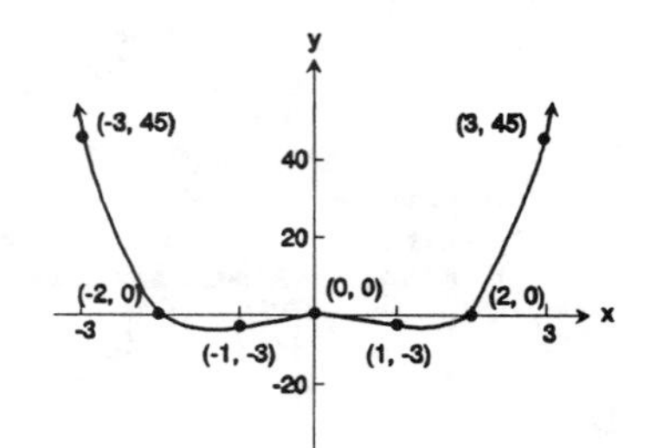

45. $f(x) = x^2(x - 2)^2$

(a) x-intercepts: 0, 2; y-intercept: 0

(b) touches at 0 and 2

(c) $y = x^4$

(d) 3

(e)

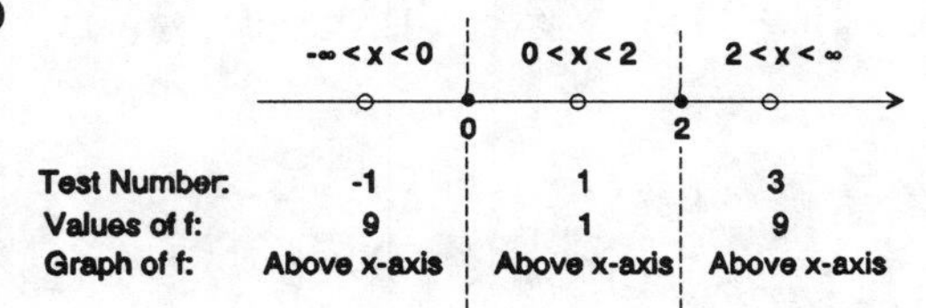

(f)

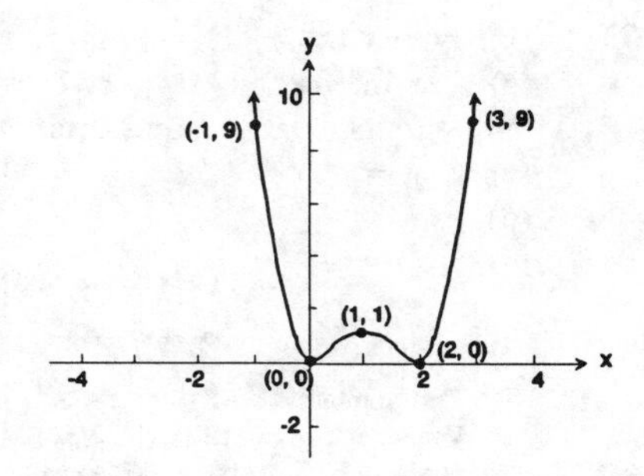

47. $f(x) = x^2(x - 3)(x + 1)$

(a) x-intercepts: $-1, 0, 3$; y-intercept: 0

(b) crosses at -1 and 3; touches at 0

(c) $y = x^4$

(d) 3

(e)

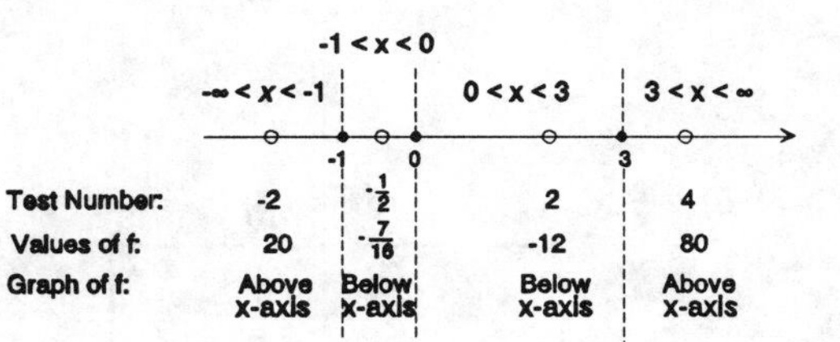

(f)

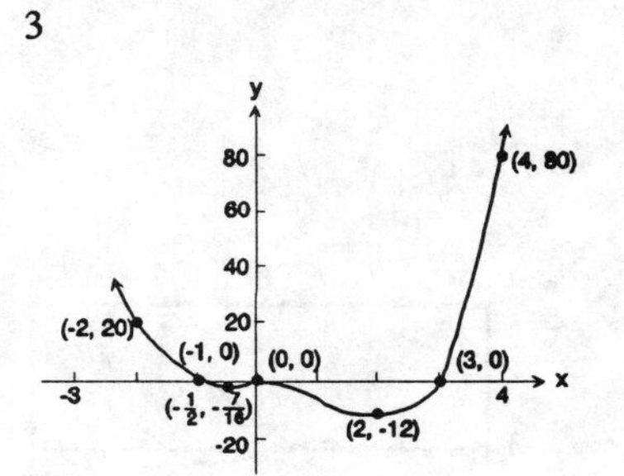

49. $f(x) = x(x + 2)(x - 4)(x - 6)$

(a) x-intercepts: $-2, 0, 4, 6$; y-intercept: 0

(b) crosses at -2, 0, 4, and 6

(c) $y = x^4$

(d) 3

(e)

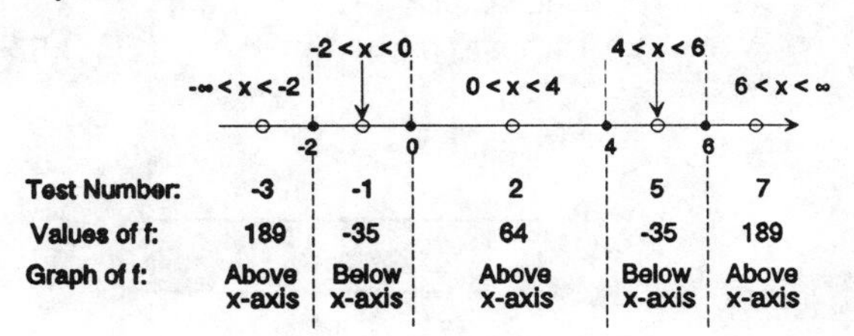

(f)

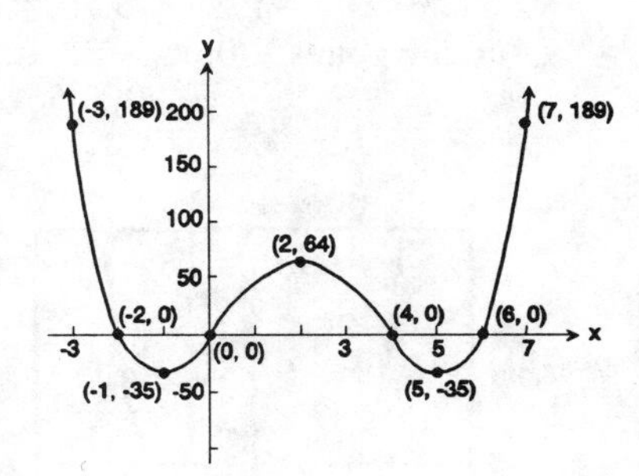

51. $f(x) = x^2(x - 2)(x^2 + 3)$

(a) x-intercepts: 0, 2; y-intercept: 0

(b) touches at 0; crosses at 2;

(c) $y = x^5$

(d) 4

(e)

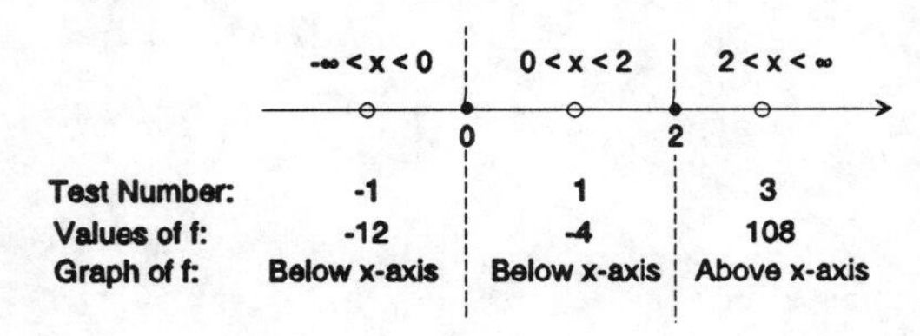

(f)

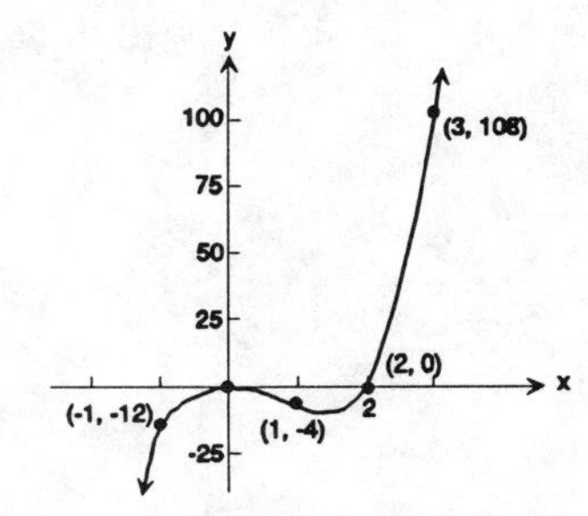

53. $f(x) = -x^2(x^2 - 1)(x + 1) = -x^2(x + 1)(x - 1)(x + 1) = -x^2(x + 1)^2(x - 1)$

(a) x-intercepts: 0, -1, 1; y-intercept: 0

(b) touches at -1 and 0; crosses at 1

(c) $y = -x^5$

(d) 4

(e)

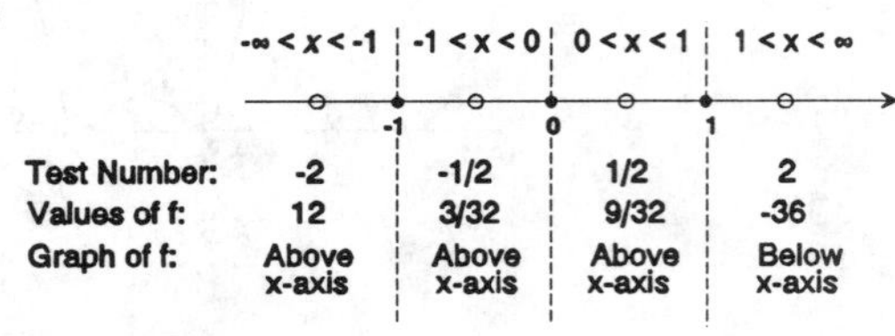

(f)

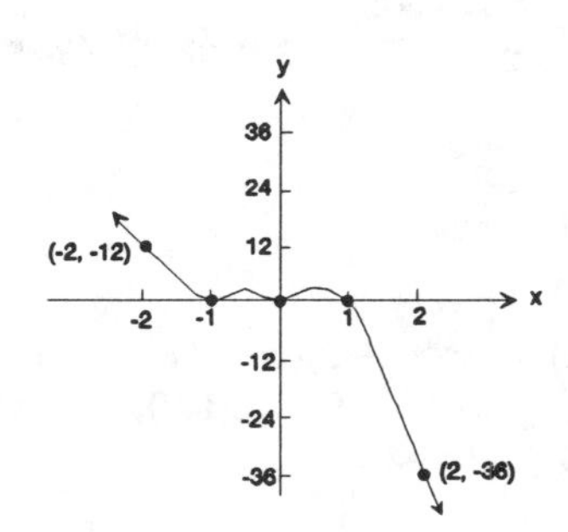

55. c, e, f

57. c, e

59.

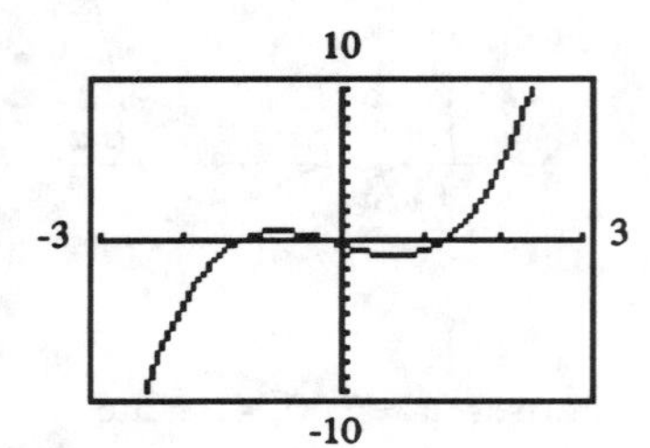

x-intercepts: -1.26, -0.20, 1.26;
Turning points: $(0.66, -0.99)$;
$(-0.79, 0.56)$

61.

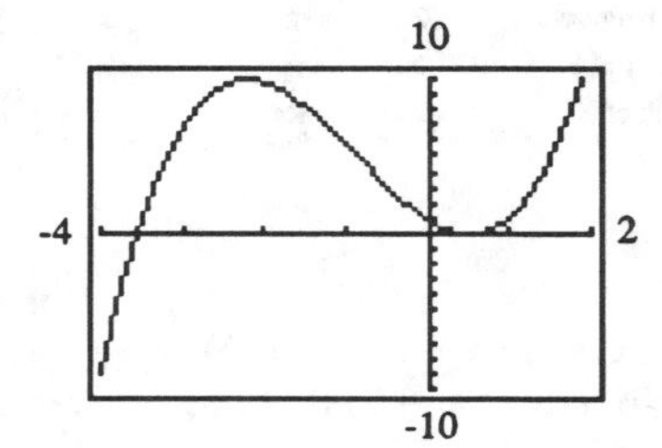

x-intercepts: -3.56, 0.50;
Turning points: $(0.50, 0)$; $(-2.20, 9.91)$

63.

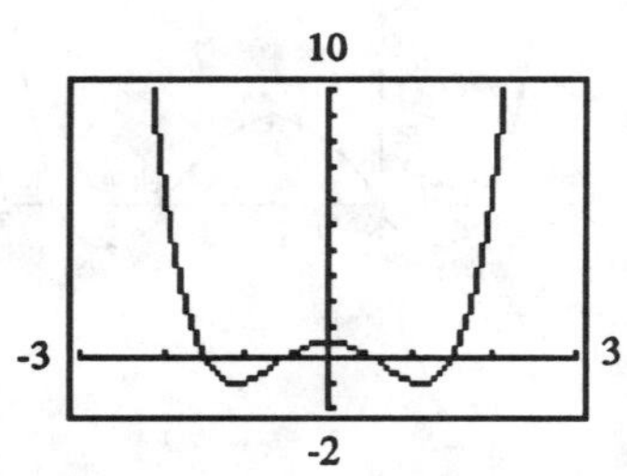

x-intercepts: -1.50, -0.50, 0.50, 1.50
Turning points: $(-1.11, -1)$, $(1.11, -1)$;
$(0, 0.5625)$

65.

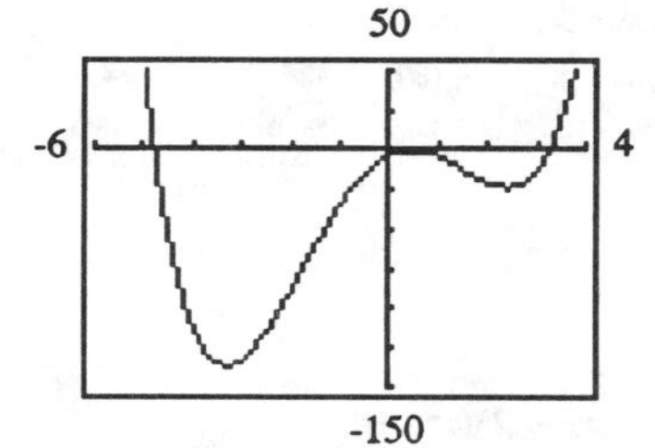

x-intercepts: -4.78, 0.45, 3.23;
Turning points: $(-3.31, -135.91)$,
$(2.37, -22.66)$; $(0.45, 0)$

67.

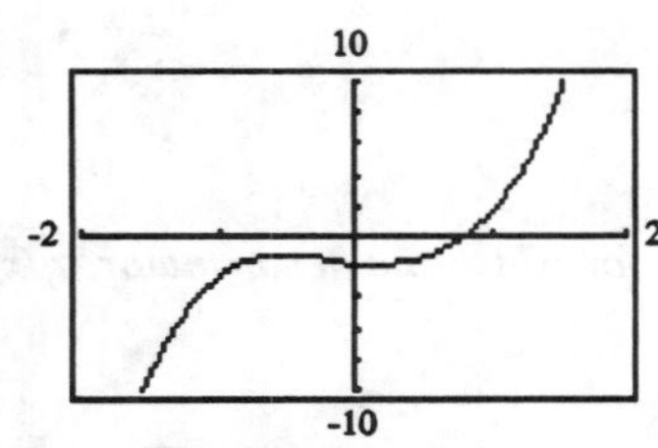

x-intercept: 0.83

Turning points: $(-0.50, -1.53)$; $(0.20, -2.11)$

69.

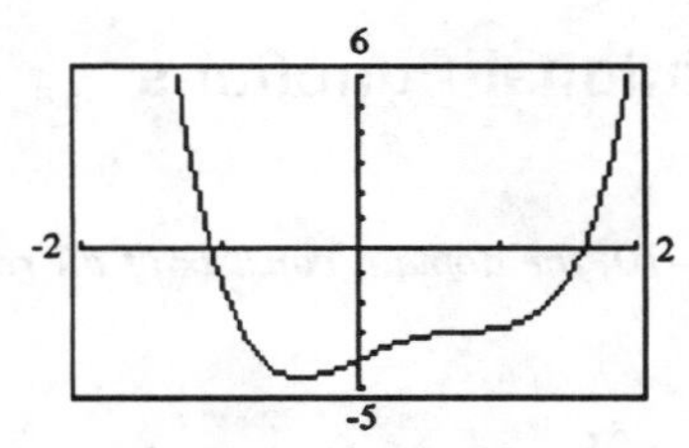

x-intercepts: $-1.06, 1.61$

Turning point: $(-0.41, -4.64)$

71.

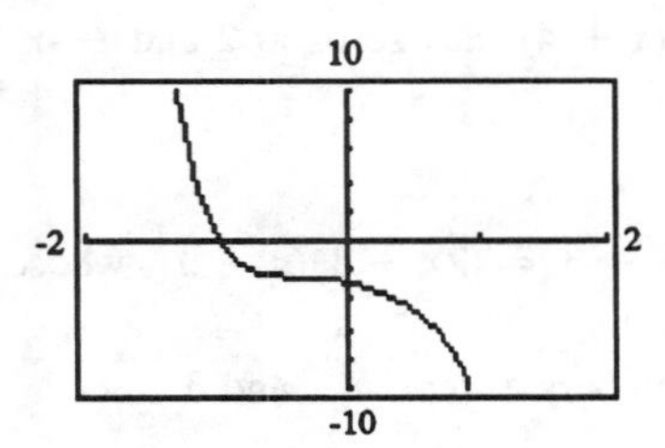

x-intercept: -0.97

No turning points

73. (a)

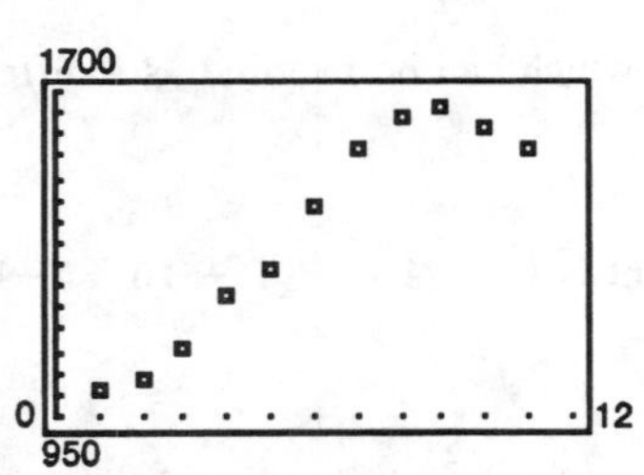

Cubic relation with $a < 0$.

(b) $M(12) = -2.4(12)^3 + 37.3(12)^2 - 70.4(12) + 1043.8$
≈ 1370 motor vehicle thefts

(c)

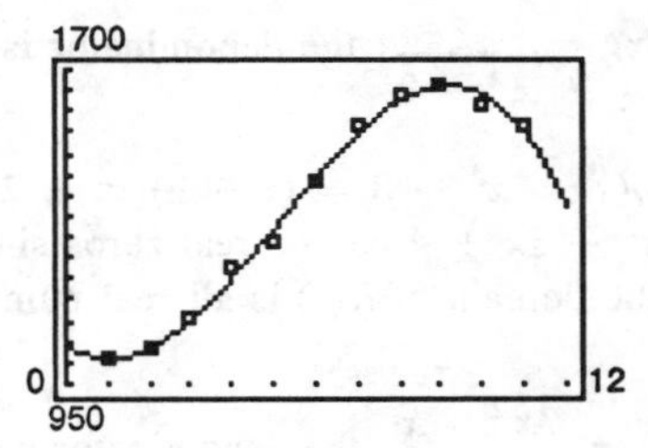

75. (a)

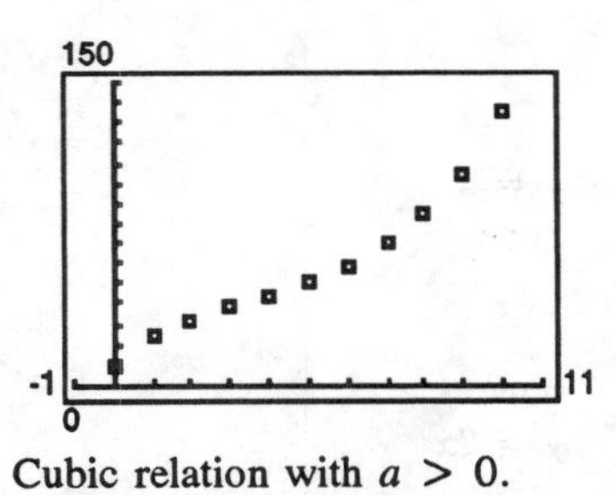

Cubic relation with $a > 0$.

(b) Average Rate of Change

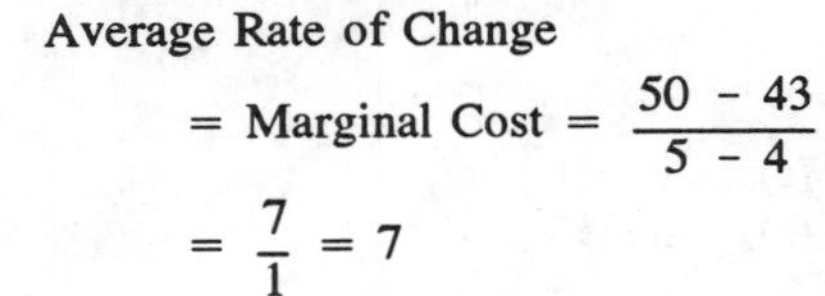

$= \text{Marginal Cost} = \dfrac{50 - 43}{5 - 4}$

$= \dfrac{7}{1} = 7$

(c) Marginal cost $= \dfrac{105 - 85}{9 - 8} = \dfrac{20}{1}$
$= 20$

(d) $C(11) = 0.2(11)^3 - 2.3(11)^2 + 14.3(11) + 10.2 \approx 155.4$

Thus, it will cost \$155,400 to produce 11 Cavaliers.

(f)

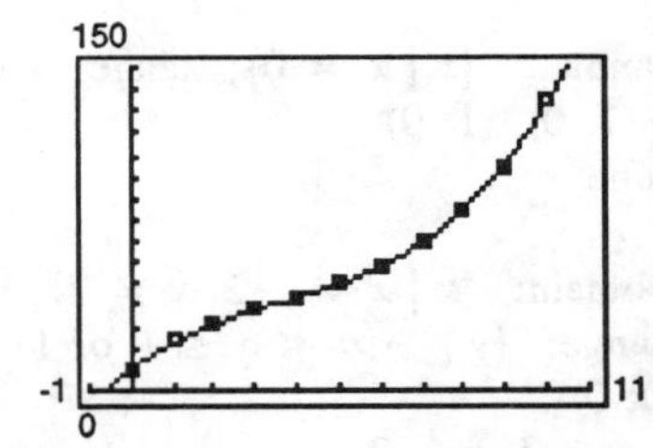

(g)

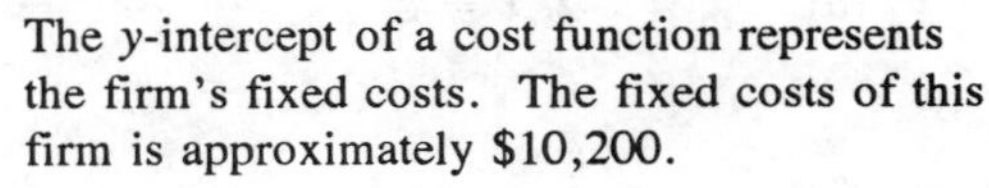

The y-intercept of a cost function represents the firm's fixed costs. The fixed costs of this firm is approximately \$10,200.

5.3 Rational Functions

In Problems 1–10, the domain consists of all real numbers except those for which the denominator, q(x), is zero.

1. $R(x) = \dfrac{4x}{x - 3}$. Here the denominator is $q(x) = x - 3$, which has 3 as its only zero. Thus, the domain of $R(x)$ consists of all real numbers except 3.

3. For $H(x) = \dfrac{-4x^2}{(x - 2)(x + 4)}$, the denominator, $q(x) = (x - 2)(x + 4)$, has zeros at 2 and -4. Thus, the domain of $H(x)$ is all real numbers except 2 and -4.

5. In $F(x) = \dfrac{3x(x - 1)}{2x^2 - 5x - 3}$, the denominator is $q(x) = 2x^2 - 5x - 3 = (2x + 1)(x - 3)$, whose zeros are $\dfrac{-1}{2}$ and 3. Thus, the domain of $F(x)$ consists of all real numbers except $\dfrac{-1}{2}$ and 3.

7. For $R(x) = \dfrac{x}{x^3 - 8}$, the denominator is $q(x) = x^3 - 8$, which can be factored as a difference of cubes:
$$q(x) = x^3 - 8 = (x - 2)(x^2 + 2x + 4)$$
Now, $x^2 + 2x + 4$ has no real zeros since its discriminant is $b^2 - 4ac = 4 - 16 = -12 < 0$. Thus, the domain of $R(x)$ is all real numbers except 2.

9. In $H(x) = \dfrac{3x^2 + x}{x^2 + 4}$, the denominator has no real zeros, so the domain is ***all*** real numbers.

11. $R(x) = \dfrac{3(x^2 - x - 6)}{4(x^2 - 9)} = \dfrac{3(x - 3)(x + 2)}{4(x + 3)(x - 3)} = \dfrac{3(x + 2)}{4(x + 3)}$, $x \neq -3$. The domain of R is $\{x \mid x \neq -3, x \neq 3\}$.

13. (a) Domain: $\{x \mid x \neq 2\}$; Range: $\{y \mid y \neq 1\}$
    (b) (0, 0) (c) $y = 1$
    (d) $x = 2$ (e) none

15. (a) Domain: $\{x \mid x \neq 0\}$; Range: all real numbers
    (b) $(-1, 0), (1, 0)$ (c) none
    (d) none (e) $y = 2x$

17. (a) Domain: $\{x \mid x \neq -2, x \neq 2\}$;
    Range: $\{y \mid -\infty < y \leq 0 \text{ or } 1 < y < \infty\}$
    (b) (0, 0) (c) $y = 1$
    (d) $x = -2, x = 2$ (e) none

Problems 19–27 are all based on the graphs of $R(x) = \frac{1}{x}$ *and* $H(x) = \frac{1}{x^2}$, *shown below.*

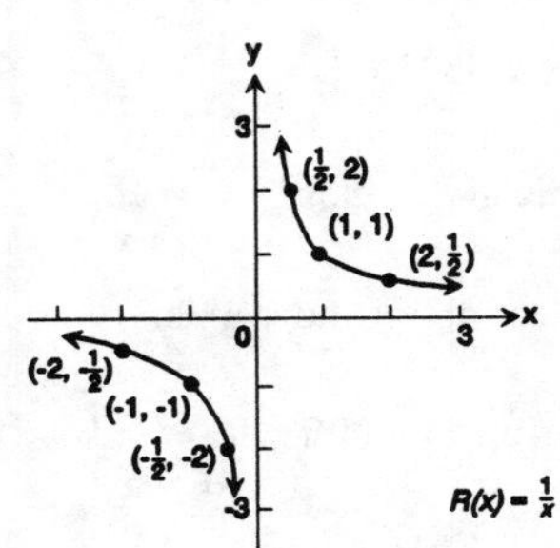

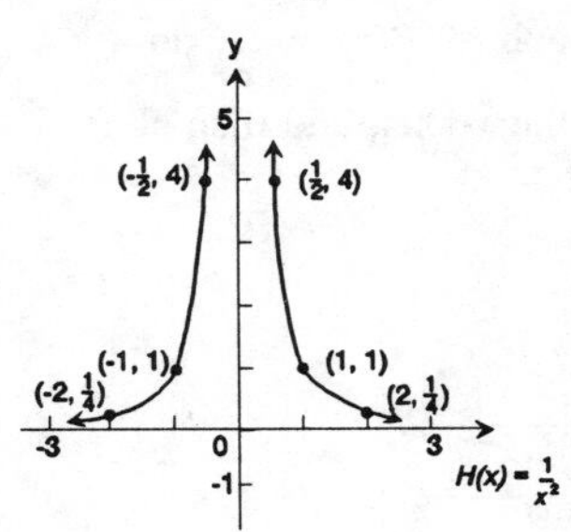

19. $R(x) = \dfrac{1}{(x - 1)^2}$. This graph can be obtained by shifting the graph of $H(x) = \dfrac{1}{x^2}$ horizontally, one unit to the right.

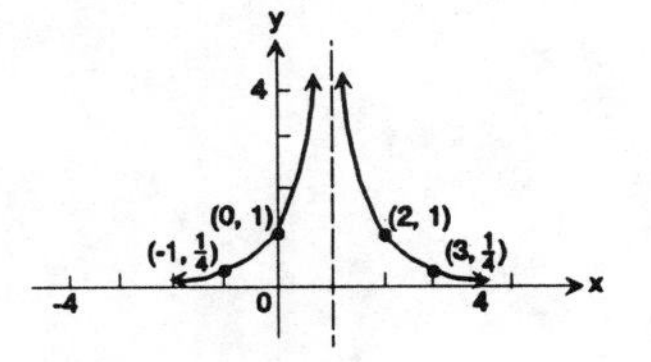

21. $H(x) = \dfrac{-2}{x + 1}$

This can be graphed in stages:

(1) Starting graph: $R(x) = \dfrac{1}{x}$

(2) Vertically stretch $y = \dfrac{2}{x}$

(3) Reflect about the x-axis: $y = \dfrac{-2}{x}$

(4) Shift one unit to the left: $y = \dfrac{-2}{x + 1}$

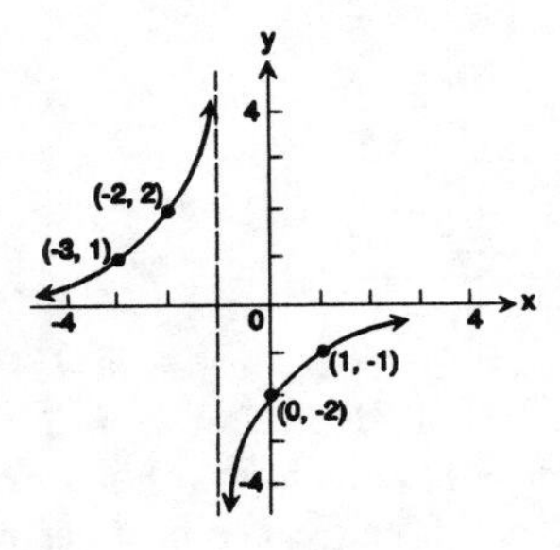

23. $R(x) = \dfrac{1}{x^2 + 4x + 4} = \dfrac{1}{(x + 2)^2}$

To obtain this graph, shift the graph of $H(x) = \dfrac{1}{x^2}$ two units to the left:

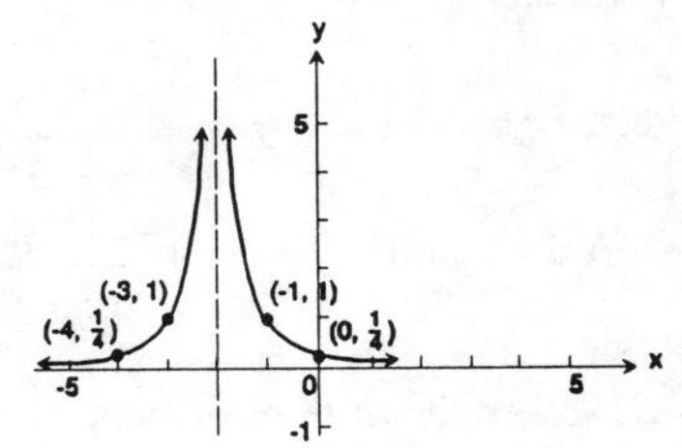

25. $F(x) = 1 - \frac{1}{x} = -\frac{1}{x} + 1$

Start with the graph of $R(x) = \frac{1}{x}$, do a reflection about the x-axis, and then shift ***upward*** one unit:

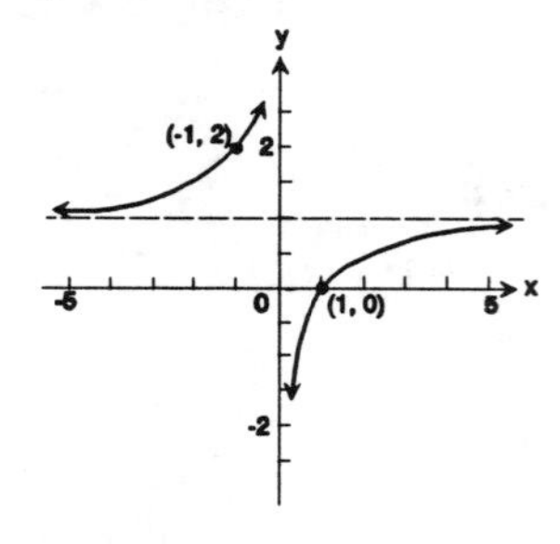

27. $R(x) = \frac{x^2 - 4}{x^2} = \frac{x^2}{x^2} - \frac{4}{x^2} = \frac{-4}{x^2} + 1$

(1) Start with $H(x) = \frac{1}{x^2}$

(2) Vertically stretch: $y = \frac{4}{x^2}$

(3) Reflect about the x-axis: $y = \frac{-4}{x^2}$

(4) Shift up one unit: $y = \frac{-4}{x^2} + 1$

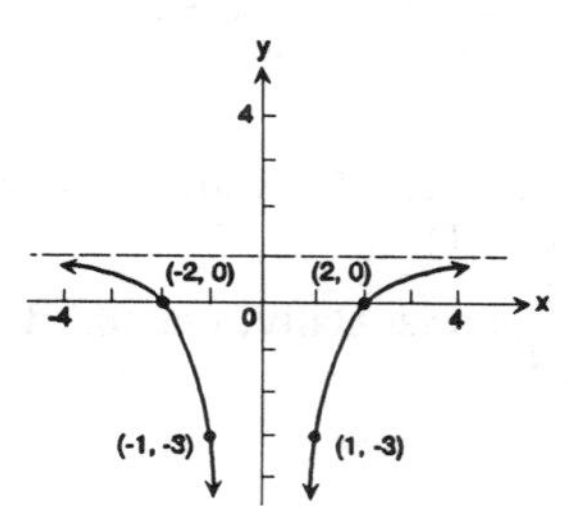

29. $G(x) = 1 + \frac{2}{(x - 3)^2}$

(1) Start with $H(x) = \frac{1}{x^2}$

(2) Vertically stretch: $y = \frac{2}{x^2}$

(3) Shift right 3 units: $y = \frac{2}{(x - 3)^2}$

(4) Shift up 1 unit: $y = 1 + \frac{2}{(x - 3)^2}$

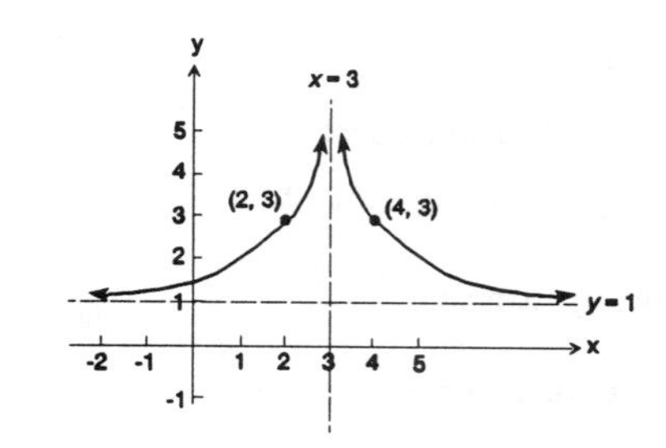

For Problems 31–39, refer to the summary concerning asymptotes.

31. $R(x) = \frac{3x}{x + 4}$

The degree of the numerator, $p(x) = 3x$, is $n = 1$. The degree of the denominator, $q(x) = x + 4$ is $m = 1$. Since $n = m$, the line $y = \frac{3}{1} = 3$ is a horizontal asymptote. The denominator is zero at $x = -4$, so $x = -4$ is a vertical asymptote.

33. $H(x) = \frac{x^4 + 2x^2 + 1}{x^2 - x + 1}$

$p(x) = x^4 + 2x^2 + 1$; degree $= n = 4$
$q(x) = x^2 - x + 1$; degree $= m = 2$
Since $n > m + 1$, $H(x)$ has no horizontal nor oblique asymptote. Since $q(x)$ has no real zeros, there is no vertical asymptote.

35. $T(x) = \dfrac{x^3}{x^4 - 1}$

$p(x) = x^3$; degree $= n = 3$
$q(x) = x^4 - 1$; degree $= m = 4$
Since $n < m$, the line $y = 0$ is a horizontal asymptote. We can factor $q(x) = x^4 - 1 = (x^2 - 1)(x^2 + 1) = (x - 1)(x + 1)(x^2 + 1)$, so the vertical asymptotes are the lines $x = -1$ and $x = 1$.

37. $Q(x) = \dfrac{5 - x^2}{3x^4}$

$p(x) = 5 - x^2$; $n = 2$
$q(x) = 3x^4$; $m = 4$
Since $n < m$, the line $y = 0$ is a horizontal asymptote. The vertical asymptote is $x = 0$.

39. $R(x) = \dfrac{3x^4 + 4}{x^3 + 3x}$

Since $n = m + 1$, we have an oblique asymptote, so it is necessary to perform long division:

$$\begin{array}{r|l} & 3x \\ \hline x^3 + 3x & 3x^4 + 0x^3 + 0x^2 + 0x + 4 \\ & 3x^4 \qquad\quad + 9x^2 \\ \hline & \qquad -9x^2 + 0x + 4 \quad \leftarrow \text{Remainder} \end{array}$$

Thus, $R(x) = \dfrac{3x^4 + 4}{x^3 + 3x} = 3x + \dfrac{-9x^2 + 4}{x^3 + 3x}$

Therefore, $y = 3x$ is an oblique asymptote for $R(x)$.
We can factor the denominator: $q(x) = x^3 + 3x = x\underbrace{(x^2 + 3)}_{\text{no real zeros}}$.

The vertical asymptote is $x = 0$.

41. $G(x) = \dfrac{x^3 - 1}{x - x^2} = \dfrac{x^3 - 1}{-x^2 + x}$

Since $n = m + 1$, we have an oblique asymptote, so it is necessary to perform long division:

$$\begin{array}{r|l} & -x - 1 \\ \hline -x^2 + x & x^3 + 0x^2 + 0x - 1 \\ & -(x^3 - x^2) \\ \hline & \qquad x^2 + 0x \\ & \quad -(x^2 - x) \\ \hline & \qquad\quad x - 1 \quad \leftarrow \text{Remainder} \end{array}$$

Thus, $G(x) = \dfrac{x^3 - 1}{-x^2 + x} = -x - 1 + \dfrac{x - 1}{-x^2 + x}$

Therefore, $y = -x - 1$ is an oblique asymptote for $G(x)$. We can factor the denominator: $q(x) = -x^2 + x = -x(x - 1)$. The vertical asymptotes are: $x = 0$ and $x = 1$.

In Problems 43–71, we will use the following terminology: $R(x) = \dfrac{p(x)}{q(x)}$, *where the degree of* $p(x) = n$ *and the degree of* $q(x) = m$. *In every problem, we will follow the steps listed below:*

Step 0: ***Preparation:***
- *(a)* *If* $n = m + 1$, *perform long division to determine the oblique asymptote.*
- *(b)* *Factor both* $p(x)$ *and* $q(x)$ *over the reals, in order to determine their zeros.*

Step 1: *Find the domain of the rational function.*

Step 2: ***Locate Intercepts:***
- *(a)* *For x-intercepts, find the zeros of* $p(x)$.
- *(b)* *For the y-intercepts, evaluate* $R(0)$.

Step 3: ***Test for Symmetry.***
Step 4: ***Locate Vertical Asymptotes:***
Find the zeros of q(x).
Step 5: ***Locate Horizontal and Oblique Asymptotes.***
Step 6: ***Determine where the Graph is above or below the x-axis:***
Use all the zeros of p(x) and q(x).
Step 7: ***Sketch the Graph.***

43. $R(x) = \dfrac{x + 1}{x(x + 4)}$ $p(x) = x + 1;\ q(x) = x(x + 4) = x^2 + 4x$
$n = 1,\ m = 2$

Step 0: This is already done, with
$p(x) = x + 1$
$q(x) = x(x + 4)$

Step 1: Domain: $\{x \mid x \neq -4, x \neq 0\}$

Step 2: (a) The x-intercept is the zero of $p(x)$: -1
(b) For y-intercept, $R(0)$ is not defined, since $q(0) = 0$, so there is no y-intercept.

Step 3: $R(-x) = \dfrac{-x + 1}{x^2 - 4x}$; this is neither $R(x)$ nor $-R(x)$, so there is no symmetry.

Step 4: The vertical asymptotes are the zeros of $q(x)$: $x = -4$ and $x = 0$.

Step 5: Since $n < m$, the line $y = 0$ is the horizontal asymptote; intersected at $(-1, 0)$.

Step 6: We have a total of three zeros, -4, -1, 0:

Interval	Test Number	$R(x)$	Graph of R
$x < -4$	$x = -5$	$\dfrac{-4}{-5(-1)} = \dfrac{-4}{5}$	Below the x-axis
$-4 < x < -1$	$x = -2$	$\dfrac{-1}{-2(2)} = \dfrac{1}{4}$	Above the x-axis
$-1 < x < 0$	$x = -\dfrac{1}{2}$	$\dfrac{\frac{1}{2}}{-\frac{1}{2}\left(\frac{7}{2}\right)} = \dfrac{-2}{7}$	Below the x-axis
$x > 0$	$x = 1$	$\dfrac{2}{1(5)} = \dfrac{2}{5}$	Above the x-axis

Step 7:

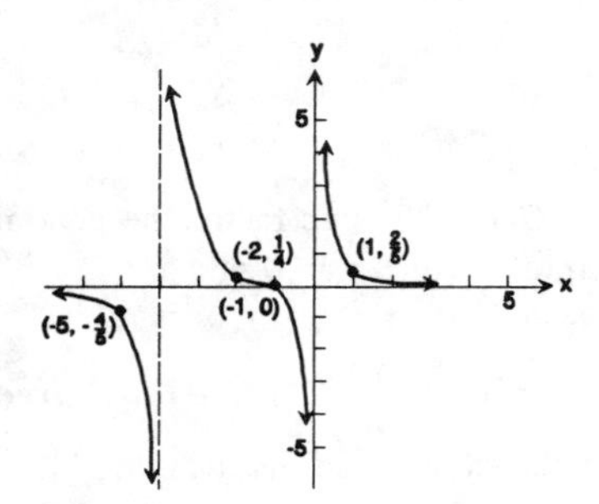

45. $R(x) = \dfrac{3x + 3}{2x + 4}$

Step 0: $p(x) = 3x + 3;\ n = 1$
$q(x) = 2x + 4;\ m = 1$

Step 1: Domain: $\{x \mid x \neq -2\}$

Step 2: (a) The x-intercept is the zero of $p(x)$: -1

(b) For y-intercept is $y = R(0) = \frac{3}{4}$

Step 3: No symmetry.

Step 4: The vertical asymptote is the zero of $q(x)$: $x = -2$.

Step 5: Since $n = m$, the horizontal asymptote is the line $y = \frac{3}{2}$; not intersected.

Step 6: We have a total of two zeros, -2, -1:

Interval	Test Number	$R(x)$	Graph of R
$x < -2$	$x = -3$	$\frac{-9 + 3}{-6 + 4} = 3$	Above the x-axis
$-2 < x < -1$	$x = \frac{-3}{2}$	$\frac{\frac{-9}{2} + 3}{-3 + 4} = \frac{-3}{2}$	Below the x-axis
$x > -1$	$x = 0$	$\frac{3}{4}$	Above the x-axis

Step 7:

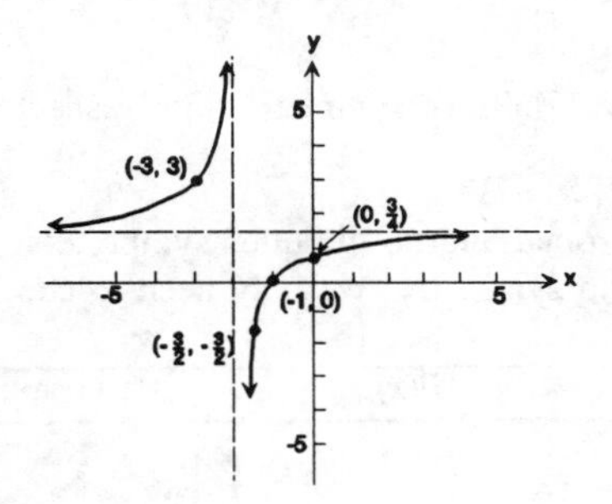

47. $R(x) = \frac{3}{x^2 - 4}$

Step 0: $p(x) = 3$; $n = 0$
$q(x) = x^2 - 4 = (x + 2)(x - 2)$; $m = 2$

Step 1: Domain: $\{x \mid x = -2, x \neq 2\}$

Step 2: (a) There are no x-intercepts, since $p(x)$ has no zeros.

(b) The y-intercept is $y = R(0) = \frac{-3}{4}$

Step 3: $R(-x) = \frac{3}{x^2 - 4} = R(x)$

Therefore, the graph of $R(x)$ is symmetric with respect to the y-axis.

Step 4: The vertical asymptotes are the zeros of $q(x)$: $x = -2$ and $x = 2$.

Step 5: Since $n < m$, the line $y = 0$ is a horizontal asymptote; not intersected.

Step 6: We have a total of two zeros, -2, 2:

Interval	Test Number	$R(x)$	Graph of R
$x < -2$	$x = -3$	$\frac{3}{9 - 4} = \frac{3}{5}$	Above the x-axis
$-2 < x < 2$	$x = 0$	$\frac{-3}{4}$	Below the x-axis
$x > 2$	$x = 3$	$\frac{3}{9 - 4} = \frac{3}{5}$	Above the x-axis

Step 7:

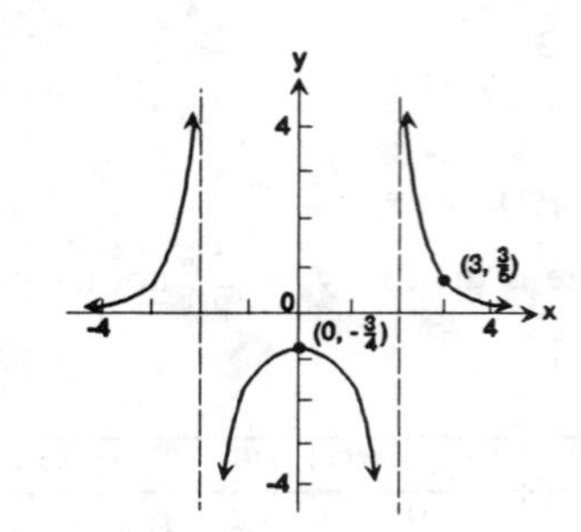

49. $P(x) = \dfrac{x^4 + x^2 + 1}{x^2 - 1}$

Step 0: $p(x) = x^4 + x^2 + 1$

$q(x) = x^2 - 1 = (x + 1)(x - 1)$

Step 1: Domain: $\{x \mid x \neq -1, x \neq 1\}$

Step 2: (a) There are no x-intercepts, since $p(x)$ has no real zeros.

(b) The y-intercept is $y = P(0) = -1$.

Step 3: $P(-x) = \dfrac{x^4 + x^2 + 1}{x^2 - 1} = P(x)$, so we do have symmetry with respect to the y-axis.

Step 4: The vertical asymptotes are $x = -1, x = 1$.

Step 5: Since $n > m + 1$, we have no horizontal and no oblique asymptote.

Step 6: We have two zeros, -1 and 1; due to symmetry, we only need to check for $x > 0$:

Interval	Test Number	$P(x)$	Graph of P
$0 < x < 1$	$x = \dfrac{1}{2}$	$\dfrac{\frac{1}{16} + \frac{1}{4} + 1}{\frac{1}{4} - 1} = \dfrac{-7}{4}$	Below the x-axis
$x > 1$	$x = 2$	$\dfrac{16 + 4 + 1}{4 - 1} = 7$	Above the x-axis

Step 7:

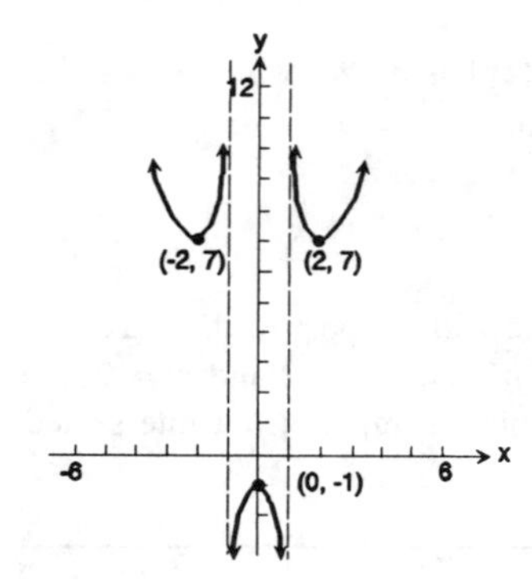

51. $H(x) = \dfrac{x^3 - 1}{x^2 - 9}$

Step 0: (a) $n = m + 1$, so we do long division:

$$\begin{array}{r} x \\ x^2 - 9 \overline{) x^3 + 0x^2 + 0x - 1} \\ \underline{x^3 - 9x } \\ 9x - 1 \end{array}$$

So $H(x) = x + \dfrac{9x - 1}{x^3 - 9}$, and we have an oblique asymptote, $y = x$.

(b) $p(x) = x^3 - 1 = (x - 1)\underbrace{(x^2 + x + 1)}_{\text{no real zeros}}$

$q(x) = x^2 - 9 = (x + 3)(x - 3)$

Step 1: Domain: $\{x \mid x \neq -3, x \neq 3\}$

Step 2: (a) The x-intercept is $x = 1$.

(b) The y-intercept is $y = H(0) = \dfrac{1}{9}$

Step 3: No symmetry.

Step 4: The vertical asymptotes are $x = -3$, $x = 3$.

Step 5: Since $n = m + 1$, we have only the oblique asymptote, $y = x$, found above; intersected at $\left(\dfrac{1}{9}, \dfrac{1}{9}\right)$.

Step 6: We have a total of three zeros, $-3, 1, 3$:

Interval	Test Number	$H(x)$	Graph of H
$x < -3$	$x = -5$	$\dfrac{-125 - 1}{25 - 9} = \dfrac{-63}{8}$	Below the x-axis
$-3 < x < 1$	$x = 0$	$\dfrac{1}{9}$	Above the x-axis
$1 < x < 3$	$x = 2$	$\dfrac{8 - 1}{4 - 9} = -\dfrac{7}{5}$	Below the x-axis
$x > 3$	$x = 5$	$\dfrac{125 - 1}{25 - 9} = \dfrac{31}{4}$	Above the x-axis

Step 7:

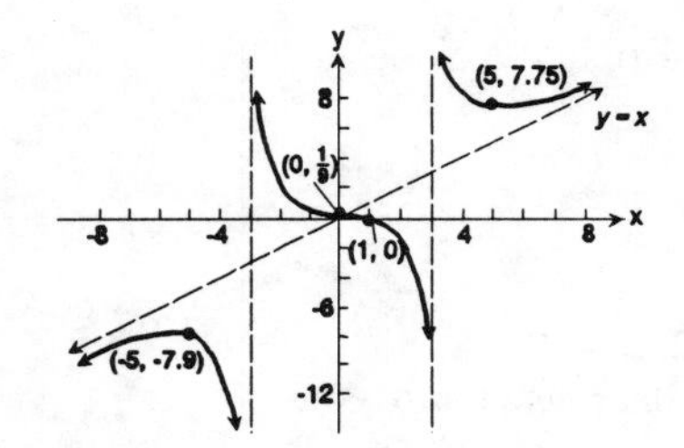

53. $R(x) = \dfrac{x^2}{x^2 + x - 6}$

Step 0: $p(x) = x^2$; $n = 2$

$q(x) = x^2 + x - 6 = (x + 3)(x - 2)$; $m = 2$

Step 1: Domain: $\{x \mid x \neq -3, x \neq 2\}$

Step 2: (a) The x-intercept is $x = 0$

(b) The y-intercept is $y = R(0) = \dfrac{0}{-6} = 0$

The intercept is (0, 0).

Step 3: No symmetry.

Step 4: Vertical asymptotes $x = -3, x = 2$.

Step 5: Since $n = m$, the line $y = \dfrac{1}{1} = 1$ is the horizontal asymptote; intersected at (6, 1).

Step 6: We have three zeros, $-3, 0, 2$:

Interval	Test Number	$R(x)$	Graph of R
$x < -3$	$x = -6$	$\dfrac{36}{36 - 6 - 6} = \dfrac{3}{2}$	Above the x-axis
$-3 < x < 0$	$x = -2$	$\dfrac{4}{4 - 2 - 6} = -1$	Below the x-axis
$0 < x < 2$	$x = 1$	$\dfrac{1}{1 + 1 - 6} = -\dfrac{1}{4}$	Below the x-axis
$x > 2$	$x = 3$	$\dfrac{9}{9 + 3 - 6} = \dfrac{3}{2}$	Above the x-axis

Step 7:

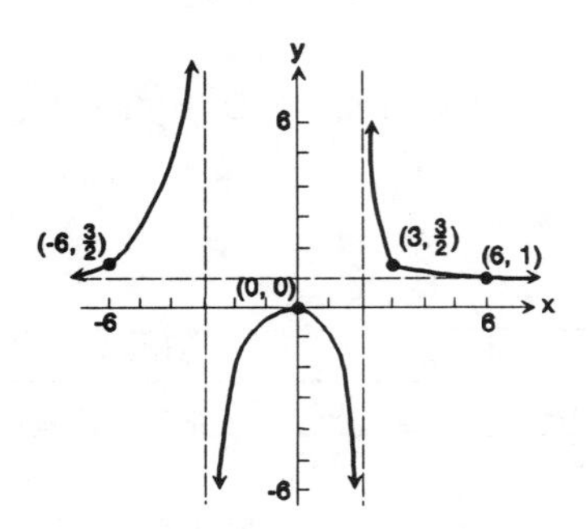

55. $G(x) = \dfrac{x}{x^2 - 4}$

Step 0: $p(x) = x;\ n = 1$

$q(x) = x^2 - 4 = (x + 2)(x - 2);\ m = 2$

Step 1: Domain: $\{x \mid x \neq -2, x \neq -2, x \neq 2\}$

Step 2: (a) The x-intercept is $x = 0$

(b) The y-intercept is $y = G(0) = 0$

The intercept is (0, 0).

Step 3: $G(-x) = \dfrac{-x}{x^2 - 4} = -G(x)$

$G(x)$ is symmetric about the origin.

Step 4: The vertical asymptotes are $x = -2, x = 2$.

Step 5: The horizontal asymptote is $y = 0$; intersected at (0, 0).

Step 6: Zeros, $-2, 0, 2$; because of symmetry, we only check $x > 0$:

Interval	Test Number	$G(x)$	Graph of G
$0 < x < 2$	$x = 1$	$\frac{1}{1-4} = -\frac{1}{3}$	Below the x-axis
$x > 2$	$x = 3$	$\frac{3}{9-4} = \frac{3}{5}$	Above the x-axis

Step 7:

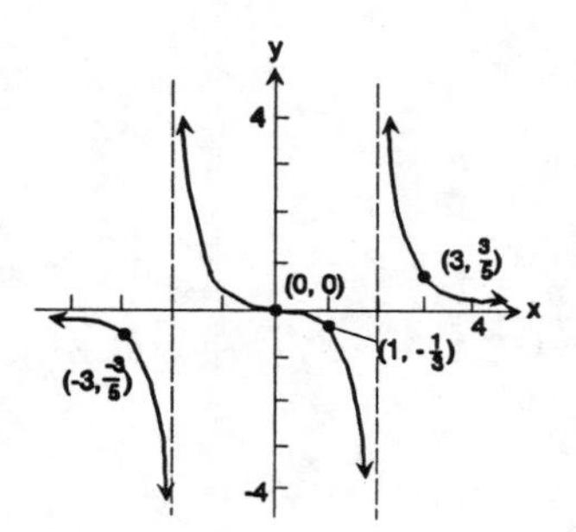

57. $R(x) = \dfrac{3}{(x-1)(x^2-4)}$

Step 0: $p(x) = 3;\ n = 0$

$q(x) = (x-1)(x^2-4) = (x-1)(x+2)(x-2);\ m = 3$

Step 1: Domain: $\{x \mid x \neq -2, x \neq 1, x \neq 2\}$

Step 2: (a) No x-intercept.

(b) y-intercept is $y = R(0) = \frac{3}{4}$

Step 3: No symmetry.

Step 4: Vertical asymptotes: $x = -2, x = 1, x = 2$

Step 5: Horizontal asymptote: $y = 0$; not intersected.

Step 6:

Interval	Test Number	$R(x)$	Graph of R
$x < -2$	$x = -3$	$\frac{3}{(-4)(5)} = \frac{-3}{20}$	Below the x-axis
$-2 < x < 1$	$x = 0$	$\frac{3}{4}$	Above the x-axis
$1 < x < 2$	$x = \frac{3}{2}$	$\frac{3}{\left(\frac{1}{2}\right)\left(\frac{-7}{4}\right)} = \frac{-24}{7}$	Below the x-axis
$x > 2$	$x = 3$	$\frac{3}{(2)(5)} = \frac{3}{10}$	Above the x-axis

Step 7:

59. $H(x) = \dfrac{4(x^2 - 1)}{x^4 - 16}$

Step 0: $p(x) = 4(x^2 - 1) = 4(x + 1)(x - 1);\ n = 2$

$q(x) = x^4 - 16 = (x^2 + 4)(x^2 - 4) = (x^2 + 4)(x + 2)(x - 2);\ m = 4$

Step 1: Domain: $\{x \mid x \neq -2, x \neq 2\}$

Step 2: (a) The x-intercepts are -1 and 1.

(b) The y-intercept is $y = H(0) = \dfrac{1}{4}$

Step 3: $H(-x) = \dfrac{4(x^2 - 1)}{x^4 - 16} = H(x)$, so the graph is symmetric with respect to the y-axis.

Step 4: The vertical asymptote are: $x = -2, x = 2$.

Step 5: The horizontal asymptote is $y = 0$; intersected at $(-1, 0)$ and $(1, 0)$.

Step 6: The zeros are $-2, -1, 1, 2$; we check the portion $x \geq 0$:

Interval	Test Number	$H(x)$	Graph of H
$-1 < x < 1$	$x = 0$	$\dfrac{1}{4}$	Above the x-axis
$1 < x < 2$	$x = \dfrac{3}{2}$	$\dfrac{4\left[\dfrac{9}{4} - 1\right]}{\dfrac{81}{16} - 16} = \dfrac{-16}{35}$	Below the x-axis
$x > 2$	$x = 3$	$\dfrac{4(9 - 1)}{81 - 16} \approx 0.49$	Above the x-axis

Step 7:

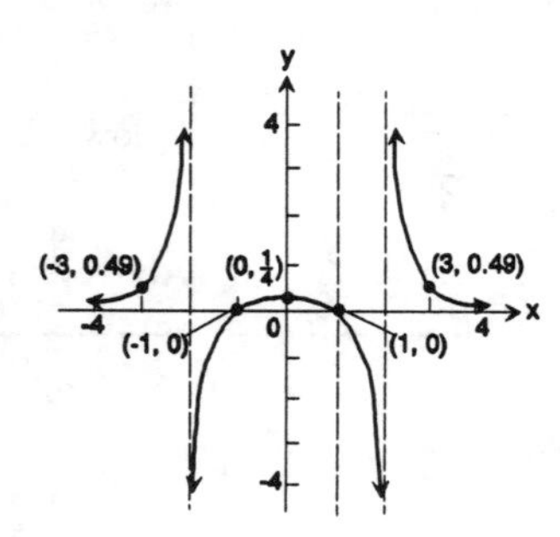

61. $F(x) = \dfrac{x^2 - 3x - 4}{x + 2}$

Step 0: (a) Since $n = m + 1$, perform division:

$$\begin{array}{r} x - 5 \\ x + 2\overline{)x^2 - 3x - 4} \\ \underline{x^2 + 2x} \\ -5x - 4 \\ \underline{-5x - 10} \\ 6 \end{array}$$

Therefore, $F(x) = x - 5 + \dfrac{6}{x + 2}$, so we have an oblique asymptote, $y = x - 5$.

(b) $p(x) = x^2 - 3x - 4 = (x + 1)(x - 4)$
$q(x) = x + 2$

Step 1: Domain: $\{x \mid x \neq -2\}$

Step 2: (a) The x-intercepts are -1, 4.
(b) The y-intercept is -2

Step 3: No symmetry.

Step 4: The vertical asymptote is $x = -2$.

Step 5: The oblique asymptote is $y = x - 5$; not intersected.

Step 6: The zeros are -2, -1, and 4.

Interval	Test Number	$F(x)$	Graph of F
$x < -2$	$x = -3$	$\dfrac{9 + 9 - 4}{-3 + 2} = -14$	Below the x-axis
$-2 < x < -1$	$x = -\dfrac{3}{2}$	$\dfrac{\frac{9}{4} + \frac{9}{2} - 4}{-\frac{3}{2} + 2} = \dfrac{11}{2}$	Above the x-axis
$-1 < x < 4$	$x = 0$	-2	Below the x-axis
$x > 4$	$x = 5$	$\dfrac{25 - 15 - 4}{5 + 2} = \dfrac{6}{7}$	Above the x-axis

Step 7:

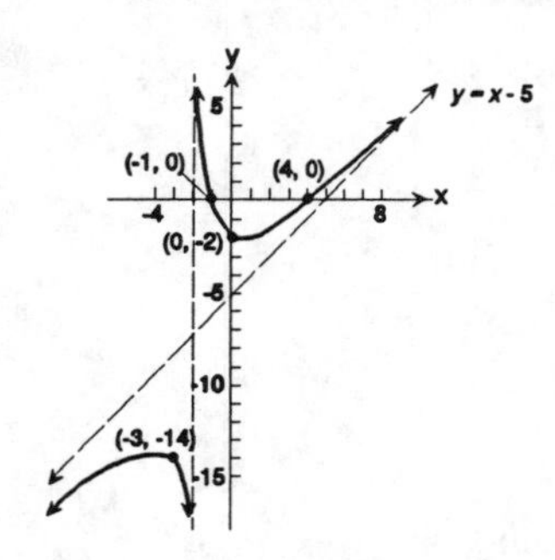

63. $R(x) = \dfrac{x^2 + x - 12}{x - 4}$

Step 0: (a) $n = m + 1$, so we do long division:

$$\begin{array}{r} x + 5 \\ x + 2\overline{)x^2 + \quad x - 12} \\ \underline{x^2 - 4x} \quad\quad \\ 5x - 12 \\ \underline{5x - 20} \\ 8 \end{array}$$

So $R(x) = x + 5 + \dfrac{8}{x - 4}$, and we have an oblique asymptote, $y = x + 5$

(b) $p(x) = x^2 + x - 12 = (x + 4)(x - 3)$
$q(x) = x - 4$

Step 1: Domain: $\{x \mid x \neq 4\}$

Step 2: (a) The x-intercepts are -4 and 3
(b) The y-intercept is $y = R(0) = 3$

Step 3: No symmetry.

Step 4: The vertical asymptote is $x = 4$.

Step 5: Since $n = m + 1$, we have only the oblique asymptote, $y = x + 5$, found above; not intersected.

Step 6: The zeros are -4, 3, and 4.

Interval	Test Number	$R(x)$	Graph of R
$x < -4$	-5	$\frac{-8}{9}$	Below x-axis
$-4 < x < 3$	1	$\frac{10}{3}$	Above x-axis
$3 < x < 4$	$\frac{7}{2}$	$\frac{-15}{2}$	Below x-axis
$x > 4$	5	18	Above x-axis

Step 7:

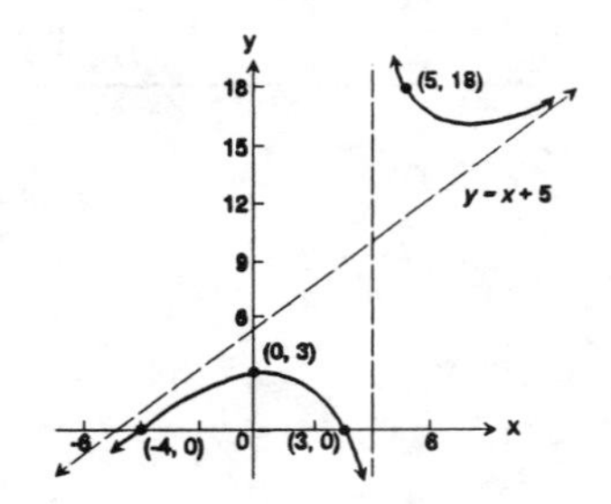

65. $F(x) = \dfrac{x^2 + x - 12}{x + 2}$

Step 0: (a) $n = m + 1$, so we do long division:

$$\begin{array}{r} x - 1 \\ x + 2\overline{)x^2 + \quad x - 12} \\ \underline{x^2 + 2x} \quad\quad \\ -x - 12 \\ \underline{-x - \;\; 2} \\ -10 \end{array}$$

So $F(x) = x - 1 + \frac{-10}{x + 2}$, and we have an oblique asymptote, $y = x - 1$

(b) $p(x) = x^2 + x - 12 = (x + 4)(x - 3)$
$q(x) = x + 2$

Step 1: Domain: $\{x \mid x \neq -2\}$

Step 2: (a) The x-intercepts are -4 and 3
(b) The y-intercept is $y = F(0) = -6$

Step 3: No symmetry

Step 4: The vertical asymptote is $x = -2$

Step 5: Since $n = m + 1$, we have only the oblique asymptote, $y = x - 1$, found above; not intersected.

Step 6: The zeros are -4, -2, and 3.

Interval	Test Number	$F(x)$	Graph of F
$x < -4$	-5	$\frac{-8}{3}$	Below x-axis
$-4 < x < -2$	-3	6	Above x-axis
$-2 < x < 3$	1	$\frac{-10}{3}$	Below x-axis
$x > 3$	4	$\frac{4}{3}$	Above x-axis

Step 7:

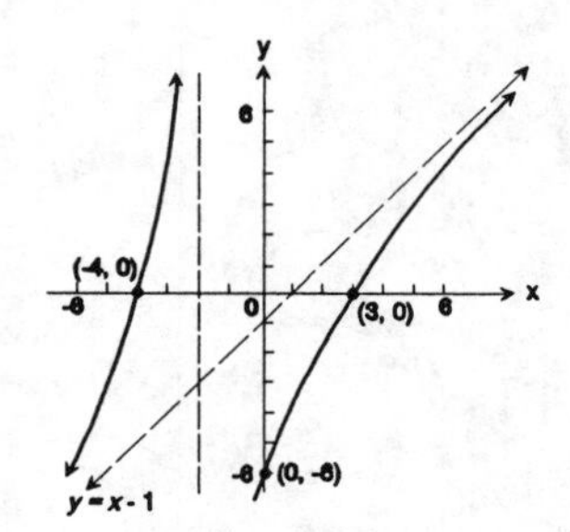

67. $R(x) = \frac{x(x - 1)^2}{(x + 3)^3}$

Step 0: $p(x) = x(x - 1)^2$; $n = 3$
$q(x) = (x + 3)^3$; $m = 3$

Step 1: Domain: $\{x \mid x \neq -3\}$

Step 2: (a) The x-intercepts are 0 and 1
(b) The y-intercept is $y = R(0) = 0$

Step 3: No symmetry

Step 4: The vertical asymptote is $x = -3$

Step 5: Since $n = m$, the horizontal asymptote is the line $y = 1$; not intersected.

Step 6: The zeros are -3, 0, and 1.

Interval	Test Number	$R(x)$	Graph of R
$x < -3$	-4	100	Above x-axis
$-3 < x < 0$	-1	$\frac{-1}{2}$	Below x-axis
$0 < x < 1$	$\frac{1}{2}$	$\frac{1}{343}$	Above x-axis
$x > 1$	2	$\frac{2}{125}$	Above x-axis

Step 7:

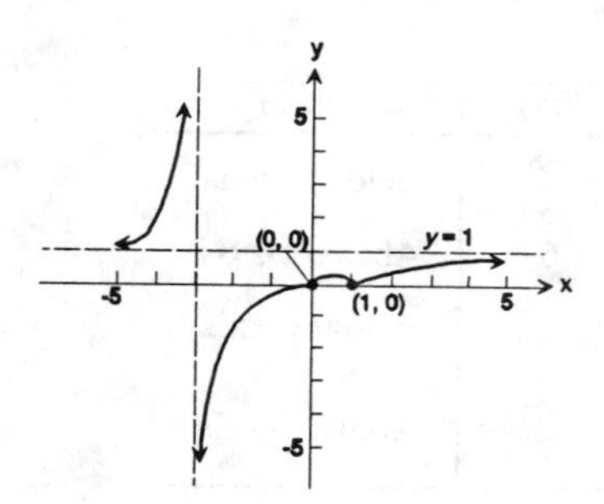

69. $R(x) = \dfrac{x^2 + x - 12}{x^2 - x - 6} = \dfrac{(x + 4)(x - 3)}{(x - 3)(x + 2)} = \dfrac{x + 4}{x + 2}, x \neq 3$

Step 0: $n = 2$

$m = 2$

Step 1: Domain: $\{x \mid x \neq -2, x \neq 3\}$

Step 2: (a) The x-intercept is -4.

(b) The y-intercept is 2.

Step 3: No symmetry.

Step 4: The vertical asymptote is $x = -2$. There is a hole at $\left(3, \dfrac{7}{5}\right)$.

Step 5: Since $n = m$, the horizontal asymptote is the line $y = 1$.

$$1 = \frac{x + 4}{x - 2}$$

$$x - 2 = x + 4$$

No solution, so horizontal asymptote not intersected.

Step 6: The zeros are -4 and -2.

Interval	Test Number	$R(x)$	Graph of R
$x < -4$	-5	$\frac{1}{3}$	Above x-axis
$-4 < x < -2$	-3	-1	Below x-axis
$x > -2$	0	2	Above x-axis

Step 7:

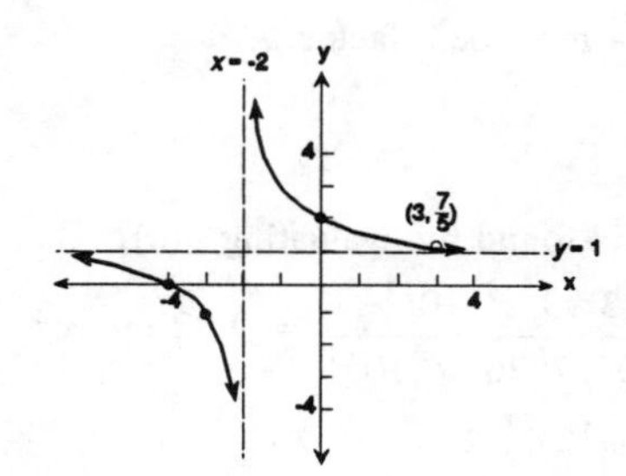

71. $R(x) = \dfrac{6x^2 - 7x - 3}{2x^2 - 7x + 6} = \dfrac{(2x - 3)(3x + 1)}{(2x - 3)(x - 2)} = \dfrac{3x + 1}{x - 2}, x \neq \dfrac{3}{2}$

Step 0: $n = 2, m = 2$

Step 1: Domain: $\{x \mid x \neq \frac{3}{2}, x \neq 2\}$

Step 2: (a) The x-intercept is $\frac{-1}{3}$.

(b) The y-intercept is $\frac{-1}{2}$.

Step 3: No symmetry.

Step 4: The vertical asymptote is $x = 2$. There is a hole at $\left(\frac{3}{2}, -11\right)$.

Step 5: Since $n = m$, the horizontal asymptote is the line

$$y = \frac{6}{2} = 3$$
$$3 = \frac{3x + 1}{x - 2}$$
$$3x - 6 = 3x + 1$$

No solution, so the horizontal asymptote is not intersected.

Step 6: The zeros are $\frac{-1}{3}$, $\frac{3}{2}$, and 2.

Interval	Test Number	$R(x)$	Graph of R
$x < \frac{-1}{3}$	-1	$\frac{2}{3}$	Above x-axis
$\frac{-1}{3} < x < \frac{3}{2}$	0	$\frac{-1}{2}$	Below x-axis
$\frac{3}{2} < x < 2$	1.75	-25	Below x-axis
$x > 2$	3	10	Above x-axis

Step 7:

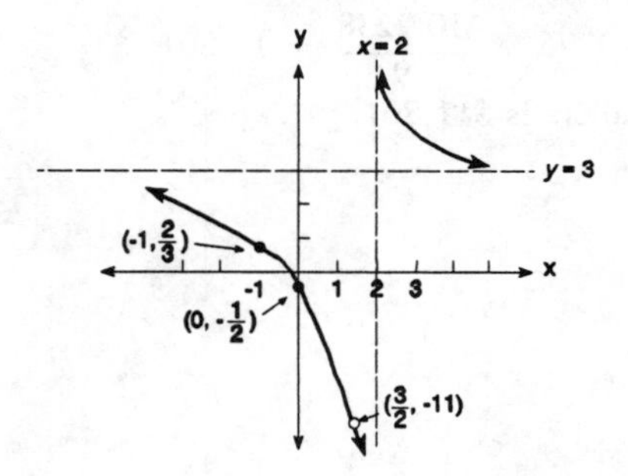

73. 4 must be a zero of the denominator; hence, $x - 4$ must be a factor.

75. c, d

77. (a) The acceleration due to gravity at sea level is found by evaluating $g(0)$.

$$g(0) = \frac{3.99 \times 10^{14}}{(6.374 \times 10^6 + 0)^2} = \frac{3.99 \times 10^{14}}{4.0627876 \times 10^{13}} = 9.82 \text{ m/sec}^2$$

(b) $$g(443) = \frac{3.99 \times 10^{14}}{(6.374 \times 10^6 + 443)^2} = 9.8195 \text{ m/sec}^2$$

(c) $$g(8848) = \frac{3.99 \times 10^{14}}{(6.374 \times 10^6 + 8848)^2} = 9.7936 \text{ m/sec}^2$$

(d) The horizontal asymptote of $g(h)$ is found by letting $h \to \infty$. As $h \to \infty$, the denominator of $g(h) \to \infty$; thus, $g(h) \to 0$. The horizontal asymptote is the h-axis.

(e)

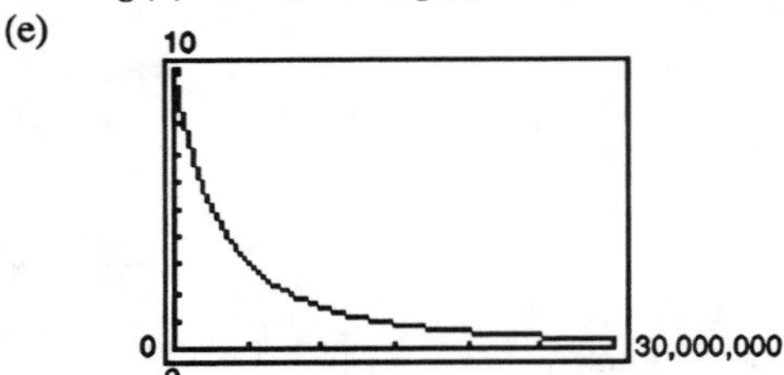

(f) $g(h) = 0$ has no solution; thus, it is never possible to escape the pull of earth's gravity.

79. (a) The horziontal asymptote of $C(t)$ is found by letting $t \to \infty$. As $t \to \infty$, $2t^2 + 1 \to \infty$, so $C(t) \to 0$. The horizontal asymptote of $C(t)$ is the t-axis.

(b)

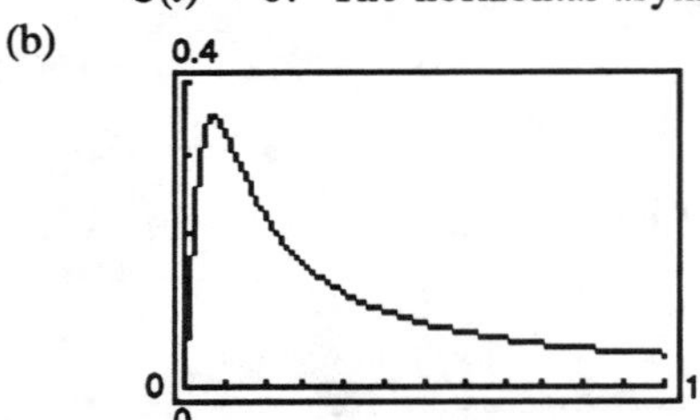

(c) Using the MAXIMUM command, $C(t)$ is highest when $t = 0.70$ hours.

81. (a) $$\bar{C}(x) = \frac{C(x)}{x} = \frac{0.2156x^3 - 2.3473x^2 + 14.3275x + 10.2238}{x}$$

$$= 0.2156x^2 - 2.3473x + 14.3275 + \frac{10.2238}{x}$$

(b) $$\bar{C}(6) = 0.2156(6)^2 - 2.3473(6) + 14.3275 + \frac{10.2238}{6} = 9.709$$

Thus, the average cost of producing 6 Cavaliers is \$9709.

(c) $$\bar{C}(9) = 0.2156(9)^2 - 2.3473(9) + 14.3275 + \frac{10.2238}{9} = 11.801$$

The average cost of producing 11 Cavaliers is \$11,801.

(d)

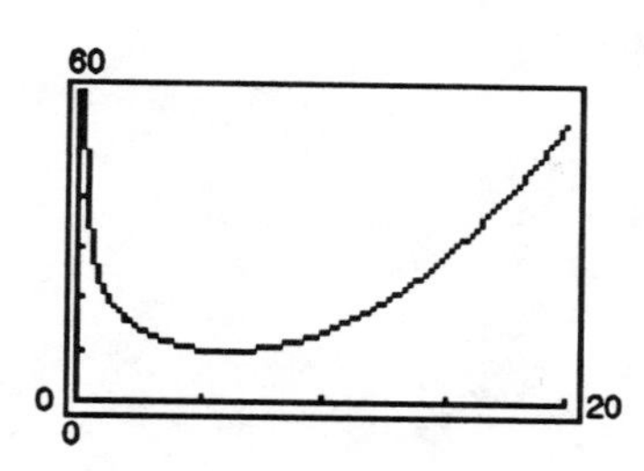

(e) Using the MINIMUM command, the number of Cavaliers to produce to minimize average cost is 6.

(f) $C(6) = \$9709$

83. (a) The surface area is the sum of the areas of the six sides of the box. The top and bottom portion of the box have surface area x^2. The four sides of the box have surface area xy, So, surface area $= 2x^2 + 4xy$.

The volume of the box is $x \cdot x \cdot y = x^2 y = 10{,}000$. Thus, $y = \dfrac{10{,}000}{x^2}$.

Substitute this for y in the expression for surface area to obtain

$$S(x) = 2x^2 + 4x\left(\frac{10{,}000}{x^2}\right) = 2x^2 + \frac{40{,}000}{x} = \frac{2x^3 + 40{,}000}{x}$$

(b)

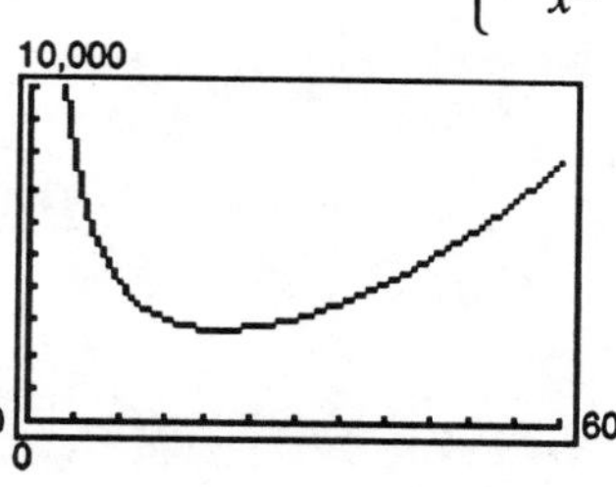

(c) Using the MINIMUM command, the smallest surface area possible is 2784.95 square inches.

(d) The value of S is minimum when $x = 21.544$. Thus,

$$y = \frac{10{,}000}{21.544^2} = 21.545$$

The dimensions of the box are 21.54" × 21.54" × 21.54".

85. (a) C = Cost of the material, r = radius, h = height

$$500 = \pi r^2 h, \quad h = \frac{500}{\pi r^2}$$

$$C(r) = 6(2\pi r^2) + 4(2\pi r h)$$

$$= 12\pi r^2 + 8\pi r\left(\frac{500}{\pi r^2}\right)$$

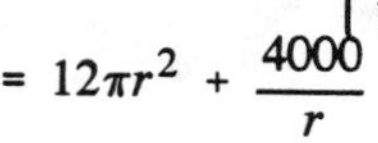

$$= 12\pi r^2 + \frac{4000}{r}$$

(b)

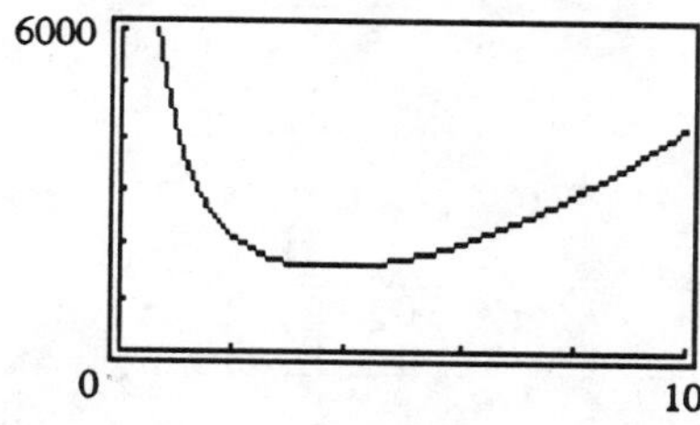

The cost is least for r about 3.75 cm.

87. No, $x = 1$ is not a vertical asymptote because each of the functions is not written in lowest terms. Each of the functions is a quotient of polynomials and is undefined for $x = 1$.

89. Minimum value: 2.00 at $x = 1.00$

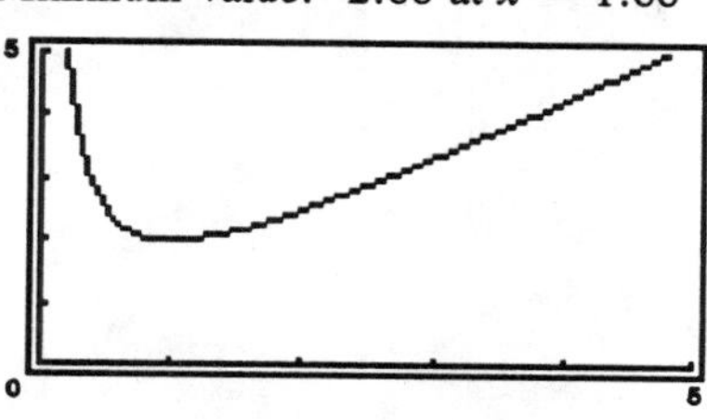

91. Minimum value: 1.88 at $x = 0.79$

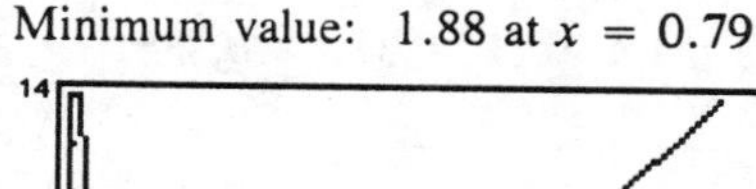

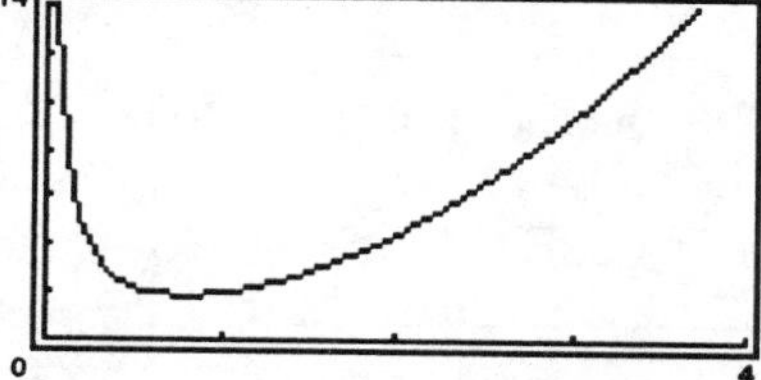

93. Minimum value: 1.75 at $x = 1.31$

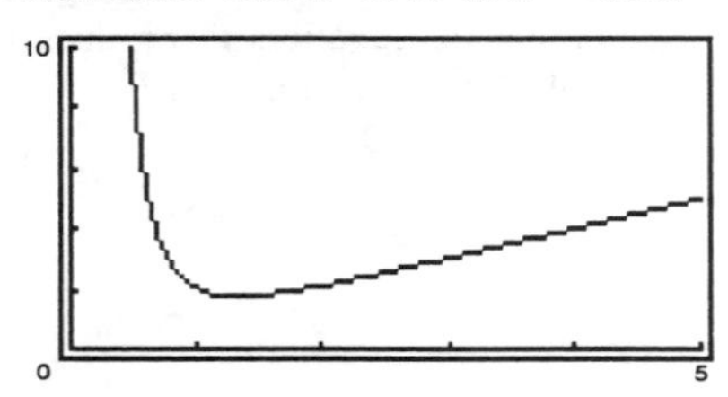

5.4 Synthetic Division

1.

$$\begin{array}{r|rrrr} 2 & 1 & -1 & 2 & 4 \\ & & 2 & 2 & 8 \\ \hline & 1 & 1 & 4 & 12 \end{array}$$

← $f(x) = x^3 - x^2 + 2x + 4$

← Remainder = 12

Quotient $= x^2 + x + 4$

3.

$$\begin{array}{r|rrrr} 3 & 3 & 2 & -1 & 3 \\ & & 9 & 33 & 96 \\ \hline & 3 & 11 & 32 & 99 \end{array}$$

← $f(x) = 3x^3 + 2x^2 - x + 3$

← Remainder = 99

Quotient $= 3x^2 + 11x + 32$

5.

$$\begin{array}{r|rrrrrr} -3 & 1 & 0 & -4 & 0 & 1 & 0 \\ & & -3 & 9 & -15 & 45 & -138 \\ \hline & 1 & -3 & 5 & -15 & 46 & -138 \end{array}$$

← $f(x) = x^5 - 4x^3 + x$

← Remainder = -138

Quotient $= x^4 - 3x^3 + 5x^2 - 15x + 46$

7.

$$\begin{array}{r|rrrrrrr} 1 & 4 & 0 & -3 & 0 & 1 & 0 & 5 \\ & & 4 & 4 & 1 & 1 & 2 & 2 \\ \hline & 4 & 4 & 1 & 1 & 2 & 2 & 7 \end{array}$$

← Remainder = 7

Quotient $= 4x^5 + 4x^4 + x^3 + x^2 + 2x + 2$

9.

$$\begin{array}{r|rrrr} -1.1 & 0.1 & 0 & 0.2 & 0 \\ & & -0.11 & 0.121 & -0.3531 \\ \hline & 0.1 & -0.11 & 0.321 & -0.3531 \end{array}$$

← Remainder = -0.3531

Quotient $= 0.1x^2 - 0.11x + 0.321$

11.

$$\begin{array}{r|rrrrrr} 1 & 1 & 0 & 0 & 0 & 0 & -1 \\ & & 1 & 1 & 1 & 1 & 1 \\ \hline & 1 & 1 & 1 & 1 & 1 & 0 \end{array}$$

← $f(x) = x^5 - 1$

← Remainder = 0

Quotient $= x^4 + x^3 + x^2 + x + 1$

In Problems 13–21, we use the following facts: the remainder in synthetic division when f(x) is divided by $x - c$, is f(c); and $x - c$ is a factor of f(x) only if $f(c) = 0$.

13. We divide by $x - 2$:

$$\begin{array}{r|rrrr} 2 & 4 & -3 & -8 & 4 \\ & & 8 & 10 & 4 \\ \hline & 4 & 5 & 2 & \boxed{8} \end{array}$$

Remainder $= 8 \neq 0$; therefore, $x - 2$ is *not* a factor of $f(x)$.

15.

$$\begin{array}{r|rrrrr} 2 & 3 & -6 & 0 & -5 & 10 \\ & & 6 & 0 & 0 & -10 \\ \hline & 3 & 0 & 0 & -5 & \boxed{0} \end{array}$$

The remainder $= 0$; therefore, $x - 2$ *is* a factor of $f(x)$.

17.

$$\begin{array}{r|rrrrrrr} -3 & 3 & 0 & 0 & 82 & 0 & 0 & 27 \\ & & -9 & 27 & -81 & -3 & 9 & -27 \\ \hline & 3 & -9 & 27 & 1 & -3 & 9 & \boxed{0} \end{array}$$

Remainder $= 0$; therefore, $x + 3$ is a factor.

19.

$$\begin{array}{r|rrrrrrr} -4 & 4 & 0 & -64 & 0 & 1 & 0 & -15 \\ & & -16 & 64 & 0 & 0 & -4 & 16 \\ \hline & 4 & -16 & 0 & 0 & 1 & -4 & \boxed{1} \end{array}$$

$x + 4$ is not a factor, since the remainder $= 1 \neq 0$.

21.

$$\begin{array}{r|rrrrr} \frac{1}{2} & 2 & -1 & 0 & 2 & -1 \\ & & 1 & 0 & 0 & 1 \\ \hline & 2 & 0 & 0 & 2 & \boxed{0} \end{array}$$

Since the remainder $= 0$; therefore, $x - \frac{1}{2}$ *is* a factor of $f(x)$.

5.5 The Real Zeros of a Polynomial Function

Historical Problems

1. $y^3 + by^2 cy + d = 0$

Using the substitution, $y = x - \frac{b}{3}$, we have:

$$\left(x - \frac{b}{3}\right)^3 + b\left(x - \frac{b}{3}\right)^2 + c\left(x - \frac{b}{3}\right) + d = 0$$

$$x^3 - 3\frac{b}{3}x^2 + 3\left(\frac{b^2}{9}\right)x - \frac{b^3}{27} + b\left(x^2 - 2\frac{b}{3}x + \frac{b^2}{9}\right) + cx - \frac{bc}{3} + d = 0$$

$$x^3 - \frac{b^2}{3}x + cx - \frac{b^3}{27} + \frac{b^3}{9} - \frac{bc}{3} + d = 0$$

$$x^3 + \left(c - \frac{b^2}{3}\right)x + \left(\frac{2b^3}{27} - \frac{bc}{3} + d\right) = 0$$

2. We start with $x^3 + px + q = 0$ and let $x = H + K$, to obtain:

$$(H + K)^3 + p(H + K) + q = 0$$
$$H^3 + 3H^2K + 3HK^2 + K^3 + pH + pK + q = 0$$
$$H^3 + K^3 + 3H^2K + 3HK^2 + pH + pK = -q \qquad (1)$$

Now we let $3HK = -p$. If we multiply both sides of $3HK = -p$ by H, we obtain $3H^2K = -pH$. If we multiply both sides by K, we obtain $3HK^2 = -pK$. This allows us to rewrite equation (1):

$$H^3 + K^3 - pH - pK + pH + pK = -q$$

or

$$H^3 + K^3 = -q, \text{ as desired.}$$

Notice we now need to find H or K such that

$$H^3 + K^3 = -q,$$

and

$$3HK = -p$$

If we can find H and K, then we will have our x, since $x = H + K$.

3. Based on Problem 2, we have two equations:

$$3HK = -p \text{ and } H^3 + K^3 = -q$$

Thus, $K = \dfrac{-p}{3H}$

so that

$$H^3 + K^3 = -q$$

$$H^3 + \left(\frac{-p}{3H}\right)^3 = -q$$

$$H^3 - \frac{p^3}{qH^3} = -q$$

$$H^6 + qH^3 - \frac{p^3}{27} = 0$$

$$H^3 = \frac{-q \pm \sqrt{q^2 + \dfrac{4p^3}{27}}}{2}; \quad \text{(choose + sign)}$$

$$H = \sqrt[3]{\frac{-q}{2} + \sqrt{\frac{q^2}{4} + \frac{p^3}{27}}}$$

4. From Problem 2 we have $H^3 + K^3 = -q$, and from Problem 3,

$$H = \sqrt[3]{\frac{-q}{2} + \sqrt{\frac{q^2}{4} + \frac{p^3}{27}}}$$

Since $H^3 + K^3 = -q$, we have

$$\frac{-q}{2} + \sqrt{\frac{q^2}{4} + \frac{p^3}{27}} + K^3 = -q$$

$$K^3 = -q + \frac{q}{2} - \sqrt{\frac{q^2}{4} + \frac{p^3}{27}}$$

$$K^3 = \frac{-q}{2} - \sqrt{\frac{q^2}{4} + \frac{p^3}{27}}$$

or $$K = \sqrt[3]{\frac{-q}{2} + \sqrt{\frac{q^2}{4} + \frac{p^3}{27}}}$$

5. From Problem 2, we know $x = H + K$

From Problem 3, $H = \sqrt[3]{\dfrac{-q}{2} + \sqrt{\dfrac{q^2}{4} + \dfrac{p^3}{27}}}$

From Problem 4, $K = \sqrt[3]{\dfrac{-q}{2} - \sqrt{\dfrac{q^2}{4} + \dfrac{p^3}{27}}}$

Therefore, $x = \sqrt[3]{\dfrac{-q}{2} + \sqrt{\dfrac{q^2}{4} + \dfrac{p^3}{27}}} + \sqrt[3]{\dfrac{-q}{2} - \sqrt{\dfrac{q^2}{4} + \dfrac{p^3}{27}}}$

6. From Problem 3, a solution of the equation $x^3 + px + q = 0$ is given by

$$x = \sqrt[3]{\frac{-q}{2} + \sqrt{\frac{q^2}{4} + \frac{p^3}{27}}} + \sqrt[3]{\frac{-q}{2} - \sqrt{\frac{q^2}{4} + \frac{p^3}{27}}}$$

We are asked to solve $x^3 - 6x - 9 = 0$
Here $p = -6$, $q = -9$

Then $\dfrac{q^2}{4} + \dfrac{p^3}{27} = \dfrac{81}{4} - \dfrac{216}{27} = \dfrac{81}{4} - 8 = \dfrac{49}{4}$

and we have $x = \sqrt[3]{\dfrac{-(-9)}{2} + \sqrt{\dfrac{49}{4}}} + \sqrt[3]{\dfrac{-(-9)}{2} - \sqrt{\dfrac{49}{4}}} = \sqrt[3]{\dfrac{9}{2} + \dfrac{7}{2}} + \sqrt[3]{\dfrac{9}{2} - \dfrac{7}{2}} = \sqrt[3]{8} + \sqrt[3]{1}$
or $x = 2 + 1 = 3$

Now we have one solution, $x = 3$. Hence $x - 3$ is a factor, and we can perform synthetic division to find the other factor:

```
-3)1   0  -6  -9
      -3  -9  -9
   1   3   3   0
```

Quotient $= x^2 + 3x + 3$

The solutions to the quadratic $x^2 + 3x + 3 = 0$ are $x = \dfrac{-3 \pm \sqrt{9 - 4(3)}}{2} = \dfrac{-3 \pm \sqrt{-3}}{2}$, which involves complex numbers (refer to Sections 1.9 and 2.4). The three solutions are:

$$x = 3, \; x = \frac{-3 + \sqrt{3}i}{2}, \; x = \frac{-3 - \sqrt{3}i}{2}$$

7. $x^3 + 3x - 14 = 0$, so $p = 3$ and $q = -14$.

$$x = H + K = \sqrt[3]{\frac{-(-14)}{2} + \sqrt{\frac{(-14)^2}{4} + \frac{3^3}{27}}} + \sqrt[3]{\frac{-(-14)}{2} - \sqrt{\frac{(-14)^2}{4} + \frac{3^3}{27}}}$$

$$= \sqrt[3]{7 + \sqrt{50}} + \sqrt[3]{7 - \sqrt{50}} = \sqrt[3]{7 + 5\sqrt{2}} + \sqrt[3]{7 - 5\sqrt{2}}$$

8. Solve $f(x) = x^3 + 3x - 14 = 0$ — There is one positive zero.
$f(-x) = -x^3 - 3x - 14$ — There are no negative zeros.

$\dfrac{p}{q}$: $\pm 1, \pm 2, \pm 7, \pm 14$ are the potential rational zeros.

$$
\begin{array}{r|rrrr} 1 & 1 & 0 & 3 & -14 \\ & & 1 & 3 & 4 \\ \hline & 1 & 1 & 4 & -10 \end{array}
\qquad
\begin{array}{r|rrrr} -1 & 1 & 0 & 3 & -14 \\ & & -1 & 1 & -4 \\ \hline & 1 & -1 & 4 & -18 \end{array}
\qquad
\begin{array}{r|rrrr} 2 & 1 & 0 & 3 & -14 \\ & & 2 & 4 & 14 \\ \hline & 2 & 2 & 7 & 0 \end{array}
$$

Thus, $x - 2$ is a factor and 2 is a zero.

$q_1(x) = 2x^2 + 2x + 7$

Thus, $f(x) = (x - 2)(2x^2 + 2x + 7) = 0$

For $2x^2 + 2x + 7$, $x = \dfrac{-2 \pm \sqrt{4 - 56}}{4} = \dfrac{-1}{2} \pm \dfrac{\sqrt{13}}{2}i$

Exercises

1. $c = 2$
   $f(2) = 4(2)^3 - 3(2)^2 - 8(2) + 4 = 8$
   Since $f(2) = 8 \neq 0$, $x - 2$ is not a factor of f.

3. $c = 2$
   $f(2) = 3(2)^4 - 6(2)^2 - 5(2) + 10 = 0$
   Since $f(2) = 0$, $x - 2$ is a factor of f.

5. $x + 3 = x - (-3)$, so $c = -3$
   $f(-3) = 3(-3)^6 + 82(-3)^3 + 27 = 0$
   Since $f(-3) = 0$, $x + 3$ is a factor of f.

7. $x + 4 = x - (-4)$, so $c = -4$
   $f(-4) = 4(-4)^6 - 64(-4)^4 + (-4)^2 - 15 = 1$
   Since $f(-4) = 1 \neq 0$, $x + 4$ is not a factor of f.

9. $c = \dfrac{1}{2}$
   $f\left(\dfrac{1}{2}\right) = 2\left(\dfrac{1}{2}\right)^4 - \left(\dfrac{1}{2}\right)^3 + 2\left(\dfrac{1}{2}\right) - 1 = 0$
   Since $f\left(\dfrac{1}{2}\right) = 0$, $x - \dfrac{1}{2}$ is a factor of f.

In Problems 11–21, we must count the number of sign changes in both $f(x)$ and $f(-x)$.

11. $f(x) = -4x^7 + x^3 - x^2 + 2$
    The maximum number of zeros is 7.

    For $f(x) = -4x^7 + x^3 - x^2 + 2$, there are three variations in sign of

    − to + + to − − to +

    the coefficients, so there will be either three positive zeros, or one positive zero.

    To find $f(-x)$, replace x in $f(x)$ by $-x$:
    $f(-x) = 4x^7 - x^3 - x^2 + 2$

    + to − − to +

    There are two variations in the sign of $f(-x)$, so there will be either two negative zeros, or none.

13. $f(x) = 2x^6 - 3x^2 - x + 1$
    The maximum number of zeros is 6.
    Here, $f(x) = 2x^6 - 3x^2 - x + 1$, and

    + to − − to +

    $f(-x) = 2x^6 - 3x^2 + x + 1$, and

    + to − − to +

    Thus, there are either two positive zeros, or none; and either two negative zeros or none.

15. $f(x) = 3x^3 - 2x^2 + x + 2$
The maximum number of zeros is 3.
For $f(x) = 3x^3 - 2x^2 + x + 2$,

$+$ to $-$ $\quad$ $-$ to $+$

there are two variations in sign of the coefficients, so there will be either two positive zeros, or none.

To find $f(-x)$, replace x in $f(x)$ by $-x$:
$$f(-x) = -3x^3 - 2x^2 - x + 2$$

$-$ to $+$

There is only one variation of sign in $f(-x)$, so there will be exactly one negative zero.

17. $f(x) = -x^4 + x^2 - 1$
The maximum number of zeros is 4.
Here, $f(x) = -x^4 + x^2 - 1$, and

$-$ to $+$ $+$ to $-$

$$f(-x) = -x^4 + x^2 - 1$$

$-$ to $+$ $+$ to $-$

Thus, there are either two positive zeros, or none; and either two negative zeros, or none.

19. $f(x) = x^5 + x^4 + x^2 + x + 1$
The maximum number of zeros is 5.
We have $f(x) = x^5 + x^4 + x^2 + x + 1$, which has ***no*** changes of signs; hence, there will be ***no*** positive zeros.

Meanwhile, $f(-x) = -x^5 + x^4 + x^2 - x + 1$ has three changes in sign; so $f(x)$ will have either three negative zeros, or one negative zero.

21. $f(x) = x^6 - 1$
The maximum number of zeros is 6.
$f(x) = x^6 - 1$; $f(-x) = x^6 - 1$
We will have exactly one positive zero and one negative zero.

For Problems 23–33, use the Rational Zeros Theorem. The possible rational (whole number or fractional) zeros must be of the form $\frac{p}{q}$, where p is a factor of the constant term, and q is a factor of the coefficient of the highest power of x.

23. For $f(x) = 3x^4 - 3x^3 + x^2 - x + 1$,
p must be a factor of $+1$: $p = +1$ or -1
q must be a factor of $+3$: $q = \pm 1$ or ± 3
So the possible zeros $\frac{p}{q}$ are $\pm 1, \pm \frac{1}{3}$

25. For $f(x) = x^5 - 6x^2 + 9x - 3$,
p must be a factor of -3: $p = \pm 1$ or ± 3
q must be a factor of $+1$: $q = \pm 1$
So the possible rational zeros are $\pm 1, \pm 3$

27. For $f(x) = -4x^3 - x^2 + x + 2$,
p must be a factor of $+2$: $p = \pm 1, \pm 2$
q must be a factor of -4: $q = \pm 1, \pm 2, \pm 4$
Hence, the possible rational zeros, $\frac{p}{q}$, are $\pm 1, \pm \frac{1}{2}, \pm \frac{1}{4}, \pm 2$

29. For $f(x) = 6x^4 - x^2 + 9$,
p must be a factor of $+9$: $p = \pm 1, \pm 3, \pm 9$
q must be a factor of $+3$: $q = \pm 1, \pm 2, \pm 3, \pm 6$
Possible rational zeros: $\pm 1, \pm \frac{1}{2}, \pm \frac{1}{3}, \pm \frac{1}{6}, \pm 3, \pm \frac{3}{2}, \pm 9, \pm \frac{9}{2}$

31. For $f(x) = 2x^5 - x^3 + 2x^2 + 12$, we have the following possibilities:

$$p = \pm 1, \pm 2, \pm 3, \pm 4, \pm 6, \pm 12; \; q = \pm 1, \pm 2$$

Rational zeros, $\frac{p}{q}$: $\pm 1, \pm\frac{1}{2}, \pm 2, \pm 3, \pm\pm\frac{3}{2}, \pm 4, \pm 6, \pm 12$

33. For $f(x) = 6x^4 + 2x^3 - x^2 + 20$, we have the following possibilities:

p: $\pm 1, \pm 2, \pm 4, \pm 5, \pm 10, \pm 20$;

q: $\pm 1, \pm 2, \pm 3, \pm 6$;

$\frac{p}{q}$: $\pm 1, \pm\frac{1}{2}, \pm\frac{1}{3}, \pm\frac{1}{6}, \pm 2, \pm\frac{2}{3}, \pm 4, \pm\frac{4}{3}, \pm 5,$

$\pm\frac{5}{2}, \pm\frac{5}{3}, \pm\frac{5}{6}, \pm 10, \pm\frac{10}{3}, \pm 20, \pm\frac{20}{3}$

35. $f(x) = x^3 + 2x^2 - 5x - 6; f(-x) = -x^3 + 2x^2 + 5x - 6$

(a) $f(x)$ has at most three zeros.

(b) $f(x)$ has one positive zero, and either two negative zeros, or none.

(c) Possible rational zeros: p: $\pm 1, \pm 2, \pm 3, \pm 6$; q: ± 1; $\frac{p}{q}$: $\pm 1, \pm 2, \pm 3, \pm 6$

(d) Synthetic Division:

$$\begin{array}{r|rrrr} -1 & 1 & 2 & -5 & -6 \\ & & -1 & -1 & 6 \\ \hline & 1 & 1 & -6 & \boxed{0} \end{array}$$

Since the remainder is 0, $x - (-1) = x + 1$ is a factor.
The other factor is the quotient: $x^2 + x - 6$.
Thus, $f(x) = (x + 1)(x^2 + x - 6) = (x + 1)(x + 3)(x - 2)$ and the zeros are -1, -3, and 2.

37. $f(x) = 2x^3 - x^2 + 2x - 1; f(-x) = -2x^3 - x^2 - 2x - 1$

(a) $f(x)$ has at most three zeros.

(b) $f(x)$ has either three positive zeros, or just one, and no negative zeros.

(c) Possible rational zeros: p: ± 1; q: $\pm 1, \pm 2$; $\frac{p}{q}$: $\pm 1, \pm\frac{1}{2}$

(d) Synthetic Division:

$$\begin{array}{r|rrrr} -1 & 2 & -1 & 2 & -1 \\ & & -2 & 3 & -5 \\ \hline & 2 & -3 & 5 & \boxed{-6} \end{array} \leftarrow (x + 1) \text{ is } \textbf{\textit{not}} \text{ a factor.}$$

So we try $x - 1$

$$\begin{array}{r|rrrr} 1 & 2 & -1 & 2 & -1 \\ & & 2 & 1 & 3 \\ \hline & 2 & 1 & 3 & \boxed{2} \end{array} \leftarrow (x - 1) \text{ is } \textbf{\textit{not}} \text{ a factor.}$$

Let's turn to fractions, say $x - \frac{1}{2}$:

$$\begin{array}{r|rrrr} \frac{1}{2} & 2 & -1 & 2 & -1 \\ & & 1 & 0 & 1 \\ \hline & 2 & 0 & 2 & \boxed{0} \end{array} \leftarrow \left(x - \frac{1}{2}\right) \text{ is a factor.}$$

Quotient $= 2x^2 + 2$

Thus, $f(x) = \left(x - \frac{1}{2}\right)(2x^2 + 2) = 2\left(x - \frac{1}{2}\right)(x^2 + 1)$

The only real zero is $\frac{1}{2}$, since $x^2 + 1$ has no real zeros.

39. $f(x) = x^4 + x^2 - 2; f(-x) = x^4 + x^2 - 2$

(a) $f(x)$ has at most four zeros.

(b) $f(x)$ has one positive zero and one negative zero. (The other two zeros must be complex numbers.)

(c) Possible rational zeros: p: $\pm 1, \pm 2$; q: ± 1; $\frac{p}{q}$: $\pm 1, \pm 2$

(d) Synthetic Division:

Try $x + 1$:
```
-1)1   0   1   0  -2
      -1   1  -2   2
   ------------------
   1  -1   2  -2  [0]  (x + 1) is a factor!
```

Quotient: $x^3 - x^2 + 2x - 2$

We have $f(x) = (x + 1)(x^3 - x^2 + 2x - 2)$

We can factor $x^3 - x^2 + 2x - 2$ by grouping terms:

$$x^3 - x^2 + 2x - 2 = x^2(x - 1) + 2(x - 1) = (x^2 + 2)(x - 1)$$

Thus, $f(x) = (x + 1)(x - 1)\underbrace{(x^2 + 2)}_{\text{no real zeros}}$

We have two real zeros: $-1, 1$, which agrees with what we discovered in part (b).

Note: $f(x) = x^4 + x^2 - 2$ could have been factored at the beginning:

$$x^4 + x^2 - 2 = (x^2 + 2)(x^2 - 1) = (x^2 + 2)(x + 1)(x - 1)$$

41. $f(x) = 4x^4 + 7x^2 - 2; f(-x) = 4x^4 + 7x^2 - 2$

(a) There are at most four zeros.

(b) There is one positive zero and one negative zero.

(c) Possible rational zeros: p: $\pm 1, \pm 2$; q: $\pm 1, \pm 2, \pm 4$; $\frac{p}{q}$: $\pm 1, \pm\frac{1}{2}, \pm\frac{1}{4}, \pm 2$

We can start by factoring:

$$f(x) = 4x^4 + 7x^2 - 2 = (4x^2 - 1)(x^2 + 2)$$

$$= 4\left(x^2 - \frac{1}{4}\right)(x^2 + 2)$$

$$= 4\left(x + \frac{1}{2}\right)\left(x - \frac{1}{2}\right)\underbrace{(x^2 + 2)}_{\text{no real zeros}}$$

Thus, $f(x)$ has the two real zeros $-\frac{1}{2}, \frac{1}{2}$.

43. $f(x) = x^4 + x^3 - 3x^2 - x + 2; f(-x) = x^4 - x^3 - 3x^2 + x + 2$

(a) There are at most four zeros.

(b) We will have either two positive zeros, or none, and either two negative zeros, or none.

(c) Possible rational zeros: p: $\pm 1, \pm 2$; q: ± 1; $\frac{p}{q}$: $\pm 1, \pm 2$.

(d) We cannot factor $f(x)$ by normal methods, so we start with synthetic division:

```
-1)1   1  -3  -1   2
      -1   0   3  -2
   ------------------
   1   0  -3   2  [0]  so x + 1 is a factor!
```

$x^3 - 3x + 2$

We now work on the depressed equation, $x^3 - 3x + 2 = 0$. This is cubic, and not easily factored. We try synthetic division again, using $x + 1$ once more (since -1 ***could*** be a root of multiplicity two).

$$\begin{array}{r|rrrr} -1 & 1 & 0 & -3 & 2 \\ & & -1 & 1 & 2 \\ \hline & 1 & -1 & -2 & \boxed{4} \end{array}$$

Therefore, $x + 1$ is not a factor of $x^3 - 3x + 2$.

But, we *do* know that there is another negative zero, and if it is rational, it must be -2. So we divide by $x - (-2) = x + 2$:

$$\begin{array}{r|rrrr} -2 & 1 & 0 & -3 & 2 \\ & & -2 & 4 & -2 \\ \hline & 1 & -2 & 1 & \boxed{0} \end{array}$$

Thus, $(x + 2)$ *is* a factor!

$x^2 - 2x + 1$

We now have:

$$\begin{aligned} f(x) &= (x + 1)(x^3 - 3x + 2) \\ &= (x + 1)(x + 2)(x^2 - 2x + 1) \\ &= (x + 1)(x + 2)(x - 1)(x - 1) \\ &= (x + 1)(x + 2)(x - 1)^2 \end{aligned}$$

and the zeros are -1, -2, 1, 1.

45. $f(x) = 4x^5 - 8x^4 - x + 2; f(-x) = -4x^5 - 8x^4 + x + 2$

(a) There are at most five zeros.

(b) There are either two or no positive zeros, and there is one negative zero.

(c) Possible rational zeros: p: $\pm 1, \pm 2$; q: $\pm 1, \pm 2, \pm 4$; $\frac{p}{q}$: $\pm 1, \pm \frac{1}{2}, \pm \frac{1}{4}, \pm 2$

(d) Here we can factor by grouping:

$$\begin{aligned} f(x) = 4x^5 - 8x^4 - x + 2 &= 4x^4(x - 2) - 1(x - 2) \\ &= (4x^4 - 1)(x - 2) \\ &= 4\left(x^4 - \frac{1}{4}\right)(x - 2) \\ &= 4\left(x^2 - \frac{1}{2}\right)\left(x^2 + \frac{1}{2}\right)(x - 2) \\ &= 4\left(x - \sqrt{\frac{1}{2}}\right)\left(x + \sqrt{\frac{1}{2}}\right)(x - 2)\left(x^2 + \frac{1}{2}\right) \end{aligned}$$

We have $f(x) = 4\left(x - \frac{1}{\sqrt{2}}\right)\left(x + \frac{1}{\sqrt{2}}\right)(x - 2)\underbrace{\left(x^2 + \frac{1}{2}\right)}_{\text{no real zeros}}$

and the real zeros of $f(x)$ are:

$\frac{1}{\sqrt{2}} = \frac{\sqrt{2}}{2}, \frac{-\sqrt{2}}{2}$, and 2

47. $x^4 - x^3 + 2x^2 - 4x - 8 = 0$

The solutions of this equation are the zeros of the polynomial function $f(x)$.

$$\begin{aligned} f(x) &= x^4 - x^3 + 2x^2 - 4x - 8 \\ f(-x) &= x^4 + x^3 + 2x^2 + 4x - 8 \end{aligned}$$

(a) There are at most four zeros.

(b) There are either three or one positive zero, and there is one negative zero.

(c) p: $\pm 1, \pm 2, \pm 4, \pm 8$; q: ± 1; $\frac{p}{q}$: $\pm 1, \pm 2, \pm 4, \pm 8$

(d) We start with synthetic division:

$$\begin{array}{r|rrrrr} -1 & 1 & -1 & 2 & -4 & -8 \\ & & -1 & 2 & -4 & 8 \\ \hline & 1 & -2 & 4 & -8 & \boxed{0} \end{array} \quad \text{Thus, } x + 1 \text{ is a factor.}$$

$$x^3 - 2x^2 + 4x - 8 = 0$$

We now have:

$$(x + 1)(x^3 - 2x^2 + 4x - 8) = 0$$
$$(x + 1)[x^2(x - 2) + 4(x - 2)] = 0$$
$$(x + 1)(x^2 + 4)(x - 2) = 0$$

no real zeros (for $x^2 + 4$)

The real zeros are $-1, 2$.

49. $3x^3 + 4x^2 - 7x + 2 = 0$

The solutions of this equation are the zeros of the polynomial function $f(x)$.

$$f(x) = 3x^3 + 4x^2 - 7x + 2$$
$$f(-x) = -3x^3 + 4x^2 + 7x + 2$$

(a) There are at most three zeros.

(b) There are two or no positive zeros, and there is one negative zero.

(c) p: $\pm1, \pm2$; q: $\pm1, \pm3$; $\frac{p}{q}$: $\pm1, \pm\frac{1}{3}, \pm2, \pm\frac{2}{3}$

(d)

$$\begin{array}{r|rrrr} -1 & 3 & 4 & -7 & 2 \\ & & -3 & -1 & 8 \\ \hline & 3 & 1 & -8 & \boxed{10} \end{array} \quad \text{so } x + 1 \text{ is not a factor.}$$

$$\begin{array}{r|rrrr} 1 & 3 & 4 & -7 & 2 \\ & & 3 & 7 & 0 \\ \hline & 3 & 7 & 0 & \boxed{2} \end{array} \quad x - 1 \text{ is not a factor.}$$

$$\begin{array}{r|rrrr} -2 & 3 & 4 & -7 & 2 \\ & & -6 & 4 & 6 \\ \hline & 3 & -2 & -3 & \boxed{8} \end{array} \quad x + 2 \text{ is not a factor.}$$

$$\begin{array}{r|rrrr} 2 & 3 & 4 & -7 & 2 \\ & & 6 & 20 & 26 \\ \hline & 3 & 10 & 13 & \boxed{28} \end{array} \quad x - 2 \text{ is not a factor.}$$

Now to the fractions:

$$\begin{array}{r|rrrr} -\frac{1}{3} & 3 & 4 & -7 & 2 \\ & & -1 & -1 & \frac{8}{3} \\ \hline & 3 & 3 & -8 & \boxed{\frac{14}{3}} \end{array} \quad x + \frac{1}{3} \text{ is not a factor.}$$

$$\begin{array}{r|rrrr} \frac{1}{3} & 3 & 4 & -7 & 2 \\ & & 1 & \frac{5}{3} & -\frac{16}{9} \\ \hline & 3 & 5 & \frac{-16}{3} & \boxed{\frac{2}{9}} \end{array} \quad x - \frac{1}{3} \text{ is not a factor.}$$

$$
\begin{array}{r|rrrr}
-\frac{2}{3} & 3 & 4 & -7 & 2 \\
 & & -2 & -\frac{4}{3} & \frac{50}{9} \\
\hline
 & 3 & 2 & \frac{-25}{3} & \boxed{\frac{68}{9}}
\end{array}
\qquad x + \frac{2}{3} \text{ is not a factor.}
$$

Sometimes you have days like this! We are down to our last possible rational root, $\frac{2}{3}$, so we try dividing by $x - \frac{2}{3}$:

$$
\begin{array}{r|rrrr}
\frac{2}{3} & 3 & 4 & -7 & 2 \\
 & & 2 & 4 & -2 \\
\hline
 & 3 & 6 & -3 & \boxed{0}
\end{array}
\qquad \text{At last! } \left(x - \frac{2}{3}\right) \text{ is a factor.}
$$

$3x^2 + 6x - 3 = 0$

$f(x) = \left[x - \frac{2}{3}\right](3x^2 + 6x - 3) = 3\left[x - \frac{2}{3}\right](x + 2x - 1)$

Now, $x^2 + 2x - 1$ cannot be factored over the integers, so we use the quadratic formula:

$$x = \frac{-2 \pm \sqrt{4 - 4(-1)}}{2} = \frac{-2 \pm \sqrt{8}}{2} = \frac{-2 \pm 2\sqrt{2}}{2}$$

The three roots are $\frac{2}{3}$, $-1 + \sqrt{2}$, $-1 - \sqrt{2}$.

Recall that once a zero, c, is known, then $x - c$ is a factor; so in factored form:

$$3\left[x - \frac{2}{3}\right]\left(x - (-1 + \sqrt{2})\right)\left(x - (-1 - \sqrt{2})\right) = 0$$

$$3\left[x - \frac{2}{3}\right]\left(x + 1 - \sqrt{2}\right)\left(x + 1 + \sqrt{2}\right) = 0$$

51. $3x^3 - x^2 - 15x + 5 = 0$

The solutions of this equation are the zeros of the polynomial function $f(x)$.

$f(x) = 3x^3 - x^2 - 15x + 5$

$f(-x) = -3x^3 - x^2 + 15x + 5$

Let's just start factoring:

$$3x^3 - x^2 - 15x + 5 = 0$$

$$x^2(3x - 1) = 5(3x - 1) = 0$$

$$(x^2 - 5)(3x - 1) = 0$$

$$\left(x + \sqrt{5}\right)\left(x - \sqrt{5}\right)(3x - 1) = 0$$

$$3\left[x - \frac{1}{3}\right]\left(x + \sqrt{5}\right)\left(x - \sqrt{5}\right) = 0$$

The zeros of $f(x)$ are $\frac{1}{3}$, $-\sqrt{5}$, $\sqrt{5}$.

53. $x^4 + 4x^3 + 2x^2 - x + 6 = 0$

The solutions of this equation are the zeros of the polynomial function $f(x)$.

$f(x) = x^4 + 4x^3 + 2x^2 - x + 6$

$f(-x) = x^4 - 4x^3 + 2x^2 + x + 6$

(a) There are at most four zeros.

(b) There are either two positive zeros or none, and either two negative zeros or none.

(c) p: $\pm 1, \pm 2, \pm 3, \pm 6$; q: ± 1; $\frac{p}{q}$: $\pm 1, \pm 2, \pm 3, \pm 6$

(d) Synthetic Division:

$$\begin{array}{r|rrrrr} -1 & 1 & 4 & 2 & -1 & 6 \\ & & -1 & -3 & 1 & 0 \\ \hline & 1 & 3 & -1 & 0 & \boxed{6} \end{array} \quad x + 1 \text{ is } \textit{not} \text{ a factor.}$$

$$\begin{array}{r|rrrrr} 1 & 1 & 4 & 2 & -1 & 6 \\ & & 1 & 5 & 7 & 6 \\ \hline & 1 & 5 & 7 & 6 & \boxed{12} \end{array} \quad x - 1 \text{ is } \textit{not} \text{ a factor.}$$

$$\begin{array}{r|rrrrr} -2 & 1 & 4 & 2 & -1 & 6 \\ & & -2 & -4 & 4 & -6 \\ \hline & 1 & 2 & -2 & 3 & \boxed{0} \end{array} \quad x + 2 \text{ } \textit{is} \text{ a factor!}$$

$x^3 + 2x^2 - 2x + 3 = 0$ ← Depressed equation

We continue with synthetic division. We do not need to try 1 or -1, since they did not work in the original polynomial, and the depressed equation is a factor of the original.

Use the Rational Zeros Theorem on the cubic polynomial $x^3 + 2x^2 - 2x + 3$:

p: $\pm 1, \pm 3$; q: ± 1; $\frac{p}{q}$: $\pm 1, \pm 3$

But we have ruled out ± 1, since they did not work in the original equation. Notice that $f(x)$ has at least one negative root ($x = -2$), since $x + 2$ is a factor. Therefore, by part (b), there must be two negative zeros. So let's try -3. We divide by $x - (-3) = x + 3$:

$$\begin{array}{r|rrrr} -3 & 1 & 2 & -2 & 3 \\ & & -3 & 3 & -3 \\ \hline & 1 & -1 & 1 & \boxed{0} \end{array} \quad \text{So, } x + 3 \text{ is a factor.}$$

We now have: $(x + 2)(x^3 + 2x^2 - 2x + 3) = 0$

$(x + 2)(x + 3)(x^2 - x + 1) = 0$

Now $x^2 - x + 1$ has no real zeros, since its discriminant is negative: $b^2 - 4ac = 1 - 4 = -3$. The real roots are -2 and -3.

55. $x^3 - \frac{2}{3}x^2 + \frac{8}{3}x + 1 = 0$

The solutions of this equation are the zeros of the polynomial function $f(x)$.

$f(x) = x^3 - \frac{2}{3}x^2 + \frac{8}{3}x + 1$

$f(-x) = -x^3 - \frac{2}{3}x^2 - \frac{8}{3}x + 1$

(a) There are at most three zeros.

(b) There are either two positive zeros, or none, and there is one negative zero.

(c) There is a trick to finding the possible rational zeros: All coefficients must be integers in order to use the Rational Zeros Theorem. But

$$x^3 - \frac{2}{3}x^2 + \frac{8}{3}x + 1 = 0$$

is equivalent to $3x^3 - 2x^2 + 8x + 3 = 0$

and we can determine the possible rational zeros:

p: $\pm 1, \pm 3$; q: $\pm 1, \pm 3$; $\frac{p}{q}$: $\pm 1, \pm \frac{1}{3}, \pm 3$

It is easy to check that 1 and -1 are not zeros of $f(x)$, by substitution.

$$f(1) = 1 - \frac{2}{3} + \frac{8}{3} + 1 \neq 0$$

$$f(-1) = -1 - \frac{2}{3} - \frac{8}{3} + 1 \neq 0$$

So let's try the fractions:

$-\frac{1}{3}$	1	$-\frac{2}{3}$	$\frac{8}{3}$	1
		$-\frac{1}{3}$	$\frac{1}{3}$	-1
	1	-1	3	0

So $\left(x + \frac{1}{3}\right)$ is a factor.

$x^2 - x + 3$

$$\left(x + \frac{1}{3}\right)(x^2 - x + 3) = 0$$

For $x^2 - x + 3$, we have: $x = \dfrac{1 \pm \sqrt{1 - 12}}{2} = \dfrac{1 \pm \sqrt{-11}}{2}$,

so we only have one real zero, $x = -\frac{1}{3}$

57. $2x^4 - 19x^3 + 57x^2 - 64x + 20 = 0$

The solutions of this equation are the zeros of the polynomial function $f(x)$.

$$f(x) = 2x^4 - 19x^3 + 57x^2 - 64x + 20$$

$$f(-x) = 2x^4 + 19x^3 + 57x^2 + 64x + 20$$

(a) There are at most four zeros.

(b) There are either 4, 2, or no positive zeros, and there are no negative zeros.

(c) p: $\pm 1, \pm 2, \pm 4, \pm 5, \pm 10, \pm 20$

q: $\pm 1, \pm 2$

$\frac{p}{q}$: $\pm 1, \pm \frac{1}{2}, \pm 2, \pm 4, \pm 5, \pm \frac{5}{2}, \pm 10, \pm 20$

(d) Try $x = 1$ as a solution:

1	2	-19	57	-64	20
		2	-17	30	-34
	2	-17	30	-34	-14

← $x - 1$ is not a factor.

Try $x = \frac{1}{2}$:

$\frac{1}{2}$	2	-19	57	-64	20
		1	-9	24	-20
	2	-18	48	-40	0

← $x - \frac{1}{2}$ is a factor.

$$f(x) = \left(x - \frac{1}{2}\right)(2x^3 - 18x^2 + 48x - 40) = 2\left(x - \frac{1}{2}\right)(x^3 - 9x^2 + 24x - 20)$$

Try $x = 2$ on depressed equation:

2	1	-9	24	-20
		2	-14	20
	1	-7	10	0

← $x - 2$ is a factor.

$$f(x) = 2\left(x - \frac{1}{2}\right)(x - 2)(x^2 - 7x + 10)$$
$$= 2\left(x - \frac{1}{2}\right)(x - 2)(x - 5)(x - 2)$$
$$= 2\left(x - \frac{1}{2}\right)(x - 2)(x - 2)^2(x - 5)$$

The solution set is $\left\{\frac{1}{2}, 2, 5\right\}$ where 2 is a zero of multiplicity 2.

59. From Problem 35:

$f(x) = x^3 + 2x^2 - 5x - 6 = (x + 1)(x + 3)(x - 2)$

(a) The y-intercept of f is $f(0) = -6$

(b) To find x-intercepts, use factored form

$f(x) = (x + 1)(x + 3)(x - 2)$

x-intercepts: $-1, -3, 2$

Use the x-intercepts to divide the x-axis into intervals:

Interval	Test Number	$f(x)$	Graph of $f(x)$
$x < -3$	-4	-18	Below the x-axis
$-3 < x < -1$	-2	4	Above the x-axis
$-1 < x < 2$	0	-6	Below the x-axis
$x > 2$	3	24	Above the x-axis

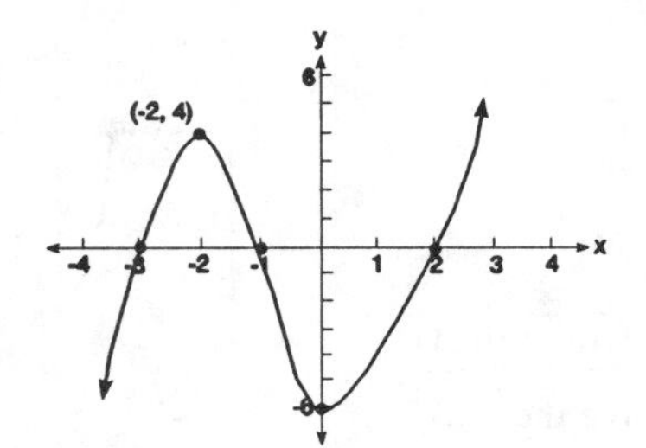

61. $f(x) = 2x^3 - x^2 + 2x - 1$

(a) To find the y-intercept, set $x = 0$:

$y = f(0) = -1$ is the y-intercept.

(b) To find x-intercepts, use the factored form (Problem 37):

$$y = f(x) = 2\left(x - \frac{1}{2}\right)(x^2 + 1)$$

Set $y = 0$ and solve for x: $2\left(x - \frac{1}{2}\right)(x^2 + 1) = 0$

So $x = \frac{1}{2}$ is the only x-intercept.

The intercepts are $\left(\frac{1}{2}, 0\right)$ and $(0, -1)$.

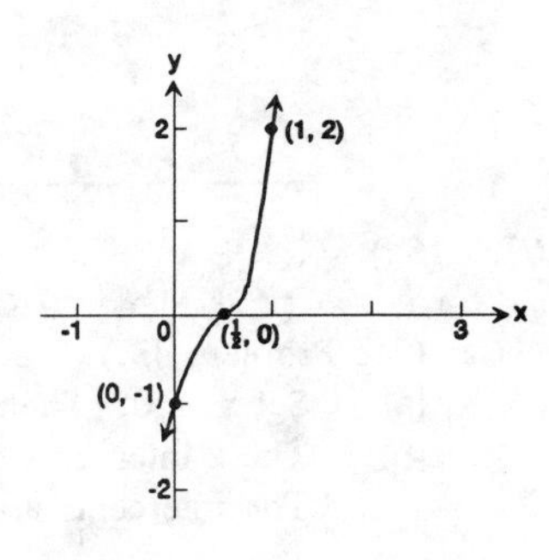

This divides the x-axis into 2 intervals:

Interval	Test Number	$f(x)$	Graph of $f(x)$
$-\infty < x < \frac{1}{2}$	$x = -1$	-6	Below the x-axis
$\frac{1}{2} < x < \infty$	$x = 1$	2	Above the x-axis

63. $f(x) = x^4 + x^2 - 2 = (x + 1)(x - 1)(x^2 + 2)$
(See Problem 39.)
(a) Set $x = 0$: $y = f(0) = -2$ is the y-intercept.
(b) The x-intercepts are the zeros of $f(x)$: $-1, 1$
The intercepts are $(-1, 0)$, $(1, 0)$, $(0, -2)$

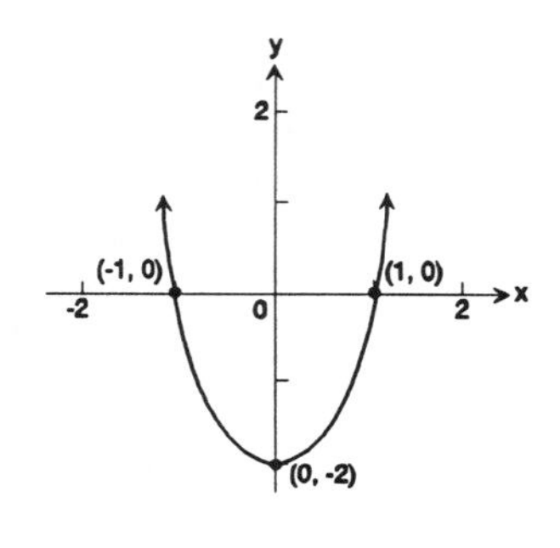

Interval	Test Number	$f(x)$	Graph of $f(x)$
$-\infty < x < -1$	$x = -2$	18	Above the x-axis
$-1 < x < 1$	$x = \frac{1}{2}$	$-\frac{27}{16}$	Below the x-axis
$1 < x < \infty$	$x = 2$	18	Above the x-axis

65. $f(x) = 4x^4 + 7x^2 - 2 = 4\left(x + \frac{1}{2}\right)\left(x - \frac{1}{2}\right)(x + 2)$
(See Problem 41.)
(a) Set $x = 0$: then $y = f(0) = -2$ is the y-intercept.
(b) The x-intercepts are the zeros: $x = -\frac{1}{2}, \frac{1}{2}$
The intercepts are $\left(-\frac{1}{2}, 0\right)$, $\left(\frac{1}{2}, 0\right)$, $(0, -2)$

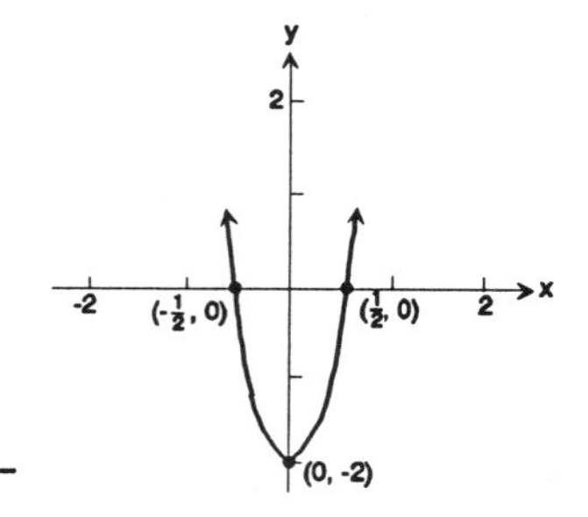

Interval	Test Number	$f(x)$	Graph of $f(x)$
$-\infty < x < -\frac{1}{2}$	$x = -1$	9	Above the x-axis
$-\frac{1}{2} < x < \frac{1}{2}$	$x = 0$	-2	Below the x-axis
$\frac{1}{2} < x < \infty$	$x = 1$	9	Above the x-axis

67. $f(x) = x^4 + x^3 - 3x^2 - x + 2$
$= (x + 1)(x + 2)(x - 1)(x - 1)$
(See Problem 43.)
(a) Set $x = 0$: Then $y = f(0) = 2$ is the y-intercept.
(b) The x-intercepts are the zeros: $-2, -1, 1$
The intercepts are $(-1, 0)$, $(-2, 0)$, $(1, 0)$, $(0, 2)$

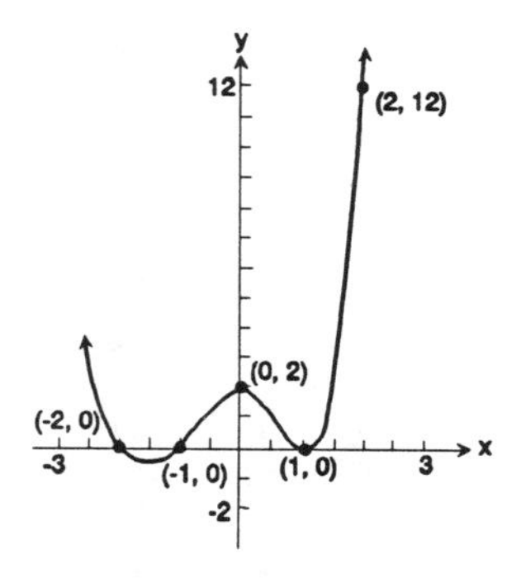

Interval	Test Number	$f(x)$	Graph of $f(x)$
$-\infty < x < -2$	$x = -3$	32	Above the x-axis
$-2 < x < -1$	$x = -\frac{3}{2}$	$-\frac{25}{16}$	Below the x-axis
$-1 < x < 1$	$x = 0$	2	Above the x-axis
$1 < x < \infty$	$x = 2$	12	Above the x-axis

69. $f(x) = 4x^5 - 8x^4 - x + 2$

$$= 4\left[x - \frac{\sqrt{2}}{2}\right]\left[x + \frac{\sqrt{2}}{2}\right](x - 2)\left[x^2 + \frac{1}{2}\right]$$

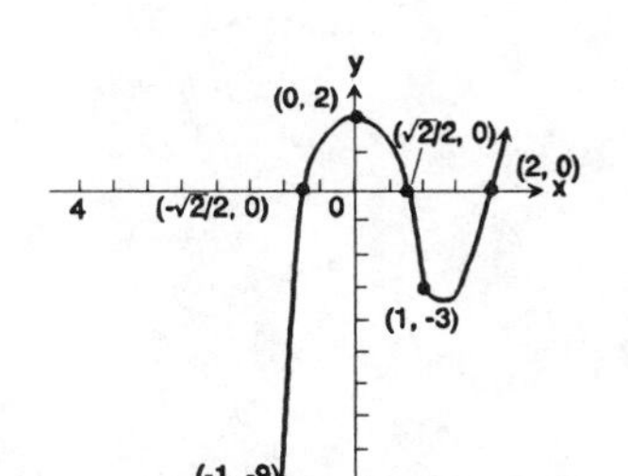

(See Problem 45.)

(a) Set $x = 0$: Then $y = f(0) = 2$ is the y-intercept.

(b) The x-intercepts are the zeros: $-\frac{\sqrt{2}}{2} \approx -.707$, $\frac{\sqrt{2}}{2} \approx .707$, 2

The intercepts are $(2, 0)$, $\left[-\frac{\sqrt{2}}{2}, 0\right]$, $\left[\frac{\sqrt{2}}{2}, 0\right]$, $(0, 2)$

Interval	Test Number	$f(x)$	Graph of $f(x)$
$-\infty < x < \frac{-\sqrt{2}}{2}$	$x = -1$	-9	Below the x-axis
$\frac{-\sqrt{2}}{2} < x < \frac{\sqrt{2}}{2}$	$x = 0$	2	Above the x-axis
$\frac{\sqrt{2}}{2} < x < 2$	$x = 1$	-3	Below the x-axis
$2 < x < \infty$	$x = 3$	323	Above the x-axis

71. $f(x) = x^4 - 3x^2 - 4$

$a_3 = 0, a_2 = -3, a_1 = 0, a_0 = -4$

$\text{Max}\{1, |-4| + |0| + |-3| + |0|\} = \text{Max}\{1, 4 + 0 + 3 + 0\} = \text{Max}\{1, 7\} = 7$

$1 + \text{Max}\{1, |-4|, |0|, |-3|, |0|\} = 1 + \text{Max}\{1, 4, 0, 3, 0\} = 1 + 4 = 5$

The smaller of the two numbers is 5. Thus, every zero of f lies between -5 and 5.

73. $f(x) = x^4 + x^3 - x - 1$

$a_3 = 1, a_2 = 0, a_1 = -1, a_0 = -1$

$\text{Max}\{1, |-1| + |-1| + |0| + |1|\} = \text{Max}\{1, 1 + 1 + 0 + 1\} = \text{Max}\{1, 3\} = 3$

$1 + \text{Max}\{1, |-1|, |-1|, |0|, |1|\} = 1 + \text{Max}\{1, 1, 1, 0, 1\} = 1 + 1 = 2$

The smaller of the two numbers is 2. Thus, every zero of f lies between -2 and 2.

75. $f(x) = 3x^4 + 3x^3 - x^2 - 12x - 12$

$$= 3\left[x^4 + 1x^3 - \frac{1}{3}x^2 - 4x - 4\right]$$ Note: Leading coefficient must be 1.

$a_3 = 1, a_2 = \frac{-1}{3}, a_1 = -4, a_0 = -4$

$\text{Max}\{1, |-4| + |-4| + |\frac{-1}{3}| + |1|\} = \text{Max}\{1, 4 + 4 + \frac{1}{3} + 1\} = \text{Max}\{1, \frac{28}{3}\}$

$= \frac{28}{3}$

$1 + \text{Max}\{1, |-4|, |-4|, |\frac{-1}{3}|, |1|\} = 1 + \text{Max}\{1, 4, 4, \frac{1}{3}, 1\} = 1 + 4 = 5$

The smaller of the two numbers is 5. Thus, every zero of f lies between -5 and 5.

77. $f(x) = 4x^5 - x^4 + 2x^3 - 2x^2 + x - 1$

$$= 4\left[x^5 - \frac{1}{4}x^4 + \frac{1}{2}x^3 - \frac{1}{2}x^2 + \frac{1}{4}x - \frac{1}{4}\right]$$ Leading coefficient must be 1.

$a_4 = \frac{1}{4}, a_3 = \frac{1}{2}, a_2 = \frac{-1}{2}, a_1 = \frac{1}{4}, a_0 = \frac{-1}{4}$

$$\text{Max}\{1, \left|\frac{-1}{4}\right| + \left|\frac{1}{2}\right| + \left|\frac{-1}{2}\right| + \left|\frac{1}{4}\right| + \frac{-1}{4}\} = \text{Max}\{1, \frac{1}{4} + \frac{1}{2} + \frac{1}{2} + \frac{1}{4} + \frac{1}{4}\}$$
$$= \text{Max}\ \{1, \frac{7}{4}\} = \frac{7}{4}$$
$$1 + \text{Max}\{1, \left|\frac{-1}{4}\right|, \left|\frac{1}{2}\right|, \left|\frac{-1}{2}\right|, \left|\frac{1}{4}\right|, \frac{-1}{4}\} = 1 + \text{Max}\{1, \frac{1}{4}, \frac{1}{2}, \frac{1}{2}, \frac{1}{4}, \frac{1}{4}\} = 1 + 1 = 2$$

The smaller of the two numbers is $\frac{7}{4}$. Thus, every zero of f lies between $\frac{-7}{4}$ and $\frac{7}{4}$.

79. $f(x) = 8x^4 - 2x^2 + 5x - 1$, $[0, 1]$
We need to evaluate $f(x)$ at the endpoints, $x = 0$ and $x = 1$. In this case, substitution is probably easiest:
$f(0) = -1$
$f(1) = 8 - 2 + 5 - 1 = 10$
Since $f(0) = -1 < 0$ and $f(1) = 10 > 0$, $f(x)$ must have a zero between 0 and 1.

81. $f(x) = 2x^3 + 6x^2 - 8x + 2$; $[-5, -4]$
In this case, synthetic division is probably the easiest way to compute $f(-5)$ and $f(-4)$.
(a) To find $f(-5)$, divide by $x - (-5) = x + 5$:

-5	2	6	-8	2
		-10	20	-60
	2	-4	12	-58

$\leftarrow f(-5) = -58$

(See the Remainder Theorem.)
(b) To find $f(-4)$, divide by $x - (-4) = x + 4$:

-4	2	6	-8	2
		-8	8	0
	2	-2	0	2

$\leftarrow f(-4) = 2$

Since $f(-5) < 0$ and $f(-4) > 0$, there must be a zero of $f(x)$ somewhere between -5 and -4.

83. $f(x) = x^5 - x^4 + 7x^3 - 7x^2 - 18x + 18$; $[1.4, 1.5]$
$f(1.4) = -0.17536$, and
$f(1.5) = 1.40625$
Since $f(1.4) < 0$ and $f(1.5) > 0$, $f(x)$ must have a zero somewhere between 1.4 and 1.5.

85. $f(x) = 8x^4 - 2x^2 + 5x - 1$
$= (8x^3 - 2x + 5)x - 1$
$= [(8x^2 - 2)x + 5]x - 1$
$f(0) = -1; f(1) = 10$
$f(0.1) = -0.5192$
$f(0.2) = -0.0672$
$f(0.3) = 0.3848$
$f(0.21) = -0.0226$
$f(0.22) = 0.0219$
Thus, the zero is 0.21 correct to two decimal places.

87. $f(x) = 2x^3 + 6x^2 - 8x + 2$
$= [(2x + 6)x - 8]x + 2$
$f-5) = -58; f(-4) = 2$
$f(-4.1) = -2.182$
$f(-4.01) = 1.598198$
$f(-4.02) = 1.192784$
$f(-4.03) = 0.783746$
$f(-4.04) = 0.371072$
$f(-4.05) = -0.04525$
Thus, the zero is -4.04 correct to two decimal places.

89. $f(x) = x^3 + x^2 + x - 4$ has exactly one positive root (because there is one change in sign of the coefficients).
We start by finding two integers which the zero must lie between:
$f(0) = -4 < 0$
$f(1) = 1 + 1 + 1 - 4 = -1 < 0$
$f(2) = 8 + 4 + 2 - 4 = 10 > 0$

Therefore, the zero lies between 1 and 2. We next divide the interval [1, 2] into ten subintervals, and evaluate $f(x)$ at each endpoint: 1.0, 1.1, 1.2, 1.3, ... , 1.9, 2.0. To do this, the nested form of $f(x)$ will be best to work with:

$$f(x) = x^3 + x^2 + x - 4$$
$$= (x^2 + x + 1)x - 4$$

or $f(x) = ([x + 1]x + 1)x - 4$

Then $f(1.0) = -1 < 0$
$f(1.1) = -0.359 < 0$
$f(1.2) = 0.368 > 0$ Stop!

The zero must lie between 1.1 and 1.2. We now evaluate $f(x)$ at 1.10, 1.11, 1.12, ... , 1.19, 1.20.

$f(1.10) = -0.359 < 0$
$f(1.11) \approx -0.290 < 0$
$f(1.12) \approx -0.221 < 0$
$f(1.13) \approx -0.150 < 0$
$f(1.14) \approx -0.079 < 0$
$f(1.15) \approx -0.007 < 0$
$f(1.16) \approx 0.066 > 0$

Therefore, the zero is 1.15 and correct to two decimal places.

Note: If we consider the possible ***rational*** zeros, we have:

$$p: \pm 1, \pm 2, \pm 4; \; q: \pm 1; \; \frac{p}{q}: \pm 1, \pm 2, \pm 4$$

The only ***possible rational*** zeros are whole numbers! Therefore, the zero that is approximately 1.15 ***must*** be irrational!

91. Here $f(x) = 2x^4 - 3x^3 - 4x^2 - 8$

To get started:

$f(0) = -8 < 0$
$f(1) = 2 - 3 - 4 - 8 = -13 < 0$
$f(2) = 32 - 24 - 16 - 8 = -16 < 0$
$f(3) = 162 - 81 - 36 - 8 = 37 > 0$

Thus, the zero lies between 2 and 3. Let's find the nested form of $f(x)$:

$$f(x) = 2x^4 - 3x^3 - 4x^2 - 8$$
$$= (2x^2 - 3x - 4)x \cdot x - 8$$

or $f(x) = ([2x - 3]x - 4)x \cdot x - 8$

Then $f(2.0) = -16 < 0$
$f(2.1) = -14.5268 < 0$
$f(2.2) = -12.4528 < 0$

Let's skip ahead to $x = 2.6$:

$f(2.6) = 3.6272 > 0$

Backup: $f(2.5) = -1.75 < 0$

Thus, the zero lies between 2.5 and 2.6.

We know: $f(2.5) = -1.75 < 0$
$f(2.6) = 3.6272 > 0$

Try: $f(2.55) \approx 0.8109 > 0$ That narrows our search to the interval 2.5 to 2.55.

Then, say $f(2.53) \approx -0.2434 < 0$
$f(2.54) \approx 0.2787 > 0$

Thus, the zero is 2.53 correct to two decimal places.

Can you see why this zero ***must*** be irrational?

93. $(x - 2)$ will be a factor of $f(x) = x^3 - kx^2 + kx + 2$ only if the remainder that results when $f(x)$ is divided by $x - 2$ is 0:

$$\begin{array}{r|rrrr} 2 & 1 & -k & k & 2 \\ & & 2 & -2k+4 & -2k+8 \\ \hline & 1 & -k+2 & -k+4 & \boxed{-2k+10} \end{array} \leftarrow \text{Remainder}$$

Therefore, we want $-2k + 10 = 0$ or $-2k = -10$, or $k = 5$

95. Either long division *or* synthetic division would be a tedious way to find the remainder in this problem, but we know by the Remainder Theorem that if $f(x)$ is divided by $x - c$, then the remainder must be $f(c)$. Here, $x - c = x - 1$, so that $c = 1$, so we want $f(1)$. This can be found by substitution:

$$f(1) = 2 - 8 + 1 - 2 = -7$$

Thus, if we divided $f(x)$ by $x - 1$, the remainder would be -7.

97. We want to prove that $x - c$ is a factor of $x^n - c^n$, for any positive integer n. By the Factor Theorem, $x - c$ will be a factor of $f(x)$ provided $f(c) = 0$. Here, $f(x) = x^n - c^n$ so that $f(c) = c^n - c^n = 0$, and therefore, $x - c$ is indeed a factor of $x^n - c^n$.

99. If 3 is a solution of $x^3 - 8x^2 + 16x - 3 = 0$, then $x - 3$ is a factor of $f(x) = x^3 - 8x^2 + 16x - 3$. Thus, using synthetic division we find:

$$\begin{array}{r|rrrr} 3 & 1 & -8 & 16 & -3 \\ & & 3 & -15 & 3 \\ \hline & 1 & -5 & 1 & 0 \end{array}$$

Hence, $x^3 - 8x^2 + 16x - 3 = (x - 3)(x^2 - 5x + 1)$

Now, we solve $x^2 - 5x + 1 = 0$

The solutions are: $\dfrac{5 \pm \sqrt{25 - 4}}{2} = \dfrac{5 \pm \sqrt{21}}{2}$

Their sum is: $\dfrac{5 + \sqrt{21}}{2} + \dfrac{5 - \sqrt{21}}{2} = 5$

101. $f(x) = 2x^3 + 3x^2 - 6x + 7$

By the Rational Zero Theorem, the only ***possible*** rational zeros of $f(x)$ are:

$$\frac{p}{q}\colon\ \pm 1,\ \pm\frac{1}{2},\ \pm 7,\ \pm\frac{7}{2}$$

Therefore, $\frac{1}{3}$ is ***not*** a zero of $f(x)$.

103. By the Rational Zero Theorem, the ***possible*** rational zeros of $f(x) = 2x^6 - 5x^4 + x^3 - x + 1$ are:

$$\frac{p}{q}\colon\ \pm 1,\ \pm\frac{1}{2},$$

so $\frac{3}{5}$ is ***not*** a zero.

105. To start with, a cube has all three dimensions of equal length.
Let x = length of each side of the original cube. After removing the slice, we will have this situation:

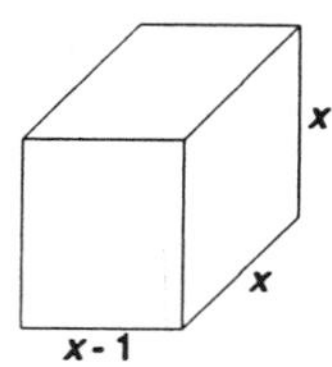

Now the volume is width times length times height, so we have

$$(x - 1)x \cdot x = 294$$
$$(x - 1)x^2 = 294$$
$$x^3 - x^2 = 294$$
$$x^3 - x^2 - 294 = 0$$

We see that there is exactly one positive zero and we can list the possibilities if we take the time to factor 294:

$$294 = 2 \cdot 147 = 2 \cdot 3 \cdot 7 \cdot 7$$

Therefore, the possibilities for p are ± 1, ± 2, ± 3, ± 6, ± 7, ± 14, ± 21, ± 42, ± 49, ± 98, ± 147, ± 294. These are also the possible rational zeros, since $q = \pm 1$. Let's try 6 (divide by $x - 6$).

$$\begin{array}{r|rrrr} 6 & 1 & -1 & 0 & -294 \\ & & 6 & 30 & 180 \\ \hline & 1 & 5 & 30 & -114 \end{array}$$

How about 7 (divide by $x - 7$)?

$$\begin{array}{r|rrrr} 7 & 1 & -1 & 0 & -294 \\ & & 7 & 42 & 294 \\ \hline & 1 & 6 & 42 & \boxed{0} \end{array}$$

Therefore, $x - 7$ is a factor.

Since $x - 7$ is a factor, 7 is a zero, and we know there is only one positive solution, by Descartes' Rule of signs.

Thus, the original length of each edge was $x = 7$ inches.

107. Since all we know about $f(x)$ is that its leading coefficient is 1, we can write:

$$f(x) = x^n = a_{n-1}x^{n-1} + \cdots + a_1 x + a_0$$

where each coefficient is an integer. (See the Rational Zeros Theorem.)

Now let r be a real zero of $f(x)$. We need to show that r is either an integer or an irrational number.

Since r is a real number, it is either rational or irrational. But if r is rational (i.e., of the form $\frac{p}{q}$, where p and q are integers), then q must be a factor of the leading coefficient, which is 1. Therefore, $q = \pm 1$, and $r = \frac{p}{q} = \pm p$, where p is a factor of a_0. Thus, if r is rational, then r is an ***integer***!

Therefore, r is either an integer, or r is irrational.

5.6 Complex Numbers; Quadratic Equations with a Negative Discriminant

1. $(2 - 3i) + (6 + 8i) = (2 + 6) + (-3 + 8)i = 8 + 5i$

3. $(-3 + 2i) - (4 - 4i) = (-3 - 4) + (2 + 4)i = -7 + 6i$

5. $(2 - 5i) - (8 + 6i) = (2 - 8) + (-5 - 6)i = -6 - 11i$

7. $3(2 - 6i) = 6 - 18i$

9. $2i(2 - 3i) = 4i - 6i^2 = 4i - 6(-1) = 6 + 4i$

11. $(3 - 4i)(2 + i) = 3(2 + i) - 4i(2 + i) = 6 + 3i - 8i - 4i^2 = 6 - 5i - 4(-1) = 10 - 5i$

13. $(-6 + i)(-6 - i) = -6(-6 - i) + i(-6 - i) = 36 + 6i - 6i - i^2 = 36 - (-1) = 37$

15. $\dfrac{10}{3 - 4i} = \dfrac{10}{3 - 4i} \cdot \dfrac{3 + 4i}{3 + 4i} = \dfrac{30 + 40i}{9 + 12i - 12i - 16i^2} = \dfrac{30 + 40i}{9 - 16(-1)} = \dfrac{30 + 40i}{25} = \dfrac{30}{25} + \dfrac{40}{25}i$

$= \dfrac{6}{5} + \dfrac{8}{5}i$

17. $\dfrac{2 + i}{i} = \dfrac{2 + i}{i} \cdot \dfrac{-i}{-i} = \dfrac{-2i - i^2}{-i^2} = -2i + 1 = 1 - 2i$

19. $\dfrac{6 - i}{1 + i} = \dfrac{6 - i}{1 + i} \cdot \dfrac{1 - i}{1 - i} = \dfrac{6 - 6i - i + i^2}{1 - i^2} = \dfrac{6 - 7i + (-1)}{1 - (-1)} = \dfrac{5 - 7i}{2} = \dfrac{5}{2} - \dfrac{7}{2}i$

21. $\left[\frac{1}{2} + \frac{\sqrt{3}}{2}i\right]^2 = \frac{1}{4} + 2\left[\frac{1}{2}\right]\left[\frac{\sqrt{3}}{2}i\right] + \frac{3}{4}i^2 = \frac{1}{4} + \frac{\sqrt{3}}{2}i + \frac{3}{4}(-1) = -\frac{1}{2} + \frac{\sqrt{3}}{2}i$

23. $(1 + i)^2 = 1 + 2i + i^2 = 1 + 2i - 1 = 2\text{i}$

25. $i^{23} = i^{22+1} = i^{22} \cdot i = (i^2)^{11} \cdot i = (-1)^{11}i = -i$

27. $i^{-15} = \frac{1}{i^{15}} = \frac{1}{i^{14+1}} = \frac{1}{i^{14}i} = \frac{1}{(i^2)^7 i} = \frac{1}{(-1)^7 i} = \frac{1}{-i} = \frac{1}{-i}\frac{i}{i} = \frac{i}{-i^2} = \frac{i}{-(-1)} = i$

29. $i^6 - 5 = (i^2)^3 - 5 = (-1)^3 - 5 = -1 - 5 = -6$

31. $6i^3 - 4i^5 = i^3(6 - 4i^2) = i^2 \cdot i(6 - 4(-1)) = -1 \cdot i(10) = -10i$

33. $(1 + i)^3 = 1^3 + 3i + 3i^2 + i^3 = 1 + 3i + 3(-1) + i^2 \cdot i = -2 + 3i + (-1)i = -2 + 2i$

35. $i^7(1 + i^2) = i^7(1 + (-1)) = i^7(0) = 0$

37. $i^6 + i^4 + i^2 + 1 = (i^2)^3 + (i^2)^2 + (-1) + 1 = (-1)^3 + (-1)^2 = -1 + 1 = 0$

39. $\sqrt{-4} = \sqrt{4}\,i = 2i$

41. $\sqrt{-25} = \sqrt{25}\,i = 5i$

43. $\sqrt{(3 + 4i)(4i - 3)} = \sqrt{12i - 9 + 16i^2 - 12i} = \sqrt{-9 - 16} = \sqrt{-25} = \sqrt{25}\,i = 5i$

For Problems 45–57 we use the quadratic formula: $x = \frac{-b \pm \sqrt{b^2 - 4ac}}{2a}$

45. $x^2 + 4 = 0$
Here $a = 1$, $b = 0$, $c = 4$, and $b^2 - 4ac = 0 - 4(1)(4) = -16$.
Then $x = \frac{-0 \pm \sqrt{-16}}{2(1)} = \frac{\pm\sqrt{16}\,i}{2} = \frac{\pm 4i}{2} = \pm 2i$
The solution set is $\{-2i, 2i\}$.

47. $x^2 - 16 = 0$
Here $a = 1$, $b = 0$, $c = -16$, and $b^2 - 4ac = 0 - 4(1)(-16) = 64$.
Then $x = \frac{-0 \pm \sqrt{64}}{2(1)} = \frac{\pm 8}{2} = \pm 4$
The solution set is $\{-4, 4\}$.

49. $x^2 - 6x + 13 = 0$
Here $a = 1$, $b = -6$, $c = 13$, and $b^2 - 4ac = (-6)^2 - 4(1)(13) = -16$.
Then $x = \frac{-(-6) \pm \sqrt{-16}}{2(1)} = \frac{6 \pm \sqrt{16}\,i}{2} = \frac{6 \pm 4i}{2} = 3 \pm 2i$
The solution set is $\{3 - 2i, 3 + 2i\}$.

51. $x^2 - 6x + 10 = 0$
Here $a = 1$, $b = -6$, $c = 10$, and $b^2 - 4ac = (-6)^2 - 4(1)(10) = -4$.
Then $x = \frac{-(-6) \pm \sqrt{-4}}{2(1)} = \frac{6 \pm \sqrt{4}i}{2} = \frac{6 \pm 2i}{2} = 3 \pm i$
The solution set is $\{3 - i, 3 + i\}$.

53. $8x^2 - 4x + 1 = 0$
Here $a = 8$, $b = -4$, $c = 1$, and $b^2 - 4ac = (-4)^2 - 4(8)(1) = -16$.

Then $x = \dfrac{-(-4) \pm \sqrt{-16}}{2(8)} = \dfrac{4 \pm \sqrt{16}\,i}{16} = \dfrac{4 \pm 4i}{16} = \dfrac{1}{4} \pm \dfrac{1}{4}i$

The solution set is $\left\{\dfrac{1}{4} - \dfrac{1}{4}i, \dfrac{1}{4} + \dfrac{1}{4}i\right\}$.

55. $5x^2 + 1 = 2x$
$5x^2 - 2x + 1 = 0$
Here $a = 5$, $b = -2$, $c = 1$, and $b^2 - 4ac = (-2)^2 - 4(5)(1) = -16$.

Then $x = \dfrac{2 \pm \sqrt{-16}}{2(5)} = \dfrac{2 \pm \sqrt{16}\,i}{10} = \dfrac{2 \pm 4i}{10} = \dfrac{1}{5} \pm \dfrac{2}{5}i$

The solution set is $\left\{\dfrac{1}{5} - \dfrac{2}{5}i, \dfrac{1}{5} + \dfrac{2}{5}i\right\}$.

57. $x^2 + x + 1 = 0$
Here $a = 1$, $b = 1$, $c = 1$, and $b^2 - 4ac = 1^2 - 4(1)(1) = -3$.

Then $x = \dfrac{-1 \pm \sqrt{-3}}{2(1)} = \dfrac{-1 \pm \sqrt{3}\,i}{2} = -\dfrac{1}{2} \pm \dfrac{\sqrt{3}}{2}i$

The solution set is $\left\{-\dfrac{1}{2} - \dfrac{\sqrt{3}}{2}i, -\dfrac{1}{2} + \dfrac{\sqrt{3}}{2}i\right\}$.

59. $x^3 - 8 = 0$
$x^3 - 8 = (x - 2)(x^2 + 2x + 4) = 0$
$x - 2 = 0 \qquad x^2 + 2x + 4 = 0$
$x = 2$
Here $a = 1$, $b = 2$, $c = 4$, and $b^2 - 4ac = 2^2 - 4(1)(4) = -12$

Then, $x = \dfrac{-2 \pm \sqrt{-12}}{2(1)} = \dfrac{-2 \pm \sqrt{12}\,i}{2} = \dfrac{-2 \pm 2\sqrt{3}\,i}{2} = -1 \pm \sqrt{3}\,i$

The solution set is $\{2, -1 - \sqrt{3}\,i, -1 + \sqrt{3}\,i\}$.

61. $x^4 - 16 = 0$
$x^4 - 16 = (x^2 - 4)(x^2 + 4) = (x - 2)(x + 2)(x^2 + 4) = 0$
$x - 2 = 0 \qquad x + 2 = 0 \qquad x^2 + 4 = 0$
$x = 2 \qquad x = -2$
Here $a = 1$, $b = 0$, $c = 4$, and $b^2 - 4ac = 0^2 - 4(1)(4) = -16$

Then, $x = \dfrac{0 \pm \sqrt{-16}}{2(1)} = \dfrac{\sqrt{16}\,i}{2} = \dfrac{\pm 4i}{2} = \pm 2i$

The solution set is $\{-2, 2, -2i, 2i\}$.

63. $x^4 + 13x^2 + 36 = 0$
$(x^2 + 9)(x^2 + 4) = 0$
$x^2 = -9 \qquad x^2 = -4$
$x^2 = 9i^2 \qquad x^2 = 4i^2$
$x = \pm 3i \qquad x = \pm 2i$
The solution set is $\{-3i, -2i, 2i, 3i\}$.

65. $3x^2 - 3x + 4 = 0$
Here $a = 3$, $b = -3$, $c = 4$,
and $b^2 - 4ac = (-3)^2 - 4(3)(4) = -39$.
Hence, this equation has two complex solutions.

67. $2x^2 + 3x - 4 = 0$
Here $a = 2$, $b = 3$, $c = -4$, and
$b^2 - 4ac = 3^2 - 4(2)(-4) = 41$.
Hence, this equation has two unequal real solutions.

69. $9x^2 - 12x + 4 = 0$
Here $a = 9$, $b = -12$, $c = 4$, and
$b^2 - 4ac = (-12)^2 - 4(9)(4) = 0$.
Hence, this equation has a repeated real solution.

71. The other solution must be the conjugate of $2 + 3i$, or $2 - 3i$.

In Problems 73–76, $z = 3 - 4i$ and $w = 8 + 3i$.

73. $z + \bar{z} = 3 - 4i + \overline{(3 - 4i)} = 3 - 4i + (3 + 4i) = (3 + 3) + (-4 + 4)i = 6$

75. $z\bar{z} = (3 - 4i)\overline{(3 - 4i)} = (3 - 4i)(3 + 4i) = 9 + 12i - 12i - 16i^2 = 9 - 16(-1) = 25$

77. For $z = a + bi$, $z + \bar{z} = a + bi + \overline{a + bi} = (a + bi) + (a - bi) = (a + a) + (b - b)i$
$= 2a + 0i = 2a$

and $z - \bar{z} = a + bi - \overline{(a + bi)} = a + bi - (a - bi) = (a - a) + (b - (-b))i$
$= 0 + 2bi = 2bi$

79. For $z = a + bi$ and $w = c + di$,
$\overline{z + w} = \overline{(a + bi) + (c + di)} = \overline{(a + c) + (b + d)i} = (a + c) - (b + d)i$
$= (a - bi) + (c - di) = \overline{a + bi} + \overline{c + di} = \bar{z} + \bar{w}$

5.7 Complex Zeros; Fundamental Theorem of Algebra

1. Since complex zeros appear as conjugate pairs, it follows that $4 + i$, the conjugate of $4 - i$, is the remaining zero of f.

3. Since complex zeros appear as conjugate pairs, it follows that $-i$, the conjugate of i, and $1 - i$, the conjugate of $1 + i$, are the remaining zeros of f.

5. Since complex zeros appear as conjugate pairs, it follows that $-i$, the conjugate of i, and $-2i$, the conjugate of $2i$, are the remaining zeros of f.

7. Since complex zeros appear as conjugate pairs, it follows that $-i$, the conjugate of i, is the remaining zero of f.

9. Since complex zeros appear as conjugate pairs, it follows that $2 - i$, the conjugate of $2 + i$, and $-3 + i$, the conjugate of $-3 - i$, are the remaining zeros of f.

For Problems 11-15, we will let the coefficient, a, of the polynomial equal 1.

11. Since $3 + 2i$ is a zero, by the Conjugate Pairs Theorem, $3 - 2i$ must also be a zero of f.
$$\begin{aligned} f(x) &= (x - 4)(x - 4)(x - (3 + 2i))(x - (3 - 2i)) \\ &= (x^2 - 8x + 16)(x^2 - (3 - 2i)x - (3 + 2i)x + (3 + 2i)(3 - 2i)) \\ &= (x^2 - 8x + 16)(x^2 - 3x + 2ix - 3x - 2ix + 9 - 6i + 6i - 4i^2) \\ &= (x^2 - 8x + 16)(x^2 - 6x + 13) \\ &= x^4 - 6x^3 + 13x^2 - 8x^3 + 48x^2 - 104x + 16x^2 - 96x + 208 \\ &= x^4 - 14x^3 + 77x^2 - 200x + 208 \end{aligned}$$

13. Since $-i$ is a zero, i is a zero; and since $1 + i$ is a zero, $1 - i$ is a zero by the Conjugate Pairs Theorem.

$$\begin{aligned} f(x) &= (x - 2)(x - i)((x + i)(x - (1 + i))(x - (1 - i)) \\ &= (x - 2)(x^2 + 1)(x^2 - (1 - i)x - (1 + i)x + (1 + i)(1 - i)) \\ &= (x^3 - 2x^2 + x - 2)(x^2 - x + ix - x - ix + 1 - i + i - i^2) \\ &= (x^3 - 2x^2 + x - 2)(x^2 - 2x + 2) \\ &= x^5 - 2x^4 + 2x^3 - 2x^4 + 4x^3 - 4x^2 + x^3 - 2x^2 + 2x - 2x^2 + 4x - 4 \\ &= x^5 - 4x^4 + 7x^3 - 8x^2 + 6x - 4 \end{aligned}$$

15. Since $-i$ is a zero, i is also a zero of f by Conjugate Pairs Theorem.

$$\begin{aligned} f(x) &= (x - 3)^2(x - i)(x + i) \\ &= (x^2 - 6x + 9)(x^2 + 1) \\ &= x^4 - 6x^3 + 9x^2 + x^2 - 6x + 9 \\ &= x^4 - 6x^3 + 10x^2 - 6x + 9 \end{aligned}$$

17. Since $2i$ is a zero of f, $-2i$ is also a zero since complex zeros appear in conjugate pairs in polynomials with real coefficients. Therefore, $(x - 2i)$ and $(x + 2i)$ are factors of f. So, $(x - 2i)(x + 2i) = x^2 + 4$ is a factor of f. We use long division to find the other factor:

$$\begin{array}{r} x - 4 \\ x^2 + 4 \overline{)\, x^3 - 4x^2 + 4x - 16} \\ \underline{-(x^3 \qquad\quad + 4x)} \\ -4x^2 \qquad - 16 \\ \underline{-(-4x^2 \qquad - 16)} \\ 0 \end{array}$$

So, $f(x) = (x^2 + 4)(x - 4)$. By the Factor Theorem, the remaining zero of f is 4. The zeros of f are $4, 2i, -2i$.

19. Since $-2i$ is a zero of f, $2i$ is also a zero of f by the Conjugate Pairs Theorem. Therefore, $x^2 + 4$ is a factor of f (See Problem 17 Solution). We use long division to find the other factor:

$$\begin{array}{r} 2x^2 + 5x - 3 \\ x^2 + 4 \overline{)\, 2x^4 + 5x^3 + 5x^2 + 20x - 12} \\ \underline{-(2x^4 \qquad\quad + 8x^2)} \\ 5x^3 - 3x^2 + 20x \\ \underline{-(5x^3 \qquad\quad + 20x)} \\ -3x^2 \qquad - 12 \\ \underline{-(-3x^2 \qquad - 12)} \\ 0 \end{array}$$

So, $f(x) = (x^2 + 4)(2x^2 + 5x - 3) = (x^2 + 4)(2x - 1)(x + 3)$. By the Factor Theorem $\frac{1}{2}$ and -3 are zeros of f. The zeros of f are $2i, -2i, -3, \frac{1}{2}$.

21. Since $3 - 2i$ is a zero of h, $3 + 2i$ is a zero of h by the Conjugate Pairs Theorem. Therefore, $(x - (3 - 2i))$ and $(x - (3 + 2i))$ are factors of h. So, $(x - (3 - 2i))(x - (3 + 2i)) = x^2 - (3 + 2i)x - (3 - 2i)x + (3 - 2i)(3 + 2i) = x^2 - 3x - 2ix - 3x + 2ix + 9 - 4i^2 = x^2 - 6x + 13$ is a factor of h. We use long division to find the other factor:

$$
\begin{array}{r l}
 & x^2 - 3x - 10 \\
x^2 - 6x + 13 \,\big) & x^4 - 9x^3 + 21x^2 + 21x - 130 \\
 & \underline{-(x^4 - 6x^3 + 13x^2)} \\
 & \quad -3x^3 + 8x^2 + 21x \\
 & \quad \underline{-(-3x^3 + 18x^2 - 39x)} \\
 & \qquad -10x^2 + 60x - 130 \\
 & \qquad \underline{-(-10x^2 + 60x - 130)} \\
 & \qquad\qquad 0
\end{array}
$$

So, $h(x) = (x^2 - 6x + 13)(x^2 - 3x - 10) = (x^2 - 6x + 13)(x - 5)(x + 2)$. By the Factor Theorem, the zeros of $h(x)$ are $3 - 2i$, $3 + 2i$, -2, 5.

23. Since $-4i$ is a zero of h, $4i$ is a zero of h by the Conjugate Pairs Theorem. Therefore, $(x + 4i)$ and $(x - 4i)$ are factors of h. So, $(x + 4i)(x - 4i) = x^2 + 16$ is a factor of h. We use long division to find the other factor:

$$
\begin{array}{r l}
 & 3x^3 + 2x^2 - 33x - 22 \\
x^2 + 16 \,\big) & 3x^5 + 2x^4 + 15x^3 + 10x^2 - 528x - 352 \\
 & \underline{-(3x^5 \qquad + 48x^3)} \\
 & \quad 2x^4 - 33x^3 + 10x^2 \\
 & \quad \underline{-(2x^4 \qquad + 32x^2)} \\
 & \qquad -33x^3 - 22x^2 - 528x \\
 & \qquad \underline{-(-33x^3 \qquad -528x)} \\
 & \qquad\qquad -22x^2 \qquad - 352 \\
 & \qquad\qquad \underline{-(-22x^2 \qquad - 352)} \\
 & \qquad\qquad\qquad 0
\end{array}
$$

So, $h(x) = (x^2 + 16)(3x^3 + 2x^2 - 33x - 22)$. The factor $3x^3 + 2x^2 - 33x - 22$ can be factored by grouping: $(3x^3 + 2x^2) + (-33x - 22) = x^2(3x + 2) - 11(3x + 2) = (3x + 2)(x^2 - 11) = (3x + 2)(x + \sqrt{11})(x - \sqrt{11})$. The zeros of h are $4i$, $-4i$, $\frac{-2}{3}$, $\sqrt{11}$, $-\sqrt{11}$.

25. $f(x) = x^3 - 1$ is a difference of cubes: $f(x) = x^3 - 1 = (x - 1)(x^2 + x + 1)$
The zeros of $x^2 + x + 1$ are:

$$x = \frac{-1 \pm \sqrt{1 - 4}}{2} = \frac{-1 \pm \sqrt{-3}}{2} = -\frac{1}{2} + \frac{\sqrt{3}}{2}i \quad or \quad -\frac{1}{2} - \frac{\sqrt{3}}{2}i$$

Thus, the zeros of $x^3 - 1$ are 1, $-\frac{1}{2} + \frac{\sqrt{3}}{2}i$, and $-\frac{1}{2} - \frac{\sqrt{3}}{2}i$.

27. $f(x) = x^3 - 8x^2 + 25x - 26$; $f(-x) = -x^3 - 8x^2 - 25x - 26$

Step 1: $f(x)$ has three zeros.
Step 2: $f(x)$ has three or one positive real zeros and zero negative real zeros.
Step 3: Possible Rational Zeros: ± 1, ± 2, ± 13, ± 26
Step 4: We check if $x = 2$ is a real zero:

$$
\begin{array}{r|rrrr}
2 & 1 & -8 & 25 & -26 \\
 & & 2 & -12 & 26 \\
\hline
 & 1 & -6 & 13 & \boxed{0}
\end{array}
\leftarrow (x - 2) \text{ is a factor}
$$

Quotient $= x^2 - 6x + 13$

To find the zeros of $x^2 - 6x + 13$, we use the quadratic formula with $a = 1$, $b = -6$, $c = 13$.

$$x = \frac{-(-6) \pm \sqrt{(-6)^2 - 4(1)(13)}}{2(1)} = \frac{6 \pm \sqrt{-16}}{2} = 3 \pm 2i$$

The complex zeros of f are 2, $3 + 2i$, and $3 - 2i$.

Step 5: No need to check Upper or Lower Bound since all the zeros were found.

29. $f(x) = x^4 + 5x^2 + 4 = (x^2 + 4)(x^2 + 1) = (x + 2i)(x - 2i)(x + i)(x - i)$
The zeros of f are $-i, i, -2i, 2i$.

31. $f(x) = x^4 + 2x^3 + 22x^2 + 50x - 75; f(-x) = x^4 - 2x^3 + 22x^2 - 50x - 75.$
Step 1: $f(x)$ has four zeros.
Step 2: $f(x)$ has one positive zero and three or one negative zeros.
Step 3: Possible Rational Zeros: $\pm 1, \pm 3, \pm 5, \pm 15, \pm 25, \pm 75$
Step 4: We check if $x = -3$ is a real zero:

$$\begin{array}{r|rrrrr} -3 & 1 & 2 & 22 & 50 & -75 \\ & & -3 & 3 & -75 & 75 \\ \hline & 1 & -1 & 25 & -25 & \boxed{0} \end{array} \leftarrow x + 3 \text{ is a factor}$$

Quotient $= x^3 - x^2 + 25x - 25$
The quotient can be factored by grouping:
$$(x^3 - x^2) + (25x - 25) = x^2(x - 1) + 25(x - 1) = (x - 1)(x^2 + 25)$$
So, $x = 1$ is a zero. The zeros of $x^2 + 25$ are:
$$x^2 + 25 = 0$$
$$x^2 = -25$$
$$x = \pm\sqrt{-25}$$
$$x = \pm 5i$$
The zeros of f are: $-3, 1, -5i, 5i$.

33. $f(x) = 3x^4 - x^3 - 9x^2 + 159x - 52; f(-x) = 3x^4 + x^3 - 9x^2 - 159x - 52$
Step 1: $f(x)$ has four zeros.
Step 2: $f(x)$ has three or one positive real zeros and one negative real zero.
Step 3: p: $\pm 1, \pm 2, \pm 4, \pm 13, \pm 26, \pm 52$; q: $\pm 1, \pm 3$

Possible Rational Zeros: $\frac{p}{q}$: $\pm 1, \pm 2, \pm 4, \pm 13, \pm 26, \pm 52,$

$$\pm\frac{1}{3}, \pm\frac{2}{3}, \pm\frac{4}{3}, \pm\frac{13}{3}, \pm\frac{26}{3}, \pm\frac{52}{3}$$

Step 4: We check if $x = -4$ is a real zero:

$$\begin{array}{r|rrrrr} -4 & 3 & -1 & -9 & 159 & -52 \\ & & -12 & 52 & -172 & 52 \\ \hline & 3 & -13 & 43 & -13 & \boxed{0} \end{array} \leftarrow x + 4 \text{ is a factor of } f$$

Quotient $= 3x^3 - 13x^2 + 43x - 13$

We check if $x = \frac{1}{3}$ is a solution of the depressed equation
$3x^3 - 13x^2 + 43x - 13 = 0$

$$\begin{array}{r|rrrr} \frac{1}{3} & 3 & -13 & 43 & -13 \\ & & 1 & -4 & 13 \\ \hline & 3 & -12 & 39 & \boxed{0} \end{array} \leftarrow x - \frac{1}{3} \text{ is a factor of } f$$

Quotient $= 3x^2 - 12x + 39 = 3(x^2 - 4x + 13)$
Find the complex zeros of $x^2 - 4x + 13$ using the quadratic formula with $a = 1$, $b = -4$, $c = 13$.

$$x = \frac{-(-4) \pm \sqrt{(-4)^2 - 4(1)(13)}}{2(1)} = \frac{4 \pm \sqrt{-36}}{2} = 2 \pm 3i$$

The complex zeros of f are: $-4, \frac{1}{3}, 2 + 3i, 2 - 3i$

35. If the coefficients are real and $2 + i$ is a zero, then $2 - i$ would also be a zero.

37. If the coefficients are real, complex zeros must come in ***pairs***; i.e., there will always be an ***even*** number of complex zeros. If the remaining zero was complex, $f(z)$ would have three complex zeros, which is impossible. Thus, the remaining zero must be real.

5 Chapter Review

1. $f(x) = (x - 2)^2 + 2$

Since the equation is in the form $f(x) = a(x - h)^2 + k$, we can read off a lot of information easily. Since $a = 1 > 0$, the parabola opens upward. Also, the vertex is at $(h, k) = (2, 2)$, and the axis of symmetry is $x = h$, or $x = 2$.

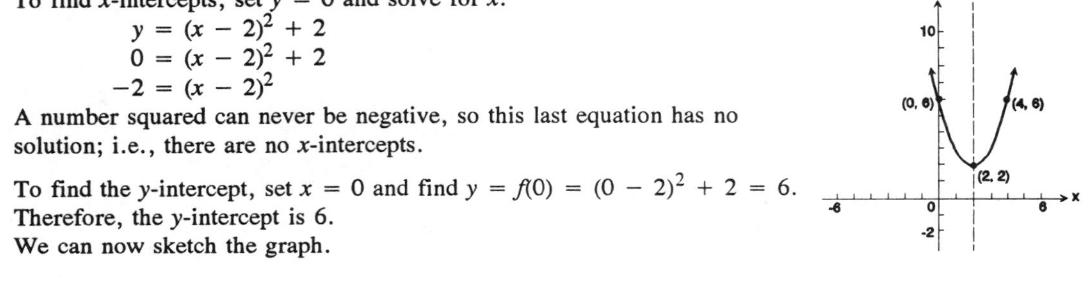

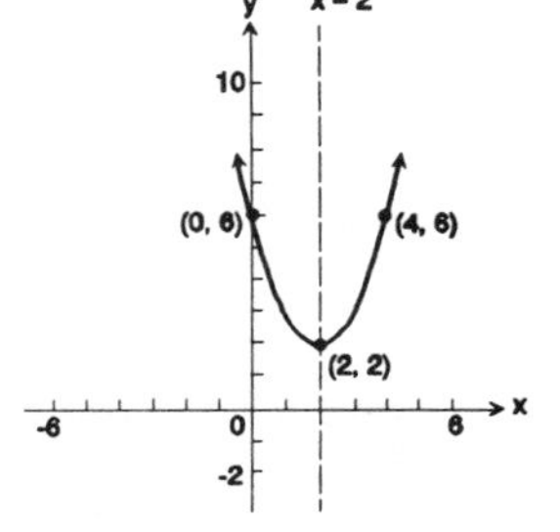

To find x-intercepts, set $y = 0$ and solve for x:

$$y = (x - 2)^2 + 2$$
$$0 = (x - 2)^2 + 2$$
$$-2 = (x - 2)^2$$

A number squared can never be negative, so this last equation has no solution; i.e., there are no x-intercepts.

To find the y-intercept, set $x = 0$ and find $y = f(0) = (0 - 2)^2 + 2 = 6$. Therefore, the y-intercept is 6. We can now sketch the graph.

3. The function $f(x) = \frac{1}{4}x^2 - 16$ is in the form $f(x) = a(x - h)^2 + k$, with $a = \frac{1}{4}$, $h = 0$, and $k = -16$. Therefore, the parabola opens upward, since $a = \frac{1}{4} > 0$, and the vertex is at $(h, k) = (0, -16)$. The axis of symmetry is $x = h$, or $x = 0$ (i.e., the y-axis). To find the x-intercepts, set $y = 0$ and solve for x:

$$y = \frac{1}{4}x^2 - 16$$
$$0 = \frac{1}{4}x^2 - 16$$
$$16 = \frac{1}{4}x^2$$
$$64 = x^2$$

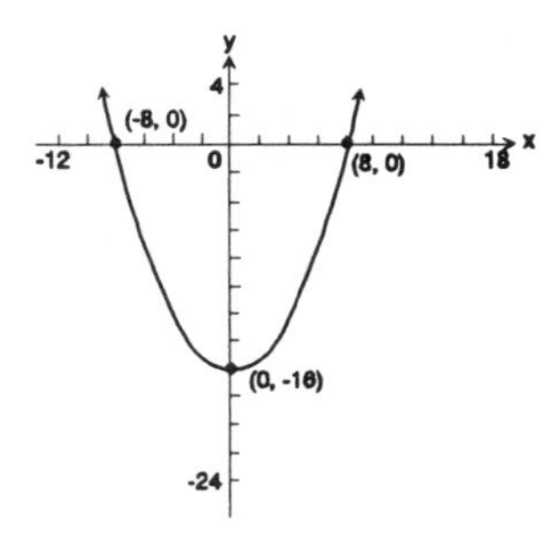

Therefore, $x = -8$ and $x = 8$ are the two x-intercepts. For the y-intercept, set $x = 0$: Then $y = f(0) = -16$. We are now ready to sketch the graph.

5. For $f(x) = -4x^2 + 4x$, since $a = -4 < 0$, the parabola opens downward. We can use the formulas in Display (2) of the text to find the vertex.

Here, $a = -4$, $b = 4$, and $c = 0$. Then the vertex is at (h, k), where

$$h = \frac{-b}{2a} = \frac{-4}{2(-4)} = \frac{1}{2} \text{ and}$$

$$k = f(h) = f\left(\frac{1}{2}\right) = -4\left(\frac{1}{2}\right)^2 + 4\left(\frac{1}{2}\right) = 1$$

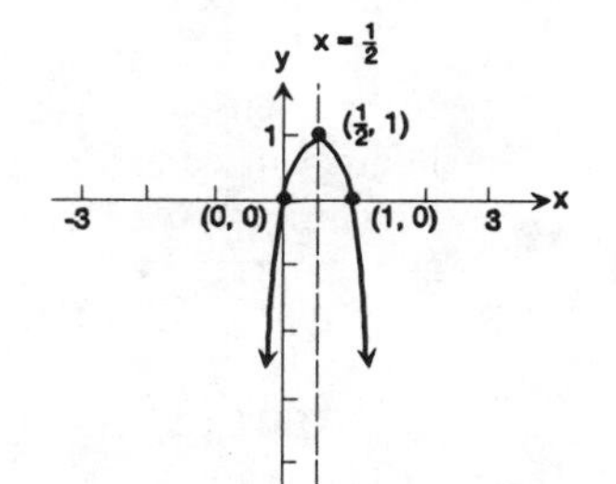

Thus, the vertex is $\left(\frac{1}{2}, 1\right)$, and the axis of symmetry is $x = \frac{1}{2}$.

The y-intercept is $y = f(0) = 0$.
To find the x-intercepts, set $y = 0$ and solve for x:

$$y = -4x^2 + 4x$$
$$0 = -4x^2 + 4x$$
$$0 = -4x(x - 1)$$

Therefore, $x = 0$ and $x = 1$ are the x-intercepts.

7. For $f(x) = \frac{9}{2}x^2 + 3x + 1$, $a = \frac{9}{2}$, $b = 3$, and $c = 1$.

The parabola opens upward, since $a = \frac{9}{2} > 0$.

We have $h = \frac{-b}{2a} = \frac{-3}{2\left(\frac{9}{2}\right)} = \frac{-3}{9} = \frac{-1}{3}$,

and $k = f(h) = f\left(-\frac{1}{3}\right) = \frac{9}{2}\left(\frac{1}{9}\right) + 3\left(-\frac{1}{3}\right) + 1 = \frac{1}{2}$.

So the vertex is at $(h, k) = \left(-\frac{1}{3}, \frac{1}{2}\right)$, and the axis of symmetry is the

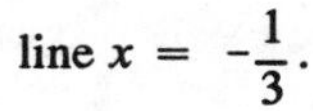
line $x = -\frac{1}{3}$.

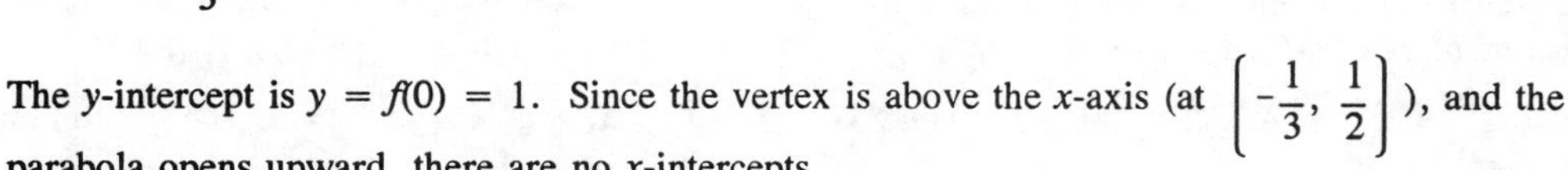
The y-intercept is $y = f(0) = 1$. Since the vertex is above the x-axis (at $\left(-\frac{1}{3}, \frac{1}{2}\right)$), and the parabola opens upward, there are no x-intercepts.

9. For $f(x) = 3x^2 + 4x - 1$, we have $a = 3$, $b = 4$, and $c = -1$. The parabola open upward, since

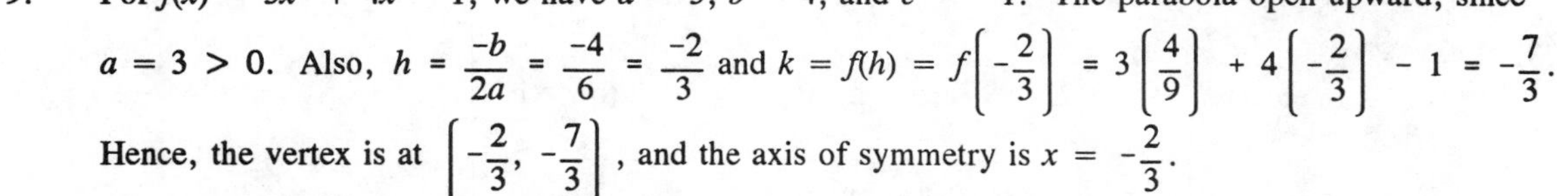
$a = 3 > 0$. Also, $h = \frac{-b}{2a} = \frac{-4}{6} = \frac{-2}{3}$ and $k = f(h) = f\left(-\frac{2}{3}\right) = 3\left(\frac{4}{9}\right) + 4\left(-\frac{2}{3}\right) - 1 = -\frac{7}{3}$.

Hence, the vertex is at $\left(-\frac{2}{3}, -\frac{7}{3}\right)$, and the axis of symmetry is $x = -\frac{2}{3}$.

The y-intercept is $y = f(0) = -1$.

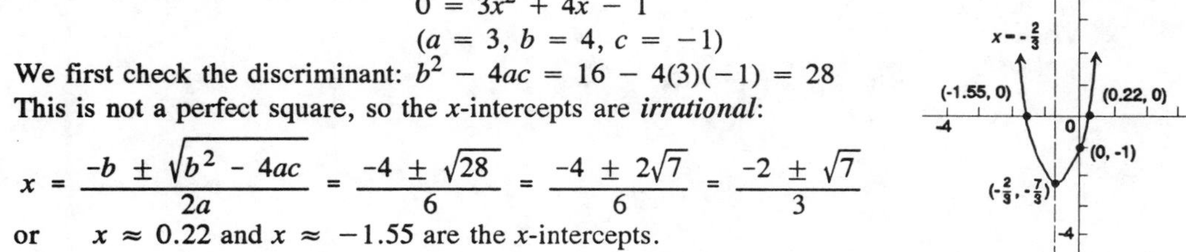
We now find the x-intercepts:

$$y = 3x^2 + 4x - 1$$
$$0 = 3x^2 + 4x - 1$$
$$(a = 3,\ b = 4,\ c = -1)$$

We first check the discriminant: $b^2 - 4ac = 16 - 4(3)(-1) = 28$
This is not a perfect square, so the x-intercepts are *irrational*:

$$x = \frac{-b \pm \sqrt{b^2 - 4ac}}{2a} = \frac{-4 \pm \sqrt{28}}{6} = \frac{-4 \pm 2\sqrt{7}}{6} = \frac{-2 \pm \sqrt{7}}{3}$$

or $\quad x \approx 0.22$ and $x \approx -1.55$ are the x-intercepts.

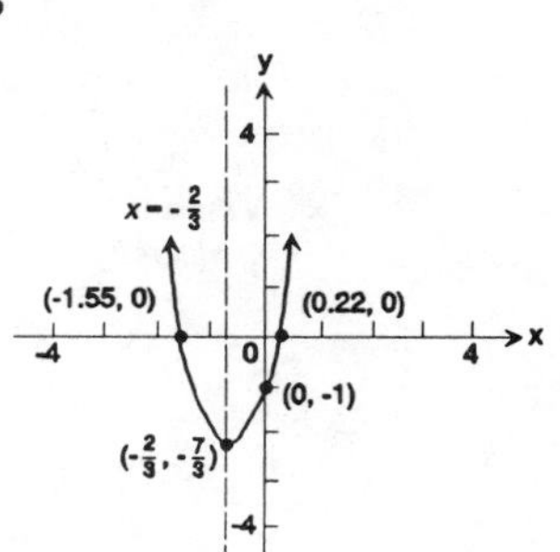

11. For $f(x) = (x + 2)^3$, start with the graph of $y = x^3$, and then perform a ***horizontal*** shift, two units to the ***left***.

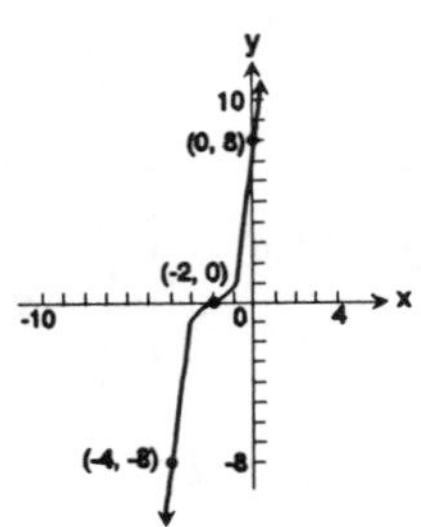

13. For $f(x) = -(x - 1)^4$, start with the graph of $y = x^4$, reflect about the x-axis (to obtain $y = -x^4$), and then shift one unit to the right.

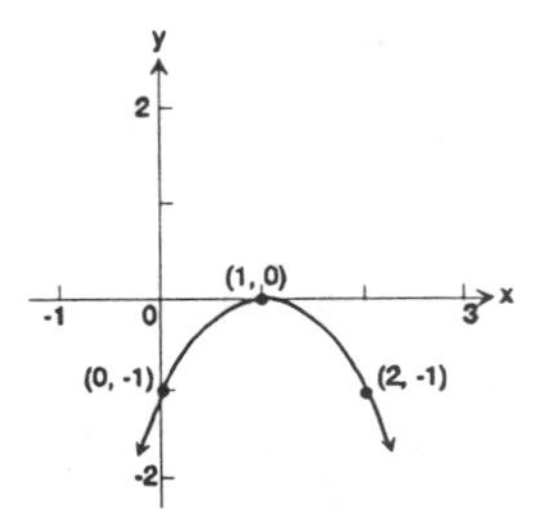

15. For $f(x) = (x - 1)^4 + 2$, start with the graph of $y = x^4$ (see Problem 13). Then shift to the ***right*** one unit, and ***up*** two units.

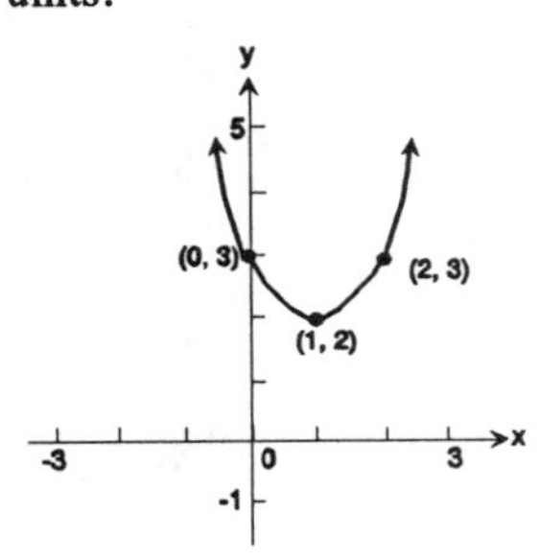

17. $f(x) = 3x^2 - 6x + 4$
Since $a = 3 > 0$, the parabola opens upward and the quadratic function has a minimum value which occurs at
$x = \dfrac{-b}{2a} = \dfrac{6}{6} = 1$. The minimum value is
$f(1) = 3(1)^2 - 6(1) + 4 = 1$.

19. $f(x) = -x^2 + 8x - 4$
Since $a = -1 < 0$, the parabola opens downward and the quadratic function has a maximum value which occurs at
$x = \dfrac{-b}{2a} = \dfrac{-8}{-2} = 4$. The maximum value is $f(4) = -(4)^2 + 8(4) - 4$
$= -16 + 32 - 4 = 12$.

21. $f(x) = -3x^2 + 12x + 4$
Since $a = -3 < 0$, the parabola opens downward and the quadratic function has a maximum value which occurs at
$x = \dfrac{-b}{2a} = \dfrac{-12}{-6} = 2$. The maximum value is $f(2) = -3(2)^2 + 12(2) + 4$
$= -12 + 24 + 4 = 16$.

23. $f(x) = x(x + 2)(x + 4)$
(a) x-intercepts: $-4, -2, 0$; y-intercept: 0
(b) crosses the x-axis at $-4, -2$, and 0
(c) $y = x^3$
(d) 2
(e)

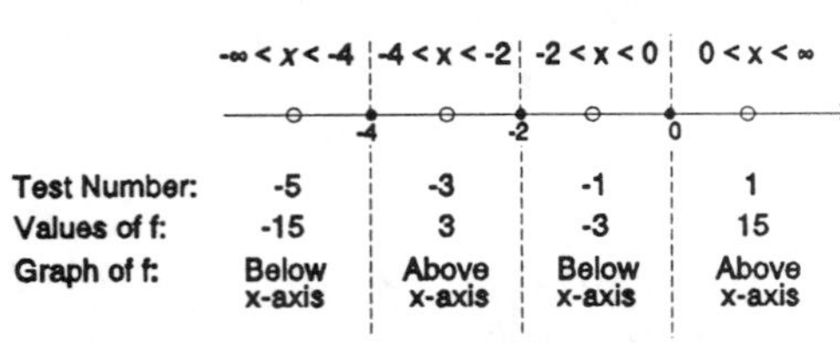

	$-\infty < x < -4$	$-4 < x < -2$	$-2 < x < 0$	$0 < x < \infty$
Test Number:	-5	-3	-1	1
Values of f:	-15	3	-3	15
Graph of f:	Below x-axis	Above x-axis	Below x-axis	Above x-axis

(f)

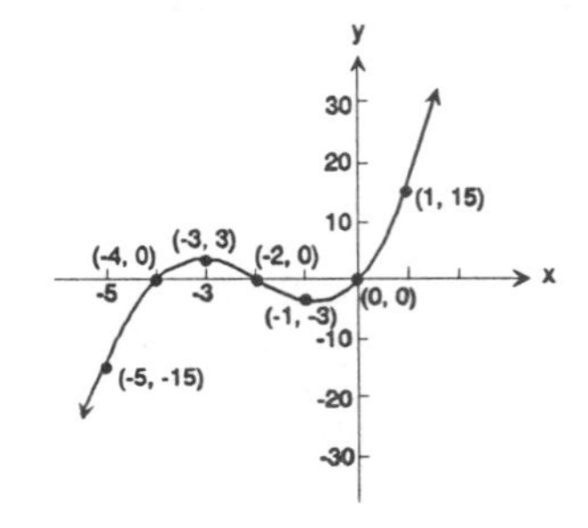

25. $f(x) = (x - 2)^2(x + 4)$

(a) x-intercepts: $-4, 2$; y-intercept: 16

(b) crosses at -4; touches at 2

(c) $y = x^3$

(d) 2

(e)

	$-\infty < x < -4$	$-4 < x < 2$	$2 < x < \infty$
Test Number:	-5	-2	3
Values of f:	-49	32	7
Graph of f:	Below x-axis	Above x-axis	Above x-axis

(f)

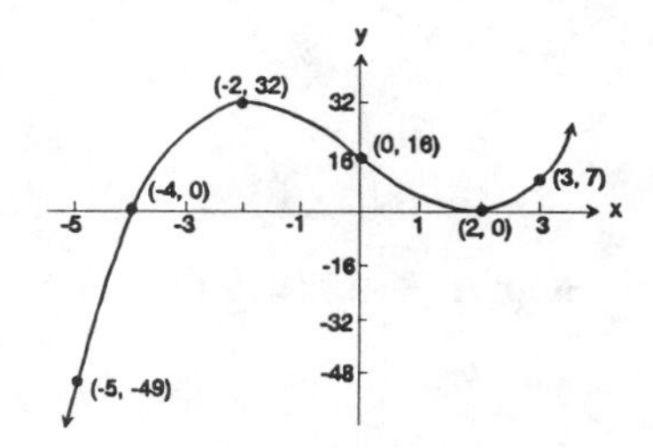

27. $f(x) = x^3 - 4x^2 = x^2(x - 4)$

(a) x-intercepts: $0, 4$; y-intercept: 0

(b) touches at 0; crosses at 4

(c) $y = x^3$

(d) 2

(e)

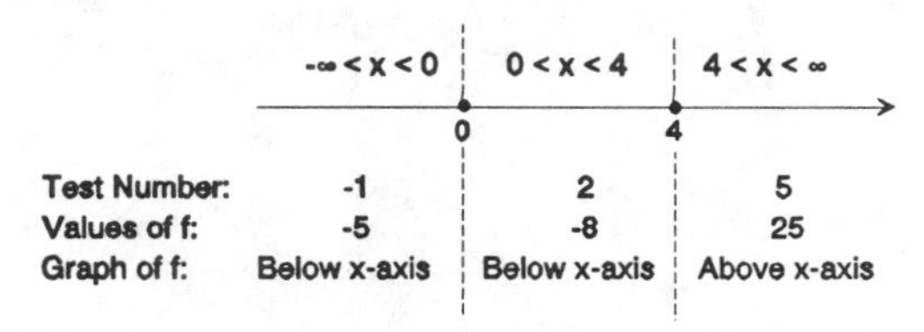

	$-\infty < x < 0$	$0 < x < 4$	$4 < x < \infty$
Test Number:	-1	2	5
Values of f:	-5	-8	25
Graph of f:	Below x-axis	Below x-axis	Above x-axis

(f)

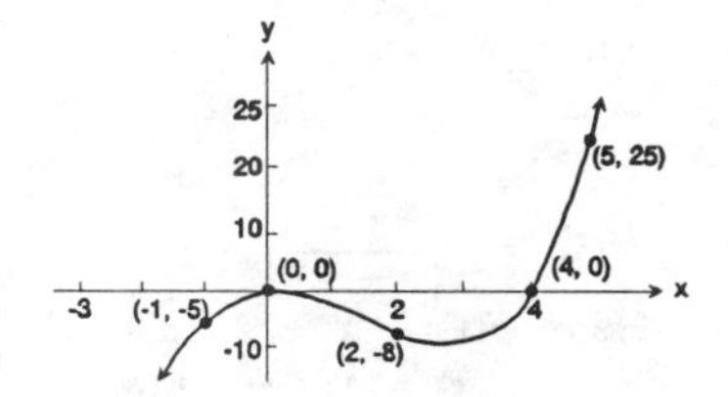

29. $f(x) = (x - 1)^2(x + 3)(x + 1)$

(a) x-intercepts: $-3, -1, 1$; y-intercept: 3

(b) crosses at -3 and -1; touches at 1

(c) $y = x^4$

(d) 3

(e)

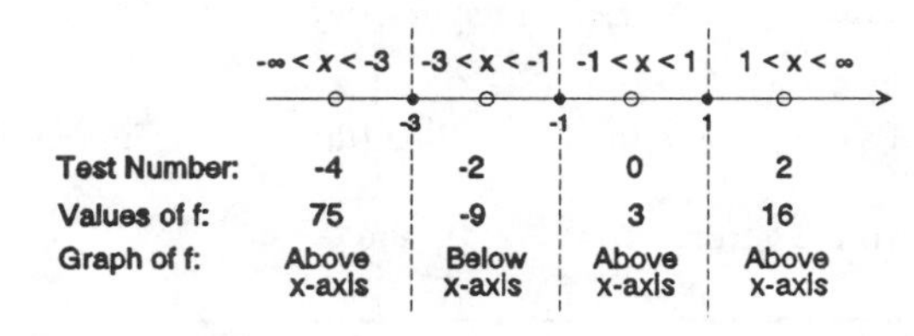

	$-\infty < x < -3$	$-3 < x < -1$	$-1 < x < 1$	$1 < x < \infty$
Test Number:	-4	-2	0	2
Values of f:	75	-9	3	16
Graph of f:	Above x-axis	Below x-axis	Above x-axis	Above x-axis

(f)

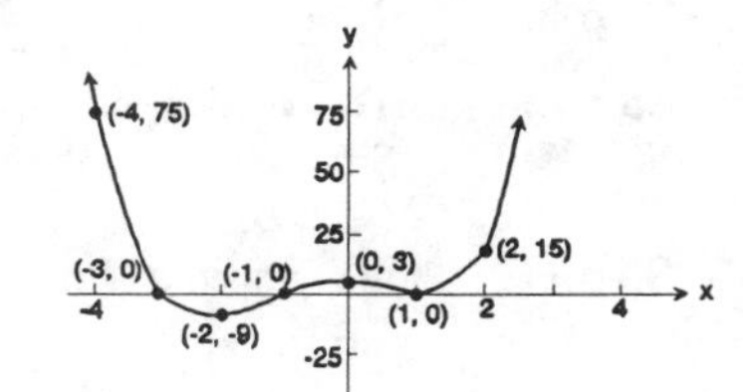

31. $R(x) = \dfrac{2x - 6}{x}$

Here, $p(x) = 2x - 6$, $n = 1$

$q(x) = x$, $m = 1$

Step 1: Domain: $\{x \mid x \neq 0\}$

Step 2: (a) The x-intercept is the zero of $p(x)$, $x = 3$.

(b) To find the y-intercept, we would let $x = 0$, but $R(0)$ is undefined. Thus, there is no y-intercept.

Step 3: $R(-x) = \dfrac{-2x - 6}{-x} = \dfrac{2x + 6}{x}$

This is neither $R(x)$ nor $-R(x)$, so there is no symmetry present.

Step 4: The vertical asymptotes are the zeros of $q(x)$: $x = 0$.

Step 5: Since $n = m$, the horizontal asymptote is the line $y = \dfrac{2}{1} = 2$; not intersected.

Step 6: The zeros of $p(x)$ and $q(x)$, $(x = 0, 3)$ divide the x-axis into three intervals:

Interval	Test Number	$R(x)$	Graph of $R(x)$
$x < 0$	$x = -2$	$\frac{-4 - 6}{-2} = 5$	Above the x-axis
$0 < x < 3$	$x = 1$	$\frac{2 - 6}{1} = -4$	Below the x-axis
$x > 3$	$x = 4$	$\frac{8 - 6}{4} = \frac{1}{2}$	Above the x-axis

Step 7: Sketch the graph:

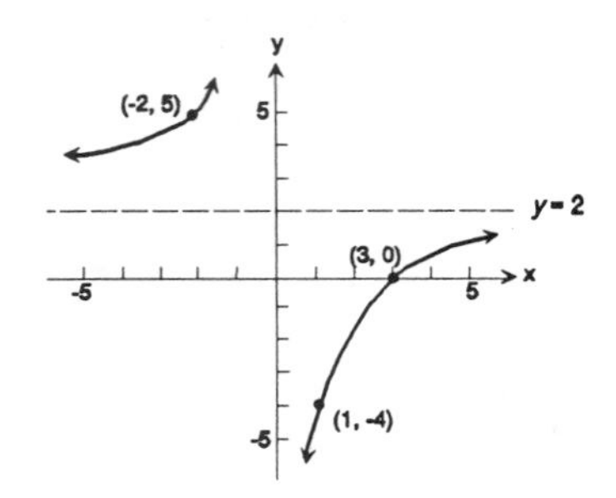

33. $H(x) = \frac{x + 2}{x(x - 2)}$

Here, $p(x) = x + 2$, $(n = 1)$
$q(x) = x(x - 2)$ $(m = 2)$

Step 1: Domain: $\{x \mid x \neq 0, x \neq 2\}$

Step 2: (a) The x-intercept is the zero of $p(x)$, $x = -2$.
(b) $H(0)$ is undefined, so there is no y-intercept.

Step 3: $H(-x) = \frac{-x + 2}{-x(-x - 2)} = \frac{-x + 2}{x(x + 2)}$ There is no symmetry.

Step 4: The vertical asymptotes are the zeros of $q(x)$: $x = 0$ and $x = 2$.

Step 5: Since $m > n$, the horizontal asymptote is the line $y = 0$ (the x-axis); intersected at $(-2, 0)$.

Step 6: $p(x)$ and $q(x)$ have a total of three zeros: $x = -2$, 0, and 2.

Interval	Test Number	$H(x)$	Graph of $H(x)$
$x < -2$	$x = -3$	$\frac{-3 + 2}{-3(-5)} = -\frac{1}{15}$	Below the x-axis
$-2 < x < 0$	$x = -1$	$\frac{-1 + 2}{-1(-3)} = \frac{1}{3}$	Above the x-axis
$0 < x < 2$	$x = 1$	$\frac{1 + 2}{1(1 - 2)} = -3$	Below the x-axis
$x > 2$	$x = 3$	$\frac{3 + 2}{3(1)} = \frac{5}{3}$	Above the x-axis

Step 7:

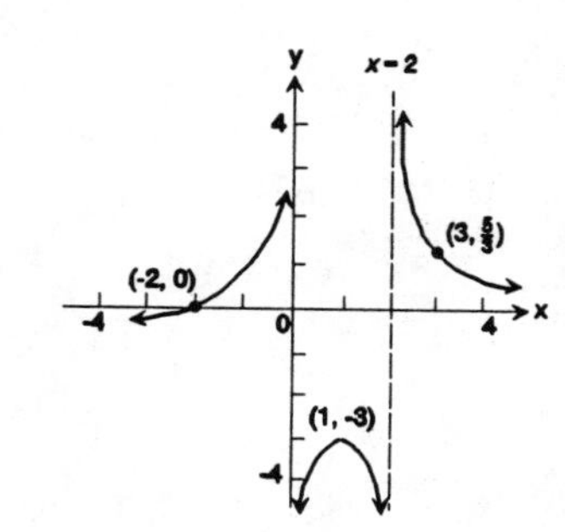

35. $R(x) = \dfrac{x^2 + x - 6}{x^2 - x - 6}$

$p(x) = x^2 + x - 6 = (x + 3)(x - 2) \qquad (n = 2)$

$q(x) = x^2 - x - 6 = (x - 3)(x + 2) \qquad (m = 2)$

Step 1: Domain: $\{x \mid x \neq -2, x \neq 3\}$

Step 2: (a) The x-intercepts are the zeros of $p(x)$, $(x + 3)(x - 2) = 0, x = -3, 2$

(b) The y-intercept is $y = R(0) = 1$.

Step 3: $R(-x) = \dfrac{x^2 - x - 6}{x^2 + x - 6}$ No symmetry present.

Step 4: The vertical asymptote is the zero of $q(x)$: $(x - 3)(x + 2) = 0; x = 3, x = -2$

Step 5: Since $m = n$, the line $y = \dfrac{1}{1} = 1$ is the horizontal asymptote; intersected at (0, 1).

Step 6: We have a total of four zeros: $x = -3, -2, 2, 3$.

Interval	Test Number	$R(x)$	Graph of $R(x)$
$x < -3$	$x = -4$	$\approx .43$	Above the x-axis
$-3 < x < -2$	$x = -2.5$	$\approx -.82$	Below the x-axis
$2 < x < 2$	$x = 0$	1	Above the x-axis
$2 < x < 3$	$x = 2.5$	≈ -1.22	Below the x-axis
$x > 3$	$x = 4$	≈ 2.33	Above the x-axis

Step 7:

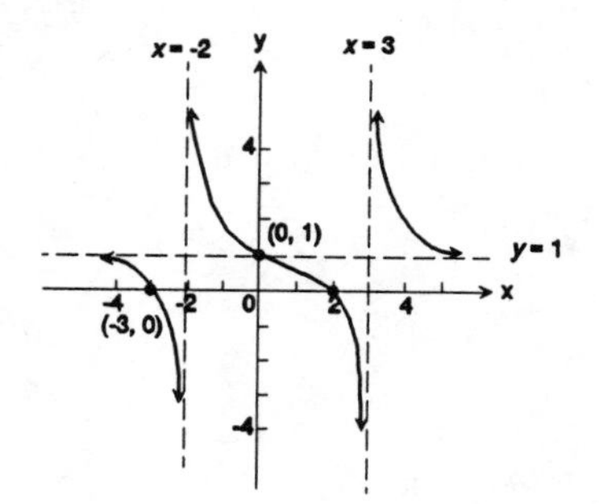

37. $F(x) = \dfrac{x^3}{x^2 - 4}$

$p(x) = x^3 \quad (n = 3)$

$q(x) = x^2 - 4 = (x + 2)(x - 2) \ (m = 2)$

Step 1: Domain: $\{x \mid x \neq -2, x \neq 2\}$

Step 2: (a) The x-intercept is $x = 0$.

(b) The y-intercept is $y = F(0) = 0$.

Step 3: $F(-x) = \dfrac{-x^3}{x^2 - 4} = -F(x)$, so the graph is symmetric about the origin.

Step 4: The vertical asymptotes are $x = -2$ and $x = 2$.

Step 5: Since $n = m + 1$, perform long division:

$$\begin{array}{r} x \\ x^2 - 4\overline{)x^3 \;\; 0x^2 \;\; 0x \;\; 0} \\ \underline{x^3 -4x} \\ 4x \end{array}$$

$$F(x) = \frac{x^3}{x^2 - 4} = x + \frac{4x}{x^2 - 4}$$

Thus, we have an oblique asymptote, $y = x$; intersected at (0, 0).

Step 6: We have a total of three zeros: $x = -2$, 0, and 2; intersected at (0, 0).

Interval	Test Number	$F(x)$	Graph of $F(x)$
$x < -2$	$x = -3$	$\frac{-27}{5}$	Below the x-axis
$-2 < x < 0$	$x = -1$	$\frac{-1}{-3} = \frac{1}{3}$	Above the x-axis
$0 < x < 2$	$x = 1$	$-\frac{1}{3}$	Below the x-axis
$x > 2$	$x = 3$	$\frac{27}{5}$	Above the x-axis

Step 7:

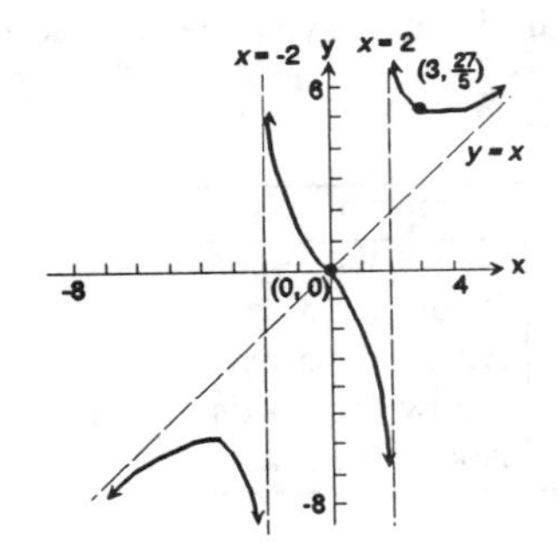

39. $R(x) = \dfrac{2x^4}{(x - 1)^2}$

$p(x) = 2x^4$ $(n = 4)$

$q(x) = (x - 1)^2$ $(m = 2)$

Step 1: Domain: $\{x \mid x \neq 1\}$

Step 2: (a) The x-intercept is $x = 0$.

(b) The y-intercept is $y = 0$.

Step 3: $R(-x) = \dfrac{2x^4}{(-x - 1)^2} = \dfrac{2x^4}{(x + 1)^2}$ No symmetry.

Step 4: The vertical asymptote is the zero of $q(x)$: $x = 1$.

Step 5: Since $n > m + 1$, there is no horizontal nor oblique asymptote.

Step 6: We have a total of two zeros: $x = 0, 1$.

Interval	Test Number	$R(x)$	Graph of $R(x)$
$x < 0$	$x = -2$	$\frac{32}{9}$	Above the x-axis
$0 < x < 1$	$x = \frac{1}{2}$	$\frac{\frac{1}{8}}{\frac{1}{4}} = \frac{1}{2}$	Above the x-axis
$x > 1$	$x = 2$	32	Above the x-axis

Step 7:

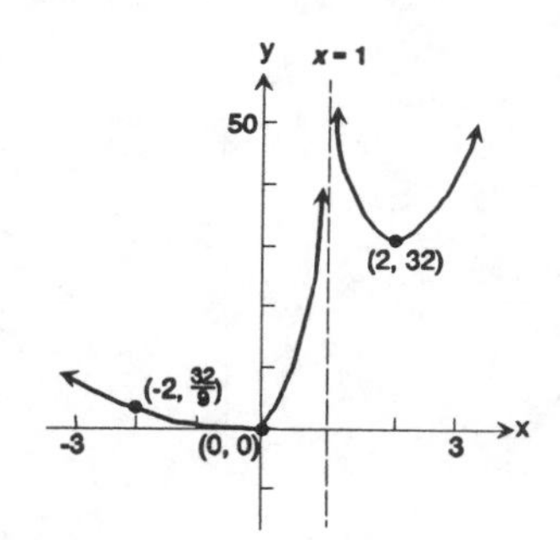

41. $G(x) = \dfrac{x^2 - 4}{x^2 - x - 2} = \dfrac{(x + 2)(x - 2)}{(x - 2)(x + 1)} = \dfrac{x + 2}{x + 1}, x \neq 2$

$p(x) = x^2 - 4 = (x + 2)(x - 2) \quad (n = 2)$

$q(x) = x^2 - x - 2 = (x - 2)(x + 1) \quad (m = 2)$

Step 1: Domain: $\{x \mid x \neq -1, x \neq 2\}$

Step 2: (a) x-intercept: -2

(b) y-intercept: 2

Step 3: $G(-x) = \dfrac{(-x)^2 - 4}{(-x)^2 - (-x) - 2} = \dfrac{x^2 - 4}{x^2 + x - 2}$ No symmetry.

Step 4: The vertical asymptote is $x = -1$. There is a hole at $\left(2, \frac{4}{3}\right)$.

Step 5: Since $n = m$, there is a horizontal asymptote at $y = 1$.

$$1 = \frac{x + 2}{x + 1}$$

$$x + 1 = x + 2$$

No solution, so $y = 1$ is not intersected.

Step 6:

Interval	Test Number	$G(x)$	Graph of $G(x)$
$x < -2$	-3	$\frac{1}{2}$	Above the x-axis
$-2 < x < -1$	-1.5	-1	Below the x-axis
$-1 < x < 2$	0	2	Above the x-axis
$x > 2$	3	$\frac{5}{4}$	Above the x-axis

Step 7:

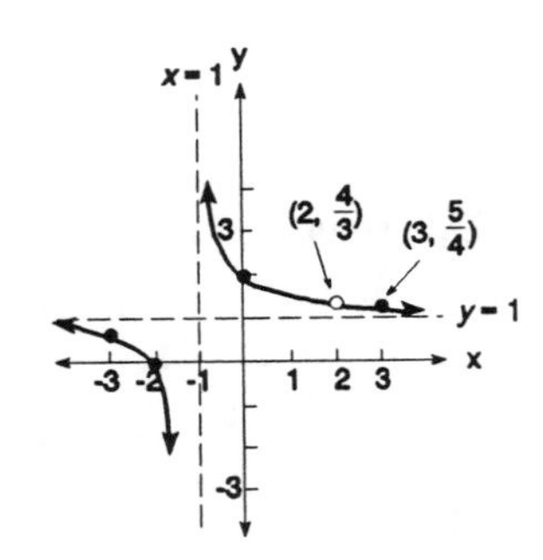

43. Divide $f(x) = 8x^3 - 3x^2 + x + 4$ by $g(x) = x - 1$:

$$\begin{array}{r|rrrr} 1 & 8 & -3 & 1 & 4 \\ & & 8 & 5 & 6 \\ \hline & 8 & 5 & 6 & \boxed{10} \end{array}$$

Remainder, $R = 10$

Quotient, $q(x) = 8x^2 + 5x + 6$

45. Divide $f(x) = x^4 - 2x^3 + x - 1$ by $g(x) = x + 2$:

$$\begin{array}{r|rrrrr} -2 & 1 & -2 & 0 & 1 & -1 \\ & & -2 & 8 & -16 & 30 \\ \hline & 1 & -4 & 8 & -15 & \boxed{29} \end{array}$$

Remainder, $R = 29$

Quotient, $q(x) = x^3 - 4x^2 + 8x - 15$

47. To find $f(4)$, we can divide $x - 4$ and find the remainder, using synthetic division:

$$\begin{array}{r|rrrrrrr} 4 & 12 & 0 & -8 & 0 & 0 & 0 & 1 \\ & & 48 & 192 & 736 & 2944 & 11776 & 47104 \\ \hline & 12 & 48 & 184 & 736 & 2944 & 11776 & \boxed{47105} \end{array}$$

$f(4)$ is the remainder: $f(4) = 47{,}105$

49. $f(x) = 12x^8 - x^7 + 8x^4 - 2x^3 + x + 3$

+ to −   − to +   + to −   − to +

$f(-x) = 12x^8 + x^7 + 8x^4 + 2x^3 - x + 3$

+ to −   − to +

We have two or no negative zeros, and either four, two, or no positive zero(s).

51. $f(x) = 12x^8 - x^7 + 6x^4 - x^3 + x - 3$

p: $\pm 1, \pm 3$

q: $\pm 1, \pm 2, \pm 3, \pm 4, \pm 6, \pm 12$

Possible rational zeros:

$$\frac{p}{q}: \pm 1, \pm\frac{1}{2}, \pm\frac{1}{3}, \pm\frac{1}{4}, \pm\frac{1}{6}, \pm\frac{1}{12}, \pm 3, \pm\frac{3}{2}, \pm\frac{3}{4}$$

53. First apply Descartes' Rule of Signs:

$f(x) = x^3 - 3x^2 - 6x + 8$

+ to −   − to +

$f(-x) = -x^3 - 3x^2 + 6x + 8$

− to +

We see that there are either two positive zeros, or none; and exactly one negative zero.

To find the possible ***rational*** zeros, p must be a factor of 8: p: $\pm1, \pm2, \pm4, \pm8$
and q must be a factor of 1: q: ±1

Therefore, the ***possible rational*** zeros are: $\frac{p}{q}$: $\pm1, \pm2, \pm4, \pm8$

We now proceed to synthetic division:

$$\begin{array}{r|rrrr} -1 & 1 & -3 & -6 & 8 \\ & & -1 & 4 & 2 \\ \hline & 1 & -4 & -2 & \boxed{10} \end{array} \leftarrow (x+1) \text{ is not a factor.}$$

$$\begin{array}{r|rrrr} 1 & 1 & -3 & -6 & 8 \\ & & 1 & -2 & -8 \\ \hline & 1 & -2 & -8 & \boxed{0} \end{array} \leftarrow (x-1) \text{ is a factor.}$$

We now have: $f(x) = x^3 - 3x^2 - 6x + 8$

$= (x - 1)(x^2 - 2x - 8)$

$= (x - 1)(x - 4)(x + 2)$

Therefore, the real zeros of $f(x)$ are 1, 4, and -2.

55. We have: $f(x) = 4x^3 + 4x^2 - 7x + 2$ (2 sign changes)
and $f(-x) = -4x^3 + 4x^2 + 7x + 2$ (1 sign change)

By Descartes' Rule of Signs, $f(x)$ must have one negative zero, and either two positive zeros or none.

Check for possible ***rational*** zeros:

p: $\pm1, \pm2$; q: $\pm1, \pm2, \pm4$; $\frac{p}{q}$: $\pm1, \pm\frac{1}{2}, \pm\frac{1}{4}, \pm2$

We begin synthetic division:

$$\begin{array}{r|rrrr} -1 & 4 & 4 & -7 & 2 \\ & & -4 & 0 & 7 \\ \hline & 4 & 0 & -7 & \boxed{9} \end{array} \leftarrow (x+1) \text{ is not a factor.}$$

$$\begin{array}{r|rrrr} 1 & 4 & 4 & -7 & 2 \\ & & 4 & 8 & 1 \\ \hline & 4 & 8 & 1 & \boxed{3} \end{array} \leftarrow (x-1) \text{ is not a factor.}$$

$$\begin{array}{r|rrrr} -2 & 4 & 4 & -7 & 2 \\ & & -8 & 8 & -2 \\ \hline & 4 & -4 & 1 & \boxed{0} \end{array} \leftarrow (x+2) \text{ \underline{is} a factor.}$$

Quotient: $q(x) = 4x^2 - 4x + 1$

We now have: $f(x) = 4x^3 + 4x^2 - 7x + 2 = (x + 2)(4x^2 - 4x + 1)$

Let's find the zeros of the depressed equation, $4x^2 - 4x + 1 = 0$, by the quadratic formula:

$$x = \frac{-b \pm \sqrt{b^2 - 4ac}}{2a} = \frac{4 \pm \sqrt{16 - 4(4)(1)}}{8} \text{ or } x = \frac{4 \pm 0}{8} = \frac{1}{2} \text{ is a double root.}$$

Thus, $\left(x - \frac{1}{2}\right)$ is a repeated factor.

Since the leading coefficient is 4, we have: $4x^2 - 4x + 1 = 4\left(x - \frac{1}{2}\right)^2$

and $f(x) = (x + 2)(4x^2 - 4x + 1) = 4(x + 2)\left(x - \frac{1}{2}\right)^2$

The real zeros of $f(x)$ are -2 and $\frac{1}{2}$ (with multiplicity 2).

57. $f(x) = x^4 - 4x^3 + 9x^2 - 20x + 20$
$f(-x) = x^4 + 4x^3 + 9x^2 + 20x + 20$
We see that we either have four positive zeros, or two positive zeros, or none, and we have no negative zeros (since there are no changes of sign in $f(-x)$).
Possible rational zeros:

$$p: \pm1, \pm2, \pm4, \pm5, \pm10, \pm20; \; q: \pm1; \; \frac{p}{q}: \pm1, \pm2, \pm4, \pm5, \pm10, \pm20$$

(But we can exclude the negative possibilities, since $f(x)$ has no negative zeros).

To see if $x = 1$ is a zero, we check to see if $(x - 1)$ is a factor:

$$\begin{array}{r|rrrrr} 1 & 1 & -4 & 9 & -20 & 20 \\ & & 4 & 0 & 9 & -11 \\ \hline & 1 & 0 & 9 & -11 & \boxed{9} \end{array} \leftarrow (x - 1) \text{ is } \textbf{\textit{not}} \text{ a factor.}$$

Now try $(x - 2)$:

$$\begin{array}{r|rrrrr} 2 & 1 & -4 & 9 & -20 & 20 \\ & & 2 & -4 & 10 & -20 \\ \hline & 1 & -2 & 5 & -10 & \boxed{0} \end{array} \leftarrow (x - 2) \text{ } \textit{is} \text{ a factor.}$$

Quotient: $q(x) = x^3 - 2x^2 + 5x - 10$

We now have: $f(x) = x^4 - 4x^3 + 9x^2 - 20x + 20$
$= (x - 2)(x^3 - 2x^2 + 5x - 10)$

factor by grouping

$= (x - 2)[x^2(x - 2) + 5(x - 2)]$
$= (x - 2)(x - 2)(x^2 + 5)$

no real zeros

$= (x - 2)^2(x^2 + 5)$

The real zero of $f(x)$ is 2 (multiplicity two).

59. $2x^4 + 2x^3 - 11x^2 + x - 6 = 0$
The solutions of this equation are the zeros of the polynomial function $f(x)$.
$f(x) = 2x^4 + 2x^3 - 11x^2 + x - 6$
$f(-x) = 2x^4 - 2x^3 - 11x^2 - x - 6$
We have either three positive zeros, or one; and exactly one negative zero.
Possible ***rational*** zeros:

$$p: \pm1, \pm2, \pm3, \pm6; \; q: \pm1, \pm2; \; \frac{p}{q}: \pm1, \pm\frac{1}{2}, \pm2, \pm3, \pm\frac{3}{2}, \pm6$$

Start synthetic division:

$$\begin{array}{r|rrrrr} -1 & 2 & 2 & -11 & 1 & -6 \\ & & -2 & 0 & 11 & -12 \\ \hline & 2 & 0 & -11 & 12 & \boxed{-18} \end{array} \leftarrow (x+1) \text{ is } \textbf{\textit{not}} \text{ a factor.}$$

$$\begin{array}{r|rrrrr} 1 & 2 & 2 & -11 & 1 & -6 \\ & & 2 & 4 & -7 & -6 \\ \hline & 2 & 4 & -7 & -6 & \boxed{-12} \end{array} \leftarrow (x-1) \text{ is not a factor.}$$

$$\begin{array}{r|rrrrr} -2 & 2 & 2 & -11 & 1 & -6 \\ & & -4 & 4 & 14 & -30 \\ \hline & 2 & -2 & -7 & 15 & \boxed{-36} \end{array} \leftarrow (x+2) \text{ is not a factor.}$$

$$\begin{array}{r|rrrrr} 2 & 2 & 2 & -11 & 1 & -6 \\ & & 4 & 12 & 2 & 6 \\ \hline & 2 & 6 & 1 & 3 & \boxed{0} \end{array} \leftarrow (x-2) \text{ } \textbf{\textit{is}} \text{ a factor.}$$

We now have: $2x^4 + 2x^3 - 11x^2 + x - 6 = 0$

$$(x - 2)(\underbrace{2x^3 + 6x^2 + x + 3}_{\text{factor by grouping}}) = 0$$

$$(x - 2)[2x^2(x + 3) + 1(x = 3)] = 0$$

$$(x - 2)(x + 3)(2x^2 + 1) = 0$$

The real zeros are -3 and 2.

61. $2x^4 + 7x^3 + x^2 - 7x - 3 = 0$

The solutions of this equation are the zeros of the polynomial function $f(x)$.

$$f(x) = 2x^4 + 7x^3 + x^2 - 7x - 3$$

$$f(-x) = 2x^4 - 7x^3 + x^2 + 7x - 3$$

We have exactly one positive zero, and either three or one negative zero(s).

For possible ***rational*** zeros: p: $\pm 1, \pm 3$; q: $\pm 1, \pm 2$; $\frac{p}{q}$: $\pm 1, \pm\frac{1}{2}, \pm 3, \pm\frac{3}{2}$

Synthetic division:

$$\begin{array}{r|rrrrr} 1 & 2 & 7 & 1 & -7 & -3 \\ & & 2 & 9 & 10 & 3 \\ \hline & 2 & 9 & 10 & 3 & \boxed{0} \end{array}$$

Therefore, $(x - 1)$ is a factor, so $x = 1$ is a zero. All other real zeros *must* be negative.

We have: $f(x) = (x - 1)(2x^3 + 9x^2 + 10x + 3)$

We now concentrate on the depressed equation, $q(x) = (2x^3 + 9x^2 + 10x + 3)$. The ***possible*** rational zeros are the same as before, but since we have found the only positive zero, we only need to consider the negative ones: $-1, -3, -\frac{1}{2}, -\frac{3}{2}$

To see if $x = -1$ is a zero, divide by $x - (-1) = x + 1$:

$$\begin{array}{r|rrrr} -1 & 2 & 9 & 10 & 3 \\ & & -2 & -7 & -3 \\ \hline & 2 & 7 & 3 & \boxed{0} \end{array}$$

So $(x + 1)$ ***is*** a factor, and we have:

$$f(x) = (x - 1)(2x^3 + 9x^2 + 10x + 3) = (x - 1)(x + 1)(2x^2 + 7x + 3)$$

Now $2x^2 + 7x + 3$ can be factored by trial and error:

$$2x^2 + 7x + 3 = (2x + 1)(x + 3)$$

We have:

$$2\left(x + \frac{1}{2}\right)$$

$$(x - 1)(x + 1)(2x + 1)(x + 3) = 0$$

$$2(x - 1)(x + 1)\left(x + \frac{1}{2}\right)(x + 3) = 0$$

The real zeros are: $-3, -1, -\frac{1}{2}$, and 1.

63. From Problem 53:

$$f(x) = x^3 - 3x^2 - 6x + 8 = (x - 1)(x - 4)(x + 2)$$

Thus, $f(x)$ has three real zeros, -2, 1, and 4, which divide the x-axis into 4 intervals. Recall that the zeros *are* the x-intercepts, and the y-intercept is $y = f(0) = 8$.

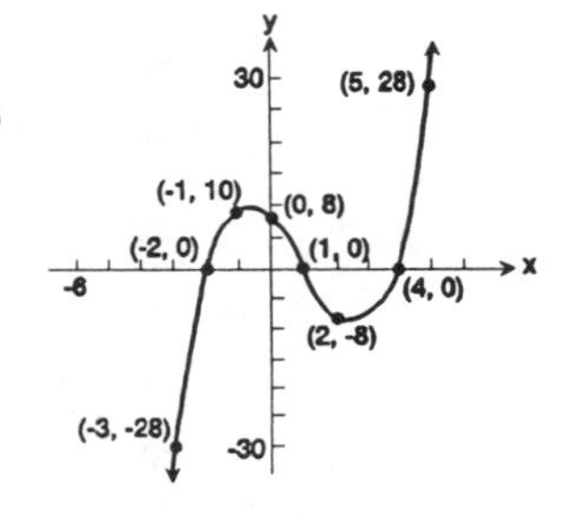

Interval	Test Number	$f(x)$	Graph of $f(x)$
$-\infty < x < -2$	$x = -3$	-28	Below the x-axis
$-2 < x < 1$	$x = -1$	10	Above the x-axis
$1 < x < 4$	$x = 2$	-8	Below the x-axis
$4 < x < \infty$	$x = 5$	28	Above the x-axis

65. From Problem 55, we have:

$$f(x) = 4x^3 + 4x^2 - 7x + 2 = 4(x + 2)\left(x - \frac{1}{2}\right)^2$$

The y-intercept is $y = f(0) = 2$.

The x-intercepts are the real zeros: $-2, \frac{1}{2}$.

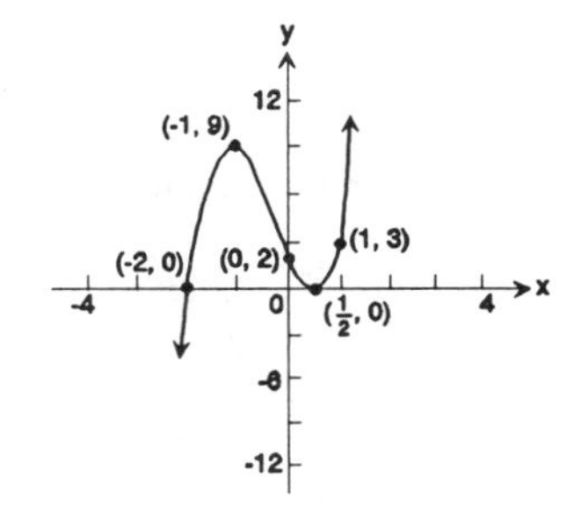

Interval	Test Number	$f(x)$	Graph of $f(x)$
$-\infty < x < -2$	$x = -3$	-49	Below the x-axis
$-2 < x < \frac{1}{2}$	$x = -1$	9	Above the x-axis
$x > \frac{1}{2}$	$x = 1$	3	Above the x-axis

67. From Problem 57,

$$f(x) = x^4 - 4x^3 + 9x^2 - 20x + 20$$
$$= (x - 2)(x - 2)(x^2 + 5)$$

The y-intercept is $y = f(0) = 20$.

The only x-intercept is the one zero of $f(x)$, $x = 2$.

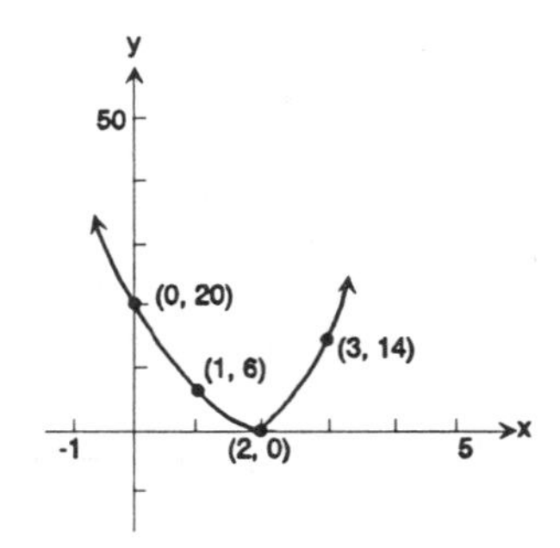

Interval	Test Number	$f(x)$	Graph of $f(x)$
$x < 2$	$x = 1$	6	Above the x-axis
$x > 2$	$x = 3$	14	Above the x-axis

69. From Problem 59,

$$f(x) = 2x^4 + 2x^3 - 11x^2 + x - 6$$
$$= (x - 2)(x + 3)(2x^2 + 1)$$

The y-intercept is $y = f(0) = -6$.
The x-intercepts are the zeros, -3 and 2.

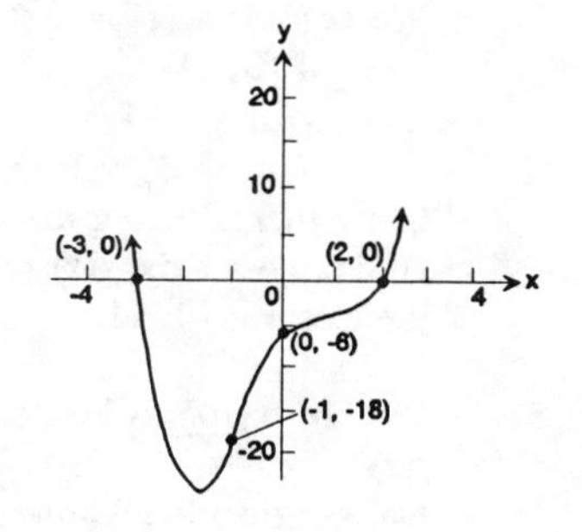

Interval	Test Number	$f(x)$	Graph of $f(x)$
$-\infty < x < -3$	$x = -4$	198	Above the x-axis
$-3 < x < 2$	$x = -1$	-18	Below the x-axis
$2 < x < \infty$	$x = 3$	171	Above the x-axis

71. From Problem 61,

$$f(x) = 2x^4 + 7x^3 + x^2 - 7x - 3$$
$$= 2(x - 1)(x + 1)\left[x + \frac{1}{2}\right](x + 3)$$

The y-intercept is $y = f(0) = -3$.

The x-intercepts are the one zeros, -3, -1, $-\frac{1}{2}$ and 1.

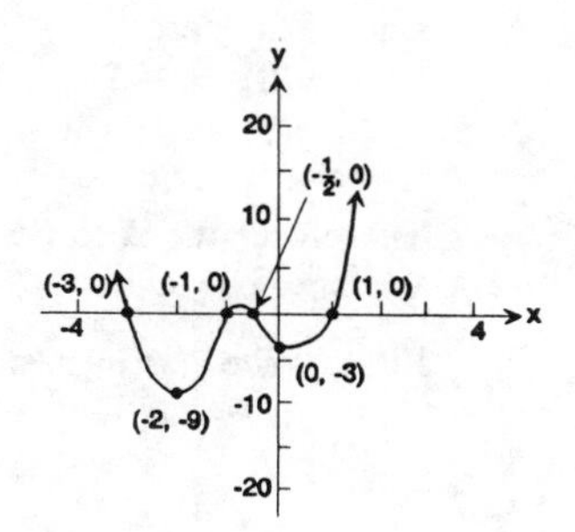

Interval	Test Number	$f(x)$	Graph of $f(x)$
$-\infty < x < -3$	$x = -4$	105	Above the x-axis
$-3 < x < -1$	$x = -2$	-9	Below the x-axis
$-1 < x < -\frac{1}{2}$	$x = -\frac{3}{4}$	$\frac{63}{128}$	Above the x-axis
$-\frac{1}{2} < x < 1$	$x = 0$	-3	Below the x-axis
$1 < x < \infty$	$x = 2$	75	Above the x-axis

73. $f(x) = x^3 - x^2 - 4x + 2$

$a_0 = 2$, $a_1 = -4$, $a_2 = -1$

Max: $\{1, |2| + |-4| + |-1|\}$ = Max: $\{1, 2 + 4 + 1\}$ = Max$\{1, 7\} = 7$

$1 +$ Max$(1, |2|, |-4|, |-1|\} = 1 +$ Max: $\{1, 2, 4, 1\} = 1 + 4 = 5$

The smaller of the two numbers is 5. Thus, the zeros of f lie between -5 and 5.

75. $f(x) = 2x^3 - 7x^2 - 10x + 35 = 2\left[x^3 - \frac{7}{2}x^2 - 5x + \frac{35}{2}\right]$ Leading coefficient must be 1.

$a_0 = \frac{35}{2}$, $a_1 = -5$, $a_2 = \frac{-7}{2}$

Max$\{1, \left|\frac{35}{2}\right| + |-5| + \left|\frac{-7}{2}\right|\}$ = Max$\{1, 26\} = 26$

$1 +$ Max$\{1, \left|\frac{35}{2}\right|, |-5|, \left|\frac{-7}{2}\right|\} = 1 + \frac{35}{2} = \frac{37}{2}$

The smaller of the two numbers is $\frac{37}{2}$. Thus, the zeros of f lie between $\frac{-37}{2}$ and $\frac{37}{2}$.

77. We consider $f(x) = 3x^3 - x - 1$ on the interval [0, 1]:

$f(0) = -1$

$f(1) = 3 - 1 - 1 = 1$

Since $f(0)$ is below the x-axis, and $f(1)$ is above the x-axis, $f(x)$ must have a zero in the interval [0, 1].

79. For $f(x) = 8x^4 - 4x^3 - 2x - 1$ on [0, 1], we have:

$f(0) = -1$

$f(1) = 8 - 4 - 2 - 1 = 1$

Therefore, $f(x)$ must have a zero between 0 and 1.

81. We are trying to locate the positive zero of $f(x) = x^3 - x - 2$. We start by finding the two consecutive whole numbers on either side of the zero:

$f(0) = -2$

$f(1) = 1 - 1 - 2 = -2 < 0$

$f(2) = 8 - 2 - 2 = 4 > 0$

Therefore, the zero lies between 1 and 2. We now check $x = 1.1, 1.2, 1.3, \ldots, 1.9$.

First, write $f(x)$ in nested form:

$$\begin{aligned} f(x) &= x^3 - x - 2 \\ &= (x^2 - 1)x - 2 \\ &= (x \cdot x - 1)x - 2 \end{aligned}$$

Then, $f(1.1) = -1.769 < 0$

$f(1.2) = -1.472 < 0$

Let's skip to: $f(1.5) = -.125 < 0$

(See how that saved some time?)

Now, $f(1.6) = .496 > 0$.

The zero must lie between 1.5 and 1.6. We continue once more, to isolate the zero to one of the intervals:

[1.5, 1.51], [1.51, 1.52], ... , [1.59, 1.6].

We have $f(1.5) = -.125 < 0$

Then $f(1.51) = -.06705 < 0$

$f(1.52) = -.00819 < 0$

$f(1.53) = .05158 > 0$

The zero is 1.52 and correct to two decimal places.

83. We write $f(x)$ in nested form:

$$\begin{aligned} f(x) &= 8x^4 - 4x^3 - 2x - 1 \\ &= (8x^3 - 4x^2 - 2)x - 1 \\ &= ([8x - 4]x \cdot x - 2)x - 1 \end{aligned}$$

We start with whole numbers:

$f(0) = -1 < 0$

$f(1) = 1 > 0$

The zero lies in the interval [0, 1].

Now proceed by tenths:

$f(0) = -1 < 0$

$f(0.1) = -1.2032 < 0$

Skip to: $f(0.5) = -2 < 0$

How about: $f(0.7) = -1.8512 < 0$

$f(0.8) = -1.3712 < 0$

$f(0.9) = -0.4672 < 0$

The zero lies between 0.9 and 1.

We go again:

$f(0.9) = -.4672 < 0$

$f(0.91) = -.34829 < 0$

⋮

$f(0.95) = 0.18655 > 0$

This time I went too far, so I back up:

$f(0.94) = .04366 > 0$

$f(0.93) = -.09301 < 0$

So the zero is 0.93 correct to two decimal places.

85. $(6 + 3i) -)2 - 4i) = (6 - 2) + (3 + 4)i = 4 + 7i$

87. $4(3 - i) + 3(-5 + 2i) = 12 - 4i - 15 + 6i = -3 + 2i$

89. $$\frac{3}{3+i} = \frac{3}{3+i} \cdot \frac{3-i}{3-i} = \frac{3(3-i)}{(3+i)(3-i)} = \frac{3(3-i)}{9-i^2} = \frac{3(3-i)}{9-(-1)} = \frac{3(3-i)}{10} = \frac{9}{10} - \frac{3}{10}i$$

91. $i^{50} = (i^2)^{25} = (-i)^{25} = -1$

93. $$\begin{aligned} (2 + 3i)^3 &= 2^3 + 3 \cdot 2^2 \cdot 3i + 3(3i)^2 \cdot 2 + (3i)^3 \\ &= 8 + 36i + 54i^2 + 27i^3 \\ &= 8 + 36i - 54 - 27i \\ &= -46 + 9i \end{aligned}$$

95. Since complex zeros appear as conjugate pairs, it follows that $4 - i$, the conjugate of $4 + i$, is the remaining zero of f.

97. $-i$, the conjugate of i, and $1 - i$, the conjugate of $1 + i$, are the remaining zeros of f.

99. $x^2 + x + 1 = 0$
Here $a = 1$, $b = 1$, $c = 1$ and $b^2 - 4ac = -3$.

Then, $x = \dfrac{-1 \pm \sqrt{-3}}{2(1)} = \dfrac{-1 \pm \sqrt{3}\,i}{2}$

The solution set is $\left\{\dfrac{-1 - \sqrt{3}i}{2}, \dfrac{-1 + \sqrt{3}i}{2}\right\}$.

101. $2x^2 + x - 2 = 0$
Here $a = 2$, $b = 1$, $c = -2$ and $b^2 - 4ac = 17$.

Then, $x = \dfrac{-1 \pm \sqrt{17}}{2(2)} = \dfrac{-1 \pm \sqrt{17}}{4}$

The solution set is $\left\{\dfrac{-1 - \sqrt{17}}{4}, \dfrac{-1 + \sqrt{17}}{4}\right\}$.

103. $x^2 + 3 = x$
$x^2 - x + 3 = 0$
Here, $a = 1$, $b = -1$, $c = 3$ and $b^2 - 4ac = -11$.

Then, $x = \dfrac{-(-1) \pm \sqrt{-11}}{2(1)} = \dfrac{1 \pm \sqrt{11}\,i}{2}$

The solution set is $\left\{\dfrac{1 - \sqrt{11}i}{2}, \dfrac{1 + \sqrt{11}i}{2}\right\}$.

105. $x(1 - x) = 6$
$x - x^2 = 6$
$-x^2 + x - 6 = 0$
$x^2 - x + 6 = 0$
Here, $a = 1$, $b = -1$, $c = 6$ and $b^2 - 4ac = -23$.

Then, $x = \dfrac{-(-1) \pm \sqrt{-23}}{2(1)} = \dfrac{1 \pm \sqrt{23}\,i}{2}$

The solution set is $\left\{\dfrac{1 - \sqrt{23}i}{2}, \dfrac{1 + \sqrt{23}i}{2}\right\}$.

107. $x^4 + 2x^2 - 8 = 0$
Let $u = x^2$ and $u^2 = x^4$.
Then $u^2 + 2u - 8 = 0$
$(u + 4)(u - 2) = 0$
$(u + 4) = 0$ or $u - 2 = 0$
$u = -4$ or $u = 2$
$x^2 = -4$ or $x^2 = 2$
$x = \pm\sqrt{-4}$ or $x = \pm\sqrt{2}$
$x = \pm\sqrt{4}\,i$
$= \pm 2i$

The solution set is $\{-\sqrt{2}, \sqrt{2}, -2i, 2i\}$.

109. $x^3 - x^2 - 8x + 12 = 0$
The solutions of this equation are the zeros of the polynomial function $f(x)$.
$f(x) = x^3 - x^2 - 8x + 12$
Determine the ***possible rational*** zeros:

p: $\pm 1, \pm 2, \pm 3, \pm 4, \pm 6, \pm 12$; q: ± 1; $\dfrac{p}{q}$: $\pm 1, \pm 2, \pm 3, \pm 4, \pm 6, \pm 12$

We will check if $x = 2$ is a solution.

$$\begin{array}{r|rrrr} 2 & 1 & -1 & -8 & 12 \\ & & 2 & 2 & -12 \\ \hline & 1 & 1 & -6 & \boxed{0} \end{array}$$
$\leftarrow (x - 2)$ is a factor.

So $x^3 - x^2 - 8x + 12 = 0$
$(x - 2)(x^2 + x - 6) = 0$
$(x - 2)(x + 3)(x - 2) = 0$
The zeros are $x = 2$ (multiplicity two), and $x = -3$.

111. $3x^4 - 4x^3 + 4x^2 + 1 = 0$

The solutions of this equation are the zeros of the polynomial function $f(x)$.

$$f(x) = 3x^4 - 4x^3 + 4x^2 - 4x + 1$$

For possible rational zeros: p: ±1; q: $\pm1, \pm3$; $\frac{p}{q}$: $\pm1, \pm\frac{1}{3}$

We will check if $x = 1$ is a solution.

$$\begin{array}{r|rrrrr} 1 & 3 & -4 & 4 & -4 & 1 \\ & & 3 & -1 & 3 & -1 \\ \hline & 3 & -1 & 3 & -1 & \boxed{0} \end{array} \leftarrow (x - 1) \textit{ is a factor.}$$

We have:

$$(x - 1)(3x^3 - x^2 + 3x - 1) = 0$$
$$(x - 1)[x^2(3x - 1) + 1(3x - 1)] = 0$$
$$(x - 1)(3x - 1)(x^2 + 1) = 0$$
$$3(x - 1)\left[x - \frac{1}{3}\right](x + i)(x - i) = 0$$

The zeros are $1, \frac{1}{3}, -i, i$.

113. Let (x, y) be any point on the line $y = x$. Then the distance from (x, y) to $(3, 1)$ is given by the distance formula.

$$d = \sqrt{(x - 3)^2 + (y - 1)^2} = \sqrt{x^2 - 6x + 9 + y^2 - 2y + 1}$$

But since (x, y) in on the line $y = x$, we can replace all y's in the above formula by x:

$$d = \sqrt{x^2 - 6x + 9 + x^2 - 2x + 1}$$

or $d = \sqrt{2x^2 - 8x + 10}$

Now d is a fairly complicated function, but notice: $d = 2x^2 - 8x + 10$

so d^2 is a parabola that opens upward, so we can find the minimum value of d^2, and then it will be easy to find the minimum value of d by taking a square root.

Consider the parabola $y = 2x^2 - 8x + 10$. We have $a = 2$, $b = -8$, $c = 10$, so that

$h = \frac{-b}{2a} = \frac{8}{4} = 2$, and $k = f(2) = 2$.

Thus, the vertex is at $(2, 2)$. That means that the minimum value of d^2 is 2, and that value occurs when $x = 2$. (So the minimum value of d is $\sqrt{2}$.) Since $y = x$, and $x = 2$, the point on the line is the point $(2, 2)$.

115. See illustration.

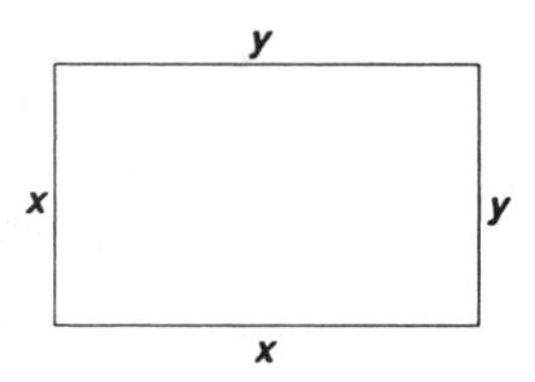

Since we have 200 feet of border, we know $2x + 2y = 200$. We want to maximize area where Area $= x \cdot y$. Solving the perimeter formula for y:

$$2x + 2y = 200$$
$$2y = 200 - 2x$$
$$y = 100 - x$$

Substitute this value into the Area formula to obtain:

$$x(100 - x) = \text{Area}$$
$$A(x) = -x^2 + 100x$$

Thie is a quadratic function with $a = -1 < 0$. The maximum value occurs at the vertex:

$$x = \frac{-b}{2a} = \frac{-100}{2(-1)} = 50$$

The pond should be 50 feet by 50 feet.

117. See illustration:

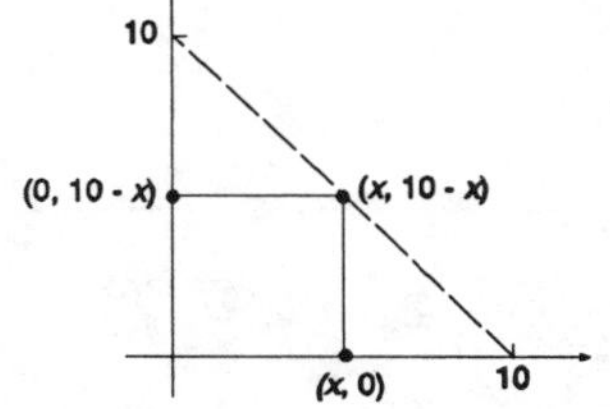

Area $= A(x) = x(10 - x) = 10x - x^2$. This is a quadratic function with $a = -1 < 0$. The vertex will be a maximum.

$$x = \frac{-b}{2a} = \frac{-10}{2(-1)} = 5$$

$A(5) = 10(5) - 5^2 = 25$ square units.

119. $C(x) = 0.003x^2 - 30x + 111{,}800$

(a) Quadratic function with $a = 0.003 > 0$. The vertex will be a minimum.

$$x = \frac{-b}{2a} = \frac{-(-30)}{2(0.003)} = 5000 \text{ books}$$

(b) $C(5000) = 0.003(5000)^2 - 30(5000) + 111{,}800 = \$36{,}800$ total cost

Cost/Book $= \$36{,}800/5000 = \$7.36$

(c) $C(1) = 0.003(1)^2 - 30(1) + 111{,}800 = \$111{,}770$

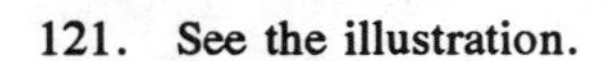

121. See the illustration.

The circumference of each semi-circle is $\pi r = \dfrac{\pi h}{2}$. Thus,

$$\frac{\pi h}{2} + \frac{\pi h}{2} + 2w = 100$$

$$\pi h + 2w = 100$$

The area of the rectangle is: Area $= wh$.

Since $\pi h + 2w = 100$, we have $h = \dfrac{100 - 2w}{\pi}$. Thus,

$A(w) = w\left[\dfrac{100 - 2w}{\pi}\right] = \dfrac{-2}{\pi}w^2 + \dfrac{100}{\pi}w$. This is a quadratic function with $a = \dfrac{-2}{\pi} < 0$. Thus, the vertex is a maximum.

$$\text{Vertex} = w = \frac{-b}{2a} = \frac{-100/\pi}{2(-2/\pi)} = 25$$

$$h = \frac{100 - 2(25)}{\pi} = \frac{50}{\pi}$$

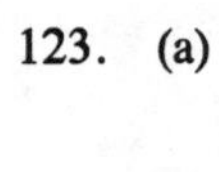

123. (a)

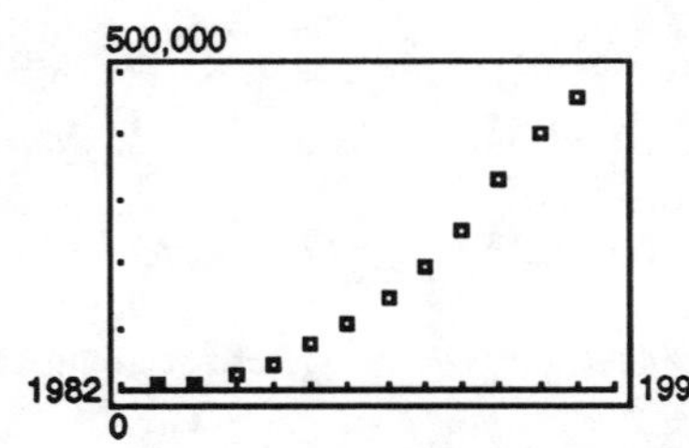

(d)

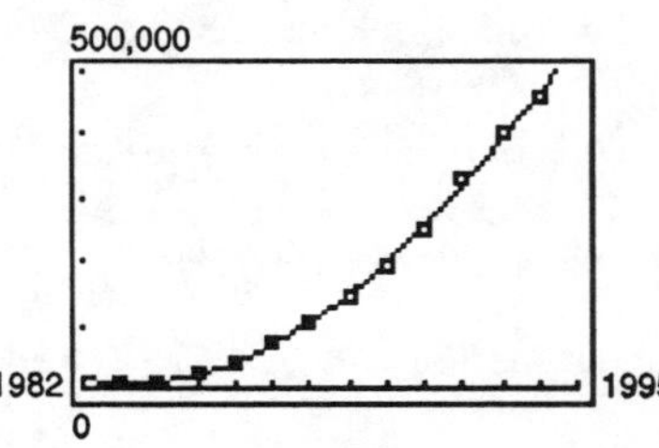

(b) $A(13) = -68.8(13)^3 + 4988.16t^2 - 12436.1t + 14611.3 = 544{,}789$

127. (a) Even

(b) Positive

(c) Even

(d) Since 0 is a zero of even multiplicity (it touches the x-axis)

(e) 8 since there are 8 zeros.

Chapter 6

EXPONENTIAL AND LOGARITHMIC FUNCTIONS

6.1 One-to-One Functions; Inverse Functions

1. (a)

Domain		Range
\$200	→	20 hours
\$300	→	25 hours
\$350	→	30 hours
\$425	→	40 hours

(b) Inverse is a function.

3. (a)

Domain		Range
\$200	→	20 hours
\$200	→	25 hours
\$350	→	30 hours
\$425	→	40 hours

(b) Inverse not a function since \$200 corresponds to two elements in the range.

5. (a) $\{(6, 2), (6, -3), (9, 4), (10, 1)\}$ (b) Inverse not a function since 6 corresponds to 2 and -3.

7. (a) $\{(0, 0), (1, 1), (16, 2), (81, 3)\}$ (b) Inverse is a function.

9. Yes, any horizontal line intersects the graph of f at most in one point. *One-to-one*.

11. No, there are horizontal lines which intersect the graph of f at more than one point. *Not one-to-one*.

13. Yes, any horizontal line intersects the graph of f at most in one point. *One-to-one*.

15. Reflect about the line $y = x$.

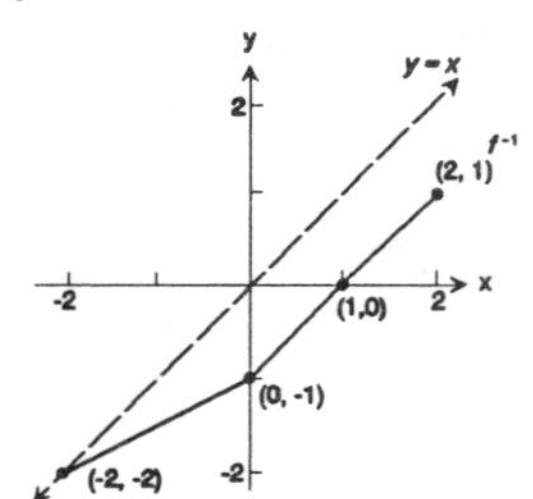

17. Reflect about the line $y = x$.

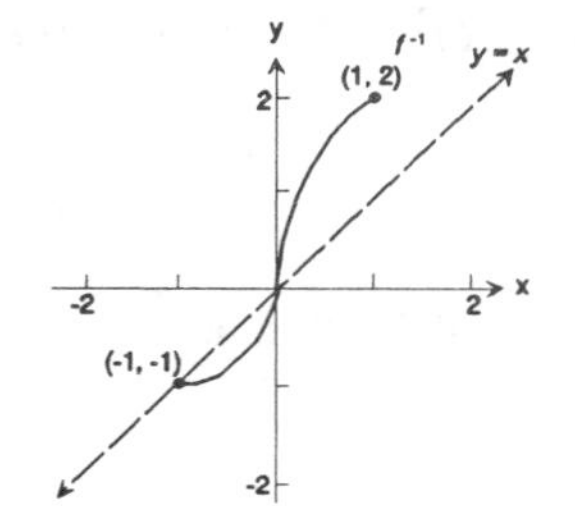

19. Reflect about the line $y = x$.

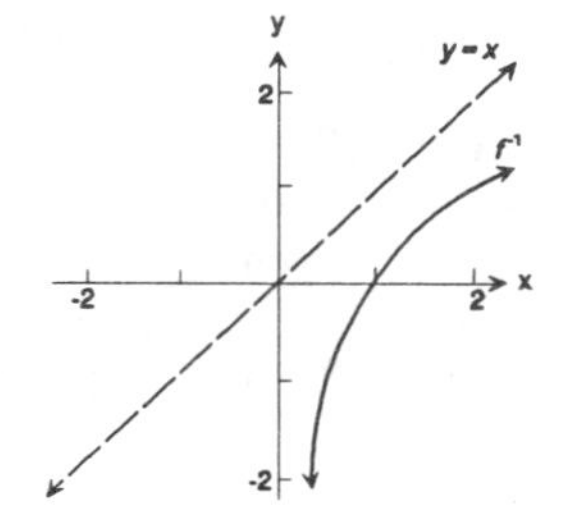

21. $f(x) = 3x + 4, \qquad g(x) = \frac{1}{3}(x - 4)$

$$f(g(x)) = f\left[\frac{1}{3}(x - 4)\right] = 3\left[\frac{1}{3}(x - 4)\right] + 4$$
$$= (x - 4) + 4 = x$$
$$g(f(x)) = g(3x + 4) = \frac{1}{3}(3x + 4 - 4) = \frac{1}{3}(3x) = x$$

23. $f(x) = 4x - 8, \qquad g(x) = \frac{x}{4} + 2$

$$f(g(x)) = f\left[\frac{x}{4} + 2\right] = 4\left[\frac{x}{4} + 2\right] - 8$$
$$= (x + 8) - 8 = x$$
$$g(f(x)) = g(4x - 8) = \frac{4x - 8}{4} + 2 = x - 2 + 2 = x$$

25. $f(x) = x^3 - 8, \qquad g(x) = \sqrt[3]{x + 8}$

$$f(g(x)) = f\left(\sqrt[3]{x + 8}\right) = \left[\sqrt[3]{x + 8}\right]^3 - 8$$
$$= (x + 8) - 8 = x$$
$$g(f(x)) = g(x^3 - 8) = \sqrt[3]{x^3 - 8 + 8} = \sqrt[3]{x^3} = x$$

27. $f(x) = \frac{1}{x}, \qquad g(x) = \frac{1}{x}$

$$f(g(x)) = f\left[\frac{1}{x}\right] = \frac{1}{\frac{1}{x}} = x$$
$$g(f(x)) = g\left[\frac{1}{x}\right] = \frac{1}{\frac{1}{x}} = x$$

29. $f(x) = \frac{2x + 3}{x + 4} \qquad g(x) = \frac{4x - 3}{2 - x}$

$$f(g(x)) = f\left[\frac{4x - 3}{2 - x}\right] = \frac{2\left[\frac{4x - 3}{2 - x}\right] + 3}{\frac{4x - 3}{2 - x} + 4} = \frac{\frac{2(4x - 3) + 3(2 - x)}{2 - x}}{\frac{(4x - 3) + 4(2 - x)}{2 - x}}$$
$$= \frac{\frac{8x - 6 + 6 - 3x}{2 - x}}{\frac{4x - 3 + 8 - 4x}{2 - x}} = \frac{\frac{5x}{2 - x}}{\frac{5}{2 - x}} = \frac{5x}{2 - x} \cdot \frac{(2 - x)}{5} = x$$
$$g(f(x)) = g\left[\frac{2x + 3}{x + 4}\right] = \frac{4\left[\frac{2x + 3}{x + 4}\right] - 3}{2 - \left[\frac{2x + 3}{x + 4}\right]} = \frac{\frac{4(2x + 3) - 3(x + 4)}{x + 4}}{\frac{2(x + 4) - (2x + 3)}{x + 4}}$$
$$= \frac{\frac{8x + 12 - 3x - 12}{x + 4}}{\frac{2x + 8 - 2x - 3}{x + 4}} = \frac{\frac{5x}{x + 4}}{\frac{5}{x + 4}} = \frac{5x}{x + 4} \cdot \frac{(x + 4)}{5} = x$$

31. $f(x) = 3x$

$y = 3x$

$x = 3y$

$y = \dfrac{x}{3}$

$f^{-1}(x) = \dfrac{x}{3}$

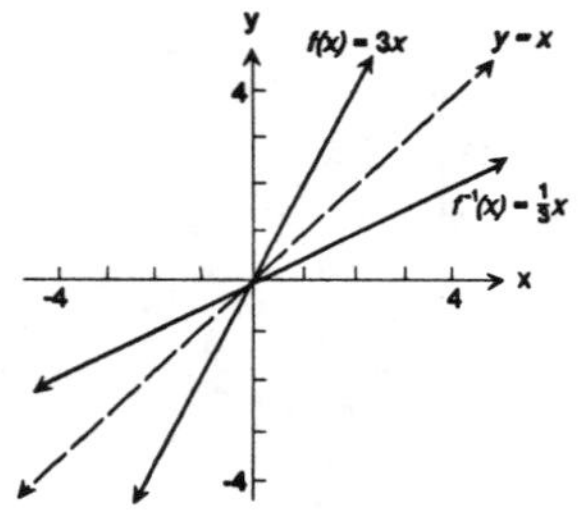

Verify: $f(f^{-1}(x)) = f\left[\dfrac{x}{3}\right] = 3\left[\dfrac{x}{3}\right] = x$

and $f^{-1}(f(x)) = f^{-1}(3x) = \dfrac{3x}{3} = x$

Domain of f = range of $f^{-1} = (-\infty, \infty)$

Range of f = domain of $f^{-1} = (-\infty, \infty)$

33. $f(x) = 4x + 2$

$y = 4x + 2$

$x = 4y + 2$

$4y = x - 2$

$y = \dfrac{x}{4} - \dfrac{1}{2}$

$f^{-1}(x) = \dfrac{x}{4} - \dfrac{1}{2}$

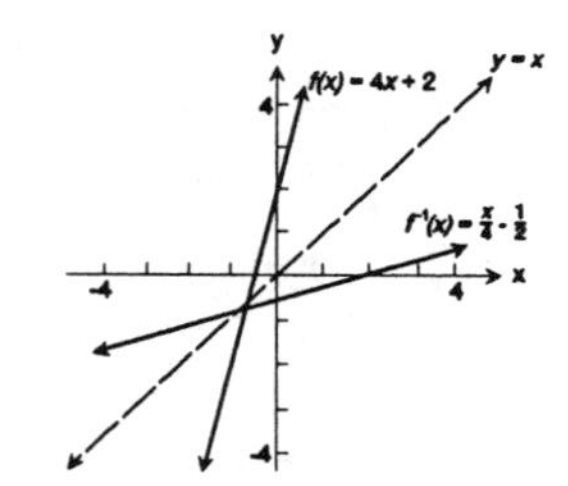

Verify:

$f(f^{-1}(x)) = f\left[\dfrac{x}{4} - \dfrac{1}{2}\right] = 4\left[\dfrac{x}{4} - \dfrac{1}{2}\right] + 2 = x - 2 + 2 = x$

$f^{-1}(f(x)) = f^{-1}(4x + 2) = \dfrac{4x + 2 - 2}{4} = x$

Domain of f = range of $f^{-1} = (-\infty, \infty)$

Range of f = domain of $f^{-1} = (-\infty, \infty)$

35. $f(x) = x^3 - 1$

$y = x^3 - 1$

$x = y^3 - 1$

$y^3 = x + 1$

$y = \sqrt[3]{x + 1}$

$f^{-1}(x) = \sqrt[3]{x + 1}$

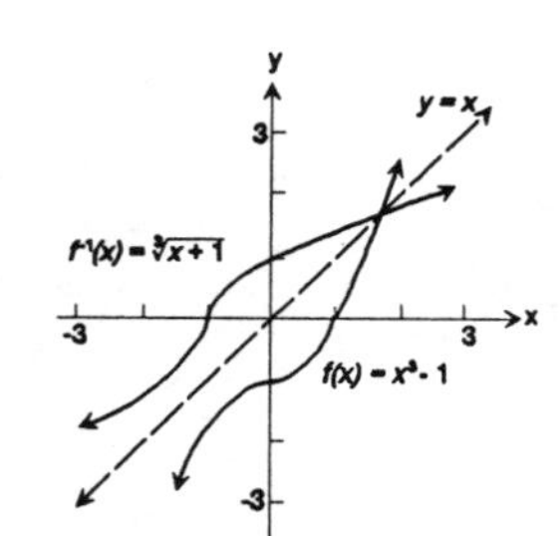

Verify:

$f(f^{-1}(x)) = f\left(\sqrt[3]{x + 1}\right) = \left(\sqrt[3]{x + 1}\right)^3 - 1 = x + 1 - 1 = x$

$f^{-1}(f(x)) = f^{-1}(x^3 - 1) = \sqrt[3]{x^3 - 1 + 1} = \sqrt[3]{x^3} = x$

Domain of f = range of $f^{-1} = (-\infty, \infty)$

Range of f = domain of $f^{-1} = (-\infty, \infty)$

37.

$$f(x) = x^2 + 4, x \ge 0$$
$$y = x^2 + 4$$
$$x = y^2 + 4$$
$$y^2 = x - 4$$
$$y = \sqrt{x - 4}$$
$$f^{-1}(x) = \sqrt{x - 4}$$

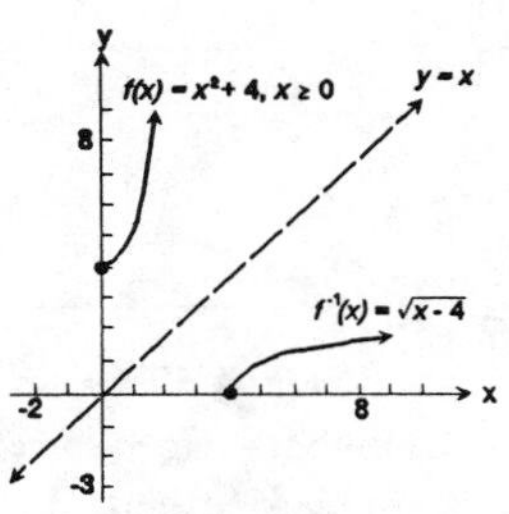

Verify:

$$f(f^{-1}(x)) = f\left(\sqrt{x - 4}\right) = \left(\sqrt{x - 4}\right)^2 + 4 = x - 4 + 4 = x$$

$$f^{-1}(f(x)) = f^{-1}(x^2 + 4) = \sqrt{x^2 + 4 - 4} = \sqrt{x^2} = |x| = x, x \ge 0$$

Domain of f = range of f^{-1} = $[0, \infty)$
Range of f = domain of f^{-1} = $[4, \infty)$

39.

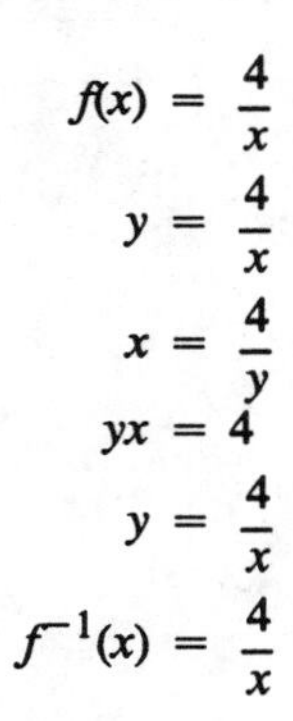

$$f(x) = \frac{4}{x}$$
$$y = \frac{4}{x}$$
$$x = \frac{4}{y}$$
$$yx = 4$$
$$y = \frac{4}{x}$$
$$f^{-1}(x) = \frac{4}{x}$$

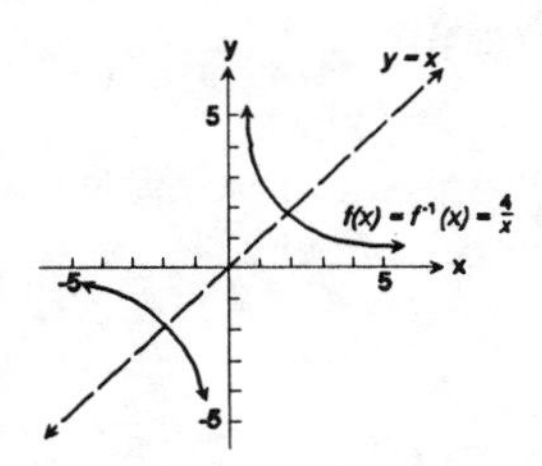

Verify: $$f(f^{-1}(x)) = f\left[\frac{4}{x}\right] = \frac{4}{\frac{4}{x}} = 4 \cdot \frac{x}{4} = x$$

$$f^{-1}(f(x)) = f^{-1}\left[\frac{4}{x}\right] = \frac{4}{\frac{4}{x}} = 4 \cdot \frac{x}{4} = x$$

Domain of f = range of f^{-1} = all real numbers except 0
Range of f = domain of f^{-1} = all real numbers except 0

41.

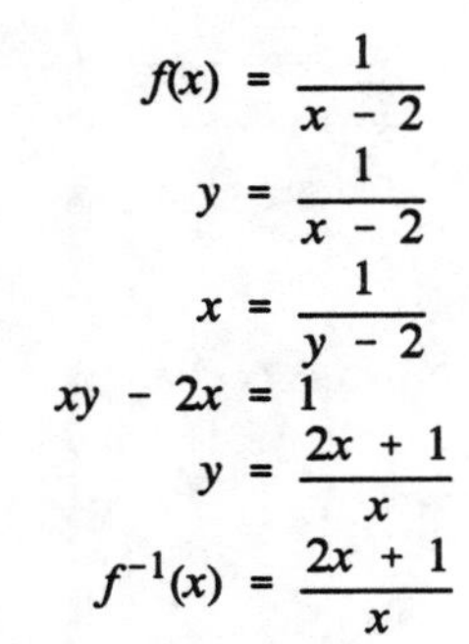

$$f(x) = \frac{1}{x - 2}$$
$$y = \frac{1}{x - 2}$$
$$x = \frac{1}{y - 2}$$
$$xy - 2x = 1$$
$$y = \frac{2x + 1}{x}$$
$$f^{-1}(x) = \frac{2x + 1}{x}$$

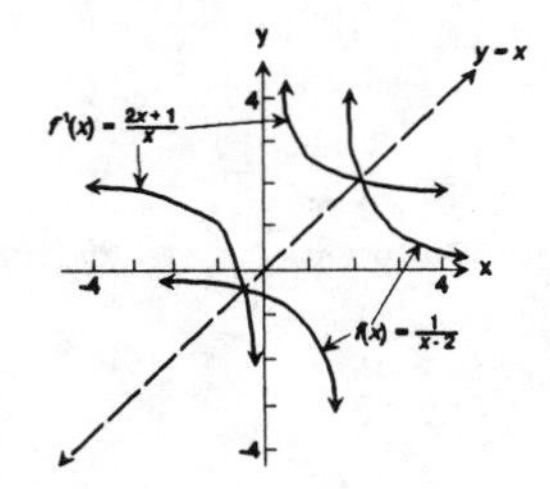

Verify: $f(f^{-1}(x)) = f\left[\dfrac{2x+1}{x}\right] = \dfrac{1}{\dfrac{2x+1}{x} - 2} = \dfrac{1}{\dfrac{2x+1-2x}{x}} = \dfrac{x}{1} = x$

$f^{-1}(f(x)) = f^{-1}\left[\dfrac{1}{x-2}\right] = \dfrac{2\left[\dfrac{1}{x-2}\right] + 1}{\dfrac{1}{x-2}} = \dfrac{\dfrac{2+(x-2)}{x-2}}{\dfrac{1}{x-2}} = x$

Domain of f = range of f^{-1} = all real numbers except 2
Range of f = domain of f^{-1} = all real numbers except 0

43.

$$f(x) = \frac{2}{3+x}$$
$$y = \frac{2}{3+x}$$
$$x = \frac{2}{3+y}$$
$$3x + xy = 2$$
$$y = \frac{2-3x}{x}$$
$$f^{-1}(x) = \frac{2-3x}{x}$$

Verify: $f(f^{-1}(x)) = f\left[\dfrac{2-3x}{x}\right] = \dfrac{2}{3 + \left[\dfrac{2-3x}{x}\right]} = \dfrac{2}{\dfrac{3x+2-3x}{x}} = x$

$f^{-1}(f(x)) = f^{-1}\left[\dfrac{2}{3+x}\right] = \dfrac{2 - 3\left[\dfrac{2}{3+x}\right]}{\dfrac{2}{3+x}} = \dfrac{\dfrac{6+2x-3}{3+x}}{\dfrac{2}{3+x}} = x$

Domain of f = range of f^{-1} = all real numbers except -3
Range of f = domain of f^{-1} = all real numbers except 0

45.

$$f(x) = (x+2)^2,\ x \geq -2$$
$$y = (x+2)^2,\ x \geq -2$$
$$x = (y+2)^2$$
$$\sqrt{x} = y + 2$$
$$y = \sqrt{x} - 2$$
$$f^{-1}(x) = \sqrt{x} - 2$$

Verify: $f(f^{-1}(x)) = f(\sqrt{x} - 2) = (\sqrt{x} - 2 + 2)^2 = \sqrt{x^2} = x$

$f^{-1}(f(x)) = f^{-1}[(x+2)^2] = \sqrt{(x+2)^2} - 2 = x + 2 - 2$
$= x$

Domain of f = range of f^{-1} = $[-2, \infty)$
Range of f = domain of f^{-1} = $[0, \infty)$

47.

$$f(x) = \frac{2x}{x-1}$$
$$y = \frac{2x}{x-1}$$
$$x = \frac{2y}{y-1}$$
$$xy - x = 2y$$
$$xy - 2y = x$$
$$y(x-2) = x$$
$$y = \frac{x}{x-2}$$
$$f^{-1}(x) = \frac{x}{x-2}$$

Verify: $f(f^{-1}(x)) = f\left[\dfrac{x}{x-2}\right] = \dfrac{2\left[\dfrac{x}{x-2}\right]}{\left[\dfrac{x}{x-2}\right] - 1} = \dfrac{\dfrac{2x}{x-2}}{\dfrac{x-(x-2)}{x-2}} = \dfrac{2x}{2} = x$

$f^{-1}(f(x)) = f^{-1}\left[\dfrac{2x}{x-1}\right] = \dfrac{\dfrac{2x}{x-1}}{\dfrac{2x}{x-1} - 2} = \dfrac{\dfrac{2x}{x-1}}{\dfrac{2x-2x+2}{x-1}} = \dfrac{2x}{2} = x$

Domain of f = range of f^{-1} = all real numbers except 1
Range of f = domain of f^{-1} = all real numbers except 2

49. $f(x) = \dfrac{3x+4}{2x-3}$

$y = \dfrac{3x+4}{2x-3}$

$x = \dfrac{3y+4}{2y-3}$

$2xy - 3x = 3y + 4$

$2xy - 3y = 3x + 4$

$y(2x-3) = 3x + 4$

$y = \dfrac{3x+4}{2x-3}$

$f^{-1}(x) = \dfrac{3x+4}{2x-3}$

Verify:

$$f(f^{-1}(x)) = f\left(\frac{3x+4}{2x-3}\right) = \frac{3\left(\frac{3x+4}{2x-3}\right)+4}{2\left(\frac{3x+4}{2x-3}\right)-3} = \frac{\frac{9x+12+8x-12}{2x-3}}{\frac{6x+8-6x+9}{2x-3}}$$

$$= \frac{17x}{2x-3} \cdot \frac{2x-3}{17} = \frac{17x}{17} = x$$

$$f^{-1}(f(x)) = f^{-1}\left(\frac{3x+4}{2x-3}\right) = \frac{3\left(\frac{3x+4}{2x-3}\right)+4}{2\left(\frac{3x+4}{2x-3}\right)-3} = \frac{\frac{9x+12+8x-12}{2x-3}}{\frac{6x+8-6x+9}{2x-3}}$$

$$= \frac{17x}{2x-3} \cdot \frac{2x-3}{17} = \frac{17x}{17} = x$$

Domain of f = range of f^{-1} = all real numbers except $\frac{3}{2}$

Range of f = domain of f^{-1} = all real numbers except $\frac{3}{2}$

51. $f(x) = \dfrac{2x+3}{x+2}$

$y = \dfrac{2x+3}{x+2}$

$x = \dfrac{2y+3}{y+2}$

$xy + 2x = 2y + 3$

$xy - 2y = 3 - 2x$

$y(x-2) = 3 - 2x$

$y = \dfrac{3-2x}{x-2}$

$f^{-1}(x) = \dfrac{3-2x}{x-2} = \dfrac{-2x+3}{x-2}$

Verify:

$$f(f^{-1}(x)) = f\left(\frac{3-2x}{x-2}\right) = \frac{2\left(\frac{3-2x}{x-2}\right)+3}{\frac{3-2x}{x-2}+2} = \frac{\frac{6-4x+3x-6}{x-2}}{\frac{3-2x+2x-4}{x-2}} = \frac{-x}{-1} = x$$

$$f^{-1}(f(x)) = f^{-1}\left(\frac{2x+3}{x+2}\right) = \frac{3-2\left(\frac{2x+3}{x+2}\right)}{\left(\frac{2x+3}{x+2}\right)-2} = \frac{\frac{3x+6-4x-6}{x+2}}{\frac{2x+3-2x-4}{x+2}} = \frac{-x}{-1} = x$$

Domain of f = range of f^{-1} = all real numbers except -2

Range of f = domain of f^{-1} = all real numbers except 2

53. $f(x) = 2\sqrt[3]{x}$

$y = 2\sqrt[3]{x}$

$x = 2\sqrt[3]{y}$

$x^3 = 8y$

$y = \dfrac{x^3}{8}$

$f^{-1}(x) = \dfrac{x^3}{8}$

Verify:

$$f(f^{-1}(x)) = f\left(\frac{x^3}{8}\right) = 2\sqrt[3]{\frac{x^3}{8}} = 2\left(\frac{x}{2}\right) = x$$

$$f^{-1}(f(x)) = f^{-1}\left(2\sqrt[3]{x}\right) = \frac{\left(2\sqrt[3]{x}\right)^3}{8} = \frac{8x}{8} = x$$

Domain of f = range of f^{-1} = $(-\infty, \infty)$

Range of f = domain of f^{-1} = $(-\infty, \infty)$

55.
$$f(x) = mx + b,\ m \neq 0$$
$$y = mx + b$$
$$x = my + b$$
$$my = x - b$$
$$y = \frac{x - b}{m}$$
$$f^{-1}(x) = \frac{x - b}{m},\ m \neq 0$$

57. No. If a function is even, $f(-x) = f(x)$. Whenever x and $-x$ are in the domain of f, two equal y values, $f(x)$ and $f(-x)$ are present.

59. f^{-1} also lies in quadrant one because whenever (a, b) is on f, then (b, a) is on f^{-1}. In quadrant one (a, b) is $(+, +)$, so (b, a) is $(+, +)$, also in quadrant one.

61. $f(x) = |x|,\ x \geq 0$ is one to one. Thus,
$$f(x) = x$$
$$f^{-1}(x) = x$$

63. $f(x) = \frac{9}{5}x + 32 \qquad g(x) = \frac{5}{9}(x - 32)$
$$f(g(x)) = f\left(\frac{5}{9}(x - 32)\right) = \frac{9}{5}\left[\frac{5}{9}(x - 32)\right] + 32 = x - 32 + 32 = x$$
$$g(f(x)) = g\left(\frac{9}{5}x + 32\right) = \frac{5}{9}\left(\frac{9}{5}x + 32 - 32\right) = x$$

65.
$$T(\ell) = 2\pi\sqrt{\frac{\ell}{g}},\ g \approx 32.2$$
$$T = 2\pi\sqrt{\frac{\ell}{g}}$$
$$\frac{T}{2\pi} = \sqrt{\frac{\ell}{g}}$$
$$\frac{T^2}{4\pi^2} = \frac{\ell}{g}$$
$$\frac{gT^2}{4\pi^2} = \ell$$
$$\ell = \frac{gT^2}{4\pi^2}$$
$$\ell(T) = \frac{gT^2}{4\pi^2}$$

67.
$$f(x) = \frac{ax + b}{cx + d}$$
$$y = \frac{ax + b}{cx + d}$$
$$x = \frac{ay + b}{cy + d}$$
$$cxy + dx = ay + b$$
$$cxy - ay = -dx + b$$
$$y(cx - a) = -dx + b$$
$$y = \frac{-dy + b}{cx - a}$$
$$f^{-1}(x) = \frac{-dx + b}{cx - a}$$
$$f = f^{-1} \text{ if } \frac{ax + b}{cx + d} = \frac{-dx + b}{cx - a}$$
This is true if $a = -d$

6.2 Exponential Functions

1. (a) $3^{2.2} \approx 11.212$ (b) $3^{2.23} \approx 11.587$ (c) $3^{2.236} \approx 11.664$ (d) $3^{\sqrt{5}} \approx 11.665$

3. (a) $2^{3.14} \approx 8.815$ (b) $2^{3.141} \approx 8.821$ (c) $2^{3.1415} \approx 8.824$ (d) $2^{\pi} \approx 8.825$

5. (a) $3.1^{2.7} \approx 21.217$ (b) $3.14^{2.71} \approx 22.217$
(c) $3.141^{2.718} \approx 22.440$ (d) $\pi^{e} \approx 22.459$

7. 3.320 9. 0.427 11. B 13. D 15. A 17. E

19. $y = e^{-x}$
Using the graph of $y = e^x$, reflect about the y-axis.

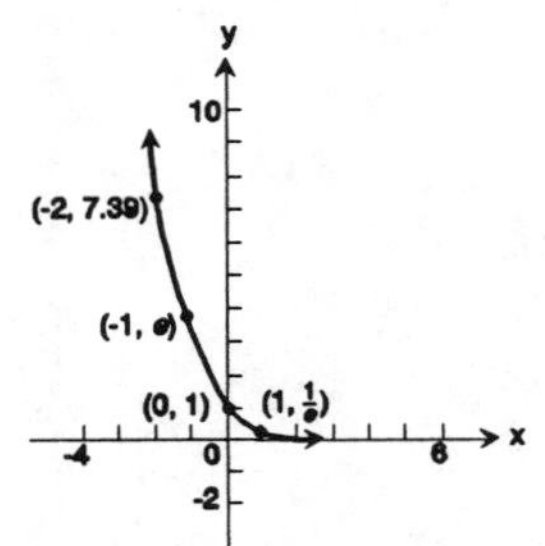

Domain: $(-\infty, \infty)$
Range: $(0, \infty)$
Horizontal Asymptote: $y = 0$

21. $y = e^{x+2}$
Using the graph of $y = e^x$, shift 2 units to the left.

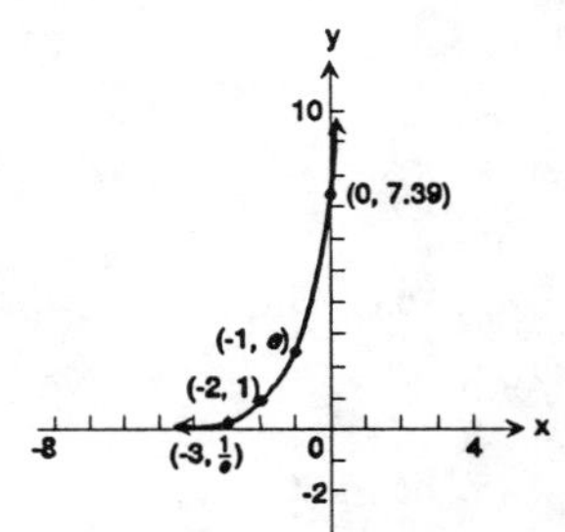

Domain: $(-\infty, \infty)$
Range: $(0, \infty)$
Horizontal Asymptote: $y = 0$

23. $y = 5 - e^{-x}$
Using the graph of $y = e^x$, reflect about the x- and y-axis, then shift up 5 units.

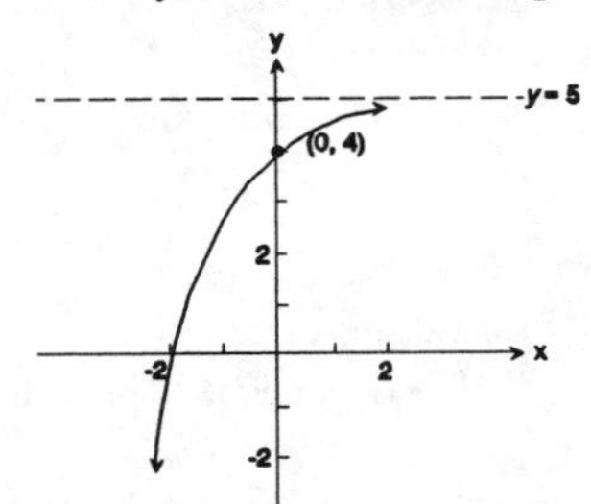

Domain: $(-\infty, \infty)$
Range: $(-\infty, 5)$
Horizontal Asymptote: $y = 5$

25. $y = 2 - e^{-x/2}$
Using the graph of $y = e^x$, vertically compress by a factor of $\frac{1}{2}$, reflect about the x-axis and y-axis, then shift up 2 units.

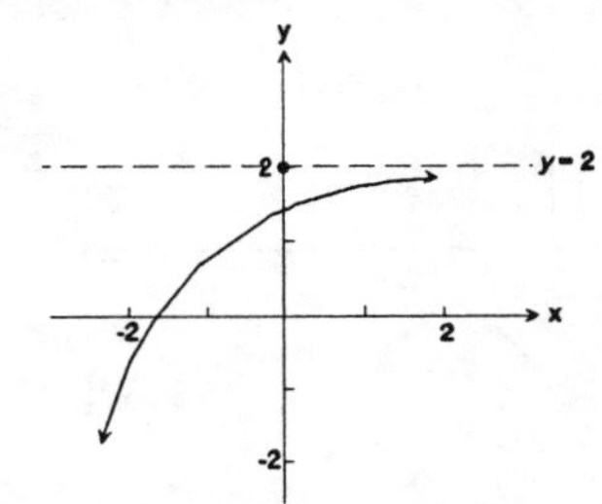

Domain: $(-\infty, \infty)$
Range: $(-\infty, 2)$
Horizontal Asymptote: $y = 2$

27.

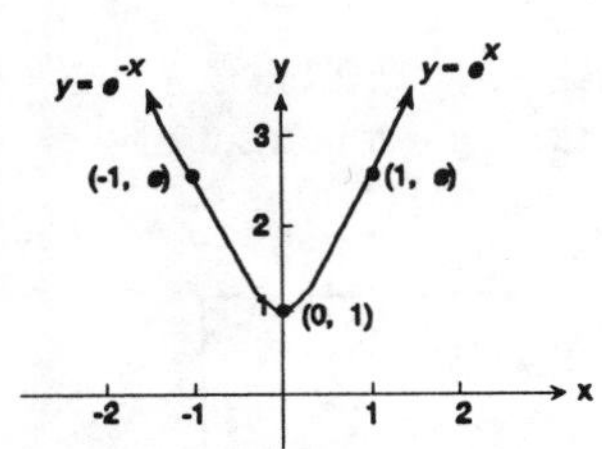

Domain: $(-\infty, \infty)$
Range: $[1, \infty)$
Intercept: $(0, 1)$

29.

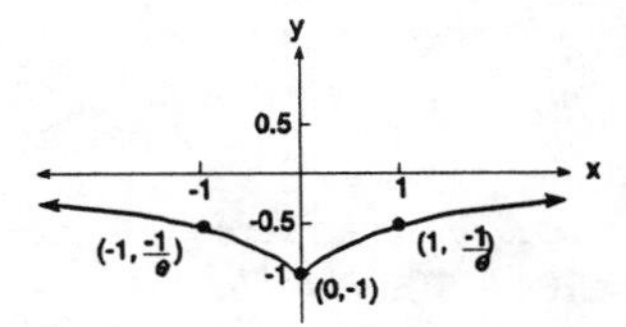

Domain: $(-\infty, \infty)$
Range: $(-1, 0]$
Intercept: $(0, -1)$

31.
$$\begin{aligned} 4^x &= 7 \\ (4^x)^{-2} &= 7^{-2} \\ 4^{-2x} &= \frac{1}{7^2} \\ 4^{-2x} &= \frac{1}{49} \end{aligned}$$

33.
$$\begin{aligned} 3^{-x} &= 2 \\ (3^{-x})^{-2} &= 2^{-2} \\ 3^{2x} &= \frac{1}{2^2} \\ 3^{2x} &= \frac{1}{4} \end{aligned}$$

35. $p = 100e^{-0.03n}$

(a)
$$\begin{aligned} p &= 100e^{-0.03(10)} \\ p &= 100e^{-0.3} \\ p &= 100(.741) \\ p &= 74\% \text{ of light} \end{aligned}$$

(b)
$$\begin{aligned} p &= 100e^{-0.03(25)} \\ p &= 100e^{-0.75} \\ p &= 100(.472) \\ p &= 47\% \text{ of light} \end{aligned}$$

37. $w = 50e^{-0.004d}$

(a)
$$\begin{aligned} w &= 50e^{-0.004(30)} \\ w &= 50e^{-0.12} \\ w &= 50(.887) \\ w &= 44 \text{ watts} \end{aligned}$$

(b)
$$\begin{aligned} w &= 50e^{-0.004(365)} \\ w &= 50e^{-1.46} \\ w &= 50(.232) \\ w &= 11.6 \text{ watts} \end{aligned}$$

39. $0 = 5e^{-0.4h}$

After 1 hour:
$$\begin{aligned} D &= 5e^{-0.4(1)} \\ D &= 5e^{-0.4} \\ D &= 5(.670) \\ D &= 3.35 \text{ milligrams} \end{aligned}$$

After 6 hours:
$$\begin{aligned} D &= 5e^{-0.4(6)} \\ D &= 5e^{-2.4} \\ D &= 5(.091) \\ D &= 0.45 \text{ milligrams} \end{aligned}$$

41. (a) $F(10) = 1 - e^{-0.1(10)} = 1 - e^{-1} = 0.63 = 63\%$

(b) $F(40) = 1 - e^{-0.1(40)} = 1 - e^{-4} = 0.98 = 98\%$

(c)

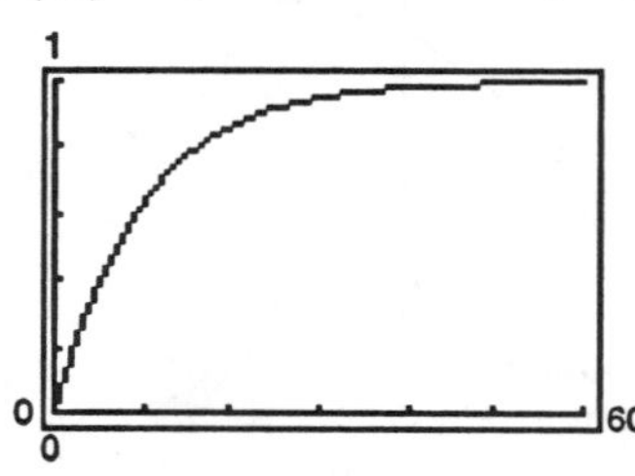

(d) Approximately 7 minutes

(e) As $t \to \infty$, $e^{-0.1t} \to 0$, so $F(t)$ approaches 1.

43. $R = 70 - 70e^{-0.2t}$

(a)
$$\begin{aligned} R &= 70 - 70e^{-0.2(10)} \\ R &= 70 - 70e^{-2} \\ R &= 70 - 70(.135) \\ R &= 60.5\% \text{ of viewers} \end{aligned}$$

(b)
$$\begin{aligned} R &= 70 - 70e^{-0.2(20)} \\ R &= 70 - 70e^{-4} \\ R &= 70 - 70(0.018) \\ &= 68.7\% \end{aligned}$$

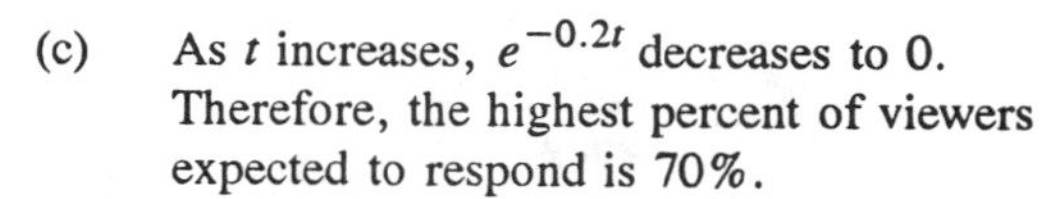

(c) As t increases, $e^{-0.2t}$ decreases to 0. Therefore, the highest percent of viewers expected to respond is 70%.

(d)

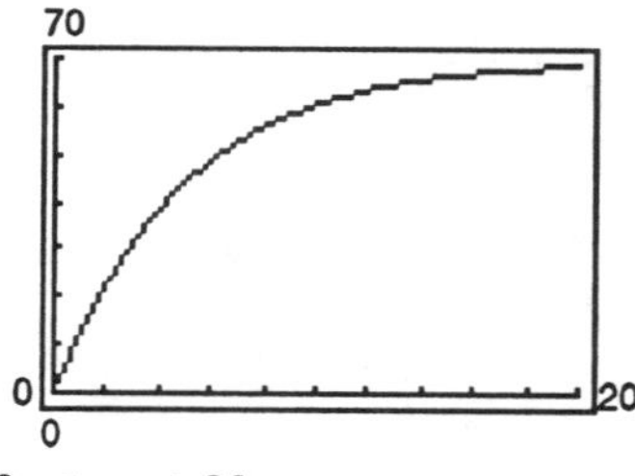

$0 \le x \le 20$
$0 \le y \le 70$

About $4\frac{1}{4}$ days.

45. $I = \frac{E}{R}[1 - e^{-(R/L)t}]$

(a) $I_1(t) = \frac{120}{10}[1 - e^{-(10/5)t}]$

After 0.3 second: $I = 5.414$ amperes

After 0.5 second: $I = 7.5854$ amperes

After 1 second: $I = 10.38$ amperes

(b) As t increases, $e^{-(10/5)t}$ goes to zero. Therefore, the maximum current is 12 amperes.

(c) See part (f)

(d) $I_2(t) = \frac{120}{5}[1 - e^{-(5/10)t}]$

After 0.3 second: $I = 3.343$ amperes

After 0.5 second: $I = 5.309$ amperes

After 1 second: $I = 9.443$ amperes

(e) The maximum current is 24 amperes.

(f)

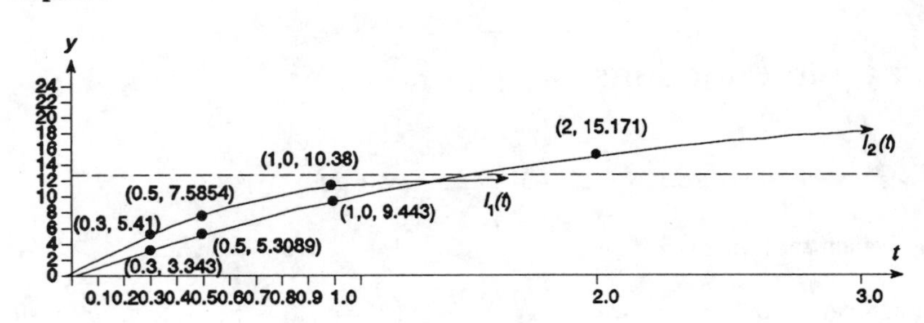

47. $y = \frac{6}{1 + e^{-(5.085 - 0.1156x)}}$

(a) $y = \frac{6}{1 + e^{-(5.085 - 0.1156(100))}}$
$= 9.23 \times 10^{-3}$

(b) $y = \frac{6}{1 + e^{-(5.085 - 0.1156(60))}} = 0.81$

(c) $y = \frac{6}{1 + e^{-(5.085 - 0.1156(30))}} = 5$

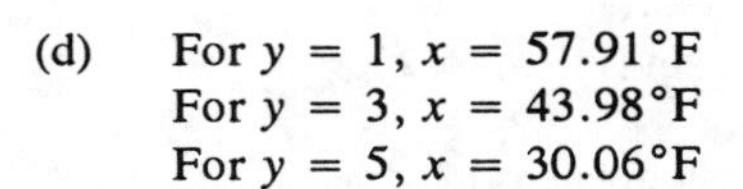
(d) For $y = 1$, $x = 57.91°F$
For $y = 3$, $x = 43.98°F$
For $y = 5$, $x = 30.06°F$

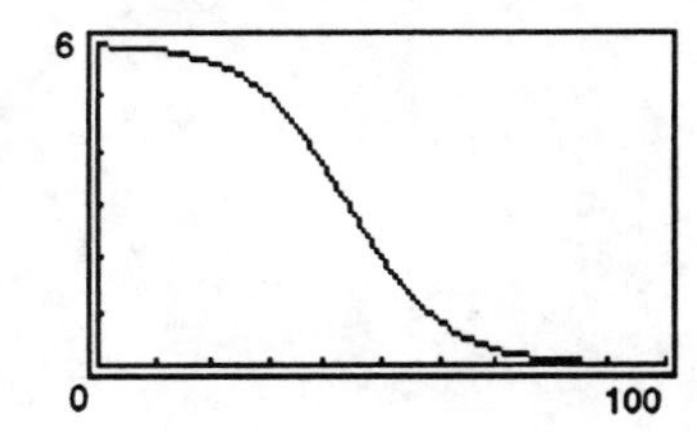

49. $2 + \frac{1}{2!} + \frac{1}{3!} + \ldots + \frac{1}{n!}$

$n = 4$, $2 + \frac{1}{2!} + \frac{1}{3!} + \frac{1}{4!} = 2.7083$

$n = 6$, $2 + \frac{1}{2!} + \frac{1}{3!} + \frac{1}{4!} + \frac{1}{5!} + \frac{1}{6!} = 2.7181$

$n = 8$, $2 + \frac{1}{2!} + \frac{1}{3!} + \frac{1}{4!} + \frac{1}{5!} + \frac{1}{6!} + \frac{1}{7!} + \frac{1}{8!} = 2.7182788$

$n = 10$, $2 + \frac{1}{2!} + \frac{1}{3!} + \frac{1}{4!} + \frac{1}{5!} + \frac{1}{6!} + \frac{1}{7!} + \frac{1}{8!} + \frac{1}{9!} + \frac{1}{10!} = 2.7182818$

$e = 2.718281828$

51. For $f(x) = a^x$, $\frac{f(x + h) - f(x)}{h} = \frac{a^{x+h} - a^x}{h} = \frac{a^x a^h - a^x}{h} = a^x\left(\frac{a^h - 1}{h}\right)$

53. For $f(x) = a^x$, $f(-x) = a^{-x} = \frac{1}{a^x} = \frac{1}{f(x)}$

55. $f(x) = 2^{(2x)} + 1$

$f(1) = 2^{(2^1)} + 1 = 2^2 + 1 = 4 + 1 = 5$

$f(2) = 2^{(2^2)} + 1 = 2^4 + 1 = 16 + 1 = 17$

$f(3) = 2^{(2^3)} + 1 = 2^8 + 1 = 256 + 1 = 257$

$f(4) = 2^{(2^4)} + 1 = 2^{16} + 1 = 65{,}536 + 1 = 65{,}537$

$f(5) = 2^{(2^5)} + 1 = 2^{32} + 1 = 4{,}294{,}967{,}296 + 1 = 4{,}294{,}967{,}297 = 641 \times 6{,}700{,}417$

6.3 Logarithmic Functions

In Problems 1–23, we use the equivalence of $a^x = M$ and $x = \log_a M$.

1. $9 = 3^2$ is equivalent to $2 = \log_3 9$

3. $a^2 = 1.6$ is equivalent to $2 = \log_a 1.6$

5. $1.1^2 = M$ is equivalent to $2 = \log_{1.1} M$

7. $2^x = 7.2$ is equivalent to $x = \log_2 7.2$

9. $x^{\sqrt{2}} = \pi$ is equivalent to $\sqrt{2} = \log_x \pi$

11. $e^x = 8$ is equivalent to $x = \ln 8$

13. $\log_2 8 = 3$ is equivalent to $2^3 = 8$

15. $\log_a 3 = 6$ is equivalent to $a^6 = 3$

17. $\log_3 2 = x$ is equivalent to $3^x = 2$

19. $\log_2 M = 1.3$ is equivalent to $2^{1.3} = M$

21. $\log_{\sqrt{2}} \pi = x$ is equivalent to $\left(\sqrt{2}\right)^x = \pi$

23. $\ln 4 = x$ is equivalent to $e^x = 4$

In Problems 25–35, we use (3)–(6):

25. $\log_2 1 = 0$

27. $\log_5 25 = \log_5 5^2 = 2$

29. $\log_{1/2} 16 = \log_{1/2} 2^4 = \log_{1/2} \left(\frac{1}{2}\right)^{-4}$
$= -4$

31. $\log_{10} \sqrt{10} = \log_{10} 10^{1/2} = \frac{1}{2}$

33. $\log_{\sqrt{2}} 4 = \log_{\sqrt{2}} \left(\sqrt{2}\right)^4$ since $4 = \left(\sqrt{2}\right)^4$
$= 4$

35. $\ln \sqrt{e} = \ln e^{1/2} = \frac{1}{2}$

37. For $f(x) = \ln(3 - x)$, the domain is all x such that

$$3 - x > 0$$
$$-x > -3$$
$$\{x \mid x < 3\}$$

y-intercept: $f(0) = \ln(3 - 0) = \ln 3$

x-intercept:
$$0 = \ln(3 - x)$$
$$3 - x = e^0$$
$$3 - x = 1$$
$$x = 2$$

39. For $F(x) = \log_2 x^2$, the domain is all x such that $x^2 > 0$, or all real numbers except zero.

y-intercept: None, zero is not in the domain of F.

x-intercept:
$$0 = \log_2 x^2$$
$$x^2 = 2^0$$
$$x^2 = 1$$
$$x = \pm 1$$

41. For $h(x) = \log_{1/2}(x^2 - 2x + 1)$, the domain is all x such that

$$x^2 - 2x + 1 > 0$$
$$(x - 1)^2 > 0$$

Domain: $\{x \mid x \neq 1\}$

y-intercept: $h(0) = \log_{1/2}(0^2 - 2(0) + 1) = \log_{1/2}1 = 0$

x-intercept: $0 = \log_{1/2}(x^2 - 2x + 1)$

$$\left(\frac{1}{2}\right)^0 = x^2 - 2x + 1$$
$$1 = x^2 - 2x + 1$$
$$0 = x^2 - 2x = x(x - 2)$$
$$x = 0, x = 2$$

43. For $g(x) = \log_5\left(\frac{x + 1}{x}\right)$, the domain is all x such that $\frac{x + 1}{x} > 0$.

	$x + 1$	x	$\frac{x + 1}{x}$
$x < -1$	$-$	$-$	$+$
$-1 < x < 0$	$+$	$-$	$-$
$x > 0$	$+$	$+$	$+$

$\frac{x + 1}{x} > 0$ when $x < -1$ or $x > 0$. The domain is $\{x \mid x < -1 \text{ or } x > 0\}$

y-intercept: None, $x = 0$ not in domain of g.

x-intercept: $0 = \log_5 \frac{x + 1}{x}$

$$5^0 = \frac{x + 1}{x}$$

$x = x + 1$ No solution, therefore no x-intercept.

45. 0.511

47. 30.099

49. For $f(x) = \log_a x$, we want to find a so that $f(2) = \log_a 2 = 2$ or $a^2 = 2$ or $a = \sqrt{2}$. Recall that $a > 0$ by definition.

51. B 53. D 55. A 57. E

59.

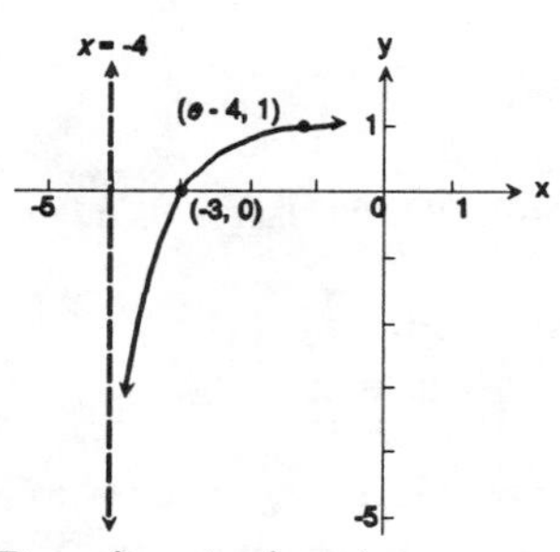

Domain: $(-4, \infty)$
Range: $(-\infty, \infty)$
Vertical Asymptote: $x = -4$

61.

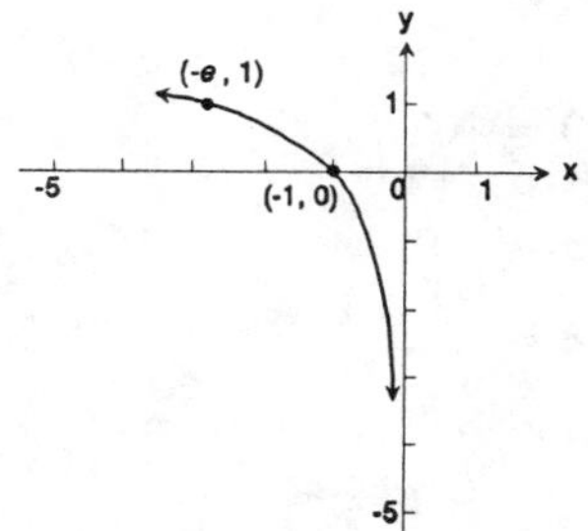

Domain: $(-\infty, 0)$
Range: $(-\infty, \infty)$
Vertical Asymptote: $x = 0$

63.

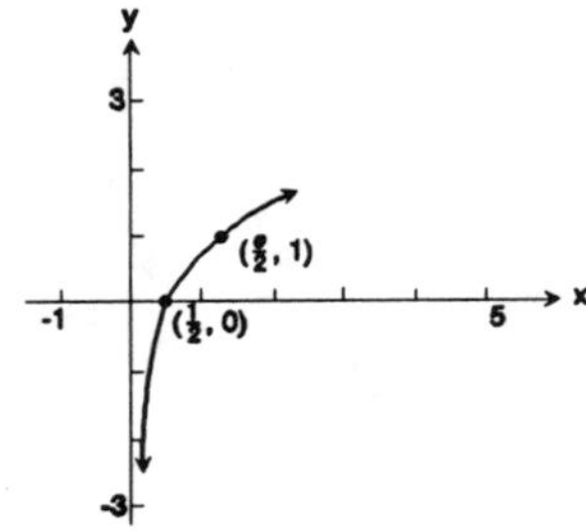

Domain: $(0, \infty)$
Range: $(-\infty, \infty)$
Vertical Asymptote: $x = 0$

65.

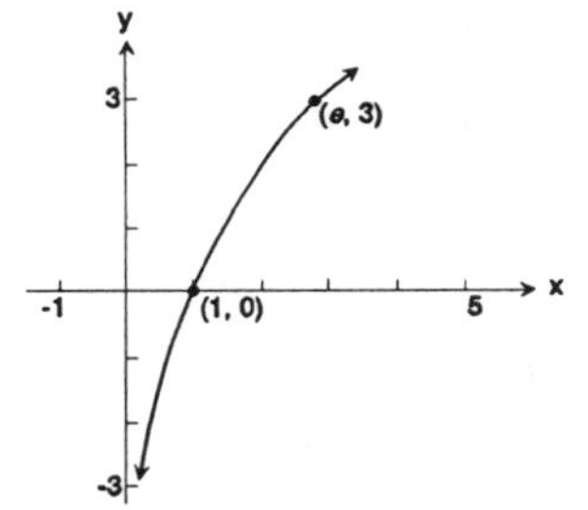

Domain: $(0, \infty)$
Range: $(-\infty, \infty)$
Vertical Asymptote: $x = 0$

67.

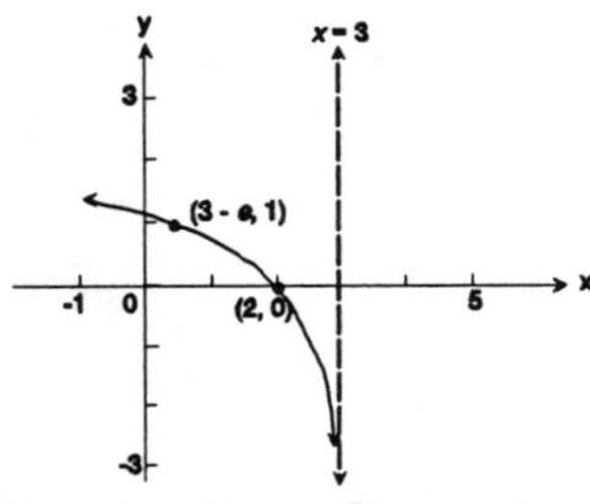

Domain: $(-\infty, 3)$
Range: $(-\infty, \infty)$
Vertical Asymptote: $x = 3$

69.

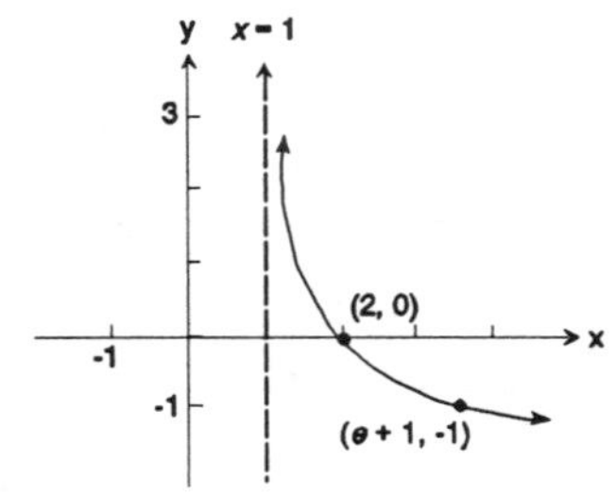

Domain: $(1, \infty)$
Range: $(-\infty, \infty)$
Vertical Asymptote: $x = 1$

71.

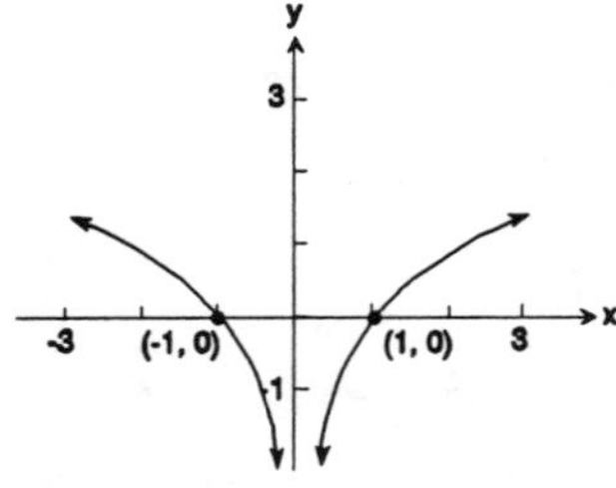

Domain: $\{x \mid x \neq 0\}$
Range: $(-\infty, \infty)$
Intercepts: $(-1, 0)$, $(1, 0)$

73.

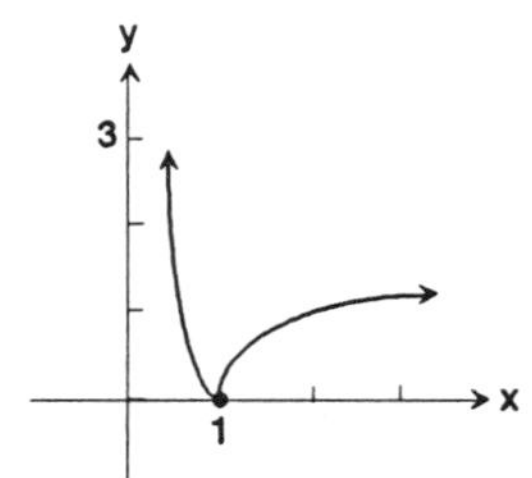

Domain: $(0, \infty)$
Range: $[0, \infty)$
Intercept: $(1, 0)$

75. $p = 100e^{-0.1n}$

(a)
$$50 = 100e^{-0.1n}$$
$$0.5 = e^{-0.1n}$$
$$\ln 0.5 = -0.1\,n$$
$$n = \frac{\ln 0.5}{-0.1}$$
$n = 6.93$, so 7 panes of glass are necessary.

(b)
$25 = 100e^{-0.1n}$ ($p = 25$ since 25% of the light passes through.)
$$0.25 = e^{-0.1n}$$
$$\ln 0.25 = -0.1\,n$$
$$n = \frac{\ln 0.25}{-0.1}$$
$n = 13.86$, so 14 panes of glass are necessary.

77. $w = 50e^{-0.004d}$

(a)
$$30 = 50e^{-0.004d}$$
$$0.6 = e^{-0.004d}$$
$$\ln 0.6 = -0.004\,d$$
$$d = \frac{\ln 0.6}{-0.004}$$
$d = 127.7$, so it takes about 128 days.

(b)
$$5 = 50e^{-0.004d}$$
$$0.1 = e^{-0.004d}$$
$$\ln 0.1 = -0.004\,d$$
$$d = \frac{\ln 0.1}{-0.004}$$
$d = 575.6$, so it takes about 576 days.

79.
$$D = 5e^{-0.4h}$$
$$2 = 5e^{-0.4h}$$
$$0.4 = e^{-0.4h}$$
$$\ln 0.4 = -0.4\,h$$
$$h = \frac{\ln 0.4}{-0.4}$$
$d = 2.29$ hours or 2 hours, 17 minutes

81. 0.5 ampere:
$$0.5 = \frac{12}{10}\left[1 - e^{-(10/5)t}\right]$$
$$0.4167 = 1 - e^{-2t}$$
$$e^{-2t} = 0.583$$
$$-2t = \ln 0.5833$$
$$t = \frac{\ln 0.5833}{-2}$$
$t = 0.2695$ second

1.0 ampere:
$$1.0 = \frac{12}{10}\left[1 - e^{-(10/5)t}\right]$$
$$0.8\overline{333} = 1 - e^{-2t}$$
$$e^{-2t} = 0.1\overline{666}$$
$$-2t = \ln 0.1\overline{666}$$
$$t = \frac{\ln 0.1\overline{666}}{-2}$$
$t = 0.8959$ second

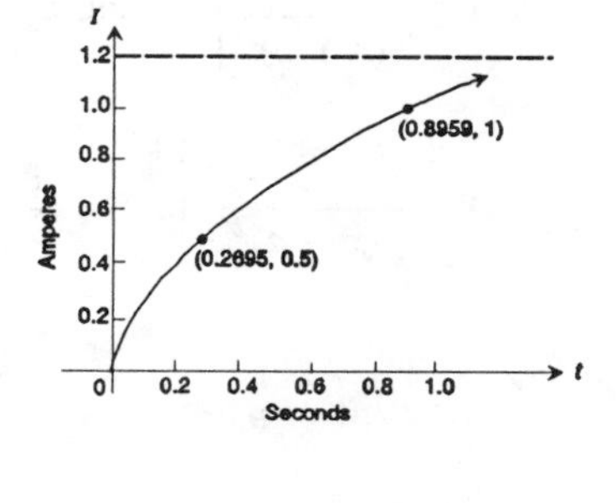

83. $r = 3e^{kx}$

(a)
$$10 = 3e^{k(0.06)}$$
$$3.\overline{333} = e^{k(0.06)}$$
$$\ln 3.\overline{3333} = k(0.06)$$
$$k = 20.07$$

(b)
$$R = 3e^{(20.07)(0.17)}$$
$$R = 3e^{3.4119}$$
$$R = 91\%$$

(c)
$$100 = 3e^{20.07x}$$
$$33.\overline{3333} = e^{20.07x}$$
$$\ln 33.\overline{3333} = 20.07x$$
$$x = 0.175$$

(d)
$$15 = 3e^{20.07x}$$
$$5 = e^{20.07x}$$
$$\ln 5 = 20.07x$$
$$x = 0.08$$

85. $y = 20e^{0.023t}$; $y = 89.2$ is predicted.

6.4 Properties of Logarithms; Curve Fitting

For Problems 1–11, we use $\ln 2 = a$ *and* $\ln 3 = b$.

1. $\ln 6 = \ln(3 \cdot 2) = \ln 3 + \ln 2 = b + a$

3. $\ln 1.5 = \ln\frac{3}{2} = \ln 3 - \ln 2 = b - a$

5. $\ln 2e = \ln 2 + \ln e = a + 1$

7. $\ln 12 = \ln(3 \cdot 4) = \ln 3 + \ln 4$
$= \ln 3 + \ln 2^2 = \ln 3 + 2\ln 2$
$= b + 2a$

9. $\ln\sqrt[5]{18} = \ln(18)^{1/5} = \frac{1}{5}\ln(2 \cdot 3^2) = \frac{1}{5}(\ln 2 + \ln 3^2) = \frac{1}{5}(\ln 2 + 2\ln 3) = \frac{1}{5}[a + 2b]$

11. $\log_2 3 = \dfrac{\log_e 3}{\log_e 2} = \dfrac{\ln 3}{\ln 2} = \dfrac{b}{a}$

13. $\ln\left[x^2\sqrt{1 - x}\right] = \ln x^2 + \ln\sqrt{1 - x} = \ln x^2 + \ln(1 - x)^{1/2} = 2\ln x + \frac{1}{2}\ln(1 - x)$

15. $\log_2\left(\dfrac{x^3}{x - 3}\right) = \log_2 x^3 - \log_2(x - 3) = 3\log_2 x - \log_2(x - 3)$

17. $\log\left[\dfrac{x(x + 2)}{(x + 3)^2}\right] = \log x(x + 2) - \log(x + 3)^2 = \log x + \log(x + 2) - 2\log(x + 3)$

19. $$\ln\left[\frac{x^2 - x - 2}{(x + 4)^2}\right]^{1/3} = \frac{1}{3}\ln\left(\frac{(x + 1)(x - 2)}{(x + 4)^2}\right) = \frac{1}{3}\left(\ln(x + 1)(x - 2) - \ln(x + 4)^2\right)$$
$$= \frac{1}{3}\left(\ln(x + 1) + \ln(x - 2) - 2\ln(x + 4)\right)$$
$$= \frac{1}{3}\ln(x + 1) + \frac{1}{3}\ln(x - 2) - \frac{2}{3}\ln(x + 4)$$

21. $$\ln\frac{5x\sqrt{1 - 3x}}{(x - 4)^3} = \ln\left(5x\sqrt{1 - 3x}\right) - \ln(x - 4)^3 = \ln 5 + \ln x + \ln\sqrt{1 - 3x} - \ln(x - 4)^3$$
$$= \ln 5 + \ln x + \ln(1 - 3x)^{1/2} - \ln(x - 4)^3$$
$$= \ln 5 + \ln x + \frac{1}{2}\ln(1 - 3x) - 3\ln(x - 4)$$

23. $3\log_5 u + 4\log_5 v = \log_5 u^3 + \log_5 v^4 = \log_5 u^3v^4$

25. $\log_{1/2}\sqrt{x} - \log_{1/2}x^3 = \log_{1/2}\dfrac{\sqrt{x}}{x^3} = \log_{1/2}x^{(1/2)-3} = \log_{1/2}x^{-5/2} = -\dfrac{5}{2}\log_{1/2}x$

27. $$\ln\left(\frac{x}{x - 1}\right) + \ln\left(\frac{x + 1}{x}\right) - \ln(x^2 - 1) = \ln\left[\frac{x}{x - 1} \cdot \frac{x + 1}{x}\right] - \ln(x^2 - 1)$$
$$= \ln\left[\frac{x + 1}{x - 1} \div (x^2 - 1)\right] = \ln\frac{1}{(x - 1)^2}$$
$$= \ln(x - 1)^{-2} = -2\ln(x - 1)$$

29. $$8\log_2\sqrt{3x - 2} - \log_2\left(\frac{4}{x}\right) + \log_2 4 = \log_2\left(\sqrt{3x - 2}\right)^8 - \log_2\frac{4}{x} + \log_2 4$$
$$= \log_2\frac{\left(\sqrt{3x - 2}\right)^8 \cdot 4}{\frac{4}{x}} = \log_2\left[x\left(\sqrt{3x - 2}\right)^8\right]$$
$$= \log_2\left[x\left((3x - 2)^{1/2}\right)^8\right] = \log_2\left[x(3x - 2)^4\right]$$

31. $2\log_a 5x^3 - \frac{1}{2}\log_a(2x+3) = \log_a(5x^3)^2 - \log_a(2x+3)^{1/2} = \log_a 25x^6 - \log_a\sqrt{2x+3}$

$$= \log_a\frac{25x^6}{\sqrt{2x+3}} = \log_a\left[\frac{25x^6}{(2x+3)^{1/2}}\right]$$

33. $3^{\log_3 5 - \log_3 4} = 3^{\log_3\left(\frac{5}{4}\right)} = \frac{5}{4}$

35. $e^{\log_{e^2} 16} = e^{\frac{\ln(16)}{\ln e^2}} = e^{\frac{\ln 16}{2\ln e}} = e^{\frac{\ln 16}{2}} = e^{\frac{1}{2}\ln 16} = e^{\ln 16^{1/2}} = e^{\ln 4} = 4$

37. $\log_3 21 = \frac{\log 21}{\log 3} = \frac{1.32222}{0.47712} = 2.771$

39. $\log_{1/3} 71 = \frac{\log 71}{\log\frac{1}{3}} = \frac{\log 7}{-\log 3} = \frac{1.85126}{-0.47712} = -3.880$

41. $\log_{\sqrt{2}} 7 = \frac{\log 7}{\log\sqrt{2}} = \frac{\log 7}{\log 2^{1/2}} = \frac{\log 7}{\frac{1}{2}\log 2} = \frac{0.84510}{0.5(0.30103)} = 5.615$

43. $\log_\pi e = \frac{\ln e}{\ln \pi} = \frac{1}{1.14473} = 0.874$

45. $\log_a\left(x + \sqrt{x^2-1}\right) + \log_a\left(x - \sqrt{x^2-1}\right)$

$$= \log_a\left(x + \sqrt{x^2-1}\right)\left(x - \sqrt{x^2-1}\right) = \log_a\left[x^2 - \left(\sqrt{x^2-1}\right)^2\right]$$

$$= \log_a\left(x^2 - (x^2-1)\right) = \log_a(x^2 - x^2 + 1) = \log_a 1 = 0$$

47. $\ln(1 + e^{2x}) = \ln[e^{2x}(e^{-2x} + 1)]$
$= \ln e^{2x} + \ln(e^{-2x} + 1)$
$= 2x + \ln(1 + e^{-2x})$

49. If $y = f(x) = \log_a x$,
$a^y = x$
$\left[\frac{1}{a}\right]^{-y} = x$
$-y = \log_{1/a} x$
Thus, $-f(x) = \log_{1/a} x$

51. If $f(x) = \log_a x$,
$f(AB) = \log_a AB$
$= \log_a A + \log_a B$
$= f(A) + f(B)$

53. $y = \log_4 x = \dfrac{\ln x}{\ln 4} = \dfrac{\log x}{\log 4}$

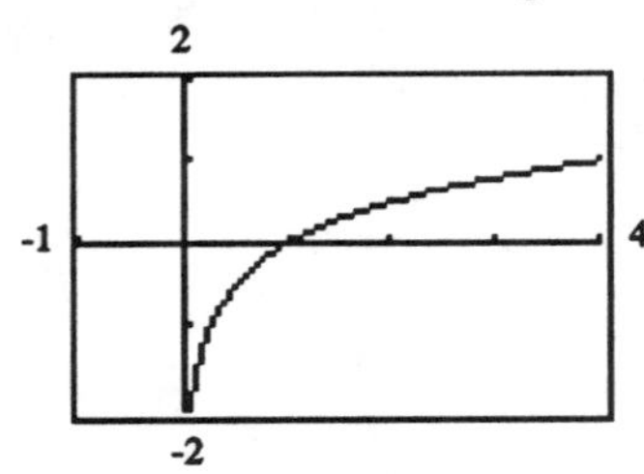

55. $y = \log_2(x + 2) = \dfrac{\ln(x + 2)}{\ln 2}$

$= \dfrac{\log(x + 2)}{\log 2}$

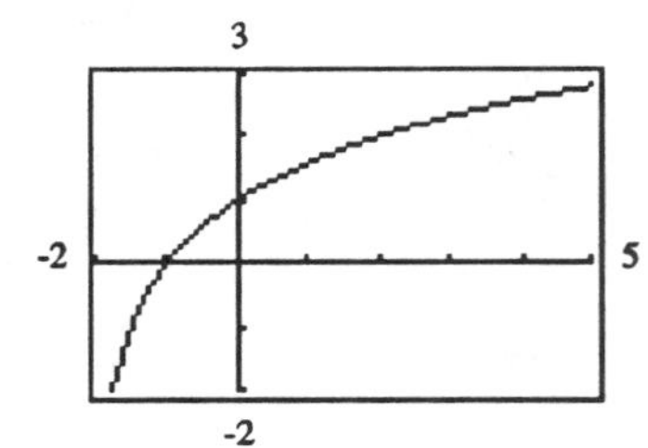

57.
$$\begin{aligned} \ln y &= \ln x + \ln C \\ \ln y &= \ln xC \\ y &= xC = Cx \end{aligned}$$

59.
$$\begin{aligned} \ln y &= \ln x + \ln(x + 1) + \ln C \\ \ln y &= \ln x(x + 1)C \\ y &= x(x + 1)C = Cx(x + 1) \end{aligned}$$

61.
$$\begin{aligned} \ln y &= 3x + \ln C \\ \ln y &= \ln e^{3x} + \ln C \\ \ln y &= \ln Ce^{3x} \\ y &= Ce^{3x} \end{aligned}$$

63.
$$\begin{aligned} \ln(y - 3) &= -4x + \ln C \\ \ln(y - 3) &= \ln e^{-4x} + \ln C \\ \ln(y - 3) &= \ln Ce^{-4x} \\ y - 3 &= Ce^{-4x} \\ y &= Ce^{-4x} + 3 \end{aligned}$$

65.
$$\begin{aligned} 3 \ln y &= \frac{1}{2} \ln(2x + 1) - \frac{1}{3} \ln(x + 4) + \ln C \\ \ln y^3 &= \ln(2x + 1)^{1/2} - \ln(x + 4)^{1/3} + \ln C \\ \ln y^3 &= \ln\left[\frac{C(2x + 1)^{1/2}}{(x + 4)^{1/3}}\right] \\ y^3 &= \left[\frac{C(2x + 1)^{1/2}}{(x + 4)^{1/3}}\right] \\ y &= \left[\frac{C(2x + 1)^{1/2}}{(x + 4)^{1/3}}\right]^{1/3} \\ &= \frac{\sqrt[3]{C}(2x + 1)^{1/6}}{(x + 4)^{1/9}} \end{aligned}$$

67. $\log_2 3 \cdot \log_3 4 \cdot \log_4 5 \cdot \log_5 6 \cdot \log_6 7 \cdot \log_7 8$

$$= \frac{\log 3}{\log 2} \cdot \frac{\log 4}{\log 3} \cdot \frac{\log 5}{\log 4} \cdot \frac{\log 6}{\log 5} \cdot \frac{\log 7}{\log 6} \cdot \frac{\log 8}{\log 7} = \frac{\log 8}{\log 2} = \frac{\log 2^3}{\log 2} = \frac{3 \log 2}{\log 2} = 3$$

69. $\log_2 3 \cdot \log_3 4 \cdot \cdots \cdot \log_n(n + 1) \cdot \log_{n+1} 2$

$$= \frac{\log 3}{\log 2} \cdot \frac{\log 4}{\log 3} \cdot \ldots \cdot \frac{\log n + 1}{\log n} \cdot \frac{\log 2}{\log n + 1} = \frac{\log 2}{\log 2} = 1$$

71. (a)

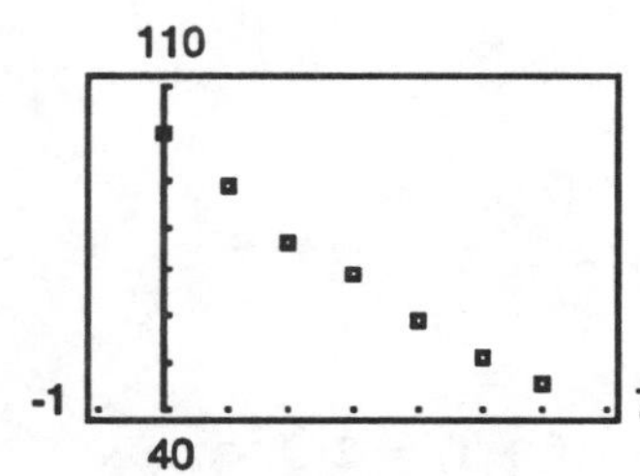

(b) $100(0.88)^t = A_0 e^{kt}$

Thus, $A_0 = 100$ and $0.88 = e^k$

$k = \ln 0.88 = -0.1278$

So, $A = 100e^{-0.1278t}$

(c) $50 = 100e^{-0.1278t}$

$\frac{1}{2} = e^{-0.1278t}$

$-0.1278t = \ln 1/2$

$t = \frac{\ln 1/2}{-0.1278} \approx 5.4$ weeks

(d) $A = 100e^{-0.1278(50)} \approx 0.17$ grams

(f)

73. (a)

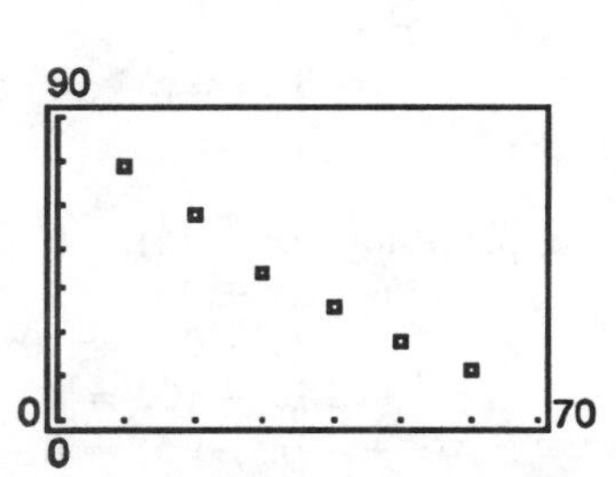

(b) $y = 96(0.98)^x$

$y = A_0 e^{kx}$

Comparing the two equations, we see

$A_0 = 96$ and $0.98^x = e^{kx}$, thus $0.98 = e^k$

$k = \ln 0.98 \approx -0.02$. Thus,

$y = 96e^{-0.02x}$

(c) $60 = 96e^{-0.02x}$ since y represents price.

$\frac{60}{96} = e^{-0.02x}$

$-0.02x = \ln(60/96)$

$x = \frac{\ln(60/96)}{-0.02} \approx 24$ shoes

(e)

75. (a)

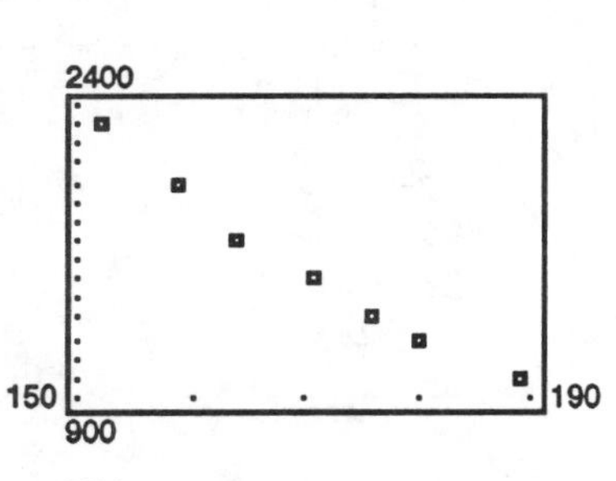

(b) $1650 = 32{,}741 - 6071 \ln x$

$-31091 = -6071 \ln x$

$\ln x = \frac{31091}{6071}$

$e^{(31091/6071)} = x$

$x \approx 168$ computers

(d)

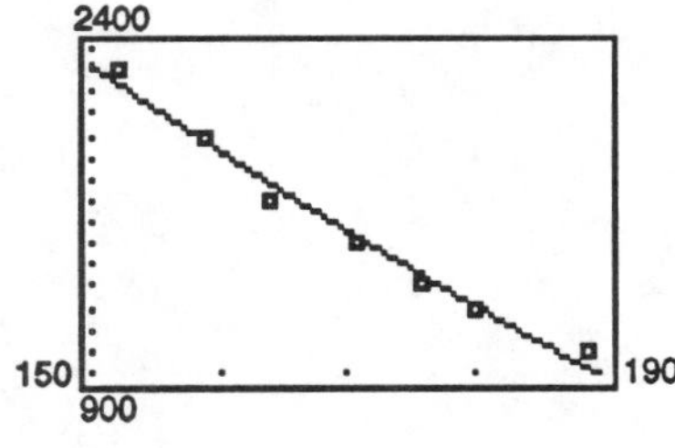

77. $\log_a\left[\frac{1}{N}\right] = \log_a N^{-1} = -\log_a N$, with $a \neq 1$

6.5 Logarithmic and Exponential Equations

1. $\log_2(2x + 1) = 3$
$2x + 1 = 2^3$
$2x + 1 = 8$
$2x = 7$
$x = \frac{7}{2}$

3. $\log_3(x^2 + 1) = 2$
$x^2 + 1 = 3^2$
$x^2 + 1 = 9$
$x^2 = 8$
$x = \pm 2\sqrt{2}$
$\{-2\sqrt{2}, 2\sqrt{2}\}$

5. $\frac{1}{2}\log_3 x = 2\log_3 2$
$\log_3 x^{1/2} = \log_3 2^2$
$x^{1/2} = 4$
$x = 16$

7. $2\log_5 x = 3\log_5 4$
$\log_5 x^2 = \log_5 4^3$
$x^2 = 64$
$x = 8$ since $x > 0$

9. $3\log_2(x - 1) + \log_2 4 = 5$
$\log_2(x - 1)^3 + \log_2 4 = 5$
$\log_2(x - 1)^3 \cdot 4 = 5$
$4(x - 1)^3 = 2^5$
$(x - 1)^3 = \frac{2^5}{4}$
$(x - 1)^3 = 8$
$x - 1 = 2$
$x = 3$

11. $\log_{10} x + \log_{10}(x + 15) = 2$
$\log_{10} x(x + 15) = 2$
$x(x + 15) = 10^2$
$x^2 + 15x - 100 = 0$
$(x + 20)(x - 5) = 0$
$x = -20$ or $x = 5$
Since $\log_{10}(-20)$ is undefined, we choose only $x = 5$.

13. $\log_x 4 = 2$
$4 = x^2$
$\pm 2 = x$
Since $\log_x(-2)$ is undefined, we have only $x = 2$.

15. $\log_3(x - 1)^2 = 2$
$(x - 1)^2 = 3^2$
$x - 1 = 3$ or $x = 1 - 3$
$x = 4$ or $x = -2$
$\{-2, 4\}$

17. $\log_{1/2}(3x + 1)^{1/3} = -2$
$(3x + 1)^{1/3} = \left[\frac{1}{2}\right]^{-2}$
$3x + 1 = \left[\frac{1}{2}\right]^{-6}$
$3x = 2^6 - 1$
$\frac{3x}{6} = \frac{63}{3}$
$x = 21$

19. $2^{2x+1} = 4$
$2^{2x+1} = 2^2$
$2x + 1 = 2$
$2x = 1$
$x = \frac{1}{2}$

21.

$$\begin{aligned} 3^{x^3} &= 9x \\ 3^{x^3} &= 3^{2x} \\ x^3 &= 2x \\ x^3 - 2x &= 0 \\ x(x^2 - 2) &= 0 \end{aligned}$$

$$x = 0 \qquad x^2 - 2 = 0$$

$$x^2 = 2$$

$$x = \pm\sqrt{2}$$

$$\{-\sqrt{2}, 0, \sqrt{2}\}$$

23.

$$\begin{aligned} 8^{x^2-2x} &= \frac{1}{2} \\ 2^{3(x^2-2x)} &= 2^{-1} \\ 3x^2 - 6x &= -1 \\ 3x^2 - 6x + 1 &= 0 \\ x &= \frac{6 \pm\sqrt{36 - 4(3)(1)}}{2(3)} \\ &= \frac{6 \pm \sqrt{24}}{6} = 1 \pm \frac{\sqrt{6}}{3} \end{aligned}$$

$$\left\{1 - \frac{\sqrt{6}}{3}, 1 + \frac{\sqrt{6}}{3}\right\}$$

25.

$$\begin{aligned} 2^x \cdot 8^{-x} &= 4^x \\ 2^x \cdot 2^{-3x} &= 2^{2x} \\ 2^{-2x} &= 2^{2x} \\ -2x &= 2x \\ x &= 0 \end{aligned}$$

27.

$$\begin{aligned} 2^{2x} + 2^x - 12 &= 0 \\ (2^x)^2 + 2^x - 12 &= 0 \\ (2^x + 4)(2^x - 3) &= 0 \end{aligned}$$

$2^x + 4 = 0$ or $2^x - 3 = 0$

$2^x = -4$ or $2^x = 3$

No solution $\qquad x = \log_2 3$

$$x = \frac{\ln 3}{\ln 2} \approx 1.585$$

29.

$$\begin{aligned} 3^{2x} + 3^{x+1} - 4 &= 0 \\ 3^{2x} + 3 \cdot 3^x - 4 &= 0 \\ (3^x + 4)(3^x - 1) &= 0 \end{aligned}$$

$3^x + 4 = 0$ or $3^x - 1 = 0$

No solution $\qquad 3^x = 1$

$$x = 0$$

31.

$$\begin{aligned} 4^x &= 8 \\ 2^{2x} &= 2^3 \\ 2x &= 3 \\ x &= \frac{3}{2} \end{aligned}$$

33.

$$\begin{aligned} 2^x &= 10 \\ \log 2^x &= \log 10 \\ x \log 2 &= 1 \\ x &= \frac{1}{\log 2} \\ &= 3.322 \end{aligned}$$

35.

$$\begin{aligned} 8^{-x} &= 1.2 \\ \log 8^{-x} &= \log 1.2 \\ -x \log 8 &= \log 1.2 \\ -x &= \frac{\log 1.2}{\log 8} \\ -x &= 0.088 \\ x &= -0.088 \end{aligned}$$

37.

$$\begin{aligned} 3^{1-2x} &= 4^x \\ \log 3^{1-2x} &= \log 4^x \\ (1 - 2x)\log 3 &= x \log 4 \\ \log 3 - 2x \log 3 &= x \log 4 \\ \log 3 &= x \log 4 + 2x \log 3 \\ \log 3 &= x (\log 4 + 2 \log 3) \\ \frac{\log 3}{\log 4 + 2 \log 3} &= x \end{aligned}$$

$$x = \frac{0.47712}{0.60206 + 2(0.47712)} = 0.307$$

39.

$$\left(\frac{3}{5}\right)^x = 7^{1-x}$$
$$\log\left(\frac{3}{5}\right)^x = \log 7^{1-x}$$
$$x \log\frac{3}{5} = (1 - x)\log 7$$
$$x \log\frac{3}{5} = \log 7 - x \log 7$$
$$x \log\frac{3}{5} + x \log 7 = \log 7$$
$$x\left[\log \frac{3}{5} + \log 7\right] = \log 7$$
$$x = \frac{\log 7}{\log \frac{3}{5} + \log 7} = \frac{\log 7}{\log 3 - \log 5 + \log 7} = 1.356$$

41.

$$1.2^x = (0.5)^{-x}$$
$$\log 1.2^x = \log (0.5)^{-x}$$
$$x \log 1.2 = -x \log 0.5$$
$$x \log 1.2 + x \log 0.5 = 0$$
$$x(\log 1.2 + \log 0.5) = 0$$
$$x = 0$$

43.

$$\pi^{1-x} = e^x$$
$$\ln \pi^{1-x} = \ln e^x$$
$$(1 - x)\ln \pi = x \ln e$$
$$\ln \pi - x \ln \pi = x(1)$$
$$\ln \pi = x + x \ln \pi$$
$$\ln \pi = x(1 + \ln \pi)$$
$$\frac{\ln \pi}{1 + \ln \pi} = x$$
$$0.534 = x$$

45.

$$5(2^{3x}) = 8$$
$$2^{3x} = \frac{8}{5}$$
$$\ln 2^{3x} = \ln \frac{8}{5}$$
$$3x \ln 2 = \ln \frac{8}{5}$$
$$x = \frac{\ln \frac{8}{5}}{3 \ln 2}$$
$$x = 0.226$$

47.

$$400e^{0.2x} = 600$$
$$e^{0.2x} = \frac{600}{400}$$
$$e^{0.2x} = \frac{3}{2}$$
$$0.2x = \ln \frac{3}{2}$$
$$x = \frac{\ln \frac{3}{2}}{0.2}$$
$$= 2.027$$

49.

$$\log_a(x - 1) - \log_a(x + 6) = \log_a(x - 2) - \log_a(x + 3)$$
$$\log_a\frac{x - 1}{x + 6} = \log_a\frac{x - 2}{x + 3}$$
$$\frac{x - 1}{x + 6} = \frac{x - 2}{x + 3}$$
$$(x - 1)(x + 3) = (x - 2)(x + 6)$$
$$x^2 + 2x - 3 = x^2 + 4x - 12$$
$$9 = 2x$$
$$\frac{9}{2} = x$$

51.

$$\log_{1/3}(x^2 + x) - \log_{1/3}(x^2 - x) = -1$$
$$\log_{1/3}\frac{x^2 + x}{x^2 - x} = -1$$
$$\frac{x^2 + x}{x^2 - x} = \left(\frac{1}{3}\right)^{-1}$$
$$\frac{x(x + 1)}{x(x - 1)} = 3$$
$$x + 1 = 3(x - 1)$$
$$x + 1 = 3x - 3$$
$$4 = 2x$$
$$2 = x$$

53.

$$\log_2 8^x = -3$$
$$x \log_2 8 = -3$$
$$x = \frac{-3}{\log_2 8}$$
$$= \frac{-3}{\log_2 2^3}$$
$$= \frac{-3}{3}$$
$$= -1$$

55.

$$\log_2(x^2+1) - \log_4 x^2 = 1$$
$$\log_2(x^2+1) - \frac{\log_2 x^2}{\log_2 4} = 1$$
$$\log_2(x^2+1) - \frac{\log_2 x^2}{2} = 1$$
$$\log_2(x^2+1) - \frac{1}{2}(\log_2 x^2) = 1$$
$$\log_2(x^2+1) - \log_2 x = 1$$
$$\log_2 \frac{x^2+1}{x} = 1$$
$$\frac{x^2+1}{x} = 2$$
$$x^2 - 2x + 1 = 0$$
$$(x-1)^2 = 0$$
$$x = 1$$

57.

$$\log_{16} x + \log_4 x + \log_2 x = 7$$
$$\frac{\log_2 x}{\log_2 16} + \frac{\log_2 x}{\log_2 4} + \log_2 x = 7$$
$$\frac{\log_2 x}{4} + \frac{\log_2 x}{2} + \log_2 x = 7$$
$$\left[\frac{1}{4} + \frac{1}{2} + 1\right] \log_2 x = 7$$
$$\frac{7}{4}\log_2 x = 7$$
$$\log_2 x^{\frac{7}{4}} = 7$$
$$x^{\frac{7}{4}} = 2^7$$
$$x^{\frac{1}{4}} = 2$$
$$x = 2^4 = 16$$

59.

$$\left(\sqrt[3]{2}\right)^{2-x} = 2^{x^2}$$
$$(2^{1/3})^{2-x} = 2^{x^2}$$
$$2^{\frac{2}{3}-\frac{1}{3}x} = 2^{x^2}$$
$$\frac{2}{3} - \frac{1}{3}x = x^2$$
$$2 - x = 3x^2$$
$$3x^2 + x - 2 = 0$$
$$(3x-2)(x+1) = 0$$
$$3x - 2 = 0 \text{ or } x + 1 = 0$$
$$x = \frac{2}{3} \text{ or } x = -1$$
$$\left\{-1, \frac{2}{3}\right\}$$

61. 1.92

63. 2.78

65. −0.56

67. −0.70

69. 0.56

71. 0.39, 1.00

73. 1.31

75. 1.30

6.6 Compound Interest

1. Here, $P = \$100$, $r = 0.04$, $n = 4$ and $t = 2$ in the formula

$$A = P\left[1 + \frac{r}{n}\right]^{nt} = 100\left[1 + \frac{0.04}{4}\right]^{(4)(2)} = \$108.29$$

3. Here, $P = \$500$, $r = 0.08$, $n = 4$ and $t = 2.5$ in

$$A = P\left[1 + \frac{r}{n}\right]^{nt} = 500\left[1 + \frac{0.08}{4}\right]^{4(2.5)} = \$609.50$$

5. Here, $P = \$600$, $r = 0.05$, $n = 365$ and $t = 3$ in

$$A = P\left[1 + \frac{r}{n}\right]^{nt} = 600\left[1 + \frac{0.05}{365}\right]^{(365)(3)} = \$697.09$$

7. Here $P = \$10$, $r = 0.11$, and $t = 2$ in

$$A = Pe^{rt} = 10e^{(0.11)(2)} = \$12.46$$

9. Here $P = \$100$, $r = 0.10$, and $t = 2.25$ in

$$A = Pe^{rt} = 100e^{(0.10)(2.25)} = \$125.23$$

11. Here, $A = \$100$, $t = 2$, $r = 0.06$, and $n = 12$ in

$$V = A\left[1 + \frac{r}{n}\right]^{-nt} = 100\left[1 + \frac{0.06}{12}\right]^{-12(2)} = \$88.72$$

13. Here, $A = \$1000$, $t = 2.5$, $r = 0.06$, and $n = 365$ in

$$V = A\left[1 + \frac{r}{n}\right]^{-nt} = 1000\left[1 + \frac{0.06}{365}\right]^{-365(2.5)} = \$860.72$$

15. Here, $A = \$600$, $t = 2$, $r = 0.04$, and $n = 4$ in

$$V = A\left[1 + \frac{r}{n}\right]^{-nt} = 600\left[1 + \frac{0.04}{4}\right]^{-4(2)} = \$554.09$$

17. Here $A = \$80$, $t = 3.25$, and $r = 0.09$ in

$$V = Ae^{-rt} = 80e^{-0.09(3.25)} = \$59.71$$

19. Here $A = \$400$, $t = 1$, and $r = 0.10$ in

$$V = Ae^{-rt} = 400e^{-0.10(1)} = \$361.93$$

21.
$$\begin{aligned} r_e &= \left[1 + \frac{.0525}{4}\right]^4 - 1 \\ &= 1.0535 - 1 \\ &= .0535 \\ &= 5.35\% \end{aligned}$$

23.
$$\begin{aligned} 2P &= P(1 + r)^3 \\ 2 &= (1 + r)^3 \\ \sqrt[3]{2} &= 1 + r \\ 1.26 &= 1 + r \\ r &= 26\% \end{aligned}$$

25. 6% compounded quarterly:

$$\begin{aligned} A &= \$10,000\left[1 + \frac{.06}{4}\right]^4 \\ &= \$10,000(1.0614) \\ &= \$10,614 \end{aligned}$$

$6\frac{1}{4}\%$ compounded annually:

$$\begin{aligned} A &= \$10,000(1 + .0625) \\ &= \$10,000(1.0625) \\ &= \$10,625 \end{aligned}$$

$6\frac{1}{4}\%$ compounded annually yields a larger amount.

27. 9% compounded monthly:

$$\begin{aligned} A &= \$10,000\left[1 + \frac{.09}{12}\right]^{12} \\ &= \$10,000(1.0938) \\ &= \$10,938 \end{aligned}$$

8.8% compounded daily:

$$\begin{aligned} A &= \$10,000\left[1 + \frac{.088}{365}\right]^{365} \\ &= \$10,000(1.0920) \\ &= \$10,920 \end{aligned}$$

9% compounded monthly yields a larger amount.

29. Compounded monthly:

$$2P = P\left[1 + \frac{.08}{12}\right]^{12t}$$
$$2 = (1.00\overline{66})^{12t}$$
$$\ln 2 = 12t \ln (1.00\overline{66})$$
$$12t = \frac{\ln 2}{\ln(1.00\overline{66})}$$
$$t = 104.32 \text{ months}$$

Compounded continuously:

$$2P = Pe^{.08t}$$
$$2 = e^{.08t}$$
$$\ln 2 = .08t$$
$$t = 8.66 \text{ years or } 103.97 \text{ months}$$

31. Compounded monthly:

$$\$150 = \$100\left[1 + \frac{.08}{12}\right]^{.12t}$$
$$1.5 = (1.00\overline{66})^{12t}$$
$$\ln 1.5 = 12t \ln (1.00\overline{66})$$
$$t = \frac{\ln 1.5}{12 \ln(1.00\overline{66})}$$
$$t = 5.0852 \text{ years or } 61.02 \text{ months}$$

Compounded continuously:

$$\$150 = \$100e^{.08t}$$
$$1.5 = e^{.08t}$$
$$\ln 1.5 = .08t$$
$$t = 5.0683 \text{ years or } 60.82 \text{ months}$$

33.

$$\$25{,}000 = \$10{,}000e^{.06t}$$
$$2.5 = e^{.06t}$$
$$\ln 2.5 = .06t$$
$$t = 15.27 \text{ years or 15 years, 4 months}$$

35.

$$A = \$90{,}000(1 + .03)^5$$
$$A = \$90{,}000(1.15927)$$
$$A \approx \$104{,}335$$

37.

$$P = \$15{,}000e^{-.05(3)}$$
$$= \$15{,}000(.86071)$$
$$= \$12{,}910.62$$

39.

$$A = \$1500(1 + .15)^5$$
$$= \$1500(2.01136)$$
$$\approx \$3017$$

41.

$$\$850{,}000 = \$650{,}000(1 + r)^3$$
$$1.3077 = (1 + r)^3$$
$$\sqrt[3]{1.3077} = 1 + r$$
$$r = .0935$$
$$r = 9.35\%$$

43. Compounded continuously:

$$A = \$1000e^{.056}$$
$$= \$1057.60$$

You do not quite have enough money.

Compounded monthly:

$$A = \$1000\left[1 + \frac{.059}{12}\right]^{12}$$
$$= \$1000(1.06062)$$
$$= \$1060.62$$

So, the second bank offers a better deal.

45. Will: $P = \$2000$, $r = 0.09$, $n = 2$, and $t = 20$ in

$$A = P\left[1 + \frac{r}{n}\right]^{nt} = 2000\left[1 + \frac{0.09}{2}\right]^{2(20)} = \$11{,}632.73$$

Henry: $P = \$2000$, $r = 0.085$, and $t = 20$ in

$$A = Pe^{rt} = 2000e^{0.085(20)} = \$10{,}947.89$$

Will has more money after 20 years.

47. Here $P = \$50{,}000$ and $t = 5$ in each of the following:

(a) Now $r = 0.12$ and $n = \dfrac{1}{5}$ (no compounding) in

$$A = P\left(1 + \frac{r}{n}\right)^{nt} = 50{,}000\left(1 + \frac{0.12}{\frac{1}{5}}\right)^{1/5(5)} = \$80{,}000.00$$

(b) Now $r = 0.115$ and $n = 12$ in

$$A = P\left(1 + \frac{r}{n}\right)^{nt} = 50{,}000\left(1 + \frac{0.115}{12}\right)^{12(5)} = \$88{,}613.59$$

(c) Now, $r = 0.1125$ in

$A = 50{,}000e^{0.1125(5)} = \$87{,}752.73$

Subtracting the original 50,000 from each, we get the interest of

(a) \$30,000.00
(b) \$38,613.59
(c) \$37,752.73

Option (a) results in the least interest.

49. (a)

$$\$10{,}000 = P\left(1 + \frac{.10}{12}\right)^{12(20)}$$
$$\$10{,}000 = P(7.328074)$$
$$P = \$1364.62$$

(b)

$$P = \$10{,}000e^{-.10(20)}$$
$$= \$1353.35$$

51.

$$\$10{,}000 = P(1 + .08)^{10}$$
$$P = \$10{,}000(1 + .08)^{-10}$$
$$= \$10{,}000(.4631935)$$
$$= \$4631.93$$

59. (a)

$$y = \frac{\ln 2}{1 \cdot \ln\left(1 + \frac{.12}{1}\right)}$$
$$y = \frac{\ln 2}{\ln(1.12)}$$
$$y = 6.1 \text{ years}$$

(b)

$$y = \frac{\ln 3}{4 \cdot \ln\left(1 + \frac{.06}{4}\right)}$$
$$y = \frac{\ln 3}{4 \cdot \ln 1.015}$$
$$y = 18.45 \text{ years}$$

(c)

$$mP = P\left(1 + \frac{r}{n}\right)^{nt}$$
$$m = \left(1 + \frac{r}{n}\right)^{nt}$$
$$\ln m = nt \ln\left(1 + \frac{r}{n}\right)$$
$$t = \frac{\ln m}{n \cdot \ln\left(1 + \frac{r}{n}\right)}$$

6.7 Growth and Decay

1. (a) For $P = 500e^{0.02t}$ we want to find t (in days)

when $P = 1000$:

$$
\begin{aligned}
1000 &= 500e^{0.02t} \\
2 &= e^{0.02t} \\
0.02t &= \ln 2 \\
t &= \frac{\ln 2}{0.02} \\
&= 34.7 \text{ days}
\end{aligned}
$$

when $P = 2000$:

$$
\begin{aligned}
2000 &= 500e^{0.02t} \\
4 &= e^{0.02t} \\
0.02t &= \ln 4 \\
t &= \frac{\ln 4}{0.02} = 69.3 \text{ days}
\end{aligned}
$$

3. (a) For $A = A_0 = e^{-0.0244t}$, the half-life is the time until $\frac{1}{2}A_0$ remains, so that

$$
\begin{aligned}
A = \frac{1}{2}A_0 &= A_0e^{-0.0244t} \\
\frac{1}{2} &= e^{-0.0244t} \\
-0.0244t &= \ln \frac{1}{2} \\
t &= \frac{-\ln 2}{-0.0244} = 28.4 \text{ years}
\end{aligned}
$$

(b)

$$
\begin{aligned}
10 &= 100e^{-0.0244t} \\
0.10 &= e^{-0.0244t} \\
\ln 0.1 &= -0.0244t \\
t &= \frac{\ln 0.1}{-0.0244} \approx 94.4 \text{ years}
\end{aligned}
$$

5. Using $N(t) = N_0e^{kt}$ where $N_0 = 1000$, $N(t) = 1800$, and $t = 1$, we get

$$
\begin{aligned}
1800 &= 1000e^{k1} \\
1.8 &= e^k \\
k &= \ln 1.8 \\
&= 0.5878
\end{aligned}
$$

Thus, $N(t) = N_0e^{0.5878t}$ and when $t = 3$

$$N(3) = 1000e^{0.5878(3)} = 1000e^{1.7634} = 5832 \text{ mosquitoes}$$

When $N(t) = 10{,}000$, we have

$$
\begin{aligned}
10{,}000 &= 1000e^{0.5878t} \\
10 &= e^{0.5878t} \\
0.5878t &= \ln 10 \\
t &= \frac{\ln 10}{0.5878} = 3.9 \text{ days}
\end{aligned}
$$

7. Using $A = A_0e^{kt}$ if after 18 months (= 1.5 years) we have $A = 2A_0$, then

$$2A_0 = A_0e^{k(1.5)}$$
$$2 = e^{1.5k}$$
$$1.5k = \ln 2$$
$$k = \frac{\ln 2}{1.5} = 0.4621$$

Here $A_0 = 10{,}000$ and when $t = 2$,

$$A = 10{,}000e^{0.4621(2)} = 10{,}000e^{0.9242} = 25{,}198 \text{ is the population 2 years from now.}$$

9. Using $A = A_0e^{kt}$ where the half-life, when $\frac{1}{2}A_0$ is present, is $t = 1690$ years, gives

$$\frac{1}{2}A_0 = A_0e^{k(1690)}$$
$$\frac{1}{2} = e^{1690k}$$
$$1690k = \ln \frac{1}{2}$$
$$k = \frac{-\ln 2}{1690} = -0.00041$$

If $A_0 = 10$, the amount present after 50 years is

$$A = 10e^{-0.00041(50)} = 10e^{-0.0205} = 9.797 \text{ grams}$$

11. (a) Using $A_0 = A_0e^{kt}$ the half-life $t = 5600$ years when $\frac{1}{2}A_0$ will be present:

$$\frac{1}{2}A_0 = A_0e^{k(5600)}$$
$$\frac{1}{2} = e^{5600k}$$
$$5600k = \ln \frac{1}{2}$$
$$k = \frac{-\ln 2}{5600} = -0.000124$$

Thus, $A = A_0e^{-0.000124t}$ and let $A_0 = 100$ so that $A = 30$:

$$30 = 100e^{-0.000124t}$$
$$0.3 = e^{-0.000124t}$$
$$-0.000124t = \ln 0.3$$
$$t = \frac{\ln 0.3}{-0.000124} = 9727 \text{ years ago}$$

13. (a) Using $u = T + (u_0 - T)e^{kt}$ where $t = 5$, $T = 70$, $u_0 = 450$, and $u = 300$:

$$300 = 70 + (450 - 70)e^{k(5)}$$
$$230 = 380e^{5k}$$
$$0.6053 = e^{5k}$$
$$5k = \ln 0.6053$$
$$k = -0.1004$$

Thus, $u = T + (u_0 - T)e^{-0.1004t} = 70 + (450 - 70)e^{-0.1004t} = 70 + 380e^{-0.1004t}$

And, when $u = 135$ (with $T = 70$, $u_0 = 450$ still)

$$135 = 70 + (450 - 70)e^{-0.1004t}$$
$$65 = 380e^{-0.1004t}$$
$$0.17105 = e^{-0.1004t}$$
$$-0.1004t = \ln 0.17105$$
$$t = 17.59 \text{ minutes past 5:00 p.m.,}$$
$$\approx 5{:}18 \text{ p.m.}$$

(b)
$$160 = 70 + 380e^{-0.1004t}$$
$$90 = 380e^{-0.1004t}$$
$$\frac{9}{38} = e^{-0.1004t}$$
$$-0.1004t = \ln\left[\frac{9}{38}\right]$$
$$t = \frac{\ln\left[\frac{9}{38}\right]}{-0.1004} = 14.3 \text{ minutes}$$

(c) As time passes, the temperature of the pizza gets closer to 70°F.

15. (a) Using $u = T + (u_0 - T)e^{kt}$ where $T = 35$, $u_0 = 8$, and $u = 15$, when $t = 3$:

$$15 = 35 + (8 - 35)e^{k(3)}$$
$$-20 = -27e^{3k}$$
$$0.7407 = e^{3k}$$
$$3k = \ln 0.7407$$
$$k = -0.100035$$

Thus, $u = T + (u_0 - T)e^{kt}$

$$= 35 + (8 - 35)e^{-0.100035t}$$

When $t = 5$, $u = 35 - 27e^{-0.100035(5)}$

$$= 18.63°\text{C}$$

When $t = 10$, $u = 35 - 27e^{-0.100035(10)}$

$$= 25.1°\text{C}$$

17. Using $A = A_0e^{kt}$ where $A_0 = 25$ and $A = 15$ when $t = 10$:

$$15 = 25e^{k(10)}$$
$$0.6 = e^{10k}$$
$$10k = \ln 0.6$$
$$k = -0.0511$$

Thus, $A = A_0e^{kt} = 25e^{-0.0511t}$

When $t = 24$, $A = 25e^{-0.0511(24)} = 7.34$ kg. remain.

When $A = \frac{1}{2}$kg, $\frac{1}{2} = 25e^{-0.0511t}$

$$.02 = e^{-0.0511t}$$
$$-0.0511t = \ln .02$$
$$t = 76.6 \text{ hours have passed.}$$

19. Using

$$A = A_0 e^{-0.087t}$$
$$.10A_0 = A_0 e^{-0.087t}$$
$$.10 = e^{-0.087t}$$
$$\ln .10 = -0.087t$$
$$t = \frac{\ln(.10)}{-0.087}$$
$$t \approx 26.5 \text{ days}$$

Farmers need to wait 26.5 days to use the hay.

21. (a) Since $t = 0$ represents 1984, evaluate $P(0)$.

$$P(0) = \frac{0.9}{1 + 6e^{-0.32(0)}} = \frac{0.9}{1 + 6} = \frac{0.9}{7} = 0.1286$$

(b) The maximum proportion of households is the carrying capacity, 0.9.

(c)

$$0.8 = \frac{0.9}{1 + 6e^{-0.32t}}$$
$$0.8(1 + 6e^{-0.32t}) = 0.9$$
$$1 + 6e^{-0.32t} = 1.125$$
$$6e^{-0.32t} = 0.125$$
$$e^{-0.32t} = 0.020833$$
$$-0.32t = \ln(0.020833)$$
$$t \approx 12.09$$

$t = 12$ corresponds to 1996. So 80% of the population will own VCRs in 1996.

23. (a) The carrying capacity is found by finding $P(t)$ as $t \to \infty$. As $t \to \infty$, $e^{-0.439t} \to 0$, so $P(t) \to \frac{1000}{1} = 1000$.

(b) The initial amount of bacteria is $P(0)$.

$$P(0) = \frac{1000}{1 + 32.33e^{-0.439(0)}} = \frac{1000}{1 + 32.33} = \frac{1000}{33.33} = 30$$

(c)

$$800 = \frac{1000}{1 + 32.33e^{-0.439t}}$$
$$1 + 32.33e^{-0.439t} = \frac{1000}{800}$$
$$32.33e^{-0.439t} = 0.25$$
$$e^{-0.439t} = 0.007733$$
$$-0.439t = \ln(0.007733)$$
$$t \approx 11.076 \text{ hours}$$

6.8 Logarithmic Scales

1.
$$L(10^{-5}) = 10 \log \frac{10^{-5}}{10^{-12}}$$
$$= 10 \log 10^7$$
$$= 10(7)$$
$$= 70 \text{ decibels}$$

3.
$$L(0.15) = 10 \log \frac{0.15}{10^{-12}}$$
$$= 10 \log (0.15(10^{12}))$$
$$= 10(\log 0.15 + \log 10^{12})$$
$$= 10(-0.8239 + 12)$$
$$= 111.76 \text{ decibels}$$

5.

$$L(x) = 10 \log \frac{x}{10^{-12}} = 130$$
$$\log (x(10^{12})) = 13$$
$$\log x + \log 10^{12} = 13$$
$$\log x + 12 = 13$$
$$\log x = 1$$
$$x = 10^1$$
$x = 10$ Watts per square meter

7. For $M(x) = \log \dfrac{x}{x_0}$ we have $x = 10.0$ and $x_0 = 10^{-3}$:

$$M(10.0) = \log \frac{10.0}{10^{-3}} = \log 10^4 = 4.0$$

on the Richter scale.

9. Using $M(x) = \log \dfrac{x}{x_0}$ we have $M(x) = 8.1$ and $x_0 = 10^{-3}$:

$$8.1 = \log \frac{x}{10^{-3}}$$
$$\frac{x}{10^{-3}} = 10^{8.1}$$
$$x = 10^{5.1} = 125{,}892.54 \text{ mm}$$

For $M(x) = 6.9 = \log \dfrac{x}{10^{-3}}$

$$\frac{x}{10^{-3}} = 10^{6.9}$$
$$x = 10^{3.9}$$

The Mexico City earthquake was $\dfrac{10^{5.1}}{10^{3.9}} = 10^{1.2} = 15.85$ times as intense as the one in San Francisco.

11. Intensity of Delta Center: $110 = 10 \log \left(\dfrac{x}{I_0}\right)$

$$11 = \log \left(\frac{x}{10^{-2}}\right)$$
$$10^{11} = \frac{x}{10^{-2}}$$
$$x = 10^{13}$$

Intensity of NBA guidelines: $95 = 10 \log \left(\dfrac{x}{I_0}\right)$

$$9.5 = \log \left(\frac{x}{10^{-2}}\right)$$
$$10^{9.5} = \frac{x}{10^{-2}}$$
$$x = 10^{11.5}$$

Ratio of Intensities: $\dfrac{10^{13}}{10^{11.5}} = 31.6$

The Delta Center crowd noise was 31.6 times as intense as NBA guidelines.

6 Chapter Review

1.
$$\begin{aligned} f(x) &= \frac{2x + 3}{5x - 2} \\ y &= \frac{2x + 3}{5x - 2} \\ x &= \frac{2y + 3}{5y - 2} \\ 5xy - 2x &= 2y + 3 \\ 5xy - 2y &= 2x + 3 \\ y(5x - 2) &= 2x + 3 \\ y &= \frac{2x + 3}{5x - 2} \\ f^{-1}(x) &= \frac{2x + 3}{5x - 2} \end{aligned}$$

Verify: $$f(f^{-1}(x)) = f\left(\frac{2x + 3}{5x - 2}\right) = \frac{2\left(\frac{2x + 3}{5x - 2}\right) + 3}{5\left(\frac{2x + 3}{5x - 2}\right) - 2} = \frac{\frac{4x + 6 + 15x - 6}{5x - 2}}{\frac{10x + 15 - 10x + 4}{5x - 2}} = \frac{19x}{19} = x$$

$$f^{-1}(f(x)) = f^{-1}\left(\frac{x + 3}{5x - 2}\right) = \frac{2\left(\frac{2x + 3}{5x - 2}\right) + 3}{5\left(\frac{2x + 3}{5x - 2}\right) - 2} = x$$

Domain of f = range of f^{-1} = all real numbers except $\frac{2}{5}$

Range of f = domain of f^{-1} = all real numbers except $\frac{2}{5}$

3.
$$\begin{aligned} f(x) &= \frac{1}{x - 1} \\ y &= \frac{1}{x - 2} \\ x &= \frac{1}{y + 1} \\ xy - x &= 1 \\ xy &= x + 1 \\ y &= x + 1 \\ f^{-1} &= \frac{x + 1}{x} \end{aligned}$$

Verify: $$f(f^{-1}(x)) = \frac{1}{\frac{x + 1}{x} + 1} = \frac{1}{\frac{x + 1 - x}{x}} = x$$

$$f^{-1}(f(x)) = \frac{\frac{1}{x - 1} + 1}{\frac{1}{x - 1}} = \frac{\frac{1 + x - 1}{x - 1}}{\frac{1}{x - 1}} = x$$

Domain of f = range of f^{-1} = all real numbers except 1

Range of f = domain of f^{-1} = all real numbers except 0

5. $$f(x) = \frac{3}{x^{1/3}} \qquad \text{Verify:} \quad f(f^{-1}(x)) = \frac{3}{\left[\frac{27}{x^3}\right]^{1/3}} = \frac{3}{\frac{3}{x}} = x$$

$$y = \frac{3}{x^{1/3}} \qquad f^{-1}(f(x)) = \frac{27}{\left[\frac{3}{x^{1/3}}\right]^3} = \frac{27}{\frac{27}{x}} = x$$

$$x = \frac{3}{y^{1/3}}$$

$$y^{1/3}x = 3$$

$$y^{1/3} = \frac{3}{x}$$

$$y = \frac{27}{x^3}$$

$$f^{-1}(x) = \frac{27}{x^3}$$

Domain of f = range of f^{-1} = all real numbers except 0
Range of f = domain of f^{-1} = all real numbers except 0

7. $\log_2 \frac{1}{8} = \log_2(2)^{-3} = -3$

9. $\ln e^{\sqrt{2}} = \sqrt{2}$

11. $2^{\log_2 0.4} = 0.4$

13. $$3 \log_4 x^2 + \frac{1}{2}\log_4\sqrt{x} = \log_4(x^2)^3 + \log_4\left(\sqrt{x}\right)^{1/2} = \log_4 x^6 + \log_4 x^{1/4} = \log_4 x^6 \cdot x^{1/4}$$
$$= \log_4 x^{25/4} = \frac{25}{4}\log_4 x$$

15. $$\ln\left[\frac{x-1}{x}\right] + \ln\left[\frac{x}{x+1}\right] - \ln(x^2-1) = \ln\frac{\frac{x-1}{x} \cdot \frac{x}{x+1}}{x^2-1} = \ln\frac{x-1}{(x^2-1)(x+1)}$$
$$= \ln\frac{1}{(x+1)^2} = \ln(x+1)^{-2} = -2\ln(x+1)$$

17. $$2\log 2 + 3\log x - \frac{1}{2}[\log(x+3) + \log(x-2)]$$
$$= \log 2^2 + \log x^3 - \frac{1}{2}[\log(x+3)(x-2)]$$
$$= \log 2^2x^3 - \log\left((x+3)(x-2)\right)^{1/2}$$
$$= \log\frac{4x^3}{\left((x+3)(x-2)\right)^{1/2}}$$

19. $$\ln y = 2x^2 + \ln C$$
$$\ln y = \ln e^{2x^2} + \ln C$$
$$\ln y = \ln Ce^{2x^2}$$
$$y = Ce^{2x^2}$$

21. $$\frac{1}{2}\ln y = 3x^2 + \ln C$$
$$\ln y^{1/2} = \ln e^{3x^2} + \ln C$$
$$\ln y^{1/2} = \ln Ce^{3x^2}$$
$$y^{1/2} = Ce^{3x^2}$$
$$y = \left(Ce^{3x^2}\right)^2$$

23. $$\ln(y-3) + \ln(y+3) = x + C$$
$$\ln(y-3)(y+3) = x + C$$
$$(y-3)(y+3) = e^{x+C}$$
$$y^2 - 9 = e^{x+C}$$
$$y^2 = e^{x+C} + 9$$
$$y = \sqrt{e^{x+C} + 9}$$

25.

$$\begin{aligned} e^{y+C} &= x^2 + 4 \\ \ln e^{y+C} &= \ln(x^2 + 4) \\ y + C &= \ln(x^2 + 4) \\ y &= \ln(x^2 + 4) - C \end{aligned}$$

27. The graph of $f(x) = e^{-x}$ is that of the reciprocal values of $y = e^x$.

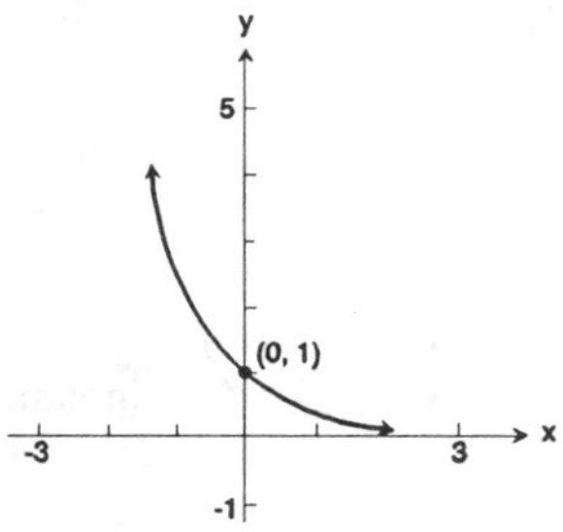

Domain: $(-\infty, \infty)$; Range: $(0, \infty)$;
Horizontal Asymptote: $y = 0$

29. The graph of $f(x) = 1 - e^x$ is that of $y = -e^x$ raised one unit. Note that the graph of $y = -e^x$ is the reflection of that of $y = e^x$ in the x-axis.

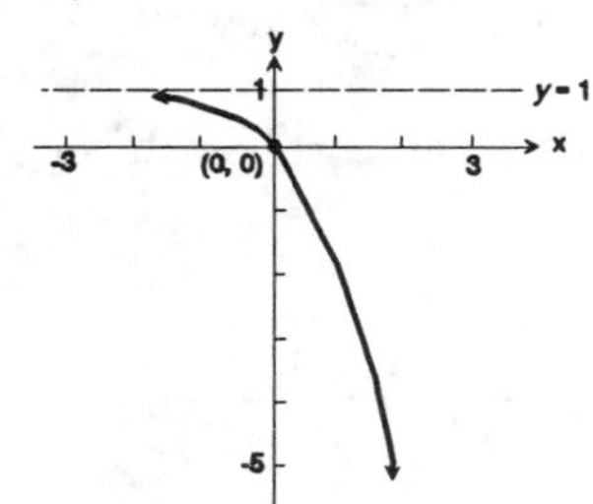

Domain: $(-\infty, \infty)$; Range: $(-\infty, 1)$;
Horizontal Asymptote: $y = 1$

31. The graph of $f(x) = 3e^x$ is that of $y = e^x$ with each y-value multiplied by 3.

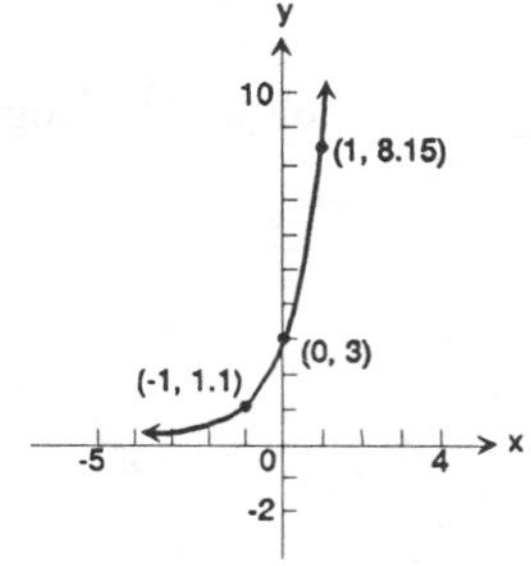

Domain: $(-\infty, \infty)$; Range: $(0, \infty)$
Horizontal Asymptote: $y = 1$

33. The graph of $f(x) = e^{|x|}$

$= \begin{cases} e^x & x \geq 0 \\ e^{-x} & x < 0 \end{cases}$ is that of $y = e^x$ in either direction from $x = 0$.

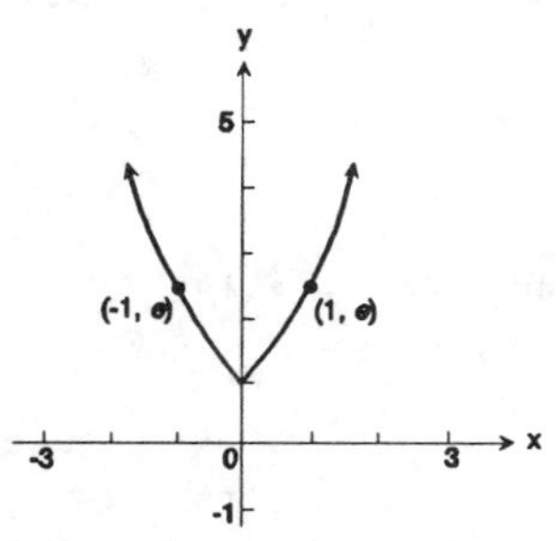

Domain: $(-\infty, \infty)$; Range; $[1, \infty)$
No Asymptotes

35. See discussion of Problem 23.

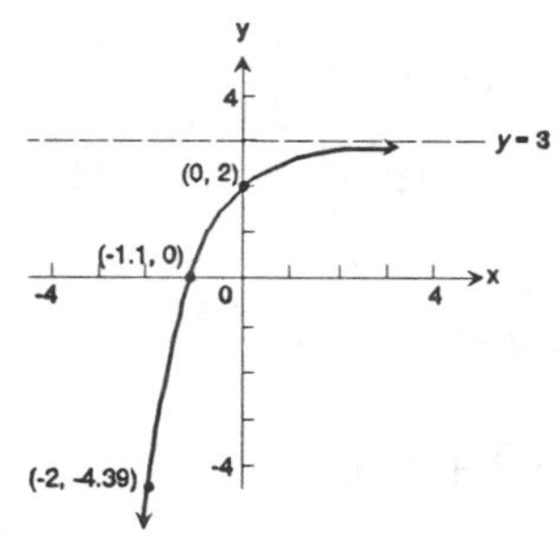

Domain: $(-\infty, \infty)$; Range: $(-\infty, 3)$
Horizontal Asymptote: $y = 3$

37.
$$\begin{aligned} 4^{1-2x} &= 2 \\ 2^{2(1-2x)} &= 2^1 \\ 2(1-2x) &= 1 \\ 2-4x &= 1 \\ -4x &= -1 \\ x &= \frac{1}{4} \end{aligned}$$

39.
$$\begin{aligned} 3^{x^2+x} &= \sqrt{3} \\ \log_3 3^{x^2+x} &= \log_3\sqrt{3} \\ (x^2+x)\log_3 3 &= \log_3 3^{1/2} \\ x^2+x &= \frac{1}{2} \\ x^2+x-\frac{1}{2} &= 0 \\ 2x^2+2x-1 &= 0 \end{aligned}$$

Here $a = 2$, $b = 2$, $c = -1$ and $b^3 - 4ac = 12$, so that

$$x = \frac{-2 \pm \sqrt{12}}{2(2)} = \frac{-2 \pm 2\sqrt{3}}{4} = \frac{-1 \pm \sqrt{3}}{2}$$

$$x = \frac{-1-\sqrt{3}}{2} \text{ or } x = \frac{-1+\sqrt{3}}{2}$$

$$\left\{\frac{-1-\sqrt{3}}{2}, \frac{-1+\sqrt{3}}{2}\right\}$$

41.
$$\begin{aligned} \log_x 64 &= -3 \\ 64 &= x^{-3} \\ 64^{-1/3} &= (x^{-3})^{-1/3} \\ \frac{1}{\sqrt[3]{64}} &= x \\ \frac{1}{4} &= x \end{aligned}$$

43.
$$\begin{aligned} 5^x &= 3^{x+2} \\ \log 5^x &= \log 3^{x+2} \\ x\log 5 &= (x+2)\log 3 \\ x\log 5 &= x\log 3 + 2\log 3 \\ x\log 5 - x\log 3 &= \log 3^2 \\ x(\log 5 - \log 3) &= \log 9 \\ x &= \frac{\log 9}{\log 5 - \log 3} = 4.301 \end{aligned}$$

45.
$$\begin{aligned} 9^{2x} &= 27^{3x-4} \\ 3^{2(2x)} &= 3^{3(3x-4)} \\ 2(2x) &= 3(3x-4) \\ 4x &= 9x - 12 \\ -5x &= -12 \\ x &= \frac{12}{5} \end{aligned}$$

47.
$$\begin{aligned} \log_3\sqrt{x-2} &= 2 \\ \sqrt{x-2} &= 3^2 \\ x-2 &= 9^2 \\ x &= 83 \end{aligned}$$

49.
$$\begin{aligned} 8 &= 4^{x^2} \cdot 2^{5x} \\ 8 &= 2^{2x^2} \cdot 2^{5x} \\ 2^3 &= 2^{2x^2+5x} \\ 3 &= 2x^2 + 5x \\ 0 &= 2x^2 + 5x - 3 \\ 0 &= (2x-1)(x+3) \end{aligned}$$

$$2x - 1 = 0 \quad \text{or} \quad x + 3 = 0$$

$$2x = 1$$

$$x = \frac{1}{2} \quad \text{or} \quad x = -3$$

$$\left\{-3, \frac{1}{2}\right\}$$

51.
$$\begin{aligned} \log_6(x+3) + \log_6(x+4) &= 1 \\ \log_6(x+3)(x+4) &= 1 \\ (x+3)(x+4) &= 6^1 \\ x^2+7x+12 &= 6 \\ x^2+7x+6 &= 0 \\ (x+1)(x+6) &= 0 \\ x = -1 \text{ or } x &= -6 \end{aligned}$$

But, -6 does not check, so $x = -1$ is the only solution.

53.
$$\begin{aligned} e^{1-x} &= 5 \\ \ln e^{1-x} &= \ln 5 \\ (1-x)\ln e &= \ln 5 \\ 1-x &= \ln 5 \\ 1-\ln 5 &= x \\ -0.609 &= x \end{aligned}$$

55.
$$\begin{aligned} 2^{3x} &= 3^{2x+1} \\ \log 2^{3x} &= \log 3^{2x+1} \\ 3x \log 2 &= (2x+1)\log 3 \\ 3x \log 2 &= 2x \log 3 + \log 3 \\ 3x \log 2 - 2x \log 3 &= \log 3 \\ x(3 \log 2) - 2 \log 3 &= \log 3 \\ x &= \frac{\log 3}{3 \log 2 - 2 \log 3} \\ x &= -9.327 \end{aligned}$$

57.
$$\begin{aligned} h &= (30 \cdot 0 + 8000)\log\left[\frac{760}{300}\right] \\ &= 8000(.403692) \\ &\approx 3229.5 \text{ meters} \end{aligned}$$

59.
$$\begin{aligned} 10{,}000 &= (30 \cdot (-100) + 8000)\log\left[\frac{760}{x}\right] \\ 10{,}000 &= 5000 \log\left[\frac{760}{x}\right] \\ 2 &= \log\left[\frac{760}{x}\right] \\ 10^2 &= \frac{760}{x} \\ x &= 7.6 \text{ mm Hg} \end{aligned}$$

61. $P = 25e^{0.1d}$

(a)
$$\begin{aligned} P &= 25e^{0.1(4)} \\ &= 37.3 \text{ watts} \end{aligned}$$

(b)
$$\begin{aligned} 50 &= 25e^{0.1d} \\ 2 &= e^{0.1d} \\ \ln 2 &= 0.1\,d \\ d &= 6.9 \text{ decibels} \end{aligned}$$

63. $P = 90 - 80\left[\frac{3}{4}\right]^t$

(a)
$$\begin{aligned} P &= 90 - 80\left[\frac{3}{4}\right]^5 \\ &= 71\% \end{aligned}$$

(b)
$$\begin{aligned} P &= 90 - 80\left[\frac{3}{4}\right]^{10} \\ &= 85.5\% \end{aligned}$$

(c) As t increases, $\left[\frac{3}{4}\right]^t$ approaches 0. Therefore, the maximum percent of purchases is 90%.

(d)
$$\begin{aligned} 40 &= 90 - 80\left[\frac{3}{4}\right]^t \\ -50 &= -80\left[\frac{3}{4}\right]^t \\ 0.625 &= 0.75t \\ \ln 0.625 &= t \ln 0.75 \\ t &= \frac{\ln 0.625}{\ln 0.75} \\ t &\approx 1.6 \text{ months} \end{aligned}$$

(e)
$$\begin{aligned} 70 &= 90 - 80\left[\frac{3}{4}\right]^t \\ -20 &= -80(0.75)^t \\ 0.25 &= 0.75t \\ \ln 0.25 &= t \ln 0.75 \\ t &= \frac{\ln 0.25}{\ln 0.75} \\ t &\approx 4.8 \text{ months} \end{aligned}$$

65. $n = \dfrac{\log_{10} s - \log_{10} i}{\log_{10}(1 - d)}$

(a) $n = \dfrac{\log_{10} 10{,}000 - \log_{10} 90{,}000}{\log_{10}(1 - 0.2)}$

$n = 9.85$ years

(b) $n = \dfrac{\log_{10} (.5i) - \log_{10}}{\log_{10}(1 - 0.15)}$

$n = \dfrac{\log_{10}\left[\dfrac{.5i}{i}\right]}{\log_{10}(.85)}$

$n = \dfrac{\log_{10}(0.5)}{\log_{10}(0.85)}$

$n = 4.27$ years

67. $\$85{,}000 = P\left[1 + \dfrac{.04}{2}\right]^{2(18)}$

$P = \$85{,}000\left[1 + \dfrac{.04}{2}\right]^{-36}$

$P = \$85{,}000(1.02)^{-36}$

$P = \$41.668.97$

The grandparents should purchase about $41,669.

69. Using $L(x) = 10 \log\dfrac{x}{I^0}$, where $I_0 = 10^{-12}$

$L(10^{-4}) = 10 \log\dfrac{10^{-4}}{10^{-12}}$

$= 10 \log 10^8$

$= 10(8)$

$= 80$ decibels

71. Using $A = A_0e^{kt}$ where $A = \dfrac{1}{2}A_0$ when $t = 5600$:

$\dfrac{1}{2}A_0 = A_0e^{k(5600)}$

$\dfrac{1}{2} = e^{5600k}$

$\ln \dfrac{1}{2} = \ln e^{5600k}$

$-\ln 2 = 5600k$

$k = \dfrac{-\ln 2}{5600} = -0.000124$

Thus, $A = A_0e^{-0.000124t}$

In this case, $A = 0.05A_0$:

$0.05A_0 = A_0e^{-0.000124t}$

$\ln 0.05 = \ln e^{-0.000124t}$

$\ln 0.05 = -0.000124t$

$t = \dfrac{\ln 0.05}{-0.000124t}$

$= 24{,}203$ years ago

73. (a)

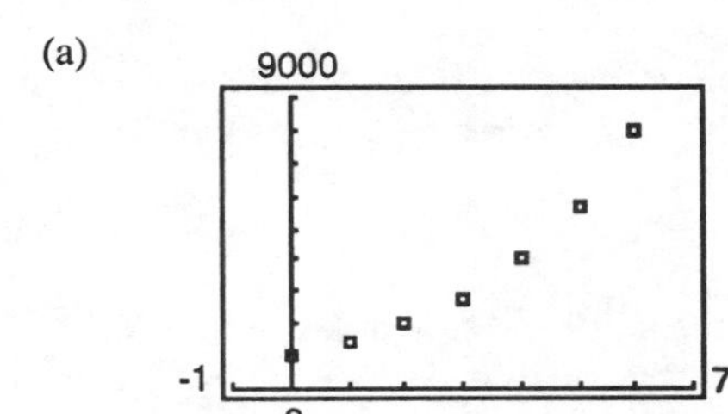

(b) $y = 1000\left(\sqrt{2}\right)^x$

$N = N_0e^{kt}$

Comparing the two equations, we have

$N_0 = 1000$ and $\sqrt{2}^{\,x} = e^{kt}$, thus

$\sqrt{2} = e^k$

$k = \ln \sqrt{2} = 0.347$. So,

$N = 1000e^{0.347t}$

(c) $N = 1000e^{0.347(7)} = 11{,}348$

(e)

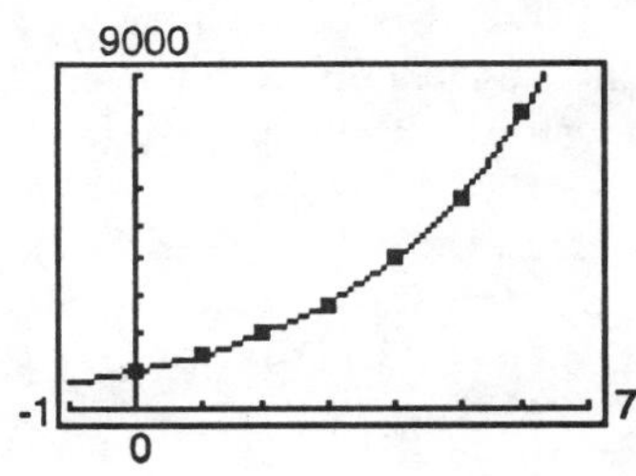

Chapter 7

THE CONICS

7.2 The Parabola

1. B 3. E 5. H 7. C

9. We want an equation of the parabola with focus at (4, 0) and vertex at (0, 0). The focus and vertex both lie on the horizontal line $y = 0$ (i.e., the x-axis). The distance, a, from (4, 0) to (0, 0) is $a = 4$. Also, the focus is to the right of the vertex, so the parabola opens to the right. Since the vertex is at (0, 0), the equation of the parabola is:

$$y^2 = 4ax \quad \text{(from Table 1)}$$

or $$y^2 = 16x$$

Letting $x = 4$, we find $y^2 = 64$ or $y = \pm 8$. The points (4, 8) and (4, −8) define the latus rectum.

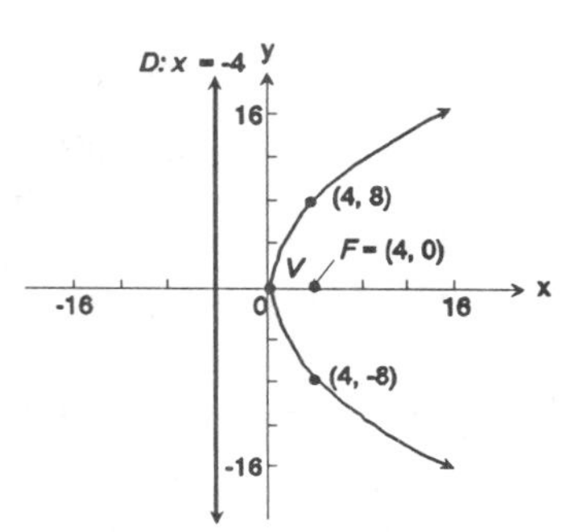

11. The focus, (0, −3), and the vertex, (0, 0), both lie on the vertical line $x = 0$ (the y-axis). We have $a = 3$, and since (0, −3) is ***below*** (0, 0), the parabola opens down. Therefore, from Table 1,

$$x^2 = -4ay$$

or $$x^2 = -12y \quad (a = 3)$$

Letting $y = -3$, we find $x^2 = 36$ or $x = \pm 6$. The points (−6, −3) and (6, −3) define the latus rectum.

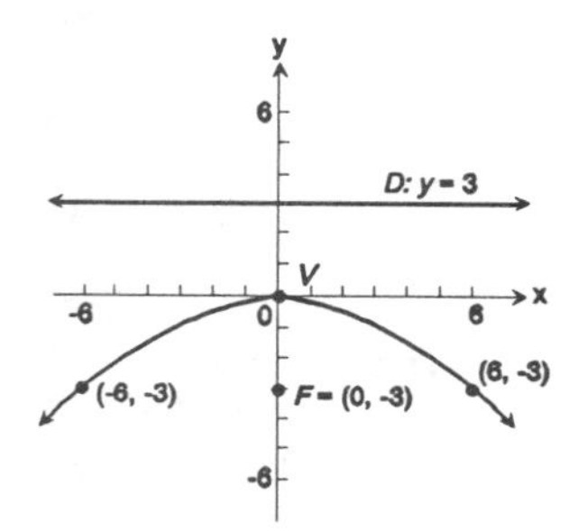

13. The vertex must be midway between the focus, (−2, 0), and the directrix, the line $x = 2$. Therefore, the vertex is the point (0, 0). The distance from the focus to the vertex is $a = 2$, and the parabola opens to the left, since (−2, 0) is to the ***left*** of the vertex. Therefore, we have:

$$y^2 = -4ax$$

or $$y^2 = -8x \quad (a = 2)$$

Letting $x = -2$, we find $y^2 = 16$ or $y = \pm 4$. The points (−2, 4) and (−2, −4) define the latus rectum.

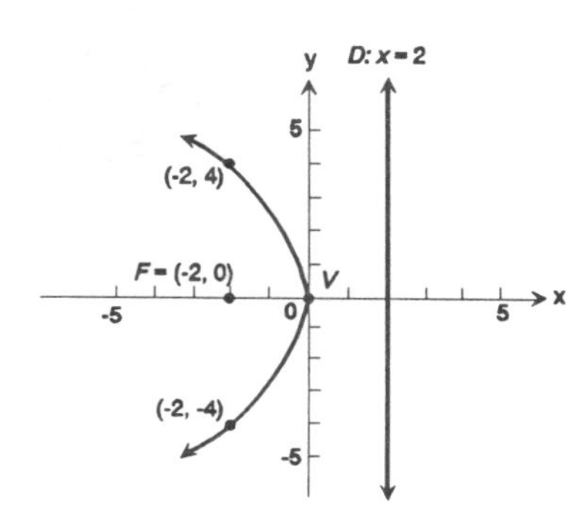

15. The directrix, $y = -\frac{1}{2}$, lies $\frac{1}{2}$ unit ***below*** the vertex, (0, 0), so the focus must be $\frac{1}{2}$ unit ***above*** the vertex, at the point $\left(0, \frac{1}{2}\right)$.

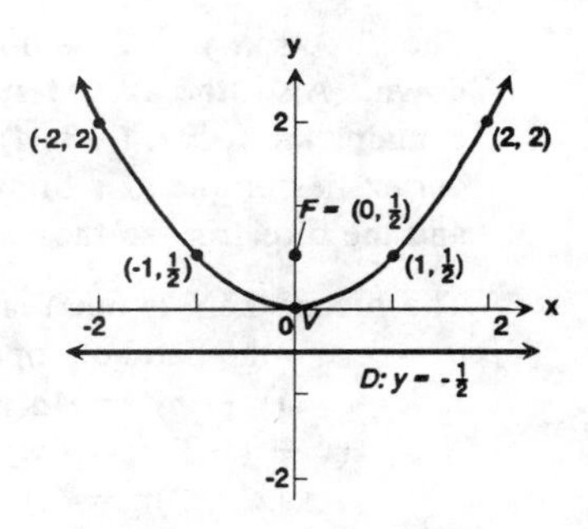

Therefore, $a = \frac{1}{2}$ and the parabola opens upward, so we have:

$$x^2 = 4ay \quad \text{or} \quad x^2 = 2y \left(a = \frac{1}{2}\right)$$

Letting $y = \frac{1}{2}$, we find $x^2 = 1$ or $x = \pm 1$.

The points $\left(1, \frac{1}{2}\right)$ and $\left(-1, \frac{1}{2}\right)$ define the latus rectum.

17. Here, the vertex, $(2, -3)$, and the focus, $(2, -5)$, both lie on the vertical line, $x = 2$. The distance between the vertex and the focus is $a = 2$, and the parabola opens down, since the focus is ***below*** the vertex. From Table 2, we have:

$$(x - h)^2 = -4a(y - k)$$
$$(x - 2)^2 = -4a(y - (-3)), \text{ since } (h, k) = (2, -3)$$
$$(x - 2)^2 = -8(y + 3), \text{ since } a = 2$$

Letting $y = -5$, we find $(x - 2)^2 = 16$ or $x - 2 = \pm 4$ so that $x = 6$ or $x = -2$. The points $(-2, -5)$ and $(6, -5)$ define the latus rectum.

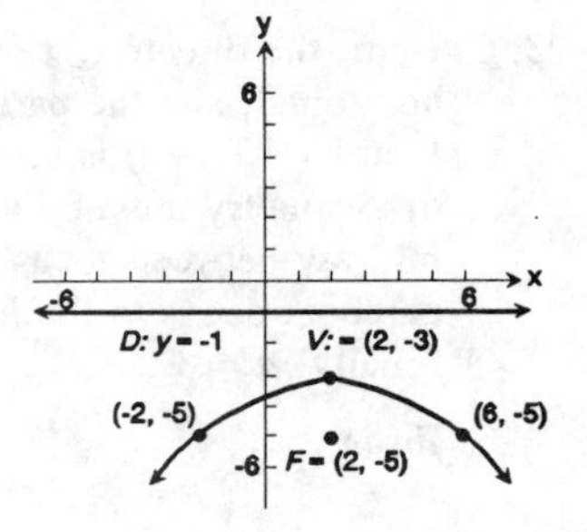

19. Since the axis of symmetry, the y-axis, is vertical, the parabola opens up or down. The point (2, 3) is ***above*** the vertex, (0, 0), so the parabola opens ***up***. Therefore, from Table 1,

$$x^2 = 4ay$$

But here we don't know what a is. But the point (2, 3) must satisfy the equation, since it lies on the graph.

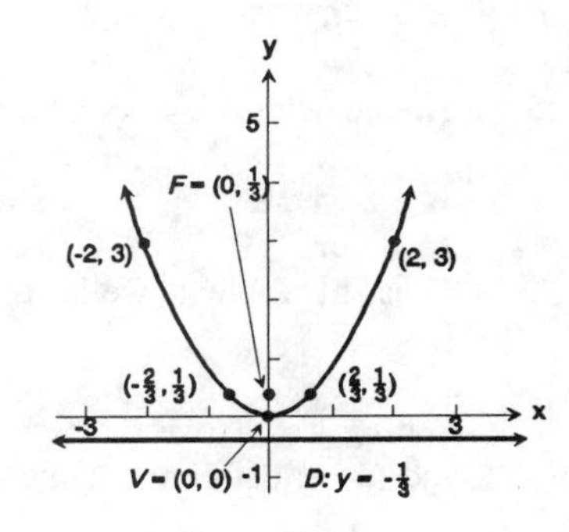

Therefore,
$$x^2 = 4ay$$
$$4 = 4a(3) \qquad \text{(using } x = 2,\ y = 3\text{)}$$
$$1 = 3a$$
$$a = \frac{1}{3}$$

The equation of the parabola is: $x^2 = 4\left(\frac{1}{3}\right)y$ or $x^2 = \frac{4}{3}y$

To help sketch the graph, note that the focus is at $(0, a) = \left(0, \frac{1}{3}\right)$. Letting $y = \frac{1}{3}$, we find $x^2 = \frac{4}{9}$ or $x = \pm\frac{2}{3}$. The points $\left(\frac{-2}{3}, \frac{1}{3}\right)$ and $\left(\frac{2}{3}, \frac{1}{3}\right)$ define the latus rectum.

21. The directrix, $y = 2$, is horizontal, so the parabola opens up or down. Also, the axis of symmetry must be vertical, and it contains the focus, $(-3, 4)$, so it must be the line $x = -3$. The vertex lies on the axis of symmetry, midway between the focus and the directrix, so the vertex must be the point $(-3, 3)$.

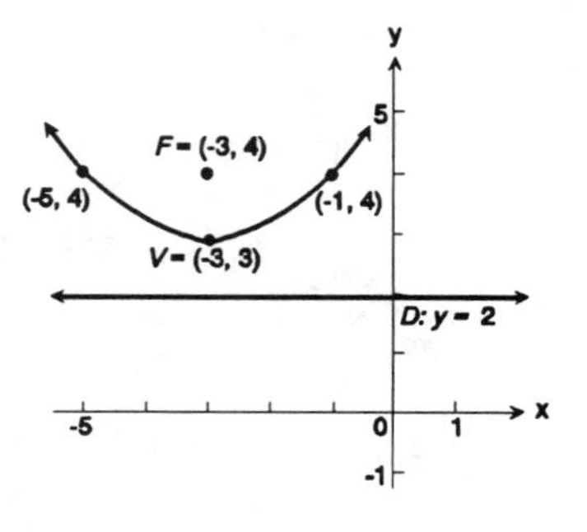

The focus, then, is one unit ***above*** the vertex, so we know that $a = 1$ and the parabola opens up. From Table 2, we have:

$$(x - h)^2 = 4a(y - k)$$
$$(x - (-3))^2 = 4a(y - 3), \text{ since } (h, k) = (-3, 3)$$
$$(x + 3)^2 = 4(y - 3), \text{ since } a = 1$$

Letting $y = 4$, we find $(x + 3)^2 = 4$ or $x + 3 = \pm 2$, so that $x = -1$ or $x = -5$. The points $(-1, 4)$ and $(-5, 4)$ define the latus rectum.

23. Here, the directrix, $x = 1$, is vertical, so the axis of symmetry is horizontal, and the parabola opens to the left or right. Since the focus, $(-3, -2)$ is on the axis of symmetry, the equation of the axis of symmetry must be $y = -2$. Again, the vertex is on the axis, midway between focus and directrix, so the vertex is $(-1, -2)$. The parabola opens to the left, since the focus is to the left of the vertex. Finally, $a = 2$.

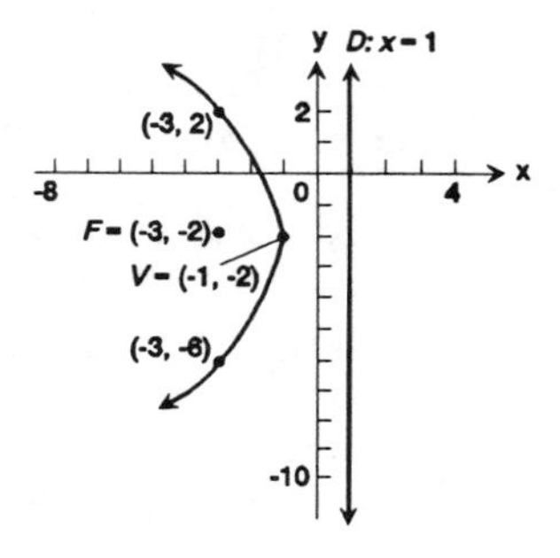

From 2, $\quad (y - k)^2 = -4a(x - h)$

or $\quad (y + 2)^2 = -8(x + 1)$

Letting $x = -3$, $(y + 2)^2 = 16$ or $y + 2 = \pm 4$ so that $y = 2$ or $y = -6$. The points $(-3, -6)$ or $(-3, 2)$ define the latus rectum.

25. The equation $x^2 = 4y$ is in the form:

$$x^2 = 4ay$$

where $\quad 4a = 4,$

or $\quad a = 1$

Thus, by Table 1, we have: Vertex: $(0, 0)$
Focus: $(0, 1)$
Directrix: $y = -1$

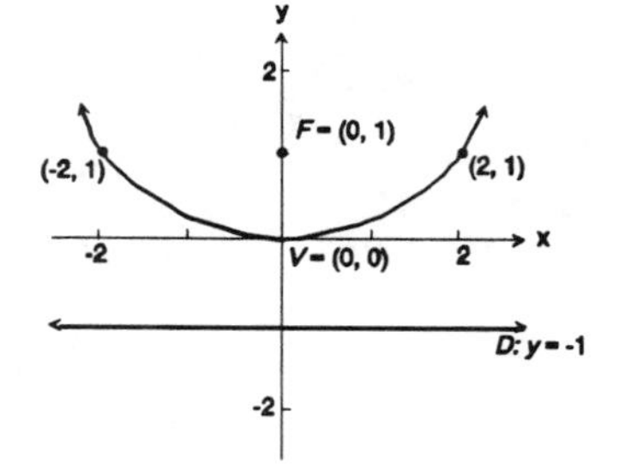

Letting $y = 1$, we find $x^2 = 4$ or $x = \pm 2$. The points $(-2, 1)$ and $(2, 1)$ define the latus rectum.

27. The equation $y^2 = -16x$ is in the form:

$$y^2 = -4ax$$

where $\quad -4a = -16$

or $\quad a = 4$

By Table 1, we have: Vertex: $(0, 0)$
Focus: $(-4, 0)$
Directrix: $x = 4$

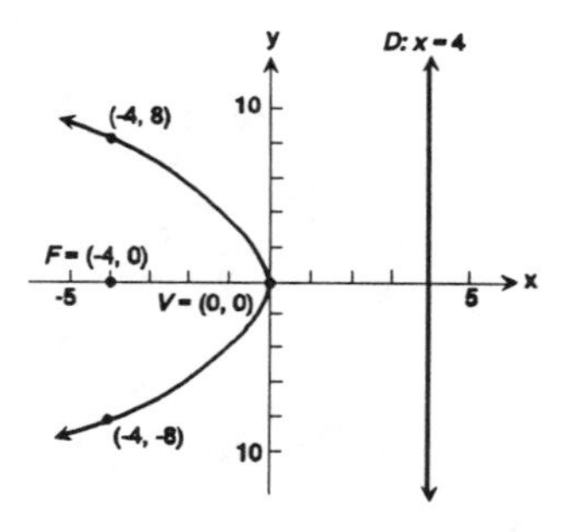

Letting $x = -4$, we have $y^2 = 64$ or $y = \pm 8$. The points $(-4, 8)$ and $(-4, -8)$ define the latus rectum.

29. The equation $(y - 2)^2 = 8(x + 1)$ is in the form:

$$(y - k)^2 = 4a(x - h),$$

where: (1) $4a = 8$, or $a = 2$

(2) $y - k = y - 2$, or $k = 2$

(3) $x - h = x + 1$, or $-h = 1$, or $h = -1$

Thus, the vertex is at $(h, k) = (-1, 2)$. From Table 2, the parabola opens to the right, so the focus is $a = 2$ units to the ***right*** of the vertex. The focus is $(1, 2)$. The directrix is a vertical line 2 units to the ***left*** of the vertex: $x = -3$.

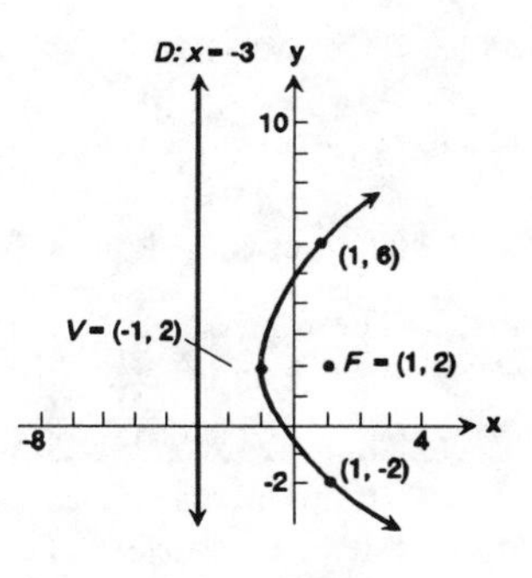

Letting $x = 1$, we have $(y - 2)^2 = 16$ or $y - 2 = \pm 4$, so that $y = 6$ or $y = -2$. The points $(1, 6)$ and $(1, -2)$ define the latus rectum.

31. The equation $(x - 3)^2 = -(y + 1)$ is in the form $(x - h)^2 = -4a(y - k)$, where:

(1) $-4a = -1$, or $a = \dfrac{1}{4}$

(2) $x - h = x - 3$

$-h = -3$

$h = 3$

and (3) $y - k = y + 1$

$-k = 1$

$k = -1$

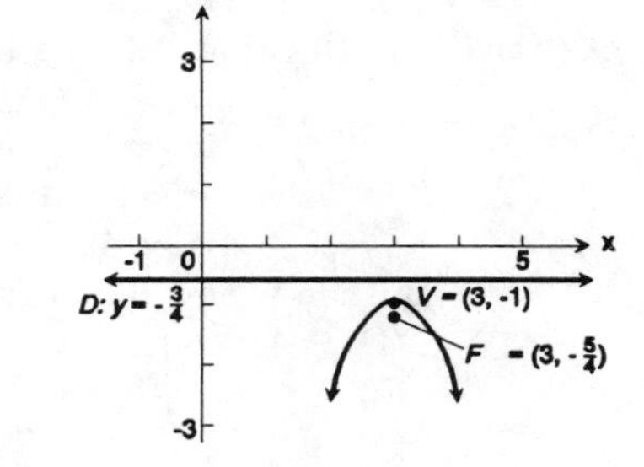

So, from Table 2, we have: Vertex: $(h, k) = (3, -1)$

The parabola opens ***down***, so the focus is ***below*** the vertex: Focus: $\left(3, -\dfrac{5}{4}\right)$

The directrix is a horizontal line $a = \dfrac{1}{4}$ unit ***above*** the vertex: Directrix: $y = -\dfrac{3}{4}$

Letting $y = \dfrac{-5}{4}$, we have $(x - 3)^2 = \dfrac{1}{4}$ or $(x - 3) = \pm\dfrac{1}{2}$ so that $x = \dfrac{7}{2}$ or $x = \dfrac{5}{2}$.

The points $\left(\dfrac{7}{2}, \dfrac{-5}{4}\right)$ and $\left(\dfrac{5}{2}, \dfrac{-5}{4}\right)$ define the latus rectum.

33. The equation $(y + 3)^2 = 8(x - 2)$ is in the form:

$$(y - k)^2 = 4a(x - h)$$

where: (1) $4a = 8$, or $a = 2$

(2) $y - k = y + 3$, or $k = -3$

(3) $x - h = x - 2$, or $h = 2$

Thus, from Table 2, we have: Vertex: $(h, k) = (2, -3)$

The parabola opens to the right, so the focus is $a = 2$ units to the right of the vertex:

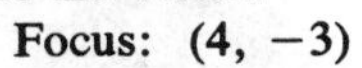

Focus: $(4, -3)$

The directrix is the vertical line 2 units to the left of the vertex:

Directrix: $x = 0$

Letting $x = 4$, we have $(y + 3)^2 = 16$ or $y + 3 = \pm 4$ so that $y = 1$ or $y = -7$. The points $(4, 1)$ and $(4, -7)$ define the latus rectum.

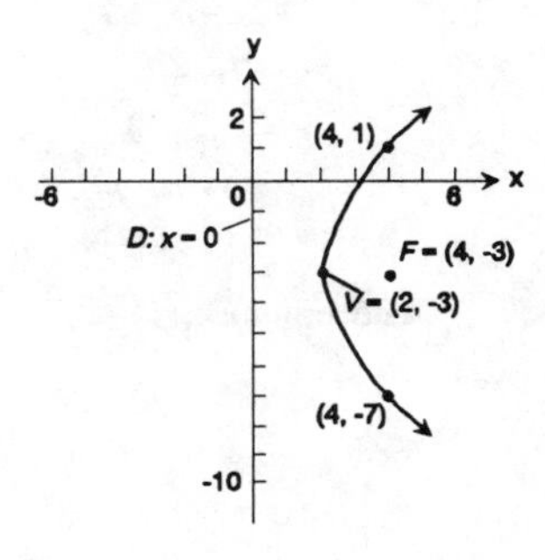

35.
$$y^2 - 4y + 4x + 4 = 0$$
$$y^2 - 4y = -4x - 4$$
$$y^2 - 4y + 4 = -4x$$
$$(y - 2)^2 = -4(x + 0)$$

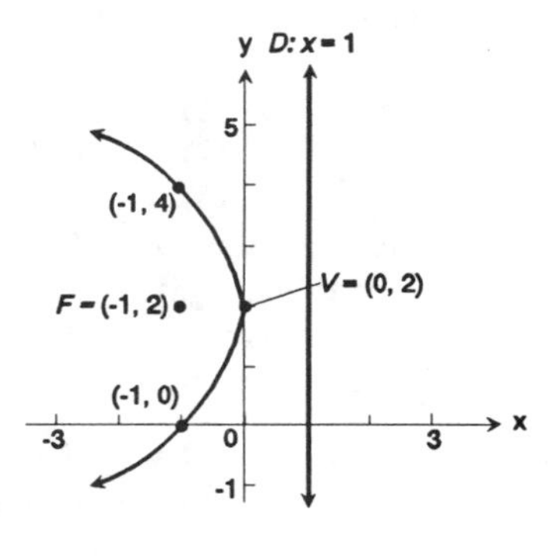

The equation is now in the form: $(y - k)^2 = -4a(x - h)$,

where: (1) $-4a = -4$, or $a = 1$

(2) $y - k = y - 2$, or $k = 2$

(3) $x - h = x + 0$, or $h = 0$

Thus, from Table 2, we have: Vertex: $(h, k) = (0, 2)$

The parabola opens to the left, so the focus is $a = 1$ unit to the left of the vertex: Focus: $(-1, 2)$

The directrix is the vertical line 1 unit to the right of the vertex:

Directrix: $x = 1$

Letting $x = -1$, we have $(y - 2)^2 = 4$ or $y - 2 = \pm 2$ so that $y = 4$ or $y = 0$. The points $(-1, 4)$ and $(-1, 0)$ define the latus rectum.

37.
$$x^2 + 8x = 4y - 8$$
$$x^2 + 8x + 16 = 4y + 8$$
$$(x + 4)^2 = 4(y + 2)$$

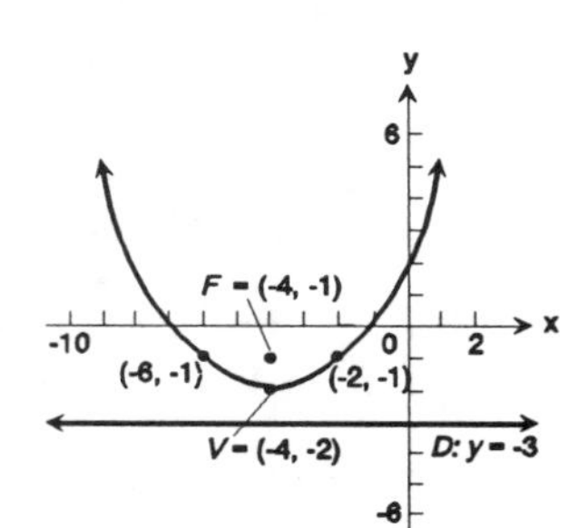

The equation is now in the form $(x - h)^2 = 4a(y - k)$, where:

(1) $4a = 4$, or $a = 1$

(2) $x - h = x + 4$, or $h = -4$

(3) $y - k = y + 2$, or $y = -2$

Thus, from Table 2, we have:

Vertex: $(h, k) = (-4, -2)$

The parabola opens up, so the focus is $a = 1$ unit above the vertex:

Focus: $(-4, -1)$

The directrix is the horizontal line 1 unit below the vertex:

Directrix: $y = -3$

Letting $y = -1$, we have $(x + 4)^2 = 4$ or $x + 4 = \pm 2$ so that $x = -2$ or $x = -6$. The points $(-2, -1)$ and $(-6, -1)$ define the latus rectum.

39. For $y^2 + 2y - x = 0$, complete the square:
$$y^2 + 2y - x = 0$$
$$y^2 + 2y + \underline{\quad} = x + \underline{\quad}$$
$$y^2 + 2y + 1 = x + 1$$
$$(y + 1)^2 = (x + 1)$$

This is in the form: $(y - k)^2 = 4a(x - h)$,

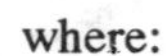

where: (1) $4a = 1$, or $a = \frac{1}{4}$

(2) $x - h = x + 1$

$h = -1$

and (3) $y - k = y + 1$

$-k = 1$

$k = -1$

From Table 2, we have: Vertex: $(h, k) = (-1, -1)$

The parabola opens to the right, and has: Focus: $\left(-\frac{3}{4}, -1\right)$

The directrix is $x = -\frac{5}{4}$

Letting $x = \frac{-3}{4}$, we have $(y + 1)^2 = \frac{1}{4}$ or $y + 1 = \pm\frac{\sqrt{1}}{2}$ so that $y = \frac{\sqrt{1}}{2} - 1 = \frac{-1}{2}$ or $y = \frac{-\sqrt{1}}{2} - 1 = \frac{-3}{2}$. The points $\left(\frac{-3}{4}, \frac{-1}{2}\right)$ and $\left(\frac{-3}{4}, \frac{3}{2}\right)$ define the latus rectum.

41. We complete the square:

$$x^2 - 4x + \underline{\quad} = y + 4 + \underline{\quad}$$
$$x^2 - 4x + 4 = y + 4 + 4$$
$$(x - 2)^2 = y + 8$$

which is in the form: $(x - h)^2 = 4a(y - k)$

where: (1) $4a = 1$, $a = \frac{1}{4}$

(2) $x - h = x - 2$
$h = 2$

and (3) $y - k = y + 8$
$k = -8$

We have: Vertex: $(2, -8)$

Focus: $\left(2, -7\frac{3}{4}\right) = \left(2, \frac{-31}{4}\right)$

Directrix: $y = -8\frac{1}{4}$ or $y = \frac{-33}{4}$

Letting $y = \frac{-31}{4}$, we have $(x - 2)^2 = \frac{1}{4}$ or $x - 2 = \pm\frac{1}{2}$ so that $x = \frac{5}{2}$ or $x = \frac{3}{2}$.

The points $\left(\frac{5}{2}, \frac{-31}{4}\right)$ and $\left(\frac{3}{2}, \frac{-31}{4}\right)$ define the latus rectum.

43.
$$(y - 1)^2 = c(x - 0)$$
$$(y - 1)^2 = cx$$
$$(2 - 1)^2 = c(1)$$
$$1 = c$$
$$(y - 1)^2 = x$$

45.
$$(y - 1)^2 = c(x - 2)$$
$$(0 - 1)^2 = c(1 - 2)$$
$$1 = c(-1)$$
$$c = -1$$
$$(y - 1)^2 = -(x - 2)$$

47.
$$(x - 0)^2 = c(y - 1)$$
$$x^2 = c(y - 1)$$
$$2^2 = c(2 - 1)$$
$$4 = c$$
$$x^2 = 4(y - 1)$$

49.
$$(y - 0)^2 = c(x - (-2))$$
$$y^2 = c(x + 2)$$
$$1^2 = c(0 + 2)$$
$$1 = 2c$$
$$\frac{1}{2} = c$$
$$y^2 = \frac{1}{2}(x + 2)$$

51. Situate the parabola so that its vertex is at (0, 0), and it opens up. Then, we know:

$$x^2 = 4ay$$

Since the parabola is 10 feet across and 4 feet deep, the points $(-5, 4)$ and $(5, 4)$ must satisfy the equation:

$$x^2 = 4ay$$
$$25 = 4a(4) \qquad (x = 5, y = 4)$$
$$25 = 16a$$
$$a = \frac{25}{16} \approx 1.5625$$

But a is, by definition, the distance from the vertex to the focus.

Therefore, the receiver (at the focus) is $\dfrac{25}{16} = 1.5625$ feet, or 18.75 inches, from the base of the dish, along the axis of symmetry.

53. The light is placed at the focus of the parabola. $x^2 = 4y$. Since $F = (0, 1)$, the bulb should be placed 1 inch from the vertex.

55. Situate the vertex at (0, 0): $x^2 = cy$

Point on parabola is (300, 80):

$$300^2 = c \cdot 80$$
$$c = 1125$$
$$x^2 = 1125y$$

When $x = 150$, $\quad (150)^2 = 1125h$

$$h = \frac{150^2}{1125} = 20 \text{ feet}$$

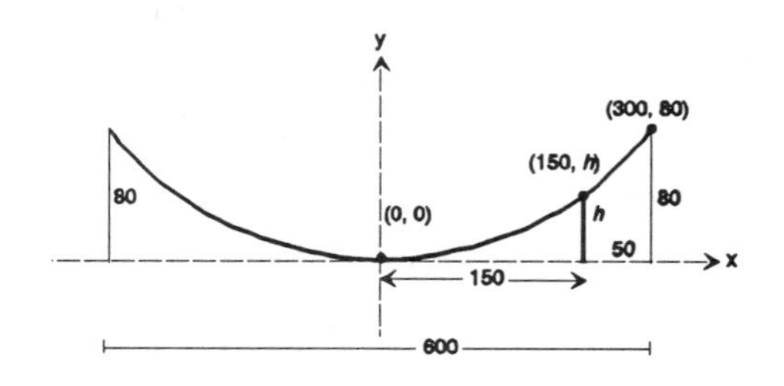

57. Situate the parabola so that its vertex is at (0, 0) and it opens up. We know:

$$x^2 = 4ay$$

The light source is located 2 feet from the base, so $a = 2$.

Since the width is 5 feet, $x = \pm 2.5$ feet. Thus, we have:

$$(2.5)^2 = 4(2)y$$
$$6.25 = 8y$$
$$y = 0.78125 \text{ feet}$$

Thus, the depth of the search light is 0.78125 feet.

59. Situate the parabola so that its vertex is at (0, 0) and it opens up. We know:

$$x^2 = 4ay$$

The parabola is 20 feet wide, so $x = \pm 10$ feet, and 6 feet deep, so $y = 6$ feet. Thus, we have:

$$(10)^2 = 4a(6)$$
$$100 = 24a$$
$$a \approx 4.17 \text{ feet}$$

Therefore, the heat source will be concentrated 4.17 feet from the base, along the axis of symmetry.

61. Situate the parabola so that its vertex is at (0, 0) and it opens down. Then, we know

$$x^2 = -4ay$$

Since the bridge has a span of 120 feet and a maximum height of 25 feet, two points on the parabola are $(-60, -25)$ and $(60, -25)$. See figure.

Since the form of the equation of this parabola is:

$$x^2 = -4ay$$

Letting $x = 60$ and $y = -25$, we can find a.

$$60^2 = -4a(-25)$$
$$a = 36$$

So, the equation of the parabola is:

$$x^2 = -144y$$

So the height of the bridge 10 feet from the center is:

$$10^2 = -144y$$
$$y = -0.69 \text{ feet}$$

Therefore, the height of the bridge 10 feet from center is $25 - 0.69 = 24.31$ feet.

The height of the bridge 30 feet from center is $25 - 6.25 = 18.75$ feet.

The height of the bridge 50 feet from center is $25 - 17.36 = 7.64$ feet.

63. $Ax^2 + Ey = 0 \quad A \neq 0, E \neq 0$

$$Ax^2 = -Ey$$

$$x^2 = \frac{-E}{A}y$$

Parabola; vertex at (0, 0) and axis of symmetry the y-axis; focus at $\left(0, \frac{-E}{4A}\right)$;

directrix the line $y = \frac{E}{4A}$. The parabola opens up if $\frac{-E}{A} > 0$ and down if $\frac{-E}{A} < 0$.

65. $Ax^2 + Dx + Ey + F = 0 \qquad A \neq 0$

(a) If $E \neq 0$, then

$$Ax^2 + Dx + Ey + F = 0$$

$$Ax^2 + Dx = -Ey - F$$

$$A\left(x^2 + \frac{D}{A}x + \frac{D^2}{4A^2}\right) = -Ey - F + \frac{D^2}{4A}$$

$$\left(x + \frac{D}{2A}\right)^2 = \frac{1}{A}\left(-Ey - F + \frac{D^2}{4A}\right)$$

$$\left(x + \frac{D}{2A^2}\right) = \frac{-E}{A}\left[y + \left(\frac{F}{E} - \frac{D^2}{4AE}\right)\right]$$

$$\left(x + \frac{D}{2A}\right)^2 = \frac{-E}{A}\left(y - \frac{D^2 - 4AF}{4AE}\right)$$

Parabola; vertex at $\left(\frac{-D}{2A}, \frac{D^2 - 4AF}{4AE}\right)$, axis of symmetry parallel to y-axis.

(b) If $E = 0$, then

$$Ax^2 + Dx + F = 0$$

$$x = \frac{-D \pm \sqrt{D^2 - 4AF}}{2A}$$

If $D^2 - 4AF = 0$, then

$$x = \frac{-D}{2A}$$

Vertical line

(c) If $E = 0$, then

$$Ax^2 + Dx + F = 0$$

$$x = \frac{-D \pm \sqrt{D^2 - 4AF}}{2A}$$

If $D^2 - 4AF > 0$, then

$$x = \frac{-D + \sqrt{D^2 - 4AF}}{2A} \text{ or}$$

$$x = \frac{-D - \sqrt{D^2 - 4AF}}{2A}$$

Two vertical lines

(d) If $E = 0$, then

$$Ax^2 + Dx + F = 0$$

$$x = \frac{-D \pm \sqrt{D^2 - 4AF}}{2A}$$

If $D^2 - 4AF < 0$, there is no real solution. Hence, the graph contains no points.

7.3 The Ellipse

1. C

3. B

In Problems 9–17, write the equation in the form shown in Table 3 in the text, so that you can identify (h, k), a, and b, and also tell whether the major axis is parallel to the x-axis or the y-axis.

5. $\dfrac{x^2}{25} + \dfrac{y^2}{4} = 1$

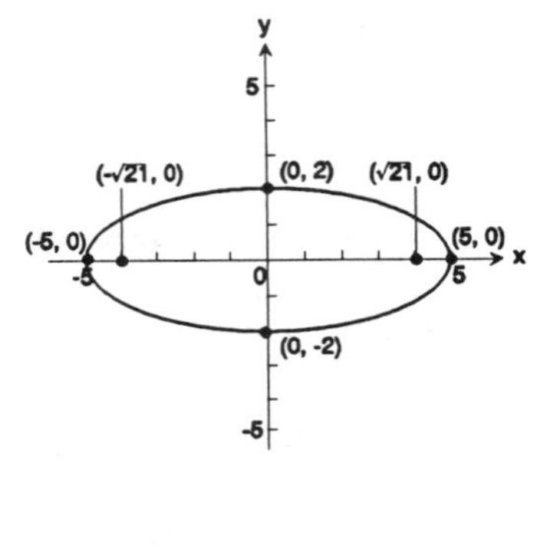

This is in the proper form, and we see that $h = 0$ and $k = 0$, so the center of the ellipse is at the origin. Since the larger denominator is associated with x^2, we see from Table 3 that the major axis is parallel to the x-axis, and:

$a^2 = 25$, so $a = 5$

$b^2 = 4$, so $b = 2$

We will be able to locate the vertices using $a = 5$, but to find the foci we will need c:

$c^2 = a^2 - b^2 = 21$

$c = \sqrt{21}$

From Table 3, the vertices are located at $(h \pm a, k) = (0 \pm 5, 0)$:

Vertices: $(-5, 0)$, $(5, 0)$

and the foci are at $(h \pm c, k)$:

Foci: $\left(-\sqrt{21}, 0\right), \left(\sqrt{21}, 0\right)$

(Notice that, since the major axis is parallel to the x-axis, we find the vertices by moving left and right of the center (h, k), a distance $a = 5$. Similarly, the foci are $c = \sqrt{21}$ units to the left and right of the center.)

Remember that we use b to locate the end-points of the ***minor axis***, which passes through the center and is perpendicular to the major axis.

7. $\dfrac{x^2}{9} + \dfrac{y^2}{25} = 1$

This is in the proper form, and the fact that the larger denominator appears in the y^2-term means that the major axis is parallel to the y-axis. By Table 3,

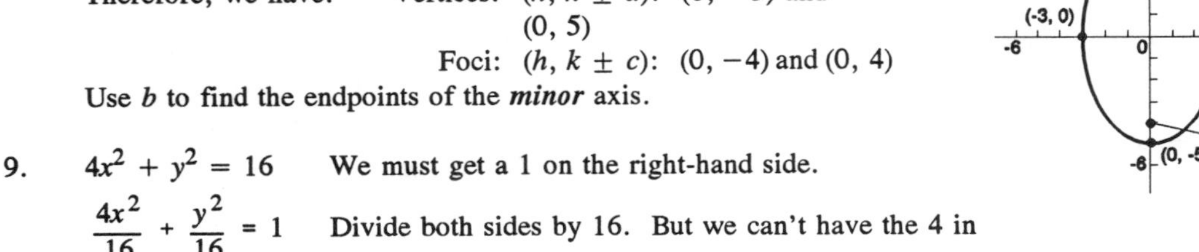

(1) $h = 0$, $k = 0$, so the center is $(0, 0)$

(2) $a^2 = 25$, so $a = 5$

(3) $b^2 = 9$, so $b = 3$, and

(4) $c^2 = a^2 - b^2 = 16$, so $c = 4$

Therefore, we have: Vertices: $(h, k \pm a)$: $(0, -5)$ and $(0, 5)$

Foci: $(h, k \pm c)$: $(0, -4)$ and $(0, 4)$

Use b to find the endpoints of the ***minor*** axis.

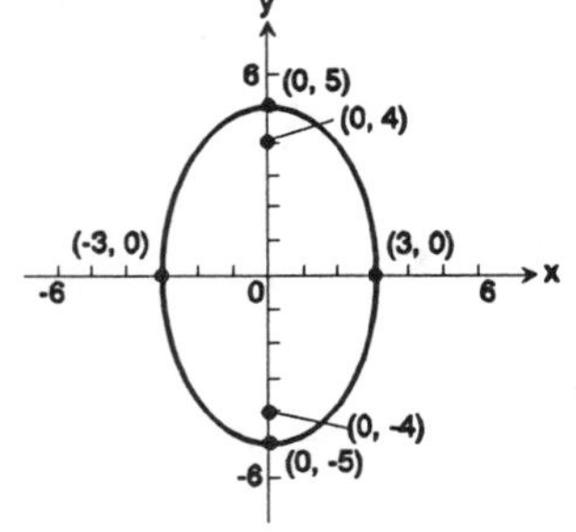

9. $4x^2 + y^2 = 16$ We must get a 1 on the right-hand side.

$\dfrac{4x^2}{16} + \dfrac{y^2}{16} = 1$ Divide both sides by 16. But we can't have the 4 in the numerator, so we simplify.

$\dfrac{x^2}{4} + \dfrac{y^2}{16} = 1$

Now we have the proper form, with $h = 0$, $k = 0$. By Table 3, the major axis is parallel to the y-axis, and:

$a^2 = 16$, $a = 4$

$b^2 = 4$, $b = 2$

$c^2 = a^2 - b^2 = 12$, $c = \sqrt{12} = 2\sqrt{3}$

Also: Foci: $(h, k \pm c)$: $\left(0, -2\sqrt{3}\right), \left(0, 2\sqrt{3}\right)$

Vertices: $(h, k \pm a)$: $(0, -4)$, $(0, 4)$

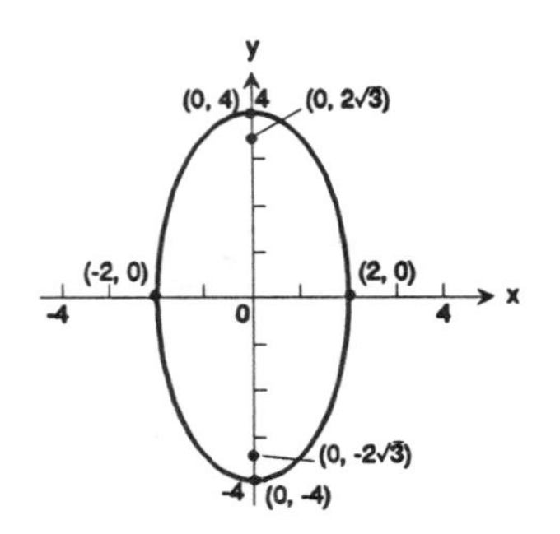

11. $4y^2 + x^2 = 8$

$$\frac{4y^2}{8} + \frac{x^2}{8} = 1$$

Obtain a 1 on the right.

$$\frac{y^2}{2} + \frac{x^2}{8} = 1$$

Simplify.

We know:
(1) Major axis is parallel to the x-axis.
(2) $a^2 = 8$, so $a = \sqrt{8} = 2\sqrt{2}$
(3) $b^2 = 2$, so $b = \sqrt{2}$
(4) $c^2 = a^2 - b^2 = 6$, $c = \sqrt{6}$
(5) $h = 0$, $k = 0$, so the center is at (0, 0)
(6) Foci: $(h \pm c, k)$: $\left(-\sqrt{6}, 0\right), \left(\sqrt{6}, 0\right)$
(7) Vertices: $(h \pm a, k)$: $\left(-2\sqrt{2}, 0\right), \left(2\sqrt{2}, 0\right)$

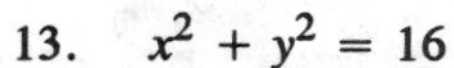

13. $x^2 + y^2 = 16$

This is the equation of a circle, with center at (0, 0) and radius = 4.

Vertices: $(\pm 4, 0)$

Foci: (0, 0)

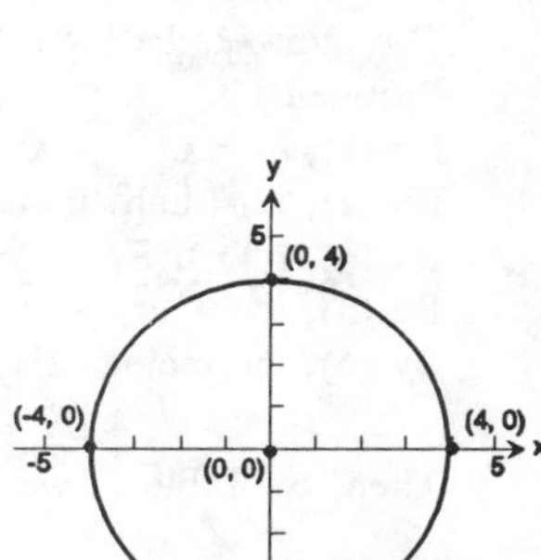

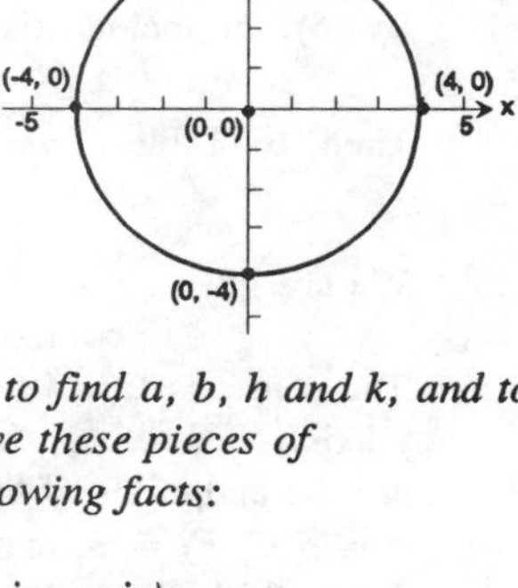

For Problems 19–28, in order to find the equation of an ellipse, it is necessary to find a, b, h and k, and to know whether the major axis is parallel to the x-axis or the y-axis. Once we have these pieces of information, we can use Table 3 to write down the equation. We will use the following facts:

(1) (h, k) are the coordinates of the center of the ellipse.

(2) a = the distance from the center to either vertex (half the length of the major axis)

(3) b = the distance from the center to either endpoint of the minor axis (half the length of the minor axis), or, if c is known, then $b^2 = a^2 - c^2$

(4) c = the distance from the center to either focus.

(5) Since $b^2 = a^2 - c^2$, if any two of the three numbers a, b, and c are known, we can find the third.

(6) The center, the foci, and the vertices all lie on the major axis.

Use these facts to find a, b, h, and k, and to determine if the major axis is vertical or horizontal.

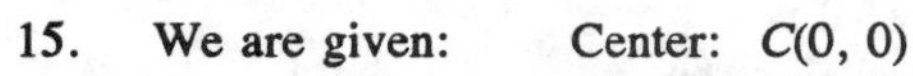

15. We are given: Center: $C(0, 0)$
Focus: $F(3, 0)$
Vertex: $V(5, 0)$

Please refer to the paragraph above.

By fact (1) above, $h = 0$, $k = 0$

By fact (2), $a = d(C, V) = 5$

By fact (4), $c = d(C, F) = 3$

By fact (3), $b^2 = a^2 - c^2 = 25 - 9 = 16$

By fact (6), since C, F and V all lie on the horizontal line, $y = 0$, the major axis is parallel to the x-axis.

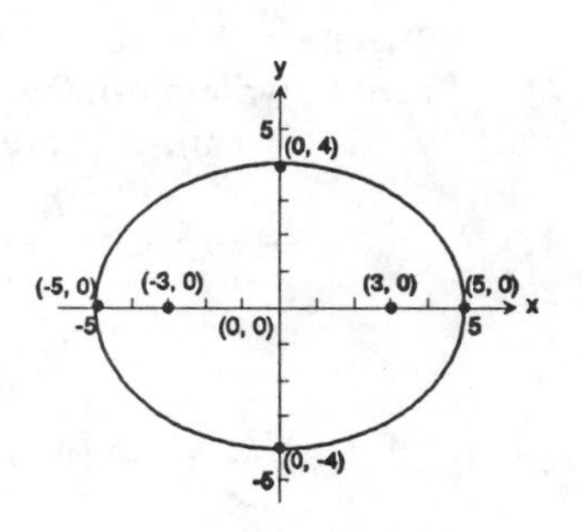

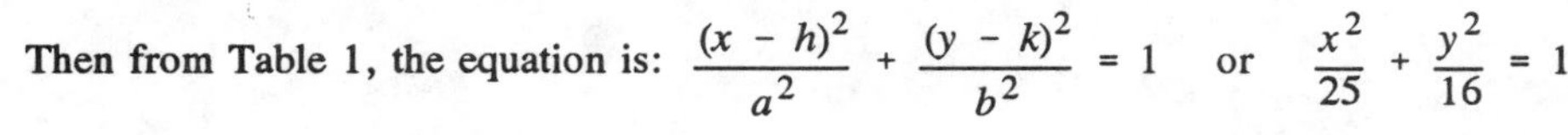

Then from Table 1, the equation is: $\dfrac{(x - h)^2}{a^2} + \dfrac{(y - k)^2}{b^2} = 1$ or $\dfrac{x^2}{25} + \dfrac{y^2}{16} = 1$

17. We are given: Center: $C(0, 0)$
Focus: $F(0, -4)$
Vertex: $V(0, 5)$

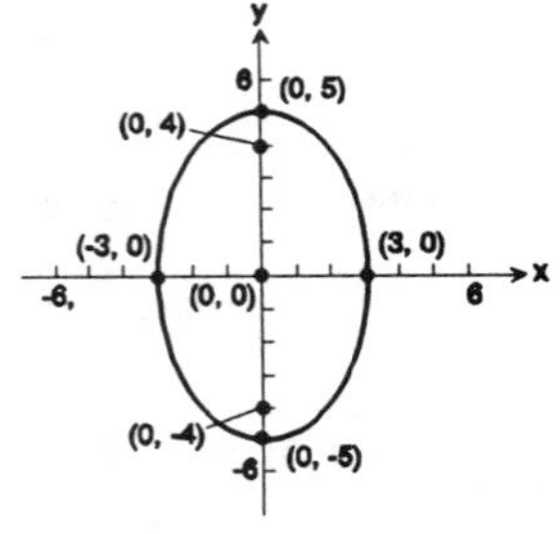

The numbering below refers to the paragraph preceding the solution to Problem 19.
By (1), $h = 0$, $k = 0$

By (2), $a = d(C, V) = 5$
By (4), $c = d(C, F) = 4$
By (3), $b^2 = a^2 - c^2 = 9$
By (6), the major axis is the vertical line $x = 0$.

From Table 1, the equation is: $\dfrac{(x - h)^2}{b^2} + \dfrac{(y - k)^2}{a^2} = 1$ or $\dfrac{x^2}{9} + \dfrac{y^2}{25} = 1$

19. We are given: Foci: $F_1(-2, 0)$ and $F_2(2, 0)$
Length of Major Axis = 6.
First of all, the center is the midpoint between F_1 and F_2: $C(0, 0)$.
The numbers below refer to the paragraph preceding the solution to Problem 19.
By (1), $h = 0$, $k = 0$
By (2), a = half the length of the major axis, or $a = 3$
By (4), $c = d(F_1, C) = 2$
By (3), $b^2 = a^2 - c^2 = 5$
By (6), the major axis is the horizontal line, $y = 0$.

Then, by Table 3, we have: $\dfrac{(x - h)^2}{a^2} + \dfrac{(y - k)^2}{b^2} = 1$ or $\dfrac{x^2}{9} + \dfrac{y^2}{5} = 1$

21. We are given: Foci: $F_1(0, -3)$ and $F_2(0, 3)$
x-intercepts: ± 2

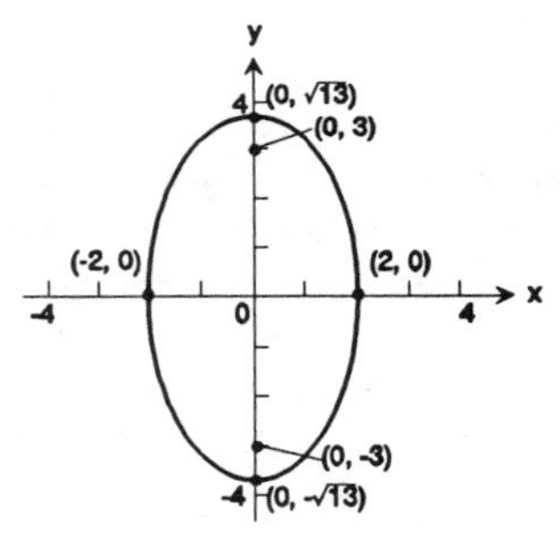

The center is $C(0, 0)$, and the major axis is the vertical line $x = 0$, (the y-axis). Therefore, the minor axis lies on the x-axis, so the x-intercepts are the endpoints of the minor axis. Thus, we have $b = 2$. We know $c = d(C, F) = 3$, and $b^2 = a^2 - c^2$, or $a^2 = b^2 + c^2 = 13$.
From Table 3, we have:

$$\frac{(x - h)^2}{b^2} + \frac{(y - k)^2}{a^2} = 1 \quad \text{or} \quad \frac{x^2}{4} + \frac{y^2}{13} = 1$$

23. Here we are given: Center: $C(0, 0)$
Vertex: $V(0, 4)$
and $b = 1$

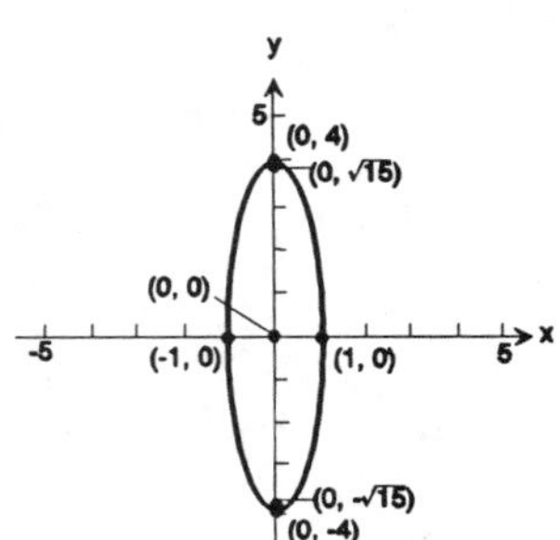

Therefore, $h = 0$, $k = 0$, and $a = d(C, V) = 4$, so we have all we need to write down the equation. Since the major axis is parallel to the y-axis ($x = 0$), we have:

$$\frac{(x - h)^2}{b^2} + \frac{(y - k)^2}{a^2} = 1 \quad \text{or} \quad \frac{x^2}{1} + \frac{y^2}{16} = 1$$

25. $\dfrac{(x + 1)^2}{4} + (y - 1)^2 = 1$

27. $(x - 1)^2 + \dfrac{y^2}{4} = 1$

29. $\dfrac{(x-3)^2}{4} + \dfrac{(y+1)^2}{9} = 1$

This is in the form: $\dfrac{(x-h)^2}{b^2} + \dfrac{(y-k)^2}{a^2} = 1$

where $h = 3$, $k = -1$, $a = 3$, $b = 2$.

Since the larger denominator belongs to the y^2-term, we see from Table 3 that the major axis is parallel to the y-axis, and we have:

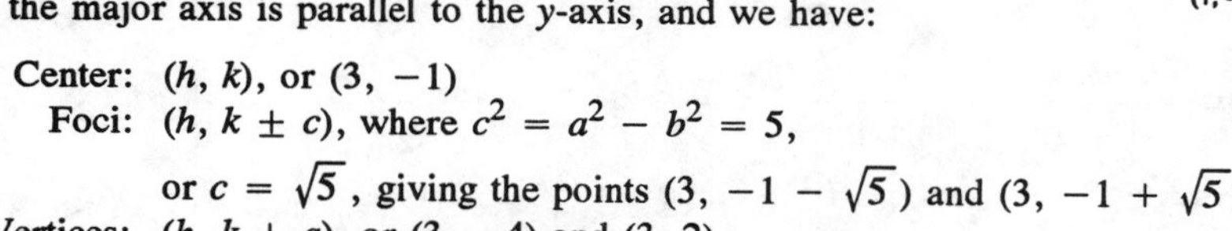

Center: (h, k), or $(3, -1)$

Foci: $(h, k \pm c)$, where $c^2 = a^2 - b^2 = 5$,

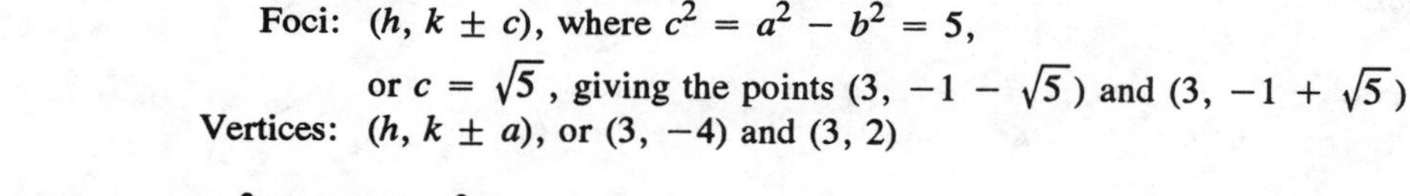

or $c = \sqrt{5}$, giving the points $(3, -1-\sqrt{5})$ and $(3, -1+\sqrt{5})$

Vertices: $(h, k \pm a)$, or $(3, -4)$ and $(3, 2)$

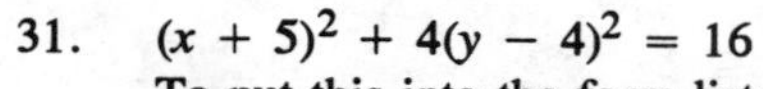

31. $(x+5)^2 + 4(y-4)^2 = 16$

To put this into the form listed in Table 3, we need to obtain a 1 on the right-hand-side:

$$\frac{(x+5)^2}{16} + \frac{4(y-4)^2}{16} = 1$$

Now get rid of the 4 in the numerator of the y^2-term (multiply top and bottom by $\frac{1}{4}$).

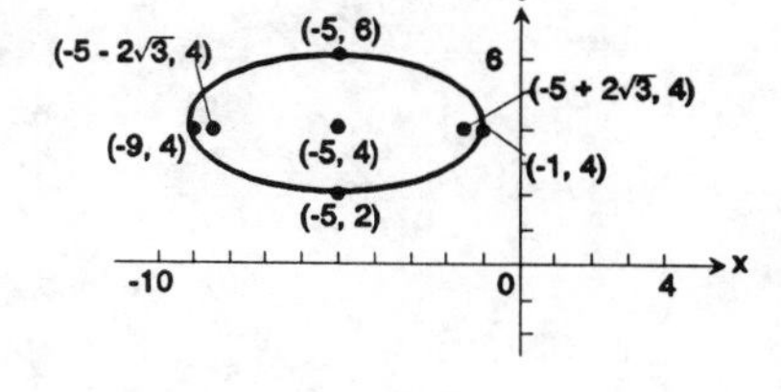

$$\frac{(x+5)^2}{16} + \frac{(y-4)^2}{4} = 1$$

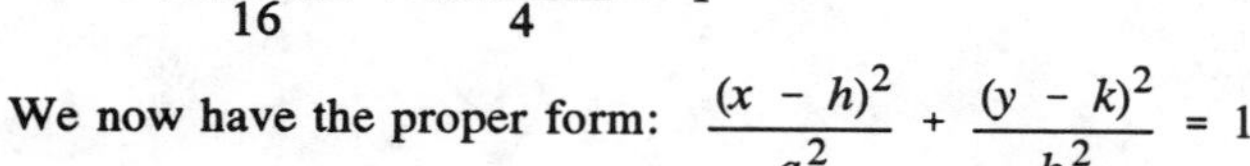

We now have the proper form: $\dfrac{(x-h)^2}{a^2} + \dfrac{(y-k)^2}{b^2} = 1$

where $h = -5$, $k = 4$, $a = 4$, $b = 2$. Also, $c^2 = a^2 - b^2 = 12$, so $c = \sqrt{12}$.

Since the larger denominator is associated with the x^2 term, the major axis is parallel to the x-axis, and, from Table 3, we have:

Center: (h, k), or $(-5, 4)$

Foci: $(h \pm c, k)$, or $\left(-5 - 2\sqrt{3},\ 4\right)$ and $\left(-5 + 2\sqrt{3},\ 4\right)$

Vertices: $(h \pm a, k)$, or $(-9, 4)$ and $(-1, 4)$

33. $x^2 + 4x + 4y^2 - 8y + 4 = 0$

Here we start by completing the square in both x and y:

$$x^2 + 4x + \underline{\quad} + 4(y^2 - 2y + \underline{\quad}) = -4$$

$$x^2 + 4x + 4 + 4(y^2 - 2y + 1) = -4 + 4 + 4$$

$$(4)(1) = 4$$

$$(x+2)^2 + 4(y-1)^2 = 4$$

$$\frac{(x+2)^2}{4} + \frac{(y-1)^2}{1} = 1$$

This is in the form: $\dfrac{(x-h)^2}{a^2} + \dfrac{(y-k)^2}{b^2} = 1$

where $h = -2$, $k = 1$, $a = 2$, $b = 1$, and $c^2 = a^2 - b^2 = 3$, so $c = \sqrt{3}$.

From Table 3, we have:

Center: (h, k), or $(-2, 1)$

Foci: $(h \pm c, k)$, or $\left(-2 - \sqrt{3},\ 1\right)$ and $\left(-2 + \sqrt{3},\ 1\right)$

Vertices: $(h \pm a, k)$, or $(-4, 1)$ and $(0, 1)$

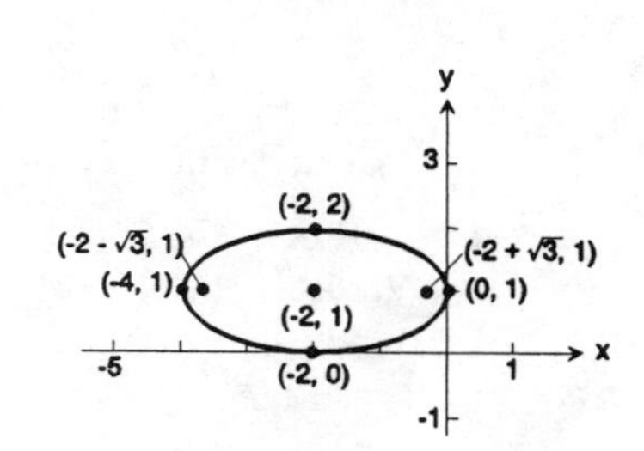

35.

$$2x^2 + 3y^2 - 8x + 6y + 5 = 0$$
$$2x^2 - 8x + 3y^2 + 6y = -5$$
$$2(x^2 - 4x + \_\_\_) + 3(y^2 + 2y + \_\_\_) = -5$$
$$2(x^2 - 4x + 4) + 3(y^2 + 2y + 1) = -5 + 8 + 3$$
$$2(x - 2)^2 + 3(y + 1)^2 = 6$$
$$\frac{(x-2)^2}{3} + \frac{(y+1)^2}{2} = 1$$

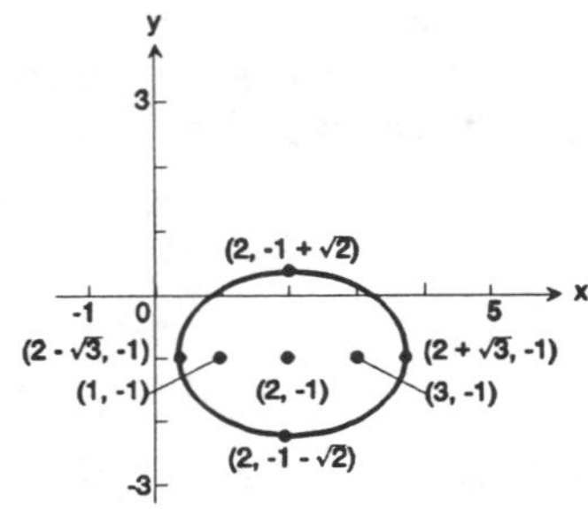
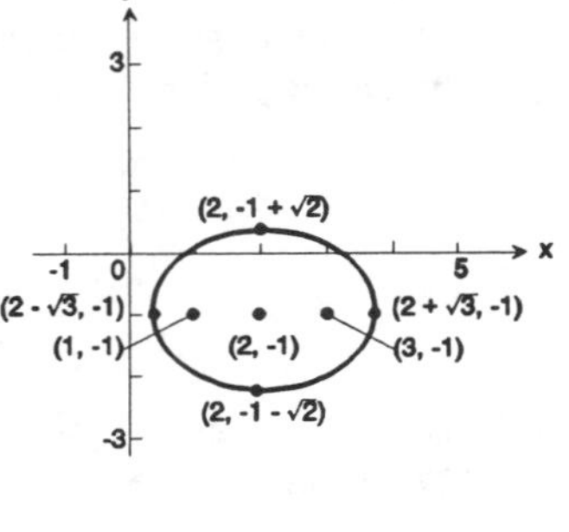

This is in the form: $\frac{(x-h)^2}{a^2} + \frac{(y-k)^2}{b^2} = 1$, where

$h = 2, k = -1, a^2 = 3$, so $a = \sqrt{3}$, $b^2 = 2$, so $b = \sqrt{2}$, and $c^2 = a^2 - b^2 = 1$, so that $c = 1$. Then we have:

Center: (h, k), or $(2, -1)$

Foci: $(h \pm c, k)$, or $(1, -1)$ and $(3, -1)$

Vertices: $(h \pm a, k)$, or $\left(2 - \sqrt{3}, -1\right)$ and $\left(2 + \sqrt{3}, -1\right)$

37.

$$9x^2 + 4y^2 - 18x + 16y - 11 = 0$$
$$9x^2 - 18x + 4y^2 + 16y = 11$$
$$9(x^2 - 2x + \_\_\_) + 4(y^2 + 4y + \_\_\_) = 11$$
$$9(x^2 - 2x + 1) + 4(y^2 + 4y + 4) = 11 + 9 + 16$$
$$9(x - 1)^2 + 4(y + 2)^2 = 36$$
$$\frac{(x-1)^2}{4} + \frac{(y+2)^2}{9} = 1$$

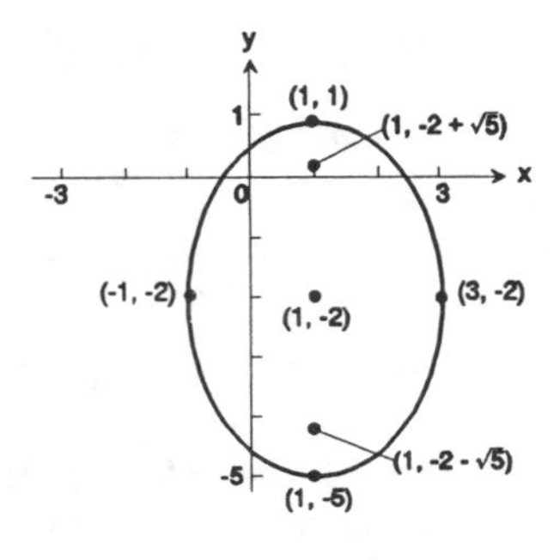

This is in the form: $\frac{(x-h)^2}{b^2} + \frac{(y-k)^2}{a^2} = 1$

where $h = 1, k = -2, a^2 = 9$, so $a = 3$, $b^2 = 4$, so $b = 2$, and

$c^2 = a^2 - b^2 = 5$, so that $c = \sqrt{5}$.

Then we have: Center: (h, k), or $(1, -2)$

Foci: $(h, k \pm c)$, or $(1, -2 + \sqrt{5})$ and $(1, -2 - \sqrt{5})$

Vertices: $(h, k \pm a)$, or $(1, 1)$ and $(1, -5)$

39.

$$4x^2 + y^2 + 4y = 0$$
$$4x^2 + (y^2 + 4y + \_\_\_) = 0$$
$$4x^2 + (y^2 + 4y + 4) = 4$$
$$4x^2 + (y + 2)^2 = 4$$
$$x^2 + \frac{(y+2)^2}{4} = 1$$

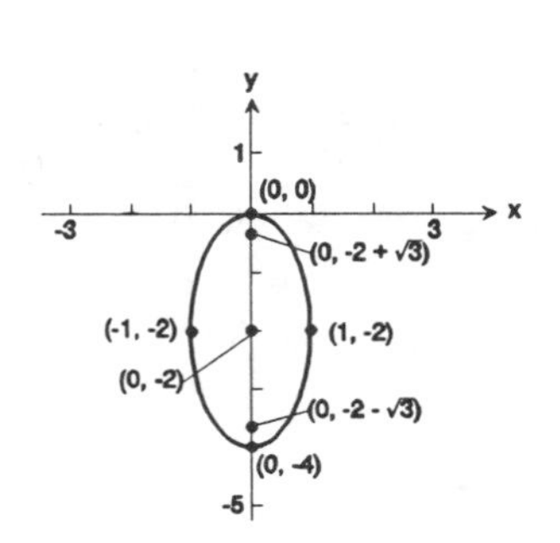

This is in the form: $\frac{(x-h)^2}{b^2} + \frac{(y-k)^2}{a^2} = 1$, where

$h = 0, k = -2, a^2 = 4$, so $a = 2$, $b^2 = 1$, so $b = 1$, and

$c^2 = a^2 - b^2 = 3$, so that $c = \sqrt{3}$. Then we have:

Center: $(0, -2)$

Foci: $(h, k \pm c)$, or $(0, -2 + \sqrt{3})$ and $(0, -2 - \sqrt{3})$

Vertices: $(h, k \pm a)$, or $(0, 0)$ and $(0, -4)$

41. We are given: Center: $C(2, -2)$
Vertex: $V(7, -2)$
Focus: $F(4, -2)$

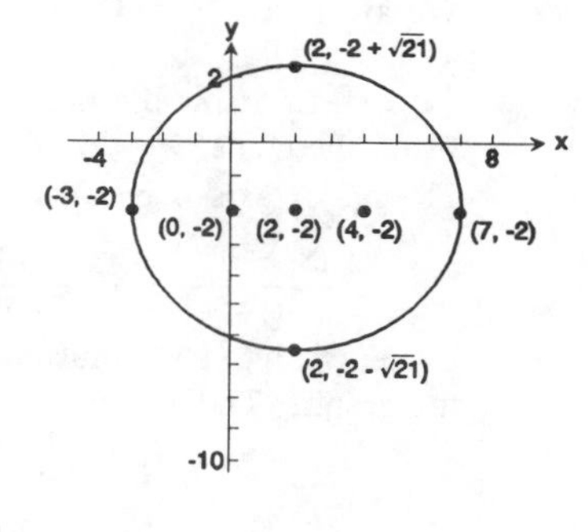

From the center, $h = 2$, $k = -2$

Also, $a = d(C, V) = 5$, and $c = d(C, F) = 2$. Therefore, $b^2 = a^2 - c^2 = 21$, so $b = \sqrt{21}$. Finally, C, V, and F all lie on the horizontal line $y = -2$, so the major axis is parallel to the x-axis.

From Table 3, the equation is of the form:

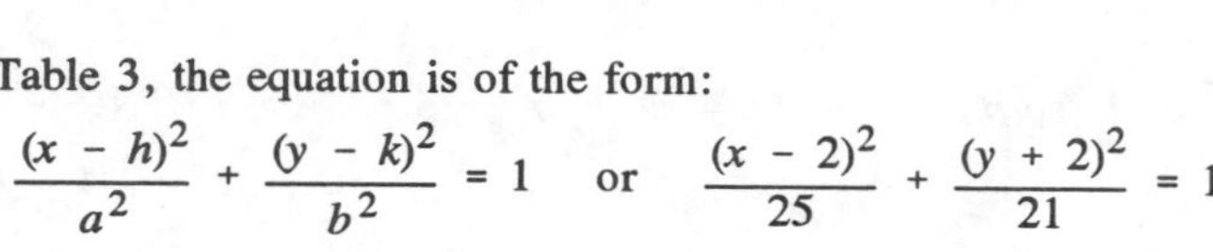

$$\frac{(x-h)^2}{a^2} + \frac{(y-k)^2}{b^2} = 1 \quad \text{or} \quad \frac{(x-2)^2}{25} + \frac{(y+2)^2}{21} = 1$$

43. We are given: Vertices: $V_1(4, 3)$ and $V_2(4, 9)$
Focus: $F(4, 8)$

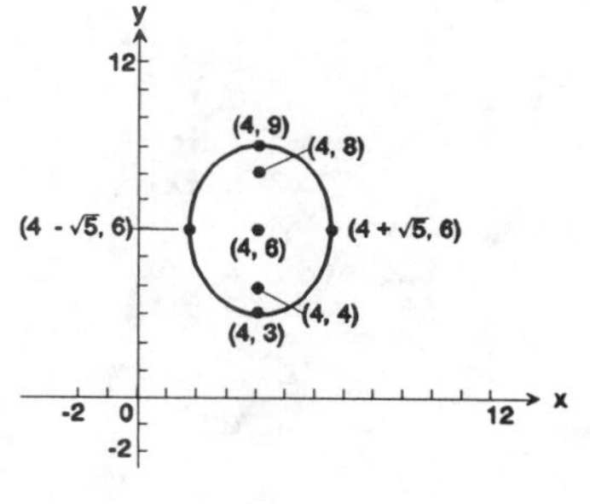

First of all, the center is the midpoint between V_1 and V_2:
$C(4, 6)$, and $h = 4$, $k = 6$.

Then, $a = d(C, V) = 3$, $c = d(C, F) = 2$

and $b^2 = a^2 - c^2 = 5$, so $b = \sqrt{5}$

Now V_1, V_2, F, and C all lie on the vertical line $x = 4$, so the major axis is parallel to the y-axis, and the equation is of the form:

$$\frac{(x-h)^2}{b^2} + \frac{(y-k)^2}{a^2} = 1 \quad \text{or} \quad \frac{(x-4)^2}{5} + \frac{(y-6)^2}{9} = 1$$

45. We are given: Foci: $F_1(5, 1)$ and $F_2(-1, 1)$
Length of major axis: 8

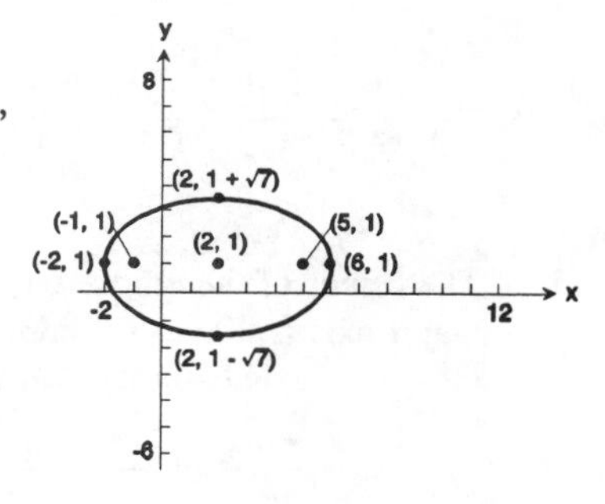

Then the center is midway between F_1 and F_2: $C(2, 1)$, so $h = 2$, $k = 1$. The length of the major axis is $2a$, so $a = 4$, and $c = d(C, F_1) = 3$. Therefore, $b^2 = a^2 - c^2 = 7$. The major axis (the line $y = 1$) is parallel to the x-axis, so we have:

$$\frac{(x-h)^2}{a^2} + \frac{(y-k)^2}{b^2} = 1 \quad \text{or} \quad \frac{(x-2)^2}{16} + \frac{(y-1)^2}{7} = 1$$

47. We are given: Center: $C(1, 2)$
Focus: $F(4, 2)$
Contains the point: $(1, 3)$

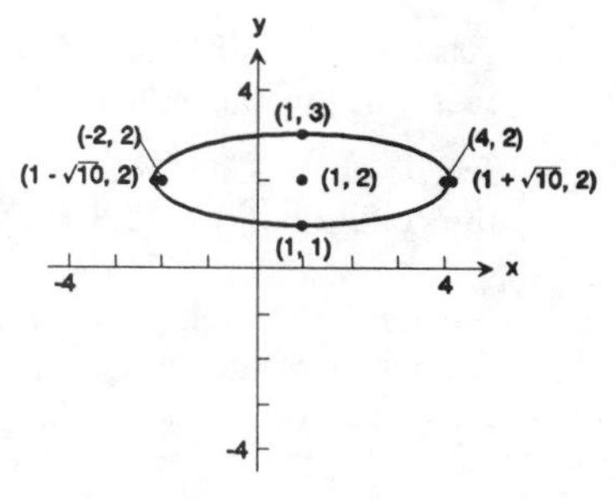

From the center $h = 1$, $k = 2$. Also, $c = d(C, F) = 3$, so $c^2 = 9$. The major axis is parallel to the x-axis, so we have:

$$\frac{(x-h)^2}{a^2} + \frac{(y-k)^2}{b^2} = 1 \quad \text{or} \quad \frac{(x-1)^2}{a^2} + \frac{(y-2)^2}{b^2} = 1$$

But the point $(1, 3)$ must satisfy the equation since it lies on the graph. Therefore:

$$\frac{0}{a^2} + \frac{1}{b^2} = 1 \text{ or } b^2 = 1, \text{ so } b = 1 \text{ and } a^2 = b^2 + c^2 = 10.$$

Thus, $\dfrac{(x-1)^2}{10} + \dfrac{(y-2)^2}{1} = 1$

49. We are given: Center: $C(1, 2)$
Vertex: $V(4, 2)$
Contains the point: $(1, 3)$

From the center $h = 1$, $k = 2$. Also, $a = d(C, V) = 3$, so $a^2 = 9$. The major axis is parallel to the x-axis, so we have:

$$\frac{(x - h)^2}{a^2} + \frac{(y - k)^2}{b^2} = 1 \quad \text{or} \quad \frac{(x - 1)^2}{a^2} + \frac{(y - 2)^2}{b^2} = 1$$

But the point $(1, 3)$ must satisfy the equation since it lies on the graph. Therefore:

$$\frac{0}{a^2} + \frac{1}{b^2} = 1 \text{ or } b^2 = 1, \text{ so } b = 1 \text{ and } c^2 = a^2 - b^2 = 8.$$

Thus, $\dfrac{(x - 1)^2}{9} + \dfrac{(y - 2)^2}{1} = 1$

51.

$$y = \sqrt{16 - 4x^2}$$
$$y^2 = 16 - 4x^2, \quad y \geq 0$$
$$4x^2 + y^2 = 16, \quad y \geq 0$$
$$\frac{x^2}{4} + \frac{y^2}{16} = 1, \quad y \geq 0$$

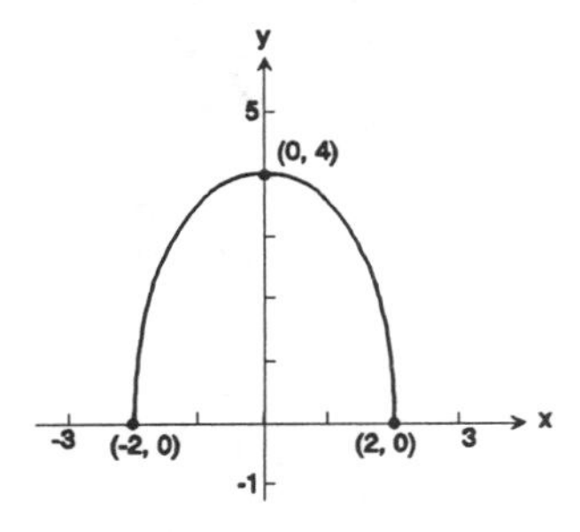

53.

$$y = -\sqrt{64 - 16x^2}$$
$$y^2 = 64 - 16x^2, \quad y \leq 0$$
$$16x^2 + y^2 = 64, \quad y \leq 0$$
$$\frac{x^2}{4} + \frac{y^2}{64} = 1, \quad y \leq 0$$

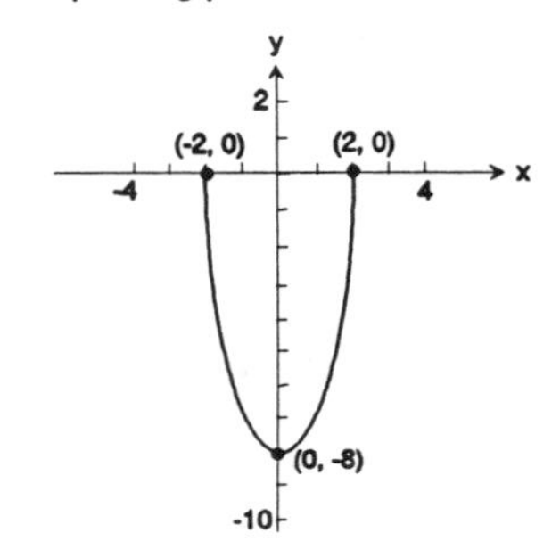

55. The center of the ellipse will be at (0, 0) due to the nice positioning of the axes. The length of the major axis is 20, so we know $a = 10$. The length of *half* the minor axis is 6, i.e., $b = 6$. Finally, the major axis is horizontal, so we have:

$$\frac{(x - h)^2}{a^2} + \frac{(y - k)^2}{b^2} = 1 \quad \text{or} \quad \frac{x^2}{100} + \frac{y^2}{36} = 1$$

57. Assume the half-ellipse formed by the whispering galley is centered at (0, 0). Since the hall is 100 feet long, we have $2a = 100$, so $a = 50$. The distance from the center of the ellipse to the foci is 25 feet, so $c = 25$. Since $b^2 = a^2 - c^2$, $b^2 = 50^2 - 25^2 = 1875$; therefore, $b = \sqrt{1875} = 43.3$. So the ceiling will be 43.3 feet high at the center.

59. Situate the semielliptical arch so that the x-axis coincides with the water and the y-axis passes through the center of the arch. Since the bridge has a span of 120 feet, the length of the major axis is 120, so $2a = 120$ or $a = 60$. The length of half the minor axis is 25, so $b = 25$. Therefore, we have:

$$\frac{x^2}{3600} + \frac{y^2}{625} = 1$$

When $x = 10$ feet from center:

$$\frac{10^2}{3600} + \frac{y^2}{625} = 1$$
$$\frac{y^2}{625} = 1 - \frac{100}{3600}$$
$$y^2 = 625 \cdot \frac{3500}{3600}$$
$$y \approx 24.65 \text{ feet}$$

When $x = 30$ feet from center:

$$\frac{30^2}{3600} + \frac{y^2}{625} = 1$$
$$y \approx 21.65 \text{ feet}$$

When $x = 50$ feet from center:

$$\frac{50^2}{3600} + \frac{y^2}{625} = 1$$
$$y \approx 13.82 \text{ feet}$$

61. First we need to find an equation for the ellipse. Let the center of the ellipse be the origin, (0, 0), with the x-axis at ground-level. Then the major axis has length 40, so we know $a = 20$, and the length of ***half*** the minor axis is 15, i.e., $b = 15$. Since the major axis is horizontal, we have:

$$\frac{x^2}{a^2} + \frac{y^2}{b^2} = 1 \text{ or } \frac{x^2}{400} + \frac{y^2}{225} = 1$$

We wish to find y when $x = 0, \pm 10, \pm 20$, so we solve for y:

$$\frac{x^2}{400} + \frac{y^2}{225} = 1$$
$$\frac{y^2}{225} = 1 - \frac{x^2}{400}$$
$$\frac{y^2}{225} = \frac{400 - x^2}{400}$$
$$y^2 = 225\left(\frac{400 - x^2}{400}\right)$$
$$y^2 = \frac{225}{400}(400 - x^2)$$
$$y = \sqrt{\frac{225}{400}}\sqrt{400 - x^2}$$

or $$y = \frac{15}{20}\sqrt{400 - x^2} = \frac{3}{4}\sqrt{400 - x^2}$$

x	$y = \frac{3}{4}\sqrt{400 - x^2}$
0	$\frac{3}{4}\sqrt{400} = \frac{3}{4}(20) = 15$
± 10	$\frac{3}{4}\sqrt{400 - 100} \approx 12.99$
± 20	$\frac{3}{4}\sqrt{400 - 400} = 0$

63. If the mean distance is 93 million miles, then $a = 93$. Therefore, the length of the major axis is 186. So, the perihelion is $186 - 94.5 = 91.5$ million miles. The distance from the center of the ellipse to the sun (the focus) is $93 - 91.5 = 1.5$ million miles; therefore, $c = 1.5$. Since $b^2 = a^2 - c^2$, we have:

$$b^2 = 93^2 - 1.5^2$$
$$b^2 = 8646.75$$

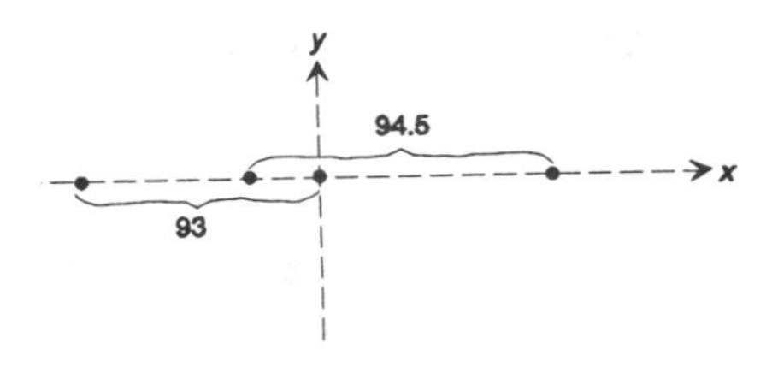

So, the equation of the earth's orbit around the sun is: $\dfrac{x^2}{93^2} + \dfrac{y^2}{8646.75} = 1$

65. The mean distance is $507 - 23.2 = 483.8$ million miles. Therefore, the perihelion is $483.3 - 23.2 = 460.6$ million miles. So, since $a = 483.8$ and $c = 23.2$, we can find b using $b^2 = a^2 - c^2$. Therefore, $b^2 = 483.8^2 - 23.2^2 = 233524.2$. So, the equation for the orbit of Jupiter is

$$\frac{x^2}{483.8^2} + \frac{y^2}{233524.2} = 1.$$

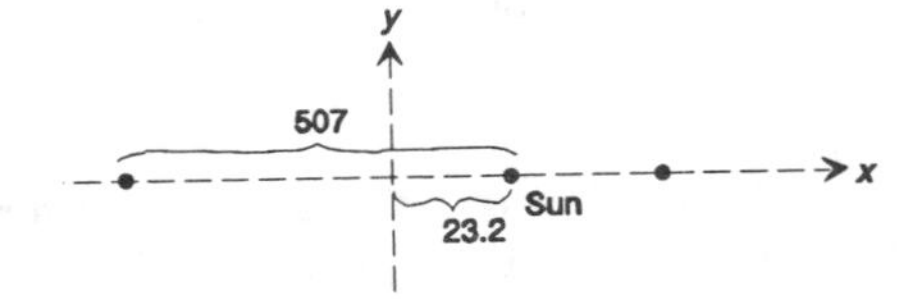

67. If the x-axis is placed along the 100 ft. portion and the y-axis along the 50 ft. portion, one equation for the ellipse is:

$$\frac{x^2}{(50)^2} + \frac{y^2}{(25)^2} = 1$$

When $x = 40$, then $\dfrac{(40)^2}{(50)^2} + \dfrac{y^2}{(25^2)} = 1$

$$\frac{y^2}{(25)^2} = 1 - \left[\frac{4}{5}\right]^2 = \frac{9}{25}$$

$$y^2 = (25)(9)$$

$$y = (5)(3) = 15$$

The width 10 feet from the side is 30 feet.

69. (a) $Ax^2 + Cy^2 + F = 0$

$$Ax^2 + Cy^2 = -F$$

If A and C are the same sign and F is of the opposite sign, then the equation takes the form $x^2/(-F/A) + y^2/(-F/C) = 1$, where $-F/A$ and $-F/C$ are positive. This is the equation of an ellipse with center at (0, 0).

(b) If $A = C$, the equation may be written as $x^2 + y^2 = -F/A$. This is the equation of a circle with center at (0, 0) and radius equal to $\sqrt{-F/A}$.

71. (a) $e = \dfrac{c}{a}$ is close to zero when c is close to zero. Since $c^2 = a^2 - b^2$, c is close to zero when $a^2 \approx b^2$. Hence, the ellipse is close to a circle.

(b) $e = \dfrac{c}{a} = \dfrac{1}{2}$ when $c = 1$ and $a = 2$. Thus, $b^2 = a^2 - c^2 = 4 - 1 = 3$. Hence, the ellipse is oval.

(c) $e = \frac{c}{a}$ is close to 1 when $c \approx a$ or $c^2 \approx a^2$. Hence, $a^2 - c^2$ will be close to zero. Since $b^2 = a^2 - c^2$, then b^2 is close to zero. Thus, the ellipse is elongated with the length of the minor axis small in comparison to the major axis.

7.4 The Hyperbola

1. B

3. A

For Problems 5–14, refer to Table 4. There we see that in order to find the equation of a hyperbola, we need to determine, h, k, a, b, and c to decide whether the transverse axis is horizontal or vertical.

We will use the following facts:

(1) (h, k) are the coordinates of the center of the hyperbola, which is midway between the vertices, and also midway between the foci.

(2) a = the distance from the center to either vertex.

(3) $b^2 = c^2 - a^2$, where c = the distance from the center to either focus.

(4) The center, the vertices and the foci all lie on the transverse axis.

5. We are given: Center: $C(0, 0)$
Focus: $F_2(3, 0)$
Vertex: $V_2(1, 0)$

Please refer to the paragraph above.
By (1) we have $h = 0, k = 0$
By (2), $a = d(C, V_2) = 1$
By (3), $c = d(C, F_2) = 3$, and

$$b^2 = c^2 - a^2 = 8, \text{ so that } b = \sqrt{8} = 2\sqrt{2}$$

By (4), since C, F_2 and V_2 lie on the horizontal line $y = 0$, the transverse axis is parallel to the x-axis.

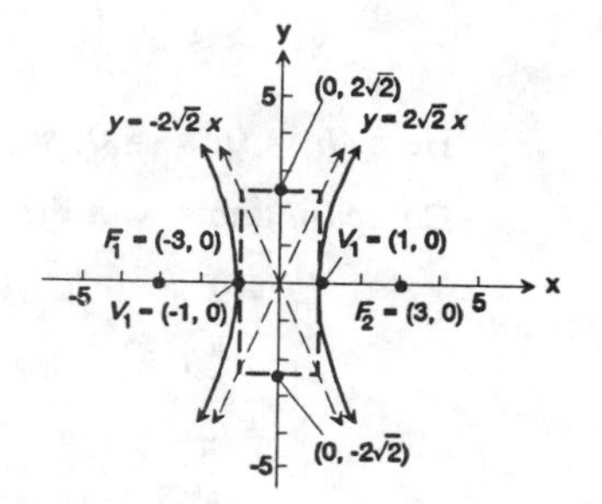

Then by Table 4, the equation is: $\frac{(x - h)^2}{a^2} - \frac{(y - k)^2}{b^2} = 1$ or $\frac{x^2}{1} - \frac{y^2}{8} = 1$

As an aid in sketching the graph, we locate the asymptotes of the hyperbola. First, plot the points that lie on the conjugate axis (perpendicular to the transverse axis) a distance b from the center: $\left(0, -2\sqrt{2}\right)$ and $\left(0, 2\sqrt{2}\right)$. These two points, together with the vertices, determine a rectangle whose diagonals are the asymptotes of the hyperbola.

7. Here we are given: Center: $C(0, 0)$
Focus: $F_1(0, -6)$
Vertex: $V_2(0, 4)$

Please refer to the paragraph before the solution to Problem 9.
By (1), $h = 0, k = 0$
By (2), $a = d(C, V_2) = 4$
By (3), $c = d(C, F_1) = 6$, and $b^2 = c^2 - a^2 = 20$, so that

$$b = \sqrt{20} = 2\sqrt{5}$$

By (4), the transverse axis is the vertical line $x = 0$.

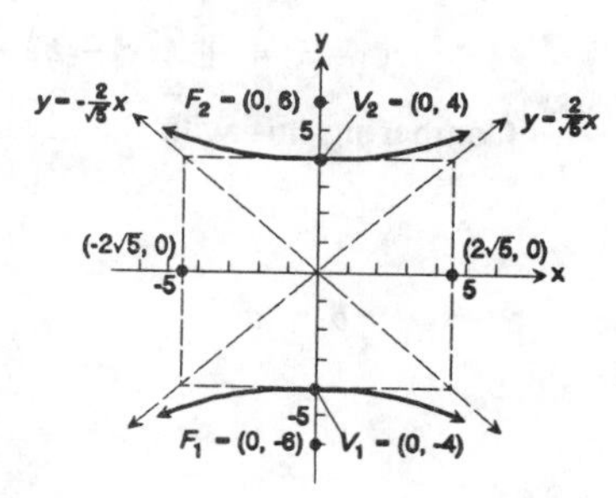

Hence, from Table 4, we have: $\frac{(y - k)^2}{a^2} - \frac{(x - h)^2}{b^2} = 1$ or $\frac{y^2}{16} - \frac{x^2}{20} = 1$

9. We are given: Foci: $F_1(-5, 0)$ and $F_2(5, 0)$
Vertex: $V_2\ (3, 0)$

Refer to the paragraph before the solution to Problem 9.

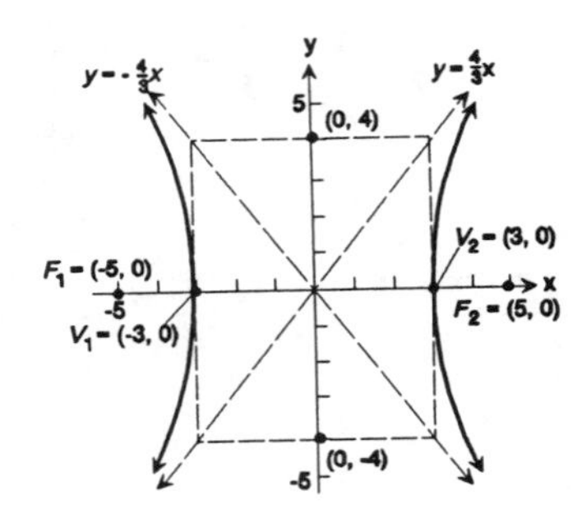

By (1), the center is the midpoint between F_1 and F_2: $C(0, 0)$, so
$h = 0, k = 0.$

By (2), $a = d(C, V_2) = 3$

By (3), $c = d(C, F_1) = 5$, and $b^2 = c^2 - a^2 = 16$
$b = 4$

By (4), the transverse axis is the horizontal line $y = 0$.

Therefore, the equation is:

$$\frac{(x - h)^2}{a^2} - \frac{(y - k)^2}{b^2} = 1 \quad \text{or} \quad \frac{x^2}{9} - \frac{y^2}{16} = 1$$

11. We are given: Vertices: $V_1(0, -6)$ and $V_2(0, 6)$
Asymptote: $y = 2x$

The center is the midpoint between V_1 and V_2: $C(0, 0)$. Then $a = d(C, V_1) = 6$. The transverse axis is the vertical line $x = 0$.
From Table 4, we see that the asymptotes of a hyperbola with a vertical transverse axis are:

$$y - k = \pm\frac{a}{b}(x - h)$$

Here, $h = 0$, $k = 0$, so one asymptote would be: $y = \frac{a}{b}x$

Comparing this with the given asymptote, $y = 2x$, we find:

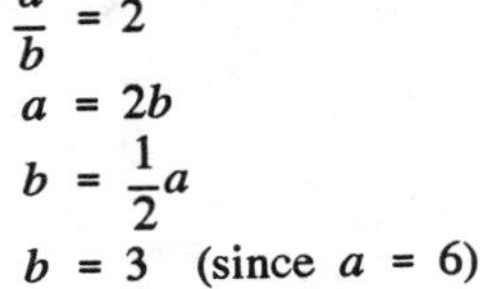

$$\frac{a}{b} = 2$$
$$a = 2b$$
$$b = \frac{1}{2}a$$
$$b = 3 \quad \text{(since } a = 6\text{)}$$

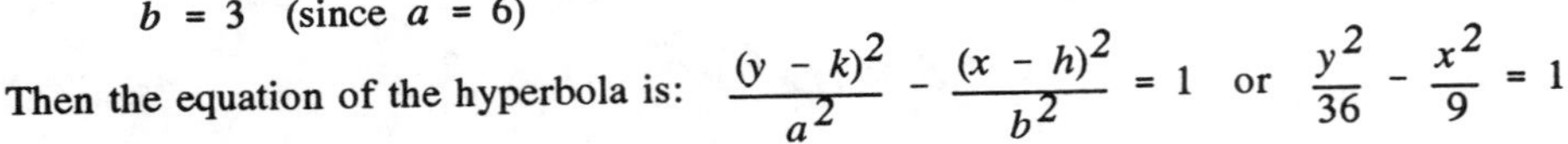

Then the equation of the hyperbola is: $\frac{(y - k)^2}{a^2} - \frac{(x - h)^2}{b^2} = 1$ or $\frac{y^2}{36} - \frac{x^2}{9} = 1$

13. Here we are given: Foci: $F_1(-4, 0)$ and $F_2(4, 0)$
Asymptote: $y = -x$

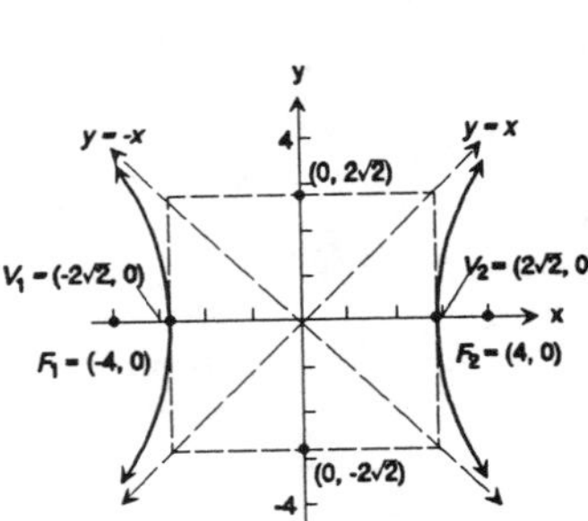

The center is the midpoint between F_1 and F_2: $C(0, 0)$. The transverse axis is the ***horizontal*** line $y = 0$.

By Table 4, the asymptotes are:

$$y - k = \pm\frac{b}{a}(x - h) \quad \text{or} \quad y = \pm\frac{b}{a}x \quad (\text{since } h = 0, k = 0)$$

Comparing this with the given asymptote, $y = -x$, we see:

$$-\frac{b}{a} = -1$$

$$b = a$$

Now $c = d(C, F_1) = 4$, and

$$\begin{aligned} b^2 &= c^2 - a^2 \\ a^2 + b^2 &= c^2 \\ a^2 + b^2 &= 16 \qquad (c = 4) \\ a^2 + a^2 &= 16 \qquad (b = a) \\ 2a^2 &= 16 \\ a^2 &= 8 \\ a &= \sqrt{8} = 2\sqrt{2} \\ b &= \sqrt{8} = 2\sqrt{2} \qquad (b = a) \end{aligned}$$

The equation is: $\dfrac{(x - h)^2}{a^2} - \dfrac{(y - k)^2}{b^2} = 1$ or $\dfrac{x^2}{8} - \dfrac{y^2}{8} = 1$

15. $\dfrac{x^2}{25} - \dfrac{y^2}{9} = 1$

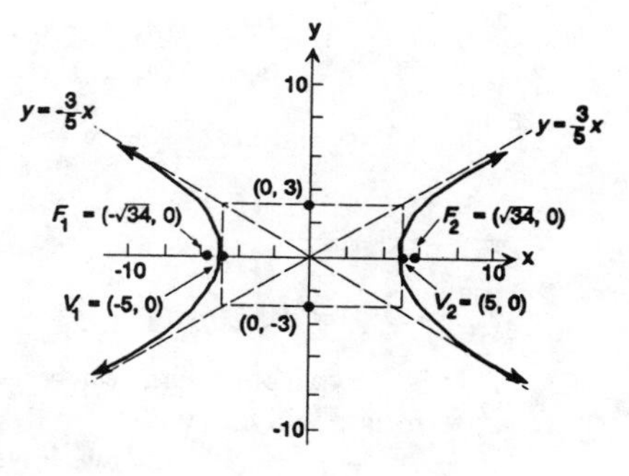

This is in the form: $\dfrac{(x - h)^2}{a^2} - \dfrac{(y - k)^2}{b^2} = 1$

with $h = 0$, $k = 0$, $a = 5$, and $b = 3$.

From Table 4, we have:

$c^2 = a^2 + b^2 = 34$, so $c = \sqrt{34}$

Center: $(h, k) = (0, 0)$

Transverse axis: Parallel to x-axis, and contains the Center: $y = 0$

Foci: $(h \pm c, k)$: $\left(-\sqrt{34}, 0\right)$ and $\left(\sqrt{34}, 0\right)$

Vertices: $(h \pm a, k)$: $(-5, 0)$ and $(5, 0)$

Asymptotes: $y - k = \pm\dfrac{b}{a}(x - h)$, or $y = \pm\dfrac{3}{5}x$

(Lines through (0, 0) with slopes $\dfrac{3}{5}$ and $\dfrac{-3}{5}$.)

17. $4x^2 - y^2 = 16$

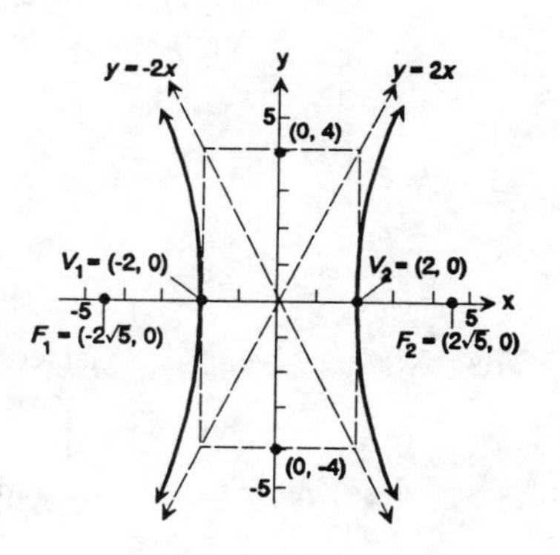

$\dfrac{4x^2}{16} - \dfrac{y^2}{16} = 1$ Obtain a 1 on the right-hand side.

$\dfrac{x^2}{4} - \dfrac{y^2}{16} = 1$ Simplify.

This is a hyperbola with transverse axis parallel to the x-axis (since the x^2-term is the positive one), with $h = 0$, $k = 0$, $a^2 = 4$ and $b^2 = 16$. Then $c^2 = a^2 + b^2 = 20$.

Therefore, $a = 2$

$b = 4$

$c = \sqrt{20} = 2\sqrt{5}$

Center: $(h, k) = (0, 0)$

Transverse axis: $y = 0$

Foci: $(h \pm c, k)$: $(-2\sqrt{5}, 0)$ and $(2\sqrt{5}, 0)$

Vertices: $(h \pm a, k)$: $(-2, 0)$ and $(2, 0)$

Asymptotes: $y - k = \pm\dfrac{b}{a}(x - h)$, or $y = \pm 2x$

(Lines through (0, 0) with slopes 2 and -2.)

19. $y^2 - 9x^2 = 9$

$$\frac{y^2}{9} - \frac{x^2}{1} = 1$$

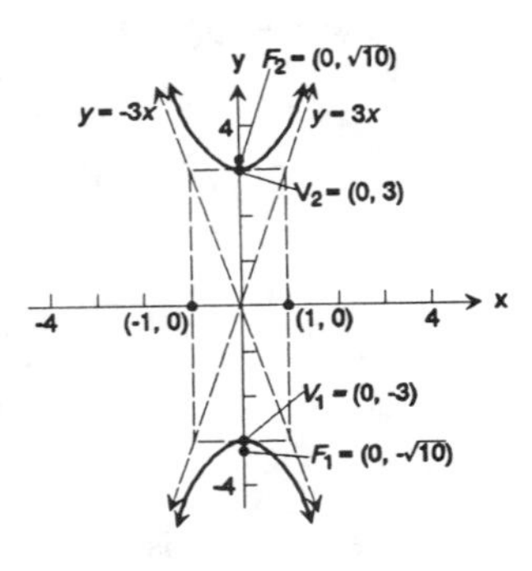

This is a hyperbola with transverse axis parallel to the y-axis (since the y^2 term is positive), with $h = 0$, $k = 0$, $a^2 = 9$ and $b^2 = 1$.

Therefore, $a = 3$

$b = 1$

$c^2 = a^2 + b^2 = 10$, so $c = \sqrt{10}$

Center: $(h, k) = (0, 0)$

Transverse axis: $x = 0$

Foci: $(h, k \pm c)$: $(0, -\sqrt{10})$ and $(0, \sqrt{10})$

Vertices: $(h, k \pm a)$: $(0, -3)$ and $(0, 3)$

Asymptotes: $y - k = \pm \frac{a}{b}(x - h)$, or $y = \pm 3x$

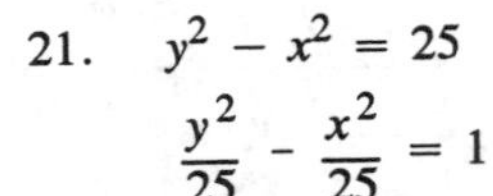

21. $y^2 - x^2 = 25$

$$\frac{y^2}{25} - \frac{x^2}{25} = 1$$

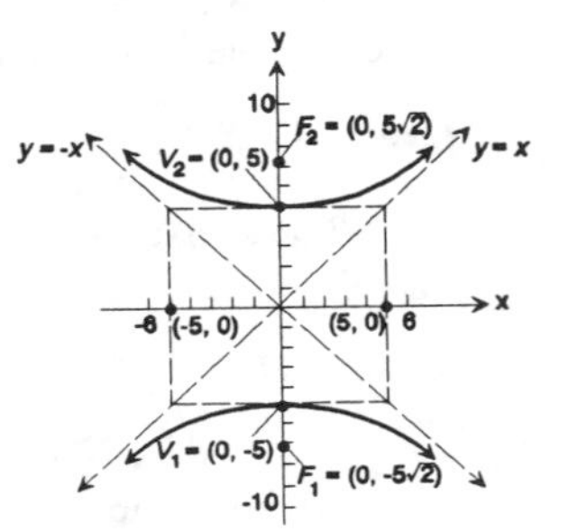

This is a hyperbola with transverse axis parallel to the y-axis (since the y^2-term is positive), with $h = 0$, $k = 0$, $a^2 = 25$, $b^2 = 25$. Therefore, $c^2 = a^2 + b^2 = 50$, and we have:

$a = 5$

$b = 5$

$c = \sqrt{50} = 5\sqrt{2}$

Center: $(h, k) = (0, 0)$

Transverse axis: $x = 0$

Foci: $(h, k \pm c)$: $(0, -5\sqrt{2})$ and $(0, 5\sqrt{2})$

Vertices: $(h, k \pm a)$: $(0, -5)$ and $(0, 5)$

Asymptotes: $y - k = \pm \frac{a}{b}(x - h)$, or $y = \pm x$

23. $x^2 - y^2 = 1$

25. $\frac{y^2}{36} - \frac{x^2}{9} = 1$

27. We are given: Center: $C(4, -1)$

Focus: $F_2(7, -1)$

Vertex: $V_2(6, -1)$

Please refer to the paragraph before the solution to Problem 9.

By (1), $h = 4$, $k = -1$

By (2), $a = d(C, V_2) = 2$

By (3), $c = d(C, F_2) = 3$, and

$b^2 = c^2 - a^2 = 5$, so that $b = \sqrt{5}$

By (4), the Transverse axis is the horizontal line $y = -1$

Then, by Table 4, we have: $\frac{(x - h)^2}{a^2} - \frac{(y - k)^2}{b^2} = 1$ or $\frac{(x - 4)^2}{4} - \frac{(y + 1)^2}{5} = 1$

29. We are given: Center: $C(-3, -4)$
Focus: $F_1(-3, -8)$
Vertex: $V_2(-3, -2)$

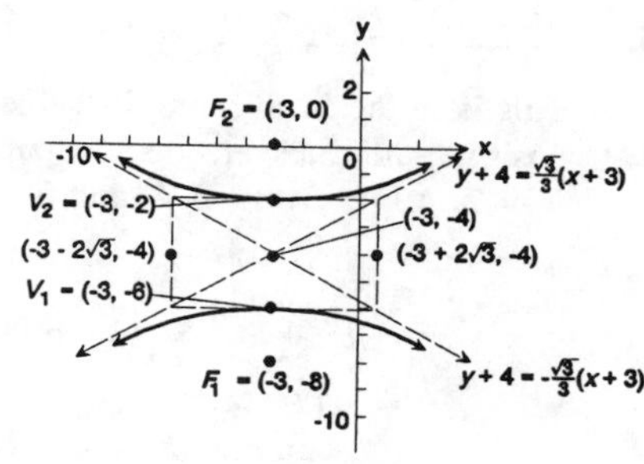

Refer to the paragraph before the solution to Problem 9.
By (1), $h = -3$, $k = -4$
By (2), $a = d(C, V_2) = 2$
By (3), $c = d(C, F_1) = 4$

$$b^2 = c^2 - a^2 = 12$$

$$b = \sqrt{12} = 2\sqrt{3}$$

By (4), the transverse axis is the vertical line $x = -3$

Then, by Table 4, we have: $$\frac{(y-k)^2}{a^2} - \frac{(x-h)^2}{b^2} = 1 \quad \text{or} \quad \frac{(y+4)^2}{4} - \frac{(x+3)^2}{12} = 1$$

31. We are given: Foci: $F_1(3, 7)$ and $F_2(7, 7)$
Vertex: $V_1(6, 7)$

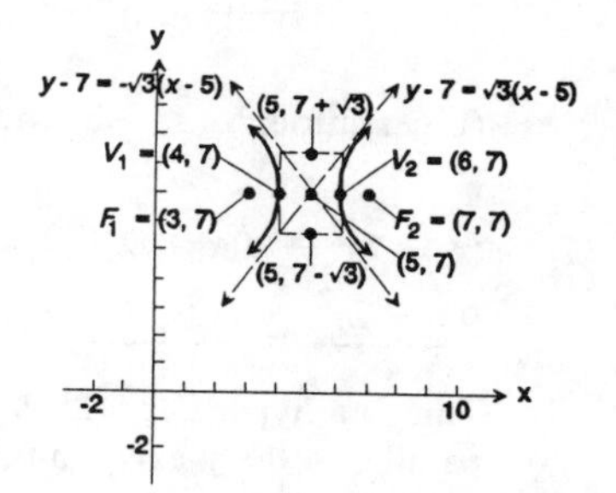

Refer to the paragraph before the solution to Problem 9.
By (1) the Center is midway between F_1 and F_2: $C(5, 7)$, so
$h = 5$, $k = 7$.
By (2), $a = d(C, V_1) = 1$
By (3), $c = d(C, F_1) = 2$

$$b^2 = c^2 - a^2 = 3$$

$$b = \sqrt{3}$$

By (4), the transverse axis is the *horizontal* line $y = 7$

By Table 4, $$\frac{(x-h)^2}{a^2} - \frac{(y-k)^2}{b^2} = 1 \quad \text{or} \quad \frac{(x-5)^2}{1} - \frac{(y-7)^2}{3} = 1$$

33. We are given: Vertices: $V_1(-1, -1)$ and $V_2(3, -1)$

Asymptote: $$\frac{(x-1)}{2} = \frac{(y+1)}{3}$$

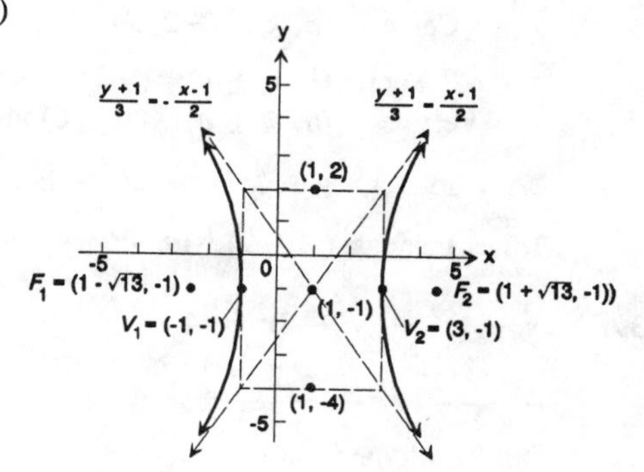

Refer to the paragraph before the solution to Problem 9.
By (1) the center is $C(1, -1)$.
By (2), $a = d(C, V_1) = 2$
By (4), the transverse axis is the *horizontal* line
$y = -1$.
We still need b, and we don't know c. From Table 4, the asymptotes would be:

$$y - k = \pm\frac{b}{a}(x - h), \text{ or } y + 1 = \pm\frac{b}{a}(x - 1)$$

Compare that formula with the *given* asymptote:

$$\frac{(x-1)}{2} = \frac{(y+1)}{3}, \text{ or } y + 1 = \frac{3}{2}(x - 1)$$

We see that $\frac{b}{a} = \frac{3}{2}$ or $b = \frac{3a}{2}$

$$b = \frac{6}{2} \quad (\text{since } a = 2)$$

$$b = 3$$

Therefore, the equation is: $$\frac{(x-h)^2}{a^2} - \frac{(y-k)^2}{b^2} = 1 \quad \text{or} \quad \frac{(x-1)^2}{4} - \frac{(y+1)^2}{9} = 1$$

35. $\dfrac{(x-2)^2}{4} - \dfrac{(y+3)^2}{9} = 1$

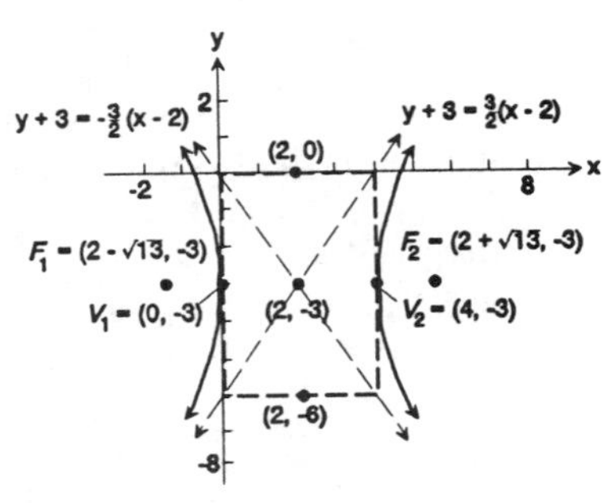

This is in the form found in Table 4. Since the x^2-term is positive, the transverse axis is parallel to the x-axis. We have $h = 2$, $k = -3$, $a^2 = 4$, and $b^2 = 9$. Therefore,

$a = 2$
$b = 3$
$c^2 = a^2 + b^2 = 13$, so that
$c = \sqrt{13}$

Center: $(h, k) = (2, -3)$

Foci: $(h \pm c, k)$: $\left(2 - \sqrt{13}, -3\right)$ and $\left(2 + \sqrt{13}, -3\right)$

Vertices: $(h \pm a, k)$: $(0, -3)$ and $(4, -3)$

Asymptotes: $y - k = \pm\dfrac{b}{a}(x - h)$, or $y + 3 = \pm\dfrac{3}{2}(x - 2)$

(Lines through $(3, -2)$ with slopes $\dfrac{3}{2}$ and $-\dfrac{3}{2}$.)

37. $(y - 2)^2 - 4(x + 2)^2 = 4$

$\dfrac{(y-2)^2}{4} - \dfrac{(x+2)^2}{1} = 1$

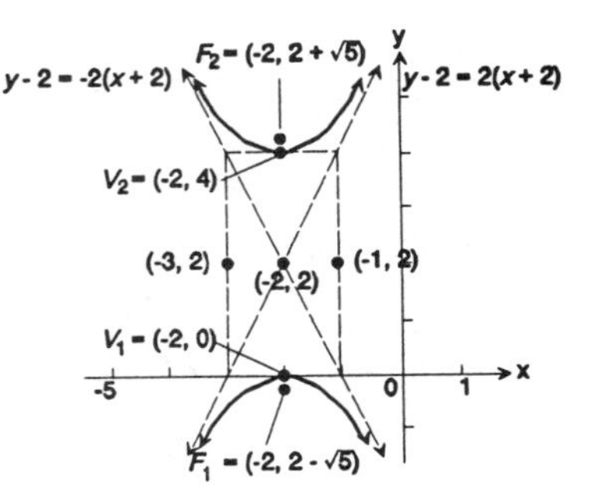

This is a hyperbola with $h = -2$, $k = 2$; the transverse axis is parallel to the y-axis, and $a^2 = 4$, $b^2 = 1$. Then:

$a = 2$
$b = 1$
$c^2 = a^2 + b^2 = 5$, so
$c = \sqrt{5}$

Center: (h, k): $(-2, 2)$

Foci: $(h, k \pm c)$: $(-2, 2 - \sqrt{5})$ and $(-2, 2 + \sqrt{5})$

Vertices: $(h, k \pm a)$: $(-2, 0)$ and $(-2, 4)$

Asymptotes: $y - k = \pm\dfrac{a}{b}(x - h)$, or $y - 2 = \pm 2(x + 2)$

(Lines through $(-2, 2)$ with slopes 2 and -2.)

39. $(x + 1)^2 - (y + 2)^2 = 4$

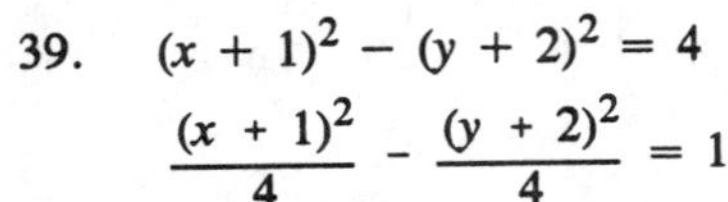

$\dfrac{(x+1)^2}{4} - \dfrac{(y+2)^2}{4} = 1$

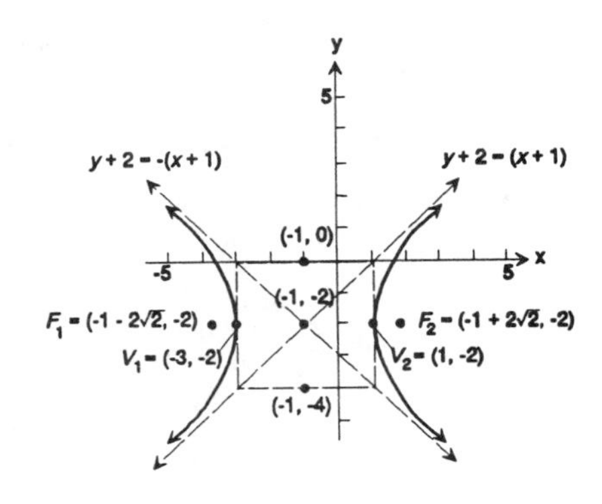

This is a hyperbola with $h = -1$, $k = -2$; the transverse axis is parallel to the x-axis, and $a^2 = 4$, $b^2 = 4$. Then:

$a = 2$
$b = 2$
$c^2 = a^2 + b^2 = 8$, so $c = 2\sqrt{2}$

Center: $(h, k) = (-1, -2)$

Foci: $(h \pm c, k)$: $\left(-1 - 2\sqrt{2}, -2\right)$ and $\left(-1 + 2\sqrt{2}, -2\right)$

Vertices: $(h \pm a, k)$: $(-3, -2)$ and $(1, -2)$

Asymptotes: $y - k = \pm\dfrac{b}{a}(x - h)$, or $y + 2 = \pm(x + 1)$

(Lines through $(-1, -2)$ with slopes 1 and -1.)

41.
$$x^2 - y^2 - 2x - 2y - 1 = 0$$
$$(x^2 - 2x + 1) - (y^2 + 2y + 1) = 1 + 1 - 1$$
$$(x - 1)^2 - (y + 1)^2 = 1$$

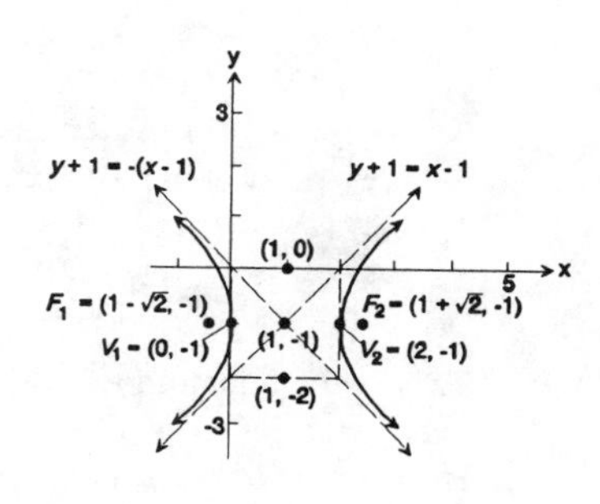

This is a hyperbola with $h = 1$, $k = -1$; the transverse axis is parallel to the x-axis, and $a^2 = 1$, $b^2 = 1$. Then:

$a = 1$

$b = 1$

$c^2 = a^2 + b^2 = 2$, so

$c = \sqrt{2}$

Center: $(h, k) = (1, -1)$

Foci: $(h \pm c, k)$: $\left(1 - \sqrt{2}, -1\right)$ and $\left(1 + \sqrt{2}, -1\right)$

Vertices: $(h \pm a, k)$: $(0, -1)$ and $(2, -1)$

Asymptotes: $y + 1 = \pm(x - 1)$

(Lines through $(1, -1)$ with slopes 1 and -1.)

43.
$$y^2 - 4x^2 - 4y - 8x - 4 = 0$$
$$(y^2 - 4y + 4) - 4(x^2 + 2x + 1) = 4 + 4 - 4$$
$$(y - 2)^2 - 4(x + 1)^2 = 4$$
$$\frac{(y - 2)^2}{4} - (x + 1)^2 = 1$$

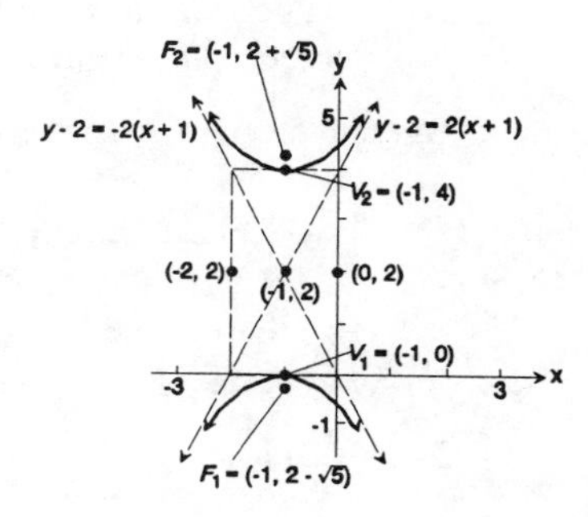

This is a hyperbola with $h = -1$, $k = 2$; the transverse axis is parallel to the y-axis, and $a^2 = 4$, $b^2 = 1$. Then:

$a = 2$

$b = 1$

$c^2 = a^2 + b^2 = 5$, so $c = \sqrt{5}$

Center: $(h, k) = (-1, 2)$

Foci: $(h, k \pm c)$: $\left(-1, 2 - \sqrt{5}\right)$ and $\left(-1, 2 + \sqrt{5}\right)$

Vertices: $(h, k \pm a)$: $(-1, 0)$ and $(-1, 4)$

Asymptotes: $y - 2 = \pm 2(x + 1)$

(Lines through $(-1, 2)$ with slopes 2 and -2.)

45.
$$4x^2 - y^2 - 24x - 4y + 16 = 0$$
$$4x^2 - 24x - y^2 - 4y = -16$$
$$4(x^2 - 6x + \underline{\quad}) - 1(y^2 + 4y + \underline{\quad}) = -16$$
$$4(x^2 - 6x + 9) - 1(y^2 + 4y + 4) = -16 + 36 - 4$$
$$4(x - 3)^2 - (y + 2)^2 = 16$$
$$\frac{(x - 3)^2}{4} - \frac{(y + 2)^2}{16} = 1$$

This is now in a form we can recognize: A hyperbola with transverse axis parallel to the x-axis (since the x^2-term is positive), with center at $C(3, -2)$, and $a^2 = 4$, and $b^2 = 16$. Then $c^2 = a^2 + b^2 = 20$, and we have:

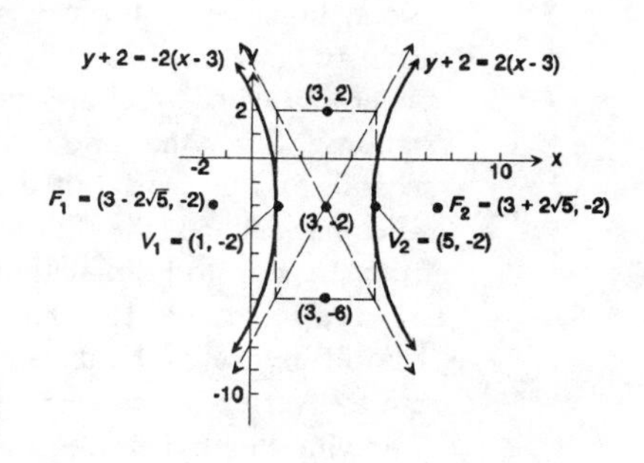

$a = 2$

$b = 4$

$c = \sqrt{20} = 2\sqrt{5}$

Center: $C(3, -2)$

Foci: $(h \pm c, k)$: $\left(3 - 2\sqrt{5}, -2\right)$ and $\left(3 + 2\sqrt{5}, -2\right)$

Vertices: $(h \pm a, k)$: $(1, -2)$ and $(5, -2)$

Asymptotes: $y - k = \pm\frac{b}{a}(x - h)$, or $y + 2 = \pm 2(x - 3)$

47.

$$y^2 - 4x^2 - 16x - 2y - 19 = 0$$
$$y^2 - 2y - 4x^2 - 16x = 19$$
$$(y^2 - 2y + \_\_\_) - 4(x^2 + 4x + \_\_\_) = 19$$
$$(y^2 - 2y + 1) - 4(x^2 + 4x + 4) = 19 + 1 - 16$$
$$(y - 1)^2 - 4(x + 2)^2 = 4$$
$$\frac{(y - 1)^2}{4} - \frac{(x + 2)^2}{1} = 1$$

This is the equation of a hyperbola with transverse axis parallel to the y-axis, with center at $C(-2, 1)$, with $a^2 = 4$ and $b^2 = 1$. Then $c^2 = a^2 + b^2 = 5$, and we have:

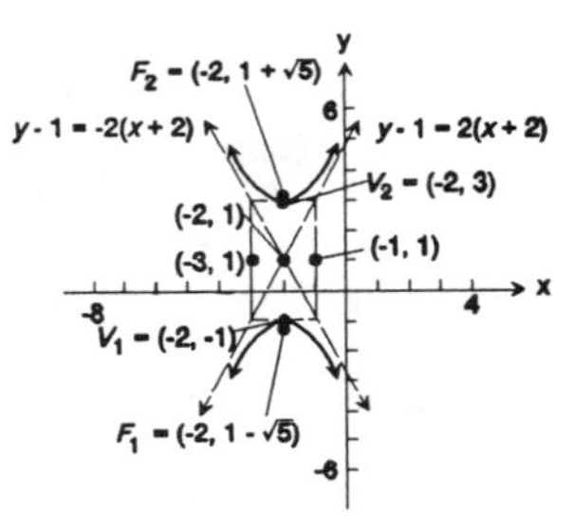

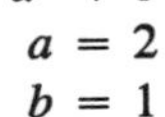

$a = 2$

$b = 1$

$c = \sqrt{5}$

Center: $C(-2, 1)$

Foci: $(h, k \pm c)$: $\left(-2,\ 1 - \sqrt{5}\right)$ and $\left(-2,\ 1 + \sqrt{5}\right)$

Vertices: $(h, k \pm a)$: $(-2, -1)$ and $(-2, 3)$

Asymptotes: $y - k = \pm\frac{a}{b}(x - h)$, or $y - 1 = \pm 2(x + 2)$

49.

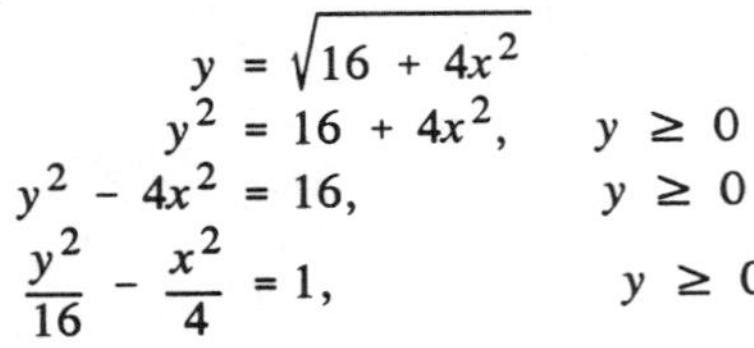

$$y = \sqrt{16 + 4x^2}$$
$$y^2 = 16 + 4x^2, \quad y \geq 0$$
$$y^2 - 4x^2 = 16, \quad y \geq 0$$
$$\frac{y^2}{16} - \frac{x^2}{4} = 1, \quad y \geq 0$$

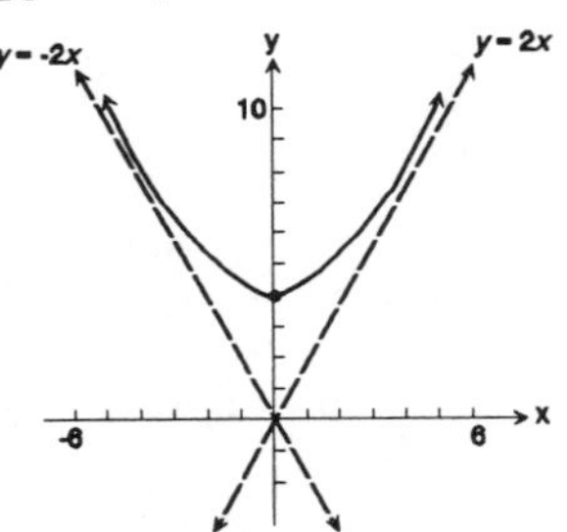

51.

$$y = -\sqrt{-25 + x^2}$$
$$y^2 = -25 + x^2, \quad y \leq 0$$
$$y^2 - x^2 = -25, \quad y \leq 0$$
$$\frac{y^2}{25} - \frac{x^2}{25} = -1, \quad y \leq 0$$
$$\frac{x^2}{25} - \frac{y^2}{25} = 1, \quad y \leq 0$$

53. (a) Set up a rectangular coordinate system so that the two stations lie on the x-axis and the origin is midway between them. The ship lies on a hyperbola whose foci are the locations of the two stations. Since the time difference is .00038 seconds and the speed of the signal is 186,000 miles per second, the difference of the distances from the ship to each station (foci) is:

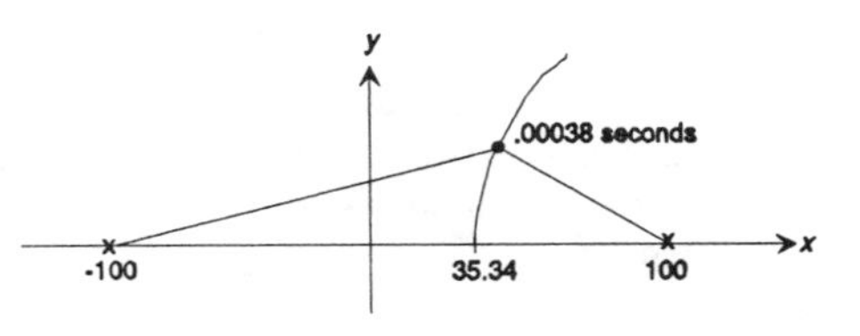

distance = (186,000)(.00038) = 70.68 miles

The difference of the distances from the ship to each station, 70.68, equals $2a$, so $a = 35.34$ and the vertex of the corresponding hyperbola is at (35.34, 0). Since the focus is at (100, 0), following this hyperbola, the ship would reach shore 64.66 miles from the master station.

(b) The ship should follow a hyperbola with vertex at (80, 0). For this hyperbola, $a = 80$, so the constant difference of the distances from the ship to each station is 160. The time difference the ship should look for is:

$$\text{time} = \frac{160}{186{,}000} = 0.00086 \text{ seconds}$$

(c) We need to find the equation of the hyperbola with vertex at (80, 0) and a focus at (100, 0). The form of the equation of this hyperbola is:

$$\frac{x^2}{a^2} - \frac{y^2}{b^2} = 1$$

where $a = 80$. Since $c = 100$ and $b^2 = c^2 - a^2$, we have $b^2 = 100^2 - 80^2 = 3600$. So, the equation of the hyperbola is:

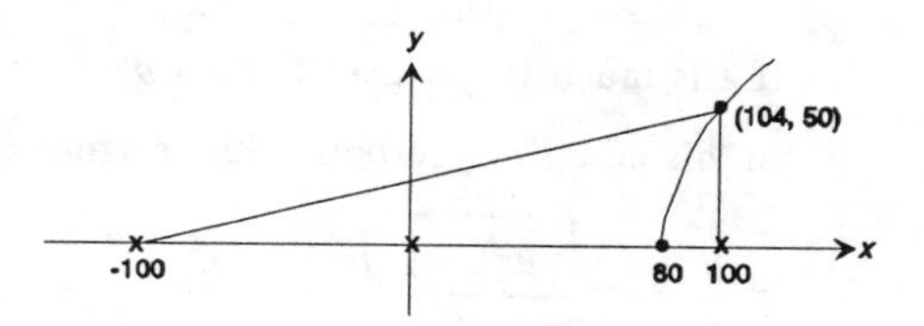

$$\frac{x^2}{6400} - \frac{y^2}{3600} = 1$$

Since the ship is 50 miles off shore, we have $y = 50$. Solve the above equation for x.

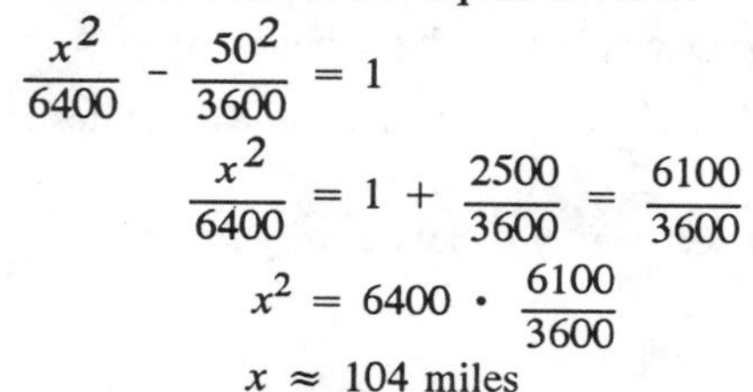

$$\frac{x^2}{6400} - \frac{50^2}{3600} = 1$$

$$\frac{x^2}{6400} = 1 + \frac{2500}{3600} = \frac{6100}{3600}$$

$$x^2 = 6400 \cdot \frac{6100}{3600}$$

$$x \approx 104 \text{ miles}$$

The ship is at the position (104, 50).

55. (a) Set up a rectangular coordinate system so the two devices lie on the x-axis and the origin is midway between them. The devices serve as foci to the hyperbola so $c = \frac{2000}{2} = 1000$. Since the explosion occurs 200 feet from point B, the vertex of the hyperbola will be (800, 0); therefore, $a = 800$. Since $b^2 = c^2 - a^2$, we have $b^2 = 1000^2 - 800^2 = 360{,}000$. Therefore, the equation of the hyperbola is:

$$\frac{x^2}{800^2} - \frac{y^2}{360{,}000} = 1$$

If $x = 1000$ feet, then we can find y.

$$\frac{1000^2}{800^2} - \frac{y^2}{360{,}000} = 1$$

$$y^2 = 360{,}000\left[\frac{360{,}000}{640{,}000}\right]$$

$$y = 450 \text{ feet}$$

Therefore, the second detonation should occur 450 feet above Point B.

57. By definition of the eccentricity, e,

$$e = \frac{c}{a}, \text{ or } c = ae$$

Therefore, if $e \approx 1$, then $c \approx a$, and $b^2 = c^2 - a^2 \approx 0$, so that b is close to 0.

Assume, for the sake of simplicity, that we have a hyperbola, centered at (0, 0), with transverse axis lying along the x-axis. The asymptotes are:

$$y = \pm\frac{b}{a}x,$$

i.e., lines through the origin with slope $\pm\frac{b}{a}$. Now, if $e \approx 1$, we have $b \approx 0$, so the slopes of the asymptotes are nearly 0. Hence, the asymptotes are nearly horizontal, so the hyperbola is very narrow.

On the other hand, if e is very large, we have:

$$c = ae \quad \text{and} \quad \begin{aligned} b^2 &= c^2 - a^2 \\ &= e^2a^2 - a^2 \\ &= (e^2 - 1)a^2, \text{ and} \end{aligned}$$

$$b = \left(\sqrt{e^2 - 1}\right)a$$

If e is much larger than 1, then b will be much larger than a.

In this case, a hyperbola with horizontal transverse axis will have asymptotes with slopes

$$\pm\frac{b}{a} = \pm\frac{\left(\sqrt{e^2 - 1}\right)a}{a} = \pm\sqrt{e^2 - 1} > 1.$$

Thus, the asymptotes will be nearly vertical, producing a ***wide*** hyperbola. As an example, look at the graph of the hyperbola in Problem 9. There, $e = \frac{c}{a} = \frac{4}{1} = 4$, and the asymptotes have slopes

$\pm\sqrt{e^2 - 1} = \pm\sqrt{15} \approx \pm 3.9$. As you can see, the hyperbola is very wide.

59. $\frac{x^2}{4} - y^2 = 1 \quad (a^2 = 4, b^2 = 1)$

is a hyperbola with ***horizontal*** transverse axis, centered at (0, 0) with asymptotes

$$y - k = \pm\frac{b}{a}(x - h), \text{ or } y = \pm\frac{1}{2}x$$

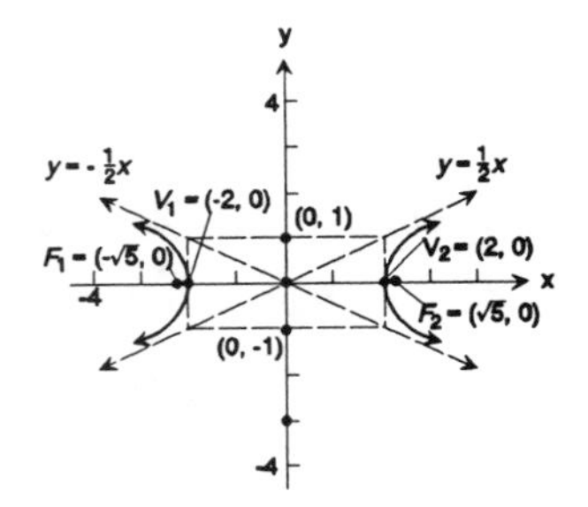

$y^2 - \frac{x^2}{4} = 1 \quad (a^2 = 1, b^2 = 4)$

is a hyperbola with ***vertical*** transverse axis, also is centered at (0, 0) and has asymptotes

$$y - k = \pm\frac{a}{b}(x - h), \text{ or } y = \pm\frac{1}{2}x$$

Since the two hyperbolas have the same asymptotes, they are conjugate.

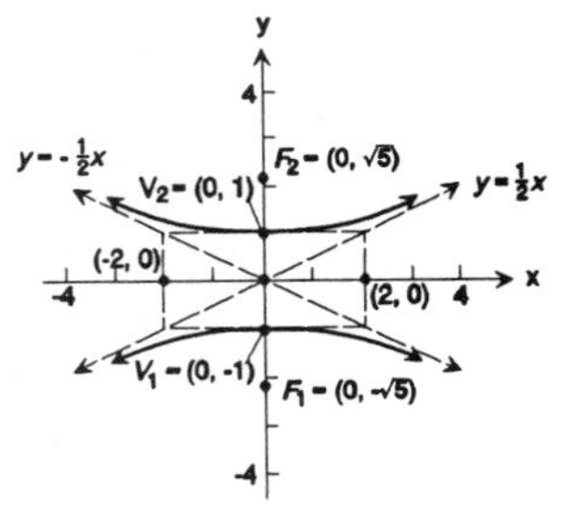

61. $Ax^2 + Cy^2 + F = 0 \qquad A \neq 0, C \neq 0, F \neq 0$

$Ax^2 + Cy^2 = -F$

Since A and C are of opposite signs and $F \neq 0$, this equation may be written as $x^2/-F/A) + y^2/(-F/C) = 1$, where $-F/A$ and $-F/C$ are opposite in sign. This is the equation of a hyperbola with center at (0, 0). The transverse axis is the x-axis if $-F/A > 0$. The transverse axis is the y-axis if $-F/A < 0$.

7 Chapter Review

For Problems 1–20, use the following rules:

I. *If only* ***one*** *variable is squared, the equation represents a parabola.*

II. *If* ***both*** *variables are squared, the equation is either an ellipse or a hyperbola.*

A. *If both the x^2-term and the y-term are positive, the graph is an ellipse.*

B. *If one of the two squared terms is negative, the graph is a hyperbola.*

1. $y^2 = -16x$.

This is a parabola, with equation of the form: $y^2 = -4ax$

where $a = 4$. By Table 3:

Vertex: $(0, 0)$

Focus: $(-4, 0)$

Directrix: $x = 4$

3. $\dfrac{x^2}{25} - y^2 = 1$

This is a hyperbola in the form: $\dfrac{(x - h)^2}{a^2} - \dfrac{(y - k)^2}{b^2} = 1$

where $a = 5$, $b = 1$, $h = 0$, $k = 0$.

By Table 4, $c^2 = a^2 + b^2 = 26$, so $c = \sqrt{26}$, and we have:

Center: $(0, 0)$

Foci: $(h \pm c, k)$: $\left(-\sqrt{26}, 0\right)$ and $\left(\sqrt{26}, 0\right)$

Vertices: $(h \pm a, k)$: $(-5, 0)$ and $(5, 0)$

Asymptotes: $y - k = \pm\dfrac{b}{a}(x - h)$ or $y = \pm\dfrac{1}{5}x$

5. $\dfrac{y^2}{25} + \dfrac{x^2}{16} = 1$

This is an ellipse since both variables are squared, and both terms are positive. The equation is already in the form:

$$\frac{(y - k)^2}{a^2} + \frac{(x - h)^2}{b^2} = 1$$

where $h = 0$, $k = 0$, $a = 5$, $b = 4$

By Table 3, we have $c^2 = a^2 - b^2 = 9$, so that $c = 3$, and:

Center: (h, k), or $(0, 0)$

Foci: $(h, k \pm c)$, or $(0, -3)$ and $(0, 3)$

Vertices: $(h, k \pm a)$, or $(0, -5)$ and $(0, 5)$

7. $x^2 + 4y = 4$ is a ***parabola***.

$$x^2 + 4y = 4$$
$$x^2 = -4y + 4$$
$$x^2 = -4(y - 1)$$

This is in the form $(x - h)^2 = -4a(y - k)$, where

(1) $a = 1$

(2) $x - h = x$
$h = 0$

(3) $y - k = y - 1$
$k = 1$

From Table 2, we have: Vertex: $(0, 1)$
Focus: $(0, 0)$
Directrix: $y = 2$

9. $4x^2 - y^2 = 8$
This is a hyperbola, since it consists of a *difference* of squared terms:
$$4x^2 - y^2 = 8$$
$$\frac{x^2}{2} - \frac{y^2}{8} = 1$$
From Table 4, $a^2 = 2$, so $a = \sqrt{2}$, and $b^2 = 8$, so $b = \sqrt{8} = 2\sqrt{2}$.
Also, $c^2 = a^2 + b^2 = 10$, so $c = \sqrt{10}$. Then we have:
Transverse axis: horizontal: $y = 0$
Center: $(0, 0)$
Foci: $\left(-\sqrt{10}, 0\right)$ and $\left(\sqrt{10}, 0\right)$
Vertices: $\left(-\sqrt{2}, 0\right)$ and $\left(\sqrt{2}, 0\right)$
Asymptotes: $y - k = \pm\frac{b}{a}(x - h)$, or $y = \pm 2x$

11. $x^2 - 4x = 2y$ is a parabola:
$$x^2 - 4x + \underline{\quad} = 2y + \underline{\quad}$$
$$x^2 - 4x + 4 = 2y + 4$$
$$(x - 2)^2 = 2(y + 2)$$
We have: (1) $4a = 2$, or $a = \frac{1}{2}$
(2) $h = 2$
(3) $k = -2$
and: Vertex: $(2, -2)$
Focus: $\left(2, -\frac{3}{2}\right)$
Directrix: $y = -\frac{5}{2}$

13. $y^2 - 4y - 4x^2 + 8x = 4$
Complete the square:
$$(y^2 - 4y + \underline{\quad}) - 4(x^2 - 2x + \underline{\quad}) = 4$$
$$(y^2 - 4y + 4) - 4(x^2 - 2x + 1) = 4 + 4 - 4$$
$$(y - 2)^2 - 4(x - 1)^2 = 4$$
$$\frac{(y - 2)^2}{4} - \frac{(x - 1)^2}{1} = 1$$
This is a hyperbola with vertical transverse axis, center at $(1, 2)$, $a^2 = 4$, $b^2 = 1$, and $c^2 = a^2 + b^2 = 5$.

Then, $a = 2$
$b = 1$
$c = \sqrt{5}$
Center: $(h, k) = (1, 2)$
Foci: $(h, k \pm c) = \left(1, 2 - \sqrt{5}\right)$ and $\left(1, 2 + \sqrt{5}\right)$
Vertices: $(h, k \pm a) = (1, 0)$ and $(1, 4)$
Asymptotes: $y - 2 = \pm 2(x - 1)$
(Lines through $(1, 2)$ with slopes 2 and -2.)

15. $4x^2 + 9y^2 - 16x - 18y = 11$
Since both variables are squared, this is either an ellipse or a hyperbola. Since both squared terms are positive, the graph *must* be an ellipse. We start by completing the square, to put the equation in a recognizable form:

$$4x^2 - 16x + 9y^2 - 18y = 11$$
$$4(x^2 - 4x + \_\_) + 9(y^2 - 2y + \_\_) = 11$$
$$4(x^2 - 4x + 4) + 9(y^2 - 2y + 1) = 11 + 16 + 9$$
$$4(x - 2)^2 + 9(y - 1)^2 = 36$$
$$\frac{4(x - 2)^2}{36} + \frac{9(y - 1)^2}{36} = 1$$
$$\frac{(x - 2)^2}{9} + \frac{(y - 1)^2}{4} = 1$$

We now have the form: $\frac{(x - h)^2}{a^2} + \frac{(y - k)^2}{b^2} = 1$

where $h = 2$, $k = 1$, $a = 3$, $b = 2$. By Table 3, $c^2 = a^2 - b^2 = 5$, so $c = \sqrt{5}$, and we have:
Center: $(h, k) = (2, 1)$

Foci: $(h \pm c, k)$: $\left(2 - \sqrt{5}, 1\right)$ and $\left(2 + \sqrt{5}, 1\right)$
Vertices: $(h \pm a, k)$: $(-1, 1)$ and $(5, 1)$

17. $4x^2 - 16x + 16y + 32 = 0$ is a ***parabola***.

$$4x^2 - 16x = -16y - 32$$
$$4(x^2 - 4x + \_\_) = -16y - 32 + \_\_$$
$$4(x^2 - 4x + 4) = -16y - 32 + 16$$
$$(4)(4) = 16$$
$$4(x - 2)^2 = -16y - 16$$
$$4(x - 2)^2 = -16(y + 1)$$
$$(x - 2)^2 = -4(y + 1)$$

We have: (1) $a = 1$
(2) $h = 2$
(3) $k = -1$

and: Vertex: $(2, -1)$
Focus: $(2, -2)$
Directrix: $y = 0$

19. $9x^2 + 4y^2 - 18x + 8y = 23$
Both variables are squared, and the squared terms are both positive, so this is an ellipse.

$$9x^2 - 18x + 4y^2 + 8y = 23$$
$$9(x^2 - 2x + \_\_) + 4(y^2 + 2y + \_\_) = 23$$
$$9(x^2 - 2x + 1) + 4(y^2 + 2y + 1) = 23 + 9 + 4$$
$$9(x - 1)^2 + 4(y + 1)^2 = 36$$
$$\frac{(x - 1)^2}{4} + \frac{(y + 1)^2}{9} = 1$$

This is in the form: $\frac{(x - h)^2}{b^2} + \frac{(y - k)^2}{a^2} = 1$
where $h = 1$, $k = -1$, $a = 3$, $b = 2$.

From Table 3, $c^2 = a^2 - b^2 = 5$, so $c = \sqrt{5}$, and we have:

Center: $(h, k) = (1, -1)$

Foci: $(h, k \pm c)$, or $\left(1, -1 - \sqrt{5}\right)$ and $\left(1, -1 + \sqrt{5}\right)$

Vertices: $(h, k \pm a)$, or $(1, -4)$ and $(1, 2)$

21. We are given: Type of graph: Parabola

Focus: $F(-2, 0)$

Directrix: $x = 2$

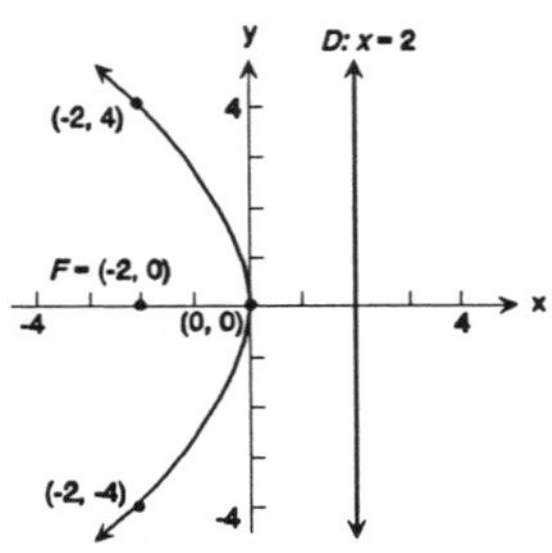

From Table 2, we see that to find the equation of a parabola, we need to know the coordinates of the vertex, (h, k), the distance from the vertex to the focus (a), and whether the parabola opens up, down, left, or right.

Here, the directrix, $x = 2$, is a vertical line, so the axis of symmetry is a horizontal line (which passes through the focus, $(-2, 0)$, i.e., $y = 0$.

The vertex is on the axis, midway between the focus and the directrix, at $V(0, 0)$. Then we have:

$a = d(V, F) = 2$

Finally, the focus is to the ***left*** of the vertex, so the parabola opens to the left. By Table 2, the parabola has an equation of the form:

$(y - k)^2 = -4a(x - h)$, or

$y^2 = -4ax$, since $h = 0, k = 0$

$y^2 = -8x$, since $a = 2$

23. We are given: Type of graph: Hyperbola

Center: $C(0, 0)$

Focus: $F_2(0, 4)$

Vertex: $V_1(0, -2)$

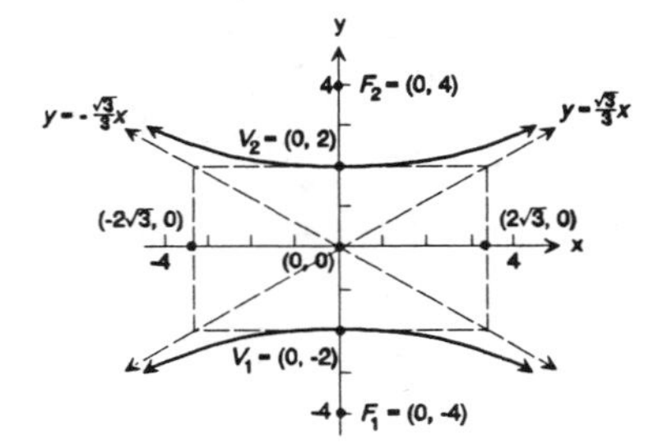

As we see from Table 4, to determine the equation of a hyperbola, we need to know the orientation of the transverse axis, the coordinates of the center, (h, k) and the constants a and b.

We know $h = 0, k = 0$

$c = d(C, F_2) = 4$

$a = d(C, V_1) = 2$

and $b^2 = c^2 - a^2$

$= 16 - 4 = 12$

or $b = \sqrt{12} = 2\sqrt{3}$

The center, focus, and vertex all lie on the ***vertical*** line $x = 0$, so the transverse axis is parallel to the y-axis. Therefore,

$$\frac{(y - k)^2}{a^2} - \frac{(x - h)^2}{b^2} = 1, \quad \text{or} \quad \frac{y^2}{4} - \frac{x^2}{12} = 1$$

As a further aid in graphing the hyperbola, the asymptotes are:

$y - k = \pm\frac{a}{b}(x - h)$, or

$y = \pm\frac{1}{\sqrt{3}}x$

$y = \pm\frac{\sqrt{3}}{3}x$

(Lines through $(0, 0)$ with slopes $\frac{\sqrt{3}}{3}$ and $\frac{-\sqrt{3}}{3}$.)

25. We are given: Type of Graph: Ellipse

Foci: $F_1(-3, 0)$, $F_2(3, 0)$

Vertex: $V_2(4, 0)$

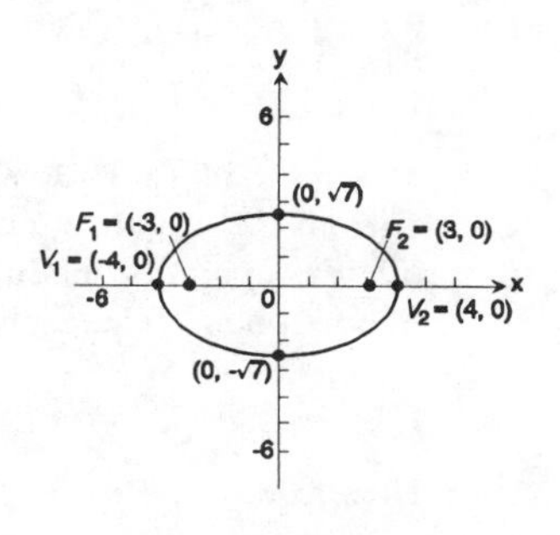

From Table 3, we see that we need the center, (h, k), a(the distance from the center to a vertex), and b, and we need to know whether the major axis is vertical or horizontal.

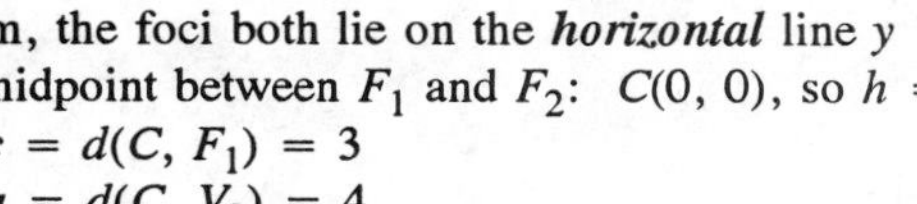

In this problem, the foci both lie on the *horizontal* line $y = 0$. The center is the midpoint between F_1 and F_2: $C(0, 0)$, so $h = 0$, $k = 0$.

Also: $c = d(C, F_1) = 3$

$a = d(C, V_2) = 4$

and $b^2 = a^2 - c^2 = 7$, so

$b = \sqrt{7}$

Therefore, the equation is of the form: $\dfrac{(x - h)^2}{a^2} + \dfrac{(y - k)^2}{b^2} = 1$, or $\dfrac{x^2}{16} + \dfrac{y^2}{7} = 1$

27. We are given: Type of graph: Parabola

Vertex: $V(2, -3)$

Focus: $F(2, -4)$

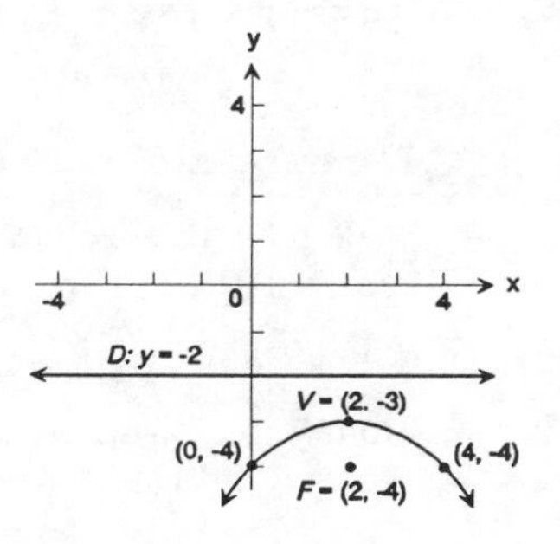

From Table 2, we need to know the vertex, $V(h, k)$; the distance, a, from the vertex to the focus; and which direction the parabola opens.

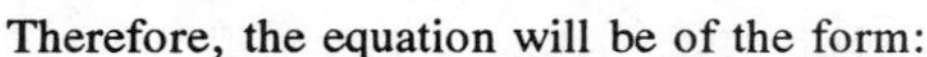

Since the focus is ***below*** the vertex, the parabola opens ***down***. Also, $a = d(V, F) = 1$.

Therefore, the equation will be of the form:

$(x - h)^2 = -4a(y - k)$

or $(x - 2)^2 = -4(y + 3)$

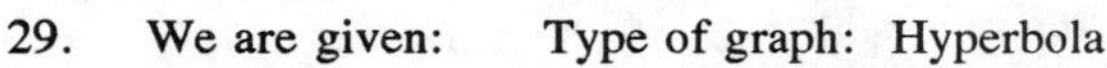

29. We are given: Type of graph: Hyperbola

Center: $C(-2, -3)$

Focus: $F_1(-4, -3)$

Vertex: $V_1(-3, -3)$

The transverse axis is the *horizontal* line $y = -3$, and:

$a = d(C, V_1) = 1$

$c = d(C, F_1) = 2$

$b^2 = c^2 - a^2 = 3$, so

$b = \sqrt{3}$

Foci: $(h \pm c, k)$: $(-4, -3)$ and $(0, -3)$

Vertices: $(h \pm a, k)$: $(-3, -3)$ and $(-1, -3)$

Asymptotes: $y - k = \pm\dfrac{b}{a}(x - h)$ or $y + 3 = \pm\sqrt{3}\,(x + 2)$

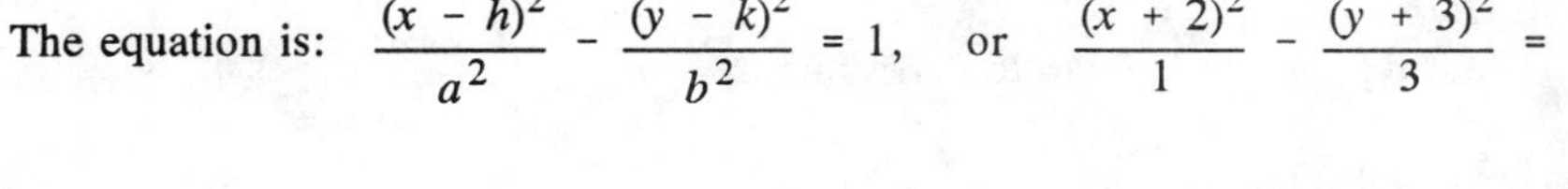

The equation is: $\dfrac{(x - h)^2}{a^2} - \dfrac{(y - k)^2}{b^2} = 1$, or $\dfrac{(x + 2)^2}{1} - \dfrac{(y + 3)^2}{3} = 1$

31. We are given: Type of graph: Ellipse
Foci: $F_1(-4, 2)$ and $F_2(-4, 8)$
Vertex: $V_2(-4, 10)$

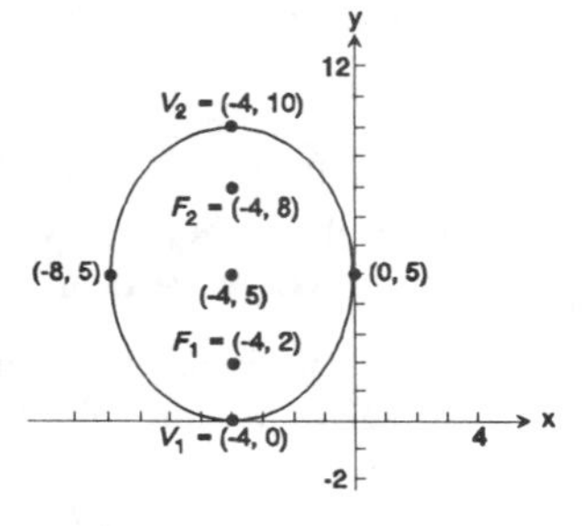

As we see in Table 3, we need to know the center, (h, k); the distance, a, from the center to either vertex; b, the distance from the center to either endpoint of the minor axis; and whether the major axis is vertical or horizontal. The center is the midpoint between F_1 and F_2:
$C(-4, 5)$.
Then $a = d(C, V_2) = 5$, and $c = d(C, F_1) = 3$.

Therefore, $b^2 = a^2 - c^2 = 16$, so that $b = 4$.

Finally, F_1, F_2 and V_2 all lie on the ***vertical*** line $x = -4$, so the major axis is parallel to the y-axis. By Table 3, the equation of the ellipse is:

$$\frac{(x - h)^2}{b^2} + \frac{(y - k)^2}{a^2} = 1, \quad \text{or} \quad \frac{(x + 4)^2}{16} + \frac{(y - 5)^2}{25} = 1$$

33. We are given: Center: $C(-1, 2)$
$a = 3$
$c = 4$

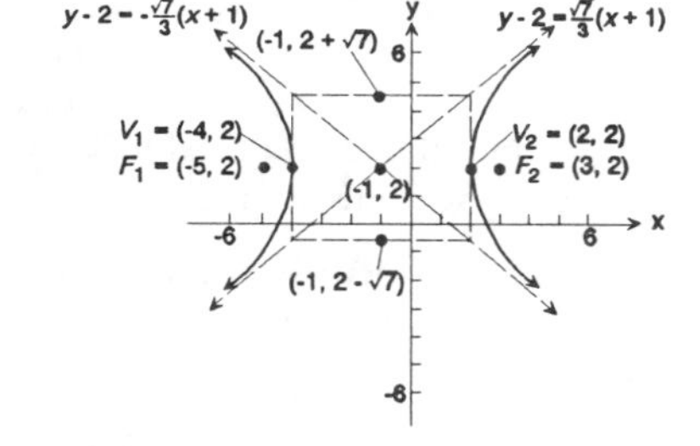

Transverse axis parallel to the x-axis.

Since the conic section has a transverse axis, it must be a hyperbola. For hyperbolas, $b^2 = c^2 - a^2$, so $b^2 = 7$, or $b = \sqrt{7}$.

From Table 4, the equation is of the form:

$$\frac{(x - h)^2}{a^2} - \frac{(y - k)^2}{b^2} = 1$$

where (h, k) are the coordinates of the center. Therefore, we have:

$$\frac{(x + 1)^2}{9} - \frac{(y - 2)^2}{7} = 1$$

As an aid to graphing, the vertices are at $(h \pm a, k)$, or $(-4, 2)$ and $(2, 2)$, and the asymptotes are:

$$y - 2 = \pm\frac{\sqrt{7}}{3}(x + 1)$$

(Lines through $(-1, 2)$ with slopes $\frac{\sqrt{7}}{3} \approx .88$ and $-\frac{\sqrt{7}}{3} \approx -.88$)

35. We are given: Vertices: $V_1(0, 1)$ and $V_2(6, 1)$
Asymptote: $3y + 2x - 9 = 0$

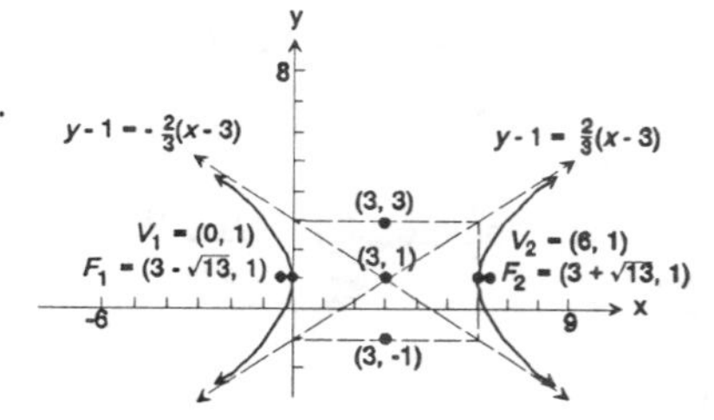

Since this conic section has asymptotes, it must be a hyperbola. The center is midway between V_1 and V_2: $C(3, 1)$, and the transverse axis is the ***horizontal*** line $y = 1$. Also, $a = d(C, V_1) = 3$. But we need to know b to determine the equation.
From Table 4, the asymptotes are:

$$y - k = \pm\frac{b}{a}(x - h), \quad \text{or} \quad y - 1 = \pm\frac{b}{a}(x - 3)$$

We must put the given asymptote into this form:

$$3y + 2x - 9 = 0$$
$$3y = -2x + 9$$
$$y = -\frac{2}{3}x + 3$$
$$y - 1 = -\frac{2}{3}x + 2$$
$$y - 1 = -\frac{2}{3}x + \frac{2 \cdot 3}{3}$$
$$y - 1 = -\frac{2}{3}(x - 3)$$

Therefore, $-\frac{b}{a} = -\frac{2}{3}$

$$3b = 2a$$
$$3b = 6 \quad (\text{since } a = 3)$$
$$b = 2$$

The equation of the hyperbola is: $\frac{(x-3)^2}{9} - \frac{(y-1)^2}{4} = 1$

37. We start with the ellipse and determine its foci and vertices:

$$4x^2 + 9y^2 = 36$$
$$\frac{x^2}{9} + \frac{y^2}{4} = 1$$

This is an ellipse centered at (0, 0). The larger denominator is associated with the x^2 term, so we know that the major axis is horizontal, and $a^2 = 9$, $b^2 = 4$, and $c^2 = a^2 - b^2 = 5$, so $c = \sqrt{5}$.

From Table 3: Vertices of the ellipse: $(h \pm a, k)$, or $(-3, 0)$ and $(3, 0)$

Foci of the ellipse: $(h \pm c, k)$, or $\left(-\sqrt{5}, 0\right)$ and $\left(\sqrt{5}, 0\right)$

Therefore, Foci of the hyperbola: $F_1(-3, 0)$ and $F_2(3, 0)$

Vertices of the hyperbola: $V_1\left(-\sqrt{5}, 0\right)$ and $V_2\left(\sqrt{5}, 0\right)$

The center of the hyperbola is midway between F_1 and F_2: $C(0, 0)$, and the transverse axis is the ***horizontal*** line, $y = 0$.

Finally,

$$a = d(C, V_1) = \sqrt{5}$$
$$c = d(C, F_1) = 3, \text{ and}$$
$$b^2 = c^2 - a^2 = 9 - 5 = 4, \text{ so}$$
$$b = 2$$

By Table 3, the equation of the ***hyperbola*** is: $\frac{(x-h)^2}{a^2} - \frac{(y-k)^2}{b^2} = 1$, or $\frac{x^2}{5} - \frac{y^2}{4} = 1$

39. Let (x, y) be any point in this collection of points.

The distance from (x, y), to $(3, 0) = \sqrt{(x - 3)^2 + y^2}$

The distance (x, y) to the line $x = \dfrac{16}{3}$ is $\left|x - \dfrac{16}{3}\right|$

Therefore, we have:

$$\sqrt{(x - 3)^2 + y^2} = \frac{3}{4}\left|x - \frac{16}{3}\right|$$

$$(x - 3)^2 + y^2 = \frac{9}{16}\left(x - \frac{16}{3}\right)^2 \quad \text{square both sides}$$

$$x^2 - 6x + 9 + y^2 = \frac{9}{16}\left(x^2 - \frac{32}{3}x + \frac{256}{9}\right)$$

$$16x^2 - 96x + 144 + 16y^2 = 9x^2 - 96x + 256$$

$$7x^2 + 16y^2 = 122$$

$$\frac{7x^2}{112} + \frac{16y^2}{112} = 1$$

$$\frac{x^2}{16} + \frac{y^2}{7} = 1$$

Thus the set of points is an ellipse.

41. Situate the parabola so that its vertex is at $(0, 0)$ and it opens up. We know

$$x^2 = 4ay$$

Since the light source is located at the focus, and the light source is one foot from the base, we have $a = 1$. Since the diameter of the opening is 2 feet across, we have $x = \dfrac{1}{2}(2) = 1$. Thus, we have:

$$x^2 = 4ay$$

$$1^2 = 4(1)y$$

$$y = 0.25 \text{ feet}$$

The mirror should be 0.25 feet or 3 inches deep.

43. Situate the ellipse so that its center is at $(0, 0)$. Since the bridge has a span of 60 feet, the length of the major axis is 60; thus, $a = 30$. The maximum height of the bridge is 20 feet; thus, $b = 20$. The equation of the ellipse is therefore:

$$\frac{x^2}{900} + \frac{y^2}{400} = 1$$

The height, y, of the arch a distance $x = 5$ feet from center is:

$$\frac{5^2}{900} + \frac{y^2}{400} = 1$$

$$y^2 = 400 \cdot \frac{875}{900}$$

$$y \approx 19.72 \text{ feet}$$

Similarly, when $x = 10$ feet, $y \approx 18.86$ feet. When $x = 20$ feet, $y \approx 14.91$ feet.

45. (a) Set up a rectangular coordinate system so that the two stations lie on the x-axis and the origin is midway between them. The ship lies on a hyperbola whose foci are the locations of the two stations. Since the time difference is .00032 seconds and the speed of the signal is 186,000 miles per second, the difference of the distances from the ship to each station is:

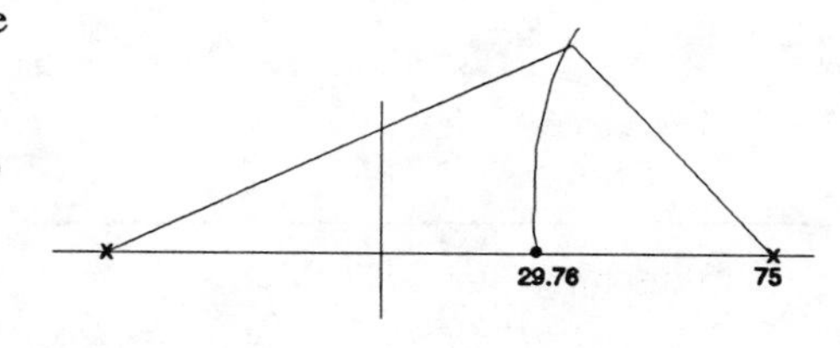

$$\text{distance} = (186{,}000)(.00032) = 59.52 \text{ miles}$$

The difference of the distances from the ship to each station, 59.52, equals $2a$, so $a = 29.76$ and the vertex of the corresponding hyperbola is at (29.76, 0). Since the focus is at (75, 0), the ship would reach shore $75 - 29.76 = 45.24$ miles from the master station.

(b) The ship should follow a hyperbola with vertex at (60, 0). For this hyperbola, $a = 60$, so the constant difference of the distances from the ship to each station is 120. The time difference the ship should look for is:

$$\text{time} = \frac{120}{186{,}000} = 0.000645 \text{ seconds}$$

(c) We need to find the equation of the hyperbola with vertex at (60, 0) and a focus at (75, 0). The form of the equation of this hyperbola is:

$$\frac{x^2}{a^2} - \frac{y^2}{b^2} = 1$$

where $a = 60$. Since $c = 75$ and $b^2 = c^2 - a^2$, we have $b^2 = 75^2 - 60^2 = 2025$. So, the equation of the hyperbola is:

$$\frac{x^2}{3600} - \frac{y^2}{2025} = 1$$

Since the ship is 20 miles off shore, we have $y = 20$. Solve the above equation for x.

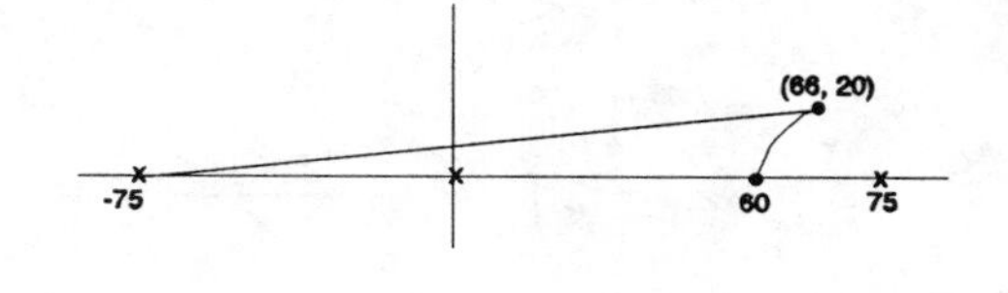

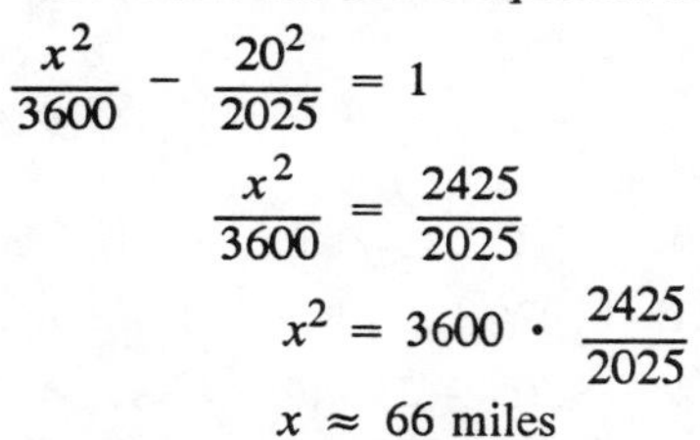

$$\frac{x^2}{3600} - \frac{20^2}{2025} = 1$$

$$\frac{x^2}{3600} = \frac{2425}{2025}$$

$$x^2 = 3600 \cdot \frac{2425}{2025}$$

$$x \approx 66 \text{ miles}$$

The ship is at the position (66, 20).

Chapter 

SYSTEMS OF EQUATIONS AND INEQUALITIES

8.1 Systems of Linear Equations: Substitution; Elimination

1. $\left.\begin{array}{r} 2x - y = 5 \\ 5x + 2y = 8 \end{array}\right\} \xrightarrow{x = 2,\ y = -1} \begin{array}{r} 2(2) - (-1) = 5 \\ 5(2) + 2(-1) = 8 \end{array} \quad \begin{array}{r} 5 = 5 \\ 8 = 8 \end{array}$

   Therefore, $x = 2$, $y = -1$ is a solution to the system of equations, since they satisfy ***both*** equations.

3. $\left.\begin{array}{r} 3x - 4y = 4 \\ \frac{1}{2}x - 3y = \frac{-1}{2} \end{array}\right\} \xrightarrow{x = 2,\ y = \frac{1}{2}} \begin{array}{r} 3(2) - 4\left(\frac{1}{2}\right) = 4 \\ \frac{1}{2}(2) - 3\left(\frac{1}{2}\right) = \frac{-1}{2} \end{array} \quad \begin{array}{r} 4 = 4 \\ \frac{-1}{2} = \frac{-1}{2} \end{array}$

   Each equation is satisfied, so $x = 2$, $y = \frac{1}{2}$ is a solution of the system.

5. $\left.\begin{array}{r} x^2 - y^2 = 3 \\ xy = 2 \end{array}\right\} \xrightarrow{x = 2,\ y = 1} \begin{array}{r} 2^2 - 1^2 = 3 \\ (2)(1) = 2 \end{array} \quad \begin{array}{r} 3 = 3 \\ 2 = 2 \end{array}$

7. $\left.\begin{array}{r} \frac{x}{1 + x} + 3y = 6 \\ x + 9y^2 = 36 \end{array}\right\} \xrightarrow{x = 0,\ y = 2} \begin{array}{r} \frac{0}{1 + 0} + 3(2) = 6 \\ 0 + 9(2)^2 = 36 \end{array} \quad \begin{array}{r} 6 = 6 \\ 36 = 36 \end{array}$

9. $\left.\begin{array}{r} 3x + 3y + 2z = 4 \\ x - y - z = 0 \\ 2y - 3z = -8 \end{array}\right\} \xrightarrow{x = 1,\ y = -1,\ z = 2} \begin{array}{r} 3(1) + 3(-1) + 2(2) = 4 \\ 1 - (-1) - 2 = 0 \\ 2(-1) - 3(2) = -8 \end{array} \quad \begin{array}{r} 4 = 4 \\ 0 = 0 \\ -8 = -8 \end{array}$

   Therefore, $x = 1$, $y = 2$, $z = 2$ is a solution to the system.

11. $\begin{cases} x + y = 8 & (1) \\ x - y = 4 & (2) \end{cases}$

    We can use substitution.

    Step 1: We can easily solve either equation for x *or* y. Let us begin by solving equation (1) for x:

    $$x + y = 8$$
    $$x = -y + 8$$

    Step 2: Now substitute $x = -y + 8$ into (2):

    $$x - y = 4$$
    $$-y + 8 - y = 4$$
    $$-2y + 8 = 4$$

Step 3: Solve the last equation for y:

$$-2y + 8 = 4$$
$$-2y = -4$$
$$y = 2$$

Step 4: We can now find x by using the fact that $x = -y + 8$ (Step 1) and $y = 2$ (Step 3):

$$x = -y + 8$$
$$x = -(2) + 8, \text{ since } y = 2$$
$$x = 6$$

The solution of the system is $x = 6$, $y = 2$.

13. $\begin{cases} 5x - y = 13 & (1) \\ 2x + 3y = 12 & (2) \end{cases}$

We use elimination.

$\begin{cases} 15x - 3y = 39 & (1) \\ 2x + 3y = 12 & (2) \end{cases}$ Multiply both sides by 3.

$\begin{cases} 15x - 3y = 39 & (1) \\ 17x = 51 & (2) \end{cases}$ Replace (2) by (1) + (2)

$\begin{cases} 15x - 3y = 39 & (1) \\ x = 3 & (2) \end{cases}$

$\begin{cases} 45 - 3y = 39 & (1) \\ x = 3 & (2) \end{cases}$ Back-substitute; $x = 3$

$\begin{cases} -3y = -6 & (1) \\ x = 3 & (2) \end{cases}$

$\begin{cases} y = 2 & (1) \\ x = 3 & (2) \end{cases}$

The solution is $x = 3$, $y = 2$.

15. $\begin{cases} 3x = 24 & (1) \\ x + 2y = 0 & (2) \end{cases}$

We use substitution.

Step 1: Since equation (1) contains only one variable, it is best to start there:

$$3x = 24$$
$$x = 8$$

Step 2: Substitute $x = 8$ into (2):

$$x + 2y = 0$$
$$8 + 2y = 0$$

Step 3: Solve for y:

$$8 + 2y = 0$$
$$2y = -8$$
$$y = -4$$

We already know x, so the solution of the system is $x = 8$, $y = -4$.

17. $\begin{cases} 3x - 6y = 2 & (1) \\ 5x + 4y = 1 & (2) \end{cases}$

We use substitution.

Step 1: It is harder to decide which equation to solve for which variable. Since the coefficient of x in equation (1) divides all other constants in equation (1), I choose to solve for x, in order to avoid fractions for as long as possible:

$$3x - 6y = 2$$
$$3x = 6y + 2$$
$$x = 2y + \frac{2}{3}$$

Step 2: Substitute $x = 2y + \frac{2}{3}$ into (2):

$$5x + 4y = 1$$
$$5\left[2y + \frac{2}{3}\right] + 4y = 1$$
$$10y + \frac{10}{3} + 4y = 1$$
$$14y + \frac{10}{3} = 1$$
$$42y + 10 = 3$$

Step 3: Solve for y:

$$42y + 10 = 3$$
$$42y = -7$$
$$y = -\frac{1}{6}$$

Step 4: Determine x:

$$x = 2y + \frac{2}{3} \quad \text{(Step 1)}$$
$$x = 2\left[-\frac{1}{6}\right] + \frac{2}{3}$$
$$x = \frac{1}{3}$$

The solution is $x = \frac{1}{3}$, $y = -\frac{1}{6}$.

19. $\begin{cases} 2x + y = 1 & (1) \\ 4x + 2y = 3 & (2) \end{cases}$

We use substitution.

Step 1: It is easiest to solve equation (1) for y:

$$2x + y = 1$$
$$y = -2x + 1$$

Step 2: Substitute $y = -2x + 1$ into (2):

$$4x + 2y = 3$$
$$4x + 2(-2x + 1) = 3$$
$$4x - 4x + 2 = 3$$
$$0x = 1$$

This has no solution, so the system is inconsistent.

21. $\begin{cases} 2x - y = 0 & (1) \\ 3x + 2y = 7 & (2) \end{cases}$

We use elimination.

$\begin{cases} 4x - 2y = 0 & (1) \\ 3x + 2y = 7 & (2) \end{cases}$ Multiply both sides by 2.

$\begin{cases} 4x - 2y = 0 & (1) \\ 7x = 7 & (2) \end{cases}$ Replace (2) by (1) + (2).

$\begin{cases} 4x - 2y = 0 & (1) \\ x = 1 & (2) \end{cases}$

$\begin{cases} 4 - 2y = 0 & (1) \\ x = 1 & (2) \end{cases}$ Back-substitute; $x = 1$.

$\begin{cases} -2y = -4 & (1) \\ x & (2) \end{cases}$

$\begin{cases} y = 2 & (1) \\ x = 1 & (2) \end{cases}$

The solution is $x = 1$, $y = 2$.

23. $\begin{cases} x + 2y = 4 & (1) \\ 2x + 4y = 8 & (2) \end{cases}$

We use substitution.

Step 1: Solve for x in (1):

$$\begin{aligned} x + 2y &= 4 \\ x &= -2y + 4 \end{aligned}$$

Step 2: Substitute $x = -2y + 4$ into (2):

$$\begin{aligned} 2x + 4y &= 8 \\ 2(-2y + 4) + 4y &= 8 \\ -4y + 8 + 4y &= 8 \\ 0y &= 0 \end{aligned}$$

This is an identity (i.e., true for ***any*** value of y), so ***any*** value of y is a solution, and we let $x = 4 - 2y$, y any real number.

25. $\begin{cases} 2x - 3y = -1 & (1) \\ 10x + y = 11 & (2) \end{cases}$

We use elimination.

$\begin{cases} -10x + 15y = 5 & (1) \\ 10x + y = 11 & (2) \end{cases}$ Multiply both sides by -5.

$\begin{cases} -10x + 15y = 5 & (1) \\ 16y = 16 & (2) \end{cases}$ Replace (2) by (1) + (2).

$\begin{cases} -10x + 15y = 5 & (1) \\ y = 1 & (2) \end{cases}$

$\begin{cases} -10x + 15(1) = 5 & (1) \\ y = 1 & (2) \end{cases}$ Back-substitute; $y = 1$.

$\begin{cases} -10x + 15 = 5 & (1) \\ y = 1 & (2) \end{cases}$

$\begin{cases} x = 1 \\ y = 1 \end{cases}$

The solution is $x = 1$, $y = 1$.

27. $\begin{cases} 2x + 3y = 6 & (1) \\ x - y = \frac{1}{2} & (2) \end{cases}$

We use substitution.

Step 1: Solve for (2) for x:

$$\begin{aligned} x - y &= \frac{1}{2} \\ x &= y + \frac{1}{2} \end{aligned}$$

Step 2: Substitute $x = y + \frac{1}{2}$ into (1):

$$\begin{aligned} 2x + 3y &= 6 \\ 2\left[y + \frac{1}{2}\right] + 3y &= 6 \\ 2y + 1 + 3y &= 6 \\ 5y + 1 &= 6 \end{aligned}$$

Step 3: Solve for y:

$$\begin{aligned} 5y + 1 &= 6 \\ 5y &= 5 \\ y &= 1 \end{aligned}$$

Step 4: Determine x:

$$x = y + \frac{1}{2} \quad \text{(Step 1)}$$
$$x = 1 + \frac{1}{2} \quad (y = 1)$$
$$x = \frac{3}{2}$$

The solution is $x = \frac{3}{2}$, $y = 1$.

29. $\begin{cases} \frac{1}{2}x + \frac{1}{3}y = 3 & (1) \\ \frac{1}{4}x - \frac{2}{3}y = -1 & (2) \end{cases}$

Multiply equation (1) by 6.
Multiply equation (2) by 12.
This eliminates the fractions.

$$\begin{cases} 3x + 2y = 18 & (1) \\ 3x - 8y = -12 & (2) \end{cases}$$

We use elimination.

$$\begin{cases} 3x + 2y = 18 \\ 10y = 30 & \text{Replace (2) by (1)} - (2). \end{cases}$$

$$\begin{cases} 3x + 2y = 18 \\ y = 3 \end{cases}$$

$$\begin{cases} 3x + 2(3) = 18 \\ y = 3 \end{cases}$$

$$\begin{cases} x = 4 \\ y = 3 \end{cases}$$

The solution is $x = 4$, $y = 3$.

31. $\begin{cases} 3x - 5y = 3 & (1) \\ 15x + 5y = 21 & (2) \end{cases}$

We use elimination.

$$\begin{cases} 3x - 5y = 3 & (1) \\ 18x = 24 & (2) \quad \text{Replace (2) by (1) + (2).} \end{cases}$$

$$\begin{cases} 3x - 5y = 3 & (1) \\ x = \frac{4}{3} & (2) \end{cases}$$

$$\begin{cases} 4 - 5y = 3 & (1) \quad \text{Back-substitution; } x = \frac{4}{3}. \\ x = \frac{4}{3} & (2) \end{cases}$$

$$\begin{cases} 5y = 1 & (1) \\ x = \frac{4}{3} & (2) \end{cases}$$

$$\begin{cases} y = \frac{1}{5} & (1) \\ x = \frac{4}{3} & (2) \end{cases}$$

The solution is $x = \frac{4}{3}$, $y = \frac{1}{5}$.

33. $\begin{cases} x - y \quad\quad = 6 & (1) \\ 2x \quad\quad - 3z = 16 & (2) \\ \quad 2y + \quad z = 4 & (3) \end{cases}$

We use substitution.

Step 1: We can solve for x in equation (1): $x - y = 6$

$$x = y + 6$$

Step 2: Substitute this expression for x into equations (2) and (3):

$$\begin{cases} 2x \quad\quad - 3z = 16 & (2) \\ \quad 2y + \quad z = 4 & (3) \end{cases}$$

$$\begin{cases} 2(y + 6) - 3z = 16 & (2) \\ \quad 2y + \quad z = 4 & (3) \end{cases}$$

$$\begin{cases} 2y - 3z + 12 = 16 & (2) \\ 2y + z \quad\quad = 4 & (3) \end{cases}$$

$$\begin{cases} 2y - 3z + \quad = 4 & (2) \\ 2y + z \quad\quad = 4 & (3) \end{cases}$$

Now we must return to Step 1 and solve (2) or (3) for y or z.

It is easiest to solve for z in equation (3): $2y + z = 4$

$$z = 4 - 2y$$

Now substitute $z = 4 - 2y$ into (2): $2y - 3z = 4$

$$2y - 3(4 - 2y) = 4$$

$$2y - 12 + 6y = 4$$

$$8y = 16$$

Step 3: $y = 2$

Step 4: From $z = 4 - 2y$, we have: $z = 4 - 2(2)$

$$z = 0,$$

and from $x = y + 6$ (Step 1)

we have $x = 2 + 6$

$$x = 8$$

The solution is: $x = 8$, $y = 2$, and $z = 0$.

35. $\begin{cases} x - 2y + 3z = 7 & (1) \\ 2x + y + z = 4 & (2) \\ -3x + 2y - 2z = -10 & (3) \end{cases}$

We use substitution.

Step 1A: Solve equation (1) for x: $x - 2y + 3z = 7$

$$x = 7 + 2y - 3z$$

Step 2A: Substitute $x = 7 + 2y - 3z$ into (2) and (3):

$$\begin{cases} 2x + y + z = 4 & (2) \\ -3x + 2y - 2z = -10 & (3) \end{cases}$$

$$\begin{cases} 2(7 + 2y - 3z) + y + z = 4 & (2) \\ -3(7 + 2y - 3z) + 2y - 2z = -10 & (3) \end{cases}$$

$$\begin{cases} 14 + 4y - 6z + y + z = 4 & (2) \\ -21 - 6y + 9z + 2y - 2z = -10 & (3) \end{cases}$$

$$\begin{cases} 5y - 5z = -10 & (2) \\ -4y + 7z = 11 & (3) \end{cases}$$

Step 1B: We must now solve for y or z in (2) or (3). If we work with (2), we can avoid fractions on this step:

$$5y - 5z = -10 \quad (2)$$

$$-5z = -5y - 10$$

$$z = y + 2$$

Step 2B: Substitute this into (3):

$$-4y + 7z = 11$$
$$-4y + 7(y + 2) = 11$$
$$-4y + 7y + 14 = 11$$
$$3y = -3$$

Step 3: $y = -1$

Step 4: Determine z: $z = y + 2$ (Step 1B)

$z = -1 + 2$ $(y = -1)$

$z = 1$

Determine x: $x = 7 + 2y - 3z$ (Step 1A)

$x = 7 + 2(-1) - 3(1)$ $(y = -1), z = 1)$

$x = 2$

The solution is $x = 2$, $y = -1$, $z = 1$.

37. $$\begin{cases} x - y - z = 1 & (1) \\ 2x + 3y + z = 2 & (2) \\ 3x + 2y \quad = 0 & (3) \end{cases}$$

We use elimination.

Since z is already missing from equation (3), we try to eliminate it from either (1) or (2).

$$\begin{cases} x - y - z = 1 & (1) \\ 3x + 2y \quad = 3 & (2) \\ 3x + 2y \quad = 0 & (3) \end{cases}$$ Replace (2) by (1) + (2).

$$\begin{cases} x - y - z = 1 & (1) \\ -3x - 2y \quad = -3 & (2) \\ 3x + 2y \quad = 0 & (3) \end{cases}$$ Multiply both sides by -1.

$$\begin{cases} x - y - z = 1 & (1) \\ -3x - 2y \quad = -3 & (2) \\ 0x + 0y \quad = -3 & (3) \end{cases}$$ Replace (3) by (2) + (3).

(3) has no solution, so the original system is inconsistent.

39. $$\begin{cases} x - y - z = 1 & (1) \\ -x + 2y - 3z = -4 & (2) \\ 3x - 2y - 7z = 0 & (3) \end{cases}$$

We use substitution.

Step 1A: Solve for x in (1): $x - y - z = 1$

$x = 1 + y + z$

Step 2A: Substitute into (2) and (3):

$$\begin{cases} -x + 2y - 3z = -4 & (2) \\ 3x - 2y - 7z = 0 & (3) \end{cases}$$

$$\begin{cases} -(1 + y + z) + 2y - 3z = -4 & (2) \\ 3(1 + y + z) - 2y - 7z = 0 & (3) \end{cases}$$

$$\begin{cases} y - 4z = -3 & (2) \\ y - 4z = -3 & (3) \end{cases}$$

Step 1B: Solve (2) for y: $y - 4z = -3$

$y = -3 + 4z$

Step 2B: Substitute this into (3): $y - 4z = -3$

$-3 + 4z - 4z = -3$

$0 \cdot z = 0$

This equation is satisfied by all values of z, so z can be any real number.

Then, from Step 1B, $y = 4z - 3$, and from Step 1A,

$$x = 1 + y + z$$
$$x = 1 + (4z - 3) + z$$
$$x = 5z - 2$$

Thus, we can write the solution as: $x = 5z - 2$; $y = 4z - 3$; where z is *any* real number.

We could also express our answer in terms of the following:

$$x = \frac{5}{4}y + \frac{7}{4};\ z = \frac{1}{4}y + \frac{3}{4},\ \text{where } y \text{ is any real number.}$$

Also, the solution is: $y = \frac{4}{5}x - \frac{7}{5};\ z = \frac{1}{5}x + \frac{2}{5}$, where x is any real number.

41. $$\begin{cases} 2x - 2y + 3z = 6 & (1) \\ 4x - 3y + 2z = 0 & (2) \\ -2x + 3y - 7z = 1 & (3) \end{cases}$$

We use elimination.

We can eliminate x from equation (3) by adding (1) and (3):

$$\begin{cases} 2x - 2y + 3z = 6 & (1) \\ 4x - 3y + 2z = 0 & (2) \\ y - 4z = 7 & (3) \quad \text{Replace (3) by (1) + 3.} \end{cases}$$

We need to eliminate the ***same*** variable, x, from either (1) or (2):

$$\begin{cases} -4x + 4y - 6z = -12 & (1) \quad \text{Multiply both sides by } -2. \\ 4x - 3y + 2z = 0 & (2) \\ y - 4z = 7 & (3) \end{cases}$$

$$\begin{cases} -4x + 4y - 6z = -12 & (1) \\ y - 4z = -12 & (2) \quad \text{Replace (2) by (1) + (2).} \\ y - 4z = 7 & (3) \end{cases}$$

We can see that (2) and (3) are contradictory: if we subtract (3) from (2), we obtain $0 = -19$.

Thus, the system is inconsistent.

43. $$\begin{cases} x + y - z = 6 & (1) \\ 3x - 2y + z = -5 & (2) \\ x + 3y - 2z = 14 & (3) \end{cases}$$

We use substitution.

Step 1A: I choose to solve equation (2) for z: $3x - 2y + z = -5$

$$z = -5 - 3x + 2y$$

Step 2A: Now substitute this expression for z into (1) and (3):

$$\begin{cases} x + y - z = 6 & (1) \\ x + 3y - 2z = 14 & (3) \end{cases}$$

$$\begin{cases} x + y - (-5 - 3x + 2y) = 6 & (1) \\ x + 3y - 2(-5 - 3x + 2y) = 14 & (3) \end{cases}$$

$$\begin{cases} 4x - y = 1 & (1) \\ 7x - y = 4 & (3) \end{cases}$$

Step 1B: Now I solve for y in (1). (Solving equation (3) for y would be equally easy.)

$$4x - y = 1$$
$$-y = 1 - 4x$$
$$y = -1 + 4x$$

Step 2B: Substitute this into (3):

$$7x - y = 4$$
$$7x - (-1 + 4x) = 4$$
$$3x + 1 = 4$$

Step 3: $x = 1$

Step 4: We now find y and z:

$$y = -1 + 4x \quad \text{(Step 1B)}$$
$$y = -1 + 4(1) \quad (x = 1)$$
$$y = 3$$

Also,

$$z = -5 - 3x + 2y \quad \text{(Step 1A)}$$
$$z = -5 - 3(1) + 2(3) \quad (x = 1, y = 3)$$
$$z = -2$$

So, the solution is: $x = 1$, $y = 3$, $z = -2$.

45. $$\begin{cases} x + 2y - z = -3 & (1) \\ 2x - 4y + z = -7 & (2) \\ -2x + 2y - 3z = 4 & (3) \end{cases}$$

We can eliminate z by adding (1) and (2):

$$\begin{cases} x + 2y - z = -3 & (1) \\ 3x - 2y = -10 & (2) \quad \text{Replace (2) by (1) + (2).} \\ -2x + 2y - 3z = 4 & (3) \end{cases}$$

Now eliminate z from (1) or (3):

$$\begin{cases} -3x - 6y + 3z = 9 & (1) \quad \text{Multiply both sides by -3.} \\ 3x - 2y = -10 & (2) \\ -2x + 2y - 3z = 4 & (3) \end{cases}$$

$$\begin{cases} -3x - 6y + 3z = 9 & (1) \\ 3x - 2y = -10 & (2) \\ -5x - 4y = 13 & (3) \quad \text{Replace (3) by (1) + (3).} \end{cases}$$

$$\begin{cases} -3x - 6y + 3z = 9 & (1) \\ -6x + 4y = 20 & (2) \quad \text{Multiply both sides by -2.} \\ -5x - 4y = 13 & (3) \end{cases}$$

$$\begin{cases} -3x - 6y + 3z = 9 & (1) \\ -6x + 4y = 20 & (2) \\ -11x = 33 & (3) \quad \text{Replace (3) by (2) + (3).} \end{cases}$$

$$\begin{cases} -3x - 6y + 3z = 9 & (1) \\ -6x + 4y = 20 & (2) \\ x = -3 & (3) \end{cases}$$

$$\begin{cases} 9 - 6y + 3z = 9 & (1) \quad \text{Back-substitute; } x = -3. \\ 18 + 4y = 20 & (2) \quad \text{Back-substitute; } x = -3. \\ x = -3 & (3) \end{cases}$$

$$\begin{cases} -6y + 3z = 0 & (1) \\ y = \frac{1}{2} & (2) \\ x = -3 & (3) \end{cases}$$

$$\begin{cases} -3 + 3z = 0 & (1) \quad \text{Back-substitute; } y = \frac{1}{2}. \\ y = \frac{1}{2} & (2) \\ x = -3 & (3) \end{cases}$$

$$\begin{cases} z = 1 & (1) \\ y = \frac{1}{2} & (2) \\ x = -3 & (3) \end{cases}$$

The solution is $x = -3$, $y = \frac{1}{2}$, $z = 1$.

47. $\begin{cases} \dfrac{1}{x} + \dfrac{1}{y} = 8 & (1) \\ \dfrac{3}{x} - \dfrac{5}{y} = 0 & (2) \end{cases}$

$$\begin{cases} \dfrac{1}{x} + \dfrac{1}{y} = 8 & (1) \\ 3\left(\dfrac{1}{x}\right) - 5\left(\dfrac{1}{y}\right) = 0 & (2) \end{cases}$$

Now let $u = \dfrac{1}{x}, v = \dfrac{1}{y}$: $\begin{cases} u + v = 8 & (1) \\ 3u - 5v = 0 & (2) \end{cases}$

Step 1: Solve (1) for v: $u + v = 8$

$v = 8 - u$

Step 2: Substitute into (2): $3u - 5v = 0$

$3u - 5(8 - u) = 0$

$3u - 40 + 5u = 0$

Step 3: Solve for u: $8u = 40$

$u = 5$

Step 4: $v = 8 - u$ (Step 1)

$v = 8 - 5$ $(u = 5)$

$v = 3$

So $u = 5, v = 3$.

But we are supposed to find x and y.

We have: $u = \dfrac{1}{x}, v = \dfrac{1}{y}$

$5 = \dfrac{1}{x}, 3 = \dfrac{1}{y}$

$x = \dfrac{1}{5}, y = \dfrac{1}{3}$

49. $\begin{cases} y_1 = \sqrt{2}x - 20\sqrt{7} \\ y_2 = -0.1x + 20 \end{cases}$

Graph y_1 and y_2.

Using the INTERSECT feature on the graphing calculator, the point of intersection is $x = 48.15, y = 15.18$.

51. $\begin{cases} \sqrt{2}x + \sqrt{3}y + \sqrt{6} = 0 \\ \sqrt{3}x - \sqrt{2}y + 60 = 0 \end{cases}$

Solve each equation for y.

$\sqrt{2}x + \sqrt{3}y + \sqrt{6} = 0$

$\sqrt{3}y = -\sqrt{2}x - \sqrt{6}$

$y_1 = \dfrac{-\sqrt{2}x - \sqrt{6}}{\sqrt{3}}$

$\sqrt{3}x - \sqrt{2}y + 60 = 0$

$-\sqrt{2}y = -\sqrt{3}x - 60$

$y_2 = \dfrac{\sqrt{3}x + 60}{\sqrt{2}}$

Graph y_1 and y_2.

Using the INTERSECT feature on the graphing calculator, the point of intersection is $x = -21.47, y = 16.12$.

53. $\begin{cases} \sqrt{3}x + \sqrt{2}y = \sqrt{0.3} \\ 100x - 95y = 20 \end{cases}$

Solve each equation for y.

$\sqrt{3}x + \sqrt{2}y = \sqrt{0.3}$

$y_1 = \dfrac{-\sqrt{3}x + \sqrt{0.3}}{\sqrt{2}}$

$100x - 95y = 20$

$y_2 = \dfrac{-100x + 20}{-95}$

Graph y_1 and y_2.

Using the INTERSECT feature on the graphing calculator, the point of intersection is $x = 0.26$, $y = 0.06$.

In Problems 55–64, start by giving variable names (x, y, etc.) to the unknowns. Then translate each statement about the unknowns into an equation involving the variables. (The number of equations and the number of variables must be equal.) Solve the equations for the unknowns. Finally, be sure you have answered the original question.

55. Let the two numbers be x and y.

$\begin{cases} x + y = 81 & \text{(Their sum is 81.)} \\ 2x - 3y = 62 & \text{(Twice one minus three times the other is 62.)} \end{cases}$

Solve the first equation for x: $x + y = 81$

$x = 81 - y$

Substitute this into the second equation: $2x - 3y = 62$

$2(81 - y) - 3y = 62$

$-5y = -100$

$y = 20$

Now find x: We know $x = 81 - y$, and $y = 20$,

so $x = 81 - 20$

$x = 61$

The two numbers are 20 and 61.

57. Denote the width and length of the rectangle by w and ℓ. Then:

(1) $2w + 2\ell = 90$ (the perimeter is 90)

(2) $\ell = 2w$ (length is twice the width)

Substitute $\ell = 2w$ into (1): $2w + 2\ell = 90$

$2w + 2(2w) = 90$

$6w = 90$

$w = 15$

Then $\ell = 2w$, or $\ell = 30$.

The room is 15 feet by 30 feet.

59. Let x = cost of one cheeseburger, in cents

y = cost of a shake, in cents

Then: $4x + 2y = 790$ (1)

$2y = x + 15$

From (2), we have: $y = \frac{1}{2}x + \frac{15}{2}$

Substitute this into equation (1): $4x + 2y = 790$

$$4x + 2\left[\frac{1}{2}x + \frac{15}{2}\right] = 790$$
$$4x + x + 15 = 790$$
$$5x = 775$$
$$x = 155$$

Then, since $y = \frac{1}{2}x + \frac{15}{2}$, we have: $y = \frac{155}{2} + \frac{15}{2}$

$y = 85$

Therefore, a cheeseburger costs \$1.55, and a shake costs \$0.85.

61. Here, we really have only one unknown:

x = Number of pounds of cashews to use

Since we are using 30 pounds of peanuts, the mixture will contain $30 + x$ pounds.

The basic formula is: Revenue = (Number of pounds) • (Price per pound)

Revenue from peanuts alone = (30)(1.50) = 45 dollars

Revenue from cashews alone = $x(5) = 5x$ dollars

Revenue from mixture = $(30 + x) \cdot 3 = 90 + 3x$ dollars

So we want: $90 + 3x = 45 + 5x$

$$45 = 2x$$
$$x = \frac{45}{2} = 22\frac{1}{2}$$

The manager should use 22.5 pounds of cashews.

63. Here, the unknown quantities are speeds.

Let x = average windspeed

y = average air speed of the Piper

both in miles per hour.

Going ***with*** the wind, the plane has a groundspeed of $x + y$. Then:

(rate) • (time) = distance

$(x + y)(3) = 600$, or

$x + y = 200$ (1)

Against the wind, the speed of the plane will be $(y - x)$.

(rate) • (time) = distance

$(y - x)(4) = 600$, or

$y - x = 150$ (2)

We have: $\begin{cases} x + y = 200 & (1) \\ y - x = 150 & (2) \end{cases}$

Solve (2) for y: $y - x = 150$

$y = 150 + x$

Then from (1): $x + y = 200$

$$x + (150 + x) = 200$$
$$2x = 50$$
$$x = 25$$

Finally, $y = 150 + x = 175$

Thus, the wind speed is $x = 25$ mph, and the air speed of the plane is $y = 175$ mph.

65. Let x = one design of a set of dishes.
and y = second design of dishes

$$\begin{aligned} x + y &= 200 \\ 25x + 45y &= 7400 \\ x &= 200 - y \\ 25(200 - y) + 45y &= 7400 \\ 5000 - 25y + 45y &= 7400 \\ 5000 + 20y &= 7400 \\ 20y &= 2400 \\ y &= 120 \\ x &= 200 - 120 = 80 \end{aligned}$$

80 \$25 sets of dishes should be ordered and 120 \$45 sets of dishes should be ordered.

67. In order to determine the size of the refund, we must determine the cost-per-package for bacon and eggs. We will use the general principle:

Total cost for an Item = (Cost per package) • (Number of Packages)

Let x = Cost-per-package of bacon
and y = Cost-per-carton of eggs

Now translate the sentences into equations:

(1) Three packages of bacon and two cartons of eggs cost \$7.45:
$$3x + 2y = 7.45 \quad (1)$$

(2) Two packages of bacon and three cartons of eggs cost \$6.45:
$$2x + 3y = 6.45 \quad (2)$$

Solve for x and y:

$$\begin{cases} 3x + 2y = 7.45 & (1) \\ 2x + 3y = 6.45 & (2) \end{cases}$$

$$\begin{cases} 6x + 4y = 14.90 & (1) \quad \text{Multiply both sides by 2.} \\ -6x - 9y = -19.35 & (2) \quad \text{Multiply both sides by } -3. \end{cases}$$

$$\begin{cases} 6x + 4y = 14.90 & (1) \\ \quad - 5y = -4.45 & (2) \quad \text{Replace (2) by (1) + (2).} \end{cases}$$

$$\begin{cases} 6x + 4y = 14.90 & (1) \\ y = .89 & (2) \end{cases}$$

$$\begin{cases} 6x + 3.56 = 14.90 & (1) \quad \text{Back-substitute; } y = .89. \\ y = .89 & (2) \end{cases}$$

$$\begin{cases} 6x = 11.34 & (1) \\ y = .89 & (2) \end{cases}$$

$$\begin{cases} x = 1.89 & (1) \\ y = .89 & (2) \end{cases}$$

We now know that bacon costs \$1.89 per package, and eggs cost \$.89 per carton. But the question is, how much money will be refunded on two packages of bacon and two cartons of eggs?

Refund = 2(1.89) + 2(.89) = \$5.56

69. Let x = cc of liquid one

y = cc of liquid two

$$.20x + .40y = 40 \text{ (Amount of Vitamin C)}$$
$$.30x + .20y = 30 \text{ (Amount of Vitamin D)}$$
$$20x + 40y = 4000$$
$$30x + 20y = 3000$$
$$40y = 4000 - 20x$$
$$y = \frac{4000}{40} - \frac{20x}{40}$$
$$y = 100 - \frac{1}{2}x$$
$$30x + 20y = 3000$$
$$30x + 20\left[100 - \frac{1}{2}x\right] = 3000$$
$$30x + 2000 - 10x = 3000$$
$$\frac{20}{20}x = \frac{1000}{20}$$
$$x = 50 \text{ cc}$$
$$y = 100 - \frac{1}{2}(50)$$
$$y = 74 \text{ cc}$$

50 cc of the liquid containing 20% Vitamin C and 30% Vitamin D, and 75cc of the liquid containing 40% Vitamin C and 20% Vitamin D should be mixed to fill the prescription.

71.
$$\begin{cases} I_2 = I_1 + I_3 \\ 5 - 3I_1 - 5I_2 = 0 \\ 10 - 5I_2 - 7I_3 = 0 \end{cases}$$

$$\begin{cases} I_2 = I_1 + I_3 \\ 5 - 3I_1 - 5(I_1 + I_3) = 0 \\ 10 - 5(I_1 + I_3) - 7I_3 = 0 \end{cases}$$

$$\begin{cases} I_2 = I_1 + I_3 \\ -8I_1 - 5I_3 + 5 = 0 \\ -5I_1 - 12I_3 + 10 = 0 \end{cases}$$

$$\begin{cases} I_2 = I_1 + I_3 \\ 40I_1 + 25I_3 = 25 \\ -40I_1 - 96I_3 = -80 \end{cases}$$

$$-71I_3 = -55$$
$$I_3 = \frac{55}{71}$$
$$-8I_1 - 5\left(\frac{55}{71}\right) + 5 = 0$$
$$-8I_1 = -5 + \frac{275}{71}$$
$$-8I_1 = \frac{-355 + 275}{71}$$
$$-8I_1 = \frac{-80}{71}$$
$$I_1 = \frac{10}{71}$$

$I_2 = I_1 + I_3$

$I_2 = \frac{10}{71} + \frac{55}{71}$

$I_2 = \frac{65}{71}$

$I_1 = \frac{10}{71},\ I_2 = \frac{65}{71},\ I_3 = \frac{55}{71}$

73. Let x = Number of orchestra seats,
y = Number of main seats,
and z = Number of balcony seats.

Then $x + y + z = 500$, since there are a total of 500 seats.

If all seats are sold, the revenue is \$17,100: $50x + 35y + 25z = 17{,}100$

Finally, if we sell only half of the orchestra seats, revenue is \$14,600:

$$50\left(\frac{1}{2}x\right) + 35y + 25z = 14{,}600$$

$$\begin{cases} x + y + z = 500 & (1) \\ 50x + 35y + 25z = 17{,}100 & (2) \\ 25x + 35y + 25z = 14{,}600 & (3) \end{cases}$$

$$\begin{cases} x + y + z = 500 & (1) & \\ 10x + 7y + 5z = 3420 & (2) & \text{Divide both sides by 5.} \\ -5x - 7y - 5z = -2920 & (3) & \text{Divide both sides by } -5. \end{cases}$$

$$\begin{cases} x + y + z = 500 & (1) & \\ 10x + 7y + 5z = 3420 & (2) & \\ 5x = 500 & (3) & \text{Replace (3) by (2) + (3).} \end{cases}$$

$$\begin{cases} x + y + z = 500 & (1) \\ 10x + 7y + 5z = 3420 & (2) \\ x = 100 & (3) \end{cases}$$

$$\begin{cases} 100 + y + z = 500 & (1) & \text{Back-substitute; } x = 100. \\ 1000 + 7y + 5z = 3420 & (2) & \text{Back-substitute; } x = 100. \\ x = 100 & (3) & \end{cases}$$

$$\begin{cases} y + z = 400 & (1) \\ 7y + 5z = 2420 & (2) \\ x = 100 & (3) \end{cases}$$

$$\begin{cases} -5y - 5z = -2000 & (1) & \text{Multiply both sides by } -5. \\ 7y + 5z = 2420 & (2) & \\ x = 100 & (3) & \end{cases}$$

$$\begin{cases} 2y = 420 & (1) & \text{Replace (1) by (1) + (2).} \\ 7y + 5z = 2420 & (2) & \\ x = 100 & (3) & \end{cases}$$

$$\begin{cases} y = 210 & (1) \\ 7y + 5z = 2420 & (2) \\ x = 100 & (3) \end{cases}$$

$$\begin{cases} y = 210 & (1) \\ 1470 + 5z = 2420 & (2) \\ x = 100 & (3) \end{cases}$$ Back-substitute; $y = 210$.

$$\begin{cases} y = 210 & (1) \\ 5z = 950 & (2) \\ x = 100 & (3) \end{cases}$$

$$\begin{cases} y = 210 & (1) \\ z = 190 & (2) \\ x = 100 & (3) \end{cases}$$

Thus, there are: $x = 100$ orchestra seats,
$y = 210$ main seats, and
$z = 190$ balcony seats.

79. We have $y = x^2 + bx + c$.

If this passes through (1, 2), then:

$$y = x^2 + bx + c$$
$$2 = 1^2 + b(1) + c \quad (x = 1, y = 2)$$
$$2 = 1 + b + c$$
$$b + c = 1 \qquad (1)$$

If $(-1, 3)$ lies on the graph, then:

$$y = x^2 + bx + c$$
$$3 = (-1)^2 + b(-1) + c$$
$$3 = 1 - b + c$$
$$-b + c = 2 \qquad (2)$$

Let's solve (2) for c:

$$-b + c = 2$$
$$c = 2 + b$$

Substitute this into (1):

$$b + c = 1$$
$$b + (2 + b) = 1$$
$$2b = -1$$
$$b = -\frac{1}{2}$$

Now find c:

$$c = 2 + b$$
$$c = 2 + \left(-\frac{1}{2}\right)$$
$$c = \frac{3}{2}$$

The solution is $b = -\frac{1}{2}$, $c = \frac{3}{2}$

81. We have $y = ax^2 + bx + c$.

From $(-1, 4)$: $4 = a(-1)^2 + b(-1) + c$
$4 = a - b + c \qquad (1)$

From (2, 3): $3 = a(2)^2 + b(2) + c$
$3 = 4a + 2b + c \qquad (2)$

From (0, 1): $1 = a(0)^2 + b(0) + c$
$c = 1 \qquad (3)$

Substitute $c = 1$ into (1) and (2):

$$\begin{cases} a - b + c = 4 & (1) \\ 4a + 2b + c = 3 & (2) \end{cases}$$

$$\begin{cases} a - b + 1 = 4 & (1) \\ 4a + 2b + 1 = 3 & (2) \end{cases}$$

$$\begin{cases} a - b = 3 & (1) \\ 4a + 2b = 2 & (2) \end{cases}$$

Now solve for (1) for a:

$$a - b = 3$$
$$a = 3 + b$$

Then, from (2):

$$4a + 2b = 2$$
$$4(3 + b) + 2b = 2$$
$$6b + 12 = 2$$
$$6b = -10$$
$$b = -\frac{10}{6}$$

or

$$b = -\frac{5}{3}$$

Finally,

$$a = 3 + b$$
$$a = 3 - \frac{5}{3}$$
$$a = \frac{4}{3}$$

The solution is $a = \frac{4}{3}$, $b = -\frac{5}{3}$, $c = 1$.

83. Solve: $\begin{cases} y = m_1x + b_1 & (1) \\ y = m_2x + b_2 & (2) \end{cases}$

From (1), $y = m_1x + b_1$. Substitute this into (2):

$$y = m_2x + b_2 \qquad (2)$$
$$m_1x + b_1 = m_2x + b_2$$
$$m_1x - m_2x = b_2 - b_1$$
$$(m_1 - m_2)x = b_2 - b_1$$
$$x = \frac{b_2 - b_1}{m_1 - m_2} = \frac{b_1 - b_2}{m_2 - m_1}$$

(Note $m_1 - m_2 \neq 0$)

Then:

$$y = m_1x + b_1$$
$$y = m_1\left[\frac{b_2 - b_1}{m_1 - m_2}\right] + b_1$$
$$y = \frac{m_1b_2 - m_1b_1}{m_1 - m_2} + \frac{b_1(m_1 - m_2)}{m_1 - m_2} \quad \text{(common denominator)}$$
$$y = \frac{m_1b_2 - m_1b_1 + m_1b_1 - m_2b_1}{m_1 - m_2}$$
$$y = \frac{m_1b_2 - m_2b_1}{m_1 - m_2} = \frac{m_2b_1 - m_1b_2}{m_2 - m_1}$$

85. We have: $\begin{cases} y = m_1x + b_1 & (1) \\ y = m_2x + b_2 & (2) \end{cases}$

Equation (1) is already solved for y, so we can substitute this into (2):

$$y = m_2x + b_2 \qquad (2)$$
$$m_1x + b_1 = m_2x + b_2$$
$$m_1x - m_2x = b_2 - b_1$$
$$mx - mx = b - b$$
$$0 \cdot x = 0$$

This is solved by *every* value of x. So the solution is:
$y = mx + b$, where x can be *any* real number.

8.2 Systems of Linear Equations: Matrices

1. $\begin{cases} x - 5y = 5 \\ 4x + 3y = 6 \end{cases}$ becomes $\left[\begin{array}{cc|c} 1 & -5 & 5 \\ 4 & 3 & 6 \end{array}\right]$

3. $\begin{cases} 2x + 3y - 6 = 0 \\ 4x - 6y + 2 = 0 \end{cases}$

First, put the constants on the right-hand side: $\begin{cases} 2x + 3y = 6 \\ 4x - 6y = -2 \end{cases}$

This can be represented as: $\left[\begin{array}{cc|c} 2 & 3 & 6 \\ 4 & -6 & -2 \end{array}\right]$

5. $\begin{cases} 0.01x - 0.03y = 0.06 \\ 0.13x + 0.10y = 0.20 \end{cases}$ becomes $\left[\begin{array}{cc|c} 0.01 & -0.03 & 0.06 \\ 0.13 & 0.10 & 0.20 \end{array}\right]$

7. $\begin{cases} x - y + z = 10 \\ 3x + 2y = 5 \\ x + y + 2z = 2 \end{cases}$ becomes $\left[\begin{array}{ccc|c} 1 & -1 & 1 & 10 \\ 3 & 2 & 0 & 5 \\ 1 & 1 & 2 & 2 \end{array}\right]$

9. $\begin{cases} x + y - z = 2 \\ 3x - 2y = 2 \end{cases}$ becomes $\left[\begin{array}{ccc|c} 1 & 1 & -1 & 2 \\ 3 & -2 & 0 & 2 \end{array}\right]$

11. $\begin{cases} x - w = 5 + y + z \\ 3x + y + w = 4 + 4z \end{cases}$

$\begin{cases} x - y - z - w = 5 \\ 3x + y - 4z + w = 4 \end{cases}$ becomes $\left[\begin{array}{cccc|c} 1 & -1 & -1 & -1 & 5 \\ 3 & -1 & -4 & 1 & 4 \end{array}\right]$

For Problems 11–20, we will show the augmented matrix that you should see on your graphing utility after each row operation is performed.

13. $\left[\begin{array}{ccc|c} 1 & -3 & -5 & -2 \\ 2 & -5 & -4 & 5 \\ -3 & 5 & 4 & 6 \end{array}\right] \rightarrow \left[\begin{array}{ccc|c} 1 & -3 & -5 & -2 \\ 0 & 1 & 6 & 9 \\ -3 & 5 & 4 & 6 \end{array}\right] \rightarrow \left[\begin{array}{ccc|c} 1 & -3 & -5 & -2 \\ 0 & 1 & 6 & 9 \\ 0 & -4 & -11 & 0 \end{array}\right] \rightarrow \left[\begin{array}{ccc|c} 1 & -3 & -5 & -2 \\ 0 & 1 & 6 & 9 \\ 0 & 0 & 13 & 36 \end{array}\right]$

(a) $R_2 = -2r_1 + r_2$ (b) $R_3 = 3r_1 + r_3$ (c) $R_3 = 4r_2 + r_3$

15. $\left[\begin{array}{ccc|c} 1 & -3 & 4 & 3 \\ 2 & -5 & 6 & 6 \\ -3 & 3 & 4 & 6 \end{array}\right] \rightarrow \left[\begin{array}{ccc|c} 1 & -3 & 4 & 3 \\ 0 & 1 & -2 & 0 \\ -3 & 3 & 4 & 6 \end{array}\right] \rightarrow \left[\begin{array}{ccc|c} 1 & -3 & 4 & 3 \\ 0 & 1 & -2 & 0 \\ 0 & -6 & 16 & 15 \end{array}\right] \rightarrow \left[\begin{array}{ccc|c} 1 & -3 & 4 & 3 \\ 0 & 1 & -2 & 0 \\ 0 & 0 & 4 & 15 \end{array}\right]$

(a) $R_2 = -2r_1 + r_2$ (b) $R_3 = 3r_1 + r_3$ (c) $R_3 = 6r_2 + r_3$

17. $$\left[\begin{array}{ccc|c} 1 & -3 & 2 & -6 \\ 2 & -5 & 3 & -4 \\ -3 & -6 & 4 & 6 \end{array}\right] \rightarrow \left[\begin{array}{ccc|c} 1 & -3 & 2 & -6 \\ 0 & 1 & -1 & 8 \\ -3 & -6 & 4 & 6 \end{array}\right] \rightarrow \left[\begin{array}{ccc|c} 1 & -3 & 2 & -6 \\ 0 & 1 & -1 & 8 \\ 0 & -15 & 10 & -12 \end{array}\right] \rightarrow \left[\begin{array}{ccc|c} 1 & -3 & 2 & -6 \\ 0 & 1 & -1 & 8 \\ 0 & 0 & -5 & 108 \end{array}\right]$$

(a) $R_2 = -2r_1 + r_2$ (b) $R_3 = 3r_1 + r_3$ (c) $R_3 = 15r_2 + r_3$

19. $$\left[\begin{array}{ccc|c} 1 & -3 & 1 & -2 \\ 2 & -5 & 6 & -2 \\ -3 & 1 & 4 & 6 \end{array}\right] \rightarrow \left[\begin{array}{ccc|c} 1 & -3 & 1 & -2 \\ 0 & 1 & 4 & 2 \\ -3 & 1 & 4 & 6 \end{array}\right] \rightarrow \left[\begin{array}{ccc|c} 1 & -3 & 1 & -2 \\ 0 & 1 & 4 & 2 \\ 0 & -8 & 7 & 0 \end{array}\right] \rightarrow \left[\begin{array}{ccc|c} 1 & -3 & 1 & -2 \\ 0 & 1 & 4 & 2 \\ 0 & 0 & 39 & 16 \end{array}\right]$$

(a) $R_2 = -2r_1 + r_2$ (b) $R_3 = 3r_1 + r_3$ (c) $R_3 = 8r_2 + r_3$

21. $$\left[\begin{array}{ccc|c} 1 & -3 & -2 & 3 \\ 2 & -5 & 2 & -1 \\ -3 & -2 & 4 & 6 \end{array}\right] \rightarrow \left[\begin{array}{ccc|c} 1 & -3 & -2 & 3 \\ 0 & 1 & 6 & -7 \\ -3 & -2 & 4 & 6 \end{array}\right] \rightarrow \left[\begin{array}{ccc|c} 1 & -3 & -2 & 3 \\ 0 & 1 & 6 & -7 \\ 0 & -11 & -2 & 15 \end{array}\right] \rightarrow \left[\begin{array}{ccc|c} 1 & -3 & -2 & 3 \\ 0 & 1 & 6 & -7 \\ 0 & 0 & 64 & -62 \end{array}\right]$$

(a) $R_2 = -2r_1 + r_2$ (b) $R_3 = 3r_1 + r_3$ (c) $R_3 = 11r_2 + r_3$

23. $\begin{cases} x = 5 \\ y = -1 \end{cases}$ consistent $x = 5$, $y = -1$

25. $\begin{cases} x = 1 \\ y = 2 \\ 0 = 3 \end{cases}$ inconsistent

27. $\begin{cases} x + 2z = -1 \\ y - 4z = -2 \\ 0 = 0 \end{cases}$ consistent $x = -1 - 2z$, $y = -2 + 4z$, z any real number

29. $\begin{cases} x_1 = 1 \\ x_2 + x_4 = 2 \\ x_3 + 2x_4 = 3 \end{cases}$ consistent $x_1 = 1, x_2 = 2 - x_4, x_3 = 3 - 2x_4$, x_4 any real number

31. $\begin{cases} x_1 + 4x_4 = 2 \\ x_2 + x_3 + 3x_4 = 3 \\ 0 = 0 \end{cases}$ consistent $x_1 = 2 - 4x_4, x_2 = 3 - x_3 - 3x_4$, x_3, x_4 any real numbers

33. $\begin{cases} x_1 + x_4 = -2 \\ x_2 + 2x_4 = 2 \\ x_3 - x_4 = 0 \\ 0 = 0 \end{cases}$ consistent $x_1 = -2 - x_4, x_2 = 2 - 2x_4, x_3 = x_4$, x_4 any real number

35. $\begin{cases} x + y = 8 \\ x - y = 4 \end{cases}$ can be represented as: $\left[\begin{array}{cc|c} 1 & 1 & 8 \\ 1 & -1 & 4 \end{array}\right]$

Since we already have a 1 in the first row, first column, we proceed to obtain 0's below it:

$$\left[\begin{array}{cc|c} 1 & 1 & 8 \\ 1 & -1 & 4 \end{array}\right] \rightarrow \left[\begin{array}{cc|c} 1 & 1 & 8 \\ 0 & -2 & -4 \end{array}\right] \rightarrow \left[\begin{array}{cc|c} 1 & 1 & 8 \\ 0 & 1 & 2 \end{array}\right]$$

$$R_2 = -1r_1 + r_2 \quad R_2 = -\frac{1}{2}r_2$$

That means: $\begin{cases} x + y = 8 \\ \quad\; y = 2 \end{cases}$

$\begin{cases} x + 2 = 8 \\ \quad\; y = 2 \end{cases}$ Back-substitute; $y = 2$.

$\begin{cases} x = 6 \\ y = 2 \end{cases}$

The solution is $x = 6$, $y = 2$.

37. $\begin{cases} x - 5y = -13 \\ 3x + 2y = 12 \end{cases}$

$$\left[\begin{array}{cc|c} 1 & -5 & -13 \\ 3 & 2 & 12 \end{array}\right] \rightarrow \left[\begin{array}{cc|c} 1 & -5 & -13 \\ 0 & 17 & 51 \end{array}\right] \rightarrow \left[\begin{array}{cc|c} 1 & -5 & -13 \\ 0 & 1 & 3 \end{array}\right]$$

$$R_2 = -3r_1 + r_2 \qquad R_2 = \frac{1}{17}r_2$$

$y = 3 \qquad x - 5y = -13$

$x = 5(3) - 13$

$x = 2$

$x = 2$, $y = 3$

39. $\begin{cases} 3x - 6y = 24 \\ 5x + 4y = 12 \end{cases}$ becomes: $\left[\begin{array}{cc|c} 3 & -6 & 24 \\ 5 & 4 & 12 \end{array}\right]$

$$\rightarrow \left[\begin{array}{cc|c} 1 & -2 & 8 \\ 5 & 4 & 12 \end{array}\right] \rightarrow \left[\begin{array}{cc|c} 1 & -2 & 8 \\ 0 & 14 & -28 \end{array}\right] \rightarrow \left[\begin{array}{cc|c} 1 & -2 & 8 \\ 0 & 1 & -2 \end{array}\right] \rightarrow \left[\begin{array}{cc|c} 1 & 0 & 4 \\ 0 & 1 & -2 \end{array}\right]$$

$$R_1 = \frac{1}{3}r_1 \qquad R_2 = -5r_1 + r_2 \quad R_2 = \frac{1}{14}r_2 \qquad R_1 = 2r_2 + r_1$$

The solution is $x = 4$, $y = -2$.

41. $\begin{cases} 2x + y = 1 \\ 4x + 2y = 6 \end{cases}$ becomes: $\left[\begin{array}{cc|c} 2 & 1 & 1 \\ 4 & 2 & 6 \end{array}\right] \rightarrow \left[\begin{array}{cc|c} 1 & \frac{1}{2} & \frac{1}{2} \\ 4 & 2 & 6 \end{array}\right] \rightarrow \left[\begin{array}{cc|c} 1 & \frac{1}{2} & \frac{1}{2} \\ 0 & 0 & 4 \end{array}\right]$

$$R_1 = \frac{1}{2}r_1 \qquad R_2 = -4r_1 + r_2$$

The system is inconsistent.

43. $\begin{cases} 2x - 4y = -2 \\ 3x + 2y = 3 \end{cases}$ becomes: $\left[\begin{array}{cc|c} 2 & -4 & -2 \\ 3 & 2 & 3 \end{array}\right]$

$$\to \left[\begin{array}{cc|c} 1 & -2 & -1 \\ 3 & 2 & 3 \end{array}\right] \to \left[\begin{array}{cc|c} 1 & -2 & -1 \\ 0 & 8 & 6 \end{array}\right] \to \left[\begin{array}{cc|c} 1 & -2 & -1 \\ 0 & 1 & \frac{3}{4} \end{array}\right] \to \left[\begin{array}{cc|c} 1 & 0 & \frac{1}{2} \\ 0 & 1 & \frac{3}{4} \end{array}\right]$$

$R_1 = \frac{1}{2}r_1$ $\quad R_2 = -3r_1 + r_2$ $\quad R_2 = \frac{1}{8}r_2$ $\quad R_1 = 2r_2 + r_1$

The solution is $x = \frac{1}{2}$, $y = \frac{3}{4}$.

45. $\begin{cases} x + 2y = 4 \\ 2x + 4y = 8 \end{cases}$ becomes: $\left[\begin{array}{cc|c} 1 & 2 & 4 \\ 2 & 4 & 8 \end{array}\right] \to \left[\begin{array}{cc|c} 1 & 2 & 4 \\ 0 & 0 & 0 \end{array}\right]$

$R_2 = -2r_1 + r_2$

Therefore, the system is equivalent to the single equation: $x + 2y = 4$

(a) In terms of y, $x = -2y + 4$
where y can be any real number.

(b) In terms of x, $2y = -x + 4$,
or $y = -\frac{1}{2}x + 2$,
where x can be any real number.

47. $\begin{cases} 2x + 3y = 6 \\ x - y = \frac{1}{2} \end{cases}$ becomes: $\left[\begin{array}{cc|c} 2 & 3 & 6 \\ 1 & -1 & \frac{1}{2} \end{array}\right]$

$$\to \left[\begin{array}{cc|c} 1 & -1 & \frac{1}{2} \\ 2 & 3 & 6 \end{array}\right] \to \left[\begin{array}{cc|c} 1 & -1 & \frac{1}{2} \\ 0 & 5 & 5 \end{array}\right] \to \left[\begin{array}{cc|c} 1 & -1 & \frac{1}{2} \\ 0 & 1 & 1 \end{array}\right] \to \left[\begin{array}{cc|c} 1 & 0 & \frac{3}{2} \\ 0 & 1 & 1 \end{array}\right]$$

Interchange r_1 and r_2 $\quad R_2 = -2r_1 + r_2$ $\quad R_2 = \frac{1}{5}r_2$ $\quad R_1 = r_2 + r_1$

The solution is $x = \frac{3}{2}$, $y = 1$.

49. $\begin{cases} 3x - 5y = 3 \\ 15x + 5y = 21 \end{cases}$ becomes: $\left[\begin{array}{cc|c} 3 & -5 & 3 \\ 15 & 5 & 21 \end{array}\right]$

$$\to \left[\begin{array}{cc|c} 1 & -\frac{5}{3} & 1 \\ 15 & 5 & 21 \end{array}\right] \to \left[\begin{array}{cc|c} 1 & -\frac{5}{3} & 1 \\ 0 & 30 & 6 \end{array}\right] \to \left[\begin{array}{cc|c} 1 & -\frac{5}{3} & 1 \\ 0 & 1 & \frac{1}{5} \end{array}\right]$$

$R_1 = \frac{1}{3}r_1$ $\quad R_2 = -15r_1 + r_2$ $\quad R_2 = \frac{1}{30}r_2$

Thus, we have:

$$\begin{cases} x - \frac{5}{3}y = 1 \\ \quad y = \frac{1}{5} \end{cases}$$

$$\begin{cases} x - \frac{5}{3}\left(\frac{1}{5}\right) = 1 \quad \text{(Back-substitution)} \\ \quad y = \frac{1}{5} \end{cases}$$

$$\begin{cases} x = \frac{4}{3} \\ y = \frac{1}{5} \end{cases}$$

The solution is $x = \frac{4}{3}, y = \frac{1}{5}$

51. $\begin{cases} x - y \quad = 6 \\ 2x \quad - 3z = 16 \\ \quad 2y + z = 4 \end{cases}$ becomes: $\left[\begin{array}{ccc|c} 1 & -1 & 0 & 6 \\ 2 & 0 & -3 & 16 \\ 0 & 2 & 1 & 4 \end{array}\right]$ (We already have a 1 in row 1, column 1.)

$$\rightarrow \left[\begin{array}{ccc|c} 1 & -1 & 0 & 6 \\ 0 & 2 & -3 & 4 \\ 0 & 2 & 1 & 4 \end{array}\right] \quad \text{(Use the 1 to get 0's below it.)}$$

$\uparrow$

$R_2 = -2r_1 + r_2$

$$\rightarrow \left[\begin{array}{ccc|c} 1 & -1 & 0 & 6 \\ 0 & 1 & -\frac{3}{2} & 2 \\ 0 & 2 & 1 & 4 \end{array}\right] \quad \text{(We need a 1 in row 2, column 2.)}$$

$\uparrow$

$R_2 = \frac{1}{2}r_2$

$$\rightarrow \left[\begin{array}{ccc|c} 1 & -1 & 0 & 6 \\ 0 & 1 & -\frac{3}{2} & 2 \\ 0 & 0 & 4 & 0 \end{array}\right] \quad \text{(Obtain a 0 below the 1 in row 2.)}$$

$\uparrow$

$R_3 = -2r_2 + r_3$

$$\rightarrow \left[\begin{array}{ccc|c} 1 & -1 & 0 & 6 \\ 0 & 1 & -\frac{3}{2} & 2 \\ 0 & 0 & 1 & 0 \end{array}\right]$$

$\uparrow$

$R_3 = \frac{1}{4}r_3$

Thus, we have:

$$\begin{cases} x - y = 6 \\ y - \frac{3}{2}z = 2 \\ z = 0 \end{cases}$$

$$\begin{cases} x - y = 6 \\ y = 2 \\ z = 0 \end{cases} \quad \text{(Back-substitution; } z = 0\text{)}$$

$$\begin{cases} x - 2 = 6 \\ y = 2 \\ z = 0 \end{cases}$$

The solution is $x = 8$, $y = 2$, $z = 0$.

53. $\begin{cases} x - 2y + 3z = 7 \\ 2x + y + z = 4 \\ -3x + 2y - 2z = -10 \end{cases}$ becomes: $\left[\begin{array}{ccc|c} 1 & -2 & 3 & 7 \\ 2 & 1 & 1 & 4 \\ -3 & 2 & -2 & -10 \end{array}\right]$

$$\rightarrow \left[\begin{array}{ccc|c} 1 & -2 & 3 & 7 \\ 0 & 5 & -5 & -10 \\ 0 & -4 & 7 & 11 \end{array}\right] \quad \text{(Use the 1 in row 1 to get 0's below it.)}$$

$R_2 = -2r_1 + r_2$
$R_3 = 3r_1 + r_3$

To obtain a 1 in row 2, column 2, we can either multiply row 2 by $\frac{1}{5}$, or add row 3 to row 2:

$$\rightarrow \left[\begin{array}{ccc|c} 1 & -2 & 3 & 7 \\ 0 & 1 & -1 & -2 \\ 0 & -4 & 7 & 11 \end{array}\right] \rightarrow \left[\begin{array}{ccc|c} 1 & 0 & 1 & 3 \\ 0 & 1 & -1 & -2 \\ 0 & 0 & 3 & 3 \end{array}\right]$$

$R_2 = \frac{1}{5}r_2$ $\qquad$ $R_1 = 2r_2 + r_1$
$R_3 = 4r_2 + r_3$

The zero ***above*** the 1 in row 2 will eliminate the need to do back-substitution at the end of the problem.

$$\rightarrow \left[\begin{array}{ccc|c} 1 & 0 & 1 & 3 \\ 0 & 1 & -1 & -2 \\ 0 & 0 & 1 & 1 \end{array}\right] \rightarrow \left[\begin{array}{ccc|c} 1 & 0 & 0 & 2 \\ 0 & 1 & 0 & -1 \\ 0 & 0 & 1 & 1 \end{array}\right]$$

$R_3 = \frac{1}{3}r_3$ $\qquad$ $R_1 = -1r_3 + r_1$
$R_2 = r_3 + r_2$

The solution is $x = 2$, $y = -1$, $z = 1$.

55. $\begin{cases} 2x - 2y - 2z = 2 \\ 2x + 3y + z = 2 \\ 3x + 2y = 0 \end{cases}$ becomes: $\left[\begin{array}{ccc|c} 2 & -2 & -2 & 2 \\ 2 & 3 & 1 & 2 \\ 3 & 2 & 0 & 0 \end{array}\right]$

$$\rightarrow \left[\begin{array}{ccc|c} 2 & -2 & -2 & 2 \\ 0 & 5 & 3 & 0 \\ 1 & 4 & 2 & -2 \end{array}\right] \rightarrow \left[\begin{array}{ccc|c} 1 & 4 & 2 & -2 \\ 0 & 5 & 3 & 0 \\ 2 & -2 & -2 & 2 \end{array}\right] \rightarrow \left[\begin{array}{ccc|c} 1 & 4 & 2 & -2 \\ 0 & 5 & 3 & 0 \\ 0 & -10 & -6 & 6 \end{array}\right]$$

$R_2 = -r_1 + r_2$, $R_3 = -r_1 + r_3$; $R_1 \leftrightarrow R_3$; $R_3 = -2r_1 + r_3$

$$\rightarrow \left[\begin{array}{ccc|c} 1 & 4 & 2 & -2 \\ 0 & 1 & \frac{3}{5} & 0 \\ 0 & -10 & -6 & 6 \end{array}\right] \rightarrow \left[\begin{array}{ccc|c} 1 & 0 & \frac{22}{5} & -2 \\ 0 & 5 & \frac{3}{5} & 0 \\ 0 & 0 & 0 & 6 \end{array}\right]$$

$R_2 = \frac{1}{5}r_2$; $R_1 = -4r_2 + r_1$, $R_3 = 10r_2 + r_3$

No solution. Inconsistent.

57. $\begin{cases} -x + y + z = -1 \\ -x + 2y - 3z = -4 \\ 3x - 2y - 7z = 0 \end{cases}$ becomes: $\left[\begin{array}{ccc|c} -1 & 1 & 1 & -1 \\ -1 & 2 & -3 & -4 \\ 3 & -2 & -7 & 0 \end{array}\right]$

$$\rightarrow \left[\begin{array}{ccc|c} 1 & -1 & -1 & 1 \\ -1 & 2 & -3 & -4 \\ 3 & -2 & -7 & 0 \end{array}\right] \rightarrow \left[\begin{array}{ccc|c} 1 & -1 & -1 & 1 \\ 0 & 1 & -4 & -3 \\ 0 & 1 & -4 & -3 \end{array}\right] \rightarrow \left[\begin{array}{ccc|c} 1 & 0 & -5 & -2 \\ 0 & 1 & -4 & -3 \\ 0 & 0 & 0 & 0 \end{array}\right] \rightarrow \begin{array}{l} x - 5z = -2 \\ y - 4z = -3 \end{array}$$

$R_1 = -r_1$; $R_2 = r_1 + r_2$, $R_3 = -3r_1 + r_3$; $R_1 = r_2 + r_1$, $R_3 = -r_2 + r_3$

Hence, $x = 5z - 2$; $y = 4z - 3$ where z is any real number.

59. $\begin{cases} 2x - 2y + 3z = 6 \\ 4x - 3y + 2z = 0 \\ -2x + 3y - 7z = 1 \end{cases}$ becomes: $\left[\begin{array}{ccc|c} 2 & -2 & 3 & 6 \\ 4 & -3 & 2 & 0 \\ -2 & 3 & -7 & 1 \end{array}\right]$

$$\rightarrow \left[\begin{array}{ccc|c} 1 & -1 & \frac{3}{2} & 3 \\ 4 & -3 & 2 & 0 \\ -2 & 3 & -7 & 1 \end{array}\right] \rightarrow \left[\begin{array}{ccc|c} 1 & -1 & \frac{3}{2} & 3 \\ 0 & 1 & -4 & -12 \\ 0 & 1 & -4 & 7 \end{array}\right] \rightarrow \left[\begin{array}{ccc|c} 1 & 0 & -\frac{5}{2} & -9 \\ 0 & 1 & -4 & -12 \\ 0 & 0 & 0 & 19 \end{array}\right]$$

$R_1 = \frac{1}{2}r_1$; $R_2 = -4r_1 + r_2$, $R_3 = 2r_1 + r_3$; $R_1 = r_2 + r_1$, $R_3 = -r_2 + r_3$

The system is inconsistent.

61. $\begin{cases} x + y - z = 6 \\ 3x - 2y + z = -5 \\ x + 3y - 2z = 14 \end{cases}$ becomes: $\left[\begin{array}{ccc|c} 1 & 1 & -1 & 6 \\ 3 & -2 & 1 & -5 \\ 1 & 3 & -2 & 14 \end{array}\right]$

$$\rightarrow \left[\begin{array}{ccc|c} 1 & 1 & -1 & 6 \\ 0 & -5 & 4 & -23 \\ 0 & 2 & -1 & 8 \end{array}\right] \rightarrow \left[\begin{array}{ccc|c} 1 & 1 & -1 & 6 \\ 0 & 1 & -\frac{4}{5} & \frac{23}{5} \\ 0 & 2 & -1 & 8 \end{array}\right] \rightarrow \left[\begin{array}{ccc|c} 1 & 0 & -\frac{1}{5} & \frac{7}{5} \\ 0 & 1 & -\frac{4}{5} & \frac{23}{5} \\ 0 & 0 & \frac{3}{5} & -\frac{6}{5} \end{array}\right] \rightarrow \left[\begin{array}{ccc|c} 1 & 0 & -\frac{1}{5} & \frac{7}{5} \\ 0 & 1 & -\frac{4}{5} & \frac{23}{5} \\ 0 & 0 & 1 & -2 \end{array}\right]$$

$R_2 = -3r_1 + r_2$, $R_3 = -r_1 + r_3$; $\quad R_2 = \frac{-1}{5}r_2$; $\quad R_1 = -r_2 + r_1$, $R_3 = -2r_2 + r_3$; $\quad R_3 = \frac{5}{3}r_3$

$$\rightarrow \left[\begin{array}{ccc|c} 1 & 0 & 0 & 1 \\ 0 & 1 & 0 & 3 \\ 0 & 0 & 1 & -2 \end{array}\right]$$

$R_1 = \frac{1}{5}r_3 + r_1$

$R_2 = \frac{4}{5}r_3 + r_2$

The solution is $x = 1$, $y = 3$, $z = -2$.

63. $\begin{cases} x + 2y - z = -3 \\ 2x - 4y + z = -7 \\ -2x + 2y - 3z = 4 \end{cases}$ becomes: $\left[\begin{array}{ccc|c} 1 & 2 & -1 & -3 \\ 2 & -4 & 1 & -7 \\ -2 & 2 & -3 & 4 \end{array}\right]$

$$\rightarrow \left[\begin{array}{ccc|c} 1 & 2 & -1 & -3 \\ 0 & -8 & 3 & -1 \\ 0 & 6 & -5 & -2 \end{array}\right] \rightarrow \left[\begin{array}{ccc|c} 1 & 2 & -1 & -3 \\ 0 & -2 & -2 & -3 \\ 0 & 6 & -5 & -2 \end{array}\right]$$

This will make the fractions in row 2 easier to work with.

$R_2 = -2r_1 + r_2$, $R_3 = 2r_1 + r_3$; $\quad R_2 = r_3 + r_2$

$$\rightarrow \left[\begin{array}{ccc|c} 1 & 2 & -1 & -3 \\ 0 & 1 & 1 & \frac{3}{2} \\ 0 & 6 & -5 & -2 \end{array}\right] \rightarrow \left[\begin{array}{ccc|c} 1 & 0 & -3 & -6 \\ 0 & 1 & 1 & \frac{3}{2} \\ 0 & 0 & -11 & -11 \end{array}\right]$$

$R_2 = -\frac{1}{2}r_2$; $\quad R_1 = -2r_2 + r_1$, $R_3 = -6r_2 + r_3$

$$\rightarrow \begin{bmatrix} 1 & 0 & -3 & | & -6 \\ 0 & 1 & 1 & | & \frac{3}{2} \\ 0 & 0 & 1 & | & 1 \end{bmatrix} \rightarrow \begin{bmatrix} 1 & 0 & 0 & | & -3 \\ 0 & 1 & 0 & | & \frac{1}{2} \\ 0 & 0 & 1 & | & 1 \end{bmatrix}$$

$R_3 = -\frac{1}{11}r_3$ $\qquad$ $R_1 = 3r_3 + r_1$

$R_2 = -1r_3 + r_2$

The solution is $x = -3$, $y = \frac{1}{2}$, $z = 1$.

65. $\begin{cases} 3x + y - z = \frac{2}{3} \\ 2x - y + z = 1 \\ 4x + 2y = \frac{8}{3} \end{cases}$ becomes: $\begin{bmatrix} 3 & 1 & -1 & | & \frac{2}{3} \\ 2 & -1 & 1 & | & 1 \\ 4 & 2 & 0 & | & \frac{8}{3} \end{bmatrix}$

$$\rightarrow \begin{bmatrix} 1 & 2 & -2 & | & -\frac{1}{3} \\ 2 & -1 & 1 & | & 1 \\ 4 & 2 & 0 & | & \frac{8}{3} \end{bmatrix} \rightarrow \begin{bmatrix} 1 & 2 & -2 & | & -\frac{1}{3} \\ 0 & -5 & 5 & | & \frac{5}{3} \\ 0 & -6 & 8 & | & \frac{12}{3} \end{bmatrix}$$

$R_1 = -1r_2 + r_1$ $\qquad$ $R_2 = -2r_1 + r_2$

$R_3 = -4r_1 + r_3$

$$\rightarrow \begin{bmatrix} 1 & 2 & -2 & | & -\frac{1}{3} \\ 0 & 1 & -1 & | & -\frac{1}{3} \\ 0 & -6 & 8 & | & 4 \end{bmatrix} \rightarrow \begin{bmatrix} 1 & 0 & 0 & | & \frac{1}{3} \\ 0 & 1 & -1 & | & -\frac{1}{3} \\ 0 & 0 & 2 & | & 2 \end{bmatrix}$$

$R_2 = -\frac{1}{5}r_2$ $\qquad$ $R_1 = -2r_2 + r_1$

$R_3 = 6r_2 + r_3$

$$\rightarrow \begin{bmatrix} 1 & 0 & 0 & | & \frac{1}{3} \\ 0 & 1 & -1 & | & -\frac{1}{3} \\ 0 & 0 & 1 & | & 1 \end{bmatrix} \rightarrow \begin{bmatrix} 1 & 0 & 0 & | & \frac{1}{3} \\ 0 & 1 & 0 & | & \frac{2}{3} \\ 0 & 0 & 1 & | & 1 \end{bmatrix}$$

$R_3 = \frac{1}{2}r_3$ $\qquad$ $R_2 = r_3 + r_2$

The solution is $x = \frac{1}{3}$, $y = \frac{2}{3}$, $z = 1$.

67. $\begin{cases} x + y + z + w = 4 \\ 2x - y + z = 0 \\ 3x + 2y + z - w = 6 \\ x - 2y - 2z + 2w = -1 \end{cases}$ becomes: $\left[\begin{array}{cccc|c} 1 & 1 & 1 & 1 & 4 \\ 2 & -1 & 1 & 0 & 0 \\ 3 & 2 & 1 & -1 & 6 \\ 1 & -2 & -2 & 2 & -1 \end{array}\right]$

$$\rightarrow \left[\begin{array}{cccc|c} 1 & 1 & 1 & 1 & 4 \\ 0 & -3 & -1 & -2 & -8 \\ 0 & -1 & -2 & -4 & -6 \\ 0 & -3 & -3 & 1 & -5 \end{array}\right] \rightarrow \left[\begin{array}{cccc|c} 1 & 1 & 1 & 1 & 4 \\ 0 & -1 & -2 & -4 & -6 \\ 0 & -3 & -1 & -2 & -8 \\ 0 & -3 & -3 & 1 & -5 \end{array}\right]$$

↑ $R_2 = -2r_1 + r2$, $R_3 = -3r_1 + r_3$, $R_4 = -1r_1 + r_4$

↑ Interchange r_2 and r_3

$$\rightarrow \left[\begin{array}{cccc|c} 1 & 1 & 1 & 1 & 4 \\ 0 & 1 & 2 & 4 & 6 \\ 0 & -3 & -1 & -2 & -8 \\ 0 & -3 & -3 & 1 & -5 \end{array}\right] \rightarrow \left[\begin{array}{cccc|c} 1 & 0 & -1 & -3 & -2 \\ 0 & 1 & 2 & 4 & 6 \\ 0 & 0 & 5 & 10 & 10 \\ 0 & 0 & 3 & 13 & 13 \end{array}\right]$$

↑ $R_2 = -1r_2$

↑ $R_1 = -1r_2 + r_1$, $R_3 = 3r_2 + r_3$, $R_4 = 3r_2 + r_4$

$$\rightarrow \left[\begin{array}{cccc|c} 1 & 0 & -1 & -3 & -2 \\ 0 & 1 & 2 & 4 & 6 \\ 0 & 0 & 1 & 2 & 2 \\ 0 & 0 & 3 & 13 & 13 \end{array}\right] \rightarrow \left[\begin{array}{cccc|c} 1 & 0 & 0 & -1 & 0 \\ 0 & 1 & 0 & 0 & 2 \\ 0 & 0 & 1 & 2 & 2 \\ 0 & 0 & 0 & 7 & 7 \end{array}\right]$$

↑ $R_3 = \frac{1}{5}r_3$

↑ $R_1 = r_3 + r_1$, $R_2 = -2r_3 + r_2$, $R_4 = -3r_3 + r_4$

$$\rightarrow \left[\begin{array}{cccc|c} 1 & 0 & 0 & -1 & 0 \\ 0 & 1 & 0 & 0 & 2 \\ 0 & 0 & 1 & 2 & 2 \\ 0 & 0 & 0 & 1 & 1 \end{array}\right] \rightarrow \left[\begin{array}{cccc|c} 1 & 0 & 0 & 0 & 0 \\ 0 & 1 & 0 & 0 & 2 \\ 0 & 0 & 1 & 0 & 0 \\ 0 & 0 & 0 & 1 & 1 \end{array}\right]$$

↑ $R_4 = \frac{1}{7}r_4$

↑ $R_1 = r_4 + r_1$, $R_3 = -2r_4 + r_3$

The solution is $x = 1, y = 2, z = 0, w = 1$.

69. $\begin{cases} x + 2y + z = 1 \\ 2x - y + 2z = 2 \\ 3x + y + 3z = 3 \end{cases}$ becomes: $\left[\begin{array}{ccc|c} 1 & 2 & 1 & 1 \\ 2 & -1 & 2 & 2 \\ 3 & 1 & 3 & 3 \end{array}\right]$

$$\rightarrow \left[\begin{array}{ccc|c} 1 & 2 & 1 & 1 \\ 0 & -5 & 0 & 0 \\ 0 & -5 & 0 & 0 \end{array}\right]$$

$\uparrow$
$R_2 = -2r_1 + r_2$
$R_3 = -3r_1 + r_3$

The system is equivalent to two equations: $\begin{cases} x + 2y + z = 1 & (1) \\ -5y = 0 & (2) \end{cases}$

From (2) we have: $y = 0$, and back-substitution into (1) yields: $x + z = 1$

We can write the solution as:
$y = 0$; $x = -z + 1$; where z is any real number,
or
$y = 0$; $z = -x + 1$; where x is any real number.

71. $\begin{cases} x - y + z = 5 \\ 3x + 2y - 2z = 0 \end{cases}$ becomes: $\left[\begin{array}{ccc|c} 1 & -1 & 1 & 5 \\ 3 & 2 & -2 & 0 \end{array}\right]$

$$\rightarrow \left[\begin{array}{ccc|c} 1 & -1 & 1 & 5 \\ 0 & 5 & -5 & -15 \end{array}\right] \rightarrow \left[\begin{array}{ccc|c} 1 & -1 & 1 & 5 \\ 0 & 1 & -1 & -3 \end{array}\right] \rightarrow \left[\begin{array}{ccc|c} 1 & 0 & 0 & 2 \\ 0 & 1 & -1 & -3 \end{array}\right]$$

$\uparrow$ $R_2 = -3r_1 + r_2$ $\quad$ $\uparrow$ $R_2 = \frac{1}{5}r_2$ $\quad$ $\uparrow$ $R_1 = r_2 + r_1$

This represents the system: $\begin{cases} x = 2 \\ y - z = -3 \end{cases}$

The solution is:
$x = 2$; $y = z - 3$; where z is any real number,
or
$x = 2$; $z = y + 3$; where y is any real number.

73. $\begin{cases} 2x + 3y - z = 3 \\ x - y - z = 0 \\ -x + y + z = 0 \\ x + y + 3z = 5 \end{cases}$ becomes: $\left[\begin{array}{ccc|c} 2 & 3 & -1 & 3 \\ 1 & -1 & -1 & 0 \\ -1 & 1 & 1 & 0 \\ 1 & 1 & 3 & 5 \end{array}\right]$

$$\rightarrow \left[\begin{array}{ccc|c} 1 & -1 & -1 & 0 \\ 2 & 3 & -1 & 3 \\ -1 & 1 & 1 & 0 \\ 1 & 1 & 3 & 5 \end{array}\right] \rightarrow \left[\begin{array}{ccc|c} 1 & -1 & -1 & 0 \\ 0 & 5 & 1 & 3 \\ 0 & 0 & 0 & 0 \\ 0 & 2 & 4 & 5 \end{array}\right] \rightarrow \left[\begin{array}{ccc|c} 1 & -1 & -1 & 0 \\ 0 & 5 & 1 & 3 \\ 0 & 2 & 4 & 5 \\ 0 & 0 & 0 & 0 \end{array}\right]$$

$\uparrow$ Interchange rows one and two $\quad$ $\uparrow$ $R_2 = -2r_1 + r_2$, $R_3 = r_1 + r_3$, $R_4 = -1r_1 + r_4$ $\quad$ $\uparrow$ Interchange r_3 and r_4

$$\to \left[\begin{array}{ccc|c} 1 & -1 & -1 & 0 \\ 0 & 1 & -7 & -7 \\ 0 & 2 & 4 & 5 \\ 0 & 0 & 0 & 0 \end{array}\right] \to \left[\begin{array}{ccc|c} 1 & 0 & -8 & -7 \\ 0 & 1 & -7 & -7 \\ 0 & 1 & 18 & 19 \\ 0 & 0 & 0 & 0 \end{array}\right] \to \left[\begin{array}{ccc|c} 1 & 0 & -8 & -7 \\ 0 & 1 & -7 & -7 \\ 0 & 0 & 1 & \frac{19}{18} \\ 0 & 0 & 0 & 0 \end{array}\right]$$

$R_2 = -2r_3 + r_2$ $\quad R_1 = r_2 + r_1$, $R_3 = -2r_2 + r_3$ $\quad R_3 = \frac{1}{18} r_3$

Because of the unusual fraction, we will do back-substitution. We have:

$$\begin{cases} x - 8z = -7 \\ y - 7z = -7 \\ z = \dfrac{19}{18} \end{cases}$$

$$\begin{cases} x - 8\left(\dfrac{19}{18}\right) = -7 \\ y - 7\left(\dfrac{19}{18}\right) = -7 \\ z = \dfrac{19}{18} \end{cases}$$

$$\begin{cases} x = \dfrac{-7 \cdot 18 + 8 \cdot 19}{18} = \dfrac{26}{18} = \dfrac{13}{9} \\ y = \dfrac{-7 \cdot 18 + 7 \cdot 19}{18} = \dfrac{7}{18} \\ z = \dfrac{19}{18} \end{cases}$$

Thus, the solution is $x = \dfrac{13}{9}$, $y = \dfrac{7}{18}$, $z = \dfrac{19}{18}$.

75. $\begin{cases} 4x + y + z - w = 4 \\ x - y + 2z + 3w = 3 \end{cases}$

$$\left[\begin{array}{cccc|c} 4 & 1 & 1 & -1 & 4 \\ 1 & -1 & 2 & 3 & 3 \end{array}\right] \to \left[\begin{array}{cccc|c} 1 & -1 & 2 & 3 & 3 \\ 4 & 1 & 1 & -1 & 4 \end{array}\right]$$

Interchange rows

$$\to \left[\begin{array}{cccc|c} 1 & -1 & 2 & 3 & 3 \\ 0 & 5 & -7 & -13 & -8 \end{array}\right]$$

$R_2 = -4r_1 + r_2$

This is equivalent to the system: $\begin{cases} x - y + 2z - 3w = 3 & (1) \\ 5y - 7z - 13w = -8 & (2) \end{cases}$

$5y = 7z + 13w - 8$

From (2): $y = \frac{7}{5}z + \frac{13}{5}w - \frac{8}{5}$

Then from (1): $x = y - 2z - 3w + 3$

or $x = \left[\frac{7}{5}z + \frac{13}{5}w - \frac{8}{5}\right] - 2z - 3w + 3$

$x = -\frac{3}{5}z - \frac{2}{5}w + \frac{7}{5}$

The solution is: $x = -\frac{3}{5}z - \frac{2}{5}w + \frac{7}{5}$

$y = \frac{7}{5}z + \frac{13}{5}w - \frac{8}{5}$

where z and w are any real numbers.

77. We have $y = ax^2 + bx + c$, and each of the three points must satisfy this equation.

(1, 2): $2 = a + b + c$
(−2, −7): $-7 = 4a - 2b + c$
(−2, −3): $-3 = 4a + 2b + c$

We have three equations in three unknowns which can be represented by:

$$\left[\begin{array}{ccc|c} 1 & 1 & 1 & 2 \\ 4 & -2 & 1 & -7 \\ 4 & 2 & 1 & -3 \end{array}\right] \to \left[\begin{array}{ccc|c} 1 & 1 & 1 & 2 \\ 0 & -6 & -3 & -15 \\ 0 & -2 & -3 & -11 \end{array}\right] \to \left[\begin{array}{ccc|c} 1 & 1 & 1 & 2 \\ 0 & 1 & \frac{1}{2} & \frac{5}{2} \\ 0 & -2 & -3 & -11 \end{array}\right]$$

$R_2 = -4r_1 + r_2$ $\quad R_2 = -\frac{1}{6}r_2$

$R_3 = -4r_1 + r_3$

$$\to \left[\begin{array}{ccc|c} 1 & 0 & \frac{1}{2} & -\frac{1}{2} \\ 0 & 1 & \frac{1}{2} & \frac{5}{2} \\ 0 & 0 & -2 & -6 \end{array}\right] \to \left[\begin{array}{ccc|c} 1 & 0 & \frac{1}{2} & -\frac{1}{2} \\ 0 & 1 & \frac{1}{2} & \frac{5}{2} \\ 0 & 0 & 1 & 3 \end{array}\right] \to \left[\begin{array}{ccc|c} 1 & 0 & 0 & -2 \\ 0 & 1 & 0 & 1 \\ 0 & 0 & 1 & 3 \end{array}\right]$$

$R_1 = -1r_2 + r_1$ $\quad R_3 = -\frac{1}{2}r_3$ $\quad R_1 = -\frac{1}{2}r_3 + r_1$

$R_3 = 2r_2 + r_3$ $\quad R_2 = -\frac{1}{2}r_3 + r_2$

The solution is $a = -2$, $b = 1$, $c = 3$, so the parabola is $y = -2x^2 + x + 3$.

You can verify that each of the given points satisfies this equation.

79. $f(x) = ax^3 + bx^2 + cx + d$

$f(-3) = -112$ implies $\quad -27a + 9b - 3c + d = -112$
$f(-1) = -2$ implies $\quad -a + b - c + d = -2$
$f(1) = 4$ implies $\quad a + b + c + d = 4$
and $f(2) = 13$ implies $\quad -8a + 4b + 2c + d = 13$

Thus, we want to find the solution to a system of four equations in four unknowns.

$$\left[\begin{array}{cccc|c} -27 & 9 & -3 & 1 & -112 \\ -1 & 1 & -1 & 1 & -2 \\ 1 & 1 & 1 & 1 & 4 \\ 8 & 4 & 2 & 1 & 13 \end{array}\right] \rightarrow \left[\begin{array}{cccc|c} 1 & 1 & 1 & 1 & 4 \\ -1 & 1 & -1 & 1 & -2 \\ -27 & 9 & -3 & 1 & -112 \\ 8 & 4 & 2 & 1 & 13 \end{array}\right] \rightarrow \left[\begin{array}{cccc|c} 1 & 1 & 1 & 1 & 4 \\ 0 & 2 & 0 & 2 & 2 \\ 0 & 36 & 24 & 28 & -4 \\ 0 & -4 & -6 & -7 & -19 \end{array}\right]$$

↑ Interchange r_3 and r_1

↑ $R_2 = r_1 + r_2$
$R_3 = 27r_1 + r_3$
$R_4 = -8r_1 + r_4$

$$\rightarrow \left[\begin{array}{cccc|c} 1 & 1 & 1 & 1 & 4 \\ 0 & 1 & 0 & 1 & 1 \\ 0 & 9 & 6 & 7 & -1 \\ 0 & -4 & -6 & -7 & -19 \end{array}\right] \rightarrow \left[\begin{array}{cccc|c} 1 & 0 & 1 & 0 & 3 \\ 0 & 1 & 0 & 1 & 1 \\ 0 & 0 & 6 & -2 & -10 \\ 0 & 0 & -6 & -3 & -15 \end{array}\right]$$

Now can we get a 1 in row 3 column 3, ***and*** avoid fractions?

↑ $R_2 = \frac{1}{2}r_2$
$R_3 = \frac{1}{4}r_3$

↑ $R_1 = -1r_2 + r_1$
$R_3 = -9r_2 + r_3$
$R_4 = 4r_2 + r_4$

$$\rightarrow \left[\begin{array}{cccc|c} 1 & 0 & 1 & 0 & 3 \\ 0 & 1 & 0 & 1 & 1 \\ 0 & 0 & 3 & -1 & -5 \\ 0 & 0 & -2 & -1 & -5 \end{array}\right] \rightarrow \left[\begin{array}{cccc|c} 1 & 0 & 1 & 0 & 3 \\ 0 & 1 & 0 & 1 & 1 \\ 0 & 0 & 1 & -2 & -10 \\ 0 & 0 & -2 & -1 & -5 \end{array}\right]$$

↑ $R_3 = \frac{1}{2}r_3$
$R_4 = \frac{1}{3}r_4$

↑ $R_3 = r_4 + r_3$

$$\rightarrow \left[\begin{array}{cccc|c} 1 & 0 & 0 & 2 & 13 \\ 0 & 1 & 0 & 1 & 1 \\ 0 & 0 & 1 & -2 & -10 \\ 0 & 0 & 0 & -5 & -25 \end{array}\right] \rightarrow \left[\begin{array}{cccc|c} 1 & 0 & 0 & 2 & 13 \\ 0 & 1 & 0 & 1 & 1 \\ 0 & 0 & 1 & -2 & -10 \\ 0 & 0 & 0 & 1 & 5 \end{array}\right] \rightarrow \left[\begin{array}{cccc|c} 1 & 0 & 0 & 0 & 3 \\ 0 & 1 & 0 & 0 & -4 \\ 0 & 0 & 1 & 0 & 0 \\ 0 & 0 & 0 & 1 & 5 \end{array}\right]$$

↑ $R_1 = -1r_3 + r_1$
$R_4 = 2r_3 + r_4$

↑ $R_4 = -\frac{1}{5}r_4$

↑ $R_1 = -2r_4 + r_1$
$R_2 = -1r_4 + r_2$
$R_3 = 2r_4 + r_3$

So we have: $a = 3, b = -4, c = 0, d = 5$.

The function is: $f(x) = 3x^3 - 4x^2 + 5$

81. Let x, y, and z represent the number of liters of 15%, 25% and 50% solutions which will be mixed. Then,

$$x + y + z = 100 \quad (1)$$

Also, in x liters of 15% solution, there will be $.15x$ liters of H_2SO_4, y liters of 25% solution contain $.25y$ liters of H_2SO_4, and the z liters contain $.50z$ liters of H_2SO_4. Meanwhile, our final 100 liter mixture is 40% H_2SO_4, so it contains $.40(100) = 40$ liters of H_2SO_4.

Thus, $.15x + .25y + .50z = 40 \quad (2)$

We have 2 equations in three unknowns:

$$\left[\begin{array}{ccc|c} 1 & 1 & 1 & 100 \\ 0.15 & 0.25 & 0.50 & 40 \end{array}\right] \rightarrow \left[\begin{array}{ccc|c} 1 & 1 & 1 & 100 \\ 0 & 0.10 & 0.35 & 25 \end{array}\right] \rightarrow \left[\begin{array}{ccc|c} 1 & 1 & 1 & 100 \\ 0 & 1 & 3.5 & 250 \end{array}\right]$$

$R_2 = -.15r_1 + r_2 \qquad R_2 = 10r_2$

$$\rightarrow \left[\begin{array}{ccc|c} 1 & 0 & -2.5 & -150 \\ 0 & 1 & 3.5 & 250 \end{array}\right]$$

$R_1 = -1r_2 + r_1$

This gives: $\begin{cases} x - 2.5z = -150 \\ y + 3.5z = 250 \end{cases}$

so $x = 2.5z - 150$

$y = -3.5z + 250$

where z can be any real number.

But, we require $x \geq 0$, $y \geq 0$, and $z \geq 0$.

Since $x \geq 0$, we have: $2.5z - 150 \geq 0$

$2.5 \geq 150$

$z \geq 60$

Also, $y \geq 0$ implies: $-3.5z + 250 \geq 0$

$-3.5 \geq -250$

$z \leq 71.43$

Some possible solutions are given below:

z (50%)	$x = 2.5z - 150$ (15%)	$y = -3.5z + 250$ (25%)	40%
60	0	40	100
64	10	26	100
68	20	12	100
70	25	5	100

83. x = price of hamburger, y = price of fries, z = price of colas

$$\begin{cases} 8x + 6y + 6z = 26.10 \\ 10x + 6y + 8z = 31.60 \end{cases}$$

$$\left[\begin{array}{ccc|c} 8 & 6 & 6 & 26.10 \\ 10 & 6 & 8 & 31.60 \end{array}\right]$$

$$\left[\begin{array}{ccc|c} 4 & 3 & 3 & 13.05 \\ 5 & 3 & 4 & 15.80 \end{array}\right]$$

$$\left[\begin{array}{ccc|c} 4 & 3 & 3 & 13.05 \\ 1 & 0 & 1 & 12.75 \end{array}\right]$$

$$\left[\begin{array}{ccc|c} 1 & 0 & 1 & 2.75 \\ 4 & 3 & 3 & 13.05 \end{array}\right]$$

$$\begin{bmatrix} 1 & 0 & 1 & | & 2.75 \\ 0 & 3 & -1 & | & 2.05 \end{bmatrix}$$

$$\begin{bmatrix} 1 & 0 & 1 & | & 2.75 \\ 0 & 1 & -\frac{1}{3} & | & \frac{2.05}{3} \end{bmatrix}$$

$x = 2.75 - z$, z any real number

$y = \frac{2.05}{3} + \frac{1}{3}z$, z any real number

$y = 0.68 + \frac{1}{3}z$, z any real number

There is not sufficient information:

x	\$2.15	\$2.00	\$1.85
y	\$0.88	\$0.93	\$0.98
z	\$0.60	\$0.75	\$0.90

85. Let x = amount in Treasury bills, y = amount in corporate bonds, z = amount in junk bonds

(a) $\begin{cases} x + y + z = 20000 \\ .07x + .09y + .11z = 2000 \end{cases}$

$$\begin{bmatrix} 1 & 1 & 1 & | & 20000 \\ .07 & .09 & .11 & | & 2000 \end{bmatrix}$$

$$\rightarrow \begin{bmatrix} 1 & 1 & 1 & | & 20{,}000 \\ 7 & 9 & 11 & | & 200{,}000 \end{bmatrix}$$

$$\rightarrow \begin{bmatrix} 1 & 1 & 1 & | & 20{,}000 \\ 0 & 2 & 4 & | & 60{,}000 \end{bmatrix}$$

$$\rightarrow \begin{bmatrix} 1 & 1 & 1 & | & 20{,}000 \\ 0 & 1 & 2 & | & 30{,}000 \end{bmatrix}$$

$$\rightarrow \begin{bmatrix} 1 & 0 & -1 & | & -10{,}000 \\ 0 & 1 & 2 & | & 30{,}000 \end{bmatrix}$$

$x = -10{,}000 + z$, $y = 30{,}000 - 2z$, z any real number

Amount Invested At		
7%	9%	11%
0	10,000	10,000
1,000	8,000	11,000
2,000	6,000	12,000
3,000	4,000	13,000
4,000	2,000	14,000
5,000	0	15,000

(b) $\begin{cases} x + y + z = 25000 \\ .07x + .09y + .11z = 2000 \end{cases}$

$$\begin{bmatrix} 1 & 1 & 1 & | & 25000 \\ .07 & .09 & .11 & | & 2000 \end{bmatrix}$$

$$\rightarrow \begin{bmatrix} 1 & 1 & 1 & | & 25{,}000 \\ 7 & 9 & 11 & | & 200{,}000 \end{bmatrix}$$

$$\rightarrow \left[\begin{array}{ccc|c} 1 & 1 & 1 & 25{,}000 \\ 0 & 2 & 4 & 25{,}000 \end{array}\right]$$

$$\rightarrow \left[\begin{array}{ccc|c} 1 & 1 & 1 & 25{,}000 \\ 0 & 1 & 2 & 12{,}500 \end{array}\right]$$

$$\rightarrow \left[\begin{array}{ccc|c} 1 & 0 & -1 & 12{,}500 \\ 0 & 1 & 2 & 12{,}500 \end{array}\right]$$

$x = 12{,}500 + z$, $y = 12{,}500 - 2z$, z any real number

Amount Invested At

7%	9%	11%
12,500	12,500	0
14,500	8,500	2,000
16,500	4,500	4,000
18,750	0	6,250

(c) $\begin{cases} x + y + z = 30{,}000 \\ .07x + .09y + .11z = 2{,}000 \end{cases}$

$$\left[\begin{array}{ccc|c} 1 & 1 & 1 & 30{,}000 \\ .07 & .09 & .11 & 2{,}000 \end{array}\right]$$

$$\rightarrow \left[\begin{array}{ccc|c} 1 & 1 & 1 & 30{,}000 \\ 7 & 9 & 11 & 200{,}000 \end{array}\right]$$

$$\rightarrow \left[\begin{array}{ccc|c} 1 & 1 & 1 & 30{,}000 \\ 0 & 2 & 4 & -10{,}000 \end{array}\right]$$

$$\rightarrow \left[\begin{array}{ccc|c} 1 & 1 & 1 & 30{,}000 \\ 0 & 1 & 2 & -5{,}000 \end{array}\right]$$

$$\rightarrow \left[\begin{array}{ccc|c} 1 & 0 & -1 & 35{,}000 \\ 0 & 1 & 2 & -5{,}000 \end{array}\right]$$

$x = 35{,}000 + z$, $y = -5{,}000 - 2z$, z any real number

All the money invested at 7% provides \$30,000(.07) = \$2100, more than what is required.

87. Let x = amount of first liquid
y = amount of second liquid
z = amount of third liquid
$.20x + .40y + .30z = 40$ (Amount of Vitamin C)
$.30x + .20y + .50z = 30$ (Amount of Vitamin D)
$20x + 40y + 30x = 4000$
$30x + 20y + 50z = 3000$

$$\left[\begin{array}{ccc|c} 20 & 40 & 30 & 4000 \\ 30 & 20 & 50 & 3000 \end{array}\right] \rightarrow \left[\begin{array}{ccc|c} 1 & 2 & \frac{3}{2} & 200 \\ 30 & 20 & 50 & 3{,}000 \end{array}\right]$$

$$\rightarrow \left[\begin{array}{ccc|c} 1 & 2 & \frac{3}{2} & 200 \\ 0 & -40 & 5 & -3{,}000 \end{array}\right]$$

$$\rightarrow \left[\begin{array}{ccc|c} 1 & 2 & \frac{3}{2} & 200 \\ 0 & 1 & \frac{-1}{8} & 75 \end{array}\right]$$

$$\rightarrow \left[\begin{array}{ccc|c} 1 & 0 & \frac{7}{4} & 50 \\ 0 & 1 & \frac{-1}{8} & 75 \end{array}\right]$$

$$x = 50 - \frac{7}{4}z,\ y = 75 + \frac{1}{8}z,\ z \text{ any real number}$$

If $z = 0$ cc, then $x = 50 - \frac{7}{4}(0) = 50$ cc, $y = 75 + \frac{1}{8}(0) = 75$ cc

If $z = 8$ cc, then $x = 50 - \frac{7}{4}(8) = 36$ cc, $y = 75 + \frac{1}{8}(8) = 76$ cc

If $z = 16$ cc, then $x = 50 - \frac{7}{4}(16) = 22$ cc, $y = 75 + \frac{1}{8}(16) = 77$ cc

If $z = 24$ cc, then $x = 50 - \frac{7}{4}(24) = 8$ cc, $y = 75 + \frac{1}{8}(24) = 78$ cc

First Liquid	Second Liquid	Third Liquid
50 cc	75 cc	0 cc
36 cc	76 cc	8 cc
22 cc	77 cc	16 cc
8 cc	78 cc	24 cc

89. $$\begin{cases} I_1 + I_2 = I_3 \\ 16 - 8 - 9I_3 - 3I_1 = 0 \\ 16 - 4 - 9I_3 - 9I_2 = 0 \\ 8 - 4 - 9I_2 + 3I_1 = 0 \end{cases}$$

$$\begin{aligned} I_1 + I_2 - I_3 &= 0 \\ -3I_1 - 9I_3 &= -8 \\ -9I_2 - 9I_3 &= -12 \\ 3I_1 - 9I_2 &= -4 \end{aligned}$$

$$\left[\begin{array}{ccc|c} 1 & 1 & -1 & 0 \\ -3 & 0 & -9 & -8 \\ 0 & -9 & -9 & -12 \\ 3 & -9 & 0 & -4 \end{array}\right]$$

$$\left[\begin{array}{ccc|c} 1 & 1 & -1 & 0 \\ 0 & 3 & -12 & -8 \\ 0 & -9 & -9 & -12 \\ 0 & -12 & 3 & -4 \end{array}\right]$$

$$\left[\begin{array}{ccc|c} 1 & 0 & 3 & \frac{8}{3} \\ 0 & 1 & -4 & -\frac{8}{3} \\ 0 & 0 & -45 & -36 \\ 0 & 0 & -45 & -36 \end{array}\right]$$

$$\left[\begin{array}{ccc|c} 1 & 0 & 3 & \frac{8}{3} \\ 0 & 1 & -4 & -\frac{8}{3} \\ 0 & 0 & 1 & \frac{36}{45} \\ 0 & 0 & 0 & 0 \end{array}\right]$$

$$I_3 = \frac{36}{45} = \frac{4}{5}$$

$$I_2 = \frac{-8}{3} + 4\left(\frac{4}{5}\right) = \frac{-40 + 48}{15} = \frac{8}{15}$$

$$I_1 = \frac{8}{3} - 3\left(\frac{4}{5}\right) = \frac{40 - 36}{15} = \frac{4}{15}$$

$$I_1 = \frac{4}{15},\ I_2 = \frac{8}{15},\ I_3 = \frac{4}{5}$$

91. $$\begin{cases} I_1 = I_3 + I_2 \\ 24 - 6I_1 - 3I_3 = 0 \\ 12 + 24 - 6I_1 - 6I_2 = 0 \end{cases}$$

$$\begin{aligned} I_1 - I_2 - I_3 &= 0 \\ -6I_1 \qquad -3I_3 &= -24 \\ 6I_1 - 6I_2 \qquad &= -36 \end{aligned}$$

$$\left[\begin{array}{ccc|c} 1 & -1 & -1 & 0 \\ -6 & 0 & -3 & -24 \\ -6 & -6 & 0 & -36 \end{array}\right]$$

$$\left[\begin{array}{ccc|c} 1 & -1 & -1 & 0 \\ 0 & -6 & -9 & -24 \\ 0 & -12 & -6 & -36 \end{array}\right]$$

$$\left[\begin{array}{ccc|c} 1 & -1 & -1 & 0 \\ 0 & 1 & \frac{3}{2} & 4 \\ 0 & -12 & -6 & -36 \end{array}\right]$$

$$\begin{bmatrix} 1 & -1 & -1 & | & 0 \\ 0 & 1 & \frac{3}{2} & | & 4 \\ 0 & 2 & 1 & | & 6 \end{bmatrix}$$

$$\begin{bmatrix} 1 & 0 & \frac{1}{2} & | & 4 \\ 0 & 1 & \frac{3}{2} & | & 4 \\ 0 & 0 & -2 & | & -2 \end{bmatrix}$$

$$\begin{bmatrix} 1 & 0 & \frac{1}{2} & | & 4 \\ 0 & 1 & \frac{3}{2} & | & 4 \\ 0 & 0 & 1 & | & 1 \end{bmatrix}$$

$$\begin{bmatrix} 1 & 0 & 0 & | & \frac{7}{2} \\ 0 & 1 & 0 & | & \frac{5}{2} \\ 0 & 0 & 1 & | & 1 \end{bmatrix}$$

$I_1 = 3.5, I_2 = 2.5, I_3 = 1$

95. $\begin{cases} a_1x + b_1y = c_1 \\ a_2x + b_2y = c_2 \end{cases}$ becomes:

$$\begin{bmatrix} a_1 & b_1 & | & c_1 \\ a_2 & b_2 & | & c_2 \end{bmatrix}$$

If $a_1 \neq 0$, we can divide row one by a_1 to obtain a 1 in the top left corner:

$$\rightarrow \begin{bmatrix} 1 & \frac{b_1}{a_1} & | & \frac{c_1}{a_1} \\ a_2 & b_2 & | & c_2 \end{bmatrix}$$

↑

$R_1 = \frac{1}{a_1}r_1$, provided $a_1 \neq 0$

Our next move would depend on whether a_2 is zero or not. If $a_2 \neq 0$, we continue:

$$\rightarrow \begin{bmatrix} 1 & \frac{b_1}{a_1} & | & \frac{c_1}{a_1} \\ 0 & \frac{-a_2b_1}{a_1} + b_2 & | & \frac{-a_2c_1}{a_1} + c_2 \end{bmatrix} \rightarrow \begin{bmatrix} 1 & \frac{b_1}{a_1} & | & \frac{c_1}{a_1} \\ 0 & \frac{a_1b_2 - a_2b_1}{a_1} & | & \frac{a_1c_2 - a_2c_1}{a_1} \end{bmatrix}$$

↑ $R_2 = -a_2r_1 + r_2$ ↑ Simplifying

Now recall that $D = a_1b_2 - a_2b_1$ so we have:

$$\begin{cases} x + \dfrac{b_1}{a_1}y = \dfrac{c_1}{a_1} & (1) \\ \dfrac{D}{a_1}y = \dfrac{a_1c_2 - a_2c_1}{a_1} & (2) \end{cases}$$

Solve for (2) for y: $y = \dfrac{a_1c_2 - a_2c_1}{D}$

Then use back-substitution to find x:

$$x + \frac{b_1}{a_1}\left[\frac{a_1c_2 - a_2c_1}{D}\right] = \frac{c_1}{a_1}$$

$$x = \frac{c_1}{a_1} - \frac{b_1(a_1c_2 - a_2c_1)}{a_1D}$$

We now need to get a common denominator (a_1D) to simplify x:

$$x = \frac{c_1D}{a_1D} - \frac{b_1a_1c_2 - b_1a_2c_1}{a_1D} = \frac{c_1(a_1b_2 - a_2b_1) - b_1a_1c_2 + b_1a_2c_1}{a_1D}$$

$$= \frac{c_1a_1b_2 - c_1a_2b_1 - b_1a_1c_2 + b_1a_2c_1}{a_1D} = \frac{c_1a_1b_2 - b_1a_1c_2}{a_1D} = \frac{c_1b_2 - b_1c_2}{D}$$

and our solution is:

$$x = \frac{1}{D}(c_1b_2 - b_1c_2)$$

$$y = \frac{1}{D}(a_1c_2 - a_2c_1), \text{ provided } a_1 \neq 0,\ a_2 \neq 0, \text{ as desired.}$$

But what if a_2 *is* zero?

Then we have: $\left[\begin{array}{cc|c} 1 & \dfrac{b_1}{a_1} & \dfrac{c_1}{a_1} \\ 0 & b_2 & c_2 \end{array}\right]$

Also, $D = a_1b_2 - a_2b_1$
$= a_1b_2$ (since $a_2 = 0$)

Therefore, b_2 ***cannot*** be 0, and we continue:

$$\rightarrow \left[\begin{array}{cc|c} 1 & \dfrac{b_1}{a_1} & \dfrac{c_1}{a_1} \\ 0 & 1 & \dfrac{c_2}{b_2} \end{array}\right] \rightarrow \left[\begin{array}{cc|c} 1 & 0 & -\dfrac{b_1c_2}{a_1b_2} + \dfrac{c_1}{a_1} \\ 0 & 1 & \dfrac{c_2}{b_2} \end{array}\right] \rightarrow \left[\begin{array}{cc|c} 1 & 0 & \dfrac{b_2c_1 - b_1c_2}{a_1b_2} \\ 0 & 1 & \dfrac{c_2}{b_2} \end{array}\right] \text{ Simplifying}$$

$$\uparrow R_2 = \frac{1}{b_2}r_2 \qquad \uparrow R_1 = \frac{-b_1}{a_1}r_2 + r_1$$

The solution is:

$$x = \frac{b_2c_1 - b_1c_2}{a_1b_2} = \frac{1}{D}(b_2c_1 - b_1c_2)$$

$$y = \frac{c_2}{b_2} = \frac{a_1c_2}{a_1b_2} = \frac{1}{D}(a_1c_2) \qquad (D = a_1b_2 \text{ if } a_2 = 0)$$

This takes care of the case $a_1 \neq 0$, $a_2 = 0$.

Finally, what if $a_1 = 0$? Then

$D = a_1b_2 - a_2b_1 = -a_2b_1$, so $a_2 \neq 0$, $b_1 \neq 0$, and we want to show:

$$x = \frac{1}{D}(c_1b_2 - c_2b_1) = \frac{1}{-a_2b_1}(c_1b_2 - c_2b_1)$$

$$y = \frac{1}{D}(a_1c_2 - a_2c_1) = \frac{1}{-a_2b_1}(-a_2c_1) = \frac{c_1}{b_1}$$

Since $a_1 = 0$, we start with:

$$\left[\begin{array}{cc|c} 0 & b_1 & c_1 \\ a_2 & b_2 & c_2 \end{array}\right] \rightarrow \left[\begin{array}{cc|c} a_2 & b_2 & c_2 \\ 0 & b_1 & c_1 \end{array}\right] \rightarrow \left[\begin{array}{cc|c} 1 & \frac{b_2}{a_2} & \frac{c_2}{a_2} \\ 0 & 1 & \frac{c_1}{b_1} \end{array}\right]$$

Interchange rows $\quad R_1 = \frac{1}{a_2}r_1 \quad$ (since $a_2 \neq 0$)

$R_2 = \frac{1}{b_1}r_2 \quad$ (since $b_1 \neq 0$)

Therefore, $y = \frac{c_1}{b_1}$, as desired, and by back-substitution,

$$x + \frac{b_2}{a_2}y = \frac{c_2}{a_2}$$

$$x + \frac{b_2}{a_2}\left(\frac{c_1}{b_1}\right) = \frac{c_2}{a_2}$$

$$x = \frac{c_2}{a_2} - \frac{c_1b_2}{a_2b_1}$$

$$x = \frac{c_2b_1 - c_1b_2}{a_2b_1}$$

or

$$x = \frac{-1}{a_2b_1}(c_1b_2 - c_2b_1)$$

as desired.

8.3 Systems of Linear Equations: Determinants

1. $\begin{vmatrix} 3 & 1 \\ 4 & 2 \end{vmatrix} = (3)(2) - (4)(1) = 2$

3. $\begin{vmatrix} 6 & 4 \\ -1 & 3 \end{vmatrix} = (6)(3) - (-1)(4) = 18 + 4 = 22$

5. $\begin{vmatrix} -3 & -1 \\ 4 & 2 \end{vmatrix} = (-3)(2) - (4)(-1) = -2$

7. $\begin{vmatrix} 3 & 4 & 2 \\ 1 & -1 & 5 \\ 1 & 2 & -2 \end{vmatrix} = 3\begin{vmatrix} -1 & 5 \\ 2 & -2 \end{vmatrix} - 4\begin{vmatrix} 1 & 5 \\ 1 & -2 \end{vmatrix} + 2\begin{vmatrix} 1 & -1 \\ 1 & 2 \end{vmatrix}$

$$\begin{aligned}
&= 3[(-1)(-2) - (2)(5)] - 4[(1)(-2) - (1)(5)] + 2[(1)(2) - (1)(-1)] \\
&= 3[2 - 10] - 4[-2 - 5] + 2[2 + 1] \\
&= (3)(-8) - (4)(-7) + (2)(3) \\
&= -24 + 28 + 6 \\
&= 10
\end{aligned}$$

9. $\begin{vmatrix} 4 & -1 & 2 \\ 6 & -1 & 0 \\ 1 & -3 & 4 \end{vmatrix} = 4\begin{vmatrix} -1 & 0 \\ -3 & 4 \end{vmatrix} - (-1)\begin{vmatrix} 6 & 0 \\ 1 & 4 \end{vmatrix} + 2\begin{vmatrix} 6 & -1 \\ 1 & -3 \end{vmatrix}$

$$\begin{aligned}
&= 4[(-1)(4) - (-3)(0)] - (-1)[(6)(4) - (1)(0)] + 2[(6)(-3) - (1)(-1)] \\
&= 4[-4] - (-1)[24] + 2[-18 + 1] \\
&= -16 + 24 - 34 \\
&= -26
\end{aligned}$$

11. $\begin{cases} x + y = 8 \\ x - y = 4 \end{cases}$

Here, $D = \begin{vmatrix} 1 & 1 \\ 1 & -1 \end{vmatrix} = -1 - 1 = -2$

Since $D \neq 0$, we proceed to find D_x and D_y.

To obtain D_x, replace the first column in D by the constants on the right-hand-side of the original system of equations:

$$D_x = \begin{vmatrix} 8 & 1 \\ 4 & -1 \end{vmatrix} = -8 - 4 = -12$$

To obtain D_y, replace the second column in D by the constants:

$$D_y = \begin{vmatrix} 1 & 8 \\ 1 & 4 \end{vmatrix} = 4 - 8 = -4$$

Then by Cramer's Rule,

$$x = \frac{D_x}{D} = \frac{-12}{-2} = 6 \qquad y = \frac{D_y}{D} = \frac{-4}{-2} = 2$$

13. $\begin{cases} 5x - y = 13 \\ 2x + 3y = 12 \end{cases}$

Here, $D = \begin{vmatrix} 5 & -1 \\ 2 & 3 \end{vmatrix} = 15 - (-2) = 17$

Since $D \neq 0$, we find D_x and D_y.

To obtain D_x, replace the first column in D by the constants:

$$D_x = \begin{vmatrix} 13 & -1 \\ 12 & 3 \end{vmatrix} = 39 - (-12) = 51$$

Similarly, $D_y = \begin{vmatrix} 5 & 13 \\ 2 & 12 \end{vmatrix} = 60 - 26 = 34$

Then by Cramer's Rule,

$$x = \frac{D_x}{D} = \frac{51}{17} = 3 \quad \text{and} \quad y = \frac{D_y}{D} = \frac{34}{17} = 2$$

15. $\begin{cases} 3x = 24 \\ x + 2y = 0 \end{cases}$

This is ***easily*** solved by inspection, but, to use Cramer's Rule:

$$D = \begin{vmatrix} 3 & 0 \\ 1 & 2 \end{vmatrix} = 6$$

$$D_x = \begin{vmatrix} 24 & 0 \\ 0 & 2 \end{vmatrix} = 48$$

and $D_y = \begin{vmatrix} 3 & 24 \\ 1 & 0 \end{vmatrix} = -24$

so that $x = \dfrac{D_x}{D} = \dfrac{48}{6} = 8$ and $y = \dfrac{D_y}{D} = \dfrac{-24}{6} = -4$

17. $\begin{cases} 3x - 6y = 24 \\ 5x + 4y = 12 \end{cases}$

Here, $D = \begin{vmatrix} 3 & -6 \\ 5 & 4 \end{vmatrix} = 12 - (-30) = 42$

$$D_x = \begin{vmatrix} 24 & -6 \\ 12 & 4 \end{vmatrix} = 96 - (-72) = 168$$

and $D_y = \begin{vmatrix} 3 & 24 \\ 5 & 12 \end{vmatrix} = 36 - 120 = -84$

Therefore,

$$x = \frac{D_x}{D} = \frac{168}{42} = 4 \quad \text{and} \quad y = \frac{D_y}{D} = \frac{-84}{42} = -2$$

19. $\begin{cases} 3x - 2y = 4 \\ 6x - 4y = 0 \end{cases}$

$D = \begin{vmatrix} 3 & -2 \\ 6 & -4 \end{vmatrix} = -12 - (-12) = 0$

Since $D = 0$, we cannot use Cramer's Rule. It is not applicable.

21. $\begin{cases} 2x - 4y = -2 \\ 3x + 2y = 3 \end{cases}$

Here, $D = \begin{vmatrix} 2 & -4 \\ 3 & 2 \end{vmatrix} = 4 - (-12) = 16$

$D_x = \begin{vmatrix} -2 & -4 \\ 3 & 2 \end{vmatrix} = -4 - (-12) = 8$

and $D_y = \begin{vmatrix} 2 & -2 \\ 3 & 3 \end{vmatrix} = 6 - (-6) = 12$

By Cramer's Rule,

$$x = \frac{D_x}{D} = \frac{8}{16} = \frac{1}{2} \quad \text{and} \quad y = \frac{D_y}{D} = \frac{12}{16} = \frac{3}{4}$$

23. $\begin{cases} 2x - 3y = -1 \\ 10x + 10y = 5 \end{cases}$

Here, $D = \begin{vmatrix} 2 & -3 \\ 10 & 10 \end{vmatrix} = 20 - (-30) = 50$

$D_x = \begin{vmatrix} -1 & -3 \\ 5 & 10 \end{vmatrix} = -10 - (-15) = 5$

and $D_y = \begin{vmatrix} 2 & -1 \\ 10 & 5 \end{vmatrix} = 10 - (-10) = 20$

By Cramer's Rule,

$$x = \frac{D_x}{D} = \frac{5}{50} = \frac{1}{10} \quad \text{and} \quad y = \frac{D_y}{D} = \frac{20}{50} = \frac{2}{5}$$

25. $\begin{cases} 2x + 3y = 6 \\ x - y = \frac{1}{2} \end{cases}$

Here, $D = \begin{vmatrix} 2 & 3 \\ 1 & -1 \end{vmatrix} = -2 - 3 = -5$

$D_x = \begin{vmatrix} 6 & 3 \\ \frac{1}{2} & -1 \end{vmatrix} = -6 - \frac{3}{2} = -\frac{15}{2}$

and $D_y = \begin{vmatrix} 2 & 6 \\ 1 & \frac{1}{2} \end{vmatrix} = 1 - 6 = -5$

By Cramer's Rule,

$$x = \frac{D_x}{D} = \frac{\frac{-15}{2}}{-5} = \frac{3}{2} \quad \text{and} \quad y = \frac{D_y}{D} = \frac{-5}{-5} = 1$$

27. $\begin{cases} 3x - 5y = 3 \\ 15x + 5y = 21 \end{cases}$

Here, $D = \begin{vmatrix} 3 & -5 \\ 15 & 5 \end{vmatrix} = 15 - (-5)(15) = 90$

$D_x = \begin{vmatrix} 3 & -5 \\ 21 & 5 \end{vmatrix} = 15 - (-105) = 120$

and $D_y = \begin{vmatrix} 3 & 3 \\ 15 & 21 \end{vmatrix} = (3)(21) - (3)(15) = (3)(21 - 15) = 18$

By Cramer's Rule,

$$x = \frac{D_x}{D} = \frac{120}{90} = \frac{12}{9} = \frac{4}{3} \quad \text{and} \quad y = \frac{D_y}{D} = \frac{18}{90} = \frac{1}{5}$$

29. $\begin{cases} x + y - z = 6 \\ 3x - 2y + z = -5 \\ x + 3y - 2z = 14 \end{cases}$

Here,

$$D = \begin{vmatrix} 1 & 1 & -1 \\ 3 & -2 & 1 \\ 1 & 3 & -2 \end{vmatrix} = 1\begin{vmatrix} -2 & 1 \\ 3 & -2 \end{vmatrix} - 1\begin{vmatrix} 3 & 1 \\ 1 & -2 \end{vmatrix} + (-1)\begin{vmatrix} 3 & -2 \\ 1 & 3 \end{vmatrix}$$
$$= 1(4 - 3) - (1)(-6 - 1) + (-1)(9 - (-2))$$
$$= 1 - (-7) + (-11)$$
$$= -3$$

To obtain D_x, replace the first column in D by the column of constants:

$$D_x = \begin{vmatrix} 6 & 1 & -1 \\ -5 & -2 & 1 \\ 14 & 3 & -2 \end{vmatrix} = 6\begin{vmatrix} -2 & 1 \\ 3 & -2 \end{vmatrix} - 1\begin{vmatrix} -5 & 1 \\ 14 & -2 \end{vmatrix} + (-1)\begin{vmatrix} -5 & -2 \\ 14 & 3 \end{vmatrix}$$
$$= 6(4 - 3) - 1(10 - 14) + (-1)(-15 - (-28))$$
$$= 6 - (-4) + (-13)$$
$$= -3$$

Similarly, $D_y = \begin{vmatrix} 1 & 6 & -1 \\ 3 & -5 & 1 \\ 1 & 14 & -2 \end{vmatrix} = 1\begin{vmatrix} -5 & 1 \\ 14 & -2 \end{vmatrix} - 6\begin{vmatrix} 3 & 1 \\ 1 & -2 \end{vmatrix} + (-1)\begin{vmatrix} 3 & -5 \\ 1 & 14 \end{vmatrix}$

$$= 1(10 - 14) - 6(-6 - 1) + (-1)(42 - (-5))$$
$$= -4 - (-42) + (-47)$$
$$= -9$$

Finally, $D_z = \begin{vmatrix} 1 & 1 & 6 \\ 3 & -2 & -5 \\ 1 & 3 & 14 \end{vmatrix} = 1\begin{vmatrix} -2 & -5 \\ 3 & 14 \end{vmatrix} - 1\begin{vmatrix} 3 & -5 \\ 1 & 14 \end{vmatrix} + 6\begin{vmatrix} 3 & -2 \\ 1 & 3 \end{vmatrix}$

$$= 1(-28 - (-15)) - 1(42 - (-5)) + 6(9 - (-2))$$
$$= -13 - 47 + 66$$
$$= 6$$

$$x = \frac{D_x}{D} = \frac{-3}{-3} = 1,\ y = \frac{D_y}{D} = \frac{-9}{-3} = 3, \text{ and } z = \frac{D_z}{D} = \frac{6}{-3} = -2$$

31. $\begin{cases} x + 2y - z = -3 \\ 2x - 4y + z = -7 \\ -2x + 2y - 3z = 4 \end{cases}$

$$D = \begin{vmatrix} 1 & 2 & -1 \\ 2 & -4 & 1 \\ -2 & 2 & -3 \end{vmatrix} = 1\begin{vmatrix} -4 & 1 \\ 2 & -3 \end{vmatrix} - 2\begin{vmatrix} 2 & 1 \\ -2 & -3 \end{vmatrix} + (-1)\begin{vmatrix} 2 & -4 \\ -2 & 2 \end{vmatrix}$$
$$= 1(12 - 2) - 2(-6 - (-2)) + (-1)(4 - 8)$$
$$= 10 - (-8) + 4$$
$$= 22$$

$$D_x = \begin{vmatrix} -3 & 2 & -1 \\ -7 & -4 & 1 \\ 4 & 2 & -3 \end{vmatrix} = -3\begin{vmatrix} -4 & 1 \\ 2 & -3 \end{vmatrix} - 2\begin{vmatrix} -7 & 1 \\ 4 & -3 \end{vmatrix} + (-1)\begin{vmatrix} -7 & -4 \\ 4 & 2 \end{vmatrix}$$

$$= -3(12 - 2) - 2(21 - 4) + (-1)(-14 - (-16))$$
$$= -30 - 34 + (-2)$$
$$= -66$$

$$D_y = \begin{vmatrix} 1 & -3 & -1 \\ 2 & -7 & 1 \\ -2 & 4 & -3 \end{vmatrix} = 1\begin{vmatrix} -7 & 1 \\ 4 & -3 \end{vmatrix} - (-3)\begin{vmatrix} 2 & 1 \\ -2 & -3 \end{vmatrix} + (-1)\begin{vmatrix} 2 & -7 \\ -2 & 4 \end{vmatrix}$$

$$= 1(21 - 4) - (-3)(-6 - (-2)) + (-1)(8 - 14)$$
$$= 17 - 12 + 6$$
$$= 11$$

and

$$D_z = \begin{vmatrix} 1 & 2 & -3 \\ 2 & -4 & -7 \\ -2 & 2 & 4 \end{vmatrix} = 1\begin{vmatrix} -4 & -7 \\ 2 & 4 \end{vmatrix} - 2\begin{vmatrix} 2 & -7 \\ -2 & 4 \end{vmatrix} + (-3)\begin{vmatrix} 2 & -4 \\ -2 & 2 \end{vmatrix}$$

$$= 1(-16 - (-14)) - 2(8 - 14) + (-3)(4 - 8)$$
$$= -2 - (-12) + 12$$
$$= 22$$

By Cramer's Rule,

$$x = \frac{D_x}{D} = \frac{-66}{22} = -3, \; y = \frac{D_y}{D} = \frac{11}{22} = \frac{1}{2}, \text{ and } z = \frac{D_z}{D} = \frac{22}{22} = 1$$

33. $\begin{cases} x - 2y + 3z = 1 \\ 3x + y - 2z = 0 \\ 2x - 4y + 6z = 2 \end{cases}$

$$D = \begin{vmatrix} 1 & -2 & 3 \\ 3 & 1 & -2 \\ 2 & -4 & 6 \end{vmatrix} = 1\begin{vmatrix} 1 & -2 \\ -4 & 6 \end{vmatrix} - (-2)\begin{vmatrix} 3 & -2 \\ 2 & 6 \end{vmatrix} + 3\begin{vmatrix} 3 & 1 \\ 2 & -4 \end{vmatrix}$$

$$= 1(-2) - (-2)(22) + 3(-14)$$
$$= 0$$

Since $D = 0$, Cramer's Rule cannot be applied.

35. $\begin{cases} x + 2y - z = 0 \\ 2x - 4y + z = 0 \\ -2x + 2y - 3z = 0 \end{cases}$

$$D = \begin{vmatrix} 1 & 2 & -1 \\ 2 & -4 & 1 \\ -2 & 2 & -3 \end{vmatrix} = 1\begin{vmatrix} -4 & 1 \\ 2 & -3 \end{vmatrix} - 2\begin{vmatrix} 2 & 1 \\ -2 & -3 \end{vmatrix} + (-1)\begin{vmatrix} 2 & -4 \\ -2 & 2 \end{vmatrix}$$

$$= 1(10) - 2(-4) + (-1)(-4)$$
$$= 22$$

$$D_x = \begin{vmatrix} 0 & 2 & -1 \\ 0 & -4 & 1 \\ 0 & 2 & -3 \end{vmatrix} = 0\begin{vmatrix} -4 & 1 \\ 2 & -3 \end{vmatrix} - 2\begin{vmatrix} 0 & -1 \\ 0 & -3 \end{vmatrix} + (-1)\begin{vmatrix} 0 & -4 \\ 0 & -3 \end{vmatrix}$$

$$= 0(12 - 2) - 2(0 - 0) + (-1)(0 - 0)$$
$$= 0$$

(We could have used (12) in the text, which states that if any row or column contains only 0's, the value of the determinant is 0.)

$$D_y = \begin{vmatrix} 1 & 0 & -1 \\ 2 & 0 & 1 \\ -2 & 0 & -3 \end{vmatrix} = 0$$

Similarly, $D_z = 0$

Therefore,

$$x = \frac{D_x}{D} = \frac{0}{22} = 0 \quad \text{and} \quad y = 0,\ z = 0$$

37. $\begin{cases} x - 2y + 3z = 0 \\ 3x + y - 2z = 0 \\ 2x - 4y + 6z = 0 \end{cases}$

$$D = \begin{vmatrix} 1 & -2 & 3 \\ 3 & 1 & -2 \\ 2 & -4 & 6 \end{vmatrix} = 1\begin{vmatrix} 1 & -2 \\ -4 & 6 \end{vmatrix} - (-2)\begin{vmatrix} 3 & -2 \\ 2 & 6 \end{vmatrix} + 3\begin{vmatrix} 3 & 1 \\ 2 & -4 \end{vmatrix}$$

$$= 1(6 - 8) - (-2)(18 - (-4)) + 3(-12 - 2)$$

$$= -2 - (-44) + (-42)$$

$$= 0$$

Since $D = 0$, Cramer's Rule cannot be applied.

39. $\begin{cases} \dfrac{1}{x} + \dfrac{1}{y} = 8 \\ \dfrac{3}{x} - \dfrac{5}{y} = 0 \end{cases}$

$$\begin{cases} \dfrac{1}{x} + \dfrac{1}{y} = 8 \\ 3\left(\dfrac{1}{x}\right) - 5\left(\dfrac{1}{y}\right) = 0 \end{cases}$$

Let $u = \dfrac{1}{x}$ and $v = \dfrac{1}{y}$. Then we have: $u + v = 8$

$3u - 5v = 0$. Now we solve for u and v:

Here, $D = \begin{vmatrix} 1 & 1 \\ 3 & -5 \end{vmatrix} = -5 - 3 = -8$

$D_u = \begin{vmatrix} 8 & 1 \\ 0 & -5 \end{vmatrix} = -40 - 0 = -40$

and $D_v = \begin{vmatrix} 1 & 8 \\ 3 & 0 \end{vmatrix} = 0 - 24 = -24$

Therefore, by Cramer's Rule,

$$u = \frac{D_u}{D} = \frac{-40}{-8} = 5 \quad \text{and} \quad v = \frac{D_v}{D} = \frac{-24}{-8} = 3$$

But, we are supposed to find x and y.

Since $u = \frac{1}{x}$, we have:

$$\begin{aligned} 5 &= \frac{1}{x} \\ 5x &= 1 \\ x &= \frac{1}{5} \end{aligned}$$

Also, $v = \frac{1}{y}$

$$\begin{aligned} 3 &= \frac{1}{y} \\ 3y &= 1 \\ y &= \frac{1}{3} \end{aligned}$$

41. Since $\begin{vmatrix} x & x \\ 4 & 3 \end{vmatrix} = 3x - 4x = -x$, we have: $-x = 5$

$$x = -5$$

43. $$\begin{vmatrix} x & 1 & 1 \\ 4 & 3 & 2 \\ -1 & 2 & 5 \end{vmatrix} = x\begin{vmatrix} 3 & 2 \\ 2 & 5 \end{vmatrix} - 1\begin{vmatrix} 4 & 2 \\ -1 & 5 \end{vmatrix} + 1\begin{vmatrix} 4 & 3 \\ -1 & 2 \end{vmatrix}$$

$$\begin{aligned} &= x(15 - 4) - 1(20 - (-2)) + 1(8 - (-3)) \\ &= 11x - 22 + 11 \\ &= 11x - 11 \end{aligned}$$

so we have: $11x - 11 = 2$

$$\begin{aligned} 11x &= 13 \\ x &= \frac{13}{11} \end{aligned}$$

45. $$\begin{vmatrix} x & 2 & 3 \\ 1 & x & 0 \\ 6 & 1 & -2 \end{vmatrix} = x\begin{vmatrix} x & 0 \\ 1 & -2 \end{vmatrix} - 2\begin{vmatrix} 1 & 0 \\ 6 & -2 \end{vmatrix} + 3\begin{vmatrix} 1 & x \\ 6 & 1 \end{vmatrix}$$

$$\begin{aligned} &= x(-2x - 0) - 2(-2 - 0) + 3(1 - 6x) \\ &= -2x^2 + 4 + 3 - 18x \\ &= -2x^2 - 18x + 7 \end{aligned}$$

so we have: $-2x^2 - 18x + 7 = 7$

$$\begin{aligned} -2x^2 - 18x &= 0 \\ -2x(x + 9) &= 0 \end{aligned}$$

so $x = 0$ or $x = -9$

47. Let $D = \begin{vmatrix} x & y & z \\ u & v & w \\ 1 & 2 & 3 \end{vmatrix} = 4$

Then, $\begin{vmatrix} 1 & 2 & 3 \\ u & v & w \\ x & y & z \end{vmatrix} = -4$, because the value of a determinant changes sign if any two rows are interchanged.

49. We try to use row operations to put $\begin{vmatrix} x & y & z \\ -3 & -6 & -9 \\ u & v & w \end{vmatrix}$ into the form $\begin{vmatrix} x & y & z \\ u & v & w \\ 1 & 2 & 3 \end{vmatrix}$

Since we know that the value of the determinant on the right is 4.

$$\begin{vmatrix} x & y & z \\ -3 & -6 & -9 \\ u & v & w \end{vmatrix} = -3\begin{vmatrix} x & y & z \\ 1 & 2 & 3 \\ u & v & w \end{vmatrix} \quad \text{by (14)}$$

$$= (-3)(-1)\begin{vmatrix} x & y & z \\ u & v & w \\ 1 & 2 & 3 \end{vmatrix} \quad \text{by (11)}$$

Therefore, $\begin{vmatrix} x & y & z \\ -3 & -6 & -9 \\ u & v & w \end{vmatrix} = (-3)(-1)(4) = 12$

51. Let $D = \begin{vmatrix} x & y & z \\ u & v & w \\ 1 & 2 & 3 \end{vmatrix} = 4$

Now, $\begin{vmatrix} 1 & 2 & 3 \\ x-3 & y-6 & z-9 \\ 2u & 2v & 2w \end{vmatrix} = 2\begin{vmatrix} 1 & 2 & 3 \\ x-3 & y-6 & z-9 \\ u & v & w \end{vmatrix}$ by (14)

$$= 2(-1)\begin{vmatrix} x-3 & y-6 & z-9 \\ 1 & 2 & 3 \\ u & v & w \end{vmatrix} \quad \text{by (11)}$$

$$= 2(-1)(-1)\begin{vmatrix} x-3 & y-6 & z-9 \\ u & v & w \\ 1 & 2 & 3 \end{vmatrix} \quad \text{by (11)}$$

Note that in this last determinant, row one can be obtained from D by the operation

$$R_1 = -3r_3 + r_1$$

By (15), that operation leaves the value of the determinant unchanged, so that

$$\begin{vmatrix} x-3 & y-6 & z-9 \\ u & v & w \\ 1 & 2 & 3 \end{vmatrix} = \begin{vmatrix} x & y & z \\ u & v & w \\ 1 & 2 & 3 \end{vmatrix} = 4$$

Therefore, $\begin{vmatrix} 1 & 2 & 3 \\ x-3 & y-6 & z-9 \\ 2u & 2v & 2w \end{vmatrix} = (2)(-1)(-1)(4) = 8$

53. Let $D = \begin{vmatrix} x & y & z \\ u & v & w \\ 1 & 2 & 3 \end{vmatrix} = 4$

$$\begin{vmatrix} 1 & 2 & 3 \\ 2x & 2y & 2z \\ u-1 & v-2 & w-3 \end{vmatrix} = 2\begin{vmatrix} 1 & 2 & 3 \\ x & y & z \\ u-1 & v-2 & w-3 \end{vmatrix} \quad \text{by (14)}$$

$$= 2(-1)\begin{vmatrix} x & y & z \\ 1 & 2 & 3 \\ u-1 & v-2 & w-3 \end{vmatrix} \quad \text{by (11)}$$

$$= 2(-1)(-1)\begin{vmatrix} x & y & z \\ u-1 & v-2 & w-3 \\ 1 & 2 & 3 \end{vmatrix}$$

Note, in this last determinant, row two can be obtained from D by the operation

$$R_2 = -1r_3 + r_2$$

which leaves the value of the determinant unchanged by (15). In other words,

$$\begin{vmatrix} x & y & z \\ u-1 & v-2 & w-3 \\ 1 & 2 & 3 \end{vmatrix} = \begin{vmatrix} x & y & z \\ u & v & w \\ 1 & 2 & 3 \end{vmatrix} = 4$$

and $\begin{vmatrix} 1 & 2 & 3 \\ 2x & 2y & 2z \\ u-1 & v-2 & w-3 \end{vmatrix} = 2(-1)(-1)4 = 8$

55. $\begin{vmatrix} x & y & 1 \\ x_1 & y_1 & 1 \\ x_2 & y_2 & 1 \end{vmatrix} = x(y_1 - y_2) - y(x_1 - x_2) + (x_1y_2 - x_2y_1) = 0$

$$\begin{aligned}
x(y_1 - y_2) + y(x_2 - x_1) &= x_2y_1 - x_1y_2 \\
y(x_2 - x_1) &= x_2y_1 - x_1y_2 + x(y_2 - y_1) \\
y(x_2 - x_1) - y_1(x_2 - x_1) &= x_2y_1 - x_1y_2 + x(y_2 - y_1) - y_1(x_2 - x_1) \\
(x_2 - x_1)(y - y_1) &= x(y_2 - y_1) + x_2y_1 - x_1y_2 - y_1(x_2 - x_1) \\
(x_2 - x_1)(y - y_1) &= (y_2 - y_1)x - (y_2 - y_1)x_1 \\
(x_2 - x_1)(y - y_1) &= (y_2 - y_1)(x - x_1) \\
(y - y_1) &= \frac{y_2 - y_1}{x_2 - x_1}(x - x_1)
\end{aligned}$$

57. $\begin{vmatrix} x^2 & x & 1 \\ y^2 & y & 1 \\ z^2 & z & 1 \end{vmatrix} = x^2\begin{vmatrix} y & 1 \\ z & 1 \end{vmatrix} - x\begin{vmatrix} y^2 & 1 \\ z^2 & 1 \end{vmatrix} + 1\begin{vmatrix} y^2 & y \\ z^2 & z \end{vmatrix}$

$$\begin{aligned}
&= x^2(y - z) - x(y^2 - z^2) + 1(y^2z - yz^2) \\
&= x^2(y - z) - x(y - z)(y + z) + yz(y - z) \\
&= (y - z)[x^2 - x(y + z) + yz] \\
&= (y - z)[x^2 - xy - xz + yz] \\
&= (y - z)[x(x - y) - z(x + y)] \\
&= (y - z)(x - y)(x - z) \qquad \text{as desired.}
\end{aligned}$$

59. Generally, below is a 3 × 3 determinant.

$$\begin{vmatrix} a_{13} & a_{12} & a_{11} \\ a_{23} & a_{22} & a_{21} \\ a_{33} & a_{32} & a_{31} \end{vmatrix} = a_{13}(a_{22}a_{31} - a_{32}a_{21}) - a_{12}(a_{23}a_{31} - a_{33}a_{21}) + a_{11}(a_{23}a_{32} - a_{33}a_{22})$$

$$= a_{13}a_{22}a_{31} - a_{13}a_{32}a_{21} - a_{12}a_{23}a_{31} + a_{12}a_{33}a_{21} + a_{11}a_{23}a_{32} - a_{11}a_{33}a_{22}$$
$$= a_{11}a_{22}a_{33} + a_{11}a_{32}a_{23} + a_{12}a_{21}a_{33} - a_{12}a_{31}a_{23} - a_{13}a_{21}a_{32} + a_{13}a_{31}a_{22}$$
$$= [a_{11}(a_{22}a_{33} - a_{32}a_{23}) - a_{12}(a_{21}a_{33} - a_{31}a_{23}) + a_{13}(a_{21}a_{32} - a_{31}a_{22})]$$

$$= - \begin{bmatrix} a_{11} & a_{12} & a_{13} \\ a_{21} & a_{22} & a_{23} \\ a_{31} & a_{32} & a_{33} \end{bmatrix}$$

As an example, let $A = \begin{vmatrix} 1 & 3 & 2 \\ -1 & 4 & -3 \\ 2 & 1 & 6 \end{vmatrix}$

Then, $A = 1\begin{vmatrix} 4 & -3 \\ 1 & 6 \end{vmatrix} - 3\begin{vmatrix} -1 & -3 \\ 2 & 6 \end{vmatrix} + 2\begin{vmatrix} -1 & 4 \\ 2 & 1 \end{vmatrix} = 1(24 - (-3)) - 3(-6 - (-6)) + 2(-1 - 8)$

$= 27 - 0 - 18 = 9$

Interchange columns one and three:

$$B = \begin{vmatrix} 2 & 3 & 1 \\ -3 & 4 & -1 \\ 6 & 1 & 2 \end{vmatrix} = 2\begin{vmatrix} 4 & -1 \\ 1 & 2 \end{vmatrix} - 3\begin{vmatrix} -3 & -1 \\ 6 & 2 \end{vmatrix} + 1\begin{vmatrix} -3 & 4 \\ 6 & 1 \end{vmatrix}$$

$= 2(8 - (-1)) - 3(-6 - (-6)) + 1(-3 - 24) = 18 - 0 - 27 = -9$

Therefore, $B = -9 = (-1)9 = (-1)A$

61. To give a proof of a theorem, we cannot simply show an example. Instead, let D represent ***any*** 3 by 3 determinant in which the entries in column one equal those in column three. Then D will be of the form:

$$D = \begin{vmatrix} a & d & a \\ b & e & b \\ c & f & c \end{vmatrix}$$

where a, b, c, d, e, f can be ***any*** real numbers.

Then, $D = \begin{vmatrix} a & d & a \\ b & e & b \\ c & f & c \end{vmatrix} = a\begin{vmatrix} e & b \\ f & c \end{vmatrix} - d\begin{vmatrix} b & b \\ c & c \end{vmatrix} + a\begin{vmatrix} b & e \\ c & f \end{vmatrix}$

$= a(ce - bf) - d(bc - bc) + a(bf - ce)$
$= a(ce - bf) - 0 + a(bf - ce)$
$= ace - abf + abf - ace$
$= 0$

8.4 Matrix Algebra

Historical Problem

1. (a) Using the correspondence $a + bi \leftrightarrow \begin{bmatrix} a & b \\ -b & a \end{bmatrix}$, we have: $2 - 5i \leftrightarrow \begin{bmatrix} 2 & -5 \\ 5 & 2 \end{bmatrix}$

$1 + 3i \leftrightarrow \begin{bmatrix} 1 & 3 \\ -3 & 1 \end{bmatrix}$

(b) $\begin{bmatrix} 2 & -5 \\ 5 & 2 \end{bmatrix} \begin{bmatrix} 1 & 3 \\ -3 & 1 \end{bmatrix} = \begin{bmatrix} 17 & 1 \\ -1 & 17 \end{bmatrix}$ (c) Now $\begin{bmatrix} 17 & 1 \\ -1 & 17 \end{bmatrix} \leftrightarrow 17 + i$

(d) On the other hand, we have $(2 - 5i)(1 + 3i) = 2 - 15i^2 + 6i - 5i = 17 + i$

Exercises

In Problems 1–16, we are using: $A = \begin{bmatrix} 0 & 3 & -5 \\ 1 & 2 & 6 \end{bmatrix}$; $B = \begin{bmatrix} 4 & 1 & 0 \\ -2 & 3 & -2 \end{bmatrix}$; $C = \begin{bmatrix} 4 & 1 \\ 6 & 2 \\ -2 & 3 \end{bmatrix}$

To answer (b), enter the matrices A, B, and C into your graphing utility and perform the indicated operation. Answer should agree with (a).

1. (a) $A + B = \begin{bmatrix} 0 & 3 & -5 \\ 1 & 2 & 6 \end{bmatrix} + \begin{bmatrix} 4 & 1 & 0 \\ -2 & 3 & -2 \end{bmatrix} = \begin{bmatrix} 0 + 4 & 3 + 1 & -5 + 0 \\ 1 + (-2) & 2 + 3 & 6 + (-2) \end{bmatrix} = \begin{bmatrix} 4 & 4 & -5 \\ -1 & 5 & 4 \end{bmatrix}$

3. (b) $4A = 4\begin{bmatrix} 0 & 3 & -5 \\ 1 & 2 & 6 \end{bmatrix} = \begin{bmatrix} 4 \cdot 0 & 4 \cdot 3 & 4(-5) \\ 4 \cdot 1 & 4 \cdot 2 & 4 \cdot 6 \end{bmatrix} = \begin{bmatrix} 0 & 12 & -20 \\ 4 & 8 & 24 \end{bmatrix}$

5. (a) $3A - 2B = \begin{bmatrix} 0 & 9 & -15 \\ 3 & 6 & 18 \end{bmatrix} - \begin{bmatrix} 8 & 2 & 0 \\ -4 & 6 & -4 \end{bmatrix} = \begin{bmatrix} -8 & 7 & -15 \\ 7 & 0 & 22 \end{bmatrix}$

7. $AC = \begin{bmatrix} 0 & 3 & -5 \\ 1 & 2 & 6 \end{bmatrix} \begin{bmatrix} 4 & 1 \\ 6 & 2 \\ -2 & 3 \end{bmatrix} = \begin{bmatrix} 0 \cdot 4 + 3 \cdot 6 + (-5)(-2) & 0 \cdot 1 + 3 \cdot 2 + (-5)3 \\ 1 \cdot 4 + 2 \cdot 6 + 6(-2) & 1 \cdot 1 + 2 \cdot 2 + 6 \cdot 3 \end{bmatrix}$

$= \begin{bmatrix} 28 & -9 \\ 4 & 23 \end{bmatrix}$

9. $CA = \begin{bmatrix} 4 & 1 \\ 6 & 2 \\ -2 & 3 \end{bmatrix} \begin{bmatrix} 0 & 3 & -5 \\ 1 & 2 & 6 \end{bmatrix}$

$$= \begin{bmatrix} 4 \cdot 0 + 1 \cdot 1 & 4 \cdot 3 + 1 \cdot 2 & 4(-5) + 1 \cdot 6 \\ 6 \cdot 0 + 2 \cdot 1 & 6 \cdot 3 + 2 \cdot 2 & 6(-5) + 2 \cdot 6 \\ (-2) \cdot 0 + 3 \cdot 1 & (-2) \cdot 3 + 3 \cdot 2 & (-2)(-5) + 3 \cdot 6 \end{bmatrix} = \begin{bmatrix} 1 & 14 & -14 \\ 2 & 22 & -18 \\ 3 & 0 & 28 \end{bmatrix}$$

11. $C(A + B) = \begin{bmatrix} 4 & 1 \\ 6 & 2 \\ -2 & 3 \end{bmatrix} \begin{bmatrix} 4 & 4 & -5 \\ -1 & 5 & 4 \end{bmatrix} = \begin{bmatrix} 15 & 21 & -16 \\ 22 & 34 & -22 \\ -11 & 7 & 22 \end{bmatrix}$

13. $AC - 3I_2 = \begin{bmatrix} 0 & 3 & -5 \\ 1 & 2 & 6 \end{bmatrix} \begin{bmatrix} 4 & 1 \\ 6 & 2 \\ -2 & 3 \end{bmatrix} -3 \underbrace{\begin{bmatrix} 1 & 0 \\ 0 & 1 \end{bmatrix}}_{I_2} = \begin{bmatrix} 28 & -9 \\ 4 & 23 \end{bmatrix} - \begin{bmatrix} 3 & 0 \\ 0 & 3 \end{bmatrix} = \begin{bmatrix} 25 & -9 \\ 4 & 20 \end{bmatrix}$

15. $CA - CB = \begin{bmatrix} 4 & 1 \\ 6 & 2 \\ -2 & 3 \end{bmatrix} \begin{bmatrix} 0 & 3 & -5 \\ 1 & 2 & 6 \end{bmatrix} - \begin{bmatrix} 4 & 1 \\ 6 & 2 \\ -2 & 3 \end{bmatrix} \begin{bmatrix} 4 & 1 & 0 \\ -2 & 3 & -2 \end{bmatrix}$

$$= \begin{bmatrix} 1 & 14 & -14 \\ 2 & 22 & -18 \\ 3 & 0 & 28 \end{bmatrix} - \begin{bmatrix} 14 & 7 & -2 \\ 20 & 12 & -4 \\ -14 & 7 & -6 \end{bmatrix} = \begin{bmatrix} -13 & 7 & -12 \\ -18 & 10 & -14 \\ 17 & -7 & 34 \end{bmatrix}$$

17. $\begin{bmatrix} 2 & -2 \\ 1 & 0 \end{bmatrix} \begin{bmatrix} 2 & 1 & 4 & 6 \\ 3 & -1 & 3 & 2 \end{bmatrix}$

$$= \begin{bmatrix} 2 \cdot 2 + (-2)3 & 2 \cdot 1 + (-2)(-1) & 2 \cdot 4 + (-2)3 & 2 \cdot 6 + (-2)2 \\ 1 \cdot 2 + 0 \cdot 3 & 1 \cdot 1 + 0(-1) & 1 \cdot 4 + 0 \cdot 3 & 1 \cdot 6 + 0 \cdot 2 \end{bmatrix}$$

$$= \begin{bmatrix} -2 & 4 & 2 & 8 \\ 2 & 1 & 4 & 6 \end{bmatrix}$$

19. $\begin{bmatrix} 1 & 0 & 1 \\ 2 & 4 & 1 \\ 3 & 6 & 1 \end{bmatrix} \begin{bmatrix} 1 & 3 \\ 6 & 2 \\ 8 & -1 \end{bmatrix} = \begin{bmatrix} 9 & 2 \\ 34 & 13 \\ 47 & 20 \end{bmatrix}$

21. $A = \begin{bmatrix} 2 & 1 \\ 1 & 1 \end{bmatrix}$

Step 1: Form $\left[\begin{array}{cc|cc} 2 & 1 & 1 & 0 \\ 1 & 1 & 0 & 1 \end{array}\right] = [A|I_2]$

Step 2:

$$\left[\begin{array}{cc|cc} 2 & 1 & 1 & 0 \\ 1 & 1 & 0 & 1 \end{array}\right] \to \left[\begin{array}{cc|cc} 1 & 1 & 0 & 1 \\ 2 & 1 & 1 & 0 \end{array}\right] \to \left[\begin{array}{cc|cc} 1 & 1 & 0 & 1 \\ 0 & -1 & 0 & -2 \end{array}\right]$$

↑ Interchange rows one and two; ↑ $R_2 = -2r_1 + r2$

$$\to \left[\begin{array}{cc|cc} 1 & 0 & 1 & -1 \\ 0 & -1 & 1 & -2 \end{array}\right] \to \left[\begin{array}{cc|cc} 1 & 0 & 1 & -1 \\ 0 & 1 & -1 & 2 \end{array}\right]$$

↑ $R_2 = r_2 + r_1$; ↑ $R_2 -(1)r_2$

Step 3: We have now achieved the form $[I_2|A^{-1}]$,

so $A^{-1} = \begin{bmatrix} 1 & -1 \\ -1 & 2 \end{bmatrix}$.

Check: $A \cdot A^{-1} = \begin{bmatrix} 2 & 1 \\ 1 & 1 \end{bmatrix} \begin{bmatrix} 1 & -1 \\ -1 & 2 \end{bmatrix} = \begin{bmatrix} 1 & 0 \\ 0 & 1 \end{bmatrix} = I_2$!

23. $A = \begin{bmatrix} 6 & 5 \\ 2 & 2 \end{bmatrix}$

Step 1: $[A|I_2] = \left[\begin{array}{cc|cc} 6 & 5 & 1 & 0 \\ 2 & 2 & 0 & 1 \end{array}\right]$

Step 2:

$$\left[\begin{array}{cc|cc} 6 & 5 & 1 & 0 \\ 2 & 2 & 0 & 1 \end{array}\right] \to \left[\begin{array}{cc|cc} 2 & 2 & 0 & 1 \\ 6 & 5 & 1 & 0 \end{array}\right] \to \left[\begin{array}{cc|cc} 2 & 2 & 0 & 1 \\ 6 & -1 & 1 & -3 \end{array}\right] \to \left[\begin{array}{cc|cc} 2 & 0 & 2 & -5 \\ 0 & -1 & 1 & -3 \end{array}\right]$$

↑ Interchange rows; ↑ $R_2 = -3r_1 + r_2$; ↑ $R_1 = 2r_2 + R_1$

↑

$$\to \left[\begin{array}{cc|cc} 1 & 0 & 1 & -\frac{5}{2} \\ 0 & 1 & -1 & 3 \end{array}\right]$$

$R_1 = \frac{1}{2}r_1$

$R_2 = (-1)r_2$

Step 3: We have: $A^{-1} = \begin{bmatrix} 1 & -\frac{5}{2} \\ -1 & 3 \end{bmatrix}$

25. $A = \begin{bmatrix} 2 & 1 \\ a & a \end{bmatrix}$, where $a \neq 0$.

$$\left[\begin{array}{cc|cc} 2 & 1 & 1 & 0 \\ a & a & 0 & 1 \end{array}\right] \rightarrow \left[\begin{array}{cc|cc} 1 & \frac{1}{2} & \frac{1}{2} & 0 \\ a & a & 0 & 1 \end{array}\right] \rightarrow \left[\begin{array}{cc|cc} 1 & \frac{1}{2} & \frac{1}{2} & 0 \\ 0 & \frac{1}{2}a & -\frac{1}{2}a & 1 \end{array}\right] \rightarrow \left[\begin{array}{cc|cc} 1 & \frac{1}{2} & \frac{1}{2} & 0 \\ 0 & 1 & -1 & \frac{2}{a} \end{array}\right] \rightarrow \left[\begin{array}{cc|cc} 1 & 0 & 1 & -\frac{1}{a} \\ 0 & 1 & -1 & \frac{2}{a} \end{array}\right]$$

$$R_1 = \frac{1}{2}r_1 \qquad R_2 = -ar_1 + r_2 \qquad R_2 = \left(\frac{2}{a}\right)r_2 \qquad R_1 = -\frac{1}{2}r_2 + r_1$$

Therefore, $A^{-1} = \begin{bmatrix} 1 & -\frac{1}{a} \\ -1 & \frac{2}{a} \end{bmatrix}$

27. $A = \begin{bmatrix} 1 & -1 & 1 \\ 0 & -2 & 1 \\ -2 & -3 & 0 \end{bmatrix}$

$$\left[\begin{array}{ccc|ccc} 1 & -1 & 1 & 1 & 0 & 0 \\ 0 & -2 & 1 & 0 & 1 & 0 \\ -2 & -3 & 0 & 0 & 0 & 1 \end{array}\right] \rightarrow \left[\begin{array}{ccc|ccc} 1 & -1 & 1 & 1 & 0 & 0 \\ 0 & -2 & 1 & 0 & 1 & 0 \\ 0 & -5 & 2 & 2 & 0 & 1 \end{array}\right] \rightarrow \left[\begin{array}{ccc|ccc} 1 & -1 & 1 & 1 & 0 & 0 \\ 0 & 1 & -\frac{1}{2} & 0 & -\frac{1}{2} & 0 \\ 0 & -5 & 2 & 2 & 0 & 1 \end{array}\right]$$

$$R_3 = 2r_1 + r_3 \qquad R_2 = -\frac{1}{2}r_2$$

$$\rightarrow \left[\begin{array}{ccc|ccc} 1 & 0 & \frac{1}{2} & 1 & -\frac{1}{2} & 0 \\ 0 & 1 & -\frac{1}{2} & 0 & -\frac{1}{2} & 0 \\ 0 & 0 & -\frac{1}{2} & 2 & -\frac{5}{2} & 1 \end{array}\right] \rightarrow \left[\begin{array}{ccc|ccc} 1 & 0 & 0 & 3 & -3 & 1 \\ 0 & 1 & 0 & -2 & 2 & -1 \\ 0 & 0 & -\frac{1}{2} & 2 & -\frac{5}{2} & 1 \end{array}\right]$$

$$R_1 = r_2 + r_1 \qquad R_1 = r_3 + r_1$$
$$R_3 = 5r_2 + r_3 \qquad R_2 = (-1)r_3 + r_2$$

$$\rightarrow \left[\begin{array}{ccc|ccc} 1 & 0 & 0 & 3 & -3 & 1 \\ 0 & 1 & 0 & -2 & 2 & -1 \\ 0 & 0 & 1 & -4 & 5 & -2 \end{array}\right]$$

$$R_3 = -2r_3$$

We have $A^{-1} = \begin{bmatrix} 3 & -3 & 1 \\ -2 & 2 & -1 \\ -4 & 5 & -2 \end{bmatrix}$

29. $A = \begin{bmatrix} 1 & 1 & 1 \\ 3 & 2 & -1 \\ 3 & 1 & 2 \end{bmatrix}$

Then, $\left[\begin{array}{rrr|rrr} 1 & 1 & 1 & 1 & 0 & 0 \\ 3 & 2 & -1 & 0 & 1 & 0 \\ 3 & 1 & 2 & 0 & 0 & 1 \end{array}\right] \rightarrow \left[\begin{array}{rrr|rrr} 1 & 1 & 1 & 1 & 0 & 0 \\ 0 & -1 & -4 & -3 & 1 & 0 \\ 0 & -2 & -1 & -3 & 0 & 1 \end{array}\right]$

$\uparrow$
$R_2 = -3r_1 + r_2$
$R_3 = -3r_1 + r_3$

$\rightarrow \left[\begin{array}{rrr|rrr} 1 & 1 & 1 & 1 & 0 & 0 \\ 0 & 1 & 4 & 3 & -1 & 0 \\ 0 & -2 & -1 & -3 & 0 & 1 \end{array}\right] \rightarrow \left[\begin{array}{rrr|rrr} 1 & 0 & -3 & -2 & 1 & 0 \\ 0 & 1 & 4 & 3 & -1 & 0 \\ 0 & 0 & 7 & 3 & -2 & 1 \end{array}\right]$

$\uparrow$ $R_2 = (-1)r_2$

$\uparrow$ $R_1 = (-1)r_2 + r_1$
$R_3 = 2r_2 + r_3$

$\rightarrow \left[\begin{array}{rrr|rrr} 1 & 0 & -3 & -2 & 1 & 0 \\ 0 & 1 & 4 & 3 & -1 & 0 \\ 0 & 0 & 1 & \frac{3}{7} & -\frac{2}{7} & \frac{1}{7} \end{array}\right] \rightarrow \left[\begin{array}{rrr|rrr} 1 & 0 & 0 & -\frac{5}{7} & \frac{1}{7} & \frac{3}{7} \\ 0 & 1 & 0 & \frac{9}{7} & \frac{1}{7} & -\frac{4}{7} \\ 0 & 0 & 1 & \frac{3}{7} & -\frac{2}{7} & \frac{1}{7} \end{array}\right]$

$\uparrow$ $R_3 = \left(\frac{1}{7}\right) r_3$

$\uparrow$ $R_1 = 3r_3 + r_1$
$R_2 = -4r_3 + r_2$

Thus, $A^{-1} = \begin{bmatrix} -\frac{5}{7} & \frac{1}{7} & \frac{3}{7} \\ \frac{9}{7} & \frac{1}{7} & -\frac{4}{7} \\ \frac{3}{7} & -\frac{2}{7} & \frac{1}{7} \end{bmatrix}$

31. Let $A = \begin{bmatrix} 2 & 1 \\ 1 & 1 \end{bmatrix}$, $X = \begin{bmatrix} x \\ y \end{bmatrix}$, $B = \begin{bmatrix} 8 \\ 5 \end{bmatrix}$.

Then $2x + y = 8$
$x + y = 5$
can be written compactly as $A \cdot X = B$.

From Problem 21, $A^{-1} = \begin{bmatrix} 1 & -1 \\ -1 & 2 \end{bmatrix}$ and $X = A^{-1}B = \begin{bmatrix} 1 & -1 \\ -1 & 2 \end{bmatrix} \begin{bmatrix} 8 \\ 5 \end{bmatrix} = \begin{bmatrix} 3 \\ 2 \end{bmatrix}$;
or in other words, $x = 3$ and $y = 2$.

33. Here, $A = \begin{bmatrix} 2 & 1 \\ 1 & 1 \end{bmatrix}$, $X = \begin{bmatrix} x \\ y \end{bmatrix}$, $B = \begin{bmatrix} 0 \\ 5 \end{bmatrix}$.

From Problem 21, $A^{-1} = \begin{bmatrix} 1 & -1 \\ -1 & 2 \end{bmatrix}$, so $X = A^{-1}B = \begin{bmatrix} 1 & -1 \\ -1 & 2 \end{bmatrix} \begin{bmatrix} 0 \\ 5 \end{bmatrix} = \begin{bmatrix} -5 \\ 10 \end{bmatrix}$; or $x = -5$ and $y = 10$.

35. $A = \begin{bmatrix} 6 & 5 \\ 2 & 2 \end{bmatrix}$, $X = \begin{bmatrix} x \\ y \end{bmatrix}$, $B = \begin{bmatrix} 7 \\ 2 \end{bmatrix}$.

From Problem 23, $A^{-1} = \begin{bmatrix} 1 & -\frac{5}{2} \\ -1 & 3 \end{bmatrix}$, so $X = A^{-1}B = \begin{bmatrix} 1 & -\frac{5}{2} \\ -1 & 3 \end{bmatrix} \begin{bmatrix} 7 \\ 2 \end{bmatrix} = \begin{bmatrix} 2 \\ -1 \end{bmatrix}$; or $x = 2$ and $y = -1$.

37. $A = \begin{bmatrix} 6 & 5 \\ 2 & 2 \end{bmatrix}$, $X = \begin{bmatrix} x \\ y \end{bmatrix}$, $B = \begin{bmatrix} 13 \\ 5 \end{bmatrix}$.

From Problem 23, $A^{-1} = \begin{bmatrix} 1 & -\frac{5}{2} \\ -1 & 3 \end{bmatrix}$, so $X = A^{-1}B = \begin{bmatrix} 1 & -\frac{5}{2} \\ -1 & 3 \end{bmatrix} \begin{bmatrix} 13 \\ 5 \end{bmatrix} = \begin{bmatrix} \frac{1}{2} \\ 2 \end{bmatrix}$; or $x = \frac{1}{2}$ and $y = 2$.

39. $A = \begin{bmatrix} 2 & 1 \\ a & a \end{bmatrix}$, $X = \begin{bmatrix} x \\ y \end{bmatrix}$, $B = \begin{bmatrix} -3 \\ -a \end{bmatrix}$, where $a \neq 0$.

From Problem 25, $A^{-1} = \begin{bmatrix} 1 & -\frac{1}{a} \\ -1 & \frac{2}{a} \end{bmatrix}$, so $X = A^{-1}B = \begin{bmatrix} 1 & -\frac{1}{a} \\ -1 & \frac{2}{a} \end{bmatrix} \begin{bmatrix} -3 \\ -a \end{bmatrix} = \begin{bmatrix} -2 \\ 1 \end{bmatrix}$; or $x = -2$ and $y = 1$.

41. $A = \begin{bmatrix} 2 & 1 \\ a & a \end{bmatrix}$, $X = \begin{bmatrix} x \\ y \end{bmatrix}$, $B = \begin{bmatrix} \frac{7}{a} \\ 5 \end{bmatrix}$.

From Problem 25, $A^{-1} = \begin{bmatrix} 1 & -\frac{1}{a} \\ -1 & \frac{2}{a} \end{bmatrix}$, so $X = A^{-1}B = \begin{bmatrix} 1 & -\frac{1}{a} \\ -1 & \frac{2}{a} \end{bmatrix} \begin{bmatrix} \frac{7}{a} \\ 5 \end{bmatrix} = \begin{bmatrix} \frac{2}{a} \\ \frac{3}{a} \end{bmatrix}$; or $x = \frac{2}{a}$ and $y = \frac{3}{a}$.

43. $A = \begin{bmatrix} 1 & -1 & 1 \\ 0 & -2 & 1 \\ -2 & -3 & 0 \end{bmatrix}$, $X = \begin{bmatrix} x \\ y \\ z \end{bmatrix}$, $B = \begin{bmatrix} 0 \\ -1 \\ -5 \end{bmatrix}$.

By Problem 27, $A^{-1} = \begin{bmatrix} 3 & -3 & 1 \\ -2 & 2 & -1 \\ -4 & 5 & -2 \end{bmatrix}$, so $X = A^{-1}B = \begin{bmatrix} 3 & -3 & 1 \\ -2 & 2 & -1 \\ -4 & 5 & -2 \end{bmatrix} \begin{bmatrix} 0 \\ -1 \\ -5 \end{bmatrix} = \begin{bmatrix} -2 \\ 3 \\ 5 \end{bmatrix}$;
or $x = -2$, $y = 3$, and $z = 5$.

45. $A = \begin{bmatrix} 1 & -1 & 1 \\ 0 & -2 & 1 \\ -2 & -3 & 0 \end{bmatrix}$, $X = \begin{bmatrix} x \\ y \\ z \end{bmatrix}$, $B = \begin{bmatrix} 2 \\ 2 \\ \frac{1}{2} \end{bmatrix}$.

By Problem 27, $A^{-1} = \begin{bmatrix} 3 & -3 & 1 \\ -2 & 2 & -1 \\ -4 & 5 & -2 \end{bmatrix}$, so $X = A^{-1}B = \begin{bmatrix} 3 & -3 & 1 \\ -2 & 2 & -1 \\ -4 & 5 & -2 \end{bmatrix} \begin{bmatrix} 2 \\ 2 \\ \frac{1}{2} \end{bmatrix} = \begin{bmatrix} \frac{1}{2} \\ -\frac{1}{2} \\ 1 \end{bmatrix}$;

or $x = \frac{1}{2}$, $y = -\frac{1}{2}$, and $z = 1$.

47. $A = \begin{bmatrix} 1 & 1 & 1 \\ 3 & 2 & -1 \\ 3 & 1 & 2 \end{bmatrix}$, $X = \begin{bmatrix} x \\ y \\ z \end{bmatrix}$, $B = \begin{bmatrix} 9 \\ 8 \\ 1 \end{bmatrix}$.

By Problem 29, $A^{-1} = \frac{1}{7}\begin{bmatrix} -5 & 1 & 3 \\ 9 & 1 & -4 \\ 3 & -2 & 1 \end{bmatrix}$, so

$$X = A^{-1}B = \frac{1}{7}\begin{bmatrix} -5 & 1 & 3 \\ 9 & 1 & -4 \\ 3 & -2 & 1 \end{bmatrix} \begin{bmatrix} 9 \\ 8 \\ 1 \end{bmatrix} = \frac{1}{7}\begin{bmatrix} -34 \\ 85 \\ 12 \end{bmatrix} = \begin{bmatrix} -\frac{34}{7} \\ \frac{85}{7} \\ \frac{12}{7} \end{bmatrix};$$

or $x = -\frac{34}{7}$, $y = \frac{85}{7}$, $z = \frac{12}{7}$

49. $A = \begin{bmatrix} 1 & 1 & 1 \\ 3 & 2 & -1 \\ 3 & 1 & 2 \end{bmatrix}$, $X = \begin{bmatrix} x \\ y \\ z \end{bmatrix}$, $B = \begin{bmatrix} 2 \\ \frac{7}{3} \\ \frac{10}{3} \end{bmatrix}$.

By Problem 29, $A^{-1} = \frac{1}{7}\begin{bmatrix} -5 & 1 & 3 \\ 9 & 1 & -4 \\ 3 & -2 & 1 \end{bmatrix}$,

so $X = A^{-1}B = \frac{1}{7}\begin{bmatrix} -5 & 1 & 3 \\ 9 & 1 & -4 \\ 3 & -2 & 1 \end{bmatrix}\begin{bmatrix} 2 \\ \frac{7}{3} \\ \frac{10}{3} \end{bmatrix} = \frac{1}{7}\begin{bmatrix} \frac{7}{3} \\ 7 \\ \frac{14}{3} \end{bmatrix} = \begin{bmatrix} \frac{1}{3} \\ 1 \\ \frac{2}{3} \end{bmatrix}$;

or $x = \frac{1}{3}$, $y = 1$, $z = \frac{2}{3}$

51. $A = \begin{bmatrix} 4 & 2 \\ 2 & 1 \end{bmatrix}$. We start with $[A \mid I_2]$ and put it into reduced echelon form. If I_2 does *not* appear to the left of the vertical bar, then A has no inverse.

$$[A \mid I_2] = \left[\begin{array}{cc|cc} 4 & 2 & 1 & 0 \\ 2 & 1 & 0 & 1 \end{array}\right] \rightarrow \left[\begin{array}{cc|cc} 4 & 2 & 1 & 0 \\ 0 & 0 & -\frac{1}{2} & 1 \end{array}\right] \rightarrow \left[\begin{array}{cc|cc} 1 & \frac{1}{2} & \frac{1}{4} & 0 \\ 0 & 0 & -\frac{1}{2} & 1 \end{array}\right]$$

$\uparrow R_2 = -\frac{1}{2}r_1 + r_2$ $\quad \uparrow R_1 = -\frac{1}{4}r_1$

This is in reduced echelon form, but the identity matrix I_2 does not appear on the left. Thus, A has no inverse.

53. $$[A \mid I_2] = \left[\begin{array}{cc|cc} 15 & 3 & 1 & 0 \\ 10 & 2 & 0 & 1 \end{array}\right] \rightarrow \left[\begin{array}{cc|cc} 1 & \frac{1}{5} & \frac{1}{15} & 0 \\ 10 & 2 & 0 & 1 \end{array}\right] \rightarrow \left[\begin{array}{cc|cc} 1 & \frac{1}{5} & \frac{1}{15} & 0 \\ 0 & 0 & -\frac{2}{3} & 1 \end{array}\right]$$

$\uparrow R_1 = \frac{1}{15}r_1$ $\quad \uparrow R_2 = -10r_1 + r_2$

We cannot obtain I_2 on the left, so A has no inverse.

55. $$\left[\begin{array}{ccc|ccc} -3 & 1 & -1 & 1 & 0 & 0 \\ 1 & -4 & -7 & 0 & 1 & 0 \\ 1 & 2 & 5 & 0 & 0 & 1 \end{array}\right] \rightarrow \left[\begin{array}{ccc|ccc} 1 & 2 & 5 & 0 & 0 & 1 \\ 1 & -4 & -7 & 0 & 1 & 0 \\ -3 & 1 & -1 & 1 & 0 & 0 \end{array}\right] \rightarrow \left[\begin{array}{ccc|ccc} 1 & 2 & 5 & 0 & 0 & 1 \\ 0 & -6 & -12 & 0 & 1 & -1 \\ 0 & 7 & 14 & 1 & 0 & 3 \end{array}\right]$$

$\uparrow$ Interchange rows one and three $\quad \uparrow R_2 = -r_1 + r_2$, $R_3 = 3r_1 + r_3$

$$\rightarrow \left[\begin{array}{ccc|ccc} 1 & 2 & 5 & 0 & 0 & 1 \\ 0 & 1 & 2 & 0 & -\frac{1}{6} & \frac{1}{6} \\ 0 & 1 & 2 & \frac{1}{7} & 0 & \frac{3}{7} \end{array}\right] \rightarrow \left[\begin{array}{ccc|ccc} 1 & 2 & 5 & 0 & 0 & 1 \\ 0 & 1 & 2 & 0 & -\frac{1}{6} & \frac{1}{6} \\ 0 & 0 & 0 & \frac{1}{7} & \frac{1}{6} & \frac{11}{42} \end{array}\right]$$

$$R_2 = -\frac{1}{6}r_2 \qquad R_3 = -r_2 + r_3$$

$$R_3 = \frac{1}{7}r_3$$

The left side is not the identity matrix, I_3, so A has no inverse.

57. $\begin{bmatrix} 0.01 & 0.05 & -0.01 \\ 0.01 & -0.02 & 0.01 \\ -0.02 & 0.01 & 0.03 \end{bmatrix}$

59. $\begin{bmatrix} 0.02 & -0.04 & -0.01 & 0.01 \\ -0.02 & 0.05 & 0.03 & -0.03 \\ -0.02 & 0.01 & -0.04 & 4.88 \\ -0.02 & 0.06 & 0.07 & 0.06 \end{bmatrix}$

61. $x = 4.56, y = -6.06, z = -22.55$

63. $x = -1.19, y = 2.46, z = 8.27$

65. (a) Since we want to use a 2 by 3 matrix to represent the data, we can let rows represent stainless steel and aluminum, respectively, while the columns can represent 10-gallon, 5-gallon, and 1-gallon containers:

	Production		
	10 g.	5 g.	1 g.
Stainless steel	500	350	400
Aluminum	700	500	850

$$\begin{bmatrix} 500 & 350 & 400 \\ 700 & 500 & 850 \end{bmatrix}$$

This could have been represented by a 3 by 2 matrix, by letting rows represent size of container, and columns represent type of material.

$$\begin{bmatrix} 500 & 700 \\ 350 & 500 \\ 400 & 850 \end{bmatrix}$$

(b) Now we are given the following information:

	Pounds of Material
10-gal.	15
5-gal.	8
1-gal.	3

$$\begin{bmatrix} 15 \\ 8 \\ 3 \end{bmatrix}$$

(c) From (a) and (b): $\begin{bmatrix} 500 & 350 & 400 \\ 700 & 500 & 850 \end{bmatrix} \begin{bmatrix} 15 \\ 8 \\ 3 \end{bmatrix} = \begin{bmatrix} 11{,}500 \\ 17{,}050 \end{bmatrix}$

The first row represents stainless steel containers (11,500 pounds), and the second row represents aluminum (17,050 pounds).

(d)

	Stainless Steel	Aluminum
Cost per pound	0.10	0.05

$\Big\} \begin{bmatrix} 0.10 & 0.05 \end{bmatrix}$

(e) $\begin{bmatrix} 0.10 & 0.05 \end{bmatrix} \begin{bmatrix} 11{,}500 \\ 17{,}050 \end{bmatrix} = \begin{bmatrix} 2002.50 \end{bmatrix}$; i.e., total cost of material $=$ \$2002.50.

Note: Since the first entry in (c) was for stainless steel, the first entry, a_{11}, in (d) had to be for stainless steel.

67. Let $A = \begin{bmatrix} a & b \\ c & d \end{bmatrix}$.

We are assuming that $D = ad - bc \neq 0$, and we wish to show that A has no inverse. We start, as usual, with

$$[A|I_2] = \left[\begin{array}{cc|cc} a & b & 1 & 0 \\ c & d & 0 & 1 \end{array}\right].$$

Our first step in putting this into echelon form would be to multiply row one by $\frac{1}{a}$, to get a 1 in the first row, first column. But that will be impossible if $a = 0$. Thus, we need to consider two cases:

Case 1: If $a \neq 0$, proceed as usual:

$$\left[\begin{array}{cc|cc} a & b & 1 & 0 \\ c & d & 0 & 1 \end{array}\right] \rightarrow \left[\begin{array}{cc|cc} 1 & \frac{b}{a} & \frac{1}{a} & 0 \\ c & d & 0 & 1 \end{array}\right] \rightarrow \left[\begin{array}{cc|cc} 1 & \frac{b}{a} & \frac{1}{a} & 0 \\ 0 & d - \frac{cb}{a} & -\frac{c}{a} & 1 \end{array}\right]$$

$$R_1 = \frac{1}{a}r_1 \qquad R_2 = -cr_1 + r_2$$

$$= \left[\begin{array}{cc|cc} 1 & \frac{b}{a} & \frac{1}{a} & 0 \\ 0 & \frac{ad - bc}{a} & -\frac{c}{a} & 1 \end{array}\right] \left(\text{since } d - \frac{cb}{a} = \frac{ad - bc}{a}\right)$$

$$\rightarrow \left[\begin{array}{cc|cc} 1 & \frac{b}{a} & \frac{1}{a} & 0 \\ 0 & 1 & \frac{-c}{ad - bc} & \frac{a}{ad - bc} \end{array}\right] \rightarrow \left[\begin{array}{cc|cc} 1 & 0 & \frac{1}{a} + \frac{bc}{a(ad - bc)} & \frac{-b}{ad - bc} \\ 0 & 1 & \frac{-c}{ad - bc} & \frac{a}{ad - bc} \end{array}\right]$$

$$R_2 = \left(\frac{a}{ad - bc}\right) r_2 \qquad R_1 = \left(\frac{-b}{a}\right) r_2 + r_1$$

Now some algebra:

$$\frac{1}{a} + \frac{bc}{a(ad - bc)} = \frac{ad - bc}{a(ad - bc)} + \frac{bc}{a(ad - bc)} = \frac{ad}{a(ad - bc)} = \frac{d}{ad - bc}$$

Thus, our inverse is: $A^{-1} = \begin{bmatrix} \frac{d}{ad - bc} & \frac{-b}{ad - bc} \\ \frac{-c}{ad - bc} & \frac{a}{ad - bc} \end{bmatrix} = \frac{1}{D}\begin{bmatrix} d & -b \\ -c & a \end{bmatrix}$, since $D = ad - bc$.

Case 2: But what if $a = 0$? Then, $ad - bc \neq 0$, so we know $b \neq 0$ and $c \neq 0$. Also, we will have:

$$[A \mid I_2] = \left[\begin{array}{cc|cc} 0 & b & 1 & 0 \\ c & d & 0 & 1 \end{array}\right]$$

So $\left[\begin{array}{cc|cc} 0 & b & 1 & 0 \\ c & d & 0 & 1 \end{array}\right] \rightarrow \left[\begin{array}{cc|cc} c & d & 0 & 1 \\ 0 & b & 1 & 0 \end{array}\right] \rightarrow \left[\begin{array}{cc|cc} 1 & \frac{d}{c} & 0 & \frac{1}{c} \\ 0 & b & 1 & 0 \end{array}\right]$

Interchange rows $R_1 = \frac{1}{c} r_1$ (since $c \neq 0$)

$$= \left[\begin{array}{cc|cc} 1 & \frac{d}{c} & 0 & \frac{1}{c} \\ 0 & 1 & \frac{1}{b} & 0 \end{array}\right] \rightarrow \left[\begin{array}{cc|cc} 1 & 0 & \frac{-d}{bc} & \frac{1}{c} \\ 0 & 1 & \frac{1}{b} & 0 \end{array}\right]$$

$$R_2 = \frac{1}{b} r_2 \text{ (since } b \neq 0\text{)} \quad R_1 = \left(-\frac{d}{c}\right) r_2 + r_1$$

Therefore, $A^{-1} = \begin{bmatrix} \frac{-d}{bc} & \frac{1}{c} \\ \frac{1}{b} & 0 \end{bmatrix}$.

That looks odd, but if we get a common denominator, we have:

$$A^{-1} = \begin{bmatrix} \frac{-d}{bc} & \frac{b}{bc} \\ \frac{c}{bc} & \frac{0}{bc} \end{bmatrix} = \frac{1}{-bc}\begin{bmatrix} d & -b \\ -c & 0 \end{bmatrix} = \frac{1}{D}\begin{bmatrix} d & -b \\ -c & a \end{bmatrix},$$

since $a = 0$, and $D = ad - bc = -bc$!

8.5 Partial Fraction Decomposition

1. The expression $\dfrac{x}{x^2 - 1}$ is proper, since the degree of the numerator is less than that of the denominator.

3. $\dfrac{x^2 + 5}{x^2 - 4}$ is improper, so we do the division:

$$\begin{array}{r l} & 1 \quad \leftarrow \text{Quotient} \\ x^2 - 4 & \overline{)x^2 + 0x + 5} \\ & \underline{x^2 \qquad - 4} \\ & \qquad 9 \quad \leftarrow \text{Remainder} \end{array}$$

Then, $\dfrac{x^2 + 5}{x^2 - 4} = 1 + \dfrac{9}{x^2 - 4}$.

5. $\dfrac{5x^3 + 2x - 1}{x^2 - 4}$ is improper, so we must do long division:

$$\begin{array}{r l} & 5x \\ x^2 - 4 & \overline{)5x^3 + 0x^2 + 2x - 1} \\ & \underline{5x^3 \qquad\quad -20x} \\ & \qquad\qquad 22x - 1 \end{array}$$

Thus, $\dfrac{5x^3 + 2x - 1}{x^2 - 4} = 5x + \dfrac{22x - 1}{x^2 - 4}$

7. Here, $\dfrac{x(x - 1)}{(x + 4)(x - 3)} = \dfrac{x^2 - x}{x^2 + x - 12}$ is improper:

$$\begin{array}{r l} & 1 \\ x^2 + x - 12 & \overline{)x^2 - x + 0} \\ & \underline{x^2 + x - 12} \\ & -2x + 12 \end{array}$$

Hence, $\dfrac{x(x - 1)}{(x + 4)(x - 3)} = 1 + \dfrac{-2x + 12}{x^2 + x - 12} = 1 + \dfrac{-2(x - 6)}{(x + 4)(x - 3)}$

In Problems 9–38, we will use the following steps:

Step 1: *Perform long division, if necessary, to obtain a* ***proper*** *fraction,* $\dfrac{p(x)}{q(x)}$, *and put it in lowest terms.*

Step 2: *Factor the denominator,* $q(x)$, *completely into linear and irreducible quadratic factors, and identify which of the four cases in the text apply.*

Step 3: *Write out the partial fraction expansion, based on the factors of* $q(x)$.

Step 4: *Solve for the unknown coefficients in Step 3.*

Step 5: *Write the final decomposition.*

9. $\dfrac{4}{x(x - 1)}$

Step 1: Already ***proper***.

Step 2: Done: $q(x) = x(x - 1)$.
This is Case 1 (only non-repeated linear factors).

Step 3: $\dfrac{4}{x(x - 1)} = \dfrac{A}{x} + \dfrac{B}{x - 1}$

Step 4: Multiply both sides by $x(x - 1)$: $4 = A(x - 1) + Bx$
Let $x = 1$: then $4 = A(0) + B$, or $B = 4$
Let $x = 0$: then $4 = A(-1) + B \cdot 0$, or $A = -4$

Step 5: Therefore, $\dfrac{4}{x(x - 1)} = \dfrac{-4}{x} + \dfrac{4}{x - 1}$

11. $\dfrac{1}{x(x^2 + 1)}$

Step 1: Already proper.

Step 2: $q(x) = x(x^2 + 1)$ — (x^2+1) cannot be factored

This is Case 1 (nonrepeated linear) and Case 3 (nonrepeated irreducible quadratic).

Step 3: $\dfrac{1}{x(x^2 + 1)} = \dfrac{A}{x} + \dfrac{Bx + C}{x^2 + 1}$

(Remember, irreducible quadratic factors in the denominator require ***first degree*** numerators in the decomposition.)

Step 4: Multiply by $x(x^2 + 1)$: $1 = A(x^2 + 1) + (Bx + C)x$

Let $x = 0$: then $1 = A$.

We still have two unknowns, B and C, so we need two equations:

Let $-x = 1$: $1 = A(1 + 1) + (B \cdot 1 + C) \cdot 1$

or $1 = 2A + B + C$

or $B + C = -1$

Let $x = -1$: $1 = 2A + (-B + C)(-1)$

or $1 = 2 + B - C$ (since $A = 1$)

or $B - C = -1$ (since $A = 1$)

We have
$$\begin{aligned} B + C &= -1 \\ B - C &= -1 \\ \text{Add: } 2B &= -2 \\ B &= -1 \end{aligned}$$

Then, from $B + C = -1$, we get:
$$\begin{aligned} -1 + C &= -1 \\ C &= 0 \end{aligned}$$

Step 5: $\dfrac{1}{x(x^2 + 1)} = \dfrac{1}{x} + \dfrac{-x}{x^2 + 1}$

13. $\dfrac{1}{(x - 1)(x - 2)}$

Step 1: This is proper.

Step 2: $q(x) = (x - 1)(x - 2)$. Case 1 only.

Step 3: $\dfrac{x}{(x - 1)(x - 2)} = \dfrac{A}{x - 1} + \dfrac{B}{x - 2}$

Step 4: Multiply by $(x - 1)(x - 2)$: $x = A(x - 2) + B(x - 1)$

Let $x = 2$: $2 = B$

Let $x = 1$: $1 = -A$, or $A = -1$

Step 5: $\dfrac{x}{(x - 1)(x - 2)} = \dfrac{-1}{x - 1} + \dfrac{2}{x - 2}$

15. $\dfrac{x^2}{(x - 1)^2(x + 1)}$

Step 1: This is proper.

Step 2: $q(x) = (x - 1)^2(x + 1)$.

This involves both Case 1 and Case 2 (a repeated linear factor).

Step 3: For the factor $(x - 1)^2$, we need two terms, $\frac{A}{x-1} + \frac{B}{(x-1)^2}$, since it is a linear factor raised to the ***second*** power.

$$\frac{x^2}{(x-1)^2(x+1)} = \frac{A}{x-1} + \frac{B}{(x-1)^2} + \frac{C}{x+1}$$

Step 4: Multiply by $(x - 1)^2(x + 1)$: $x^2 = A(x - 1)(x + 1) + B(x + 1) + C(x - 1)^2$

Since we have three unknowns, we need three equations, so we will choose three values of x starting with the ***zeros*** of $q(x)$:

Let $x = 1$: $1 = 2B$, or $B = \frac{1}{2}$

Let $x = -1$: $1 = 4C$, or $C = \frac{1}{4}$

Now pick a third value for x:

Let $x = 0$: $0 = -A + B + C$

$0 = -A + \frac{1}{2} + \frac{1}{4}$, since we know B and C.

$A = \frac{3}{4}$

Step 5:

$$\frac{x^2}{(x-1)^2(x+1)} = \frac{\frac{3}{4}}{x-1} + \frac{\frac{1}{2}}{(x-1)^2} + \frac{\frac{1}{4}}{x+1}$$

17. $\frac{1}{x^3 - 8}$

Step 1: This is proper.

Step 2: We need to do some factoring:

$q(x) = (x - 2)(x^2 + 2x + 4)$

Can this be factored?

To determine if $x^2 + 2x + 4$ can be factored or if it is irreducible, check its discriminant:

$b^2 - 4ac = 4 - 4(4) = 4 - 16 = -12 < 0$

Therefore, $x^2 + 2x + 4$ has no real zeros; i.e., it is irreducible over the reals.

We have Case 1 and Case 3.

Step 3:

$$\frac{1}{x^3 - 8} = \frac{1}{(x-2)(x^2 + 2x + 4)} = \frac{A}{x-2} + \frac{Bx + C}{x^2 + 2x + 4}$$

Step 4: Multiply both sides by $x^3 - 8 = (x - 2)(x^2 + 2x + 4)$:

$1 = A(x^2 + 2x + 4) + (Bx + C)(x - 2) \quad (1)$

We only have one zero for $q(x)$, $x = 2$:

Let $x = 2$: $1 = 12A$, or $A = \frac{1}{12}$

But we need three equations. Another approach is to expand (1) and collect like terms:

$1 = Ax^2 + 2Ax + 4A + Bx^2 - 2Bx + Cx - 2C$

or

$1 = (A + B)x^2 + (2A - 2B + C)x + (4A - 2C)$

There is no x^2 on the left-hand side, so we must have $A + B = 0$. Also, there is no x-term on the left, so

$2A - 2B + C = 0$

Finally, the constant terms must be equal:

$1 = 4A - 2C$

But, we already know $A = \frac{1}{12}$, so from $A + B = 0$, we have $B = -\frac{1}{12}$, and from $1 = 4A - 2C$ we have:

$$2C = 4A - 1$$
$$2C = \frac{4}{12} - 1$$
$$2C = \frac{-8}{12}$$
$$C = \frac{-4}{12} = \frac{-1}{3}$$

Step 5: $$\frac{1}{x^3 - 8} = \frac{\frac{1}{12}}{x - 2} + \frac{\frac{-1}{12}x - \frac{1}{3}}{x^2 + 2x + 4} = \frac{\frac{1}{12}}{x - 2} + \frac{\frac{-1}{12}(x + 4)}{x^2 + 2x + 4}$$

19. $\dfrac{x^2}{(x - 1)^2(x + 1)^2}$

Step 1: This is proper.

Step 2: $q(x) = (x - 1)^2(x + 1)^2$. This is case 2.

Step 3: $$\frac{x^2}{(x - 1)^2(x + 1)^2} = \frac{A}{x - 1} + \frac{B}{(x - 1)^2} + \frac{C}{x + 1} + \frac{D}{(x + 1)^2}$$

Step 4: Multiply by $(x - 1)^2(x + 1)^2$:

$$x^2 = A(x - 1)(x + 1)^2 + B(x + 1)^2 + C(x - 1)^2(x + 1) + D(x - 1)^2$$

Let $x = 1$: $\quad 1 = 4B,\ B = \frac{1}{4}$

Let $x = -1$: $\quad 1 = 4D,\ D = \frac{1}{4}$

Let $x = 0$: $\quad 0 = -A + B + C + D$

or $\quad A - C = \frac{1}{2}$

Let $x = 2$: $\quad 4 = 9A + 9B + 3C + D$

or $\quad -9A - 3C = -\frac{3}{2}$

or $\quad 3A + C = \frac{1}{2}$

We have two equations in two unknowns:

$$A - C = \frac{1}{2}$$
$$3A + C = \frac{1}{2}$$

Add: $4A = 1$

$$A = \frac{1}{4}$$

and $$A - C = \frac{1}{2}$$
$$\frac{1}{4} - C = \frac{1}{2}$$
$$C = -\frac{1}{4}$$

Step 5: $$\frac{x^2}{(x - 1)^2(x + 1)^2} = \frac{\frac{1}{4}}{x - 1} + \frac{\frac{1}{4}}{(x - 1)^2} + \frac{\frac{-1}{4}}{x + 1} + \frac{\frac{1}{4}}{(x + 1)^2}$$

21. $\dfrac{x - 3}{(x + 2)(x + 1)^2}$

Steps 1 and 2: Proper, with $q(x) = (x + 2)(x + 1)^2$, Case 1 and Case 2.

Step 3: $\dfrac{x - 3}{(x + 2)(x + 1)^2} = \dfrac{A}{x + 2} + \dfrac{B}{(x + 1)} + \dfrac{C}{(x + 1)^2}$

Step 4: Clear of fractions:

$$x - 3 = A(x + 1)^2 + B(x + 2)(x + 1) + C(x + 2)$$

Let $x = -1$: $\quad -4 = C$

Let $x = -2$: $\quad -5 = A$

Let $x = 0$: $\quad -3 = A + 2B + 2C$

or $\quad -3 = -5 + 2B - 8$

$10 = 2B$

$B = 5$

Step 5: $\dfrac{x - 3}{(x + 2)(x + 1)^2} = \dfrac{-5}{x + 2} + \dfrac{5}{x + 1} + \dfrac{-4}{(x + 1)^2}$

23. $\dfrac{x + 4}{x^2(x^2 + 4)}$

Step 1: This is proper.

Step 2: $q(x) = x^2(x^2 + 4)$.

Be careful to note the difference between x^2 and $x^2 + 4$:

The factor x^2 is a linear factor (x), repeated:

$x^2 = x \cdot x$, so it is Case 2.

The factor $x^2 + 4$ is a quadratic that cannot be factored as a product of linear factors (Case 3).

Step 3: $\dfrac{x + 4}{x^2(x^2 + 4)} = \dfrac{A}{x} + \dfrac{B}{x^2} + \dfrac{Cx + D}{x^2 + 4}$

Step 4: $x + 4 = Ax(x^2 + 4) + B(x^2 + 4) + (Cx + D)x^2$. (1)

There is only one zero, $x = 0$:

Let $x = 0$: $4 = 4B$, or $B = 1$.

Let's expand equation (1) and combine like terms:

$$x + 4 = Ax^3 + 4Ax + Bx^2 + 4B + Cx^3 + Dx^2$$
$$= (A + C)x^3 = (B + D)x^2 + (4A)x + 4B$$

Now equate coefficients of like powers of x on the left and right:

For x^3: $\quad 0 = A + C$

For x^2: $\quad 0 = B + D$

For x^1: $\quad 1 = 4\text{A}$, or $A = \dfrac{1}{4}$

For x^0: $\quad 4 = 4B$, or $B = 1$

Then: $\quad A + C = 0 \qquad B + D = 0$

$\dfrac{1}{4} + C = 0 \qquad 1 + D = 0$

$C = \dfrac{-1}{4} \qquad D = -1$

Step 5: $\dfrac{x + 4}{x^2(x^2 + 4)} = \dfrac{\frac{1}{4}}{x} + \dfrac{1}{x^2} + \dfrac{-\frac{1}{4}x - 1}{x^2 + 4} = \dfrac{\frac{1}{4}}{x} + \dfrac{1}{x^2} - \dfrac{\frac{1}{4}(x + 4)}{x^2 + 4}$

25. $\dfrac{x^2 + 2x + 3}{(x + 1)(x^2 + 2x + 4)}$

Step 1: This is proper.

Step 3: $q(x) = (x + 1)(x^2 + 2x + 4)$

$$b^2 - 4ac = 4 - 16 = -12 < 0$$

We have Case 1 and Case 3.

Step 3: $$\frac{x^2 + 2x + 3}{(x + 1)(x^2 + 2x + 4)} = \frac{A}{x + 1} = \frac{Bx + C}{x^2 + 2x + 4}$$

Step 4: $x^2 + 2x + 3 = A(x^2 + 2x + 4) + (Bx + C)(x + 1)$

Let $x = -1$: $\quad 2 = 3A;\ A = \frac{2}{3}$

Let $x = 0$: $\quad 3 = 4A + C;\ C = 3 - 4A = 3 - \frac{8}{3} = \frac{1}{3}$

Let $x = 1$: $\quad 6 = 7A + 2B + 2C$

$2B = 6 - 7A - 2C$

$2B = 6 - \frac{14}{3} - \frac{2}{3} = \frac{2}{3};\ B = \frac{1}{3}$

Step 5: $$\frac{x^2 + 2x + 3}{(x + 1)(x^2 + 2x + 4)} = \frac{\frac{2}{3}}{x + 1} + \frac{\frac{1}{3}(x + 1)}{x^2 + 2x + 4}$$

27. $$\frac{x}{(3x - 2)(2x + 1)}$$

Step 1: This is proper.

Step 2: $q(x) = (3x - 2)(2x + 1)$, Case 1.

Step 3: $$\frac{x}{(3x - 2)(2x + 1)} = \frac{A}{3x - 2} + \frac{B}{2x + 1}$$

Step 4: $x = A(2x + 1) + B(3x - 2)$

The two zeros are $x = -\frac{1}{2}$ and $x = \frac{2}{3}$

Let $x = -\frac{1}{2}$: $\quad -\frac{1}{2} = B\left[-\frac{7}{2}\right]$

$1 = 7B$

$B = \frac{1}{7}$

Let $x = \frac{2}{3}$: $\quad \frac{2}{3} = A\left[\frac{7}{3}\right]$

$2 = 7A$

$A = \frac{2}{7}$

Step 5: $$\frac{x}{(3x - 2)(2x + 1)} = \frac{\frac{2}{7}}{3x - 2} + \frac{\frac{1}{7}}{2x + 1}$$

29. $$\frac{x}{x^2 + 2x - 3}$$

Step 1: Already proper.

Step 2: $q(x) = x^2 + 2x - 3 = (x + 3)(x - 1)$, Case 1.

Step 3: $$\frac{x}{x^2 + 2x - 3} = \frac{A}{x + 3} + \frac{B}{x - 1}$$

Step 4: $x = A(x - 1) + B(x + 3)$

Let $x = 1$: $1 = 4B$, $B = \frac{1}{4}$

Let $x = -3$: $-3 = -4A$, $A = \frac{3}{4}$

Step 5: $$\frac{x}{(x^2 + 2x - 3)} = \frac{\frac{3}{4}}{x + 3} + \frac{\frac{1}{4}}{x - 1}$$

31. $$\frac{x^2 + 2x + 3}{(x^2 + 4)^2}$$

Steps 1 and 2: This is proper, and falls under Case 4 (repeated irreducible quadratic factor).

Step 3: $$\frac{x^2 + 2x + 3}{(x^2 + 4)^2} = \frac{Ax + B}{x^2 + 4} + \frac{Cx + D}{(x^2 + 4)^2}$$

Step 4: Clear of fractions:

$$x^2 + 2x + 3 = (Ax + B)(x^2 + 4) + (Cx + D)$$

We have no zeros of $q(x)$, so expand the right-hand side:

$$x^2 + 2x + 3 = Ax^3 + Bx^2 + 4Ax + 4B + Cx + D$$

$$x^2 + 2x + 3 = Ax^3 + Bx^2 + (4A + C)x + (4B + D)$$

x^3: $0 = A$

x^2: $1 = B$

x: $2 = 4A + C$, $C = 2$ (since $A = 0$)

Constant: $3 = 4B + D$, $D = 3 - 4B = -1$

Step 5: $$\frac{x^2 + 2x + 3}{(x^2 + 4)^2} = \frac{1}{x^2 + 4} + \frac{2x - 1}{(x^2 + 4)^2}$$

33. $$\frac{7x + 3}{x^3 - 2x^2 - 3x}$$

Step 1: This is proper.

Step 2: $q(x) = x^3 - 2x^2 - 3x = x(x^2 - 2x - 3) = x(x - 3)(x + 1)$

This is Case 1.

Step 3: $$\frac{7x + 3}{x^3 - 2x^2 - 3x)} = \frac{A}{x} = \frac{B}{x - 3} + \frac{C}{x + 1}$$

Step 4: Multiply both sides by $x^3 - 2x^2 - 3x = x(x - 3)(x + 1)$:

$$7x + 3 = A(x - 3)(x + 1) + Bx(x + 1) + Cx(x - 3)$$

Let $x = 0$: $3 = -3A$; $A = -1$

Let $x = 3$: $24 = 12B$, $B = 2$

Let $x = -1$: $-4 = 4C$, $C = -1$

Step 5: $$\frac{7x + 3}{x^3 - 2x^2 - 3x} = \frac{-1}{x} + \frac{2}{x - 3} + \frac{-1}{x + 1}$$

35. $$\frac{x^2}{x^3 - 4x^2 + 5x - 2}$$

Step 1: This is proper.

Step 2: Try to find a zero by synthetic division:

```
1)1  -4   5   -2
      1  -5   10
  1  -5  10  -12
```

```
-1)1  -4   5   -2
      -1   3   -2
   1  -3   2   [0]
   x^2 - 3x + 2
```

Thus, $x - 1$ is a factor:

$$q(x) = x^3 - 4x^2 + 5x - 2 = (x - 1)(x^2 - 3x + 2) = (x - 1)(x - 1)(x - 2) = (x - 1)^2(x - 2)$$

This involves Case 1 and Case 2.

Step 3: $$\frac{x^2}{x^3 - 4x^2 + 5x - 2} = \frac{A}{x - 1} + \frac{B}{(x - 1)^2} + \frac{C}{x - 2}$$

Step 4: Multiply by $x^3 - 4x^2 + 5x - 2 = (x - 1)^2(x - 2)$:

$$x^2 = A(x - 1)(x - 2) + B(x - 2) + C(x - 1)^2$$

Let $x = 1$: $1 = -B$, $B = -1$

Let $x = 2$: $4 = C$

Let $x = 0$: $0 = 2A - 2B + C$

$2A = 2B - C$

$2A = -2 - 4$

$A = -3$

Step 5: $$\frac{x^2}{x^3 - 4x^2 + 5x - 2} = \frac{-3}{x - 1} + \frac{-1}{(x - 1)^2} + \frac{4}{x - 2}$$

37. $\dfrac{x^3}{(x^2 + 16)^3}$

Step 1 and 2: Proper, Case 4

Step 3: $$\frac{x^3}{(x^2 + 16)^3} = \frac{Ax + B}{x^2 + 16} + \frac{Cx + D}{(x^2 + 16)^2} + \frac{Ex + F}{(x^2 + 16)^3}$$

Step 4: $x^3 = (Ax + B)(x^2 + 16)^2 + (Cx + D)(x^2 + 16) + Ex + F$ or

$$x^3 = (Ax + B)(x^4 + 32x^2 + 256) + Cx^3 + Dx^2 + 16Cx + 16D + Ex + F$$

$$x^3 = Ax^5 + 32Ax^3 + 256Ax + Bx^4 + 32Bx^2 + 256B + Cx^3 + Dx^2 + 16Cx + 16D + Ex + F$$

$$x^3 = Ax^5 + Bx^4 + (32A + C)x^3 + (32B + D)x^2 + (256A + 16C + E)x + (256B + 16D + F)$$

Now we equate coefficients of like powers of x to obtain six equations in six unknowns.

x^5: $0 = A$

x^4: $0 = B$

x^3: $1 = 32A + C$, $C = 1$

x^2: $0 = 32B + D$, $D = 0$

x^1: $0 = 256A + 16C + E$, $E = -16C = -16$

x^0: $0 = 256B + 16D + F$, $F = 0$

Step 5: $$\frac{x^3}{(x^2 + 16)^3} = \frac{x}{(x^2 + 16)^2} + \frac{-16x}{(x^2 + 16)^3}$$

39. $\dfrac{4}{2x^2 - 5x - 3}$

Step 1: This is proper.

Step 2: $q(x) = 2x^2 - 5x - 3 = (2x + 1)(x - 3)$, Case 1.

Step 3: $$\frac{4}{2x^2 - 5x - 3} = \frac{A}{2x + 1} + \frac{B}{x - 3}$$

Step 4: $4 = A(x - 3) + B(2x + 1)$

The two zeros are $x = 3$ and $x = \dfrac{-1}{2}$

Let $x = 3$: $4 = 7B$

$B = \frac{4}{7}$

Let $x = \frac{-1}{2}$: $4 = \frac{-7}{2}A$

$A = \frac{-8}{7}$

Step 5: $$\frac{4}{2x^2 - 5x - 3} = \frac{-\frac{8}{7}}{2x + 1} + \frac{\frac{4}{7}}{x - 3}$$

41. $\frac{2x + 3}{x^4 - 9x^2}$

Step 1: This is proper.

Step 2: $q(x) = x^4 - 9x^2 = x^2(x - 3)(x + 3)$, Case 1 and 2.

Step 3: $$\frac{2x + 3}{x^4 - 9x^2} = \frac{A}{x} + \frac{B}{x^2} + \frac{C}{x - 3} + \frac{D}{x + 3}$$

Step 4: $2x + 3 = Ax(x - 3)(x + 3) + B(x - 3)(x + 3) + Cx^2(x - 3) + Dx^2(x - 3)$

Let $x = 0$: $3 = B(-9)$

$B = \frac{-1}{3}$

Let $x = 3$: $9 = C(9)(6)$

$C = \frac{1}{6}$

Let $x = -3$: $-3 = D(9)(-6)$

$D = \frac{1}{18}$

$2x + 3 = Ax(x - 3)(x + 3) - \frac{1}{3}(x - 3)(x + 3) + \frac{1}{6}x^2(x + 3) + \frac{1}{18}x^2(x - 3)$

Let $x = 1$: $5 = A(-2)(4) - \frac{1}{3}(-2)(4) + \frac{1}{6}(4) + \frac{1}{18}(-2)$

$5 = A(-8) + \frac{8}{3} + \frac{2}{3} - \frac{1}{9}$

$\frac{45}{9} - \frac{30}{9} + \frac{1}{9} = A(-8)$

$\frac{16}{9} = A(-8)$

$A = \frac{-2}{9}$

Step 5: $$\frac{2x + 3}{x^4 - 9x^2} = \frac{-\frac{2}{9}}{x} - \frac{\frac{1}{3}}{x^2} + \frac{\frac{1}{6}}{x - 3} + \frac{\frac{1}{18}}{x + 3}$$

8.6 Systems of Nonlinear Equations

Historical Problem

1. We wish to solve

$$\begin{cases} x^2 + y^2 = 100 & (1) \\ x = \frac{3}{4}y & (2) \end{cases}$$

This is readily done by substitution, since we know from (2) that $x = \frac{3}{4}y$. Substituting this expression for x into (1) produces:

$$x^2 + y^2 = 100$$
$$\left[\frac{3}{4}y\right]^2 + y^2 = 100$$
$$\frac{9}{16}y^2 + y^2 = 100$$
$$\frac{25}{16}y^2 = 100$$
$$y^2 = \frac{16}{25}(100)$$
$$y = \pm\sqrt{\frac{16(100)}{25}}$$
$$y = \pm\frac{(4)(10)}{5}, \quad \text{or} \quad y = \pm 8$$

Now determine x from (2): $x = \frac{3}{4}y$

If $y = -8$, $x = \frac{3}{4}(-8) = -6$

If $y = 8$, $x = \frac{3}{4}(8) = 6$

Thus, we have two solutions: $x = -6$ and $y = -8$;
$x = 6$ and $y = 8$

Each solution checks, as you can verify.

Exercises

1. $\begin{cases} y = x^2 + 1 \\ y = x + 1 \end{cases}$

$$x^2 + 1 = x + 1$$
$$x^2 - x = 0$$
$$x(x - 1) = 0$$
$$x = 0, x = 1$$
$$x = 0, y = 1; x = 1, y = 2$$

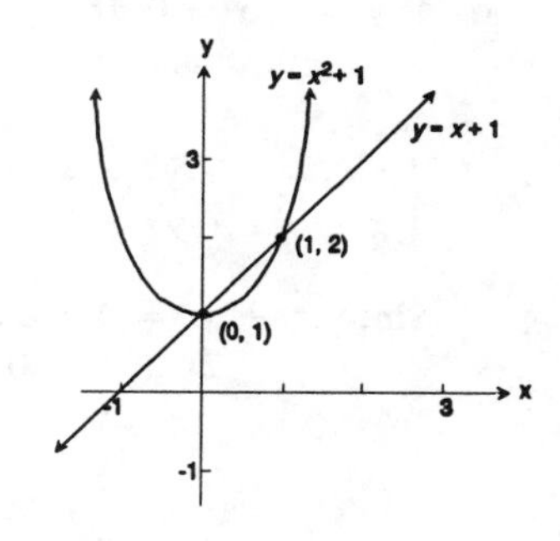

3. $\begin{cases} y = \sqrt{36 - x^2} \\ y = 8 - x \end{cases}$

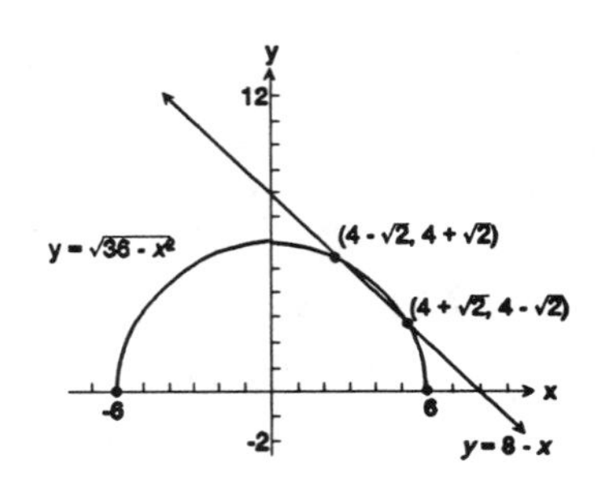

$$\sqrt{36 - x^2} = 8 - x$$
$$36 - x^2 = 64 - 16x + x^2$$
$$2x^2 - 16x + 28 = 0$$
$$x^2 - 8x + 14 = 0$$
$$x = \frac{8 \pm \sqrt{64 - 56}}{2}$$
$$x = \frac{8 \pm 2\sqrt{2}}{2}$$
$$x = 4 \pm \sqrt{2}$$
$$x = 8 - x$$
$$x = 8 - (4 \pm \sqrt{2})$$
$$y = 4 \mp \sqrt{2}$$

$\left(4 + \sqrt{2}, 4 - \sqrt{2}\right)$ and $\left(4 - \sqrt{2}, 4 + \sqrt{2}\right)$

5. $\begin{cases} y = \sqrt{x} \\ y = 2 - x \end{cases}$

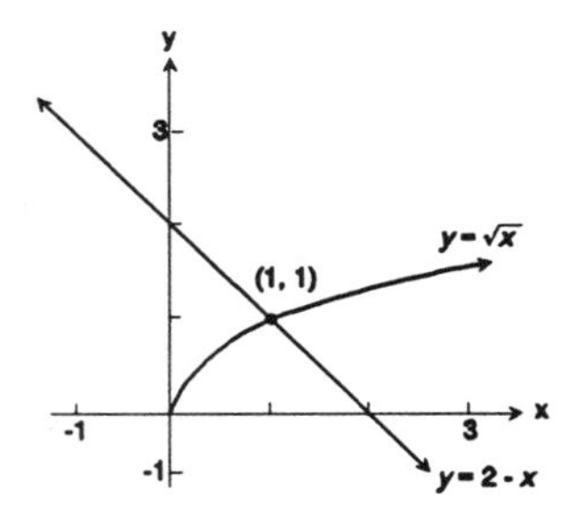

$$\sqrt{x} = 2 - x$$
$$x = 4 - 4x + x^2$$
$$x^2 - 5x + 4 = 0$$
$$(x - 1)(x - 4) = 0$$

If $x = 1$, then $y = 1$.

$(1, 1)$

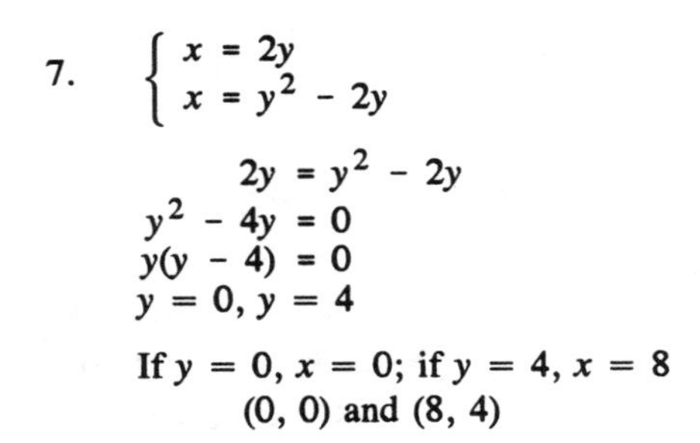

7. $\begin{cases} x = 2y \\ x = y^2 - 2y \end{cases}$

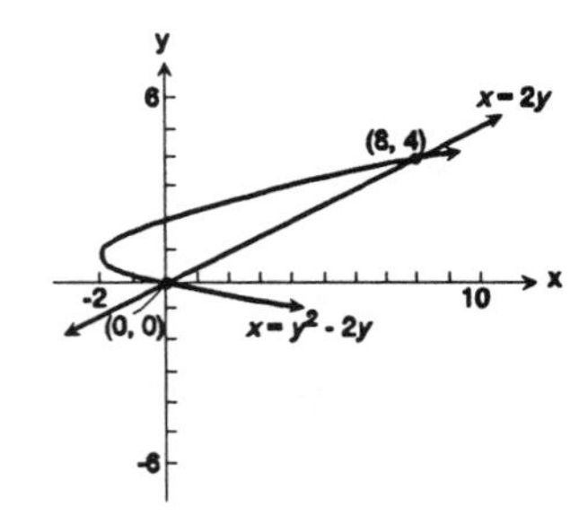

$$2y = y^2 - 2y$$
$$y^2 - 4y = 0$$
$$y(y - 4) = 0$$
$$y = 0, y = 4$$

If $y = 0$, $x = 0$; if $y = 4$, $x = 8$

$(0, 0)$ and $(8, 4)$

9. $\begin{cases} x^2 + y^2 = 4 \\ x^2 + 2x + y^2 = 0 \end{cases}$

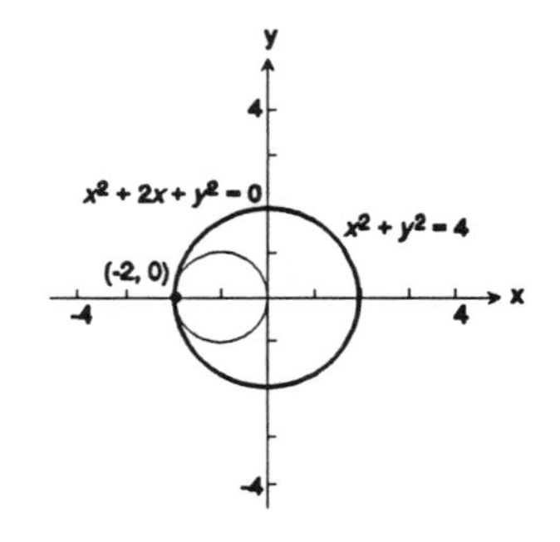

Since $x^2 + y^2 = 4$, we use substitution:

$$2x + 4 = 0$$
$$2x = -4$$
$$x = -2$$
$$x^2 + y^2 = 4$$
$$(-2)^2 + y^2 = 4$$
$$4 + y^2 = 4$$
$$y^2 = 0$$
$$y = 0$$

11. $\begin{cases} y = 3x - 5 & (1) \\ x^2 + y^2 = 5 & (2) \end{cases}$

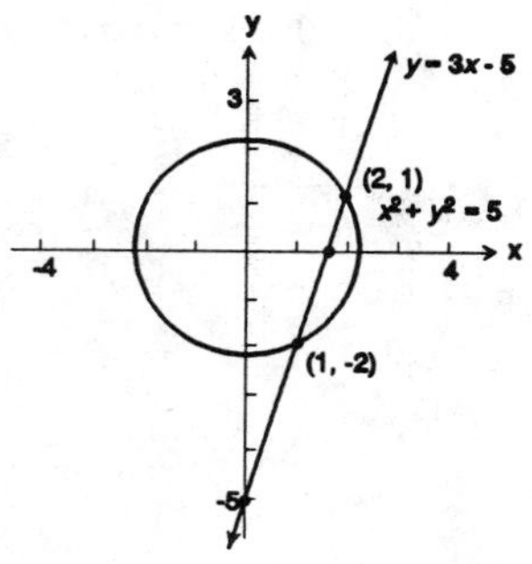

(a) First we graph (1): $y = 3x - 5$

Since both x and y appear only to the first power, this represents a straight line. An easy way to graph a line is to find the x-intercept and y-intercept, and then draw a straight line.

For the x-intercept, set $y = 0$:

$$0 = 3x - 5$$
$$5x = 3x$$
$$x = \frac{5}{3}$$

For the y-intercept, set $x = 0$:

$$y = 3(0) - 5$$
$$y = -5$$

(b) We now graph equation (2): $x^2 + y^2 = 5$

This is the equation of a circle, centered at (0, 0), with $r^2 = 5$, so that $r = \sqrt{5} \approx 2.24$.

(c) Now solve the system (i.e., find the points of intersection). We cannot use elimination, since both x and y are ***squared*** in one equation, but not in the other. Instead, we use substitution:

$$y = 3x - 5 \qquad (1)$$

Now substitute this expression into (2):

$$x^2 + y^2 = 5$$
$$x^2 + (3x - 5)^2 = 5$$
$$x^2 + 9x^2 - 30x + 25 = 5$$
$$10x^2 - 30x + 20 = 0$$
$$10(x^2 - 3x + 2) = 0$$
$$10(x - 1)(x - 2) = 0$$
$$x - 1 = 0 \qquad x - 2 = 0$$
$$x = 1 \qquad x = 2$$

Now $y = 3x - 5$, so we have:

$$y = 3(1) - 5 \quad \text{or} \quad y = 3(2) - 5$$
$$y = -2 \quad \text{or} \quad y = 1$$

We have two possible solutions: $x = 1, x = -2$

and $x = 2, y = 1$

Both solutions check, as you can verify.

13. $\begin{cases} x^2 + y^2 = 4 & (1) \\ y^2 - x = 4 & (2) \end{cases}$

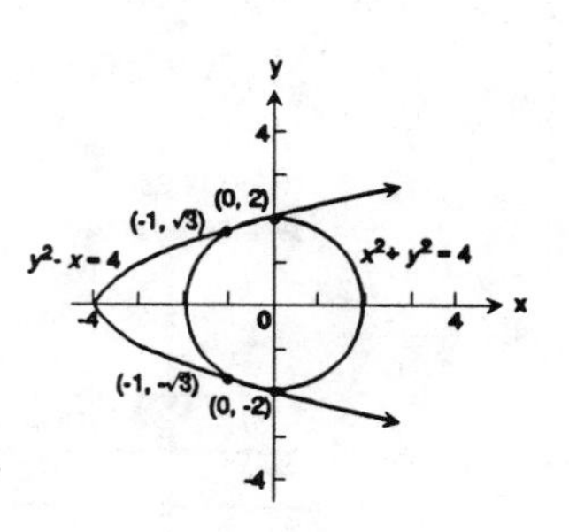

(a) To graph (1): $x^2 + y^2 = 4$

Notice that it is a circle of radius 2, centered at (0, 0).

(b) Now graph (2): $y^2 - x = 4$

$$-x = -y^2 + 4$$
$$x = y^2 - 4$$

This may be graphed by finding points and connecting them with a smooth curve:

y	$x = y^2 - 4$	Point (x, y)
−2	0	(0, −2)
−1	−3	(−3, −1)
0	−4	(−4, 0)
1	−3	(−3, 1)
2	0	(0, 2)

(c) Now solve the system:

$$\begin{cases} x^2 + y^2 = 4 & (1) \\ y^2 - x = 4 & (2) \end{cases}$$

$$\begin{cases} x^2 + y^2 = 4 & (1) \\ -y^2 + x = -4 & (2) \quad \text{Multiply by } -1. \end{cases}$$

$$\begin{cases} x^2 + y^2 = 4 & (1) \\ x^2 + x = 0 & (2) \quad \text{Replace (2) by (1) + (2).} \end{cases}$$

Now solve (2) for x:

$$x^2 + x = 0 \quad (2)$$
$$x(x + 1) = 0$$
$$x = 0 \text{ or } x + 1 = 0$$
$$x = -1$$

Now use (1) to find y: If $x = 0$, then

$$x^2 + y^2 = 4 \quad (1)$$
$$0 + y^2 = 4$$
$$y^2 = 4$$
$$y = \pm 2$$

If $x = -1$, then

$$x^2 + y^2 = 4 \quad (1)$$
$$1 + y^2 = 4$$
$$y^2 = 3$$
$$y = \pm\sqrt{3}$$

We have four possible solutions:

$x = 0$ and $y = -2$;
$x = 0$ and $y = 2$;
$x = -1$ and $y = -\sqrt{3}$;
$x = -1$ and $y = \sqrt{3}$

Let's check these:

For $x = 0$, $y = \pm 2$:

$$0^2 + (\pm 2)^2 = 0 + 4 = 4 \quad (1)$$
$$(\pm 2)^2 - 0 = 4 - 0 = 4 \quad (2)$$

For $x = -1$, $y = \pm\sqrt{3}$:

$$(-1)^2 + (\pm\sqrt{3})^2 = 1 + 3 = 4 \quad (1)$$
$$(\pm\sqrt{3})^2 - (-1) = 3 + 1 = 4 \quad (2)$$

Thus, all four possibilities check.

15. $\begin{cases} xy = 4 & (1) \\ x^2 + y^2 = 8 & (2) \end{cases}$

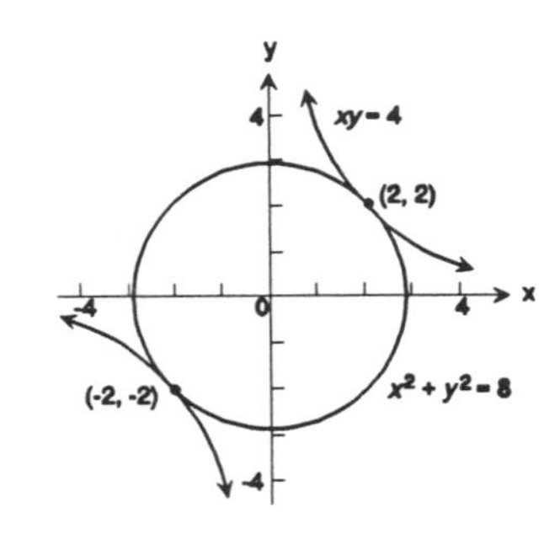

(a) To graph (1):

$$xy = 4$$
$$y = \frac{4}{x}$$

Make a table:

x	$y = \frac{4}{x}$	Point (x, y)
0	Undef.	no y-intercept
-2	-2	$(-2, -2)$
-1	-4	$(-1, -4)$
1	4	$(1, 4)$
2	2	$(2, 2)$

(b) For (2), $x^2 + y^2 = 8$ is a circle centered at $(0, 0)$ with radius $r = \sqrt{8} \approx 2.83$.

(c) Solve the system: $\begin{cases} xy = 4 & (1) \\ x^2 + y^2 = 8 & (2) \end{cases}$

Solve (1) for y: $xy = 4$ (1)

$$y = \frac{4}{x}$$

Substitute this into (2): $x^2 + y^2 = 8$

$$x^2 + \left[\frac{4}{x}\right]^2 = 8$$

$$x^2 + \frac{16}{x^2} = 8$$

$$x^4 + 16 = 8x^2 \qquad \text{Multiply both sides by } x^2.$$

$$x^4 - 8x^2 + 16 = 0 \qquad \text{This is a disguised quadratic.}$$

Let $u = x^2$ (so that $u^2 = x^4$): $u^2 - 8u + 16 = 0$

$$(u - 4)(u - 4) = 0$$

Therefore, $u = 4$, or $x^2 = 4$

$$x = \pm 2$$

If $x = 2$, then: $y = \frac{4}{x}$ (from (1))

$$y = \frac{4}{x} = 2$$

If $x = -2$, then: $y = \frac{4}{x}$

$$y = -2$$

We have two possible solutions: $x = 2$ and $y = 2$;
$x = -2$ and $y = -2$

both of which check.

17. $\begin{cases} x^2 + y^2 = 4 & (1) \\ y = x^2 - 9 & (2) \end{cases}$

(a) To graph (1), $x^2 + y^2 = 4$, we have a circle of radius 2, centered at (0, 0).

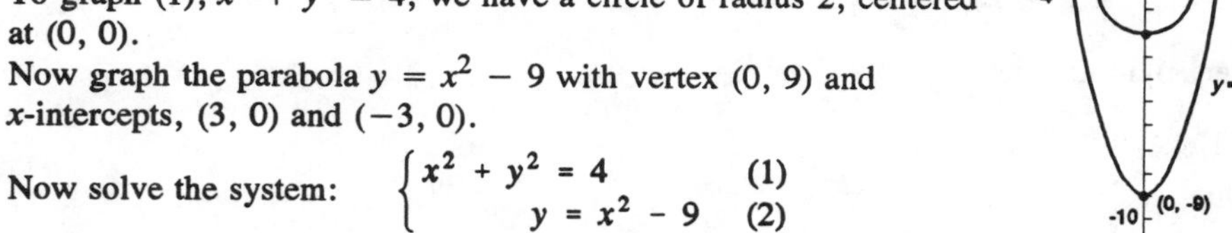

(b) Now graph the parabola $y = x^2 - 9$ with vertex (0, 9) and x-intercepts, (3, 0) and (−3, 0).

(c) Now solve the system: $\begin{cases} x^2 + y^2 = 4 & (1) \\ y = x^2 - 9 & (2) \end{cases}$

Let's substitute expression (2) into (1):

$$x^2 + (x^2 - 9)^2 = 4$$

$$x^2 + x^4 - 18x^2 + 81 = 4$$

$$x^4 - 17x^2 + 77 = 0 \qquad \text{This is quadratic.}$$

Use the quadratic formula: $x^2 = \dfrac{17 \pm \sqrt{289 - 4(77)}}{2}$

$$x^2 = \frac{17 \pm \sqrt{289 - 308}}{2} = \frac{17 \pm \sqrt{-19}}{2}$$

No real solution. Inconsistent.

19. 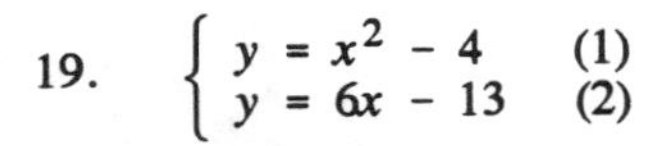

$$\begin{cases} y = x^2 - 4 & (1) \\ y = 6x - 13 & (2) \end{cases}$$

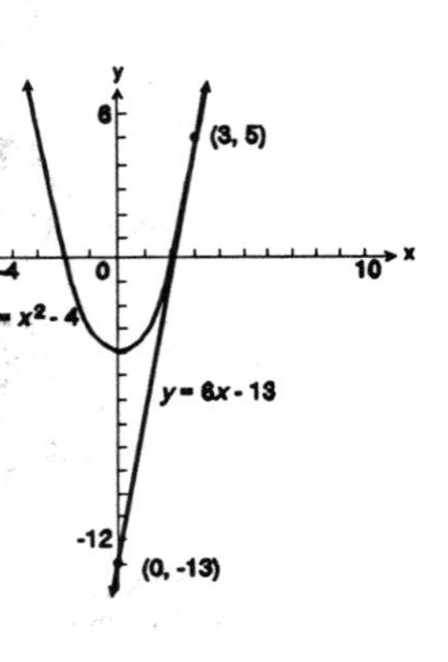

(a) Graph (1): $y = x^2 - 4$
This is a parabola with y-intercept (vertex), $(0, -4)$ and x-intercepts, $(-2, 0)$, $(2, 0)$.

(b) Now graph the line $y = 6x - 13(2)$. First find the intercepts and then draw a straight line through them.

For the x-intercept, set $y = 0$:
$$0 = 6x - 13$$
$$13 = 6x$$
$$\frac{13}{6} = x$$

For the y-intercept, set $x = 0$:
$$y = 6(0) - 13$$
$$y = -13$$

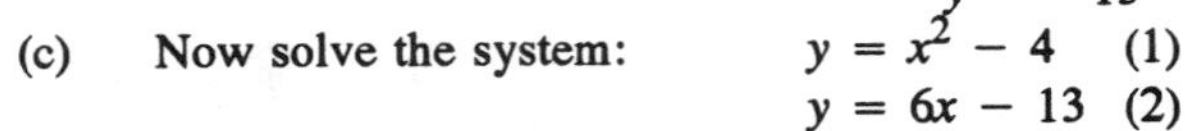

(c) Now solve the system:
$$y = x^2 - 4 \quad (1)$$
$$y = 6x - 13 \quad (2)$$

Substituting equation (2) into equation (1):
$$6x - 13 = x^2 - 4$$
$$x^2 - 6x + 9 = 0$$
$$(x - 3)^2 = 0$$
$$x = 3$$

Now, back-substitute this result into (2):
$$y = 6(3) - 13$$
$$y = 5$$

We have ***one*** solution: $x = 3$ and $y = 5$

Now to check:
$$\begin{cases} 5 = (3)^2 - 4 = 5 & (1) \\ 5 = 6(3) - 13 = 5 & (2) \end{cases}$$

The solution checks.

21.
$$\begin{cases} 2x^2 + y^2 = 18 & (1) \\ xy = 4 & (2) \end{cases}$$

Solve equation (2) for y:
$$xy = 4$$
$$y = \frac{4}{x}$$

Substitute this into (1):
$$2x^2 + y^2 = 18$$
$$2x^2 + \left[\frac{4}{x}\right]^2 = 18$$
$$2x^2 + \frac{16}{x^2} = 18$$
$$2x^4 + 16 = 18x^2 \qquad \text{Multiply both sides by } x^2.$$
$$x^4 + 8 = 9x^2 \qquad \text{Divide all terms by 2.}$$
$$x^4 - 9x^2 + 8 = 0 \qquad \text{This is a disguised quadratic.}$$

Let $u = x^2$:
$$u^2 - 9u + 8 = 0$$
$$(u - 1)(u - 8) = 0$$

Therefore, $u = 1$ or $u = 8$, so that $x^2 = 1$ or $x^2 = 8$, and
$$x = -1, 1, -\sqrt{8}, \sqrt{8}$$

If $x = -1$, then $y = \dfrac{4}{x} = -4$

If $x = 1$, then $y = \dfrac{4}{x} = 4$

If $x = -\sqrt{8} = -2\sqrt{2}$, then $y = \dfrac{4}{x} = \dfrac{4}{-2\sqrt{2}} = -\sqrt{2}$

If $x = \sqrt{8} = 2\sqrt{2}$, then $y = \sqrt{2}$

We have four *possible* solutions: $x = -1$ and $y = -4$;
$x = 1$ and $y = 4$;
$x = -2\sqrt{2}$ and $y = -\sqrt{2}$;
and $x = 2\sqrt{2}$ and $y = \sqrt{2}$.

All four check, as you may verify.

23. $\begin{cases} y = 2x + 1 & (1) \\ 2x^2 + y^2 = 1 & (2) \end{cases}$

Solve for y in (1): $y = 2x + 1$

Substitute this expression for y into (2):

$$\begin{aligned} 2x^2 + y^2 &= 1 \\ 2x^2 + (2x + 1)^2 &= 1 \\ 2x^2 + 4x^2 + 4x + 1 &= 1 \\ 6x^2 + 4x &= 0 \\ 2x + (3x + 2) &= 0 \end{aligned}$$

So we have $x = 0$ or $x = -\frac{2}{3}$.

If $x = 0$, then:

$$\begin{aligned} y &= 2x + 1 \quad \text{from (1)} \\ y &= 2(0) + 1 \\ y &= 1 \end{aligned}$$

If $x = -\frac{2}{3}$, then:

$$\begin{aligned} y &= 2x + 1 \\ y &= 2\left(-\frac{2}{3}\right) + 1 \\ y &= -\frac{1}{3} \end{aligned}$$

We have two possible solutions: $x = 0$ and $y = 1$
and $x = -\frac{2}{3}$ and $y = -\frac{1}{3}$

Let's check these:

For $x = 0$, $y = 1$: $\begin{cases} 1 = 2(0)+1 & (1) \\ 2(0)^2 + (1) = 1 & (2) \end{cases}$

For $x = -\frac{2}{3}$ and $y = -\frac{1}{3}$: $-\frac{1}{3} = 2\left(-\frac{2}{3}\right) + 1$ (1)

$2\left(-\frac{2}{5}\right)^2 + \left(-\frac{1}{3}\right)^2 = \frac{8}{9} + \frac{1}{9} = 1$ (2)

Both solutions check.

25. $\begin{cases} x + y + 1 = 0 & (1) \\ x^2 + y^2 + 6y - x = -5 & (2) \end{cases}$

We could add (1) and (2) and eliminate the x-terms, but we would still have an x^2-term. Instead, let's solve (1) for x:

$$\begin{aligned} x + y + 1 &= 0 \\ x &= -y - 1 \end{aligned}$$

Now substitute this expression for x into (2):

$$\begin{aligned} x^2 + y^2 + 6y - x &= -5 \\ (-y - 1)^2 + y^2 + 6y - (-y - 1) &= -5 \\ y^2 + 2y + 1 + y^2 + 6y + y + 1 &= -5 \quad \text{This is a quadratic.} \\ 2y^2 + 9y + 7 &= 0 \\ (2y + 7)(y + 1) &= 0 \end{aligned}$$

So we have $y = \frac{-7}{2}$ or $y = -1$.

If $y = \frac{-7}{2}$, then: $x = -y - 1$ from (1)

$$x = \frac{7}{2} - 1 = \frac{5}{2}$$

If $y = -1$, then: $x = -y - 1$

$$x = 1 - 1 = 0$$

We have two possible solutions: $x = \frac{5}{2}$ and $y = \frac{-7}{2}$; $x = 0$ and $y = -1$

You can verify that both of these solutions check in the original system of equations.

27. $\begin{cases} 4x^2 - 3xy + 9y^2 = 15 & (1) \\ 2x + 3y = 5 & (2) \end{cases}$

Solve for x in (2):

$$\begin{aligned} 2x + 3y &= 5 \\ 2x &= -3y + 5 \\ x &= \frac{-3}{2}y + \frac{5}{2} \end{aligned}$$

Substitute this expression into (1):

$$\begin{aligned} 4x^2 - 3xy + 9y^2 &= 15 \\ 4\left[\frac{-3}{2}y + \frac{5}{2}\right]^2 - 3\left[\frac{-3}{2}y + \frac{5}{2}\right]y + 9y^2 &= 15 \\ 4\left[\frac{9}{4}y^2 - \frac{15}{2}y + \frac{25}{4}\right] - 3\left[\frac{-3}{2}y^2 + \frac{5}{2}y\right] + 9y^2 &= 15 \\ 9y^2 - 30y + 25 + \frac{9}{2}y^2 - \frac{15}{2}y + 9y^2 &= 15 \\ 18y^2 - 60y + 50 + 9y^2 - 15y + 18y^2 &= 30 \\ 45y^2 - 75y + 20 &= 0 \\ 9y^2 - 15y + 4 &= 0 \\ (3y - 1)(3y - 4) &= 0 \end{aligned}$$

So $y = \frac{1}{3}$ or $y = \frac{4}{3}$

If $y = \frac{1}{3}$, then: $x = \frac{-3}{2}y + \frac{5}{2}$ from (2)

$$x = \frac{-3}{2}\left[\frac{1}{3}\right] + \frac{5}{2} = 2$$

If $y = \frac{4}{3}$, then: $x = \frac{-3}{2}y + \frac{5}{2}$

$$x = \frac{-3}{2}\left[\frac{4}{3}\right] + \frac{5}{2} = \frac{1}{2}$$

The two possible solutions are: $x = 2$ and $y = \frac{1}{3}$; $x = \frac{1}{2}$ and $y = \frac{4}{3}$

Both of these solutions check.

29. $\begin{cases} x^2 - 4y^2 + 7 = 0 & (1) \\ 3x^2 + y^2 - 31 = 0 & (2) \end{cases}$

Here we can use the method of elimination:

$$\begin{cases} x^2 - 4y^2 + 7 = 0 & (1) \\ 12x^2 + 4y^2 - 124 = 0 & (2) \text{ Multiply both sides by 4.} \end{cases}$$

$$\begin{cases} x^2 - 4y^2 + 7 = 0 & (1) \\ 13x^2 - 117 = 0 & (2) \end{cases}$$ Replace (2) by (1) + (2).

Now solve (2) for x:

$$13x^2 - 117 = 0$$
$$13x^2 = 117$$
$$x^2 = 9$$

so, $x = 3$ or $x = -3$

If $x = 3$, then:

$$x^2 - 4y^2 + 7 = 0 \quad (1)$$
$$9 - 4y^2 + 7 = 0$$
$$-4y^2 = -16$$
$$y^2 = 4$$
$$y = \pm 2$$

If $x = -3$, then:

$$x^2 - 4y^2 + 7 = 0$$
$$9 - 4y^2 + 7 = 0$$
$$-4y^2 = -16$$
$$y^2 = 4$$
$$y = \pm 2$$

We have four possible solutions: $x = 3, y = 2$; $x = 3, y = -2$; $x = -3, y = 2$; $x = -3, y = -2$

We can check all four at once. If $x = \pm 3$, and $y = \pm 2$:

$$\begin{cases} x^2 - 4y^2 + 7 = 9 - 16 + 7 = 0 & (1) \\ 3x^2 + y^2 - 31 = 27 + 4 - 31 = 0 & (2) \end{cases}$$

All four solutions check.

31. $$\begin{cases} 7x^2 - 3y^2 + 5 = 0 & (1) \\ 3x^2 + 5y^2 - 12 = 0 & (2) \end{cases}$$

Let's solve (2) for x^2:

$$3x^2 + 5y^2 - 12 = 0$$
$$3x^2 = -5y^2 + 12$$
$$x^2 = \frac{-5}{3}y^2 + 4$$

Substitute this expression for x^2 into (1):

$$7x^2 - 3y^2 + 5 = 0$$
$$7\left[\frac{-5}{3}y^2 + 4\right] - 3y^2 + 5 = 0$$
$$\frac{-35}{3}y^2 + 28 - 3y^2 + 5 = 0$$
$$-35y^2 + 84 - 9y^2 + 15 = 0 \quad \text{Multiply by 3.}$$
$$-44y^2 = -99$$
$$y^2 = \frac{99}{44} = \frac{9}{4}$$
$$y = \pm\frac{3}{2}$$

If $y = \frac{3}{2}$, then:

$$x^2 = \frac{-5}{3}y^2 + 4 \quad \text{from (2)}$$
$$x^2 = \frac{-5}{3}\left[\frac{9}{4}\right] + 4$$
$$x^2 = \frac{1}{4}$$
$$x = \pm\frac{1}{2}$$

If $y = \frac{-3}{2}$, then: $x^2 = \frac{-5}{3}y^2 + 4$

$$x^2 = \frac{-5}{3}\left[\frac{9}{4}\right] + 4, \text{ as above}$$

$$x = \pm\frac{1}{2}$$

We have four solutions: $x = \frac{1}{2}, y = \frac{3}{2}; x = -\frac{1}{2}, y = \frac{3}{2};$

$x = \frac{1}{2}, y = \frac{-3}{2}; x = \frac{-1}{2}, y = \frac{-3}{2}$

(All of them check.)

33. $\begin{cases} x^2 + 2xy = 10 & (1) \\ 3x^2 - xy = 2 & (2) \end{cases}$

$\begin{cases} x^2 + 2xy = 10 & (1) \\ 6x^2 - 2xy = 4 & (2) \end{cases}$ Multiply by 2.

$\begin{cases} x^2 + 2xy = 10 & (1) \\ 7x^2 = 14 & (2) \end{cases}$ Replace (2) by (1) + (2).

From (2): $x^2 = 2$

$$x = \pm\sqrt{2}$$

If $x = \sqrt{2}$, then: $x^2 + 2xy = 10$ (1)

$$2 + 2\sqrt{2}\,y = 10$$

$$2\sqrt{2}\,y = 8$$

$$y = \frac{8}{2\sqrt{2}} = 2\sqrt{2}$$

If $x = -\sqrt{2}$, then: $x^2 + 2xy = 10$

$$2 - 2\sqrt{2}\,y = 10$$

$$-2\sqrt{2}\,y = 8$$

$$y = \frac{8}{-2\sqrt{2}} = -2\sqrt{2}$$

The two possible solutions are: $x = \sqrt{2}, y = 2\sqrt{2}$ and $x = -\sqrt{2}, y = -2\sqrt{2}$
Both solutions check.

35. $\begin{cases} 2x^2 + y^2 = 2 & (1) \\ x^2 - 2y^2 + 8 = 0 & (2) \end{cases}$

$\begin{cases} 4x^2 + 2y^2 = 4 & (1) \\ x^2 - 2y^2 = -8 & (2) \end{cases}$ Multiply equation (1) by 2.

Adding equation (1) and (2): $5x^2 = -4$

$$x^2 = \frac{-4}{5}$$

No real solution. The system is inconsistent.

37. $\begin{cases} x^2 + 2y^2 = 16 & (1) \\ 4x^2 - y^2 = 24 & (2) \end{cases}$

$\begin{cases} x^2 + 2y^2 = 16 & (1) \\ 8x^2 - 2y^2 = 48 & (2) \end{cases}$ Multiply by 2.

$\begin{cases} x^2 + 2y^2 = 16 & (1) \\ 9x^2 = 64 & (2) \end{cases}$ Replace (2) by (1) + (2).

From (2): $x^2 = \frac{64}{9}$

$x = \pm\frac{8}{3}$

If $x = \pm\frac{8}{3}$, then:

$$x^2 + 2y^2 = 16 \quad (1)$$
$$\frac{64}{9} + 2y^2 = 16$$
$$2y^2 = \frac{144}{9} - \frac{64}{9} = \frac{80}{9}$$
$$y^2 = \frac{40}{9}$$

and $y = \pm\sqrt{\frac{40}{9}} = \pm\frac{-2\sqrt{10}}{3}$

We have four solutions, all of which check:

$$x = \frac{8}{3},\ y = \frac{2\sqrt{10}}{3};\ x = \frac{8}{3},\ y = \frac{-2\sqrt{10}}{3};$$
$$x = \frac{-8}{3},\ y = \frac{2\sqrt{10}}{3};\ x = \frac{-8}{3},\ y = \frac{-2\sqrt{10}}{3}$$

39. $\begin{cases} \frac{5}{x^2} - \frac{2}{y^2} + 3 = 0 & (1) \\ \frac{3}{x^2} + \frac{1}{y^2} - 7 = 0 & (2) \end{cases}$

$\begin{cases} \frac{5}{x^2} - \frac{2}{y^2} + 3 = 0 & (1) \\ \frac{6}{x^2} + \frac{2}{y^2} - 14 = 0 & (2) \end{cases}$ Multiply both sides by 2.

$\begin{cases} \frac{5}{x^2} - \frac{2}{y^2} + 3 = 0 & (1) \\ \frac{11}{x^2} - 11 = 0 & (2) \end{cases}$ Replace (2) by (1) + (2).

Now, from (2): $\frac{11}{x^2} = 11$

$$11 = 11x^2$$
$$x^2 = 1$$
$$x = \pm 1$$

If $x = 1$ or $x = -1$, we have:

$$\frac{5}{x^2} - \frac{2}{y^2} + 3 = 0 \quad (9((1)$$

$$\frac{5}{1} - \frac{2}{y^2} + 3 = 0 \quad \text{(since } x = \pm 1)$$

$$8 = \frac{2}{y^2}$$

$$8y^2 = 2$$

$$y^2 = \frac{1}{4}$$

$$y = \pm\frac{1}{2}$$

We have four solutions, all of which can be checked:

$$x = 1, y = \frac{1}{2};\ x = 1, y = \frac{-1}{2};\ x = -1, y = \frac{1}{2};\ x = -1, y = \frac{-1}{2}$$

41. $$\begin{cases} \dfrac{1}{x^4} + \dfrac{6}{y^4} = 6 & (1) \\ \dfrac{2}{x^4} - \dfrac{2}{y^4} = 19 & (2) \end{cases}$$

Let's make a substitution: let $u = \dfrac{1}{x^4}$ and $v = \dfrac{1}{y^4}$. Then we have:

$$\begin{cases} u + 6v = 6 & (1) \\ 2u - 2v = 19 & (2) \end{cases}$$

$$\begin{cases} u + 6v = 6 & (1) \\ 6u - 6v = 57 & (2) \end{cases} \quad \text{Multiply each term by 3.}$$

$$\begin{cases} u + 6v = 6 & (1) \\ 7u = 63 & (2) \end{cases} \quad \text{Replace (2) by (1) + (2).}$$

From (2): $7u = 63$

$u = 9$

Then, from (1): $u + 6v = 6$

$9 + 6v = 6$

$6v = -3$

$v = \dfrac{-1}{2}$

Recall that $u = \dfrac{1}{x^4}$ and $v = \dfrac{1}{y^4}$. Thus: $\dfrac{1}{y^4} = v = \dfrac{-1}{2}$

$2 = -y^4$

$y^4 = -2$

This is impossible, since an even power can never be negative, so we have no real solution. The system is inconsistent.

43. $$\begin{cases} x^2 - 3xy + 2y^2 = 0 & (1) \\ x^2 + xy = 6 & (2) \end{cases}$$

Since both x and y appear in two terms of equation (1), we cannot use elimination. We can either solve for y in (2), or factor (1). Let's try the latter approach:

$$x^2 - 3xy + 2y^2 = 0 \qquad (1)$$

$$(x - y)(x - 2y) = 0$$

Therefore, either $x = y$ or $x = 2y$.

If $x = y$, then from (2):

$$x^2 + xy = 6$$
$$y^2 + y^2 = 6 \quad (x = y)$$
$$y^2 = 3$$
$$y = \pm\sqrt{3}$$

We know $x = y$, so we have two possible solutions: $x = \sqrt{3}, y = \sqrt{3}$ and $x = -\sqrt{3}, y = -\sqrt{3}$.

If $x = 2y$, then:

$$x^2 + xy = 6 \quad (2)$$
$$4y^2 + 2y^2 = 6 \quad (x = 2y)$$
$$y^2 = 1$$
$$y = \pm 1$$

Then, from $x = 2y$, we have two more possible solutions: $x = 2, y = 1$ and $x = -2, y = -1$

You can check that all four solutions are valid.

45. $\begin{cases} xy - x^2 = -3 & (1) \\ 3xy - 4y^2 = 2 & (2) \end{cases}$

Solve equation (1) for y:

$$xy - x^2 = -3 \quad (1)$$
$$xy = x^2 - 3$$
$$y = \frac{x^2 - 3}{x}, \text{ provided } x \neq 0.$$

We will check the possibility $x = 0$ later.

Substitute $y = \dfrac{x^2 - 3}{x}$ into (2):

$$3xy - 4y^2 = 2 \quad (2)$$
$$3x\left[\frac{x^2 - 3}{x}\right] - 4\left[\frac{x^2 - 3}{x}\right]^2 = 2$$
$$3x^2 - 9 - \frac{4(x^2 - 3)^2}{x^2} = 2$$
$$3x^4 - 9x^2 - 4(x^4 - 6x^2 + 9) = 2x^2 \quad \text{Multiply by } x^2.$$
$$3x^4 - 9x^2 - 4x^4 + 24x^2 - 36 - 2x^2 = 0$$
$$-x^4 + 13x^2 - 36 = 0$$
$$x^4 - 13x^2 + 36 = 0$$
$$(x^2 - 4)(x^2 - 9) = 0$$
$$x^2 = 4 \quad \text{or} \quad x^2 = 9$$
$$x = \pm 2 \quad \text{or} \quad x = \pm 3$$

If $x = 2$, then $y = \dfrac{x^2 - 3}{x}$ or $y = \dfrac{4 - 3}{2} = \dfrac{1}{2}$

If $x = -2$, then $y = \dfrac{x^2 - 3}{x}$ or $y = \dfrac{4 - 3}{-2} = \dfrac{-1}{2}$

If $x = 3$, then $y = \dfrac{x^2 - 3}{x}$ or $y = \dfrac{9 - 3}{3} = 2$

If $x = -3$, then $y = \dfrac{x^2 - 3}{x}$ or $y = \dfrac{9 - 3}{-3} = -2$

This gives us four possible solutions. What about the possibility $x = 0$? From (1) we would have $0 = -3$. Thus, there are no solutions for which $x = 0$. Each of the four solutions above checks:

$$x = 2, y = \frac{1}{2};\ x = -2, y = \frac{-1}{2};\ x = 3, y = 2;\ x = -3, y = -2$$

47. $\begin{cases} x^3 - y^3 = 26 & (1) \\ x - y = 2 & (2) \end{cases}$

Solve (2) for x: $\quad x - y = 2$

$$x = y + 2$$

and substitute this into (1):

$$\begin{aligned} x^3 - y^3 &= 26 \\ (y + 2)^3 - y^3 &= 26 \\ y^3 + 6y^2 + 12y + 8 - y^3 &= 26 \\ 6y^2 + 12y - 18 &= 0 \\ y^2 + 2y - 3 &= 0 \\ (y + 3)(y - 1) &= 0 \end{aligned}$$

so $y = -3$ or $y = 1$.

If $y = -3$, then $x = y + 2 = -1$

If $y = 1$, then $x = y + 2 = 3$.

The two solutions (both of which check) are $x = -1, y = -3$, and $x = 3, y = 1$.

49. $\begin{cases} y^2 + y + x^2 - x - 2 = 0 & (1) \\ y + 1 + \dfrac{x - 2}{y} = 0 & (2) \end{cases}$

$\begin{cases} y^2 + y + x^2 - x - 2 = 0 & (1) \\ -y^2 - y - x + 2 = 0 & (2) \end{cases}$ Multiply both sides by $-y$.

$\begin{cases} y^2 + y + x^2 - x - 2 = 0 & (1) \\ x^2 - 2x = 0 & (2) \end{cases}$ Replace (2) by (1) + (2).

From (2):

$$\begin{aligned} x - 2x &= 0 \\ x(x - 2) &= 0 \\ x = 0 \ \text{ or } \ x &= 2 \end{aligned}$$

Now use (1):

If $x = 0$:

$$\begin{aligned} y^2 + y + x^2 - x - 2 &= 0 \qquad (1) \\ y^2 + y - 2 &= 0 \\ (y + 2)(y - 1) &= 0 \end{aligned}$$

and $\quad y = -2 \ \text{ or } y = 1$

If $x = 2$:

$$\begin{aligned} y^2 + y + x^2 - x - 2 &= 0 \qquad (1) \\ y^2 + y + 4 - 2 - 2 &= 0 \\ y^2 + y &= 0 \\ y(y + 1) &= 0 \end{aligned}$$

and $\quad y = 0 \ \text{ or } y = -1$

Thus, we have four possible solutions:

$x = 0, y = -2$; $x = 0, y = 1$; $x = 2, y = 0$; and $x = 2, y = -1$

We see from equation (2) that y cannot be zero. That eliminates the third solution above. The others can be checked:

$x = 0, y = -2$

$x = 0, y = 1$

$x = 2, y = -1$

51. $\begin{cases} \log_x y = 3 \\ \log_x (4y) = 5 \end{cases}$

$\begin{cases} y = x^3 \\ 4y = x^5 \end{cases}$

$$4x^3 = x^5$$
$$x^5 - 4x^3 = 0$$
$$x^3(x^2 - 4) = 0$$
$$x = 0, x = -2, x = 2$$

0 and -2 are extraneous (the base of a logarithm must be positive). Thus, $x = 2$ and $y = 2^3 = 8$.

53. $\begin{cases} \ln x = 4 \ln y \\ \log_3 x = 2 + 2 \log_3 y \end{cases}$

$\begin{cases} \ln x - 4 \ln y = 0 \\ \log_3 x - 2 + 2 \log_3 y = 2 \end{cases}$

$\begin{cases} \ln \dfrac{x}{y^4} = 0 \\ \log_3 \dfrac{x}{y^2} = 2 \end{cases}$

$\begin{cases} e^0 = \dfrac{x}{y^4} \\ 3^2 = \dfrac{x}{y^2} \end{cases}$

$\begin{cases} 1 = \dfrac{x}{y^4} \\ 9 = \dfrac{x}{y^2} \end{cases}$

$\begin{cases} y^4 = x \\ 9y^2 = x \end{cases}$

$$y^4 = 9y^2$$
$$y^4 - 9y^2 = 0$$
$$y^2(y^2 - 9) = 0$$
$$y^2(y - 3)(y + 3) = 0$$

$x = 81, y = 3$

55. $\begin{cases} x + 2y = 0 & (1) \\ (x - 1)^2 + (y - 1)^2 = 5 & (2) \end{cases}$

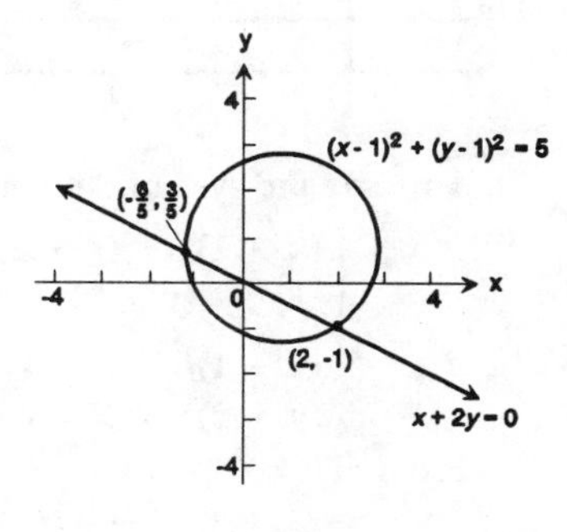

The line $x + 2y = 0$ can be rewritten as $y = \dfrac{-1}{2}x$,

so we see it is a line with slope $m = \dfrac{-1}{2}$ which passes through $(0, 0)$.

Also, $(x - 1)^2 + (y - 1)^2 = 5$ is a circle with center $C(1, 1)$ and radius $r = \sqrt{5} \approx 2.24$.

To find the points of intersection, substitute $y = \dfrac{-1}{2}x$ (from (1)) into (2):

$$(x - 1)^2 + (y - 1)^2 = 5 \quad (2)$$

$$(x - 1)^2 + \left[\frac{-1}{2}x - 1\right]^2 = 5$$

$$x^2 - 2x + 1 + \frac{1}{4}x^2 + x + 1 = 5$$

$$\frac{5}{4}x^2 - x - 3 = 0$$

$$5x^2 - 4x - 12 = 0 \quad \text{Multiply by 4.}$$

$$(5x + 6)(x - 2) = 0$$

$$\text{so } x = \frac{-6}{5}, \text{ or } x = 2$$

If $x = \frac{-6}{5}$, then $y = \frac{-1}{2}x = \frac{3}{5}$

If $x = 2$, then $y = \frac{-1}{2}x = -1$

Both solutions check, so we have two points of intersection: $\left(\frac{-6}{5}, \frac{3}{5}\right)$ and $(2, -1)$

57. $\begin{cases} (x - 1)^2 + (y + 2)^2 = 4 & (1) \\ y^2 + 4y - x + 1 = 0 & (2) \end{cases}$

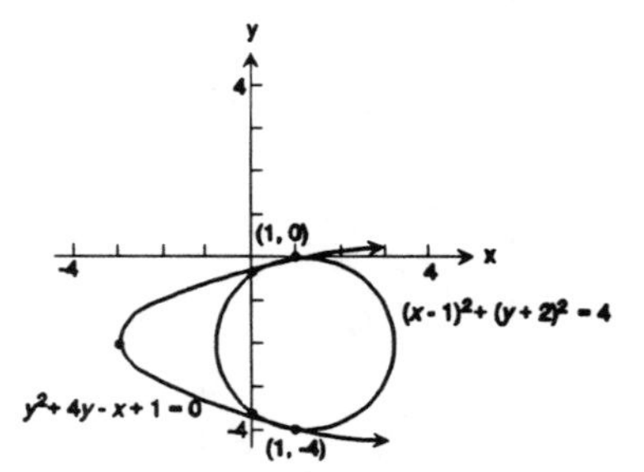

Equation (1) is a circle with center at $C(1, -2)$ and radius $r = 2$. To graph the parabola, (2), let's complete the square and then plot points:

$$y^2 + 4y + \underline{\quad} = x - 1 + \underline{\quad}$$

$$y^2 + 4y + 4 = x - 1 + 4$$

$$(y + 2)^2 = x + 3$$

$$x = (y + 2)^2 - 3$$

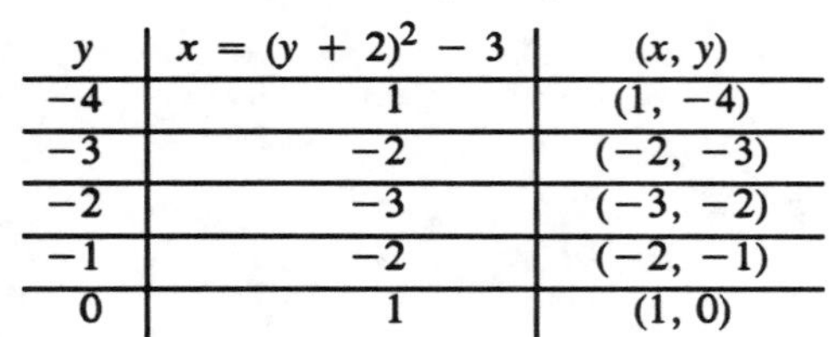

y	$x = (y + 2)^2 - 3$	(x, y)
-4	1	$(1, -4)$
-3	-2	$(-2, -3)$
-2	-3	$(-3, -2)$
-1	-2	$(-2, -1)$
0	1	$(1, 0)$

To solve the system, let's use the standard form we obtained for the parabola:

$$\begin{cases} (x - 1)^2 + (y + 2)^2 = 4 & (1) \\ (y + 2)^2 - 3 = x & (2) \end{cases}$$

$$\begin{cases} (x - 1)^2 + 3 = 4 - x & (1) \quad \text{Replace (1) by (1) - (2)} \\ (y + 2)^2 - 3 = x & (2) \end{cases}$$

From (1):

$$(x - 1)^2 + 3 = 4 - x$$

$$x - 2x + 1 + 3 = 4 - x$$

$$x^2 - x = 0$$

$$x(x - 1) = 0$$

$$x = 0 \text{ or } x = 1$$

If $x = 0$, then by (2):

$$(y + 2)^2 - 3 = x$$

$$(y + 2)^2 - 3 = 0$$

$$(y + 2)^2 = 3$$

$$y + 2 = \pm\sqrt{3}$$

$$y = -2 \pm \sqrt{3}$$

If $x = 1$, then:

$$\begin{aligned} (y + 2)^2 - 3 &= 1 \\ (y + 2)^2 &= 4 \\ y + 2 &= \pm 2 \\ y &= -2 \pm 2 \end{aligned}$$

i.e., $y = 0$ or $y = -4$

We have ***four*** points of intersection: $(0, -2 - \sqrt{3})$, $(0, -2 + \sqrt{3})$, $(1, 0)$, and $(1, -4)$

59. 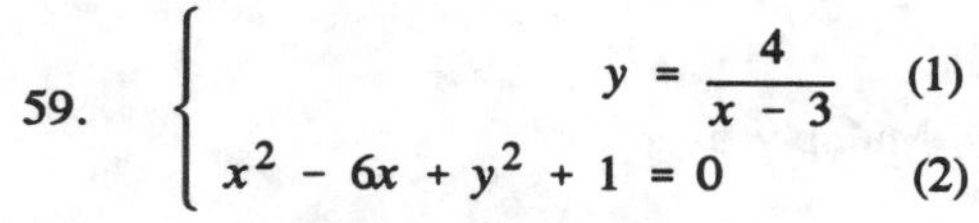

$$\begin{cases} y = \dfrac{4}{x - 3} & (1) \\ x^2 - 6x + y^2 + 1 = 0 & (2) \end{cases}$$

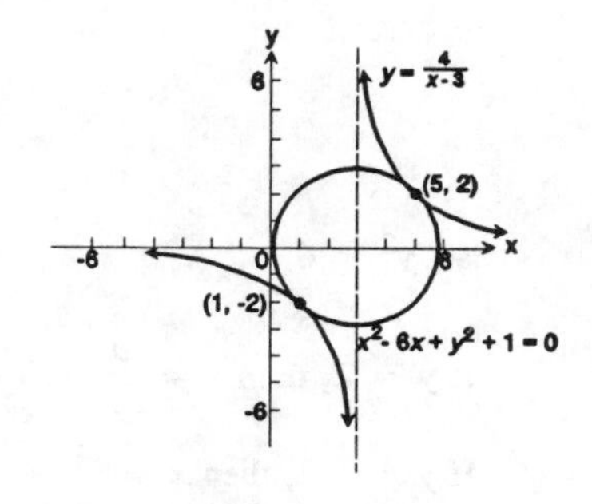

To graph (1), plot points:

x	$y = \dfrac{4}{x - 3}$	(x, y)
6	$\dfrac{4}{3}$	$\left(6, \dfrac{4}{3}\right)$
4	$\dfrac{4}{1}$	(4, 4)
3	Undefined	
2	$\dfrac{4}{-1}$	(2, −4)
0	$\dfrac{4}{-3}$	$\left(0, -\dfrac{4}{3}\right)$

To graph (2), complete the square:

$$\begin{aligned} x^2 - 6x + \_\_\_ + y^2 &= -1 + \_\_\_ \\ x^2 - 6x + 9 + y^2 &= -1 + 9 \\ (x - 3)^2 + y^2 &= 8 \end{aligned}$$

This is a circle with center at $C(3, 0)$ and radius $r = \sqrt{8} \approx 2.83$.

Now solve the system:

$$\begin{cases} y = \dfrac{4}{x - 3} & (1) \\ x^2 - 6x + y^2 + 1 = 0 & (2) \end{cases}$$

We could substitute $y = \dfrac{4}{x - 3}$ into (2), but then we obtain a complicated denominator, $(x - 3)^2 = x^2 - 6x + 9$.

Instead, let's solve (1) for x: $y(x - 3) = 4$

$$\begin{aligned} c - 3 &= \frac{4}{y} \\ x &= \frac{4}{y} + 3 \end{aligned}$$

and substitute this into (2):

$$x^2 - 6x + y^2 + 1 = 0$$
$$\left(\frac{4}{y} + 3\right)^2 - 6\left(\frac{4}{y} + 3\right) + y^2 + 1 = 0$$
$$\frac{16}{y^2} + \frac{24}{y} + 9 - \frac{24}{y} - 18 + y^2 + 1 = 0$$
$$\frac{16}{y^2} + y^2 - 8 = 0$$
$$16 + y^4 - 8y^2 = 0 \quad \text{Multiply by } y^2.$$
$$y^4 - 8y^2 + 16 = 0$$
$$(y^2 - 4)(y^2 - 4) = 0$$

so $y^2 = 4$

$y = \pm 2$

If $y = 2$, then $x = \frac{4}{y} + 3 = 5$.

If $y = -2$, then $x = \frac{4}{y} + 3 = 1$.

We have two points of intersection: (5, 2) and (1, −2)

61. Graph $y_1 = x^{2/3}$ and $y_2 = e^{-x}$. The only intersection point is (0.48, 0.61).

63. Graph $y_1 = \sqrt[3]{2 - x^2}$ and $y_2 = \frac{4}{x^3}$. The only intersection point is (−1.64, −0.89).

65. Graph $y_1 = \sqrt[4]{12 - x^4} = (12 - x^4)^{1/4}$, $y_2 = -(12 - x^4)^{1/4}$, $y_3 = \sqrt{\frac{2}{x}}$, and $y_4 = -\sqrt{\frac{2}{x}}$.
The intersection points are (0.58, 1.85); (1.81, 1.05); (0.58, −1.85) and (1.81, −1.05).

67. Graph $y_1 = \frac{2}{x}$ and $y_2 = \ln x$. The only intersection point is (2.34, 0.85).

69. Let x and y be the two numbers. Then we have:

$$\begin{cases} x - y = 2 & (1) \\ x^2 + y^2 = 10 & (2) \end{cases}$$

Solve (1) for x: $x - y = 2$

$x = y + 2$

Substitute this into (2):

$$(y + 2)^2 + y^2 = 10 \ (2)$$
$$y^2 + 4y + 4 + y^2 = 10$$
$$2y^2 + 4y - 6 = 0$$
$$2(^2 + 2y - 3) = 0$$
$$2(y + 3)(y - 1) = 0$$
$$y = -3 \text{ or } y = 1$$

If $y = -3$, then $x = y + 2 = -3 + 2 = -1$

If $y = 1$, then $x = y + 2 = 1 + 2 = 3$

The two numbers are 3 and 1 or −3 and −1.

$3^2 + 1^2 = 10$ and $(-3)^2 + (-1)^2 = 10$

so both solutions check.

71. Let x and y be the two numbers. Then we have:

$$\begin{cases} xy = 4 & (1) \\ x^2 + y^2 = 8 & (2) \end{cases}$$

Solve for y in (1): $y = \dfrac{4}{x}$

Then from (2):

$$\begin{aligned} x^2 + y^2 &= 8 \\ x^2 + \left[\frac{4}{x}\right]^2 &= 8 \\ x^2 + \frac{16}{x^2} &= 8 \\ x^4 + 16 &= 8x^2 \qquad \text{Multiply by } x^2. \\ x^4 - 8x^2 + 16 &= 0 \\ (x^2 - 4)(x^2 - 4) &= 0 \\ x^2 &= 4 \\ x &= \pm 2 \end{aligned}$$

Recall that $y = \dfrac{4}{x}$.

If $x = 2$, then $y = 2$. If $x = -2$, then $y = -2$.

Thus, we have ***two*** pairs of numbers: 2 and 2, or -2 and -2. Both pairs solve the original problem.

73. Let x and y be the two numbers. Then:

$$\begin{cases} x - y = xy & (1) \\ \dfrac{1}{x} + \dfrac{1}{y} = 5 & (2) \end{cases}$$

$$\begin{cases} x - y = xy & (1) \\ y + x = 5xy & (2) \end{cases} \qquad \text{Multiply by } xy.$$

$$\begin{cases} x - y = xy & (1) \\ 2x = 6xy & (2) \end{cases} \qquad \text{Replace (2) by (1) + (2).}$$

From (2) we have:

$$\begin{aligned} 2x - 6xy &= 0 \\ x - 3xy &= 0 \\ x(1 - 3y) &= 0 \end{aligned}$$

So either $x = 0$ or $y = \dfrac{1}{3}$. But from the original equation (2), we see that $x = 0$ is impossible.

Thus, $y = \dfrac{1}{3}$.

From (1): $x - y = xy$ (1)

$$\begin{aligned} x - \frac{1}{3} &= \frac{1}{3}x \qquad \left[y = \frac{1}{3}\right] \\ \frac{2}{3}x &= \frac{1}{3} \\ x &= \frac{1}{2} \end{aligned}$$

Thus, the two numbers are $\dfrac{1}{2}$ and $\dfrac{1}{3}$.

75. $\begin{cases} \dfrac{a}{b} = \dfrac{2}{3} \\ a + b = 10 \end{cases}$

$$\frac{10 - b}{b} = \frac{2}{3}$$
$$3(10 - b) = 2b$$
$$30 = 5b$$
$$b = 6$$
$$a = 4$$

$a + b = 10$, $b - a = 2$

Ratio of $a + b$ to $b - a$ is $\dfrac{10}{2} = 5$.

77. ℓ = length, w = width

$\begin{cases} 2\ell + 2w = 16 & (1) \quad \text{(Perimeter)} \\ \ell w = 15 & (2) \quad \text{(Area)} \end{cases}$

$2\ell + 2\left[\dfrac{15}{\ell}\right] = 16$ Substituting $w = \dfrac{15}{\ell}$ from (2) into equation (1).

$2\ell^2 - 16\ell + 30 = 0$ Multiply by ℓ.

$2(\ell^2 - 8\ell + 15) = 0$

$2(\ell - 5)(\ell - 3) = 0$

$\ell = 5$ $\quad\quad$ $\ell = 3$

$w = \dfrac{15}{\ell} = 3$ $\quad\quad$ $w = \dfrac{15}{3} = 5$

The dimensions are 3 inches × 5 inches, or 5 inches × 3 inches.

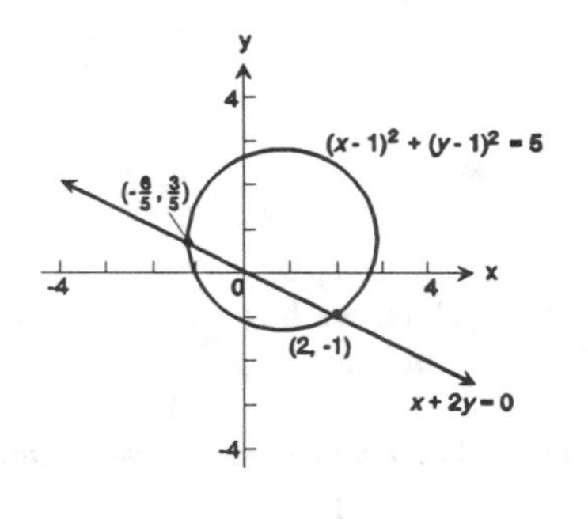

79. $\begin{cases} 2\pi r_1 + 2\pi r_2 = 12\pi \\ \pi r_1^2 + \pi r_2^2 = 20\pi \end{cases}$

$2\pi(r_1 + r_2) = 12\pi$ $\quad$ (1)

$$r_1 + r_2 = \frac{12\pi}{2\pi}$$
$$r_1 + r_2 = 6$$
$$r_1 = 6 - r_2$$

$\pi(r_1^2 + r_2^2) = 20\pi$ $\quad$ (2)

$$r_1^2 + r_2^2 = \frac{20\pi}{\pi}$$
$$r_1^2 + r_2^2 = 20$$

We substitute equation (1) into (2):

$$r_1^2 + r_2^2 = 20$$
$$(6 - r_2)^2 + r_2^2 = 20$$
$$36 - 12r_2 + r_2^2 + r_2^2 = 20$$
$$2r_2^2 - 12r_2 + 16 = 0$$
$$2(r_2 - 6r_2 + 8) = 0$$
$$2(r_2 - 2)(r_2 - 4) = 0$$

$r_2 - 2 = 0$ or $r_2 - 4 = 0$

$r_2 = 2$ or $r_2 = 4$

$r_1 = 6 - r_2$ $\quad\quad$ $r_1 = 6 - r_2$

$r_1 = 6 - 2$ $\quad\quad$ $r_1 = 6 - 4$

$r_1 = 4$ $\quad\quad$ $r_1 = 2$

The radius of each circle is 2 cm and 4 cm.

81. We know that rate × time = distance

The tortoise takes a total of $9 + 3 = 12$ ***minutes*** longer to complete the 21 meter race. In hours, 12 minutes is $\frac{12}{60}$ hour which is $\frac{1}{5}$ hour. The hare runs at a speed 0.5 meter per hour faster, which means the tortoise runs 0.5 meter per hour slower than the hare. Hence,

$$(r - 0.5)\left[t + \frac{1}{5}\right] = 21 \leftrightarrow \text{Tortoise}$$

$$rt = 21 \leftrightarrow \text{Hare}$$

$$r = \frac{21}{t}$$

$$\left[r - \frac{1}{2}\right]\left[t + \frac{1}{5}\right] = 21$$

$$\left[\frac{21}{t} - \frac{1}{2}\right]\left[t + \frac{1}{5}\right] = 21$$

$$21 + \frac{21}{5t} - \frac{t}{2} - \frac{1}{10} = 21$$

$$210t + 42 - 5t^2 - t = 210t$$

$$-5t^2 - t + 42 = 0$$

$$5t^2 + t - 42 = 0$$

$$(5t - 14)(t + 3) = 0$$

$$5t - 14 = 0 \qquad t = -3$$

$$t = \frac{14}{5} \qquad \text{Ignore because time cannot be negative}$$

$$= 2\frac{4}{5} \text{ hours}$$

$$r = \frac{21}{t} = \frac{21}{\frac{14}{5}} = \frac{105}{14} = 7.5\frac{\text{meters}}{\text{hour}}$$

$$r - 0.5 = 7.5 - 0.5 = 7\frac{\text{meters}}{\text{hour}}$$

The tortoise runs at an average speed of 7 meters per hour. The hare runs at an average speed of 7.5 meters per hour.

83. Let the piece of cardboard have width x, and length y. Then its area is $A = xy$. From the drawing, we see that the width of the box will be $w = x - 4$, its length will be $\ell = y - 4$, and its height will be $h = 2$. Therefore, its volume is

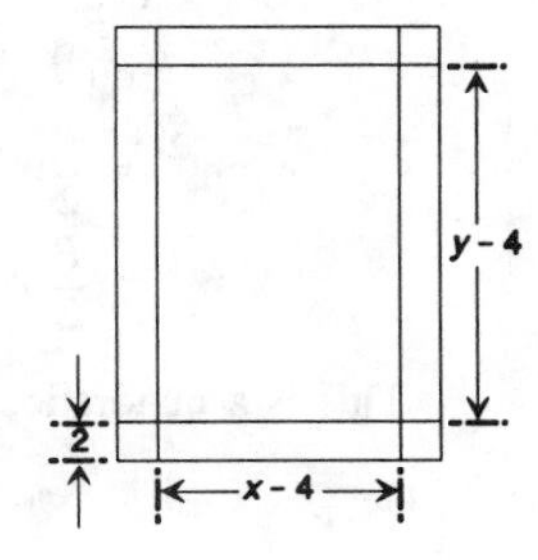

$$V = \ell \cdot w \cdot h = 2(x - 4)(y - 4)$$

We have: $\begin{cases} xy = 216 & (1) \\ 2(x - 4)(y - 4) = 224 & (2) \end{cases}$

Solve (1) for y: $y = \frac{216}{x}$

Then from (2): $2(x - 4)(y - 4) = 224$

$$(2x - 8)\left[\frac{216}{x} - 4\right] = 224$$

$$432 - 8x - \frac{1728}{x} + 32 = 224$$

$$432x - 8x^2 - 1728 + 32x = 224x \qquad \text{Multiply by } x.$$

$$-8x^2 + 240x - 1728 = 0$$

$$x^2 - 30x + 216 = 0 \qquad \text{Divide by } -8.$$

With some trial and error, this can be factored: $(x - 12)(x - 18) = 0$

so $x = 12$ or $x = 18$

If $x = 12$, then $y = \dfrac{216}{x} = 18$

If $x = 18$, then $y = \dfrac{216}{x} = 12$

Thus, the cardboard should be 12 centimeters by 18 centimeters.

85.

$$\begin{aligned}
x^2 + y^2 &= 4500 \\
3x + 3y + (x - y) &= 300 \\
4x + 2y &= 300 \\
2y &= 300 - 4x \\
y &= 150 - 2x \\
x^2 + (150 - 2x)^2 &= 4500 \\
x^2 + 22{,}500 - 600x + 4x^2 &= 4500 \\
5x^2 - 600x + 22{,}500 &= 4500 \\
5x^2 - 600x + 18{,}000 &= 0 \\
5(x^2 - 120x + 3600) &= 0 \\
5(x - 60)^2 &= 0 \\
x &= 60 \text{ ft.} \\
y &= 150 - 2x \\
y &= 150 - 2(60) \\
y &= 30 \text{ ft.}
\end{aligned}$$

87. Let ℓ and w be the length and width of a rectangle. Then the area, A, and perimeter, P, are given by:

$$\begin{cases} A = \ell w & (1) \\ P = 2\ell + 2w & (2) \end{cases}$$

We solve for ℓ and w, treating A and P as constants:

From (2), $\quad 2\ell = P - 2w$

$$\ell = \frac{P}{2} - w$$

Then from (1),

$$\begin{aligned}
\ell &= A \qquad (1) \\
\left(\frac{P}{2} - 3\right) w &= A \\
\frac{P}{2}w - w^2 &= A \\
-w^2 + \frac{P}{2}w - A &= 0 \\
w^2 - \frac{P}{2}w + A &= 0
\end{aligned}$$

This is a quadratic, in w, with $a = 1$, $b = \dfrac{-P}{2}$ and $c = A$.

$$\text{Then, } w = \frac{-b \pm \sqrt{b^2 - 4ac}}{2a} = \frac{\dfrac{P}{2} \pm \sqrt{\dfrac{P^2}{4} - 4A}}{2} = \frac{\dfrac{P}{2} \pm \sqrt{\dfrac{P^2 - 16A}{4}}}{2} = \frac{\dfrac{P}{2} - \dfrac{\sqrt{P^2 - 16A}}{2}}{2}$$

or $\quad w = \dfrac{P \pm \sqrt{P^2 - 16A}}{4}$ Multiplying top and bottom by 2.

Recall that $\ell = \dfrac{P}{2} - w$.

If $w = \dfrac{P \pm \sqrt{P^2 - 16A}}{4}$, then we have

$$\ell = \frac{P}{2} - \frac{P}{4} + \frac{\sqrt{P^2 - 16A}}{4} \text{ or } \ell = \frac{P}{4} - \frac{\sqrt{P^2 - 16A}}{4} = \frac{P - \sqrt{P^2 - 16A}}{4}$$

This gives a length ***smaller*** than the width, so we reject this solution.

If $w = \dfrac{P - \sqrt{P^2 - 16A}}{4}$, then we obtain $\ell = \dfrac{P + \sqrt{P^2 - 16A}}{4}$, and these are the proper formulas.

89.
$$\begin{aligned} M^2 - 4(2M - 4) &= 0 \\ M^2 - 8M + 16 &= 0 \\ (M - 4)^2 &= 0 \\ M &= 4 \end{aligned}$$

Now, using the point-slope equation with slope 4 and the point, (2, 4),

$$\begin{aligned} y - 4 &= 4(x - 2) \\ y &= 4x - 8 + 4 \\ y &= 4x - 4 \end{aligned}$$

91. Refer to Problem 89 for the method to be used. We want the system

$$\begin{cases} y = x^2 + 2 & (1) \\ y = mx + b & (2) \end{cases}$$

to have one solution. Substitute $y = mx + b$ into the first equation:

$$\begin{aligned} mx + b &= x^2 + 2 \\ -x^2 + mx + b - 2 &= 0 \\ x^2 - mx + 2 - b &= 0 \qquad \text{(a quadratic)} \end{aligned}$$

Here $A = 1$, $B = -m$ and $C = 2 - b$. The quadratic formula will produce just one solution provided $B^2 - 4AC = 0$, i.e.,

$$\begin{aligned} (-m)^2 - 4(1)(2 - b) &= 0 \\ m^2 - 8 + 4b &= 0 \qquad (1) \end{aligned}$$

We also want (1, 3) to lie on the line $y = mx + b$, so that $3 = m(1) + b$. We have two equations:

$$\begin{cases} m^2 - 8 + 4b = 0 & (1) \\ 3 = m + b & (2) \end{cases}$$

From (2), $b = 3 - m$, so, by (1):

$$\begin{aligned} m^2 - 8 + 4b &= 0 \qquad (1) \\ m^2 - 8 + 4(3 - m) &= 0 \\ m^2 - 8 + 12 - 4m &= 0 \\ m^2 - 4m + 4 &= 0 \\ (m - 2)(m - 2) &= 0 \end{aligned}$$

so that $m = 2$. Then $b = 3 - m - 1$, and the equation of the tangent line is: $y = mx + b$, or

$$y = 2x + 1$$

93. Refer to Problems 89 and 91 to see the method used. The system

$$\begin{cases} 2x^2 + 3y^2 = 14 & (1) \\ y = mx + b & (2) \end{cases}$$

is to have just one solution. Substitute $y = mx + b$ into the first equation:

$$\begin{aligned} 2x^2 + 3y^2 &= 14 \\ 2x^2 + 3(mx + b)^2 &= 14 \\ 2x^2 + 3(m^2x^2 + 2mbx + b^2) &= 14 \\ 2x^2 + 3m^2x^2 + 6mbx + 3b^2 - 14) &= 0 \\ (2 + 3m^2)x^2 + (6mb)x + (3b^2 - 14) &= 0 \qquad \text{(a quadratic)} \end{aligned}$$

Here $A = 2 + 3m^2$, $B = 6mb$ and $C = 3b^2 - 14$.
There will be ***one*** solution to the quadratic if $B^2 - 4AC = 0$, i.e.,

$$(6mb)^2 - 4(2 + 3m^2)(3b^2 - 14) = 0$$
$$36m^2b^2 - 4(6b^2 - 28 + 9m^2b^2 - 42m^2) = 0$$
$$36m^2b^2 - 24b^2 + 112 - 36m^2b^2 + 168m^2 = 0$$
$$-24b^2 + 112 + 168m^2 = 0$$
$$3b^2 - 14 - 21m^2 = 0 \quad \text{(1) Divide by } -8.$$

We also want (1, 2) to lie on the line $y = mx + b$, i.e, we need

$$2 = m + b \quad (2)$$

We have two equations to solve:

$$\begin{cases} 3b^2 - 14 - 21m^2 = 0 & (1) \\ 2 = m + b & (2) \end{cases}$$

From (2), $b = 2 - m$, and by (1),

$$3b^2 - 14 - 21m^2 = 0$$
$$3(2 - m)^2 - 14 - 21m^2 = 0$$
$$3(4 - 4m + m^2) - 14 - 21m^2 = 0$$
$$12 - 12m + 3m^2 - 14 - 21m^2 = 0$$
$$-18m^2 - 12m - 2 = 0$$
$$9m^2 + 6m + 1 = 0$$
$$(3m + 1)(3m + 1) = 0$$

so $m = \frac{-1}{3}$ and $b = 2 - m = \frac{7}{3}$, and the equation of the tangent line is

$$y = mx + b, \text{ or}$$
$$y = \frac{-1}{3}x + \frac{7}{3}$$

95. We want the system

$$\begin{cases} x^2 - y^2 = 3 \\ y = mx + b \end{cases}$$

to have just ***one*** solution.
Substitute $y = mx + b$ into the first equation:

$$x^2 - y^2 = 3$$
$$x^2 - (mx + b)^2 = 3$$
$$x^2 - (m^2x^2 + 2mbx - b^2) = 3$$
$$x^2 - m^2x^2 + 2mbx - b^2 = 3$$
$$(1 - m^2)x^2 + (-2mb)x + (-b^2 - 3) = 0$$

Let $A = 1 - m^2$, $B = -2mb$, $C = -b^2 - 3$
We want $B^2 - 4AC = 0$, or

$$(-2mb)^2 - 4(1 - m^2)(-b^2 - 3) = 0$$
$$4m^2b^2 - 4(-b^2 - 3 + m^2b^2 + 3m^2) = 0$$
$$4b^2 + 12 - 12m^2 = 0$$
$$b^2 + 3 - 3m^2 = 0 \quad (1)$$

We want (2, 1) to lie on the line $y = mx + b$: $1 = 2m + b$ (2)
This gives us the system:

$$\begin{cases} b^2 + 3 - 3m^2 = 0 & (1) \\ 1 = 2m + b & (2) \end{cases}$$

From (2): $b = 1 - 2m$, and we have

$$
\begin{aligned}
b^2 + 3 - 3m^2 &= 0 \qquad (1)\\
(1 - 2m)^2 + 3 - 3m^2 &= 0\\
1 - 4m + 4m^2 + 3 - 3m^2 &= 0\\
m^2 - 4m + 4 &= 0\\
(m - 2)(m - 2) &= 0
\end{aligned}
$$

so $m = 2$, $b = 1 - 2m = -3$, and the tangent line is

$y = mx + b$, or

$y = 2x - 3$

97. Solve for r_1 and r_2:

$$
\begin{cases}
r_1 + r_2 = \dfrac{-b}{a} & (1)\\[2ex]
r_1 r_2 = \dfrac{c}{a} & (2)
\end{cases}
$$

From (1),

$$r_1 = -r_2 - \frac{b}{a}, \text{ and}$$

$$r_1 r_2 = \frac{c}{a} \quad (2)$$

$$
\begin{aligned}
\left[-r_2 - \frac{b}{a}\right] r_2 &= \frac{c}{a}\\
-r_2^2 - \frac{b}{a} r_2 - \frac{c}{a} &= 0\\
a r_2^2 + b r_2 + c &= 0 \quad \text{Multiply by } -a.
\end{aligned}
$$

By the quadratic formula,

$$r_2 = \frac{-b \pm \sqrt{b^2 - 4ac}}{2a}$$

Then,

$$r_1 = -r_2 - \frac{b}{a} = -\left[\frac{-b \pm \sqrt{b^2 - 4ac}}{2a}\right] - \frac{2b}{2a} = \frac{b \mp \sqrt{b^2 - 4ac} - 2b}{2a}$$

$$= \frac{-b \mp \sqrt{b^2 - 4ac}}{2a}$$

Thus, we have one pair of numbers:

$$\frac{-b + \sqrt{b^2 - 4ac}}{2a} \text{ and } \frac{-b - \sqrt{b^2 - 4ac}}{2a}$$

99. (a)

$V = x^2 y = 9$

$A = (x + 2y)^2 = 100$

Solving for y, we get $y = \dfrac{9}{x^2}$.

Substituting into Area equation, we get

$$100 = x^2 + 4xy + 4y^2$$

$$100 = x^2 + 4x\left[\frac{9}{x^2}\right] + 4\left[\frac{81}{x^4}\right]$$

$$100 = x^2 + \frac{36}{x} + \frac{324}{x^4}$$

$$100x^4 = x^6 + 36x^3 + 324$$

$$0 = (x^6 + 36x^3 + 324) - 100x^4$$

$$0 = (x^3 + 18 - 10x^2)(x^3 + 18 - 10x^2)$$

$$x + 2y = 3.5$$

3.5 ft. by 3.5 ft.

8.7 Systems of Inequalities

1. $x \geq 0$:

Step 1: Graph the line $x = 0$ (a vertical line through the origin, i.e., the y-axis). Use a solid line, since the inequality is nonstrict.

Step 2: Choose a test point not on the line, say, the point (2, 0), which is to the right of the line.

Step 3: For this point we have $x = 2 \geq 0$, so that the inequality ***is*** satisfied.

Step 4: Therefore, we shade to the right of the line $x = 0$.

3. $x \geq 4$:

Step 1: Graph the line $x = 4$, using a solid line (for a nonstrict inequality). The x-intercept is 4.

Step 2: Choose a test point say, (5, 0), which is to the right of the line.

Step 3: For this point (5, 0), we have: $5 \geq 4$, so the inequality is satisfied.

Step 4: Therefore, we shade the region to the right of the line $x = 4$.

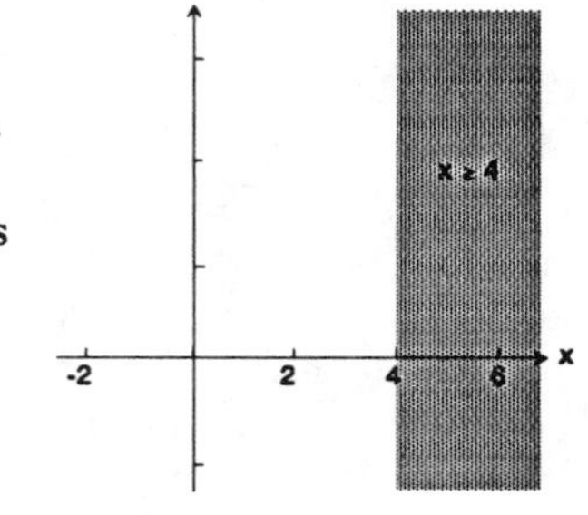

5. $2x + y \geq 6$:

Step 1: Graph $2x + y = 6$, using a ***solid*** line. The x-intercept is $x = 3$, and the y-intercept is $y = 6$.

Step 2: Choose a test point, say, (0, 0), which is below the line.

Step 3: For this point (0, 0), we have: $2x + y = 0 < 6$, so the point does ***not*** satisfy the inequality.

Step 4: Therefore, we shade the region ***above*** the line $2x + y = 6$.

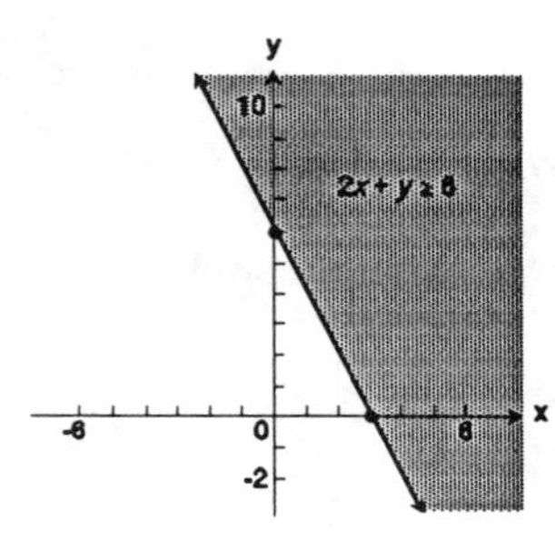

7.

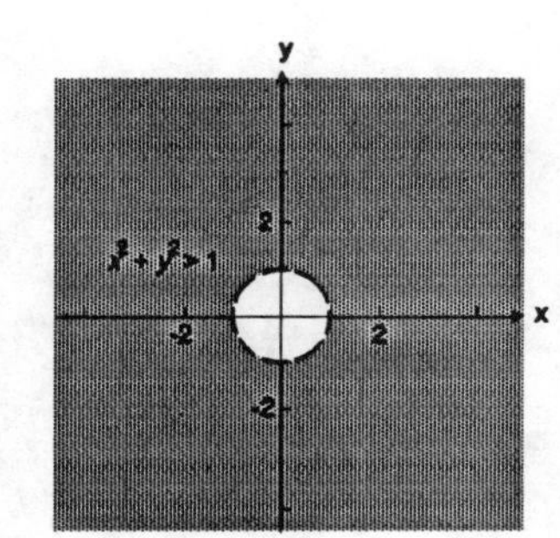

9.

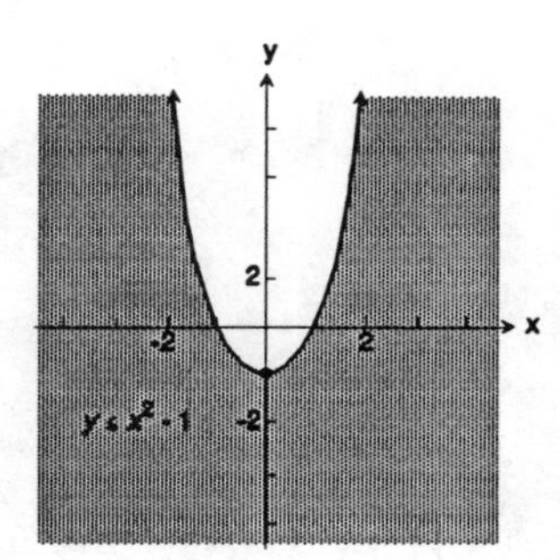

11.

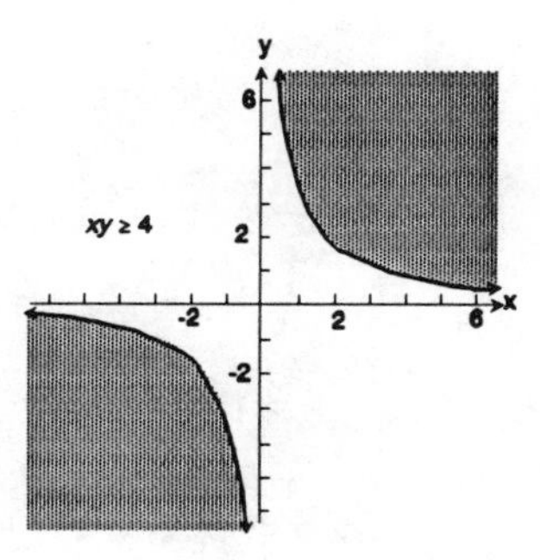

13. $\begin{cases} x + y \le 2 & (1) \\ 2x + y \ge 4 & (2) \end{cases}$

(a) $x + y \le 2$:

Step 1: Graph $x + y = 2$ with a solid line. The x-intercept is $x = 2$, and the y-intercept is $y = 2$.

Step 2: We will use (0, 0), below the line, as a test point.

Step 3: At (0, 0), $x + y = 0 < 2$, so the inequality is satisfied.

Step 4: Shade the region below the line $x + y = 2$.

(b) $2x + y \ge 4$:

Step 1: Graph $2x + y = 4$ with a solid line. The x-intercept is $x = 2$, and the y-intercept is $y = 4$.

Step 2: Use (0, 0), below the line, as a test point.

Step 3: At (0, 0), $2x + y = 0 < 4$, so the inequality is *not* satisfied.

Step 4: Shade the region above the line.

(c) Where the two shaded regions overlap is the graph of the system.

15. $\begin{cases} 2x - y \le 4 & (1) \\ 3x + 2y \ge -6 & (2) \end{cases}$

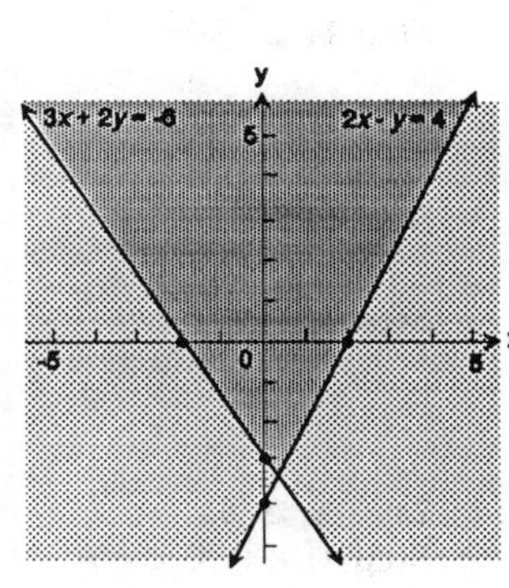

(a) $2x - y \le 4$:

Step 1: Graph $2x - y = 4$ with a solid line. The x-intercept is $x = 2$, and the y-intercept is $y = -4$.

Step 2: We will use (0, 0) as a test point (above the line).

Step 3: At (0, 0), $2x - y = 0 < 4$.

Step 4: Shade the region above the line $2x - y = 4$.

(b) $3x + 2y \ge -6$:

Step 1: Graph $3x + 2y = -6$ with a solid line. The x-intercept is $x = -2$, and the y-intercept is $y = -3$.

Step 2: Use (0, 0), above the line.

Step 3: At (0, 0), $3x + 2y = 0 > -6$.

Step 4: Shade the region above the line.

17. $\begin{cases} 2x - 3y \le 0 & (1) \\ 3x + 2y \le 6 & (2) \end{cases}$

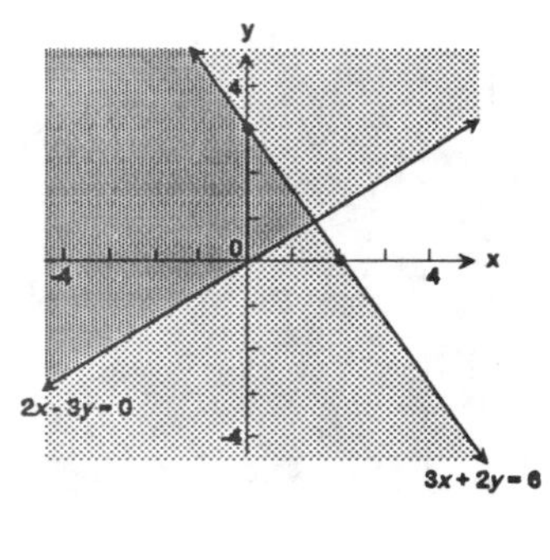

(a) $2x - 3y \le 0$:

Step 1: Graph $2x - 3y = 0$, a line through (0, 0) with slope $= \frac{2}{3}$.

Step 2: Let's use (1, 0) below the line.

Step 3: At (1, 0), $2x - 3y = 2 > 0$, so the point does not satisfy the inequality.

Step 4: Shade the region ***above*** the line $2x - 3y = 0$.

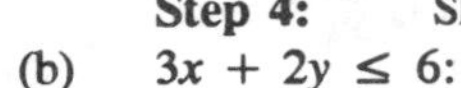

(b) $3x + 2y \le 6$:

Step 1: Graph $3x + 2y = 6$ with a solid line. The x-intercept is $x = 2$, and the y-intercept is $y = 3$.

Step 2: We can use (0, 0), below the line.

Step 3: At (0, 0), $3x + 2y = 0 < 6$.

Step 4: Shade the region ***below*** the line $3x + 2y = 6$.

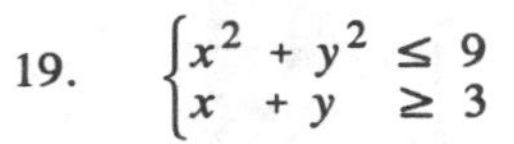

19. $\begin{cases} x^2 + y^2 \le 9 \\ x + y \ge 3 \end{cases}$

$x^2 + y^2 = 9$

$x + y = 3$

(a) $x^2 + y^2 \le 9$:

Step 1: Graph circle $x^2 + y^2 = 9$ with radius $= 3$.

Step 2: Use (0, 0) inside the circle.

Step 3: At (0, 0), $0 \le 9$, so the point does satisfy the inequality.

Step 4: Shade the region ***inside*** circle $x^2 + y^2 \le 9$.

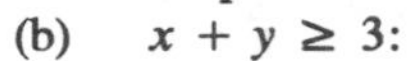

(b) $x + y \ge 3$:

Step 1: Graph solid line $x + y = 3$. The x-intercept is $x = 3$ and the y-intercept is $y = 3$.

Step 2: Use (3, 2) above the line.

Step 3: At (3, 2), $5 \ge 3$, so the point does satisfy the inequality.

Step 4: Shade the region ***above*** the line $x + y = 3$.

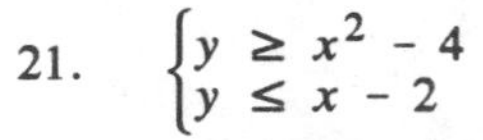

21. $\begin{cases} y \ge x^2 - 4 \\ y \le x - 2 \end{cases}$

$y = x^2 - 4$

$y = x - 2$

(a) $y \ge x^2 - 4$:

Step 1: Graph parabola $y = x^2 - 4$ with vertex (0, −4) and going upward.

Step 2: Use (0, 2) inside the parabola.

Step 3: At (0, 2), $2 \ge -4$ so the point satisfies the inequality.

Step 4: Shade the region ***inside*** the prarabola $y = x^2 - 4$

(b) $y \le x - 2$:

Step 1: Graph the line $y = x - 2$ with x-intercept $x = 2$ and y-intercept $y = -2$.

Step 2: Use (0, −3) below the line.

Step 3: At (0, −3), $-3 \le -2$ so the point satisfies the inequality.

Step 4: Shade the region ***below*** the line $y = x - 2$.

23. 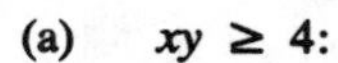
$$\begin{cases} xy \geq 4 \\ y \geq x^2 + 1 \end{cases}$$

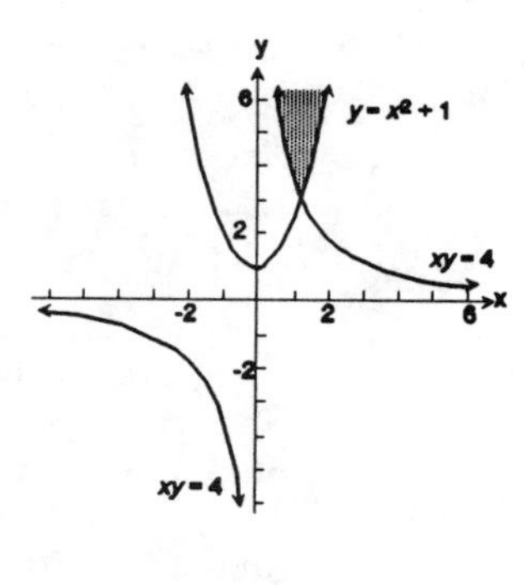

(a) $xy \geq 4$:

Step 1: Graph $xy = 4$ using solid lines.
Step 2: Use (0, 0) between parts of the graph.
Step 3: At (0, 0), $0 \geq 4$ is false.
Step 4: Shade the region ***outside*** the parts of the graph $xy = 4$.

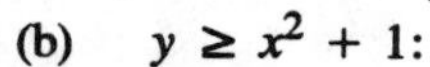
(b) $y \geq x^2 + 1$:

Step 1: Graph $y = x^2 + 1$. It is an upward parabola with vertex (0, 1).
Step 2: Use (0, 2) inside the parabola.
Step 3: At (0, 2) $2 \geq 1$ is true.
Step 4: Shade the region ***inside*** the parabola $y = x^2 + 1$.

25. 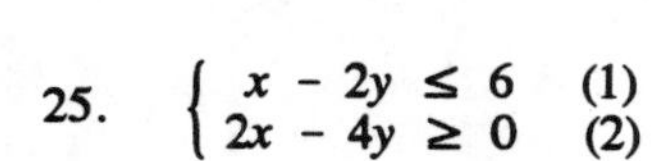
$$\begin{cases} x - 2y \leq 6 & (1) \\ 2x - 4y \geq 0 & (2) \end{cases}$$

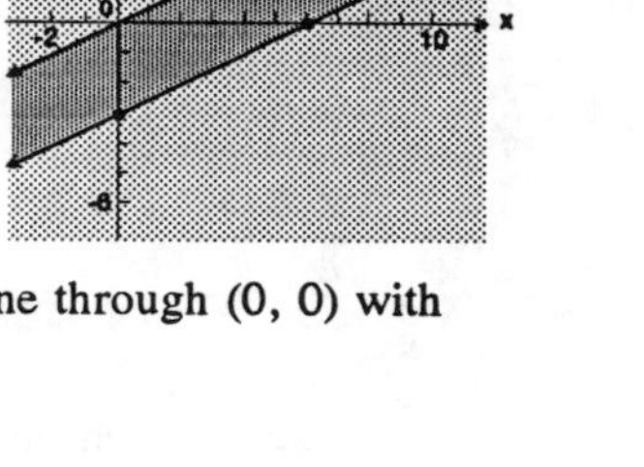

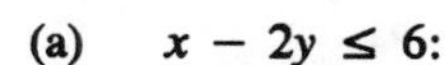
(a) $x - 2y \leq 6$:

Step 1: Graph $x - 2y = 6$, using a solid line. The x-intercept is $x = 6$; the y-intercept is $y = -3$.
Step 2: Use (0, 0) above the line.
Step 3: At (0, 0), $x - 2y = 0 < 6$.
Step 4: Shade the region ***above*** the line $x - 2y = 6$.

(b) $2x - 4y \geq 0$:

Step 1: Graph $2x - 4y = 0$ with a solid line. This is a line through (0, 0) with slope $= \frac{1}{2}$.
Step 2: Let's use (1, 0) below the line.
Step 3: At (1, 0), $2x - 4y = 2 > 0$.
Step 4: Shade the region ***below*** $2x - 4y = 0$.

27. $$\begin{cases} 2x + y \geq -2 & (1) \\ 2x + y \geq 2 & (2) \end{cases}$$

(a) $2x + y \geq -2$:

Step 1: Graph $2x + y = -2$, with a solid line. The x-intercept is $x = -1$; the y-intercept is $y = -2$.
Step 2: Use (0, 0) above the line.
Step 3: At (0, 0), $2x + y = 0 > -2$, so the inequality is satisfied.
Step 4: Shade the region above the line $2x + y = -2$.

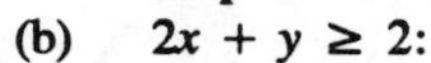
(b) $2x + y \geq 2$:

Step 1: Graph $2x + y = 2$ with a solid line. The x-intercept is $x = 1$; the y-intercept is $y = 2$.
Step 2: Use (0, 0) ***below*** the line.
Step 3: At (0, 0), $2x + y = 0 < 2$, so (0, 0) does ***not*** satisfy the inequality.
Step 4: Therefore, shade the region ***above*** the line $2x + y = 2$.

29. $\begin{cases} 2x + 3y \geq 6 & (1) \\ 2x + 3y \leq 0 & (2) \end{cases}$

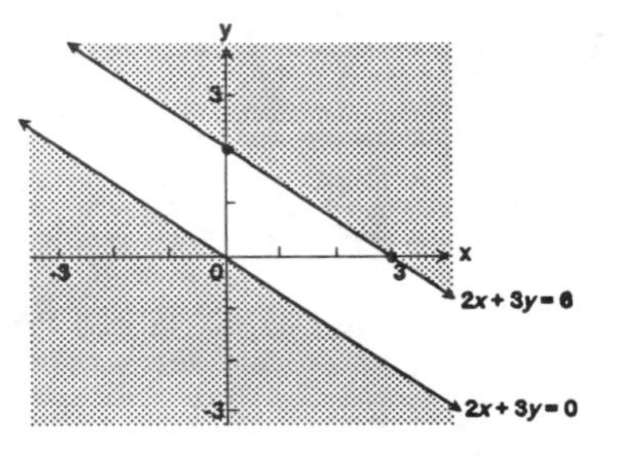

This system has no solution, because $2x + 3y$ cannot be greater than 6 ***and*** less than 0 at the same time.

(a) $2x + 3y \geq 6$:

Step 1: Graph $2x + 3y = 6$, with a solid line. The x-intercept is $x = 3$; the y-intercept is $y = 2$.

Step 2: Use (0, 0) below the line as a test point.

Step 3: At (0, 0), $2x + 3y = 0 < 6$, so the inequality is ***not*** satisfied.

Step 4: Shade the region ***above*** the line $2x + 3y = 6$.

(b) $2x + 3y \leq 0$:

Step 1: Graph $2x + 3y = 0$ with a solid line. This is a line through (0, 0) with slope $= \dfrac{-2}{3}$.

Step 2: Use (1, 0) above the line as a test point.

Step 3: At (1, 0), $2x + 3y = 2 > 0$.

Step 4: Therefore, we shade the region ***below*** the line $2x + 3y = 0$.

(c) Notice that the two shaded regions do ***not*** overlap. Thus, the system of inequalities has no solution.

31. $\begin{cases} x \geq 0 & (1) \\ y \geq 0 & (2) \\ 2x + y \leq 6 & (3) \\ x + 2y \leq 6 & (4) \end{cases}$

(a) $x \geq 0$; $y \geq 0$:

These two inequalities require that our shaded region must be restricted to quadrant I.

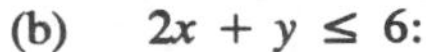

(b) $2x + y \leq 6$:

Step 1: Graph $2x + y = 6$ with a solid line. The x-intercept is $x = 3$; the y-intercept is $y = 6$.

Step 2: Use (0, 0) below the line as a test point.

Step 3: At (0, 0), $2x + y = 0 < 6$, so the inequality is satisfied.

Step 4: Shade the region below the line $2x + y = 6$.

(c) $x + 2y \leq 6$:

Step 1: Graph $x + 2y = 6$ with a solid line. The x-intercept is $x = 6$; the y-intercept is $y = 3$.

Step 2: Let's use (0, 0) ***below*** the line as a test point.

Step 3: At (0, 0), $x + 2y = 0 < 6$.

Step 4: Therefore, shade the region ***below*** the line $x + 2y = 6$.

(d) The graph is bounded.

(e) To list the vertices, consult the graph. We see that there are four vertices:

(1) Intersection of x-axis and y-axis: (0, 0).

(2) Intersection of $x + 2y = 6$ and y-axis: (0, 3).

(3) Intersection of $x + 2y = 6$ and $2x + y = 6$. To find this point of intersection, it will be necessary to solve a system of equations:

$$\begin{cases} x + 2y = 6 & (1) \\ 2x + y = 6 & (2) \end{cases}$$

$$\begin{cases} -2x - 4y = -12 & (1) \quad \text{Multiply by } -2. \\ 2x + y = 6 & (2) \end{cases}$$

$$\begin{cases} -3y = -6 & (1) \quad \text{Replace (1) by (1) + (2).} \\ 2x + y = 6 & (2) \end{cases}$$

$$\begin{cases} y = 2 & (1) \quad \text{Divide by } -3. \\ 2x + y = 6 & (2) \end{cases}$$

$$\begin{cases} y = 2 & (1) \\ 2x + 2 = 6 & (2) \quad \text{Back-substitution: } y = 2. \end{cases}$$

$$\begin{cases} y = 2 & (1) \\ x = 2 & (2) \end{cases}$$

Thus, the point of intersection is (2, 2).

(4) The intersection of $2x + y = 6$ and the x-axis: (3, 0)

So, the four vertices are: (0, 0), (0, 3), (2, 2), and (3, 0).

33.
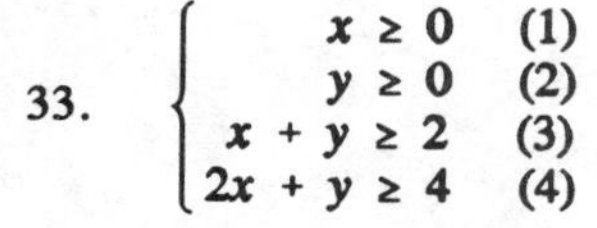

$$\begin{cases} x \geq 0 & (1) \\ y \geq 0 & (2) \\ x + y \geq 2 & (3) \\ 2x + y \geq 4 & (4) \end{cases}$$

(a) 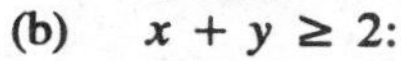
$x \geq 0; y \geq 0$:
These inequalities require that the final shaded area be in quadrant I.

(b) $x + y \geq 2$:

Step 1: Graph $x + y = 2$ with a solid line. The x-intercept is $x = 2$; the y-intercept is $y = 2$.

Step 2: Use (0, 0) below the line as a test point.

Step 3: 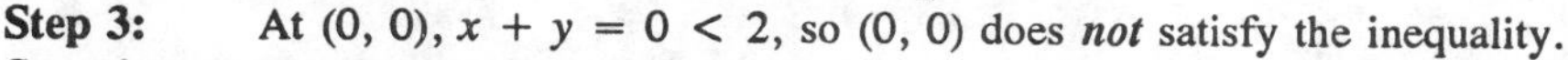
At (0, 0), $x + y = 0 < 2$, so (0, 0) does ***not*** satisfy the inequality.

Step 4: Shade the region ***above*** the line $x + y = 2$.

(c) 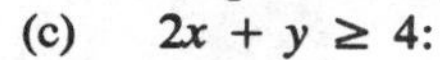
$2x + y \geq 4$:

Step 1: Graph $2x + y = 4$. The x-intercept is 2; the y-intercept is 4.

Step 2: Use (0, 0) ***below*** the line as a test point.

Step 3: At (0, 0), $2x + y = 0 < 4$.

Step 4: Therefore, we shade the region ***above*** the line $2x + y = 4$.

(d) The graph is ***unbounded*** since it extends forever toward the upper right.

(e) We only have two vertices:

(1) The intersection of $2x + y = 4$ and the y-axis: (0, 4).

(2) The intersection of either $x + y = 2$ or $2x + y = 4$ and the x-axis: (2, 0).

35.
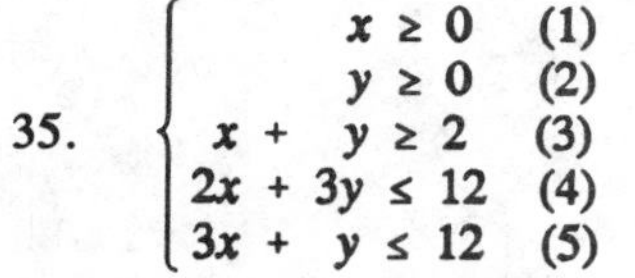

$$\begin{cases} x \geq 0 & (1) \\ y \geq 0 & (2) \\ x + y \geq 2 & (3) \\ 2x + 3y \leq 12 & (4) \\ 3x + y \leq 12 & (5) \end{cases}$$

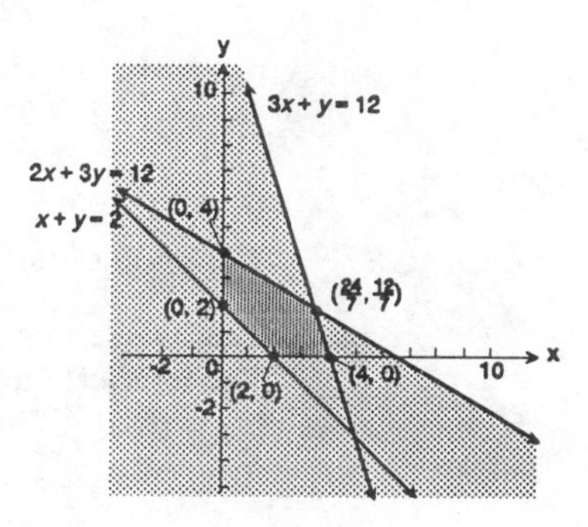

(a) 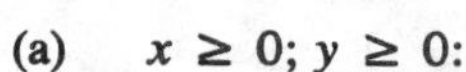
$x \geq 0; y \geq 0$:
Our graph will lie in quadrant I.

(b) 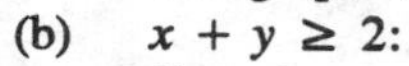
$x + y \geq 2$:

Step 1: Graph $x + y = 2$. The x-intercept is 2; the y-intercept is 2.

Step 2: Use (0, 0) below the line

Step 3: At (0, 0), $x + y = 0 < 2$.

Step 4: Therefore, we shade the region ***above*** the line $x + y = 2$.

(c) 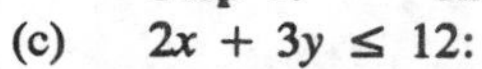
$2x + 3y \leq 12$:

Step 1: Graph $2x + 3y = 12$. The x-intercept is = 6; the y-intercept is 4.

Step 2: Use (0, 0) below the line.

Step 3: At (0, 0), $2x + 3y = 0 < 12$.

Step 4: Therefore, shade ***below*** the line $2x + 3y = 12$.

(d) $3x + y \le 12$:

Step 1: Graph $3x + y = 12$. The x-intercept is $x = 4$; the y-intercept is 12.

Step 2: Use (0, 0) ***below*** the line as a test point.

Step 3: At (0, 0), $3x + y = 0 < 12$.

Step 4: Shade the region ***below*** the line $3x + y = 12$.

(e) We see that the graph is bounded.

(f) We have five vertices:

(1) Intersection of $x + y = 2$ and the y-axis: (0, 2).

(2) Intersection of $2x + 3y = 12$ and the y-axis: (0, 4).

(3) Intersection of $2x + 3y = 12$ and $3x + y = 12$:

$$\begin{cases} 2x + 3y = 12 & (1) \\ 3x + y = 12 & (2) \end{cases}$$

$$\begin{cases} 2x + 3y = 12 & (1) \\ -9x - 3y = -36 & (2) \quad \text{Multiply by } -3. \end{cases}$$

$$\begin{cases} 2x + 3y = 12 & (1) \\ -7x = -24 & (2) \quad \text{Replace (2) by (1) + (2).} \end{cases}$$

$$\begin{cases} 2x + 3y = 12 & (1) \\ x = \dfrac{24}{7} & (2) \quad \text{Divide by } -7. \end{cases}$$

$$\begin{cases} \dfrac{48}{7} + 3y = 12 & (1) \quad \text{Back-substitution: } x = \dfrac{24}{7} \\ x = \dfrac{24}{7} & (2) \end{cases}$$

$$\begin{cases} 3y = \dfrac{84}{7} - \dfrac{48}{7} & (1) \\ x = \dfrac{24}{7} & (2) \end{cases}$$

$$\begin{cases} y = \dfrac{12}{7} & (1) \\ x = \dfrac{24}{7} & (2) \end{cases}$$

The point of intersection is $\left(\dfrac{24}{7}, \dfrac{12}{7}\right)$

(4) The intersection of $3x + y = 12$ and the x-axis: (4, 0)

(5) The intersection of $x + y = 2$ and the x-axis: (2, 0)

37. $\begin{cases} x \geq 0 & (1) \\ y \geq 0 & (2) \\ x + y \geq 2 & (3) \\ x + y \leq 8 & (4) \\ 2x + y \leq 10 & (5) \end{cases}$

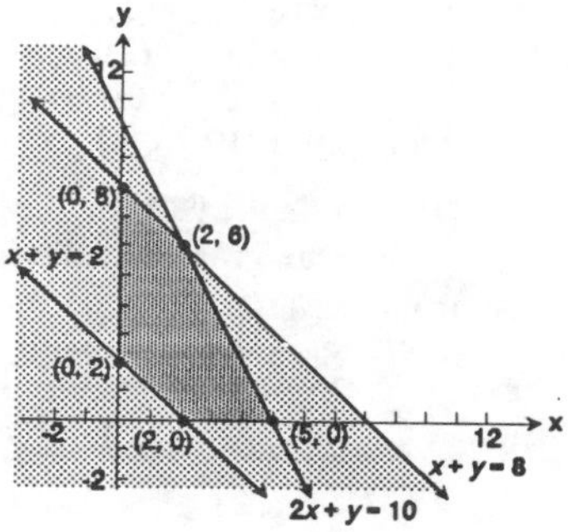

(a) $x \geq 0$; $y \geq 0$:
This places our final graph in quadrant I.

(b) $x + y \geq 2$:
Step 1: Graph $x + y = 2$. The x-intercept is 2; the y-intercept is 2.
Step 2: Use (0, 0) ***below*** the line, as a test point.
Step 3: At (0, 0), $x + y = 0 < 2$, so the inequality is ***not*** satisfied.
Step 4: Therefore, shade ***above*** the line $x + y = 2$.

(c) $x + y \leq 8$:
Step 1: Graph $x + y = 8$. The x-intercept is 8; the y-intercept is 8.
Step 2: Use (0, 0) below the line.
Step 3: At (0, 0), $x + y = 0 < 8$.
Step 4: Therefore, shade ***below*** the line $x + y = 8$.

(d) $2x + y \leq 10$:
Step 1: Graph $2x + y = 10$. The x-intercept is $x = 5$; the y-intercept is $y = 10$.
Step 2: Use (0, 0) below the line as a test point.
Step 3: At (0, 0), $2x + y = 0 < 10$.
Step 4: Shade the region below the line $2x + y = 10$.

(e) The graph is bounded.

(f) We have five vertices:

(1) Intersection of $x + y = 2$ and the y-axis: (0, 2).

(2) Intersection of $x + y = 8$ and the y-axis: (0, 8).

(3) Intersection of $x + y = 8$ and $2x + y = 10$:

$$\begin{cases} x + y = 8 & (1) \\ 2x + y = 10 & (2) \end{cases}$$

$$\begin{cases} x + y = 8 & (1) \\ -2x - y = -10 & (2) \end{cases}$$

$$\begin{cases} x + y = 8 & (1) \\ -x = -2 & (2) \end{cases} \quad \text{Replace (2) by (1) + (2).}$$

$$\begin{cases} x + y = 8 & (1) \\ x = 2 & (2) \end{cases}$$

$$\begin{cases} y = 6 & (1) \\ x = 2 & (2) \end{cases} \quad \text{Back-substitution: } x = 2.$$

The vertex is (2, 6).

(4) The intersection of $2x + y = 10$ and the x-axis: (5, 0)

(5) The intersection of $x + y = 2$ and the x-axis: (2, 0)

39.

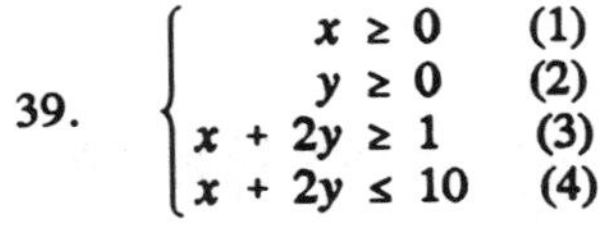

$$\begin{cases} x \geq 0 & (1) \\ y \geq 0 & (2) \\ x + 2y \geq 1 & (3) \\ x + 2y \leq 10 & (4) \end{cases}$$

(a) 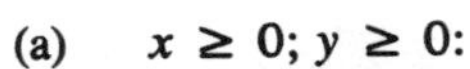

$x \geq 0$; $y \geq 0$:

Our graph will lie in quadrant I.

(b) $x + 2y \geq 1$:

Step 1: Graph $x + 2y = 1$. The x-intercept is 1; the y-intercept is $\frac{1}{2}$.

Step 2: Use (0, 0) below the line as a test point.

Step 3: At (0, 0), $x + 2y = 0 < 1$.

Step 4: Therefore, we shade the region ***above*** the line $x + 2y = 1$.

(c) $x + 2y \leq 10$:

Step 1: Graph $x + 2y = 10$. The x-intercept is $= 10$; the y-intercept is 5.

Step 2: Use (0, 0) below the line.

Step 3: At (0, 0), $x + 2y = 0 < 10$.

Step 4: Therefore, we shade the region ***below*** the line $x + 2y = 10$.

(d) We see that the region is bounded.

(e) From the graph, there are four vertices:

(1) The intersection of $x + 2y = 1$ and the y-axis: $\left(0, \frac{1}{2}\right)$.

(2) The intersection of $x + 2y = 10$ and the y-axis: (0, 5).

(3) The intersection of $x + 2y = 10$ and the x-axis: (10, 0).

(4) The intersection of $x + 2y = 1$ and the x-axis: (1, 0)

41. $\begin{cases} x \leq 4 \\ x + y \leq 6 \\ x \geq 0,\ y \geq 0 \end{cases}$

43. $\begin{cases} x \leq 20 \\ y \geq 15 \\ x + y \leq 50 \\ x \leq y \\ x \geq 0 \end{cases}$

45. (a) $\begin{cases} x + y \leq 50{,}000 \\ x \geq 35{,}000 \\ y \leq 10{,}000 \\ x \geq 0 \\ y \geq 0 \end{cases}$

(b)

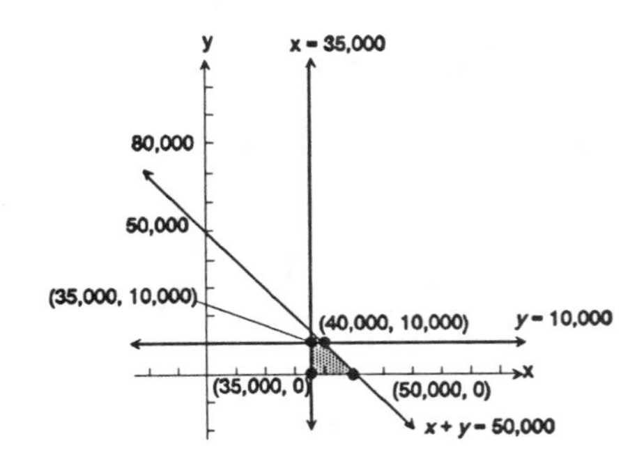

47. (a) x denotes the number of packages of the economy blend.
y denotes the number of packages of the superior blend.

$4x + 8y \leq 75(16)$ denotes the equation which states that the blends cannot exceed 75 pounds of A grade coffee or $(76)(16) = 1200$ ounces.

$12x + 8y \leq 120(16)$ denotes the equation which states that the blends cannot exceed 120 pounds of B grade coffee or $(120)(16) = 1920$ ounces.

We also must denote that the number of packages of both blends must be non-negative.

The following equations are simplified:

from $4x + 8y \leq 1200$
to $x + 2y \leq 300$

and from $12x + 8y \leq 1920$
to $3x + 2y \leq 480$

We have the following system:

$$x \geq 0,\ y \geq 0$$
$$x + 2y \leq 300$$
$$3x + 2y \leq 480$$

(b)

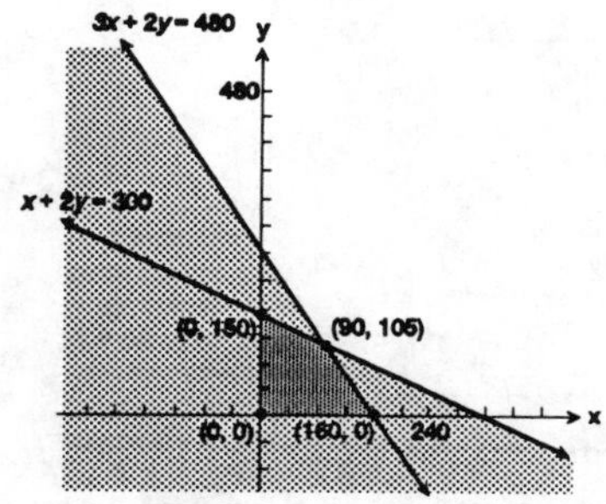

The vertices are (0, 0), (0, 150), (90, 105), and (160, 0).

49. (a)

$$\begin{cases} 30x + 20y \leq 1600 \\ 2x + 3y \leq 150 \\ x \geq 0 \\ y \geq 0 \end{cases}$$

(b)

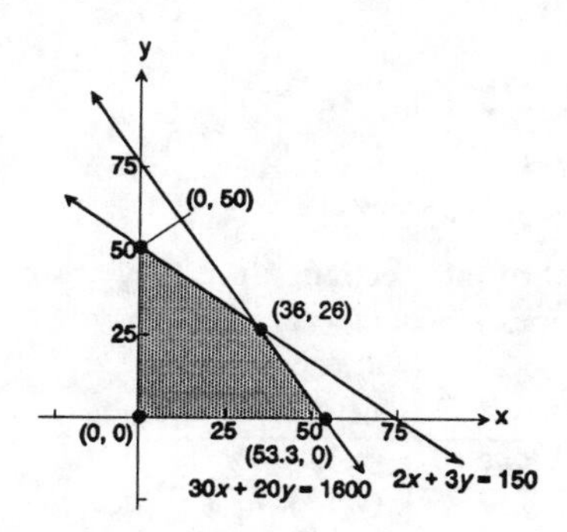

8.8 Linear Programming

In Problems 1-6, we have the same set of vertices: (0, 3), (0, 6), (5, 6), (5, 2), and (4, 0).

1. $z = x + y$

Vertex	Value of Objective Function ($z = x + y$)
(0, 3)	$z = 0 + 3 = 3$
(0, 6)	$z = 0 + 6 = 6$
(5, 6)	$z = 5 + 6 = 11$
(5, 2)	$z = 5 + 2 = 7$
(4, 0)	$z = 4 + 0 = 4$

The maximum value is 11, at (5, 6), and the minimum value is 3, at (0, 3).

3.

Vertex	Value of $z = x + 10y$
(0, 3)	$z = 0 + 10 \cdot 3 = 30$
(0, 6)	$z = 0 + 10 \cdot 6 = 60$
(5, 6)	$z = 5 + 10 \cdot 6 = 65$ ← Maximum value
(5, 2)	$z = 5 + 10 \cdot 2 = 25$
(4, 0)	$z = 4 + 10 \cdot 0 = 4$ ← Minimum value

5.

Vertex	Value of Objective Function ($z = 5x + 7y$)
(0, 3)	$z = 5 \cdot 0 + 7 \cdot 3 = 21$
(0, 6)	$z = 5 \cdot 0 + 7 \cdot 6 = 42$
(5, 6)	$z = 5 \cdot 5 + 7 \cdot 6 = 67$ ← Maximum value
(5, 2)	$z = 5 \cdot 5 + 7 \cdot 2 = 39$
(4, 0)	$z = 5 \cdot 4 + 7 \cdot 0 = 20$ ← Minimum value

7. Maximize $z = 2x + y$

Subject to $x \geq 0, y \geq 0, x + y \leq 6, x + y \geq 1$

(a) We have two lines:
$y = -x + 6$ and $y = -x + 1$.

There is no point of intersection, since they are parallel lines. (Their slopes are both -1).

(b) Now evaluate $z = 2x + y$ at each vertex:

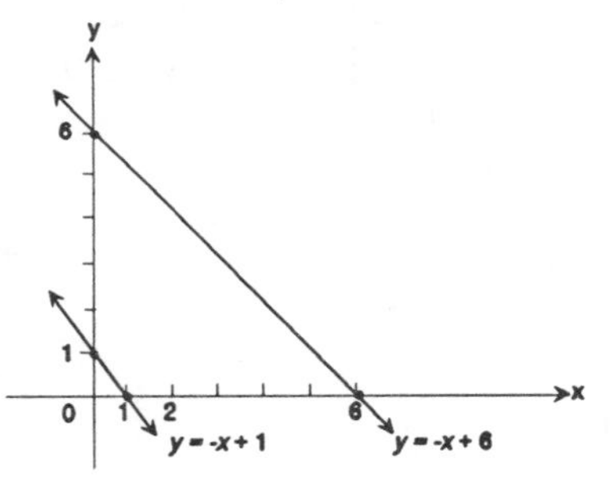

Vertex	Value of $z = 2x + y$
(0, 1)	$z = 2 \cdot 0 + 1 = 1$
(0, 6)	$z = 2 \cdot 0 + 6 = 6$
(6, 0)	$z = 2 \cdot 6 + 0 = 12$
(1, 0)	$z = 2 \cdot 1 + 0 = 2$

The maximum possible value for z is 12, at the point (6, 0).

9. Minimize $z = 2x + 5y$

Subject to $x \geq 0, y \geq 0, x + y \geq 2, x \leq 5, y \leq 3$.

(a) Graph the constraints:

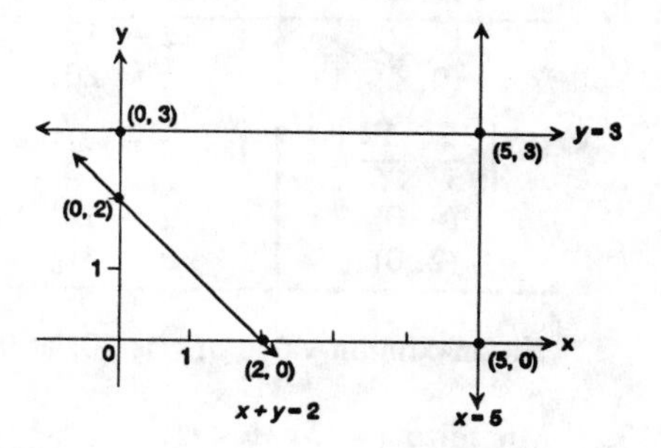

Vertex	Value of $z = 2x + 5y$
(0, 2)	$z = 10$
(0, 3)	$z = 15$
(5, 3)	$z = 25$
(5, 0)	$z = 10$
(2, 0)	$z = 4$

The minimum value of z is 4, at the point (2, 0).

11. $z = 3x + 5y$

Subject to $x \geq 0, y \geq 0, x + y \geq 2, 2x + 3y \leq 12,$
$3x + 2y \leq 12$.

We have three lines:

(1) $y = -x + 2$

(2) $y = -\frac{2}{3}x + 4$

(3) $y = -\frac{3}{2}x + 6$

Let's find the points of intersection:

(1) and (2):

$$-x + 2 = -\frac{2}{3}x + 4$$
$$-3x + 6 = -2x + 12$$
$$-x = 6$$
$$x = -6 \text{ and } y = -x + 2$$
$$y = 8$$

But, $(-6, 8)$ is not in the feasible region ($x \geq 0$).

(1) and (3):

$$-x + 2 = -\frac{3}{2}x + 6$$
$$-2x + 4 = -3x + 12$$
$$x = 8$$

and

$$y = -x + 2 = -6$$

But $(8, -6)$ is not in the feasible region.

(2) and (3):

$$-\frac{2}{3}x + 4 = -\frac{3}{2}x + 6$$
$$-4x + 24 = -9x + 36$$
$$5x = 12$$
$$x = \frac{12}{5}$$

and

$$y = -\frac{3}{2}x + 6$$
$$y = -\frac{3}{2}\left(\frac{12}{5}\right) + 6 = \frac{12}{5}$$

This gives the point $\left(\frac{12}{5}, \frac{12}{5}\right)$.

We now graph the constraints:

Vertex	Value of $z = 3x + 5y$
(0, 2)	$z = 10$
(0, 4)	$z = 20$
$\left(\frac{12}{5}, \frac{12}{5}\right)$	$z = \frac{96}{5} = 19.2$
(4, 0)	$z = 12$
(2, 0)	$z = 6$

The maximum value of z is 20, at the point (0, 4).

13. Minimize $z = 5x + 4y$

Subject to $x \geq 0$, $y \geq 0$, $x + y \geq 2$, $2x + 3y \leq 12$, $3x + y \leq 12$.

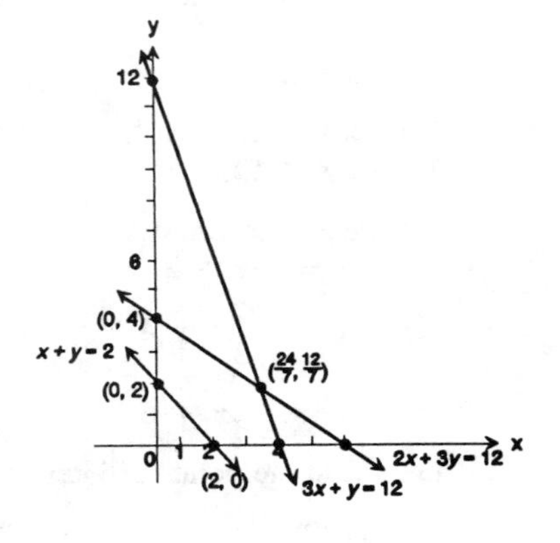

We have three lines to graph:

(1) $y = -x + 2$;
y-intercept $= 2$; x-intercept $= 2$

(2) $y = -\frac{2}{3}x + 4$;
y-intercept $= 4$; x-intercept $= 6$

(3) $y = -3x + 12$;
y-intercept $= 12$; x-intercept $= 4$

The only intersection we need is:

(2) and (3):
$$-\frac{2}{3}x + 4 = -3x + 12$$
$$-2x + 12 = -9x + 36$$
$$7x = 24$$
$$x = \frac{24}{7}$$

and $y = -3x + 12$; $y = \frac{12}{7}$

$$\left(\frac{24}{7}, \frac{12}{7}\right)$$

Vertex	Value of $z = 5x + 4y$
(0, 2)	$z = 8$
(0, 4)	$z = 16$
$\left(\frac{24}{7}, \frac{12}{7}\right)$	$z = \frac{168}{7} = 24$
(4, 0)	$z = 20$
(2, 0)	$z = 10$

The minimum value of z is 8, at the point (0, 2).

15. Maximize $z = 5x + 2y$

Subject to $x \geq 0, y \geq 0, x + y \leq 10, 2x + y \geq 10,$ $x + 2y \geq 10.$

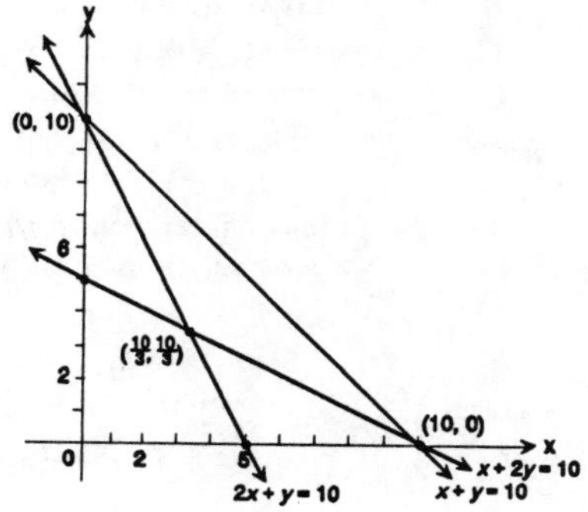

We have:

(1) $y = -x + 10;$
y-intercept $= 10$; x-intercept $= 10$

(2) $y = -2x + 10;$
y-intercept $= 10$; x-intercept $= 5$

(3) $y = -\frac{1}{2}x + 5;$
y-intercept $= 5$; x-intercept $= 10$

To find the intersection of (2) and (3):

$$-2x + 10 = -\frac{1}{2}x + 5$$
$$-4x + 20 = -x + 10$$
$$-3x = -10$$
$$x = \frac{10}{3}$$

and
$$y = -2x + 10$$
$$y = -2\left(\frac{10}{3}\right) + 10$$
$$y = \frac{10}{3}; \left(\frac{10}{3}, \frac{10}{3}\right)$$

Vertex	Value of $z = 5x + 2y$
(0, 10)	$z = 20$
(10, 0)	$z = 50$
$\left(\frac{10}{3}, \frac{10}{3}\right)$	$z = \frac{70}{3} = 23\frac{1}{3}$

The maximum value of z is 50, at the point (10, 0).

17. Let x = Number of Downhill skis produced
y = Number of Cross-country skis produced

We want to maximize profit (which is \$70 per Downhill ski, \$50 per Cross-Country ski). Thus, total profit is:

$P = 70x + 50y$

This is our objective function. Now we need to determine our constraints:

$x \geq 0, y \geq 0$ Nonnegative constraints

We only have 40 hours manufacturing time available:

$2x + y \leq 40$ Manufacturing time

Also, $x + y \leq 32$ Finishing time constraint

So we have two lines to graph:

(1) $y = -2x + 40$;
y-intercept $= 40$; x-intercept $= 20$

(2) $y = -x + 32$;
y-intercept $= 32$; x-intercept $= 32$

For the point of intersection of (1) and (2):

$$-2x + 40 = -x + 32$$
$$-x = -8$$
$$x = 8$$

and
$$y = -x + 32$$
$$y = 24; (8, 24)$$

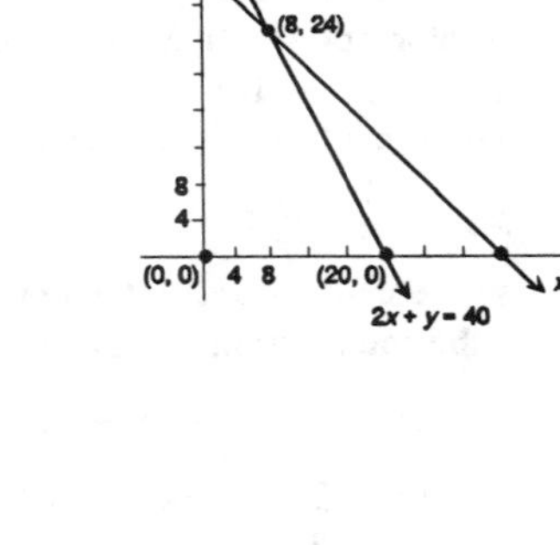

Vertex	Value of Profit: $P = 70x + 50y$
(0, 0)	$P = 0$
(0, 32)	$P = 50 \cdot 32 = 1600$
(8, 24)	$P = 70 \cdot 8 + 50 \cdot 24 = 1760$
(20, 0)	$P = 70 \cdot 20 = 1400$

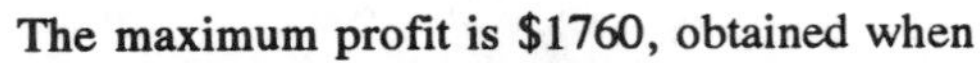

The maximum profit is $1760, obtained when

$x = 8$ Downhill skis
$y = 24$ Cross-country skis

are produced.

With 48 hours of manufacturing, we have $2x + y \leq 48$. So, $y = -2x + 48$ has x-intercept of 24. $2x + y = 48$ and $x + y = 32$ intersect at (16, 16).

Vertex	Value of Profit
(0, 0)	$P = 0$
(0, 32)	$P = 70 \cdot 0 + 50 \cdot 32 = 1600$
(16, 16)	$P = 70 \cdot 16 + 50 \cdot 16 = 1920$
(24, 0)	$P = 70 \cdot 24 = 1680$

The maximum profit is $1920.

19. Maximize profit $P = 200x + 250y$

Let x = acres of soybeans
y = acres of corn

$x \geq 0, y \geq 0$ Nonnegative constraints

$40x + 60y \leq 1800$ Cultivation Cost
$60x + 60y \leq 2400$ Labor Cost

For the point of intersection,

$$\frac{1800 - 40x}{60} = \frac{2400 - 60x}{60}$$
$$30 - \frac{2}{3}x = 40 - x$$
$$\frac{1}{3}x = 10$$
$$x = 30$$
$$y = \frac{1800 - 40(30)}{60}$$
$$y = 10$$

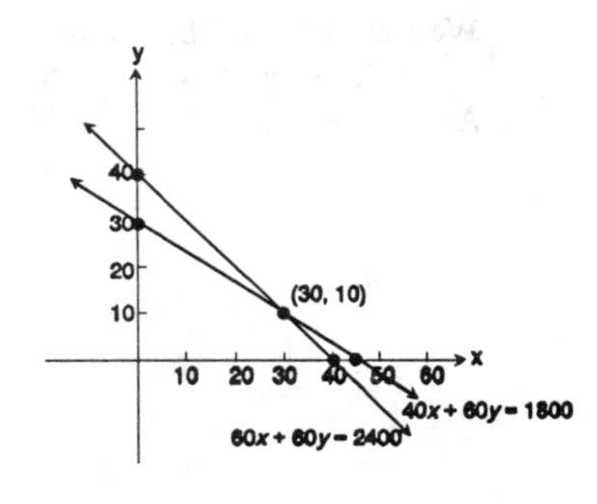

Vertex	Value of Profit $P = 200x + 250y$
(0, 0)	$P = 0$
(0, 30)	$P = 0 + 250(30) = 7500$
(30, 10)	$P = 200(30) + 250(10) = 8500$
(40, 0)	$P = 200(40) + 250(0) = 8000$

The maximum profit is $8500 obtained with 30 acres of soybeans and 10 acres of corn.

21. Let x = Machine I
y = Machine II

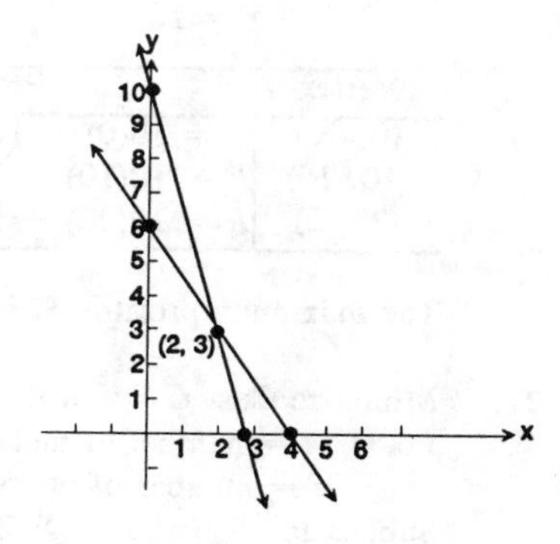

Minimize $C = 50x + 30y$

Subject to

$x \geq 0, y \geq 0$
$60x + 40y \geq 240$
$70x + 20y \geq 140$

Vertex	Cost $C = 50x + 30y$
(0, 7)	$C = 50(0) + 30(7) = \$210$
$\left[\frac{1}{2}, \frac{21}{4}\right]$	$C = 50\left[\frac{1}{2}\right] + 30\left[\frac{21}{4}\right] = \$182.5$
(4, 0)	$C = 50(4) + 30(0) = \$200$

The minimum cost is $182.50 with $\frac{1}{2}$ hour on Machine I and 5.25 hours on Machine II.

23. Let x = number of pounds of ground beef.
Let y = number of pounds of ground pork.

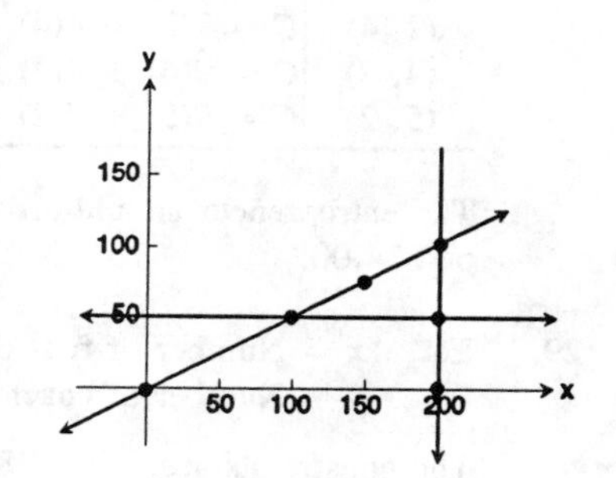

Minimize $C = .75x + .45y$

Subject to

$x \geq 0, y \geq 0$
$.75x + .60y \geq .70(x + y)$
$.75x + .60y \geq .70x + .70y$
$.05x - .10y \geq 0$
$x \leq 200$
$y \geq 50$
$.05x \geq .10y$

Vertex	Cost $C = .75x + .45y$
(100, 50)	$C = .75(100) + .45(50) = \$97.50$
(200, 100)	$C = .75(200) + .45(100) = \$195.00$
(200, 50)	$C = .75(200) + .45(50) = \$172.50$

The minimum cost is $97.50 obtained with 100 lbs. of ground beef and 50 lbs of pork.

25. Maximize Profit $P = \$10r + \$12f$

Let r = number of racing skates
f = number of figure skates

Subject to

$r \geq 0, f \geq 0$

$$6r + 4f \le 120$$
$$1r + 2f \le 40$$
$$\frac{120 - 4f}{6} = 40 - 2f$$
$$120 - 4f = 240 - 12f$$
$$8f = 120$$
$$f = 15$$
$$r = 40 - 2(15)$$
$$r = 10$$

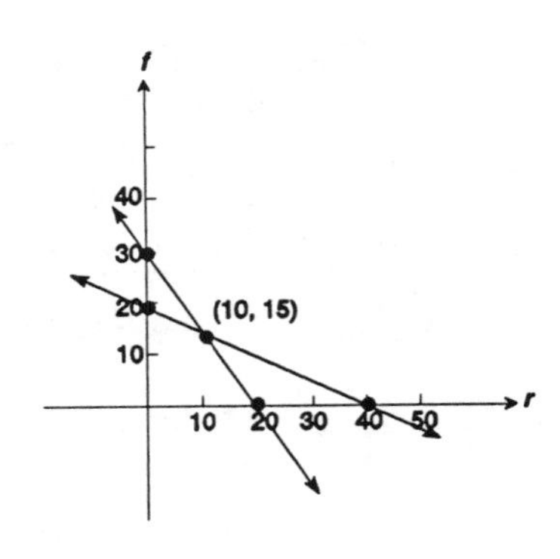

Vertex	Profit
(0, 20)	$P = 10(0) + 12(20) = \$240$
(10, 15)	$P = 10(10) + 12(15) = \$280$
(20, 0)	$P = 10(20) + 12(0) = \$200$

The maximum profit is \$280 which occurs when manufacturing 10 racing skates and 15 figure skates.

27. Minimize Cost $C = 9m + 4p$
Let m = number of metal fasteners
p = number of plastic fasteners
Subject to:
$$m \ge 2$$
$$p \ge 2$$
$$m + 2 \ge 6$$
$$4m + 2p \le 24$$

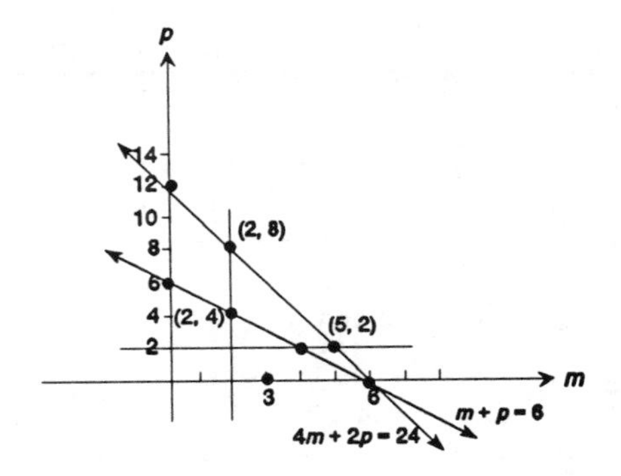

Vertex	Profit
(2, 8)	$C = 9(2) + 4(8) = \$50.00$
(2, 4)	$C = 9(2) + 4(4) = \$34.00$
(4, 2)	$C = 9(4) + 4(2) = \$44.00$
(5, 2)	$C = 9(5) + 4(2) = \$53.00$

The entrepreneur should order 2 metal fasteners and 4 plastic fasteners. The cost of the samples will be \$34.00.

29. Let x = Number of first class seats
y = Number of coach seats

The constraints are:
$$8 \le x \le 16$$
$$80 \le y \le 120$$

(a) If the ratio of first class to coach cannot exceed $\frac{1}{12}$, then $\frac{x}{y} \le \frac{1}{12}$.

Let $y = 120$. Then $\frac{x}{120} \le \frac{1}{12}$.
$$12x \le 120$$
or $x \le 10$

This satisfies the first constraint. Thus, the number of seats that maximizes revenue is 10 first class and 120 coach.

(b) If the ratio of first class to coach cannot exceed $\frac{1}{8}$, then $\frac{x}{y} \le \frac{1}{8}$.

Let $y = 120$. Then $\frac{x}{120} \le \frac{1}{8}$.

$8x \le 120$ or $x \le 15$ which satisifes the first constraint. Thus, the number of seats that maximizes revenue is 15 first class and 120 coach.

8 Chapter Review

1. $\begin{cases} 2x - y = 5 & (1) \\ 5x + 2y = 8 & (2) \end{cases}$

Step 1: It is easiest to solve (1) for y:

$$2x - y = 5$$
$$-y = 5 - 2x$$
$$y = -5 + 2x$$

Step 2: Substitute $y = -5 + 2x$ into (2):

$$5x + 2y = 8$$
$$5x + 2(-5 + 2x) = 8$$
$$5x - 10 + 4x = 8$$
$$9x = 18$$

Step 3: $x = 2$

Step 4: Determine y:

$$y = -5 + 2x \text{ (Step 1)}$$
$$y = -5 + 2(2) \qquad (x = 2)$$
$$y = -1$$

The solution is $x = 2$, $y = -1$.

3. $\begin{cases} 3x - 4y = 4 & (1) \\ x - 3y = \frac{1}{2} & (2) \end{cases}$

Step 1: Let's solve for x in (2):

$$x - 3y = \frac{1}{2}$$
$$x = \frac{1}{2} + 3y$$

Step 2: Substitute $x = \frac{1}{2} + 3y$ into (1):

$$3x - 4y = 4$$
$$3\left[\frac{1}{2} + 3y\right] - 4y = 4$$
$$\frac{3}{2} + 9y - 4y = 4$$
$$3 + 18y - 8y = 8 \quad \text{Multiply by 2.}$$
$$10y = 5$$

Step 3: $y = \frac{5}{10}$

$y = \frac{1}{2}$

Step 4: Determine x: $x = \frac{1}{2} + 3y$ (Step 1)

$$x = \frac{1}{2} + 3\left[\frac{1}{2}\right] \qquad \left[y = \frac{1}{2}\right]$$
$$x = 2$$

The solution is $x = 2$, $y = \frac{1}{2}$.

5. $\begin{cases} x - 2y - 4 = 0 & (1) \\ 3x + 2y - 4 = 0 & (2) \end{cases}$

or

$\begin{cases} x - 2y = 4 & (1) \\ 3x + 2y = 4 & (2) \end{cases}$

Step 1: It is easiest to solve for x in (1):

$$x - 2y = 4$$
$$x = 4 + 2y$$

Step 2: Substitute $x = 4 + 2y$ into (2):

$$3x + 2y = 4$$
$$3(4 + 2y) + 2y = 4$$
$$8y = -8$$

Step 3: $y = -1$

Step 4: Determine x:

$$x = 4 + 2y \quad \text{(Step 1)}$$
$$x = 4 + 2(-1)$$
$$x = 2$$

The solution is $x = 2$, $y = -1$.

7. $\begin{cases} y = 2x - 5 & (1) \\ x = 3y + 4 & (2) \end{cases}$

We can substitute $y = 2x - 5$ into (2):

$$x = 3y + 4$$
$$x = 3(2x - 5) + 4$$
$$x = 6x - 11$$
$$-5x = -11$$
$$x = \frac{11}{5}$$

Then we have:

$$y = 2x - 5$$
$$y = \frac{22}{5} - 5$$
$$y = \frac{-3}{5}$$

The solution is $x = \frac{11}{5}$, $y = \frac{-3}{5}$.

9. $\begin{cases} x - y + 4 = 0 & (1) \\ \frac{1}{2}x + \frac{1}{6}y + \frac{2}{5} = 0 & (2) \end{cases}$

or

$\begin{cases} x - y = 4 & (1) \\ 15x + 5y = -12 & (2) \end{cases}$ Multiply by 30.

Step 1: Solve equation (1) for x:

$$x - y = -4$$
$$x = -4 + y$$

Step 2: Substitute $x = -4 + y$ into (2):

$$15x + 5y = -12$$
$$15(-4 + y) + 5y = -12$$
$$-60 + 20y = -12$$
$$20y = 48$$
$$y = \frac{48}{20}$$

Step 3: $y = \frac{12}{5}$

Step 4: Determine x:

$$x = -4 + y \quad \text{(Step 1)}$$
$$x = -4 + \frac{12}{5}$$
$$x = \frac{-20}{5} + \frac{12}{5}$$
$$x = \frac{-8}{5}$$

The solution is $x = \frac{-8}{5}$, $y = \frac{12}{5}$.

11. $\begin{cases} x - 2y - 8 = 0 & (1) \\ 2x + 2y - 10 = 0 & (2) \end{cases}$

$\begin{cases} x - 2y - 8 = 0 & (1) \\ 3x \quad - 18 = 0 & (2) \end{cases}$ Replace (2) by (1) + (2).

$\begin{cases} x - 2y - 8 = 0 & (1) \\ x = 6 & (2) \end{cases}$

$\begin{cases} 6 - 2y - 8 = 0 & (1) \\ x = 6 & (2) \end{cases}$ Back-substitute; $x = 6$.

$\begin{cases} -2y = 2 & (1) \\ x = 6 & (2) \end{cases}$

$\begin{cases} y = -1 & (1) \\ x = 6 & (2) \end{cases}$

The solution is $x = 6$; $y = -1$.

13. $\begin{cases} y - 2x = 11 & (1) \\ 2y - 3x = 18 & (2) \end{cases}$

$\begin{cases} -2y + 4x = -22 & (1) \\ 2y - 3x = 18 & (2) \end{cases}$ Multiply both sides by -2.

$\begin{cases} -2y + 4x = -22 & (1) \\ x = -4 & (2) \end{cases}$ Replace (2) by (1) + (2).

$\begin{cases} -2y - 16 = -22 & (1) \\ x = -4 & (2) \end{cases}$ Back-substitute; $x = -4$.

$\begin{cases} -2y = -6 & (1) \\ x = -4 & (2) \end{cases}$

$\begin{cases} y = 3 & (1) \\ x = -4 & (2) \end{cases}$

The solution is $x = -4$; $y = 3$.

15. $\begin{cases} 2x + 3y - 13 = 0 & (1) \\ 3x - 2y = 0 & (2) \end{cases}$

$\begin{cases} 6x + 9y - 39 = 0 & (1) \\ -6x + 4y = 0 & (2) \end{cases}$ Multiply both sides by 3. Multiply both sides by -2.

$\begin{cases} 6x + 9y - 39 = 0 & (1) \\ 13y - 39 = 0 & (2) \end{cases}$ Replace (2) by (1) + (2).

$\begin{cases} 6x + 9y - 39 = 0 & (1) \\ y = 3 & (2) \end{cases}$

$\begin{cases} 6x + 27 - 39 = 0 & (1) \\ y = 3 & (2) \end{cases}$ Back-substitute; $y = 3$.

$\begin{cases} x = 2 & (1) \\ y = 3 & (2) \end{cases}$

The solution is $x = 2$, $y = 3$.

17. $\begin{cases} 3x - 2y = 8 & (1) \\ x - \frac{2}{3}y = 12 & (2) \end{cases}$

$\begin{cases} 3x - 2y = 8 & (1) \\ -3x + 2y = -36 & (2) \end{cases}$ Multiply both sides by -3.

$\begin{cases} 3x - 2y = 8 & (1) \\ 0x + 0y = -28 & (2) \end{cases}$ Replace (2) by (1) + (2).

The system is inconsistent.

19. $$\begin{cases} x + 2y - z = 6 & (1) \\ 2x - y + 3z = -13 & (2) \\ 3x - 2y + 3z = -16 & (3) \end{cases}$$

We can eliminate the variable y by adding (1) and (3):

$$\begin{cases} x + 2y - z = 6 & (1) \\ 2x - y + 3z = -13 & (2) \\ 4x + 2z = -10 & (3) \quad \text{Replace (3) by (1) + (3).} \end{cases}$$

Now we must eliminate the same variable, y, from either (1) or (2):

$$\begin{cases} x + 2y - z = 6 & (1) \\ 4x - 2y + 6z = -26 & (2) \quad \text{Multiply both sides by 2.} \\ 4x + 2z = -10 & (3) \end{cases}$$

$$\begin{cases} x + 2y - z = 6 & (1) \\ 5x + 5z = -20 & (2) \quad \text{Replace (2) by (1) + (2).} \\ 4x + 2z = -10 & (3) \end{cases}$$

Now attack (2) and (3):

$$\begin{cases} x + 2y - z = 6 & (1) \\ 20x + 20z = -80 & (2) \quad \text{Multiply both sides by 4.} \\ -20x - 10z = 50 & (3) \quad \text{Mutliply both sides by -5.} \end{cases}$$

$$\begin{cases} x + 2y - z = 6 & (1) \\ 20x + 20z = -80 & (2) \\ 10z = -30 & (3) \quad \text{Replace (3) by (2) + (3).} \end{cases}$$

$$\begin{cases} x + 2y - z = 6 & (1) \\ 20x + 20z = -80 & (2) \\ z = -3 & (3) \end{cases}$$

$$\begin{cases} x + 2y + 3 = 6 & (1) \quad \text{Back-substitute: } z = -3. \\ 20x - 60 = -80 & (2) \quad \text{Back-substitute; } z = -3. \\ z = -3 & (3) \end{cases}$$

$$\begin{cases} x + 2y = 3 & (1) \\ x = -1 & (2) \\ z = -3 & (3) \end{cases}$$

$$\begin{cases} -1 + 2y = 3 & (1) \quad \text{Back-substitute; } x = -1. \\ x = -1 & (2) \\ z = -3 & (3) \end{cases}$$

$$\begin{cases} y = 2 & (1) \\ x = -1 & (2) \\ z = -3 & (3) \end{cases}$$

The solution is $x = -1, y = 2, z = -3$.

21. $A + C = \begin{bmatrix} 1 & 0 \\ 2 & 4 \\ -1 & 2 \end{bmatrix} + \begin{bmatrix} 3 & -4 \\ 1 & 5 \\ 5 & -2 \end{bmatrix} = \begin{bmatrix} 4 & -4 \\ 3 & 9 \\ 4 & 0 \end{bmatrix}$

23. $6A = 6\begin{bmatrix} 1 & 0 \\ 2 & 4 \\ -1 & 2 \end{bmatrix} = \begin{bmatrix} 6 & 0 \\ 12 & 24 \\ -6 & 12 \end{bmatrix}$

25. $AB = \begin{bmatrix} 1 & 0 \\ 2 & 4 \\ -1 & 2 \end{bmatrix} \begin{bmatrix} 4 & -3 & 0 \\ 1 & 1 & -2 \end{bmatrix} = \begin{bmatrix} 4+0 & -3+0 & 0+0 \\ 8+4 & -6+4 & 0+(-8) \\ -4+2 & 3+2 & 0+(-4) \end{bmatrix} = \begin{bmatrix} 4 & -3 & 0 \\ 12 & -2 & -8 \\ -2 & 5 & -4 \end{bmatrix}$

27. $CB = \begin{bmatrix} 3 & -4 \\ 1 & 5 \\ 5 & -2 \end{bmatrix} \begin{bmatrix} 4 & -3 & 0 \\ 1 & 1 & -2 \end{bmatrix} = \begin{bmatrix} 8 & -13 & 8 \\ 9 & 2 & -10 \\ 18 & -17 & 4 \end{bmatrix}$

29. $A = \begin{bmatrix} 4 & 6 \\ 1 & 3 \end{bmatrix}$. We start with $[A|I_2]$, and proceed to put it into reduced echelon form.

$$[A|I_2] = \left[\begin{array}{cc|cc} 4 & 6 & 1 & 0 \\ 1 & 3 & 0 & 1 \end{array}\right] \rightarrow \left[\begin{array}{cc|cc} 1 & 3 & 0 & 1 \\ 4 & 6 & 1 & 0 \end{array}\right]$$

Interchange rows

$$\rightarrow \left[\begin{array}{cc|cc} 1 & 3 & 0 & 1 \\ 0 & -6 & 1 & -4 \end{array}\right] \rightarrow \left[\begin{array}{cc|cc} 1 & 0 & \frac{1}{2} & -1 \\ 1 & -6 & 1 & -4 \end{array}\right] \rightarrow \left[\begin{array}{cc|cc} 1 & 0 & \frac{1}{2} & -1 \\ 0 & 1 & -\frac{1}{6} & \frac{2}{3} \end{array}\right]$$

$R_2 = -4r_1 + r_2$ $\quad R_1 = \frac{1}{2}r_2 + r_1$ $\quad R_2 = -\frac{1}{6}r_2$

Therefore, $A^{-1} = \begin{bmatrix} \frac{1}{2} & -1 \\ -\frac{1}{6} & \frac{2}{3} \end{bmatrix}$

31. $A = \begin{bmatrix} 1 & 3 & 3 \\ 1 & 2 & 1 \\ 1 & -1 & 2 \end{bmatrix}$

$$\left[\begin{array}{ccc|ccc} 1 & 3 & 3 & 1 & 0 & 0 \\ 1 & 2 & 1 & 0 & 1 & 0 \\ 1 & -1 & 2 & 0 & 0 & 1 \end{array}\right] \rightarrow \left[\begin{array}{ccc|ccc} 1 & 3 & 3 & 1 & 0 & 0 \\ 0 & -1 & -2 & -1 & 1 & 0 \\ 0 & -4 & -1 & -1 & 0 & 1 \end{array}\right] \rightarrow \left[\begin{array}{ccc|ccc} 1 & 3 & 3 & 1 & 0 & 0 \\ 0 & 1 & 2 & 1 & -1 & 0 \\ 0 & -4 & -1 & -1 & 0 & 1 \end{array}\right]$$

$R_2 = -1r_1 + r_2$
$R_3 = -1r_1 + r_3$ $\quad R_2 = -1r_2$

$$\to\left[\begin{array}{ccc|ccc}1&0&-3&-2&3&0\\0&1&2&1&-1&0\\0&0&7&3&-4&1\end{array}\right]\to\left[\begin{array}{ccc|ccc}1&0&-3&-2&3&0\\0&1&2&1&-1&0\\0&0&1&\frac{3}{7}&-\frac{4}{7}&\frac{1}{7}\end{array}\right]\to\left[\begin{array}{ccc|ccc}1&0&0&-\frac{5}{7}&\frac{9}{7}&\frac{3}{7}\\0&1&0&\frac{1}{7}&\frac{1}{7}&-\frac{2}{7}\\0&0&1&\frac{3}{7}&-\frac{4}{7}&\frac{1}{7}\end{array}\right]$$

$R_1 = -3r_2 + r_1$, $R_3 = 4r_2 + r_3$; $R_3 = \frac{1}{7}r_3$; $R_1 = 3r_3 + r_1$, $R_2 = -2r_3 + r_2$

Therefore, $A^{-1} = \begin{bmatrix}-\frac{5}{7}&\frac{9}{7}&\frac{3}{7}\\\frac{1}{7}&\frac{1}{7}&-\frac{2}{7}\\\frac{3}{7}&-\frac{4}{7}&\frac{1}{7}\end{bmatrix} = \frac{1}{7}\begin{bmatrix}-5&9&3\\1&1&-2\\3&-4&1\end{bmatrix}$

33. $A = \begin{bmatrix}4&-8\\-1&2\end{bmatrix}$

$$\left[\begin{array}{cc|cc}4&-8&1&0\\-1&2&0&1\end{array}\right]\to\left[\begin{array}{cc|cc}1&-2&\frac{1}{4}&0\\-1&2&0&1\end{array}\right]\to\left[\begin{array}{cc|cc}1&-2&\frac{1}{4}&0\\0&0&\frac{1}{4}&1\end{array}\right]$$

$R_1 = \frac{1}{4}r_1$; $R_2 = r_1 + r_2$

This did not take the form $[I_2|B]$, so A has no inverse; i.e., A is singular.

35. $\begin{cases}3x - 2y = 1\\10x + 10y = 5\end{cases}$ becomes: $\left[\begin{array}{cc|c}3&-2&1\\10&10&5\end{array}\right]$

In two steps, we can get a 1 in row 1, column 1, *and* avoid fractions:

$$\to\left[\begin{array}{cc|c}3&-2&1\\1&16&2\end{array}\right]\to\left[\begin{array}{cc|c}1&16&2\\3&-2&1\end{array}\right]\to\left[\begin{array}{cc|c}1&16&2\\0&-50&-5\end{array}\right]$$

$R_2 = -3r_1 + r_2$; Interchange rows; $R_2 = -3r_1 + r_2$

$$\to\left[\begin{array}{cc|c}1&16&2\\0&1&\frac{1}{10}\end{array}\right]\to\left[\begin{array}{cc|c}1&0&\frac{20}{10}-\frac{16}{10}\\0&1&\frac{1}{10}\end{array}\right]$$

$R_2 = -\frac{1}{50}r_2$; $R_1 = -16r_2 + r_1$

The solution is $x = \frac{2}{5}$, $y = \frac{1}{10}$.

37. $\begin{cases} 5x + 6y - 3z = 6 \\ 4x - 7y - 2z = -3 \\ 3x + y - 7z = 1 \end{cases}$ becomes: $\left[\begin{array}{ccc|c} 5 & 6 & -3 & 6 \\ 4 & -7 & -2 & -3 \\ 3 & 1 & -7 & 1 \end{array}\right]$

$$\rightarrow \left[\begin{array}{ccc|c} 1 & 13 & -1 & 9 \\ 4 & -7 & -2 & -3 \\ 3 & 1 & -7 & 1 \end{array}\right] \rightarrow \left[\begin{array}{ccc|c} 1 & 13 & -1 & 9 \\ 0 & -59 & 2 & -39 \\ -1 & -38 & -4 & -26 \end{array}\right]$$

Let's get some smaller numbers:

$R_1 = -1r_2 + r_1$ $\quad R_2 = -4r_1 + r_2$, $R_3 = -3r_1 + r_3$

$$\rightarrow \left[\begin{array}{ccc|c} 1 & 13 & -1 & 9 \\ 0 & -59 & 2 & -39 \\ 0 & 19 & 2 & 13 \end{array}\right] \rightarrow \left[\begin{array}{ccc|c} 1 & 13 & -1 & 9 \\ 0 & -2 & 8 & 0 \\ 0 & 19 & 2 & 13 \end{array}\right]$$

$R_3 = -\frac{1}{2}r_3$ $\quad R_2 = 3r_3 + r_2$

$$\rightarrow \left[\begin{array}{ccc|c} 1 & 13 & -1 & 9 \\ 0 & 1 & -4 & 0 \\ 0 & 19 & 2 & 13 \end{array}\right]$$

We got a 1 in row 2, column 2, ***and*** avoided fractions, by taking a couple of extra steps!

$R_2 = -\frac{1}{2}r_2$

$$\rightarrow \left[\begin{array}{ccc|c} 1 & 0 & 51 & 9 \\ 0 & 1 & -4 & 0 \\ 0 & 0 & 78 & 13 \end{array}\right] \rightarrow \left[\begin{array}{ccc|c} 1 & 0 & 51 & 9 \\ 0 & 1 & -4 & 0 \\ 0 & 0 & 1 & \frac{13}{78} = \frac{1}{6} \end{array}\right] \rightarrow \left[\begin{array}{ccc|c} 1 & 0 & 0 & \frac{-51}{6} + \frac{54}{6} \\ 0 & 1 & 0 & \frac{4}{6} \\ 0 & 0 & 1 & \frac{1}{6} \end{array}\right]$$

$R_1 = -13r_2 + r_1$, $R_3 = -19r_2 + r_3$ $\quad R_3 = \frac{1}{78}r_3$ $\quad R_1 = -51r_3 + r_1$, $R_2 = 4r_3 + r_2$

The solution is $x = \frac{1}{2}, y = \frac{2}{3}, z = \frac{1}{6}$.

39. $\begin{cases} x - 2z = 1 \\ 2x + 3y = -3 \\ 4x - 3y - 4z = 3 \end{cases}$ becomes: $\left[\begin{array}{ccc|c} 1 & 0 & -2 & 1 \\ 2 & 3 & 0 & -3 \\ 4 & -3 & -4 & 3 \end{array}\right]$

$$\rightarrow \left[\begin{array}{ccc|c} 1 & 0 & -2 & 1 \\ 0 & 3 & 4 & -5 \\ 0 & -3 & 4 & -1 \end{array}\right] \rightarrow \left[\begin{array}{ccc|c} 1 & 0 & -2 & 1 \\ 0 & 1 & \frac{4}{3} & -\frac{5}{3} \\ 0 & -3 & 4 & -1 \end{array}\right] \rightarrow \left[\begin{array}{ccc|c} 1 & 0 & -2 & 1 \\ 0 & 1 & \frac{4}{3} & -\frac{5}{3} \\ 0 & 0 & 8 & -6 \end{array}\right]$$

$R_2 = -2r_1 + r_2$, $R_3 = -4r_1 + r_3$ $\quad R_2 = \frac{1}{3}r_2$ $\quad R_3 = 3r_2 + r_3$

$$\rightarrow \left[\begin{array}{ccc|c} 1 & 0 & -2 & 1 \\ 0 & 1 & \frac{4}{3} & -\frac{5}{3} \\ 0 & 0 & 1 & -\frac{3}{4} \end{array}\right] \rightarrow \left[\begin{array}{ccc|c} 1 & 0 & 0 & -\frac{1}{2} \\ 0 & 1 & 0 & -\frac{2}{3} \\ 0 & 0 & 1 & -\frac{3}{4} \end{array}\right]$$

$$R_3 = -\frac{1}{8}r_3 \qquad R_1 = 2r_3 + r_1, \quad R_2 = -\frac{4}{3}r_3 + r_2$$

The solution is $x = -\frac{1}{2}$, $y = -\frac{2}{3}$, $z = -\frac{3}{4}$.

41. $\begin{cases} x - y + z = 0 \\ x - y - 5z = 6 \\ 2x - 2y + z = 1 \end{cases}$ (Get the constants on the right-hand-side.)

$$\left[\begin{array}{ccc|c} 1 & -1 & 1 & 0 \\ 1 & -1 & -5 & 6 \\ 2 & -2 & 1 & 1 \end{array}\right] \rightarrow \left[\begin{array}{ccc|c} 1 & -1 & 1 & 0 \\ 0 & 0 & -6 & 6 \\ 0 & 0 & -1 & 1 \end{array}\right]$$

$$R_2 = -1r_1 + r_2, \quad R_3 = -2r_1 + r_3$$

It is impossible to obtain a 1 in row 2 column 2 (we cannot use row 1, since that would mess up the 0's we have in column 1).

So we focus on row 2, column 3:

$$\rightarrow \left[\begin{array}{ccc|c} 1 & -1 & 1 & 0 \\ 0 & 0 & 1 & -1 \\ 0 & 0 & -1 & 1 \end{array}\right] \rightarrow \left[\begin{array}{ccc|c} 1 & -1 & 0 & 1 \\ 0 & 0 & 1 & -1 \\ 0 & 0 & 0 & 0 \end{array}\right]$$

$$R_2 = -\frac{1}{6}r_2 \qquad R_1 = -r_2 + r_1, \quad R_3 = r_2 + r_3$$

Thus, we have: $x - y = 1$, $z = -1$

We can write the solution as:

$z = -1$, $x = y + 1$, where y can be any real number, or

$z = -1$, $y = x - 1$, where x can be any real number.

43. $\begin{cases} x - y - z - t = 1 \\ 2x + y + z + 2t = 3 \\ x - 2y - 2z - 3t = 0 \\ 3x - 4y + z + 5t = -3 \end{cases}$ becomes: $\left[\begin{array}{cccc|c} 1 & -1 & -1 & -1 & 1 \\ 2 & 1 & 1 & 2 & 3 \\ 1 & -2 & -2 & -3 & 0 \\ 3 & -4 & 1 & 5 & -3 \end{array}\right]$

$$\rightarrow \left[\begin{array}{cccc|c} 1 & -1 & -1 & -1 & 1 \\ 0 & 3 & 3 & 4 & 1 \\ 0 & -1 & -1 & -2 & -1 \\ 0 & -1 & 4 & 8 & -6 \end{array}\right] \rightarrow \left[\begin{array}{cccc|c} 1 & -1 & -1 & -1 & 1 \\ 0 & 1 & 1 & 0 & -1 \\ 0 & -1 & -1 & -2 & -1 \\ 0 & -1 & 4 & 8 & -6 \end{array}\right]$$

↑ $R_2 = -2r_1 + r_2$, $R_3 = -1r_2 + r_3$, $R_4 = -3r_1 + r_4$ ↑ $R_2 = 2r_3 + r_2$

$$\rightarrow \left[\begin{array}{cccc|c} 1 & 0 & 0 & -1 & 0 \\ 0 & 1 & 1 & 0 & -1 \\ 0 & 0 & 0 & -2 & -2 \\ 0 & 0 & 5 & 8 & -7 \end{array}\right]$$

Now we need a 1 in row 3, column 3.

↑ $R_1 = r_2 + r_1$, $R_3 = r_2 + r_3$, $R_4 = r_2 + r_4$

$$\rightarrow \left[\begin{array}{cccc|c} 1 & 0 & 0 & -1 & 0 \\ 0 & 1 & 1 & 0 & -1 \\ 0 & 0 & 5 & 8 & -7 \\ 0 & 0 & 0 & -2 & -2 \end{array}\right] \rightarrow \left[\begin{array}{cccc|c} 1 & 0 & 0 & -1 & 0 \\ 0 & 1 & 1 & 0 & -1 \\ 0 & 0 & 1 & \frac{8}{5} & -\frac{7}{5} \\ 0 & 0 & 0 & 1 & 1 \end{array}\right]$$

↑ Interchange r_3 and r_4 ↑ $R_3 = \frac{1}{5}r_3$, $R_4 = -\frac{1}{2}r_4$

$$\rightarrow \left[\begin{array}{cccc|c} 1 & 0 & 0 & -1 & 0 \\ 0 & 1 & 0 & -\frac{8}{5} & \frac{2}{5} \\ 0 & 0 & 1 & \frac{8}{5} & -\frac{7}{5} \\ 0 & 0 & 0 & 1 & 1 \end{array}\right] \rightarrow \left[\begin{array}{cccc|c} 1 & 0 & 0 & 0 & 1 \\ 0 & 1 & 0 & 0 & 2 \\ 0 & 0 & 1 & 0 & -3 \\ 0 & 0 & 0 & 1 & 1 \end{array}\right]$$

↑ $R_2 = -1r_3 + r_2$ ↑ $R_1 = r_4 + r_1$, $R_2 = \frac{8}{5}r_4 + r_2$, $R_3 = -\frac{8}{5}r_4 + r_3$

The solution is: $x = 1, y = 2, z = -3, t = 1$

45. $\begin{vmatrix} 3 & 4 \\ 1 & 3 \end{vmatrix} = (3)(3) - (1)(4) = 9 - 4 = 5$

47. $$\begin{vmatrix} 1 & 4 & 0 \\ -1 & 2 & 6 \\ 4 & 1 & 3 \end{vmatrix} = 1\begin{vmatrix} 2 & 6 \\ 1 & 3 \end{vmatrix} - 4\begin{vmatrix} -1 & 6 \\ 4 & 3 \end{vmatrix} + 0\begin{vmatrix} -1 & 2 \\ 4 & 1 \end{vmatrix}$$
$$= 1(6 - 6) - 4(-3 - 24) + 0$$
$$= 0 - 4(-27) + 0$$
$$= 108$$

49. $$\begin{vmatrix} 2 & 1 & -3 \\ 5 & 0 & 1 \\ 2 & 6 & 0 \end{vmatrix} = 2\begin{vmatrix} 0 & 1 \\ 6 & 0 \end{vmatrix} - 1\begin{vmatrix} 5 & 1 \\ 2 & 0 \end{vmatrix} + (-3)\begin{vmatrix} 5 & 0 \\ 2 & 6 \end{vmatrix}$$
$$= 2(0 - 6) - 1(0 - 2) + (-3)(30 - 0)$$
$$= -12 + 2 - 90$$
$$= -100$$

51. $$\begin{cases} x - 2y = 4 \\ 3x + 2y = 4 \end{cases}$$

Here, $D = \begin{vmatrix} 1 & -2 \\ 3 & 2 \end{vmatrix} = 2 - (-6) = 8$

To obtain D_x, replace the first column in D by the column of constants.

$$D_x = \begin{vmatrix} 4 & -2 \\ 4 & 2 \end{vmatrix} = 8 - (-8) = 16$$

To obtain D_y, replace the second column of D by the column of constants.

$$D_y = \begin{vmatrix} 1 & 4 \\ 3 & 4 \end{vmatrix} = 4 - 12 = -8$$

Then, $x = \dfrac{D_x}{D} = \dfrac{16}{8} = 2$ and $y = \dfrac{D_y}{D} = \dfrac{-8}{8} = -1$

53. $$\begin{cases} 2x + 3y - 13 = 0 \\ 3x - 2y = 0 \end{cases}$$
$$\begin{cases} 2x + 3y = 13 \\ 3x - 2y = 0 \end{cases}$$

Here, $D = \begin{vmatrix} 2 & 3 \\ 3 & -2 \end{vmatrix} = -4 - 9 = -13$

$$D_x = \begin{vmatrix} 13 & 3 \\ 0 & -2 \end{vmatrix} = -26 - 0 = -26$$

$$D_y = \begin{vmatrix} 2 & 13 \\ 3 & 0 \end{vmatrix} = 0 - 39 = -39$$

By Cramer's Rule, $x = \dfrac{D_x}{D} = \dfrac{-26}{-13} = 2$ and $y = \dfrac{D_y}{D} = \dfrac{-39}{-13} = 3$

55. $\begin{cases} x + 2y - z = 6 \\ 2x - y + 3z = -13 \\ 3x - 2y + 3z = -16 \end{cases}$

$$D = \begin{vmatrix} 1 & 2 & -1 \\ 2 & -1 & 3 \\ 3 & -2 & 3 \end{vmatrix} = 1\begin{vmatrix} -1 & 3 \\ -2 & 3 \end{vmatrix} - 2\begin{vmatrix} 2 & 3 \\ 3 & 3 \end{vmatrix} + (-1)\begin{vmatrix} 2 & -1 \\ 3 & -2 \end{vmatrix}$$

$$= 1(-3 + 6) - 2(6 - 9) + (-1)(-4 + 3)$$
$$= 3 + 6 + 1$$
$$= 10$$

$$D_x = \begin{vmatrix} 6 & 2 & -1 \\ -13 & -1 & 3 \\ -16 & -2 & 3 \end{vmatrix} = 6\begin{vmatrix} -1 & 3 \\ -2 & 3 \end{vmatrix} - 2\begin{vmatrix} -13 & 3 \\ -16 & 3 \end{vmatrix} + (-1)\begin{vmatrix} -13 & -1 \\ -16 & -2 \end{vmatrix}$$

$$= 6(-3 + 6) - 2(-39 + 48) + (-1)(26 - 16)$$
$$= 18 - 18 - 10$$
$$= -10$$

$$D_y = \begin{vmatrix} 1 & 6 & -1 \\ 2 & -13 & 3 \\ 3 & -16 & 3 \end{vmatrix} = 1\begin{vmatrix} -13 & 3 \\ -16 & 3 \end{vmatrix} - 6\begin{vmatrix} 2 & 3 \\ 3 & 3 \end{vmatrix} + (-1)\begin{vmatrix} 2 & -13 \\ 3 & -16 \end{vmatrix}$$

$$= 1(-39 + 48) - 6(6 - 9) + (-1)(-32 + 39)$$
$$= 9 + 18 - 7$$
$$= 20$$

$$D_z = \begin{vmatrix} 1 & 2 & 6 \\ 2 & -1 & -13 \\ 3 & -2 & -16 \end{vmatrix} = 1\begin{vmatrix} -1 & -13 \\ -2 & -16 \end{vmatrix} - 2\begin{vmatrix} 2 & -13 \\ 3 & -16 \end{vmatrix} + 6\begin{vmatrix} 2 & -1 \\ 3 & -2 \end{vmatrix}$$

$$= 1(16 - 26) - 2(-32 + 39) + 6(-4 + 3)$$
$$= -10 - 14 - 6$$
$$= -30$$

By Cramer's Rule, $x = \dfrac{D_x}{D} = \dfrac{-10}{10} = -1$, $y = \dfrac{D_y}{D} = \dfrac{20}{10} = 2$, and $z = \dfrac{D_z}{D} = \dfrac{-30}{10} = -3$

57. $\dfrac{6}{x(x - 4)}$

Step 1: This is a proper fraction.

Step 2: $q(x) = x(x - 4)$
This is Case 1 (nonrepeated linear factors).

Step 3: $\dfrac{6}{x(x - 4)} = \dfrac{A}{x} + \dfrac{B}{x - 4}$

Step 4: Multiply both sides by $x(x - 4)$: $6 = A(x - 4) + Bx$

Let $x = 0$: $6 = -4A$, or $A = -\dfrac{6}{4} = -\dfrac{3}{2}$

Let $x = 4$: $6 = 4B$, or $B = \dfrac{6}{4} = \dfrac{3}{2}$

Step 5: $\dfrac{6}{x(x - 4)} = \dfrac{-\frac{3}{2}}{x} + \dfrac{\frac{3}{2}}{x - 4}$

59. $\dfrac{x - 4}{x^2(x - 1)}$

Step 1: This is a proper fraction.

Step 2: $q(x) = x^2(x - 1)$

Note that x^2 is a ***repeated linear*** factor: $x^2 = (x)(x)$.

Step 3: $\dfrac{x - 4}{x^2(x - 1)} = \dfrac{A}{x} + \dfrac{B}{x^2} + \dfrac{C}{x - 1}$

Step 4: Multiply by $x^2(x - 1)$: $x - 4 = Ax(x - 1) + B(x - 1) + Cx^2$

Let $x = 0$: $-4 = -B$, or $B = 4$

Let $x = 1$: $-3 = C$

Now choose any value of x, say

$$x = 2: \quad -2 = 2A + B + 4C$$
$$-2 = 2A + 4 - 12$$
$$2A = -2 - 4 + 12$$
$$A = 3$$

Step 5: $\dfrac{x - 4}{x^2(x - 1)} = \dfrac{3}{x} + \dfrac{4}{x^2} - \dfrac{3}{x - 1}$

61. $\dfrac{x}{(x^2 + 9)(x + 1)}$

Step 1: This is proper.

Step 2: $q(x) = (x^2 + 9)(x + 1)$

$(x^2 + 9)$ cannot be factored

This is Case 1 and Case 3.

Step 3: $\dfrac{x}{(x^2 + 9)(x + 1)} = \dfrac{Ax + B}{x^2 + 9} + \dfrac{C}{x + 1}$

Step 4: $x = (Ax + B)(x + 1) + C(x^2 + 9) \qquad (1)$

Now $q(x)$ only has one zero, $x = -1$:

Let $x = -1$: $\quad -1 = 10C = -\dfrac{1}{10}$

To find the coefficients A and B, let's expand (1):

$$x = Ax^2 + Ax + Bx + B + Cx^2 + 9C$$
$$x = (A + C)x^2 + (A + B)x + (B + 9C)$$

Now equate coefficients of like powers of x:

For x^2: $\quad 0 = A + C$

For x: $\quad 1 = A + B$

For the constant: $\quad 0 = B + 9C$

But we know $C = -\dfrac{1}{10}$, so

$$0 = A + C$$
$$0 = A - \frac{1}{10}$$
$$A = \frac{1}{10}$$

and

$$0 = B + 9C$$
$$0 = B - \frac{9}{10}$$
$$B = \frac{9}{10}$$

Step 5: $$\frac{x}{(x^2 + 9)(x + 1)} = \frac{\frac{1}{10}x + \frac{9}{10}}{x^2 + 9} = \frac{-\frac{1}{10}}{x + 1}$$

63. $$\frac{x^3}{(x^2 + 4)^2}$$

Step 1: This is a proper fraction.

Step 2: $q(x) = (x^2 + 4)^2$

This is Case 4 (a ***repeated*** irreducible quadratic factor).

Step 3: $$\frac{x^3}{(x^2 + 4)^2} = \frac{Ax + B}{x^2 + 4} + \frac{Cx + D}{(x^2 + 4)^2}$$

Step 4: Clear of fractions:

$$x^3 = (Ax + B)(x^2 + 4) + Cx + D$$

Since $q(x)$ has no real zeros, we choose to expand the right-hand-side and equate coefficients of like powers of x:

$$x^3 = Ax^3 + Bx^2 + 4Ax + 4B + Cx + D$$
$$x^3 = Ax^3 + Bx^2 + (4A + C)x + (4B + D)$$

Coefficient of x^3: $1 = A$

Coefficient of x^2: $0 = B$

Coefficient of x: $0 = 4A + C$

Constant term: $0 = 4B + D$

Since $A = 1$, we have:

$$0 = 4A + C$$
$$0 = 4 + C$$
$$C = -4$$

Since $B = 0$, we have:

$$0 = 4B + D$$
$$0 = 0 + D$$
$$D = 0$$

Step 5: $$\frac{x^3}{(x^2 + 4)^2} = \frac{x}{x^2 + 4} + \frac{-4x}{(x^2 + 4)^2}$$

65. $$\frac{x^2}{(x^2 + 1)(x^2 - 1)}$$

Step 1: This fraction is proper.

Step 2: $q(x) = (x^2 + 1)(x^2 - 1) = (x^2 + 1)(x + 1)(x - 1)$

This is Case 1 and Case 3.

Step 3: $$\frac{x^2}{(x^2 + 1)(x^2 - 1)} = \frac{Ax + B}{x^2 + 1} + \frac{C}{x + 1} + \frac{D}{x - 1}$$

Step 4: $x^2 = (Ax + B)(x + 1)(x - 1) + C(x^2 + 1)(x - 1) + D(x^2 + 1)(x + 1)$

Let $x = 1$: $1 = 4D$ *or* $D = \frac{1}{4}$

Let $x = -1$: $1 = 4C$ *or* $C = -\frac{1}{4}$

Now we need two more equations:

Let $x = 0$: $0 = -B - C + D$

$$0 = -B + \frac{1}{4} + \frac{1}{4}$$
$$B = \frac{1}{2}$$

Finally, let $x = 2$:

$$4 = (2A + B)(3) + C(5) + D(15)$$
$$4 = 6A + 3B + 5C + 15D$$
$$4 = 6A + 3\left[\frac{1}{2}\right] + 5\left[-\frac{1}{4}\right] + 15\left[\frac{1}{4}\right]$$
$$4 = 6A + \frac{6}{4} - \frac{5}{4} + \frac{15}{4}$$
$$4 = 6A + \frac{16}{4}$$
$$0 = 6A$$
$$A = 0$$

Step 5: $$\frac{x^2}{(x^2 + 1)(x^2 - 1)} = \frac{\frac{1}{2}}{x^2 + 1} + \frac{-\frac{1}{4}}{x + 1} + \frac{\frac{1}{4}}{x - 1}$$

67. $$\begin{cases} 2x + y + 3 = 0 & (1) \\ x^2 + y^2 = 5 & (2) \end{cases}$$

We solve (1) for y: $$2x + y + 3 = 0$$
$$y = -2x - 3$$

Substitute this expression for y into (2): $$x^2 + y^2 = 5$$
$$x^2 + (-2x - 3)^2 = 5$$
$$x^2 + 4x^2 + 12x + 9 = 5$$
$$5x^2 + 12x + 4 = 0$$
$$(5x + 2)(x + 2) = 0$$
$$\text{so } x = \frac{-2}{5} \text{ or } x = -2$$

If $x = \frac{-2}{5}$, then, from (1): $$y = -2x - 3$$
$$y = \frac{4}{5} - 3 = \frac{-11}{5}$$

If $x = -2$, then: $$y = -2x - 3 = 1$$

Thus, we have two possible solutions: $x = \frac{-2}{5}$, $y = \frac{-11}{5}$ and $x = -2$, $y = 1$

Both of these check.

69. $$\begin{cases} 2xy + y^2 = 10 & (1) \\ 3y^2 - xy = 2 & (2) \end{cases}$$

$$\begin{cases} 2xy + y^2 = 10 & (1) \\ -2xy + 6y^2 = 4 & (2) \end{cases}$$ Multiply by 2.

$$\begin{cases} 2xy + y^2 = 10 & (1) \\ 7y^2 = 14 & (2) \end{cases}$$ Replace (2) by (1) + (2).

From (2): $$y^2 = 2$$
$$y = \pm\sqrt{2}$$

If $y = \sqrt{2}$, then:

$$\begin{aligned} 2xy + y^2 &= 10 \quad (1) \\ 2\sqrt{2}x + 2 &= 10 \\ 2\sqrt{2}x &= 8 \\ x &= \frac{8}{2\sqrt{2}} = 2\sqrt{2} \end{aligned}$$

If $y = -\sqrt{2}$, then:

$$\begin{aligned} -2\sqrt{2}x + 2 &= 10 \\ -2\sqrt{2}x &= 8 \\ x &= -2\sqrt{2} \end{aligned}$$

We have two possible solutions: $x = 2\sqrt{2}$, $y = \sqrt{2}$; and $x = -2\sqrt{2}$, $y = -\sqrt{2}$ and they both check.

71. $\begin{cases} x^2 + y^2 = 6y & (1) \\ x^2 = 3y & (2) \end{cases}$

From (2): $x^2 = 3y$

Then, by (1):

$$\begin{aligned} 3y + y^2 &= 6y \\ y^2 - 3y &= 0 \\ y(y - 3) &= 0 \\ y &= 0 \quad \text{or } y = 3 \end{aligned}$$

If $y = 0$, $x^2 = 3(0) = 0$, so $x = 0$

If $y = 3$, $x^2 = 3(3) = 9$, so $x = \pm 3$

All of these solutions check: $x = 0, y = 0$; $x = 3, y = 3$; and $x = -3, y = 3$

73. $\begin{cases} 3x^2 + 4xy + 5y^2 = 8 & (1) \\ x^2 + 3xy + 2y^2 = 0 & (2) \end{cases}$

Equation (2) can be factored:

$$\begin{aligned} x^2 + 3xy + 2y^2 &= 0 \\ (x + y)(x + 2y) &= 0 \end{aligned}$$

So $x = -y$ or $x = -2y$.

From (1), if $x = -y$, then:

$$\begin{aligned} 3(-y)^2 + 4(-y)y + 5y^2 &= 8 \\ 3y^2 - 4y^2 + 5y^2 &= 8 \\ 4y^2 &= 8 \\ y^2 &= 2 \\ y &= \pm\sqrt{2} \end{aligned}$$

Since $x = -y$, if $y = \sqrt{2}$, $x = -\sqrt{2}$, and if $y = -\sqrt{2}$, $x = \sqrt{2}$.

This gives two possible solutions: $x = \sqrt{2}$, $y = -\sqrt{2}$; $x = -\sqrt{2}$, $y = \sqrt{2}$, both of which check.

On the other hand, if $x = -2y$, then from (1):

$$\begin{aligned} 3(-2y)^2 + 4(-2y)y + 5y^2 &= 8 \\ 12y^2 - 8y^2 + 5y^2 &= 8 \\ 9y^2 &= 8 \\ y^2 &= \frac{8}{9} \\ y &= \pm\sqrt{\frac{8}{9}} = \pm\frac{2\sqrt{2}}{3} \end{aligned}$$

Then, since $x = -2y$, we have:

If $y = \frac{2\sqrt{2}}{3}$, then $x = \frac{-4\sqrt{2}}{3}$

If $y = \frac{-2\sqrt{2}}{3}$, then $x = \frac{4\sqrt{2}}{3}$

Let's check these two possible solutions:

For $x = \frac{4\sqrt{2}}{3}$, $y = \frac{-2\sqrt{2}}{3}$

$$\begin{cases} 3x^2 + 4xy + 5y^2 = 3\left(\frac{32}{9}\right) + 4\left(\frac{-16}{9}\right) + 5\left(\frac{8}{9}\right) = \frac{72}{9} = 8 \\ x^2 + 3xy + 2y^2 = \left(\frac{32}{9}\right) + 3\left(\frac{-16}{9}\right) + 2\left(\frac{8}{9}\right) = \frac{0}{9} = 0 \end{cases}$$

The solution $x = \frac{-4\sqrt{2}}{3}$, $y = \frac{2\sqrt{2}}{3}$ also checks.

Thus, we have four solutions: $x = \sqrt{2}, y = -\sqrt{2}$; $x = -\sqrt{2}, y = \sqrt{2}$;

$x = \frac{4\sqrt{2}}{3}, y = \frac{-2\sqrt{2}}{3}$; and $x = \frac{-4\sqrt{2}}{3}, y = \frac{2\sqrt{2}}{3}$

75. $$\begin{cases} x^2 - 3x + y^2 + y = -2 & (1) \\ \dfrac{x^2 - x}{y} + y + 1 = 0 & (2) \end{cases}$$

$$\begin{cases} x^2 - 3x + y^2 + y + 2 = 0 & (1) \\ x^2 - x + y^2 + y = 0 & (2) \text{ Multiply by } y. \end{cases}$$

We can now eliminate y:

$$\begin{cases} x^2 - 3x + y^2 + y + 2 = 0 & (1) \\ -x^2 + x - y^2 - y = 0 & (2) \text{ Multiply by } -1. \end{cases}$$

$$\begin{cases} x^2 - 3x + y^2 + y + 2 = 0 & (1) \\ -2x + 2 = 0 & (2) \text{ Replace (2) by (1) + (2).} \end{cases}$$

From (2): $x = 1$. Then from (1):

$$\begin{aligned} x^2 - 3x + y^2 + y + 2 &= 0 \\ 1 - 3 + y^2 + y + 2 &= 0 \\ y^2 + y &= 0 \\ y(y + 1) &= 0 \end{aligned}$$

So, $y = 0$ or $y = -1$.

Thus, we have two possible solutions: $x = 1, y = 0$ or $x = 1, y = -1$.

From the original equation (2), we see that y *cannot* be 0. But the other solution checks. Thus, we have just one solution:

$x = 1, y = -1$.

77. $6\begin{cases} -2x + y \le 2 & (1) \\ x + y \ge 2 & (2) \end{cases}$

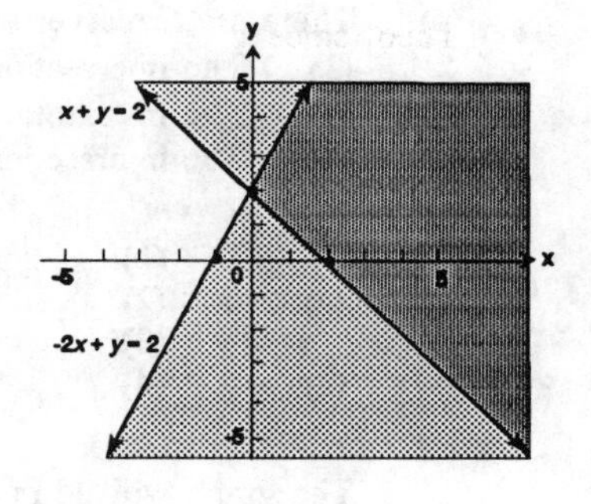

(a) $-2x + y \le 2$:

Step 1: Graph $-2x + y = 2$ with a solid line (since the inequality is nonstrict). The x-intercept is $x = -1$; the y-intercept is $y = 2$.

Step 2: Let's use (0, 0), which lies ***below*** the line, as a test point.

Step 3: At (0, 0), $-2x + y = 0 \le 2$, so the inequality is ***satisfied***.

Step 4: Therefore, we shade the region ***below*** the line $-2x + y = 2$.

(b) $x + y \ge 2$:

Step 1: Graph $x + y = 2$ with a solid line. The x-intercept is 2; the y-intercept is 2.

Step 2: Use (0, 0) ***below*** the line as a test point.

Step 3: At (0, 0), $x + y = 0 < 2$, so the inequality is ***not*** satisfied.

Step 4: Shade the region ***above*** the line $x + y = 2$.

(c) We see that the overlapping shaded region is unbounded.

(d) There is only one vertex, the intersection of $-2x + y = 2$ and $x + y = 2$:

$$\begin{cases} -2x + y = 2 & (1) \\ x + y = 2 & (2) \end{cases}$$

$$\begin{cases} -2x + y = 2 & (1) \\ -x - y = -2 & (2) \end{cases} \quad \text{Multiply by } -1.$$

$$\begin{cases} -2x + y = 2 & (1) \\ -3x = 0 & (2) \end{cases} \quad \text{Replace (2) by (1) + (2).}$$

$$\begin{cases} -2x + y = 2 & (1) \\ x = 0 & (2) \end{cases}$$

$$\begin{cases} y = 2 & (1) \\ x = 0 & (2) \end{cases}$$

The vertex is (0, 2).

79. $\begin{cases} x \ge 0 \\ y \ge 0 \\ x + y \le 4 \\ 2x + 3y \le 6 \end{cases}$

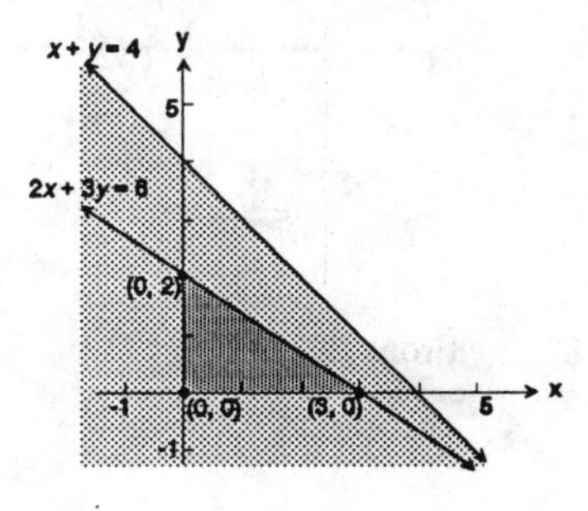

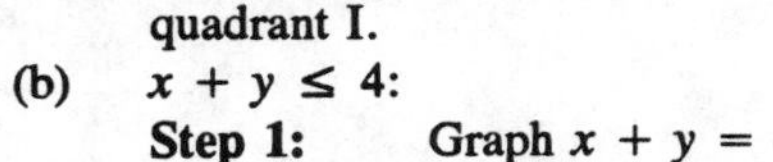

(a) $x \ge 0$; $y \ge 0$:

These inequalities require that our graph be located in quadrant I.

(b) $x + y \le 4$:

Step 1: Graph $x + y = 4$. The x-intercept is $x = 4$; the y-intercept is $y = 4$.

Step 2: Use (0, 0) below the line as a test point.

Step 3: At (0, 0), $x + y = 0 < 4$.

Step 4: Therefore, we shade ***below*** the line $x + y = 4$.

(c) $2x + 3y \le 6$:

Step 1: Graph $2x + 3y = 6$. The x-intercept is 3; the y-intercept is 2.

Step 2: Use (0, 0) below the line as a test point.

Step 3: At (0, 0), $2x + 3y = 0 < 6$.

Step 4: Shade ***below*** the line $2x + 3y = 6$.

(d) We see that the graph is bounded.

(e) There are ***three*** vertices:

(1) The intersection of $x = 0$ and $y = 0$: $(0, 0)$.

(2) The intersection of $2x + 3y = 6$ and the y-axis: $(0, 2)$.

(3) The intersection of $2x + 3y = 6$ and the x-axis: $(3, 0)$.

81. $\begin{cases} x \geq 0 & (1) \\ y \geq 0 & (2) \\ 2x + y \leq 8 & (3) \\ x + 2y \geq 2 & (4) \end{cases}$

(a) $x \geq 0$; $y \geq 0$:
The graph will lie in quadrant I.

(b) $2x + y \leq 8$:

Step 1: Graph $2x + y = 8$. The x-intercept is $x = 4$; the y-intercept is 8.

Step 2: Use $(0, 0)$ below the line as a test point.

Step 3: At $(0, 0)$, $2x + y = 0 < 8$.

Step 4: Therefore, we shade ***below*** the line $2x + y = 8$.

(c) $x + 2y \geq 2$:

Step 1: Graph $x + 2y = 2$. The x-intercept is 2; the y-intercept is 1.

Step 2: Use $(0, 0)$ below the line as a test point.

Step 3: At $(0, 0)$, $x + 2y = 0 < 2$, so the inequality is ***not*** satisfied.

Step 4: Therefore, we shade ***above*** the line $x + 2y = 2$.

(d) The graph is bounded.

(e) There are ***four*** vertices, all are either x-intercepts or y-intercepts.
$(0, 1)$, $(0, 8)$, $(4, 0)$, and $(2, 0)$.

83.

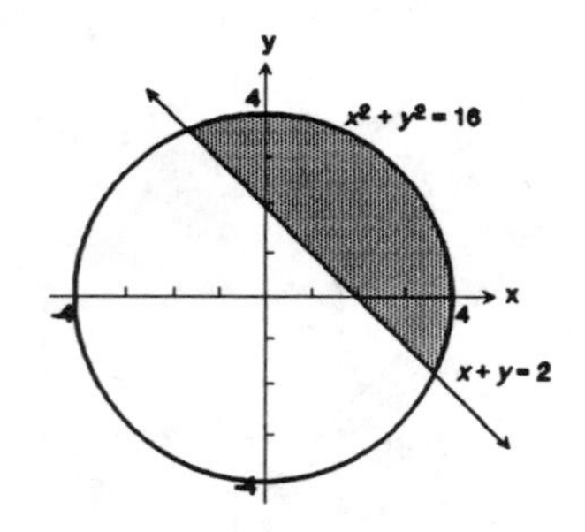

85.

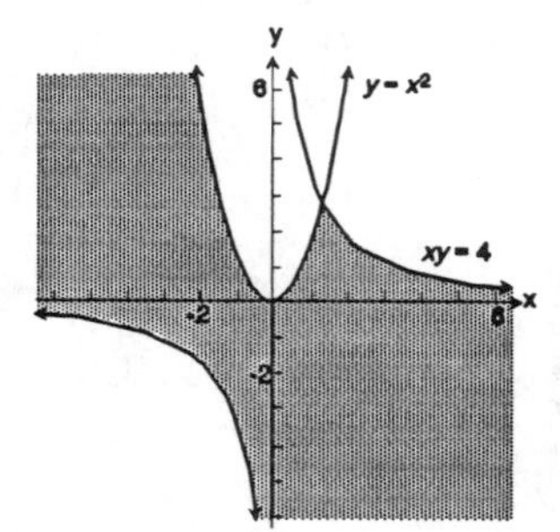

87. Maximize $z = 3x + 4y$
Subject to $x \geq 0$, $y \geq 0$, $3x + 2y \geq 6$, $x + y \leq 8$.
We graph the two lines:

(1) $y = -\frac{3}{2}x + 3$; y-intercept $= 3$; x-intercept $= 2$;

(2) $y = -x + 8$; y-intercept $= 8$; x-intercept $= 8$.

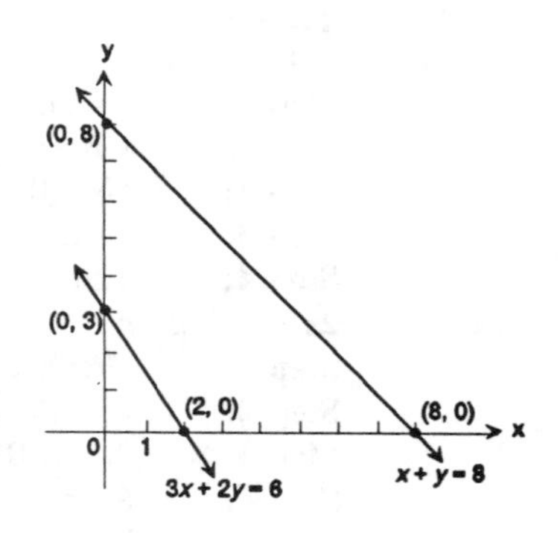

Vertex	Value of $z = 3x + 4y$
$(0, 3)$	$z = 12$
$(0, 8)$	$z = 32$
$(8, 0)$	$z = 24$
$(2, 0)$	$z = 6$

The maximum value of z is 32, at the point $(0, 8)$.

89. Minimize $z = 3x + 5y$
Subject to $x \geq 0, y \geq 0, x + y \geq 1, 3x + 2y \leq 12, x + 3y \leq 12$.
We graph the three lines:
(1) $y = -x + 1$; y-intercept = 1; x-intercept = 1
(2) $y = -\frac{3}{2}x + 6$; y-intercept = 6; x-intercept = 4
(3) $y = -\frac{1}{3}x + 4$; y-intercept = 4; x-intercept = 12

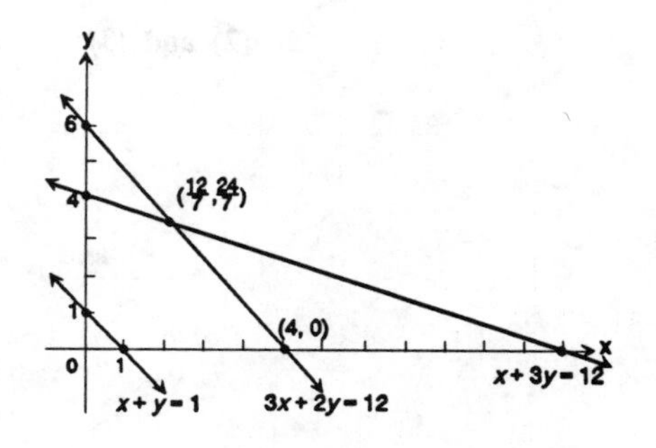

We need the point of intersection of (2) and (3):

$$-\frac{3}{2}x + 6 = -\frac{1}{3}x + 4$$
$$-9x + 36 = -2x + 24$$
$$-7x = -12$$
$$x = \frac{12}{7}$$

and $y = -\frac{1}{3}x + 4 = -\frac{4}{7} + \frac{28}{7} = \frac{24}{7}$; $\left(\frac{12}{7}, \frac{24}{7}\right)$

Vertex	Value of $z = 3x + 5y$
(0, 1)	$z = 5$
(0, 4)	$z = 20$
$\left(\frac{12}{7}, \frac{24}{7}\right)$	$z = \frac{36}{7} + \frac{120}{7} = \frac{156}{7} \approx 22.3$
(4, 0)	$z = 12$
(1, 0)	$z = 3$

The minimum value of z is 3, at (1, 0).

91. Maximize $z = 5x + 4y$
Subject to $x \geq 0, y \geq 0, x + 2y \geq 2, 3x + 4y \leq 12, y \geq x$
We have three lines:
(1) $y = -\frac{1}{2}x + 1$; y-intercept = 1; x-intercept = 2
(2) $y = -\frac{3}{4}x + 3$; y-intercept = 3; x-intercept = 4
(3) $y = x$ We need the following points of intersection:

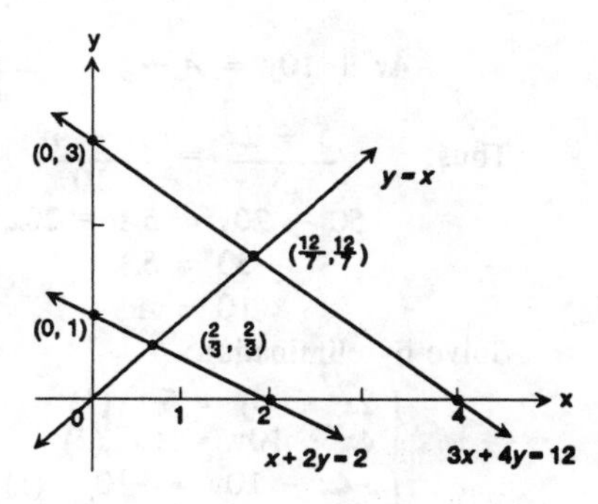

(1) and (3):
$$x = -\frac{1}{2}x + 1$$
$$2x = -x + 2$$
$$3x = 2$$
$$x = \frac{2}{3}$$
and
$$y = x$$
$$y = \frac{2}{3}$$
$$\left(\frac{2}{3}, \frac{2}{3}\right)$$

(2) and (3):

$$x = -\frac{3}{4}x + 3$$
$$4x = -3x + 12$$
$$7x = 12$$
$$x = \frac{12}{7}$$

and

$$y = x$$
$$y = \frac{12}{7}$$
$$\left(\frac{12}{7}, \frac{12}{7}\right)$$

Vertex	Value of $z = 5x + 4y$
(0, 1)	$z = 4$
(0, 3)	$z = 12$
$\left(\frac{12}{7}, \frac{12}{7}\right)$	$z = \frac{108}{7}$
$\left(\frac{2}{3}, \frac{2}{3}\right)$	$z = \frac{18}{3} = 6$

The maximum value of z is $\frac{108}{7}$, when $x = \frac{12}{7}$, and $y = \frac{12}{7}$.

93. $\begin{cases} 2x + 5y = 5 & (1) \\ 4x + 10y = A & (2) \end{cases}$

For the system of equations to have infinitely many solutions, all solutions must satisfy both equations. This means that both equations must be equal.

$$2x + 5y = 5 \rightarrow y = \frac{5 - 2x}{5}$$
$$4x + 10y = A \rightarrow y = \frac{A - 4x}{10}$$

Thus, $\frac{5 - 2x}{5} = \frac{A - 4x}{10}$

$$50 - 20x = 5A - 20x$$
$$50 = 5A$$
$$10 = A$$

Solve by elimination:

$\begin{cases} 2x + 5y = 5 & (1) \\ 4x + 10y = A & (2) \end{cases}$

$\begin{cases} -4x - 10y = -10 & (1) \quad \text{Multiply by } -2. \\ 4x + 10y = A & (2) \end{cases}$

$\begin{cases} -4x - 10y = -10 & (1) \\ 0 = A - 10 & (2) \quad \text{Replace (2) by (1) + (2).} \end{cases}$

If $A - 10 = 0$, the system will have infinitely many solutions of the form:

$$-4x - 10y = -10$$
$$-4x = 10y - 10$$
$$x = \frac{-5}{2}y + \frac{5}{2}$$

where y is any real number.

Therefore, we need $A = 10$.

95. We are given $y = ax^2 + bx + c$. If (0, 1) satisfies this equation, then:

$$1 = 0a + 0b + c,$$

or $\quad 1 = c$

From (1, 0), we have: $\quad 0 = a + b + c$

or $\quad a + b = -c$

$a + b = -1 \quad$ since $c = 1$

From (−2, 1): $\quad 1 = 4a - 2b + c$

$4a - 2b = 1 - c$

$4a - 2b = 0 \quad$ since $c = 1$

We now have two equations in the two unknowns a and b.

$$\begin{cases} a + b = -1 & (1) \\ 4a - 2b = 0 & (2) \end{cases}$$

$$\begin{cases} 2a - 2b = -2 & (1) \quad \text{Multiply by 2.} \\ 4a - 2b = 0 & (2) \end{cases}$$

$$\begin{cases} 2a + 2b = -2 & (1) \\ 6a = -2 & (2) \quad \text{Replace (2) by (1) + (2).} \end{cases}$$

From (2): $\quad a = \frac{-1}{3}$

Then from (1),

$$2a + 2b = -2$$

$$2\left(\frac{-1}{3}\right) + 2b = -2$$

$$2b = -2 + \frac{2}{3}$$

$$b = -1 + \frac{1}{3}$$

$$b = -\frac{2}{3}$$

Therefore, $y = ax^2 + bx + c = -\frac{1}{3}x^2 - \frac{2}{3}x + 1.$

97. Let $\quad x =$ blend of \$3.00 coffee

$y =$ blend of \$6.00 coffee

$$\begin{cases} x + y = 100 \\ 3.00x + 6.00y = 3.90(x + y) \end{cases}$$

$$\begin{cases} x + y = 100 \\ 3x + 6y = 3.9x + 3.9y \end{cases}$$

$$\begin{cases} x + y = 100 \\ .9x - 2.1y = 0 \end{cases}$$

$$\begin{cases} x + y = 100 \\ 9x - 21y = 0 \end{cases}$$

$$\begin{cases} y = 100 - x \\ 9x - 21(100 - x) = 0 \end{cases}$$

$$\begin{cases} y = 100 - x \\ 9x - 2100 + 21x) = 0 \end{cases}$$

$$\begin{cases} y = 100 - x \\ 30x = 2100 \end{cases}$$

$$\begin{cases} y = 100 - x \\ x = 70 \end{cases}$$

$$\begin{cases} y = 30 \\ x = 70 \end{cases}$$

The desired blend consists of 70 lbs of \$3.00 coffee and 30 lbs of \$6.00 coffee.

99. Let x = number of small boxes
y = number of medium boxes
z = number of large boxes

(1) $x + 2y + 2z = 15$
(2) $x + y + 2z = 10$
(3) $y + 3z = 11$
(1)–(2) $y = 5$
(3) $5 + 3z = 11$
$3z = 6$
$z = 2$
(1) $x + 2(5) + 2(2) = 15$
$x + 10 + 4 = 15$
$x = 1$

You should buy 1 small box, 5 medium boxes, and 2 large boxes.

101. Let w = width of the plot
ℓ = length of the plot
d = length of the diagonal
p = perimeter of the rectangle.

Now w, ℓ, d form a right triangle with hypotenuse = d, so we have
$$w^2 + \ell^2 = d^2 \quad (1)$$
Also
$$2w + 2\ell = p \quad (2)$$

So we have two equations:
$$\begin{cases} w^2 + \ell^2 = (26)^2 & (1) \\ 2w + 2\ell = 68 & (2) \end{cases}$$
$$\begin{cases} w^2 + \ell^2 = 676 & (1) \\ w + \ell = 34 & (2) \quad \text{Divide by 2.} \end{cases}$$

Solve for ℓ in (2): $w + \ell = 34$
$\ell = 34 - w$

Substitute this into (1):
$$w^2 + \ell^2 = 676$$
$$w^2 + (34 - w)^2 = 676$$
$$w^2 + 1156 - 68w + w^2 = 676$$
$$2w^2 - 68w + 480 = 0$$
$$w^2 - 34w + 240 = 0$$
$$(w - 24)(w - 10) = 0$$

so $w = 24$ or $w = 10$

Finally, from (2), $\ell = 34 - w$, so: If $w = 24$, then $\ell = 10$;
If $w = 10$, then $\ell = 24$

Thus, the rectangle is 10 feet by 24 feet.

103. Let x and y be the legs of the triangle. We have:
$$\begin{cases} x + y + 6 = 14 & (1) \text{ The perimeter.} \\ x^2 + y^2 = 36 & (2) \text{ Pythagorean Theorem.} \end{cases}$$

From (1): $x + y = 8$
$y = -x + 8$

Then, by (2):
$$x^2 + y^2 = 36$$
$$x^2 + (-x + 8)^2 = 36$$
$$x^2 + x^2 - 16x + 64 = 36$$
$$2x^2 - 16x + 28 = 0$$
$$x^2 - 8x + 14 = 0$$

so $x = \dfrac{8 \pm \sqrt{64 - 4(14)}}{2} = \dfrac{8 \pm \sqrt{8}}{2}$ or $x = 4 \pm \sqrt{2}$

If $x = 4 + \sqrt{2}$, then $y = 8 - x = 4 - \sqrt{2}$

If $x = 4 - \sqrt{2}$, then $y = 8 - x = 4 + \sqrt{2}$

In either case, the lengths of the legs are $4 + \sqrt{2}$ inches and $4 - \sqrt{2}$ inches.

105.

s 2s

$A = s^2$ $\quad A = (2s)^2$

$P = 4s$ $\quad P = 4(2s) = 8s$

$$
\begin{aligned}
s^2 + (2s)^2 &= 5000 \\
s^2 + 4s^2 &= 5000 \\
5s^2 &= 5000 \\
s^2 &= 1000 \\
s &= 10\sqrt{10} \\
2s + 8s &= \text{Fencing needed} \\
10s &= 12\left(10\sqrt{10}\right) \\
&= 100\sqrt{10} \text{ feet}
\end{aligned}
$$

107. Let $x =$ the amount Katy receives.
$y =$ the amount Mike receives.
$z =$ the amount Danny receives.
and $w =$ the amount Colleen receives.

We are told: $x + y + z + w = 45$, and

$y = 2x$

$w = x$

$z = \frac{x}{2}$

Therefore, by substitution:

$$
\begin{aligned}
x + y + z + w &= 45 \\
x + 2x + \frac{x}{2} + x &= 45 \\
2x + 4x + x + 2x &= 90 \quad \text{Multiply by 2.} \\
9x &= 90 \\
x &= 10
\end{aligned}
$$

Katy receives $x = 10$ dollars.

Mike receives $y = 2x = 20$ dollars.

Danny receives $z = \frac{x}{2} = 5$ dollars.

Colleen receives $w = x = 10$ dollars.

109. Let $x =$ # of hours for Bruce to do the job alone,
$y =$ # of hours for Bryce to do the job alone,
$z =$ # of hours for Marty to do the job alone.

We have:

	Fraction of the job done in one hour
Bruce	$\frac{1}{x}$
Bryce	$\frac{1}{y}$
Marty	$\frac{1}{z}$
Bruce & Bryce	$\frac{1}{x} + \frac{1}{y}$
Bryce & Marty	$\frac{1}{y} + \frac{1}{z}$
Marty & Bruce	$\frac{1}{x} + \frac{1}{z}$

But we are told that, working together, Bruce and Bryce can do the job in 1 hour and 20 minutes $\left(1\frac{2}{6} = \frac{8}{6} = \frac{4}{3} \text{ hours}\right)$.

Therefore, together, they complete $\frac{1}{\left(\frac{4}{3}\right)} = \frac{3}{4}$ of the job per hour:

$$\frac{1}{x} + \frac{1}{y} = \frac{3}{4} \qquad (1)$$

Similarly, Bryce and Marty do the job in 1 hour and 36 minutes

$$\left(1\frac{36}{60} = 1\frac{6}{10} = \frac{16}{10} = \frac{8}{5} \text{ hr.}\right)$$

Therefore, $$\frac{1}{y} + \frac{1}{z} = \frac{1}{\left(\frac{8}{5}\right)} = \frac{5}{8} \qquad (2)$$

Finally, Marty and Bruce take 2 hours 40 minutes $\left(2\frac{4}{6} = \frac{16}{6} = \frac{8}{3} \text{ hr.}\right)$ to do the job:

$$\frac{1}{x} + \frac{1}{z} = \frac{1}{\left(\frac{8}{3}\right)} = \frac{3}{8}$$

We have: $$\begin{cases} \frac{1}{x} + \frac{1}{y} = \frac{3}{4} & (1) \\ \frac{1}{y} + \frac{1}{z} = \frac{5}{8} & (2) \\ \frac{1}{x} + \frac{1}{z} = \frac{3}{8} & (3) \end{cases}$$

Let $u = \frac{1}{x}$, $v = \frac{1}{y}$, $w = \frac{1}{z}$:

$$\begin{cases} u + v = \frac{3}{4} & (1) \\ v + w = \frac{5}{8} & (2) \\ u + w = \frac{3}{8} & (3) \end{cases}$$

$$\begin{cases} u + v = \frac{3}{4} & (1) \\ v + w = \frac{5}{8} & (2) \\ -v + w = -\frac{3}{8} & (3) \quad \text{Replace (3) by (3) - (1).} \end{cases}$$

$$\begin{cases} u + v = \frac{3}{4} & (1) \\ v + w = \frac{5}{8} & (2) \\ 2w = \frac{1}{4} & (3) \quad \text{Replace (3) by (3) + (2).} \end{cases}$$

$$\begin{cases} u + v = \frac{3}{4} & (1) \\ v + w = \frac{5}{8} & (2) \\ w = \frac{1}{8} & (3) \quad \text{Divide by 2.} \end{cases}$$

$$\begin{cases} u + v = \frac{3}{4} & (1) \\ v + \frac{1}{8} = \frac{5}{8} & (2) \quad \text{Back-substitution; } w = \frac{1}{8}. \\ w = \frac{1}{8} & (3) \end{cases}$$

$$\begin{cases} u + v = \frac{3}{4} & (1) \\ v = \frac{1}{2} & (2) \\ w = \frac{1}{8} & (3) \end{cases}$$

$$\begin{cases} u = \frac{3}{4} & (1) \quad \text{Back-substitution; } v = \frac{1}{2}. \\ v = \frac{1}{2} & (2) \\ w = \frac{1}{8} & (3) \end{cases}$$

Finally, $\frac{1}{x} = u = \frac{1}{4}$, so $x = 4$; $\frac{1}{y} = v = \frac{1}{2}$, so $y = 2$; and $\frac{1}{z} = w = \frac{1}{8}$, so $z = 8$.

Bruce can do the job in $x = 4$ hours;
Bryce can do it in $y = 2$ hours;
Marty can do it in $z = 8$ hours.

111. Let x = Number of gasoline engines produced per week, and
y = Number of diesel engines produced per week

Then the objective function, or ***cost***, is given by:

$C = 450x + 550y$ (in dollars)

Our constraints are:

(1) $x \geq 0, y \geq 0$

(2) $x \geq 20$ (must deliver at least 20 gasoline engines)

(3) $y \geq 15$ (must deliver at least 15 diesel engines)

(4) $x \leq 60$ (factory cannot make more than 60 gasoline engines)

(5) $x \leq 40$ (factory cannot make more than 40 diesel engines)

(6) $x + y \geq 50$ (must produce a total of 50 engines)

Equations (2)–(5) bound a rectangle with vertices (20, 15), (20, 40), (60, 40), and (60, 15).

We will need two points of intersection:

(6) and (2): $\begin{cases} x + y = 50 \\ x = 20 \end{cases}$

Then $y = 30$

The point (20, 30) ***is*** in the feasible region.

(6) and (3): $\begin{cases} x + y = 50 \\ x = 15 \end{cases}$

Then $x = 35$

The point (35, 15) is also in the feasible region.

Vertex	Value of Cost: $C = 450x + 550y$
(20, 30)	$C = 25{,}500$
(20, 40)	$C = 31{,}000$
(60, 40)	$C = 49{,}000$
(60, 15)	$C = 35{,}250$
(35, 15)	$C = 24{,}000$

The minimum possible cost is \$24,000, obtained by producing $x = 35$ gasoline engines, and $y = 15$ diesel engines.

Since the factory is ***obligated*** to deliver 20 gasoline engines and 15 diesel engines, their excess capacity is 15 gasoline engines (and no excess diesel engines).

Chapter 9

SEQUENCES; INDUCTION; COUNTING; PROBABILITY

9.1 Sequences

1. $a_1 = 1$
$a_2 = 2$
$a_3 = 3$
$a_4 = 4$
$a_5 = 5$

3. $a_1 = \frac{1}{1+2} = \frac{1}{3}$
$a_2 = \frac{2}{2+2} = \frac{2}{4} = \frac{1}{2}$
$a_3 = \frac{3}{3+2} = \frac{3}{5}$
$a_4 = \frac{4}{4+2} = \frac{4}{6} = \frac{2}{3}$
$a_5 = \frac{5}{5+2} = \frac{5}{7}$

5. $a_1 = (1)(1^2) = 1$
$a_2 = (-1)(2^2) = -4$
$a_3 = (1)(3^2) = 9$
$a_4 = (-1)(4^2) = -16$
$a_5 = (1)(5^2) = 25$

7. $a_1 = \frac{3}{3} = 1$
$a_2 = \frac{3^2}{2^2+1} = \frac{9}{5}$
$a_3 = \frac{3^3}{2^3+1} = \frac{27}{9} = 3$
$a_4 = \frac{3^4}{2^4+1} = \frac{81}{17}$
$a_5 = \frac{3^5}{2^5+1} = \frac{243}{33} = \frac{81}{11}$

9. $a_1 = \frac{-1}{(2)(3)} = \frac{-1}{6}$
$a_2 = \frac{1}{(3)(4)} = \frac{1}{12}$
$a_3 = \frac{-1}{(4)(5)} = \frac{-1}{20}$
$a_4 = \frac{1}{(5)(6)} = \frac{1}{30}$
$a_5 = \frac{-1}{(6)(7)} = \frac{-1}{42}$

11. $a_1 = \frac{1}{e}$
$a_2 = \frac{2}{e^2}$
$a_3 = \frac{3}{e^3}$
$a_4 = \frac{4}{e^4}$
$a_5 = \frac{5}{e^5}$

13. $\frac{n}{n+1}$

15. $\frac{1}{2^{n-1}}$

17. $(-1)^{n+1}$

19. $(-1)^{n+1}n$

21. $a_1 = 3$
$a_2 = 2 + 3 = 5$
$a_3 = 2 + 5 = 7$
$a_4 = 2 + 7 = 9$
$a_5 = 2 + 9 = 11$

23. $a_1 = -2$
$a_2 = 1 + -2 = -1$
$a_3 = 2 + -1 = 1$
$a_4 = 3 + 1 = 4$
$a_5 = 4 + 4 = 8$

25. $a_1 = 5$
$a_2 = 2 \cdot 5 = 10$
$a_3 = 2 \cdot 10 = 20$
$a_4 = 2 \cdot 20 = 40$
$a_5 = 2 \cdot 40 = 80$

27. $a_1 = 3$
$a_2 = \frac{3}{1} = 3$
$a_3 = \frac{3}{2}$
$a_4 = \frac{\frac{3}{2}}{3} = \frac{1}{2}$
$a_5 = \frac{\frac{1}{2}}{4} = \frac{1}{8}$

29. $a_1 = 1$
$a_2 = 2$
$a_3 = 1 \cdot 2 = 2$
$a_4 = 2 \cdot 2 = 4$
$a_5 = 2 \cdot 4 = 8$

31. $a_1 = A$
$a_2 = A + d$
$a_3 = A + 2d$
$a_4 = A + 3d$
$a_5 = A + 4d$

33. $a_1 = \sqrt{2}$
$a_2 = \sqrt{2 + \sqrt{2}}$
$a_3 = \sqrt{2 + \sqrt{2 + \sqrt{2}}}$
$a_4 = \sqrt{2 + \sqrt{2 + \sqrt{2 + \sqrt{2}}}}$
$a_5 = \sqrt{2 + \sqrt{2 + \sqrt{2 + \sqrt{2 + \sqrt{2}}}}}$

35. $\sum_{k=1}^{10} 5 = 10(5) = 50$

37. $\sum_{k=1}^{6} k = 1 + 2 + 3 + 4 + 5 + 6 = 21$

39. $\sum_{k=1}^{5} (5k + 3) = \sum_{k=1}^{5} 5k + \sum_{k=1}^{5} 3 = 5\sum_{k=1}^{5} k + 5(3) = 5(1 + 2 + 3 + 4 + 5) + 15 = 5(15) + 15$
$= 75 + 15 = 90$

41. $\sum_{k=1}^{3} (k^2 + 4) = \sum_{k=1}^{3} k^2 + \sum_{k=1}^{3} 4 = 1^2 + 2^2 + 3^2 + 3(4) = 1 + 4 + 9 + 12 = 26$

43. $\sum_{k=1}^{6} (-1)^k 2^k = (-1)^1 2^1 + (-1)^2 2^2 + (-1)^3 2^3 + (-1)^4 2^4 + (-1)^5 2^5 + (-1)^6 \cdot 2^6$
$= -1 \cdot 2 + 1 \cdot 4 + -1 \cdot 8 + 1 \cdot 16 + -1 \cdot 32 + 1 \cdot 64$
$= -2 + 4 - 8 + 16 - 32 + 64 = 42$

45. $\sum_{k=1}^{4} (k^3 - 1) = \sum_{k=1}^{4} k^3 - \sum_{k=1}^{4} 1 = 1^3 + 2^3 + 3^3 + 4^3 - 4(1) = 1 + 8 + 27 + 64 - 4 = 96$

47. $\sum_{k=1}^{n} (k + 3) = 4 + 5 + 6 + \cdots + (n + 3)$

49. $\sum_{k=1}^{n} \frac{k^2}{2} = \frac{1}{2} + 2 + \frac{9}{2} + \cdots + \frac{n^2}{2}$

51. $\sum_{k=0}^{n} \frac{1}{3^k} = 1 + \frac{1}{3} + \frac{1}{9} + \cdots + \frac{1}{3^n}$

53. $\sum_{k=0}^{n-1} \frac{1}{3^{k+1}} = \frac{1}{3} + \frac{1}{9} + \frac{1}{27} + \cdots + \frac{1}{3^n}$

55. $\sum_{k=2}^{n} (-1)^k \ln k = \ln 2 - \ln 3 + \ln 4 - \cdots + (-1)^n \ln n$

57. $1 + 2 + 3 + \cdots + 20 = \sum_{k=1}^{20} k$

59. $\frac{1}{2} + \frac{2}{3} + \frac{3}{4} + \cdots + \frac{13}{13 + 1} = \sum_{k=1}^{13} \frac{k}{k + 1}$

61. $1 - \frac{1}{3} + \frac{1}{9} - \frac{1}{27} + \cdots + (-1)^6 \frac{1}{3^6} = \sum_{k=0}^{6} (-1)^k \left[\frac{1}{3^k}\right]$

63. $3 + \frac{3^2}{2} + \frac{3^3}{3} + \cdots + \frac{3^n}{n} = \sum_{k=1}^{n} \frac{3k}{k}$

65. $a + (a + d) + (a + 2d) + \cdots + (a + nd) = \sum_{k=1}^{n} a + (k - 1)d = \sum_{k=0}^{n} (a + kd)$

67. $a_1 = 1, a_2 = 1, a_3 = 2, a_4 = 3, a_5 = 5, a_6 = 8, a_7 = 13, a_8 = 21, a_{n+2} = a_{n+1} + a_n$
$a_8 = a_6 + a_7$
$a_8 = 8 + 13$
$a_8 = 21$

71. 1, 1, 2, 3, 5, 8, 13
The Fibonacci sequence.

9.2 Arithmetic Sequences

1. $d = a_{n+1} - a_n$
$d = (n + 5) - (n + 4)$
$d = n + 5 - n - 4$
$d = 1$
$a_1 = 5$
$a_2 = 6$
$a_3 = 7$
$a_4 = 8$

3. $d = [2(n + 1) - 5] - (2n - 5)$
$d = 2n + 2 - 5 - 2n + 5$
$d = 2$
$a_1 = -3$
$a_2 = -1$
$a_3 = 1$
$a_4 = 3$

5. $d = [6 - 2(n + 1)] - (6 - 2n)$
$d = 6 - 2n - 2 - 6 + 2n$
$d = -2$
$a_1 = 4$
$a_2 = 2$
$a_3 = 0$
$a_4 = -2$

7. $d = \left[\frac{1}{2} - \frac{1}{3}(n + 1)\right] - \frac{1}{2} - \frac{1}{3}n$
$d = \frac{1}{2} - \frac{1}{3}n - \frac{1}{3} - \frac{1}{2} + \frac{1}{3}n$
$d = -\frac{1}{3}$
$a_1 = \frac{1}{6}$
$a_2 = -\frac{1}{6}$
$a_3 = -\frac{1}{2}$
$a_4 = -\frac{5}{6}$

9. $d = (\ln 3^{n+1}) - (\ln 3^n)$
$d = (n + 1)\ln 3 - n(\ln 3)$
$d = \ln 3(n + 1 - n)$
$d = \ln 3$
$a_1 = \ln 3$
$a_2 = 2 \ln 3$
$a_3 = 3 \ln 3$
$a_4 = 4 \ln 3$

11. $a_n = 3 + (n - 1)2$
$a_n = 3 + 2n - 2$
$a_n = 2n + 1$
$a_5 = 2(5) + 1 = 11$

13. $a_n = 5 + (n - 1)(-3)$
$a_n = 5 - 3n + 3$
$a_n = 8 - 3n$
$a_5 = 5 + 4(-3) = -7$

15. $a_n = 0 + (n - 1)\frac{1}{2}$
$a_n = \frac{1}{2}n - \frac{1}{2}$
$a_n = \frac{1}{2}(n - 1)$
$a_5 = 4\left(\frac{1}{2}\right) = 2$

17. $a_n = \sqrt{2} + (n - 1)\sqrt{2}$
$a_n = \sqrt{2}(1 + n - 1)$
$a_n = n\sqrt{2} = \sqrt{2}n$
$a_5 = 5\sqrt{2}$

19. The first term of the arithmetic sequence is $a = 2$, and the common difference is 2. The nth term obeys the formula:
$a_n = 2 + (n - 1)2$
Hence, the 11th term is:
$a_{11} = 2 + (11 - 1)2$
$a_{11} = 2 + 20$
$a_{11} = 22$

21. $a = 1, d = -3$
$a_n = 1 + (n - 1)(-3)$
$a_n = 1 - 3n + 3$
$a_n = 4 - 3n$
$a_{10} = 1 + 9(-3)$
$a_{10} = 1 - 27 = -26$
Also, $a_{10} = 4 - 3(10)$
$a_{10} = 4 - 30$
$a_{10} = -26$

23. $a = a, d = b$
$a_n = a + (n - 1)b$
$a_8 = a + 7b$

25. $a_8 = a + 7d = 8$
$a_{21} = a + 20d = 47$
$-13d = -39$
$d = 3$
$a = 8 - 7d$
$a_1 = 8 - 7(3)$
$a_1 = 8 - 21$
$a_1 = -13$
$a_{n+1} = a_n + 3$

27. $a_9 = a + 8d = -5$
$a_{15} = a + 14d = 31$
$-6d = -36$
$d = 6$
$a = -5 - 8d$
$a_1 = -5 - 8(6)$
$a_1 = -53$
$a_{n+1} = a_n + 6$

29. $a_{15} = a + 14d = 0$
$a_{40} = a + 39d = -50$
$-25d = 50$
$d = -2$
$a = -14d$
$a_1 = -14(-2)$
$a_1 = 28$
$a_{n+1} = a_n - 2$

31. $a_{14} = a + 13d = -1$
$a_{18} = a + 17d = -9$
$-4d = 8$
$d = -2$
$a = -1 - 13d$
$a_1 = -1 - 13(-2)$
$a_1 = 25$
$a_{n+1} = a_n - 2$

33. $S_n = \frac{n}{2}(1 + (2n - 1))$
$= \frac{n}{2} \cdot 2n$
$= n^2$

35. $S_n = \frac{n}{2}(7 + (2 + 5n))$
$= \frac{n}{2}(9 + 5n)$

37. $S_{35} = \frac{35}{2}(2 + 70)$
$= \frac{35}{2} \cdot 72$
$= 1260$

39. $S_{11} = \frac{12}{2}(5 + 49)$
$= \frac{12}{2} \cdot 54$
$= 324$

41.
```
sum seq(3.45N+4.
12,N,1,20,1)
           806.9
```

43. $d = 5.2 - 2.8 = 2.4$
$a = 2.8$
$a_n = 2.8 + (n - 1)2.4$
$= 2.8 + 2.4n - 2.4$
$= 2.4n + 0.4$

```
sum seq(2.4N+.4,
N,1,15,1)
             294
```

45. $d = 7.48 - 4.9 = 2.58$
$a = 4.9$
$a_n = 4.9 + (n - 1)2.58$
$= 4.9 + 2.58n - 2.58$
$= 2.58n + 2.32$

```
sum seq(2.58N+2.
32,N,1,25,1)
           896.5
```

47. $2x + 1 - (x + 3) = d$
$5x + 2 - (2x + 1) = d$
$x - 2 = d$
$3x + 1 = d$
$-2x - 3 = 0$
$-2x = 3$
$x = \frac{-3}{2}$

49. The total number of seats, s, is
$S = 25 + 26 + 27 + \cdots$

This is the sum of an arithmetic sequence with $d = 1$ and $n = 30$.

$S_n = \frac{n}{2}[2a + (n - 1)d]$

$S_{30} = \frac{30}{2}[2(25) + 29(1)]$

$S_{30} = 15(50 + 29) = 1185$ seats

51. The lighter colored tile has 20 tiles in the bottom row and 1 tile at the top. The number of tiles decreases by one as we move up the triangle. This is an arithmetic sequence with $a = 20$ and $d = -1$. The number of terms to be added is 20.

$S = \frac{20}{2}(40 + 19 \cdot (-1)) = \frac{20}{2} \cdot 21 = 210$ tiles

The darker tile has 19 tiles in the bottom row and 1 tile at the top. The number of tiles decreases by one as we move up the triangle. This is an arithmetic sequence with $a_1 = 19$ and $d = -1$. The number of terms to be added is 19.

$$S = \frac{19}{2}(38 + 18(-1)) = \frac{19}{2} \cdot 20 = 190 \text{ tiles}$$

53. $2S = n(n - 1)$

$S = \frac{1}{2}n(n - 1)$

9.3 Geometric Sequences; Geometric Series

1. $r = \dfrac{3^{n+1}}{3^n}$

$r = 3^{n+1-n}$

$r = 3$

$a_1 = 3$

$a_2 = 9$

$a_3 = 27$

$a_4 = 81$

3. $r = \dfrac{-3\left(\frac{1}{2}\right)^{n+1}}{-3\left(\frac{1}{2}\right)^{n}}$

$r = \left(\frac{1}{2}\right)^{n+1-n}$

$r = \frac{1}{2}$

$a_1 = -\frac{3}{2}$

$a_2 = -\frac{3}{4}$

$a_3 = -\frac{3}{8}$

$a_4 = -\frac{3}{16}$

5. $r = \dfrac{\frac{2^{(n+1)-1}}{4}}{\frac{2^{n-1}}{4}}$

$r = \dfrac{2^n}{2^{n-1}}$

$r = 2^{n-(n-1)}$

$r = 2$

$a_1 = \frac{1}{4}$

$a_2 = \frac{1}{2}$

$a_3 = 1$

$a_4 = 2$

7. $r = \dfrac{2^{\frac{n+1}{3}}}{2^{\frac{n}{3}}}$

$r = 2^{\frac{n+1}{3} - \frac{n}{3}}$

$r = 2^{1/3}$

$a_1 = 2^{1/3}$

$a_2 = 2^{2/3}$

$a_3 = 2$

$a_4 = 2^{4/3}$

9. $r = \dfrac{\frac{3^{(n-1)+1}}{2^{n+1}}}{\frac{3^{n-1}}{2^n}}$

$r = \dfrac{3^n}{2(3^{n-1})}$

$r = \frac{1}{2}3^{n-(n-1)}$

$r = \frac{3}{2}$

$a_1 = \frac{1}{2}$

$a_2 = \frac{3}{4}$

$a_3 = \frac{9}{8}$

$a_4 = \frac{27}{16}$

11. Arithmetic

$d = [(n + 1) + 4] - (n + 4)$

$d = n + 5 - n - 4$

$d = 1$

13. Neither

15. Arithmetic

$$d = \left[3 - \frac{2}{3}(n + 1)\right] - \left[3 - \frac{2}{3}n\right]$$
$$d = 3 - \frac{2}{3}n - \frac{2}{3} - 3 + \frac{2}{3}n$$
$$d = \frac{-2}{3}$$

17. Neither

19. Geometric

$$r = \frac{\left(\frac{2}{3}\right)^{n+1}}{\left(\frac{2}{3}\right)^{n}}$$
$$r = \frac{2^{n+1-n}}{3}$$
$$r = \frac{2}{3}$$

21. Geometric

$$r = \frac{-\left(2^{(n+1)-1}\right)}{-\left(2^{n-1}\right)}$$
$$r = 2^{n-(n-1)}$$
$$r = 2$$

23. Geometric

$$r = \frac{3^{\frac{n+1}{2}}}{3^{\frac{n}{2}}}$$
$$r = 3^{\frac{n+1}{2} - \frac{n}{2}}$$
$$r = 3$$
$$r = 3^{1/2}$$

25. $a_n = ar^{n-1}$
$a_5 = 3(2)^4 = 48$
$a_n = 3 \cdot 2^{n-1}$

27. $a_5 = -1^4(5) = 1(5) = 5$
$a_n = (-1)^{n-1}(5)$

29.
$$a_5 = \left(\frac{1}{2}\right)^4(0) = 0$$
$$a_n = \left(\frac{1}{2}\right)^{n-1}(0)$$
$$a_n = 0$$

31.
$$a_5 = \sqrt{2}^4\left(\sqrt{2}\right) = \sqrt{2}^5 = 4\sqrt{2}$$
$$a_n = \sqrt{2}^{n-1}\sqrt{2}$$
$$a_n = \left(\sqrt{2}\right)^n$$

33. The first term of this geometric sequence is $a = 1$, and the common ratio is $\frac{1}{2}$. $\frac{\frac{1}{2}}{1} = \frac{1}{2}$.

The nth term obeys the formula:
$$a_n = \left(\frac{1}{2}\right)^{n-1}(1) = \left(\frac{1}{2}\right)^{n-1}$$
$$a_7 = \left(\frac{1}{2}\right)^6 = \frac{1}{64}$$

35. $a = 1,\ r = -1$
$a_n = (-1)^{n-1}(1)$
$a_n = (-1)^{n-1}$
$a_9 = (-1)^8 = 1$

37. $a = 0.4,\ r = 0.1$
$a_n = (0.1)^{n-1}(0.4)$
$a_8 = (0.1)^7(0.4)$
$a_8 = (0.0000001)(0.4)$
$a_8 = 0.00000004$

39. The sequence $\frac{2^{n-1}}{4}$ is a geometric sequence with $a = \frac{1}{4}$ and $r = 2$.

$$S_n = \sum_{k=1}^{n} a_k = a_1 + a_2 + \cdots + a_n = a\left(\frac{1 - r^n}{1 - r}\right)$$

In this sequence, $S_n = \sum_{k=1}^{n} \frac{2^{k-1}}{4} = \frac{1}{4} + \frac{2}{4} + \frac{2^2}{4} + \frac{2^3}{4} + \cdots + \frac{2^{n-1}}{4} = \frac{1}{4}\left[\frac{1 - 2^n}{1 - 2}\right] = -\frac{1}{4}(1 - 2^n)$

41. Geometric sequence: $a = \frac{2}{3}, r = \frac{2}{3}$

$$S_n = \sum_{k=1}^{n} \left(\frac{2}{3}\right)^k = \frac{2}{3} + \left(\frac{2}{3}\right)^2 + \left(\frac{2}{3}\right)^3 + \cdots + \left(\frac{2}{3}\right)^n$$

$$= \frac{2}{3}\left[\frac{1 - \left(\frac{2}{3}\right)^n}{1 - \left(\frac{2}{3}\right)}\right] = \frac{2}{3}\left[\frac{1 - \left(\frac{2}{3}\right)^n}{\frac{1}{3}}\right] = 2\left[1 - \left(\frac{2}{3}\right)^n\right]$$

43. Geometric sequence: $a = -1, r = 2$

$$S_n = \sum_{k=1}^{n} -(2^{k-1}) = -1 - 2 - 4 - 8 - \cdots - (2^{n-1}) = -1\left(\frac{1 - 2^n}{1 - 2}\right) = 1 - 2^n$$

45.
```
(1/4)sum seq(2^N
,N,0,14,1)
         8191.75
```

47.
```
sum seq((2/3)^N,
N,1,15,1)
     1.995432683
```

49.
```
-1sum seq(2^N,N,
0,14,1)
          -32767
```

51. This geometric series has first term $a = 1$ and common ratio $r = \frac{1}{4}$. Since $|r| < 1$, its sum is

$$1 + \frac{1}{4} + \frac{1}{16} + \cdots = \frac{1}{1 - \frac{1}{4}} = \frac{4}{3}$$

53. This geometric series has first terms $a = 8$ and common ratio $r = \frac{1}{2}$. Since $|r| < 1$, its sum is:

$$8 + 4 + 2 + \cdots = \frac{8}{1 - \frac{1}{2}} = 16$$

55. This geometric series has first terms $a = 2$ and common ratio $r = \frac{-1}{4}$. Since $|r| < 1$, its sum is:

$$2 - \frac{1}{2} + \frac{1}{8} - \frac{1}{32} + \cdots = \frac{2}{1 + \frac{1}{4}} = \frac{8}{5}$$

57. This geometric series has first terms $a = 5$ and common ratio $r = \frac{1}{4}$. Since $|r| < 1$, its sum is:

$$\sum_{k=1}^{\infty} 5\left(\frac{1}{4}\right)^{k-1} = \frac{5}{1 - \frac{1}{4}} = \frac{20}{3}$$

59. This geometric series has first term $a = 6$ and common ratio $r = \frac{-2}{3}$. Since $|r| < 1$, its sum is:

$$\sum_{k=1}^{\infty} 6\left(\frac{-2}{3}\right)^{k-1} = \frac{6}{1 + \frac{2}{3}} = \frac{18}{5}$$

61. The ratio of successive terms must be the same. Thus,

$$\frac{x + 2}{x} = \frac{x + 3}{x + 2}$$
$$x^2 + 4x + 4 = x^2 + 3x$$
$$x = -4$$

63. This is an ordinary annuity with 12 deposits per year of $P = \$100$ for 30 years. Thus, there are $n = 30(12) = 360$ payment periods. The rate of interest per payment period is $i = \dfrac{0.12}{12} = 0.01$. Thus,

$$A = 100\left[\frac{(1 + 0.01)^{360} - 1}{0.01}\right] = 100(3494.964133) = \$349{,}496.41$$

65. This is an ordinary annuity with four deposits per year of $P = \$500$ for 20 years. Thus, there are $n = 20(4) = 80$ payment periods. The rate of interest per payment period is $i = \dfrac{0.08}{4} = 0.02$. Thus,

$$A = 500\left[\frac{(1 + 0.02)^{80} - 1}{0.02}\right] = 500(193.7719578) = \$96{,}885.98$$

67. We wish to find the payment, P, necessary to achieve a value A of \$50,000. Scott and Alice are making 12 deposits a year for 10 years, so $n = 12(10) = 120$. The rate of interest per period is $i = \dfrac{0.06}{12} = 0.005$. Thus,

$$50{,}000 = P\left[\frac{(1 + 0.005)^{120} - 1}{0.005}\right]$$

$$50{,}000 = P(163.8793468)$$

$$P = \frac{50{,}000}{163.8793468} = \$305.10$$

69. (a) $a = 2$, geometric sequence

$a_{10} = (.9)^9(2) = 0.775$ feet

(b) $2(.9)^{n-1} < 1$ since $a_1 = 2$

$$.9^{n-1} < .5$$

$$(n - 1)\log .9 < \log .5$$

$$n - 1 < \frac{\log .5}{\log .9}$$

$$n - 1 < 6.58$$

$$n < 7.58$$

So, on the 8th swing of the pendulum, the arc is less than 1 foot.

(c) The sum of a geometric sequence with $n = 15$, $a = 2$, and $d = .9$ is

$$S_{15} = 2\left(\frac{1 - .9^{15}}{1 - .9}\right) = 15.88 \text{ feet}$$

(d) We must find the sum of an infinite geometric sequence with $a = 2$ and $d = 0.9$.

$$S_n = \frac{2}{1 - .9} = 20 \text{ feet}$$

71. **Begin:** 18,000 **After 1 year:** $18000 + .05(18000)$ **After 2 years:** $18000(1 + .05) + 18000(1 + .05)(.05) = 18000(1 + .05)^2$

After 4 years, the salary is $18{,}000(1 + .05)^4 = \$21{,}879.11$

73. With option 1, your total salary is \$2,000,000(7) + \$100,000(7) = \$14,700,000.

With option 2, your total salary is the sum of a geometric sequence with $a = 2{,}000{,}000$, $r = 1.045$ and $n = 7$.

$$S_7 = 2{,}000{,}000\left(\frac{1 - (1.045)^7}{1 - 1.045}\right) = \$16{,}038{,}304$$

With option 3, your total salary is the sum of an arithmetic sequence with $a = 2{,}000{,}000$, $d = 95{,}000$ and $n = 7$, so

$$S_7 = \frac{7}{2}(4{,}000{,}000 + 6(95{,}000)) = \$15{,}995{,}000$$

So option 2 provides the most moeny over the 7-year period, and option 1 provides the least.

75. The total number grains required would be the sum of a geometric sequence with $a = 1$, $r = 2$, and $n = 64$.

$$S_{64} = 1\left(\frac{1 - 2^{64}}{1 - 2}\right) = 1.845 \times 10^{19} \text{ grains}$$

79.

x	$\frac{1}{1 - x}$	n
0.1	1.111	3
0.25	1.333	5
0.5	2.000	9
0.75	4.00	24
0.9	10	73

We want the error to be less than .01. So, we solve the following inequality for n.

$$\left|\frac{1 - x^n}{1 - x} - \frac{1}{1 - x}\right| < 0.01$$

$$\left|\frac{-x^n}{1 - x}\right| < 0.01$$

$$\left|\frac{x^n}{1 - x}\right| < 0.01$$

Therefore, the expansion requires 3 terms. Similar calculations are used for $x = 0.25$, $x = 0.5$, $x = 0.75$, and $x = 0.9$.

81. Both options are geometric sequences.

A: After the 5th year, your salary would be $a_5 = \$20{,}000(1.06)^4 = \$25{,}250$

Your salary over the 5 years would be $S_5 = \$20{,}000\left(\frac{1 - 1.06^5}{1 - 1.06}\right) = \$112{,}472$

B: After the 5th year, your salary would be $a_5 = \$22{,}000(1.03)^4 = \$24{,}761$

Your salary over the 5 years would be $S_5 = \$22{,}000\left(\frac{1 - 1.03^5}{1 - 1.03}\right) = \$116{,}801$

Option B is better than A.

83. Multiplier $= \frac{1}{1 - 0.9} = \frac{1}{0.1} = 10$

85.
$$\begin{aligned}\text{Price} &= 4 + 4\left[\frac{1+0.03}{1+0.09}\right] + 4\left[\frac{1+0.03}{1+0.09}\right]^2 + 4\left[\frac{1+0.03}{1+0.09}\right]^3 + \cdots \\ &= 4 + 4\left[\frac{1.03}{1.09}\right] + 4\left[\frac{1.03}{1.09}\right]^2 + 4\left[\frac{1.03}{1.09}\right]^3 + \cdots \\ &= \frac{4}{1 - \frac{1.03}{1.09}} \\ &= 72\frac{2}{3} \approx \$72.67\end{aligned}$$

9.4 Mathematical Induction

1. (I) $n = 1$: $2 \cdot 1 = 2$ and $1(1 + 1) = 2$

(II) If $2 + 4 + 6 + \cdots + 2k = k(k + 1)$,

then $2 + 4 + 6 + \cdots + 2k + 2(k + 1)$

$$= [2 + 4 + 6 + \cdots + 2k] + 2(k + 1) = k(k + 1) + 2(k + 1)$$
$$= k^2 + k + 2k + 2 = k^2 + 3k + 2 = (k + 1)(k + 2)$$

3. (I) $n = 1$: $1 + 2 = 3$ and $\frac{1}{2}(1)(1 + 5) = \frac{1}{2}(6) = 3$

(II) If $3 + 4 + 5 + \cdots + (k + 2) = \frac{1}{2}k(k + 5)$,

then $3 + 4 + 5 + \cdots + (k + 2) + [(k + 1) + 2]$

$$= [3 + 4 + 5 + \cdots + (k + 2)] + (k + 3)$$
$$= \frac{1}{2}k(k + 5) + k + 3 = \frac{1}{2}k^2 + \frac{5}{2}k + k + 3$$
$$= \frac{1}{2}k^2 + \frac{7}{2}k + 3 = \frac{1}{2}(k^2 + 7k + 6) = \frac{1}{2}(k + 1)(k + 6)$$

5. (I) $n = 1$: $3 \cdot 1 - 1 = 2$ and $\frac{1}{2}(1)[3(1) + 1] = \frac{1}{2}(4) = 2$

(II) If $2 + 5 + 8 + \cdots + (3k - 1) = \frac{1}{2}k(3k + 1)$,

then $2 + 5 + 8 + \cdots + (3k - 1) + [3(k + 1) - 1]$

$$= [2 + 5 + 8 + \cdots + (3k - 1)] + 3k + 2$$
$$= \frac{1}{2}k(3k + 1) + (3k + 2) = \frac{3}{2}k^2 + \frac{1}{2}k + 3k + 2$$
$$= \frac{3}{2}k^2 + \frac{7}{2}k + 2 = \frac{1}{2}(3k^2 + 7k + 4) = \frac{1}{2}(k + 1)(3k + 4)$$

7. (I) $n = 1$: $2^{1-1} = 1$ and $2^1 - 1 = 1$

(II) If $1 + 2 + 2^2 + \cdots + 2^{k-1} = 2^k - 1$,

then $1 + 2 + 2^2 + \cdots + 2^{k-1} + 2^{(k+1)} - 1$

$$= (1 + 2 + 2^2 + \cdots + 2^{k-1}) + 2^k = 2k - 1 + 2^k = 2(2^k) - 1$$
$$= 2^{k+1} - 1$$

9. (I) $n = 1$: $4^{1-1} = 1$ and $\frac{1}{3}(4^1 - 1) = \frac{1}{3}(3) = 1$

(II) If $1 + 4 + 4^2 + \cdots + 4^{k-1} = \frac{1}{3}(4^k - 1)$,

then $1 + 4 + 4^2 + \cdots + 4^{k-1} + 4^{(k+1)-1}$

$$= (1 + 4 + 4^2 + \cdots + 4^{k-1}) + 4^k$$

$$= \frac{1}{3}(4^k - 1) + 4^k = \frac{1}{3}[4^k - 1 + 3(4^k)] = \frac{1}{3}[4(4^k) - 1]$$

$$= \frac{1}{3}(4^{k+1} - 1)$$

11. (I) $n = 1$: $\frac{1}{1 \cdot 2} = \frac{1}{2}$ and $\frac{1}{1+1} = \frac{1}{2}$

(II) If $\frac{1}{1 \cdot 2} + \frac{1}{2 \cdot 3} + \frac{1}{3 \cdot 4} + \cdots + \frac{1}{k(k+1)} = \frac{k}{k+1}$,

then $\frac{1}{1 \cdot 2} + \frac{1}{2 \cdot 3} + \frac{1}{3 \cdot 4} + \cdots + \frac{1}{k(k+1)} + \frac{1}{(k+1)[(k+1)+1]}$

$$= \left[\frac{1}{1 \cdot 2} + \frac{1}{2 \cdot 3} + \frac{1}{3 \cdot 4} + \cdots + \frac{1}{k(k+1)}\right] + \frac{1}{(k+1)(k+2)}$$

$$= \frac{k}{k+1} + \frac{1}{(k+1)(k+2)} = \frac{k(k+2)+1}{(k+1)(k+2)} = \frac{k^2+2k+1}{(k+1)(k+2)}$$

$$= \frac{(k+1)(k+1)}{(k+1)(k+2)} = \frac{k+1}{k+2}$$

13. (I) $n = 1$: $1^2 = 1$ and $\frac{1}{6} \cdot 1 \cdot 2 \cdot 3 = 1$

(II) If $1^2 + 2^2 + 3^2 + \cdots + k^2 = \frac{1}{6}k(k+1)(2k+1)$,

then $1^2 + 2^2 + 3^2 + \cdots + k^2 + (k+1)^2$

$$= (1^2 + 2^2 + 3^2 + \cdots + k^2) + (k+1)^2$$

$$= \frac{1}{6}k(k+1)(2k+1) + (k+1)^2 = \frac{1}{6}k(2k^2 + 3k + 1) + (k^2 + 2k + 1)$$

$$= \frac{1}{3}k^3 + \frac{1}{2}k^2 + \frac{1}{6}k + k^2 + 2k + 1 = \frac{1}{3}k^3 + \frac{3}{2}k^2 + \frac{13}{6}k + 1$$

$$= \frac{1}{6}(2k^3 + 9k^2 + 13k + 6) = \frac{1}{6}(k+1)(2k^2 + 7k + 6)$$

$$= \frac{1}{6}(k+1)(k+2)(2k+3)$$

15. (I) $n = 1$: $5 - 1 = 4$ and $\frac{1}{2}(9 - 1) = \frac{1}{2} \cdot 8 = 4$

(II) If $4 + 3 + 2 + \cdots + (5 - \mathrm{k}) = \frac{1}{2}\mathrm{k}(9 - \mathrm{k})$,

then $4 + 3 + 2 + \cdots + (5 - \mathrm{k}) = [5 - (\mathrm{k} + 1)] = [4 + 3 + 2 + \cdots + (5 - \mathrm{k})] + [5 - (\mathrm{k} + 1)]$

$$= \frac{1}{2}k(9 - k) + 4 - k = \frac{9}{2}k - \frac{1}{2}k^2 + 4 - k = -\frac{1}{2}k^2 + \frac{7}{2}k + 4$$

$$= \frac{1}{2}(-k^2 + 7k + 8) = \frac{1}{2}(k+1)(8-k) = \frac{1}{2}(k+1)[9 - (k+1)]$$

17. (I) $n = 1$: $1 \cdot (1 + 1) = 2$ and $\frac{1}{3} \cdot 1 \cdot 2 \cdot 3 = 2$

(II) If $1 \cdot 2 + 2 \cdot 3 + 3 \cdot 4 + \cdots + k(k+1) = \frac{1}{3}k(k+1)(k+2)$,

then $1 \cdot 2 + 2 \cdot 3 + 3 \cdot 4 + \cdots + k(k+1) + (k+1)(k+1+1)$

$$= [1 \cdot 2 + 2 \cdot 3 + 3 \cdot 4 + \cdots + k(k+1)] + (k+1)(k+2)$$

$$= \frac{1}{3}k(k+1)(k+2) + (k+1)(k+2) = (k+1)(k+2)\left[\frac{1}{3}k + 1\right]$$

$$= (k+1)(k+2)\frac{1}{3}(k+3) = \frac{1}{3}(k+1)(k+2)(k+3)$$

19. (I) $n = 1$: $1^2 + 1 = 2$ is divisible by 2.

(II) If $k^2 + k$ is divisible by 2,

then $(k+1)^2 + (k+1) = k^2 + 2k + 1 + k + 1 = (k^2 + k) + (2k + 2)$

Since $k^2 + k$ is divisible by 2 and $2k + 2$ is divisible by 2, then $(k+1)^2 + k + 1$ is divisible by 2.

21. (I) $n = 1$: $1^2 - 1 + 2 = 2$ is divisible by 2.

(II) If $k^2 - k + 2$ is divisible by 2,

then $(k+1)^2 - (k+1) + 2$

$$= k^2 + 2k + 1 - k - 1 + 2 = (k^2 - k + 2) + 2k + 1 - 1$$

$$= (k^2 - k + 2) + 2k$$

Since $k^2 - k + 2$ is divisible by 2 and $2k$ is divisible by 2, then $(k+1)^2 - (k+1) + 2$ is divisible by 2.

23. (I) $n = 1$: If $x > 1$, then $x^1 = x > 1$.

(II) Assume, for any natural number k, that if $x > 1$, then $x^k > 1$. Show that if $x^k > 1$, then $x^{k+1} > 1$:

$$x^{k+1} = x^k \cdot x > 1 \cdot x = x > 1$$

↑

$x^k > 1$

25. (I) $n = 1$: $a - b$ is a factor of $a^1 - b^1 = a - b$.

(II) If $a - b$ is a factor of $a^k - b^k$, show that $a - b$ is a factor of $a^{k+1} - b^{k+1}$

$= a(a^k - b^k) + b^k(a - b)$.

Since $a - b$ is a factor of $a^k - b^k$ and $a - b$ is a factor of $a - b$, then $a - b$ is a factor of $a^{k+1} - b^{k+1}$.

27. $n = 1$: $1^2 - 1 + 41 = 41$ is a prime number.

$n = 41$: $41^2 - 41 + 41 = 1681 = 41^2$ is not prime.

29. (I) $n = 1$: $ar^{1-1} = a \cdot 1 = a$ and $a \cdot \frac{1 - r^1}{1 - r} = a$ because $r \neq 1$.

(II) If $a + ar + ar + ar^2 + \cdots + ar^{k-1} = a \cdot \left(\frac{1 - r^k}{1 - r}\right)$

then, $a + ar + ar^2 + \cdots + ar^{k-1} + ar^{(k+1)-1}$

$$= (a + ar + ar^2 + \cdots + ar^{k-1}) + ar^k$$

$$= a \cdot \left(\frac{1 - r^k}{1 - r}\right) + ar^k = \frac{a(1 - r^k) + ar^k(1 - r)}{1 - r}$$

$$= \frac{a - ar^k + ar^k - ar^{k+1}}{1 - r} = a \cdot \left(\frac{1 - r^{k+1}}{1 - r}\right)$$

31. (I) $n = 3$: The sum of the angles of a triangle is $(3 - 2) \cdot 180° = 180°$.

(II) Assume that for any integer k the sum of the angles of a convex polygon of k sides is $(k - 2) \cdot 180°$. A convex polygon of $k + 1$ sides consists of a convex polygon k sides plus a triangle. See the illustration. The sum of the angles is $(k - 2)180° + 180° = (k - 1)180°$. Since Conditions I and II have been met, the result follows.

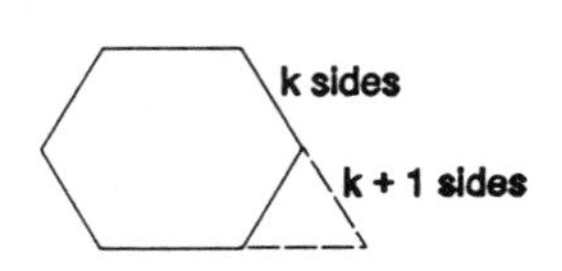

9.5 The Binomial Theorem

1. $\binom{5}{2} = \frac{5!}{2!3!} = \frac{5 \cdot 4}{2} = 10$

3. $\binom{7}{5} = \frac{7!}{5!2!} = \frac{7 \cdot 6}{2} = 21$

5. $\binom{50}{49} = \frac{50!}{49!1!} = 50$

7. $\binom{1000}{1000} = 1$

9. $\binom{55}{23} = 1.866 \times 10^{15}$

11. $\binom{47}{25} = 1.483 \times 10^{13}$

13. $(x + 1)^4 = \binom{4}{0}x^4 + \binom{4}{1}x^3 + \binom{4}{2}x^2 + \binom{4}{3}x + \binom{4}{4} = x^4 + 4x^3 + 6x^2 + 4x + 1$

15. $(x - 2)^5 = \binom{5}{0}x^5 + \binom{5}{1}(-2)x^4 + \binom{5}{2}(-2)^2x^3 + \binom{5}{3}(-2)^3x^2 + \binom{5}{4}(-2)^4x + \binom{5}{5}(-2)^5$

$= x^5 + 5(-2x^4) + 10(4x^3) + 10(-8x^2) + 5(16x) + (-32)$

$= x^5 - 10x^4 + 40x^3 - 80x^2 + 80x - 32$

17. $(3x + 1)^4 = \binom{4}{0}(3x)^4 + \binom{4}{1}(3x)^3 + \binom{4}{2}(3x)^2 + \binom{4}{3}(3x) + \binom{4}{4}$

$= 81x^4 + 4 \cdot 27x^3 + 6 \cdot 9x^2 + 4 \cdot 3x + 1 = 81x^4 + 108x^3 + 54x^2 + 12x + 1$

19. $(x^2 + y^2)^5 = \binom{5}{0}(x^2)^5 + \binom{5}{1}(y^2)(x^2)^4 + \binom{5}{2}(y^2)^2(x^2)^3 + \binom{5}{3}(y^2)^3(x^2)^2$

$+ \binom{5}{4}(y^2)^4(x^2) + \binom{5}{5}(y^2)^5$

$= x^{10} + 5y^2x^8 + 10y^4x^6 + 10y^6x^4 + 5y^8x^2 + y^{10}$

21. $(\sqrt{x} + \sqrt{2})^6 = \begin{bmatrix}6\\0\end{bmatrix}(\sqrt{x})^6 + \begin{bmatrix}6\\1\end{bmatrix}\sqrt{2}(\sqrt{x})^5 + \begin{bmatrix}6\\2\end{bmatrix}(\sqrt{2})^2(\sqrt{x})^4 + \begin{bmatrix}6\\3\end{bmatrix}(\sqrt{2})^3(\sqrt{x})^3$

$+ \begin{bmatrix}6\\4\end{bmatrix}(\sqrt{2})^4(\sqrt{x})^2 + \begin{bmatrix}6\\5\end{bmatrix}(\sqrt{2})^5(\sqrt{x}) + \begin{bmatrix}6\\6\end{bmatrix}(\sqrt{2})^6$

$= x^3 + 6\sqrt{2}(\sqrt{x})^5 + 15(\sqrt{2})^2(\sqrt{x})^4 + 20(\sqrt{2})^3(\sqrt{x})^3 + 15(\sqrt{2})^4(\sqrt{x})^2 + 6(\sqrt{2})^5\sqrt{x} + (\sqrt{2})^6$

$= x^3 + 6\sqrt{2}x^{5/2} + 30x^2 + 40\sqrt{2}x^{3/2} + 60x + 24\sqrt{2}x^{1/2} + 8$

23. $(ax + by)^5 = \begin{bmatrix}5\\0\end{bmatrix}(ax)^5 + \begin{bmatrix}5\\1\end{bmatrix}(by)(ax)^4 + \begin{bmatrix}5\\2\end{bmatrix}(by)^2(ax)^3 + \begin{bmatrix}5\\3\end{bmatrix}(by)^3(ax)^2$

$+ \begin{bmatrix}5\\4\end{bmatrix}(by)^4(ax) + \begin{bmatrix}5\\5\end{bmatrix}(by)^5$

$= (ax)^5 + 5by(ax)^4 + 10(by)^2(ax)^3 + 10(by)^3(ax)^2 + 5(by)^4(ax) + (by)^5$

25. $n = 10$, $a = 3$, $x = x$, and $j = 4$

$\begin{bmatrix}10\\10-4\end{bmatrix}3^{10-4}x^4 = \begin{bmatrix}10\\6\end{bmatrix}3^6 \cdot x^4 = \frac{10!}{6!4!} \cdot 729 \cdot x^4 = \frac{10 \cdot 9 \cdot 8 \cdot 7}{4 \cdot 3 \cdot 2} \cdot 729x^4 = 153{,}090x^4$

The coefficient of x^4 is 153,090.

27. $n = 12$, $a = -1$, $x = 2x$, and $j = 7$

$\begin{bmatrix}12\\12-7\end{bmatrix}(-1)^{12-7}(2x)^7 = \begin{bmatrix}12\\5\end{bmatrix}(-1)^5 \cdot 128 \cdot x^7 = \frac{-12!}{5!7!} \cdot 128 \cdot x^7 = -101{,}376x^7$

The coefficient of x^7 is $-101{,}376$.

29. $n = 9$, $a = 3$, $x = 2x$, and $j = 7$

$\begin{bmatrix}9\\9-7\end{bmatrix}(3)^{9-7}(2x)^7 = \begin{bmatrix}9\\2\end{bmatrix}3^2 \cdot 128 \cdot x^7 = \frac{9!}{2!7!} \cdot 9 \cdot 128 \cdot x^7 = 41{,}472x^7$

The coefficient of x^7 is 41,472.

31. The fifth term contains x^3. $n = 7$, $a = 3$, $x = x$, and $j = 3$

$\begin{bmatrix}7\\7-3\end{bmatrix}(3)^{7-3}x^3 = \begin{bmatrix}7\\4\end{bmatrix}3^4 \cdot x^3 = \frac{7!}{4!3!} \cdot 81 \cdot x^3 = 2835x^3$

33. The third term contains x^7. $n = 9$, $a = -2$, $x = 3x$, and $j = 7$

$\begin{bmatrix}9\\9-7\end{bmatrix}(-2)^{9-7}(3x)^7 = \begin{bmatrix}9\\2\end{bmatrix}(-2)^2(3x)^7 = \frac{9!}{2!7!} \cdot 4 \cdot 2187 \cdot x^7 = 314{,}928x^7$

35. $\left[x^2 + \frac{1}{x}\right]^{12}$

Constant term $= \begin{bmatrix}12\\12-j\end{bmatrix}\left[\frac{1}{x}\right]^{12-j}(x^2)^j$ where $2j = 12 - j$ or $j = 4$

Constant term $= \begin{bmatrix}12\\8\end{bmatrix}\left[\frac{1}{x}\right]^8(x^2)^4 = \begin{bmatrix}12\\8\end{bmatrix} = \frac{12!}{8!4!} = \frac{12 \cdot 11 \cdot 10 \cdot 9}{4 \cdot 3 \cdot 2} = 495$

37. $\left(x - \frac{2}{\sqrt{x}}\right)^{10}$

x^4 term $= \binom{10}{10-j}\left(\frac{-2}{\sqrt{10}}\right)^{10-j} x^j$ where $j - \frac{10-j}{2} = 4$ or $\frac{3}{2}j = 9$ or $j = 6$

x^4 term $= \binom{10}{4}\left(\frac{-2}{\sqrt{10}}\right)^4 x^6 = 16\binom{10}{6}\left(\frac{x^6}{x^2}\right) = \frac{16 \cdot 10 \cdot 9 \cdot 8 \cdot 7}{4 \cdot 3 \cdot 2} = 3360$

39. $(1.001)^5 = (1 + 10^{-3})^5 = 1 + \binom{5}{1} \cdot 10^{-3} + \binom{5}{2} \cdot (10^{-3})^2 + \binom{5}{3}(10^{-3})^3 + \cdots$

$= 1 + .005 + 10(.000001) + 10(.000000001) + \cdots$

$= 1 + .005 + .00001 + \cdots$

≈ 1.00501 correct to five decimal places

41. $\binom{n}{n} = 1$

$\frac{n!}{n!(n-n)!} = \frac{n!}{n!0!} = \frac{n!}{n! \cdot 1} = \frac{n!}{n!} = 1$

43. $\binom{n}{0} + \binom{n}{1} + \cdots + \binom{n}{n} = 2^n$

$\binom{n}{0} + \binom{n}{1} + \cdots + \binom{n}{n} = (1 + 1)^n$

$= \binom{n}{0}1^n + \binom{n}{1}(1)^1(1)^{n-1} + \binom{n}{2}(1)^2(1)^{n-2} + \cdots + \binom{n}{n}(1)^n(1)^{n-n}$

$= \binom{n}{0} + \binom{n}{1} + \cdots + \binom{n}{n}$

45. $\binom{5}{0}\left(\frac{1}{4}\right)^5 + \binom{5}{1}\left(\frac{1}{4}\right)^4\left(\frac{3}{4}\right) + \binom{5}{2}\left(\frac{1}{4}\right)^3\left(\frac{3}{4}\right)^2 + \binom{5}{3}\left(\frac{1}{4}\right)^2\left(\frac{3}{4}\right)^3$

$+ \binom{5}{4}\left(\frac{1}{4}\right)\left(\frac{3}{4}\right)^4 + \binom{5}{5}\left(\frac{3}{4}\right)^5$

$= 1 \cdot \frac{1}{4^5} + 5 \cdot \frac{1}{4^4} \cdot \frac{3}{4} + 10 \cdot \frac{1}{4^3} \cdot \frac{9}{4^2} + 10 \cdot \frac{1}{4^2} + \frac{27}{4^3} + 5 \cdot \frac{1}{4} \cdot \frac{81}{4^4} + \frac{243}{4^5}$

$= \frac{1 + 15 + 90 + 270 + 405 + 243}{4^5} = \frac{1024}{4^5} = \frac{1024}{1024} = 1$

9.6 Sets and Counting

In Problems 1–9, we have $A = \{1, 3, 5, 7, 9\}$, $B = \{1, 5, 6, 7\}$, and $C = \{1, 2, 4, 6, 8, 9\}$:

1. $A \cup B = \{1, 3, 5, 7, 9\} \cup \{1, 5, 6, 7\} = \{1, 3, 5, 6, 7, 9\}$

3. $A \cap B = \{1, 3, 5, 7, 9\} \cap \{1, 5, 6, 7\} = \{1, 5, 7\}$

5. $(A \cup B) \cap C = (\{1, 3, 5, 7, 9\} \cup \{1, 5, 6, 7\}) \cap \{1, 2, 4, 6, 8, 9\}$
$= \{1, 3, 5, 6, 7, 9\} \cap \{1, 2, 4, 6, 8, 9\}$
$= \{1, 6, 9\}$

7. $(A \cap B) \cup C = (\{1, 3, 5, 7, 9\} \cap \{1, 5, 6, 7\}) \cup \{1, 2, 4, 6, 8, 9\}$
$= \{1, 5, 7\} \cup \{1, 2, 4, 6, 8, 9\}$
$= \{1, 2, 4, 5, 6, 7, 8, 9\}$

9. $(A \cup C) \cap (B \cup C) = (\{1, 3, 5, 7, 9\} \cup \{1, 2, 4, 6, 8, 9\}) \cap (\{1, 5, 6, 7\} \cup \{1, 2, 4, 6, 8, 9\})$
$= \{1, 2, 3, 4, 5, 6, 7, 8, 9\} \cap \{1, 2, 4, 5, 6, 7, 8, 9\}$
$= \{1, 2, 4, 5, 6, 7, 8, 9\}$

In Problems 11–19, we have U = *Universal set* = $\{0, 1, 2, 3, 4, 5, 6, 7, 8, 9\}$, $A = \{1, 3, 4, 5, 9\}$, $B = \{2, 4, 6, 7, 8\}$, *and* $C = \{1, 3, 4, 6\}$:

11. $A' = \{1, 3, 4, 5, 9\}'$
$= \{0, 2, 6, 7, 8\}$

13. $(A \cap B)' = (\{1, 3, 4, 5, 9\} \cap \{2, 4, 6, 7, 8\})'$
$= \{4\}'$
$= \{0, 1, 2, 3, 5, 6, 7, 8, 9\}$

15. $A' \cup B'$
$= \{1, 3, 4, 5, 9\}' \cup \{2, 4, 6, 7, 8\}'$
$= \{0, 2, 6, 7, 8\} \cup \{0, 1, 3, 5, 9\}$
$= \{0, 1, 2, 3, 5, 6, 7, 8, 9\}$

17. $(A \cap C')'$
$= (\{1, 3, 4, 5, 9\} \cap \{1, 3, 4, 6\}')'$
$= (\{1, 3, 4, 5, 9\} \cap \{0, 2, 5, 7, 8, 9\})'$
$= \{5, 9\}'$
$= \{0, 1, 2, 3, 4, 6, 7, 8\}$

19. $(A \cup B \cup C)' = (\{1, 3, 4, 5, 9\} \cup \{2, 4, 6, 7, 8\} \cup \{1, 3, 4, 6\})'$
$= \{1, 2, 3, 4, 5, 6, 7, 8, 9\}' = \{0\}$

21. Subsets of $\{a, b, c, d)$ with
zero elements: $\varnothing$
one element: $\{a\}, \{b\}, \{c\}, \{d\}$
two elements: $\{a, b\}, \{a, c\}, \{a, d\}, \{b, c\}, \{b, d\}, \{c, d\}$
three elements: $\{a, b, c\}, \{a, b, d\}, \{a, c, d\}, \{b, c, d\}$
four elements: $\{a, b, c, d\}$

23. For $n(A) = 25$, $n(B) = 30$, $n(A \cap B) = 15$, we have by (1) that
$n(A \cup B) = n(A) + n(B) - n(A \cap B) = 25 + 30 - 15 = 40$

25. For $n(A \cup B) = 50$, $n(A \cap B) = 10$, $n(B) = 20$, we have that
$n(A \cup B) = n(A) + n(B) - n(A \cap B)$
$50 = n(A) + 20 - 10$
$40 = n(A)$

27. From figure,
$n(A) = 15 + 3 + 5 + 2 = 25$

29. From figure,
$n(A \text{ or } B) = n(A \cup B)$
$= n(A) + n(B) - n(A \cap B)$
$= 25 + 20 - 8 = 37$

31. From figure,
$n(A \text{ but not } C) = n(A) - n(A \cap C)$
$= 25 - 7 = 18$

33. From figure,
$n(A \text{ and } B \text{ and } C)$
$= n(A \cap B \cap C) = 5$

35. Let A = {those who will purchase a major appliance}
B = {those who will buy a car}
Note that $n(A) = 200$, $n(B) = 150$, $n(A \cap B) = 25$
Using (1), $n(A \cup B) = n(A) + n(B) - n(A \cap B) = 200 + 150 - 25 = 325$

Since a total of 500 were asked the number that will purchase neither, $n((A \cup B)') = 500 - 325 = 175$. The number who will purchase only a car is:
$n(B \text{ but not } A) = n(B \cap A')n(B) - n(A \cap B) = 150 - 25 = 125.$

This problem can be solved by Venn Diagram as well.

37. We fill in a Venn Diagram, reading the list from the bottom up:
(a) 15
(b) 15
(c) 15
(d) 25
(e) 40

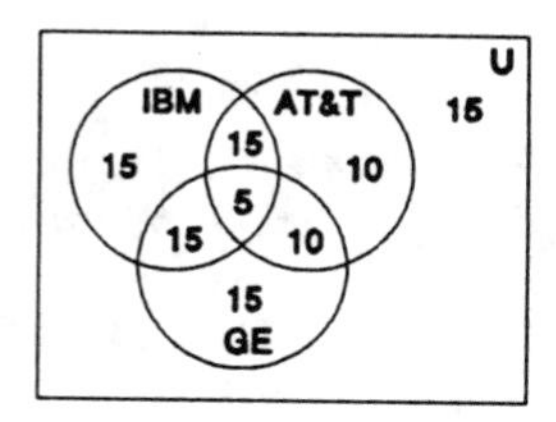

9.7 Permutations and Combinations

Use $P(n, r) = \dfrac{n!}{(n - r)!}$ *for Problems 1–7:*

1. $P(6, 2) = \dfrac{6!}{(6 - 2)!} = \dfrac{6!}{4!} = \dfrac{6 \cdot 5 \cdot 4!}{4!} = 30$

3. $P(5, 5) = \dfrac{5!}{(5 - 5)!} = \dfrac{5!}{0!} = \dfrac{5 \cdot 4 \cdot 3 \cdot 2 \cdot 1}{1} = 120$

5. $P(8, 0) = \dfrac{8!}{(8 - 0)!} = \dfrac{8!}{8!} = 1$

7. $P(8, 4) = \dfrac{8!}{(8 - 4)!} = \dfrac{8!}{4!} = \dfrac{8 \cdot 7 \cdot 6 \cdot 5 \cdot 4!}{4!} = 1680$

In Problems 9–15, we use $C = \dfrac{n!}{(n - r)!r!}$:

9. $C(8, 2) = \dfrac{8!}{(8 - 2)!2!} = \dfrac{8!}{6!2!} = \dfrac{8 \cdot 7 \cdot 6!}{2 \cdot 1 \cdot 6!} = 28$

11. $C(6, 4) = \dfrac{6!}{(6 - 4)!4!} = \dfrac{6!}{2!4!} = \dfrac{6 \cdot 5 \cdot 4!}{2 \cdot 1 \cdot 4!} = 15$

13. $C(15, 15) = \dfrac{15!}{(15 - 15)!15!} = \dfrac{15!}{0!15!} = \dfrac{15!}{1 \cdot 15!} = \dfrac{15!}{15!} = 1$

15. $C(26, 13) = \dfrac{26!}{(26 - 13)!13!} = \dfrac{26!}{13!13!}$

$$= \frac{26 \cdot 25 \cdot 24 \cdot 23 \cdot 22 \cdot 21 \cdot 20 \cdot 19 \cdot 18 \cdot 17 \cdot 16 \cdot 15 \cdot 14 \cdot 13!}{13 \cdot 12 \cdot 11 \cdot 10 \cdot 9 \cdot 8 \cdot 7 \cdot 6 \cdot 5 \cdot 4 \cdot 3 \cdot 2 \cdot 1 \cdot 13!}$$
$= 10{,}400{,}600$

17. {*abc, abd, abe, acb, acd, ace, adb, adc, ade, aeb, aec, aed, bac, bad, bae, bca, bcd, bce, bda, bde, bea, bec, bed, cab, cad, cae, cba, cbd, cde, cda, cdb, cde, cea, ceb, ced, dab, dac, dae, dba, dbc, dbe, dca, dcb, dce, dea, deb, dec, eab, eac, ead, eba, ebc, ebd, eca, ecb, ecd, eda, edb, edc*}

$$P(5, 3) = \frac{5!}{(5-3)!} = \frac{5!}{2!} = \frac{5 \cdot 4 \cdot 3 \cdot 2 \cdot 1}{2 \cdot 1} = 60.$$

19. {123, 124, 132, 134, 142, 143, 213, 214, 231, 234, 241, 243, 312, 314, 321, 324, 341, 342, 412, 413, 421, 423, 431, 432}

$$P(4, 3) = \frac{4!}{(4-3)!} = \frac{4!}{1!} = 4 \cdot 3 \cdot 2 \cdot 1 = 24$$

21. Combinations of a, b, c, d, e taken 3 at a time are:

(*abc*), (*abd*), (*abe*), (*acd*), (*ace*), (*ade*), (*bcd*), (*bce*), (*bde*), (*cde*)

Thus, $C(5, 3) = 10$.

23. Combinations of 1, 2, 3, 4 taken 3 at a time are:

(123), (124), (134), (234)

Thus, $C(4, 3) = 4$.

25. There are 6 choices and 4 choices or $6 \cdot 4 = 24$ combinations.

27. There are $4 \cdot 4 = 16$ ways since there are four choices for each of two positions.

29. For a 3-digit number using 0 and 1, we have two choices in each of three positions or $2 \cdot 2 \cdot 2 = 8$ numbers.

31. To line 4 people up, we have 4 choices for the first position, 3 for the second, 2 for the third, and only 1 for the fourth position; i.e., $4 \cdot 3 \cdot 2 \cdot 1 = 4! = 24$ arrangements.

33. Since no letter can be repeated, we have 5 choices for the first position, 4 choices for the second, and 3 choices for the third; i.e., $5 \cdot 4 \cdot 3 = 60$ codes.

35. We have 5 letters and use each only once so that there are 5 choices for the first position, 4 for the second, 3 for the third, 2 for the second, and only 1 for the last position, or $5 \cdot 3 \cdot 4 \cdot 2 \cdot 1 = 5! = 120$ arrangements.

37. To form a committee of 3 from a total of 7 students is given by:

$$C(7, 3) = \frac{7}{(7-3)!3!} = \frac{7!}{4!3!} = \frac{7 \cdot 6 \cdot 5 \cdot 4!}{3 \cdot 2 \cdot 1 \cdot 4!} = 35$$

39. There are 2 possibilities for each of the ten questions; therefore, there are $2^{10} = 1024$ different possible arrangements.

41. There are 9 possible choices for the first digit. There are 10 possible choices for the remaining 3 digits. Therefore, there are $9 \cdot 10 \cdot 10 \cdot 10 = 9000$ possible four-digit numbers.

43. There are 5 positions available for the first book, 4 positions available for the second book, 3 for the third, 2 for the fourth and 1 for the fifth. Therefore, there are $5 \cdot 4 \cdot 3 \cdot 2 \cdot 1 = 120$ different arrangements.

45. $\begin{bmatrix} 8 \\ 1 \end{bmatrix} \begin{bmatrix} 15 \\ 1 \end{bmatrix} \begin{bmatrix} 4 \\ 1 \end{bmatrix} = 8 \cdot 15 \cdot 4 = 480$ different portfolios.

47. Choosing 3 boys from 6 can be done $C(6, 3)$ ways and choosing 3 girls from 8 can be done $C(8, 3)$ ways, giving a total of

$$C(6, 3) \times C(8, 3) = \frac{6!}{(6-3)!3!} \cdot \frac{8!}{(8-3)!3!} = \frac{6!}{3!3!} \cdot \frac{8!}{5!3!}$$
$$= \frac{6 \cdot 5 \cdot 4 \cdot 3!}{3 \cdot 2 \cdot 1 \cdot 3!} \cdot \frac{8 \cdot 7 \cdot 6 \cdot 5!}{3!5!} = 20 \cdot 56 = 1120$$

There are 1120 ways of choosing a committee.

49. The committee is formed as follows:

$$\frac{\text{Administrators}}{\text{2 from 4}} \times \frac{\text{faculty}}{\text{3 from 8}} \times \frac{\text{students}}{\text{5 from 20}}$$
$$C(4, 2) \times C(8, 3) \times C(20, 5)$$
$$\frac{4!}{(4-2)!2!} \times \frac{8!}{(8-3)!3!} \times \frac{20!}{(20-5)!15!}$$
$$\frac{4!}{2!2!} \times \frac{8!}{5!3!} \times \frac{20!}{15!5!}$$
$$\frac{4 \cdot 3!}{2 \cdot 1 \cdot 2 \cdot 1} \times \frac{8 \cdot 7 \cdot 6 \cdot 5!}{3!5!} \times \frac{20 \cdot 19 \cdot 18 \cdot 17 \cdot 16 \cdot 15!}{5 \cdot 4 \cdot 3 \cdot 2 \cdot 1 \cdot 15!}$$
$$= 5{,}209{,}344 \text{ different committees}$$

51. There are 9 choices for the first position, 8 for the second, 7 for the third, etc., or $9 \cdot 8 \cdot 7 \cdot 6 \cdot 5 \cdot 4 \cdot 3 \cdot 2 \cdot 1 = 9! = 362{,}880$ possible batting orders.

53. $\frac{9!}{2!2!} = 90{,}720$ different words

55. $C(100, 22)$ ways to form the first committee. After choosing the first committee, there are $100 - 22 = 78$ senators left. Thus, there are $C(78, 13)$ ways to form the second committee. There are $C(65, 10)$ ways to form the third, $C(55, 5)$ ways to form the fourth, $C(50, 16)$ ways to form the fifth, $C(34, 17)$ ways to form the sixth, and $C(17, 17)$ ways to form the seventh committee.

$$\text{Total number of committees} = C(100, 22) \cdot C(78, 13) \cdot C(65, 10) \cdot C(55, 5) \cdot C(50, 16) \cdot C(34, 17) \cdot C(17, 17)$$
$$= 1.157 \times 10^{76}$$

57. We have 6 players and we must choose 2 of them. Therefore, there are $C(6, 2) = 15$ different teams possible.

59. (a) $C(7, 2) \cdot C(3, 1) = 21 \cdot 3 = 63$ (b) $C(7, 3) = 35$ (c) $C(3, 3) = 1$

9.8 Probability

1. The sample space, all of the logical possibilities that can occur, is $S = \{HH, HT, TH, TT\}$

No one outcome is more likely than another, so that using $P(E) = \frac{n(E)}{n(S)}$

The probabilities are $P(HH) = \frac{1}{4}$, $P(HT) = \frac{1}{4}$, $P(TH) = \frac{1}{4}$, and $P(TT) = \frac{1}{4}$

3. The sample space, generated by combining each of the outcomes of the two coins as in Problems 1 and 2 with each face of the die, is:

S = {$HH1$, $HH2$, $HH3$, $HH4$, $HH5$, $HH6$, $HT1$, $HT2$, $HT3$, $HT4$, $HT5$, $HT6$, $TH1$, $TH2$, $TH3$, $TH4$, $TH5$, $TH6$, $TT1$, $TT2$, $TT3$, $TT4$, $TT5$, $TT6$}

There are 24 equally likely outcomes, and the probability of each is $\frac{1}{24}$.

5. The sample space for three coins builds on that for two coins:

S = {HHH, HHT, HTH, HTT, THH, THT, TTH, TTT}.

Since there are 8 equally likely outcomes, the probability of each is $\frac{1}{8}$.

In Problems 7–11, we use the figures in the text. Note that the sample space for Spinner I is S_1 = {1, 2, 3, 4}, and the sample space for Spinner II is S_2 = {Yellow, Red, Green}, and the sample space for Spinner III is S_3 = {Forward, Backward}.

7. The sample space for spinning Spinner I and then Spinner II is:

S = {1 Yellow, 1 Red, 1 Green, 2 Yellow, 2 Red, 2 Green, 3 Yellow, 3 Red, 3 Green, 4 Yellow, 4 Red, 4 Green},

or 12 equally likely outcomes, and the probability of each one is $\frac{1}{12}$. The probability of getting a 2 or a 4, followed by Red is P(2 Red) + P(4 Red) $= \frac{1}{12} + \frac{1}{12} = \frac{1}{6}$

9. The sample space for spinning Spinner I, then Spinner II, and then Spinner III is

S = {1 Yellow Forward, 1 Yellow Backward, 1 Red Forward, 1 Red Backward, 1 Green Forward, 1 Green Backward, 2 Yellow Forward, 2 Yellow Backward, 2 Red Forward, 2 Red Backward, 2 Green Forward, 2 Green Backward, 3 Yellow Forward, 3 Yellow Backward, 3 Red Forward, 3 Red Backward, 3 Green Forward, 3 Green Backward, 4 Yellow Forward, 4 Yellow Backward, 4 Red Forward, 4 Red Backward, 4 Green Forward, 4 Green Backward}

or 24 equally likely outcomes, and the probability of each is $\frac{1}{24}$.

The probability of getting a 1, followed by Red or Green, followed by Backward is:

P(1 Red Backward) + P(1 Green Backward) $= \frac{1}{24} + \frac{1}{24} = \frac{1}{12}$

11. The sample space for spinning Spinner I twice and then Spinner II is:

S = {11 Yellow, 11 Red, 11 Green, 12 Yellow, 12 Red, 12 Green, 13 Yellow, 13 Red, 13 Green, 14 Yellow, 14 Red, 14 Green, 21 Yellow, 21 Red, 21 Green, 22 Yellow, 22 Red, 22 Green, 23 Yellow, 23 Red, 23 Green, 24 Yellow, 24 Red, 24 Green, 31 Yellow, 31 Red, 31 Green, 32 Yellow, 32 Red, 32 Green, 33 Yellow, 33 Red, 33 Green, 34 Yellow, 34 Red, 34 Green, 41 Yellow, 41 Red, 41 Green, 42 Yellow, 42 Red, 42 Green, 43 Yellow, 43 Red, 43 Green, 44 Yellow, 44 Red, 44 Green}

or 48 equally likely outcomes, and the probability of each is $\frac{1}{48}$.

The probability of getting a 2, followed by a 2 or a 4, followed by a Red or Green is:

P(2 2 Red) + P(2 2 Green) + P(2 4 Red) + P(2 4 Green) $= \frac{1}{48} + \frac{1}{48} + \frac{1}{48} + \frac{1}{48} = \frac{1}{12}$

13. A, B, C, F

15. B

17. P(heads) $= \frac{4}{5}$; P(tails) $= \frac{1}{5}$

19. $P(1) = P(3) = P(5) = \frac{2}{9}$; $P(2) = P(4) = P(6) = \frac{1}{9}$

21. $P(A \cup B) = P(A) + P(B)$
$= 0.25 + 0.45 = 0.70$

23. $P(A \cup B) = P(A) + P(B) - P(A \cap B)$
$= 0.25 + 0.45 - 0.15 = 0.55$

In Problems 25 and 27 a golf ball is selected from a container with 9 white balls, 8 green balls, and 3 orange balls.

25. The sample sapce, the total number of golf balls, contains 20 equally likely outcomes There are 9 white balls. Thus, the probability of getting a white ball is:

$$P(E) = \frac{n(E)}{n(S)} = \frac{n(\text{white balls})}{n(\text{golf balls present})} = \frac{9}{20}$$

27. The sample space, the total number of golf balls, contains 20 equally likely outcomes. There are 9 white balls and 8 green ones. Thus, the probability of getting a white or a green one is

$$P(E) = \frac{n(E)}{n(S)} = \frac{n(\text{white or green or balls})}{n(\text{golf balls present})} = \frac{17}{20}$$

29. There are 36 different possible rolls. The number of ways to get a 6 is 5 and the number of ways to get an 8 is 5. Therefore,

$$P(6 \text{ or } 8) = P(6) + P(8) = \frac{n(6) + n(8)}{36} = \frac{5 + 5}{36} = \frac{10}{36} = \frac{5}{18}$$

In Problems 31 and 33, we are using the table given in the text. The sample space is 100 households in each case.

31. From the table there are 30 households total out of 100 which have an income in excess of \$30,000. Thus, the probability that a household has an annual income in excess of \$30,000 is

$$P(E) = \frac{n(E)}{n(S)} = \frac{n(\text{in excess of \$30,000})}{n(\text{total households })} = \frac{30}{100} = \frac{3}{10}$$

33. From the table there are 35 households total out of 100 which have an income less than \$20,000. Thus, the probability that a household has an annual income less than \$20,000 is

$$P(E) = \frac{n(E)}{n(S)} = \frac{n(\text{less than \$20,000})}{n(\text{total households })} = \frac{35}{100} = \frac{7}{20}$$

35. (a) $P(1 \text{ or } 2) = P(1) + P(2) = 0.24 + 0.33 = 0.57$

(b) $P(1 \text{ or more}) = P(1) + P(2) + P(3) + P(4 \text{ or more})$
$= 0.24 + 0.33 + 0.21 + 0.17$
$= 0.95$

(c) $P(3 \text{ or fewer}) = P(0) + P(1) + P(2) + P(3)$
$= 0.05 + 0.24 + 0.331 + 0.21$
$= 0.83$

(d) $P(3 \text{ or more}) = P(3) + P(4 \text{ or more})$
$= 0.21 + 0.17$
$= 0.38$

(e) $P(\text{less than } 2) = P(0) + P(1)$
$= 0.05 + 0.24$
$= 0.29$

(f) $P(\text{less than } 1) = P(0) = 0.05$

(g) $P(1,2, \text{ or } 3) = P(1) + P(2) + P(3)$
$= 0.24 + 0.33 + 0.21$
$= 0.78$

(h) $P(2 \text{ or more}) = P(2) + P(3) + P(4 \text{ or more})$
$= 0.33 + 0.21 + 0.17$
$= 0.71$

37. The sample space for picking 5 numbers from a total of 10 numbers contains

$$P(10, 5) = \frac{10!}{(10-5)!} = \frac{10!}{5!} = \frac{10 \cdot 9 \cdot 8 \cdot 7 \cdot 6 \cdot 5!}{5!} = 30{,}240$$

possible outcomes. Only one of these outcomes is the desired event. Thus, the probability of winning is:

$$P(E) = \frac{n(E)}{n(S)} = \frac{n(\text{drawn number})}{n(\text{total possible outcomes})} = \frac{1}{30{,}240} = 0.000033068$$

39. (a) $P(3 \text{ heads}) = C(5, 3) \cdot \left(\frac{1}{2}\right)^5 = \frac{10}{32}$

(b) $P(0 \text{ heads}) = C(5, 0) \cdot \left(\frac{1}{2}\right)^5 = \frac{1}{32}$

41. (a) $P(\text{Sum} = 7 \text{ three times}) = P(\text{sum} = 7) \cdot P(\text{sum} = 7) \cdot P(\text{sum} = 7)$

$$= \frac{1}{6} \cdot \frac{1}{6} \cdot \frac{1}{6} = \frac{1}{216} = .00463$$

(b) $P(\text{sum} = 7 \text{ or } 11 \text{ at least twice}) = P(\text{sum} = 7 \text{ or } 11) \cdot P(\text{sum} = 7 \text{ or } 11)$
$\cdot\, P(\text{sum} \neq 7 \text{ or } 11) + P(\text{sum} = 7 \text{ or } 11) \cdot P(\text{sum} = 7 \text{ or } 11) \cdot P(\text{sum} = 7 \text{ or } 11)$

$$= \frac{8}{36} \cdot \frac{8}{36} \cdot \frac{28}{36} + \frac{8}{36} \cdot \frac{8}{36} \cdot \frac{8}{36} = .049$$

43. $P(\text{all five defective}) = \frac{1}{C(30, 5)} = 7.02 \times 10^{-6}$

$P(\text{at least two defective})$

$= P(2 \text{ def}) + P(3 \text{ def}) + P(4 \text{ def}) + P(5 \text{ def})$

$$= \frac{C(5, 2) \cdot C(25, 3)}{C(30, 5)} + \frac{C(5, 3) \cdot C(25, 2)}{C(30, 5)} + \frac{C(5, 4) \cdot C(25, 1)}{C(30, 5)} + \frac{(C(5, 5) \cdot C(25, 0)}{(30, 5)}$$

$= 0.183$

45. $P(\text{one of the 5 coins is the one valued at more than } \$10{,}000) = \frac{C(49, 4) \times C(1, 1)}{C(50, 5)} = 0.1$

9 Chapter Review

1. $5! = 5 \cdot 4 \cdot 3 \cdot 2 \cdot 1 = 120$

3. $\binom{5}{3} = \frac{5!}{3!2!} = \frac{5 \cdot 4}{2} = 10$

5. $P(8, 3) = \frac{8!}{(8-3)!} = \frac{8!}{5!} = \frac{8 \cdot 7 \cdot 6 \cdot 5!}{5!} = 336$

7. $C(8, 3) = \frac{8!}{(8-3)!3!} = \frac{8!}{5!3!} = \frac{8 \cdot 7 \cdot 6 \cdot 5!}{3 \cdot 2 \cdot 1 \cdot 5!} = 56$

9. $a_1 = (-1)^1\frac{1+3}{1+2} = \frac{-4}{3}$

$a_2 = (-1)^2\frac{2+3}{2+2} = \frac{5}{4}$

$a_3 = (-1)^3\frac{3+3}{3+2} = \frac{-6}{5}$

$a_4 = (-1)^4\frac{4+3}{4+2} = \frac{7}{6}$

$a_5 = (-1)^5\frac{5+3}{5+2} = \frac{-8}{7}$

11. $a_1 = \frac{2^1}{1^2} = 2$

$a_2 = \frac{2^2}{2^2} = 1$

$a_3 = \frac{2^3}{3^2} = \frac{8}{9}$

$a_4 = \frac{2^4}{4^2} = 1$

$a_5 = \frac{2^5}{5^2} = \frac{32}{25}$

13. $a_1 = 6$

$a_2 = \frac{2}{3}(6) = 4$

$a_3 = \frac{2}{3}(4) = \frac{8}{3}$

$a_4 = \frac{2}{3}\left(\frac{8}{3}\right) = \frac{16}{9}$

$a_5 = \frac{2}{3}\left(\frac{16}{9}\right) = \frac{32}{27}$

15. $a_1 = 2$

$a_2 = 2 - 2 = 0$

$a_3 = 2 - 0 = 2$

$a_4 = 2 - 2 = 0$

$a_7 = 2 - 0 = 2$

17. Arithmetic

$a_{n+1} - a_n = d$

$n + 6 - (n + 5) = d = 1$

$a = 6,\ a_n = (n + 5)$

$S_n = \frac{n}{2}[6 + (n + 5)] = \frac{n}{2}(n + 11)$

19. Neither

21. Geometric

$r = \frac{a_{n+1}}{a_n}$

$r = \frac{64}{8} = 8$

$S_r = \sum_{k=1}^{n} 2^{3k} = 8 + 64 + 512 + \ldots + 2^{3n}$

$= 8\left(\frac{1 - 8^n}{1 - 8}\right) = \frac{-8}{7}(1 - 8^n)$

$= \frac{-8}{7} + \frac{8^{n+1}}{7} = \frac{8}{7}(8^n - 1)$

23. Arithmetic

$d = 4 - 0$

$d = 4$

$a = 0,\ a_n = 4(n - 1)$

$S_n = \frac{n}{2}[0 + 4(n-1)]$

$= \frac{n}{2}[4(n - 1)]$

$= 2n(n - 1)$

25. Geometric

$r = \frac{\frac{3}{2}}{3} = \frac{1}{2}$

$S_n = \sum_{k=1}^{n} 3\left(\frac{1}{2}\right)^{k-1} = 3 + \frac{3}{2} + \frac{3}{4} + \ldots + 3\left(\frac{1}{2}\right)^{n-1}$

$= 3\left(\frac{1 - \left(\frac{1}{2}\right)^n}{1 - \frac{1}{2}}\right) = 3\frac{\left[1 - \left(\frac{1}{2}\right)^n\right]}{\frac{1}{2}} = 6\left[1 - \left(\frac{1}{2}\right)^n\right]$

27. Neither

29. $\sum_{k=1}^{5}(k^2 + 12) = \sum_{k=1}^{5}k^2 + \sum_{k=1}^{5}12 = 1^2 + 2^2 + 3^2 + 4^2 + 5^2 + 5(12)$

$= 1 + 4 + 9 + 16 + 25 + 60 = 115$

31. $\sum_{k=1}^{10}(3k - 9) = \sum_{k=1}^{10} 3k - \sum_{k=1}^{10} 9 = 3\sum_{k=1}^{10} k - 10(9)$
$= 3(1 + 2 + 3 + 4 + 5 + 6 + 7 + 8 + 9 + 10) - 90 = 3(55) - 90$
$= 165 - 90 = 75$

33. $\sum_{k=1}^{7}\left(\frac{1}{3}\right)^k = \left(\frac{1}{3}\right)^1 + \left(\frac{1}{3}\right)^2 + \left(\frac{1}{3}\right)^3 + \left(\frac{1}{3}\right)^4 + \left(\frac{1}{3}\right)^5 + \left(\frac{1}{3}\right)^6 + \left(\frac{1}{3}\right)^7$
$= \frac{1}{3} + \frac{1}{9} + \frac{1}{27} + \frac{1}{81} + \frac{1}{243} + \frac{1}{729} + \frac{1}{2187} \approx 0.49977$

35. Arithmetic: $d = 4, a_1 = 3$
$a_9 = 3 + 8(4)$
$a_9 = 35$

37. Geometric: $r = \frac{1}{10}, a_1 = 1$
$a_{11} = \left(\frac{1}{10}\right)^{10}(1)$
$a_{11} = \left(\frac{1}{10}\right)^{10}$

39. Arithmetic: $d = \sqrt{2}$,
$a_1 = \sqrt{2}$
$a_9 = \sqrt{2} + 8\sqrt{2}$
$a_9 = 9\sqrt{2}$

41. $a_7 = a + 6d = 31$
$a_{20} = a + 19d = 96$
$13d = 65$
$= 5$
$a = 31 - 6d$
$a = 31 - 6(5)$
$a = 1$
$a_n = 1 + (n - 1)5$
$a_n = 1 + 5n - 5$
$a_n = 5n - 4$

43. $a_{10} = a + 9d = 0$
$a_{18} = a + 17d = 8$
$8d = 8$
$d = 1$
$a = -9d$
$a = -9$
$a_n = -9 + (n - 1)1$
$a_n = -9 + n - 1$
$a_n = n - 10$

45. $a = 3, r = \frac{1}{3}$
$3 + 1 + \frac{1}{3} + \frac{1}{9} + \cdots = \frac{3}{1 - \frac{1}{3}} = \frac{9}{2}$

47. $a = 2, r = \frac{-1}{2}$
$2 - 1 + \frac{1}{2} - \frac{1}{4} + \cdots = \frac{2}{1 + \frac{1}{2}} = \frac{4}{3}$

49. $a = 4, r = \frac{1}{2}$
$\sum_{k=1}^{\infty} 4\left(\frac{1}{2}\right)^{k-1} = \frac{4}{1 - \frac{1}{2}} = 8$

51. (I) $n = 1$: $3 \cdot 1 = 3$ and $\frac{3 \cdot 1}{2}(2) = 3$

(II) If $3 + 6 + 9 + \cdots + 3k = \frac{3k}{2}(k + 1)$,

then $3 + 6 + 9 + \cdots + 3k + 3(k + 1)$

$$= (3 + 6 + 9 + \cdots + 3k) + (3k + 3) = \frac{3k^2}{2}(k + 1) + (3k + 3)$$

$$= \frac{3k^2}{2} + \frac{3k}{2} + 3k + 3 = \frac{3k^2}{2} + \frac{3k}{2} + \frac{6k}{2} + \frac{6}{2}$$

$$= \frac{3k^2}{2} + \frac{9k}{2} + \frac{6}{2} = \frac{3}{2}(k^2 + 3k + 2) = \frac{3}{2}(k + 1)(k + 2)$$

53. (I) $n = 1$: $2 \cdot 3^{1-1} = 2$ and $3^1 - 1 = 2$

(II) If $2 + 6 + 18 + \cdots + 2 \cdot 3^{k-1} = 3^k - 1$,

then $2 + 6 + 18 + \cdots + 2 \cdot 3^{k-1} + 2 \cdot 3^{(k+1)-1}$

$$= (2 + 6 + 18 + \cdots + 2 \cdot 3^{k-1}) + 2 \cdot 3^k = 3^k - 1 + 2 \cdot 3^k$$

$$= 3^k(1 + 2) - 1 = 3 \cdot 3^k - 1 = 3^{k+1} - 1$$

55. (I) $n = 1$: $1^2 = 1$ and $\frac{1}{2}(6 - 3 - 1) = \frac{1}{2}(2) = 1$

(II) If $1^2 + 4^2 + 7^2 + \cdots + (3k - 2)^2 = \frac{1}{2}k(6k^2 - 3k - 1)$

then $1^2 + 4^2 + 7^2 + \cdots + (3k - 2)^2 + [3(k + 1) - 2]^2$

$$= [1^2 + 4^2 + 7^2 + \cdots + (3k - 2)^2] + (3k + 1)^2$$

$$= \frac{1}{2}k(6k^2 - 3k - 1) + (3k + 1)^2 = \frac{1}{2}(6k^3 + 15k^2 + 11k + 2)$$

$$= \frac{1}{2}(k + 1)[6k^2 + 12k - 3k + 6 - 3 - 1]$$

$$= \frac{1}{2}(k + 1)[6k^2 + 12k + 6 - 3k - 3 - 1]$$

$$= \frac{1}{2}(k + 1)[6(k^2 + 2k + 1) - 3(k + 1) - 1]$$

$$= \frac{1}{2}(k + 1)[6(k + 1)^2 - 3(k + 1) - 1]$$

57. $(x + 3)^5 = \binom{5}{0}x^5 + \binom{5}{1}3 \cdot x^4 + \binom{5}{2}3^2x^3 + \binom{5}{3}3^3 \cdot x^2 + \binom{5}{4}3^4 \cdot x + \binom{5}{5}2^5$

$$= x^5 + 5 \cdot 3 \cdot x^4 + 10 \cdot 9 \cdot x^3 + 10 \cdot 27 \cdot x^2 + 5 \cdot 81 \cdot x + 32$$

$$= x^5 + 15x^4 + 90x^3 + 270x^2 + 405x + 32$$

59. $(2x + 3)^5 = \binom{5}{0}(2x)^5 + \binom{5}{1}3(2x)^4 + \binom{5}{2}3^2(2x)^3 + \binom{5}{3}3^3(2x^2) + \binom{5}{4}3^4(2x) + \binom{5}{5}3^5$

$$= 32x^5 + 5 \cdot 3 \cdot 16x^4 + 10 \cdot 9 \cdot 8x^3 + 10 \cdot 27 \cdot 4x^2 + 5 \cdot 81 \cdot 2x + 243$$

$$= 32x^5 + 240x^4 + 720x^3 + 1080x^2 + 810x + 243$$

61. $n = 9, a = 2, x = x$, and $j = 7$

$$\binom{9}{9-7}2^{9-7}x^7 = \binom{9}{2} \cdot 2^2x^7$$
$$= \frac{9!}{2!7!} \cdot 4 \cdot x^7$$
$$= \frac{9 \cdot 8}{2} \cdot 4 \cdot x^7$$
$$= 144x^7$$

The coefficient of x^7 is 144.

63. $n = 7, a = 1, x = 2x$, and $j = 3$

$$\binom{7}{7-3}1^{7-3}(2x)^3 = \binom{7}{4}1^4 \cdot 8x^3$$
$$= \frac{7!}{4!3!} \cdot 1 \cdot 8 \cdot x^3$$
$$= 280x^3$$

The coefficient of x^2 is 280.

In Problems 65–71, U = Universal set = {1, 2, 3, 4, 5, 6, 7, 8, 9}, A = {1, 3, 5, 7}, B = {3 5, 6, 7, 8}, C = {2, 3, 7, 8, 9}.

65. $A \cup B = \{1, 3, 5, 7\} \cup \{3, 5, 6, 7, 8\}$
$= \{1, 3, 5, 6, 7, 8\}$

67. $A \cap C = \{1, 3, 5, 7\} \cap \{2, 3, 7, 8, 9\}$
$= \{3, 7\}$

69. $A' \cup B' = \{1, 3, 5, 7\}' \cup \{3, 5, 6, 7, 8\}'$
$= \{2, 4, 6, 8, 9\} \cup \{1, 2, 4, 9\}$
$= \{1, 2, 4, 6, 8, 9\}$

71. $(B \cap C)'$
$= (\{3, 5, 6, 7, 8\} \cap \{2, 3, 7, 8, 9\})'$
$= (\{3, 7, 8\})' = \{1, 2, 4, 5, 6, 9\}$

73. For $n(A) = 8$, $n(B) = 12$, and $n(A \cap B) = 3$,
$n(A \cup B) = n(A) + n(B) - n(A \cap B)$
$= 8 + 12 - 3 = 17$

75. From the figure $n(A) = 20 + 2 + 6 + 1$
$= 29$

77. From the figure,
$n(A \text{ and } C) = n(A \cap C)$
$= 6 + 1 = 7$

79. From the figure,
n(neither in A nor in C)
$= n((A \cup C)') = 25$

81. We have 2 choices of material, 3 choices of color, and 10 choices of size, or $2 \times 3 \times 10 = 60$ choices for a complete assortment.

83. There are two possible outcomes for each game, or

$$2 \times 2 \times 2 \times 2 \times 2 \times 2 \times 2 = 2^7 = 128 \text{ outcomes for 7 games.}$$

85. Since order is significant, we have

$$P(9, 4) = \frac{9!}{(9-4)!} = \frac{9!}{5!} = \frac{9 \cdot 8 \cdot 7 \cdot 6 \cdot 5!}{5!} = 3024 \text{ ways to seat 4 people in 9 seats.}$$

87. Choosing 4 runners from 8 where order is not significant gives

$$C(8, 4) = \frac{8!}{(8-4)!4!} = \frac{8!}{4!4!} = \frac{8 \cdot 7 \cdot 6 \cdot 5 \cdot 4!}{4 \cdot 3 \cdot 2 \cdot 1 \cdot 4!} = 70 \text{ ways a squad can be chosen.}$$

89. We have 14 teams to be taken 2 at a time, or

$$C(14, 2) = \frac{14!}{(14-2)!2!} = \frac{14!}{12!2!} = \frac{14 \cdot 13 \cdot 12!}{2 \cdot 1 \cdot 12!} = 91 \text{ ways to pair 14 teams.}$$

91. There are $8 \cdot 10 \cdot 10 \cdot 10 \cdot 10 \cdot 10 \cdot 2 = 1{,}600{,}000$ possible phone numbers.

93. There are $24 \cdot 9 \cdot 10 \cdot 10 \cdot 10 = 216{,}000$ possible license plates possible.

95. $\dfrac{7!}{2!2!} = 1260$ different words.

97. (a) $\begin{bmatrix} 9 \\ 4 \end{bmatrix} \cdot \begin{bmatrix} 9 \\ 3 \end{bmatrix} \cdot \begin{bmatrix} 9 \\ 2 \end{bmatrix} = 126 \cdot 84 \cdot 36 = 381{,}024$

(b) $\begin{bmatrix} 9 \\ 4 \end{bmatrix} \cdot \begin{bmatrix} 5 \\ 3 \end{bmatrix} \cdot \begin{bmatrix} 2 \\ 2 \end{bmatrix} = 126 \cdot 10 \cdot 1 = 1{,}260$

99. The sample space contains $3 + 6 + 11 = 20$ elements. The probability of choosing a 40-watt bulb is:

$$P(E) = \frac{n(E)}{n(S)} = \frac{3}{20}$$

The probability of not choosing a 75 watt bulb is:

$$P(E) = \frac{n(E)}{n(S)} = \frac{3+6}{20} = \frac{9}{20}$$

101. The sample space for dealing out 4 cards contains $P(4, 4) = 4! = 24$ possible outcomes. Only one of these outcomes fulfills the desired event of spelling out ROSE so that the probability of this event is:

$$P(E) = \frac{n(E)}{n(S)} = \frac{1}{24}$$

103. (a) $P(\text{all 3 are Merlot}) = \dfrac{5}{12} \cdot \dfrac{4}{11} \cdot \dfrac{3}{10} = 0.045$

(b) $P(\text{2 are Merlot}) = \dfrac{5}{12} \cdot \dfrac{4}{11} \cdot \dfrac{7}{10} + \dfrac{7}{12} \cdot \dfrac{5}{11} \cdot \dfrac{4}{10} + \dfrac{5}{12} \cdot \dfrac{7}{11} \cdot \dfrac{4}{10} = 0.318$

(c) $P(\text{none Merlot}) = P(\text{all Cabernet}) = \dfrac{7}{12} \cdot \dfrac{6}{11} \cdot \dfrac{5}{10} = 0.159$

105. The bottom step requires 80 bricks and each successive step requires 3 less bricks than the prior step. This is an arithmetic sequence with $a_1 = 80$ and $d = -3$.

(a) $a_{25} = a + (n - 1)d = 80 + (24)(-3) = 8$ bricks

(b) The total number of bricks required is:

$$S = 80 + 77 + \cdots + 8$$

This is the sum of an arithmetic sequence. The number of terms to be added is 25.

$$S = \frac{25}{2}(80 + 8) = 1100$$

Thus, 1100 bricks will be needed.

107. This is an ordinary annuity with 12 deposits per year of $P = \$200$ for 20 years. Thus, there are $n = 20(12) = 240$ payment periods. The rate of interest per payment period is $i = \dfrac{0.10}{12}$. Thus,

$$A = \$200\left[\frac{\left(1 + \frac{0.10}{12}\right)^{240} - 1}{\frac{0.10}{12}}\right] = \$200(759.368836) = \$151{,}873.77$$

109. This is an example of a geometric series, with $r = \frac{3}{4}$.

(a) After striking the third time, the height is $20\left[\frac{3}{4}\right]^3 = \frac{135}{16}$ feet.

(b) After striking the nth time, the height is $20\left[\frac{3}{4}\right]^n$ feet.

(c) If the height is 6 inches (0.5 feet), then

$$0.5 = 20\left[\frac{3}{4}\right]^n$$

$$0.025 = \left[\frac{3}{4}\right]^n$$

$$\log 0.025 = n \log\left[\frac{3}{4}\right]$$

$$n = \frac{\log 0.025}{\log\left[\frac{3}{4}\right]} = 12.82$$

The height is less than 6 inches after the 13th strike.

(d) $$\text{Distance} = 20 + 20\left[\frac{3}{4}\right] + 20\left[\frac{3}{4}\right] + 20\left[\frac{3}{4}\right]^2 + 20\left[\frac{3}{4}\right]^2 + \cdots + 20\left[\frac{3}{4}\right]^n + 20\left[\frac{3}{4}\right]^n + \cdots$$

(first $20\left[\frac{3}{4}\right]$: Ball Going Up; second $20\left[\frac{3}{4}\right]$: Ball Coming Down)

$$= 20 + 40\left[\frac{3}{4}\right] + 40\left[\frac{3}{4}\right]^2 + \cdots + 40\left[\frac{3}{4}\right]^n + \cdots$$

$$= 20 + 40\left[\frac{3}{4}\right]\left[1 + \left[\frac{3}{4}\right] + \cdots\right] = 20 + \frac{40\left[\frac{3}{4}\right]}{1 - \frac{3}{4}}$$

$$= 20 + 120 = 140 \text{ feet}$$

111. (a) $P(\text{tune-up or brake}) = P(\text{tune-up}) + P(\text{brake}) - P(\text{tune-up and brake})$
$= 0.6 + 0.1 - 0.02$
$= 0.68$

(b) $P(\text{tune-up and brake}') = P(\text{tune-up}) - P(\text{tune-up and brake})$
$= 0.6 - 0.02$
$= 0.58$

(c) $P(\text{tune-up and brake})' = 1 - P(\text{tune-up or brake})$
$= 1 - 0.68$
$= 0.32$

APPENDIX A Graphing Utilities

A.1 The Viewing Rectangle

1. (−1, 4)

3. (3, 1)

5. Xmin = −11
Xmax = 5
Xscl = 1
Ymin = −3
Ymax = 6
Yscl = 1

7. Xmin = −30
Xmax = 50
Xscl = 10
Ymin = −90
Ymax = 50
Yscl = 10

9. Xmin = −10
Xmax = 110
Xscl = 10
Ymin = −10
Ymax = 160
Yscl = 10

11. Xmin = −6
Xmax = 6
Xscl = 2
Ymin = −4
Ymax = 4
Yscl = 2

13. Xmin = −9
Xmax = 9
Xscl = 3
Ymin = −4
Ymax = 4
Yscl = 2

15. Xmin = −6
Xmax = 6
Xscl = 1
Ymin = −8
Ymax = 8
Yscl = 2

17. Xmin = −6
Xmax = 6
Xscl = 2
Ymin = −1
Ymax = 3
Yscl = 1

19. Xmin = 3
Xmax = 9
Xscl = 1
Ymin = 2
Ymax = 10
Yscl = 2

A.2 Using a Graphing Utility to Graph Equations

1. (a)

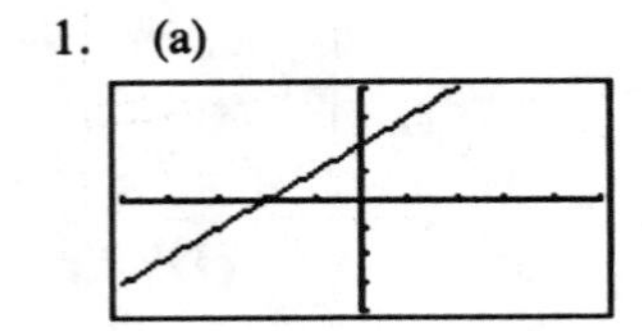

(b)

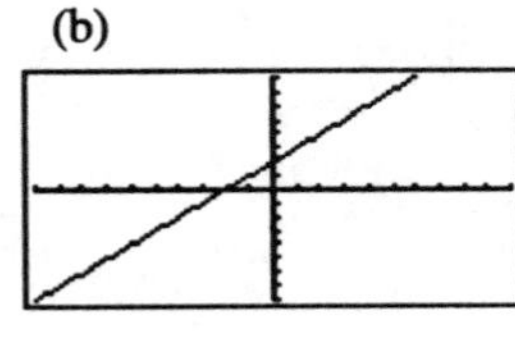

(c)

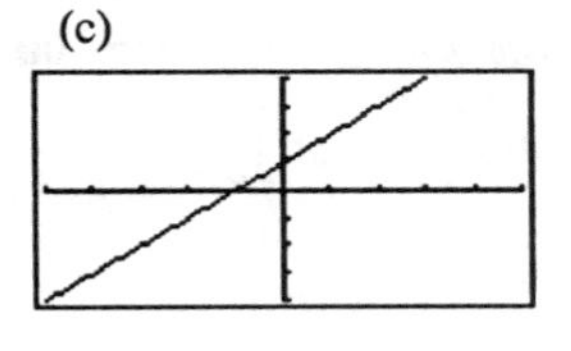

(d)

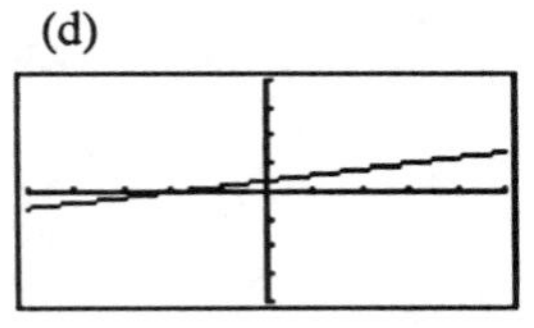

3. (a)

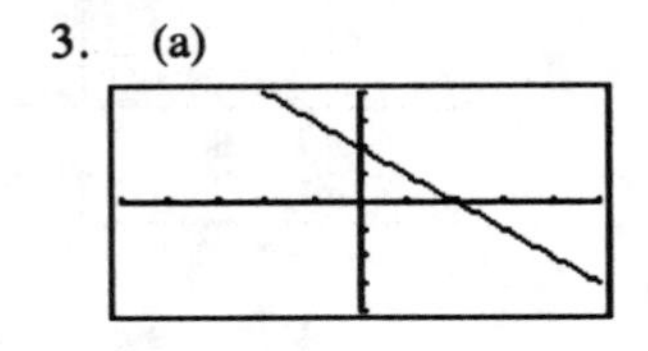

(b)

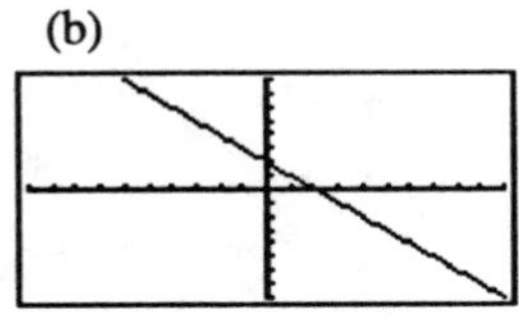

(c)

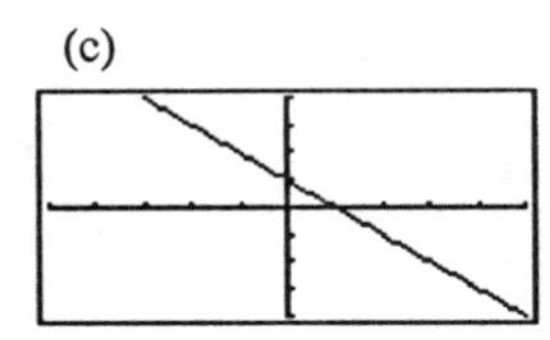

(d)

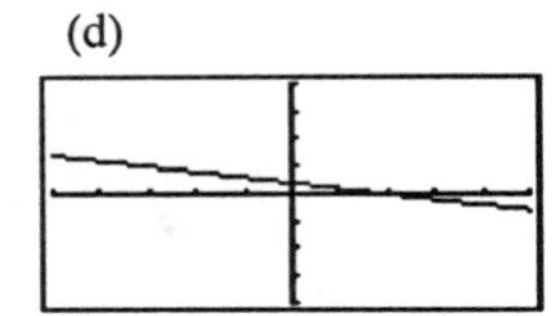

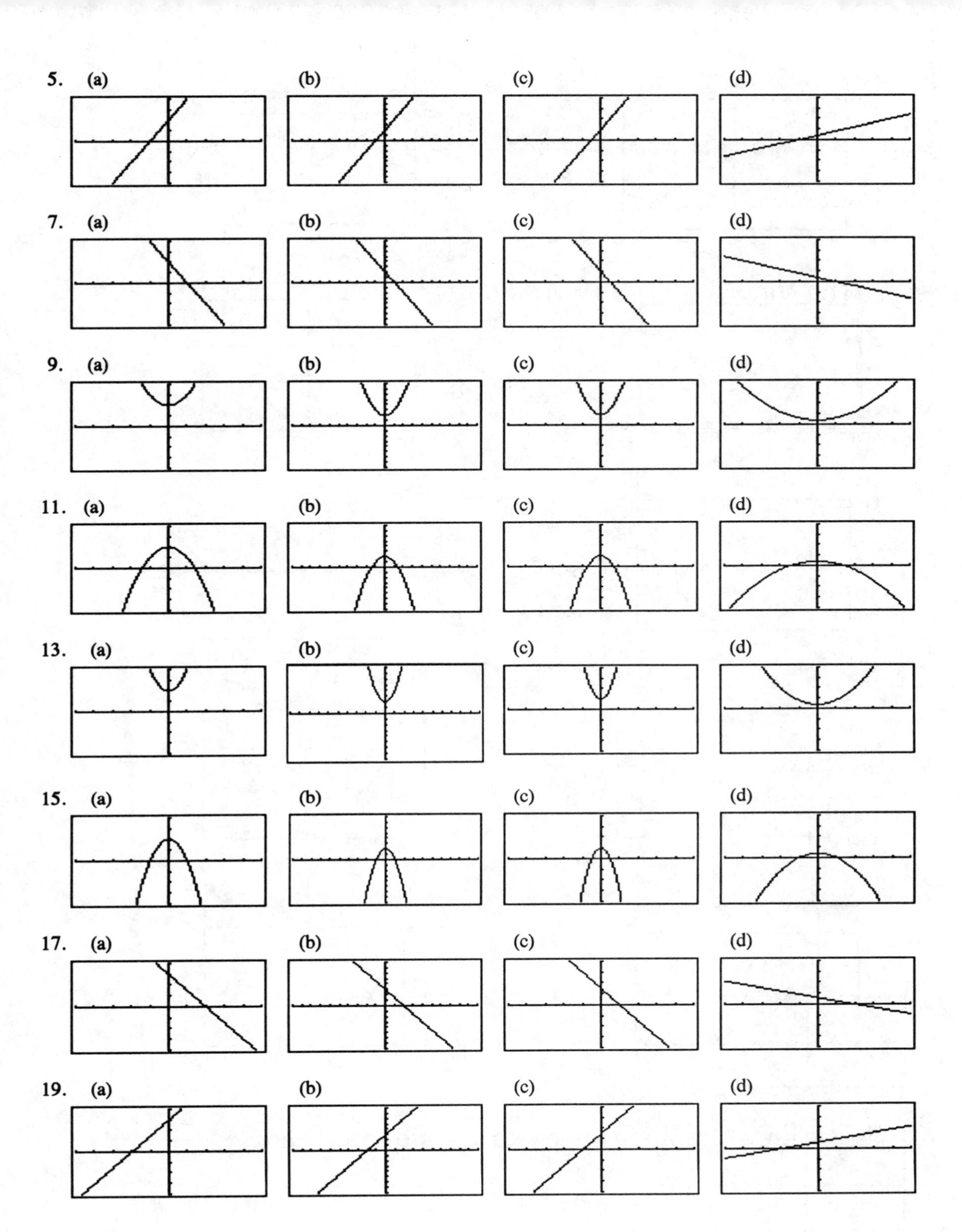
5. (a)
(b)
(c)
(d)
7. (a)
(b)
(c)
(d)
9. (a)
(b)
(c)
(d)
11. (a)
(b)
(c)
(d)
13. (a)
(b)
(c)
(d)
15. (a)
(b)
(c)
(d)
17. (a)
(b)
(c)
(d)
19. (a)
(b)
(c)
(d)

21. 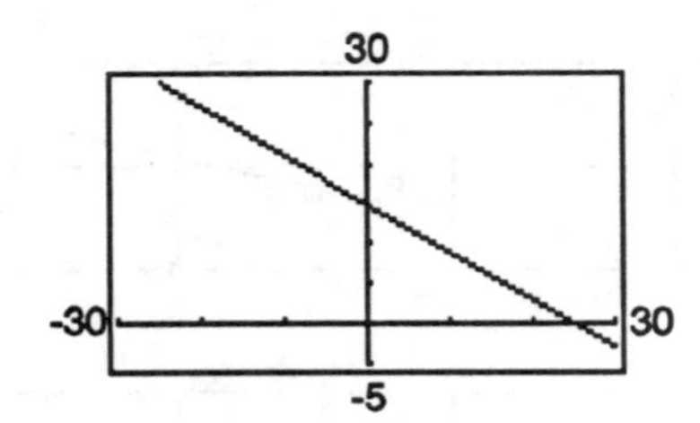

23. 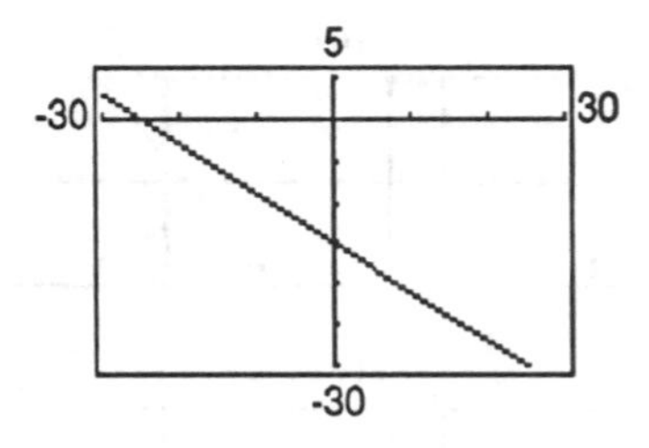

25. 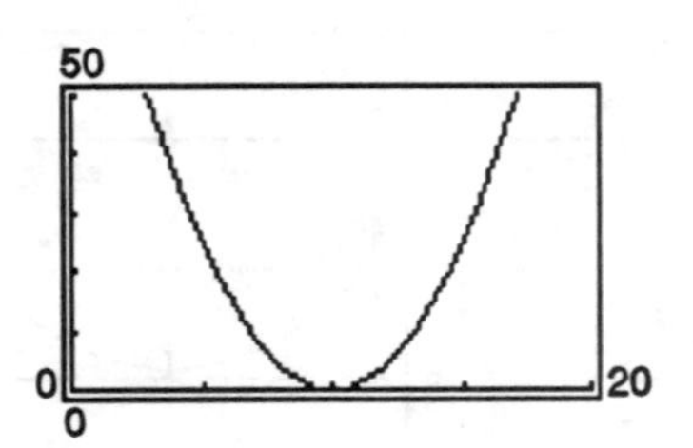

27. 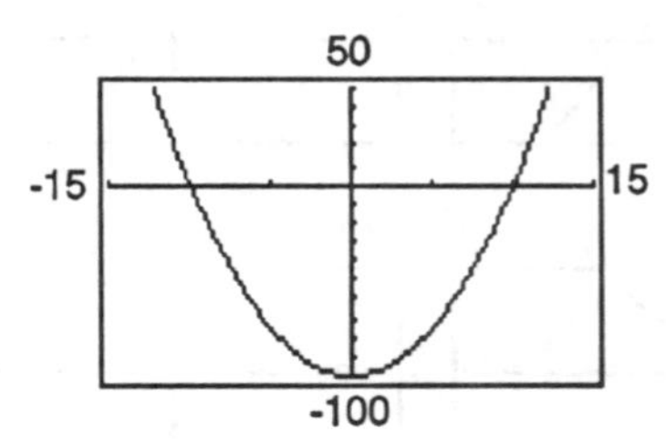

29. 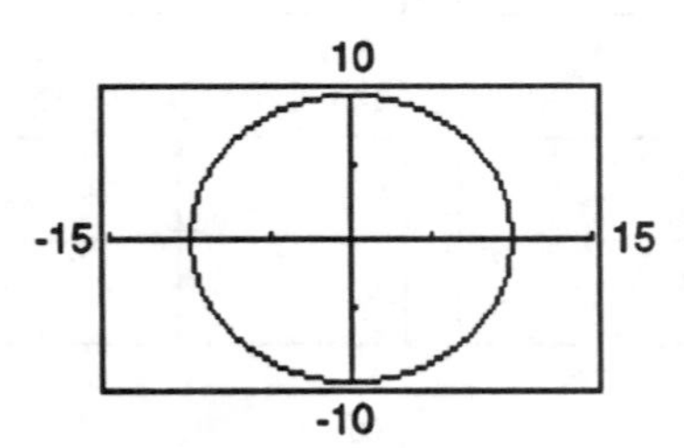

31. 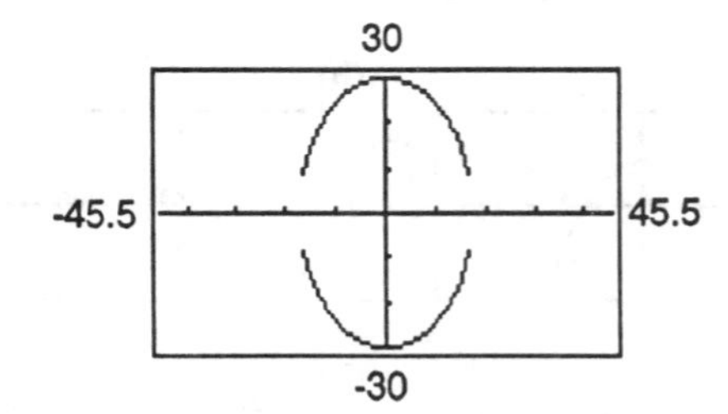

33. 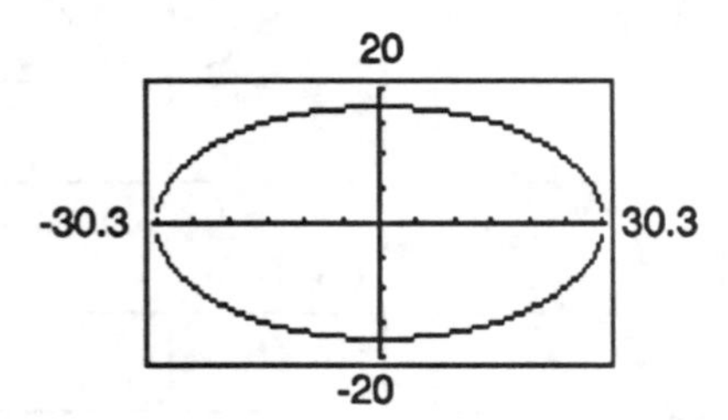

35. 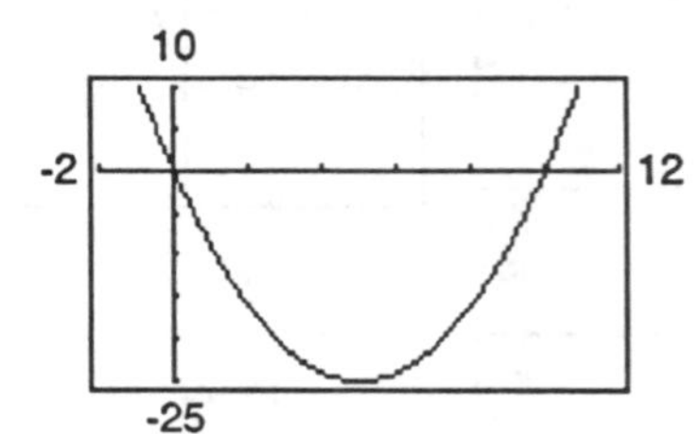

37. 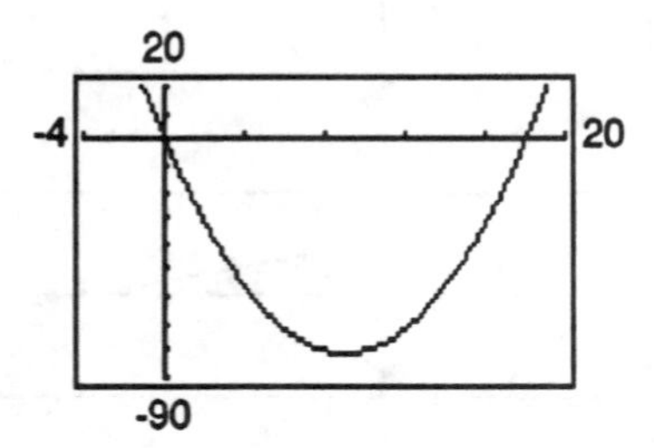

39. 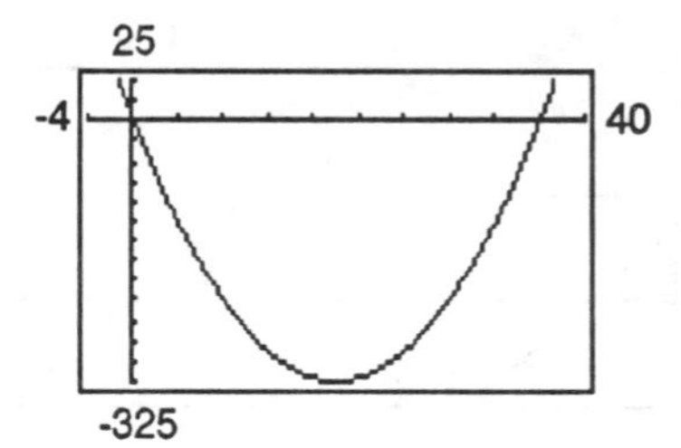

A.3 Using a Graphing Utility to Locate Intercepts

1. 0.42

3. 2.23

5. 1.25

7. Graph $y_1 = x^2 + 4x + 2$ and use ZERO or ROOT.

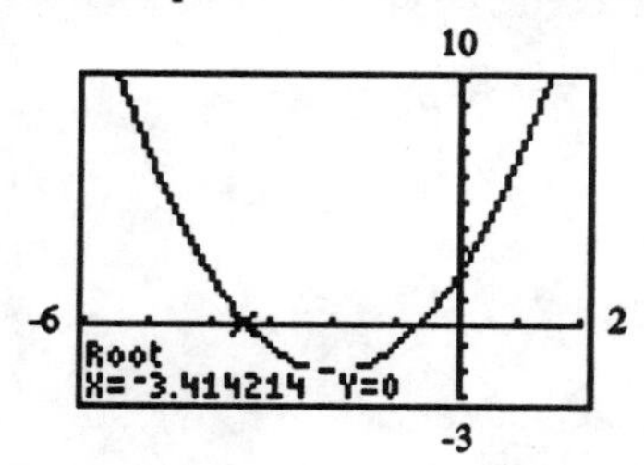

The smaller solution is -3.41 correct to two decimal places.

9. Graph $y_1 = 2x^2 + 4x + 1$ and use ZERO or ROOT.

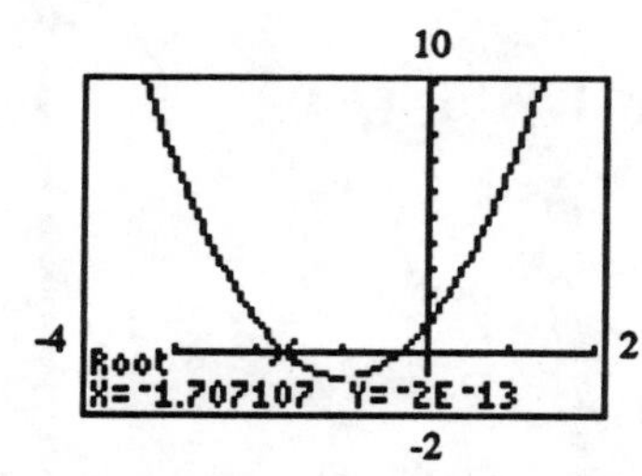

The smaller solution is -1.70 correct to two decimal places.

11. Graph $y_1 = 2x^2 - 3x - 1$ and use ZERO or ROOT.

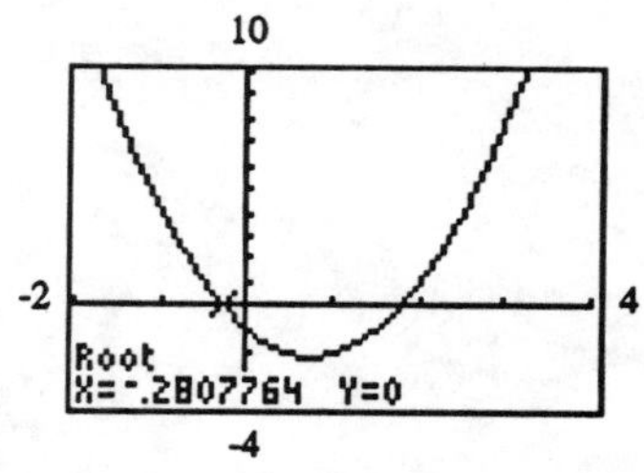

The smaller solution is -0.28 correct to two decimal places.

13. Graph $y_1 = x^3 + 3.2x^2 - 16.83x - 5.31$ and use ZERO or ROOT.

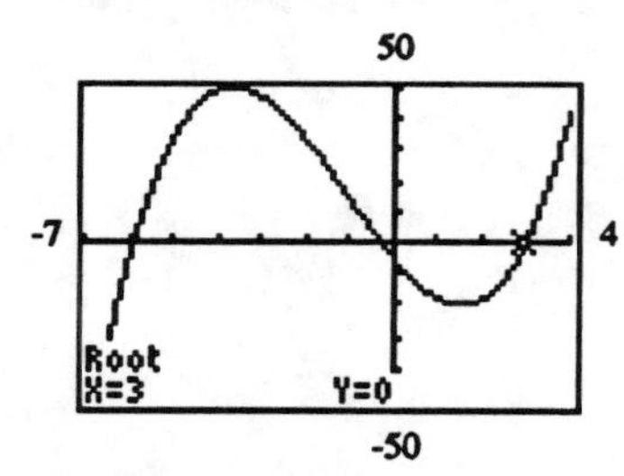

The positive solution is 3.

15. Graph $y_1 = x^4 - 1.4x^3 - 33.71x^2 + 23.94x + 292.41$ and use ZERO or ROOT.

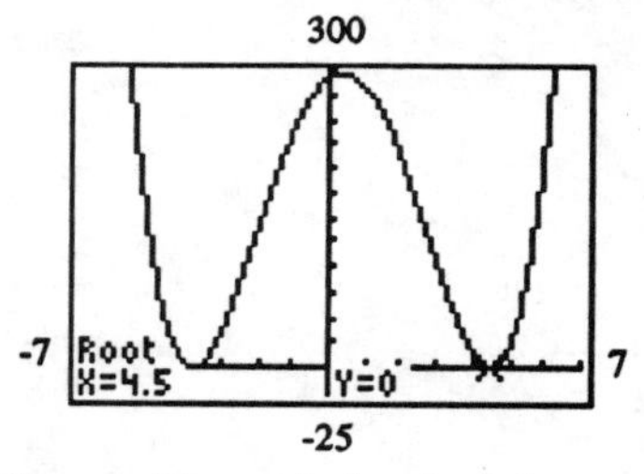

The positive solution is 4.5.

17. Graph $y_1 = \pi x^3 - (5.63\pi + 2)x^2 - (108.392\pi - 11.26)x + 216.784$ and use ZERO or ROOT.

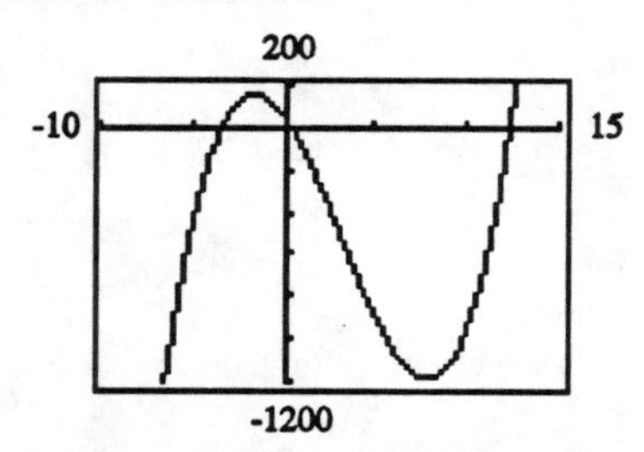

The two positive solutions are 0.31 and 12.30 correct to two decimal places.

19. Graph $y_1 = x^3 + 19.5x^2 - 1021x + 1000.5$ and use ZERO or ROOT.

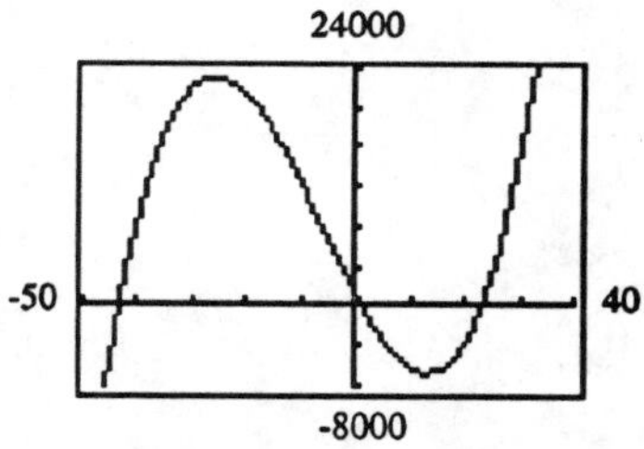

The two positive solutions are 1.00 and 23.00 correct to two decimal places.

A.4 Square Screens

1. Yes 3. Yes 5. No 7. Yes

9. ymin = 4
ymax = 12
yscl = 1

Note: Other answers are possible (ymax − ymin must equal 8).